AF783594

Engineering Materials

This series provides topical information on innovative, structural and functional materials and composites with applications in optical, electrical, mechanical, civil, aeronautical, medical, bio- and nano-engineering. The individual volumes are complete, comprehensive monographs covering the structure, properties, manufacturing process and applications of these materials. This multidisciplinary series is devoted to professionals, students and all those interested in the latest developments in the Materials Science field, that look for a carefully selected collection of high quality review articles on their respective field of expertise.

Indexed at Compendex (2021) and Scopus (2022)

Srikanta Moharana · Ram Naresh Mahaling
Editors

Surface Engineering of MXenes

Springer

Editors
Srikanta Moharana
Department of Chemistry
School of Applied Sciences
Centurion University of Technology
and Management
Bhubaneswar, Odisha, India

Ram Naresh Mahaling
School of Chemistry
Sambalpur University
Jyoti Vihar, Burla, Sambalpur
Odisha, India

ISSN 1612-1317 ISSN 1868-1212 (electronic)
Engineering Materials
ISBN 978-981-96-3639-6 ISBN 978-981-96-3640-2 (eBook)
https://doi.org/10.1007/978-981-96-3640-2

© The Editor(s) (if applicable) and The Author(s), under exclusive license to Springer Nature Singapore Pte Ltd. 2025

This work is subject to copyright. All rights are solely and exclusively licensed by the Publisher, whether the whole or part of the material is concerned, specifically the rights of translation, reprinting, reuse of illustrations, recitation, broadcasting, reproduction on microfilms or in any other physical way, and transmission or information storage and retrieval, electronic adaptation, computer software, or by similar or dissimilar methodology now known or hereafter developed.
The use of general descriptive names, registered names, trademarks, service marks, etc. in this publication does not imply, even in the absence of a specific statement, that such names are exempt from the relevant protective laws and regulations and therefore free for general use.
The publisher, the authors and the editors are safe to assume that the advice and information in this book are believed to be true and accurate at the date of publication. Neither the publisher nor the authors or the editors give a warranty, expressed or implied, with respect to the material contained herein or for any errors or omissions that may have been made. The publisher remains neutral with regard to jurisdictional claims in published maps and institutional affiliations.

This Springer imprint is published by the registered company Springer Nature Singapore Pte Ltd.
The registered company address is: 152 Beach Road, #21-01/04 Gateway East, Singapore 189721, Singapore

If disposing of this product, please recycle the paper.

Preface

The discovery of MXenes in 2011 marked a watershed moment in materials science, introducing a new family of two-dimensional transition metal carbides and nitrides that would revolutionize multiple fields of technology. As we delve deeper into the potential of these remarkable materials, it has become increasingly clear that surface engineering holds the key to unlocking their full capabilities.

The book *Surface Engineering of MXenes* emerges at a crucial time when the global scientific community is intensively exploring ways to tailor MXene properties for specific applications. Surface engineering of MXenes represents a powerful approach to modifying their chemical, electronic, and physical properties, opening new avenues for their implementation in energy storage, electromagnetic interference shielding, catalysis, sensors, and numerous other applications. The motivation for this volume stems from the rapid acceleration of research in MXene surface modification and the need to consolidate our current understanding of these techniques. We aim to provide researchers, engineers, and students with a comprehensive resource that bridges the gap between fundamental surface chemistry and practical applications of modified MXenes.

The surface engineering of MXenes is covered in 18 well-organized chapters. First Chapter, "Present Status and Prospects of MXene Research," provides a comprehensive overview of the current state of MXene research, tracing its evolution from discovery to present-day applications. It examines breakthroughs, current challenges, and emerging trends in MXene development. The chapter concludes with a detailed analysis of future research directions and potential applications. Second Chapter, "Synthesis and Fabrication Techniques of MXene," delves into a detailed exploration of various synthesis methodologies and fabrication techniques for MXenes, including selective etching, delamination processes, and advanced manufacturing methods. The chapter covers both traditional and emerging fabrication approaches, discussing their advantages, limitations, and scalability potential. Third Chapter, "Surface Characterization Techniques of MXene Films and Powders," examines analytical tools and techniques used for characterizing MXene surfaces in depth. It covers advanced microscopy techniques (SEM, TEM, AFM), spectroscopic methods (XPS, Raman), and other analytical approaches essential for understanding surface morphology, composition, and properties. Fourth Chapter, "Surface Functionalization Strategies of MXene: A Critical Overview of Its Improved Physical, Chemical, and Electrochemical Properties," critically reviews various surface modification approaches and their impact on MXene properties. It examines chemical functionalization methods, surface treatments, and modification strategies, providing a detailed analysis of how these changes affect physical, chemical, and electrochemical properties. The relationship between surface chemistry and material performance is thoroughly explored. Fifth Chapter, "MXene-Based Advanced Materials: Mechanical and Catalytic Properties and Their Key Applications," provides a comprehensive examination of the mechanical properties and catalytic applications of MXenes. The chapter discusses strength, flexibility, durability characteristics, catalytic mechanisms, and performance in various key applications. Sixth Chapter, "MXene: Antiviral and Antibacterial Properties," explores the emerging role of MXenes in antiviral and antimicrobial applications. It covers mechanisms of action against various pathogens, including viruses and bacteria, and examines factors affecting antimicrobial efficacy. Seventh Chapter, "MXene and Its Derivatives for Energy Storage Applications," covers fundamental principles of energy storage using MXenes and their derivatives, examining various architectures and compositions for optimal performance. Eighth Chapter, "Recent Progress in MXene Research for Sensor and Environmental Applications," examines the latest developments in MXene-based sensors and environmental technologies. It covers synthesis techniques, various properties of MXene-based sensors, and environmental sensing applications. Ninth Chapter, "MXene and Its Modification for Supercapacitor Application," explores MXene-based supercapacitor technologies, including electrode design, electrolyte selection, and device fabrication. The chapter discusses various modification strategies to enhance capacitance, cycling stability, and energy density. Additionally, the chapter highlights future prospects and examines the challenges associated with these technologies. Tenth Chapter, "Application of MXenes for Perovskite Solar Cells," focuses on the integration of MXenes in perovskite solar cell technology. It covers interface engineering, charge transport enhancement, and stability improvement strategies. Recent advances in efficiency and durability are discussed, along with future prospects. Eleventh Chapter, "MXene and Its Composites for Battery Applications," describes a comprehensive examination of MXene applications in various battery technologies, including lithium-ion, sodium-ion, and emerging battery systems. The chapter covers electrode design, composite formation, and performance optimization strategies followed by a summary and future prospects. Twelfth Chapter, "MXene as Nanofillers for Polymer Composites," explores the use of MXenes as reinforcing agents in polymer matrices. It covers processing techniques, interface engineering, and mechanisms for property enhancement. It provides an in-depth discussion of the advanced applications of MXene-based polymer composites. Thirteenth Chapter, "MXene and Its Composites for Biomedical Applications," provides a detailed examination of the use of MXenes in (bio)sensors for biomedical applications. It also offers an overview of recent innovations in the field. Additionally, the chapter highlights the challenges and opportunities associated with the use of these materials in biomedical contexts. Fourteenth

Chapter, "MXene and Its Composites for Industrial Applications," explores the practical implementations of MXene-based materials in various industrial settings. It covers a wide range of applications, including Li-ion batteries, energy storage components, supercapacitors, transparent conducting electrodes, gas sensors, neural electrodes, photonic diodes, electrochromic devices, triboelectric nanogenerators, composites, photocatalysts for dye degradation, and water purification. The chapter concludes with a perspective on future developments and the potential of these materials in industrial applications. Fifteenth Chapter, "Surface-Functionalized MXenes for Adsorptive Wastewater Treatment Applications," covers a comprehensive analysis of MXene applications in water purification and treatment. The chapter covers adsorption mechanisms, surface modification strategies for enhanced pollutant removal, and practical implementation in water treatment systems. Sixteenth Chapter, "Surface Engineering and Processing via MXene for Corrosion Protection Coatings," examines the use of MXenes in protective coating applications. It covers coating formulation, application techniques, and performance evaluation methods. Long-term stability, adhesion mechanisms, and practical implementation strategies are thoroughly discussed. Seventeenth Chapter, "Progress of MXene-Based Materials: Commercialization, Shortcomings, and Multifunctional Applications," explores the current progress of MXene-based materials in terms of commercialization, highlighting both their potential and existing shortcomings. Additionally, it delves into the multifunctional applications of MXenes in various industries, showcasing their versatility and promising future prospects. Eighteenth Chapter, "MXenes and MXene Based Composite Materials for Tissue Engineering and Cancer Nanotheranostics," explores the potential applications of MXenes and their composites in tissue engineering and cancer nanotheranostics. The chapter discusses the recent advancements in utilizing MXenes and their composites for targeted drug delivery, imaging, and regenerative medicine in the fields of tissue engineering and cancer nanotheranostics.

This book explores the versatility of surface-engineered MXenes, discussing synthesis, fabrication techniques, advanced surface characterization, and surface functionalization strategies. It discusses their potential in energy storage and biomedical applications, including antiviral and antibacterial properties. The book also discusses environmental applications like sensors, remediation, and wastewater treatment, as well as industrial perspectives like polymer composites and corrosion protection coatings.

This book represents a collaborative effort from leading experts in the field, each contributing their unique expertise and insights. Our goal is to provide researchers, engineers, and students with a comprehensive resource that not only presents the current state of knowledge but also identifies future opportunities and challenges in MXene surface engineering. With the rapid advancements in MXene research and development, we hope this book will inspire new ideas, foster collaboration, and accelerate the journey from fundamental research to cutting-edge applications. The future of MXene technology is bright, and we are confident that the surface engineering strategies and applications discussed in this volume will contribute significantly to this exciting field.

We are deeply grateful to all contributors who have shared their expertise and recent findings, making this book a valuable resource for the materials science community. Their collective wisdom has helped create a volume that we hope will serve as both a practical guide and an inspiration for future innovations in MXene surface engineering.

As we stand at the frontier of this exciting field, we invite readers to join us in exploring the enormous potential of surface-engineered MXenes. Whether you are an experienced researcher or a newcomer to the field, we believe this book will provide you with valuable insights and tools to advance your work in this dynamic area of materials science.

Bhubaneswar, India Srikanta Moharana, Ph.D.
Burla, India Ram Naresh Mahaling, Ph.D.

Contents

Editors and Contributors

About the Editors

Dr. Srikanta Moharana is presently working as an Associate Professor in the Department of Chemistry, School of Applied Sciences, Centurion University of Technology and Management, Odisha, India. He obtained his Ph.D. and M.Phil. degrees in Chemistry from the School of Chemistry, Sambalpur University, India. He received his M.Sc. degree in Chemistry from the National Institute of Technology, Rourkela. He has published more than 72 research papers in international peer-reviewed journals, 80 book chapters, and many more counting with renowned international publications such as the RSC, Elsevier, Wiley, Springer, and Taylor & Francis. Besides, he has edited 12 books under the banner of Elsevier, Springer Nature, Royal Society of Chemistry and CRC Taylor & Francis Publications. Under his guidance, two Ph.D students have successfully received their Ph.D. degrees. Moreover, he has actively participated in and presented his research work at several international conferences and has been a reviewer for reputed international journals. He has been awarded **Prof. G. B. Behera's Best Ph.D. Thesis Award** under the banner of Orissa Chemical Society, Odisha, India. He has more than 5 years of teaching and research experience. His research experience and interests lie in carbon nanomaterials (graphene, carbon nanotubes) and functional ceramic-based polymer nanocomposite synthesis and characterization for advanced energy storage applications. He is a life member of the Orissa Chemical Society.

Professor Ram Naresh Mahaling at present is working in the School of Chemistry, Jyoti Vihar, Sambalpur University, Odisha, India (BHARAT). He has served at Sambalpur University in the capacity of Assistant Professor and Associate Professor prior to joining as Full Professor. He has more than 14 years of teaching and 20 years of research experience and has published more than 80 research papers in peer-reviewed journals. In addition to this, he has contributed seven book chapters and author of two edited books. His current areas of research include the synthesis of perovskite-type metal oxide ($BaTiO_3$, $BiFeO_3$, $CaTiO_3$, etc.)-based polymer composite materials for electronic applications and energy storage, nanocomposites, and their structure-properties relationship. He has visited the Institute of Polymer Research Dresden, Germany, and the University of Orleans, France, for two and a half years as a Postdoctoral Fellow. He has received research grants from different funding agencies, such as UGC, New Delhi; DST SERB, New Delhi, and the Government of Odisha. So far, he has guided 14 M.Phil. and 3 Ph.D. students. Currently, 4 students are working under his guidance for their Ph.D. He has delivered ten invited talks at different national and international conferences.

Contributors

Arnet Maria Antony Centre for Nano and Material Sciences, Jain (Deemed-to-be University), Jain Global Campus, Kanakapura, Bangalore, Karnataka, India

M. A. Anu Center for Advanced Materials Research, Department of Physics, Government College for Women, Thiruvananthapuram, Kerala, India

Shahid Anwar Materials Chemistry and Interfacial Engineering Department, CSIR-Institute of Minerals and Materials Technology, Bhubaneswar, India; Academy of Scientific and Innovative Research (AcSIR) Ghaziabad, Ghaziabad, India

Sharmistha Anwar Academy of Scientific and Innovative Research (AcSIR) Ghaziabad, Ghaziabad, India;

Pyrometallurgy and Materials Engineering Department, CSIR-Institute of Minerals and Materials Technology, Bhubaneswar, India

N. Anwesha Department of Chemistry, School of Applied Sciences, Centurion University of Technology and Management, Bhubaneswar, Odisha, India

Pankaj Arora Smart Materials & Sensing Devices (SMSD) Lab, Department of Electrical and Electronics Engineering, Birla Institute of Technology and Science Pilani, Pilani, Rajasthan, India

Adarsh Kumar Arya Department of Chemical Engineering, Harcourt Butler Technical University, Kanpur, Uttar Pradesh, India

Ramesh Banoth Energy Materials and Photovoltaics Laboratory, Department of Physics, National Institute of Technology, Warangal, Telangana, India

Aruna Kumar Barick Department of Chemistry, Veer Surendra Sai University of Technology, Siddhi Vihar, Burla, Sambalpur, Odisha, India

Ajaya K. Behera Department of Chemistry, Utkal University, Bhubaneswar, India

Shreelata Behera Department of Chemistry, Veer Surendra Sai University of Technology, Siddhi Vihar, Burla, Sambalpur, Odisha, India

Jayesh Cherusseri Department of Chemistry (BK21 FOUR), Research Institute of Advanced Chemistry, Gyeongsang National University, Jinju, Republic of Korea

Nigamananda Das Department of Chemistry, Utkal University, Bhubaneswar, India

K. Sumiran Dash Materials Chemistry and Interfacial Engineering Department, CSIR-Institute of Minerals and Materials Technology, Bhubaneswar, India;
Academy of Scientific and Innovative Research (AcSIR) Ghaziabad, Ghaziabad, India

Ana Paula de Oliveira Lopes Antunes Laboratório de Eletroespectro Analítica (LEEA), Escola de Química e Alimentos, Universidade Federal do Rio Grande, Rio Grande, RS, Brazil

Daiane Dias Laboratório de Eletroespectro Analítica (LEEA), Escola de Química e Alimentos, Universidade Federal do Rio Grande, Rio Grande, RS, Brazil

Isaac dos Santos Nunes Laboratório de Eletroespectro Analítica (LEEA), Escola de Química e Alimentos, Universidade Federal do Rio Grande, Rio Grande, RS, Brazil;
Federal Institute of Education, Science and Technology of São Paulo, Suzano Campus, Suzano, SP, Brazil

Kathellen Pintado Esteves Laboratório de Eletroespectro Analítica (LEEA), Escola de Química e Alimentos, Universidade Federal do Rio Grande, Rio Grande, RS, Brazil

Lenon Medina Ferreira Laboratório de Eletroespectro Analítica (LEEA), Escola de Química e Alimentos, Universidade Federal do Rio Grande, Rio Grande, RS, Brazil

Ananda Fagundes Guarda Laboratório de Eletroespectro Analítica (LEEA), Escola de Química e Alimentos, Universidade Federal do Rio Grande, Rio Grande, RS, Brazil

Bigyan Ranjan Jali Department of Chemistry, Veer Surendra Sai University of Technology, Siddhi Vihar, Burla, Sambalpur, Odisha, India

Jyoti Jangra Academy of Scientific and Innovative Research (AcSIR), Ghaziabad, India;
CSIR-Central Scientific Instruments Organisation, Chandigarh, India

Akshay Kapoor Department of Chemical Engineering, Harcourt Butler Technical University, Kanpur, Uttar Pradesh, India

Ashish Kapoor Department of Chemical Engineering, Harcourt Butler Technical University, Kanpur, Uttar Pradesh, India

Asheesh Kumar Department of Mechanical Engineering, Mahatma Gandhi Institute of Technology, Hyderabad, Telangana, India

Maruboina Hemanth Kumar Centre for Nano and Material Sciences, Jain (Deemed-to-be University), Jain Global Campus, Kanakapura, Bangalore, Karnataka, India

Khusboo Kumari Department of Chemistry and Chemical Biology, IIT (ISM) Dhanbad, Dhanbad, India

Kusum Kumari Energy Materials and Photovoltaics Laboratory, Department of Physics, National Institute of Technology, Warangal, Telangana, India

Ram Naresh Mahaling Laboratory of Polymeric and Materials Chemistry, School of Chemistry, Sambalpur University, Jyoti Vihar, Burla, Odisha, India

Chandan Kumar Maity Department of Chemistry and Chemical Biology, IIT (ISM) Dhanbad, Dhanbad, India

Annada Mishra Department of Chemistry, Veer Surendra Sai University of Technology, Siddhi Vihar, Burla, Sambalpur, Odisha, India

Priyanka P. Mishra Department of Chemistry, Utkal University, Bhubaneswar, India

Pooja Mohapatra Department of Chemistry, Veer Surendra Sai University of Technology, Siddhi Vihar, Burla, Sambalpur, Odisha, India

Priyaranjan Mohapatra Department of Chemistry, Veer Surendra Sai University of Technology, Siddhi Vihar, Burla, Sambalpur, Odisha, India

Srikanta Moharana Department of Chemistry, School of Applied Sciences, Centurion University of Technology and Management, Bhubaneswar, Odisha, India

Drisya Mottammal Department of Physics, Government College for Women (Affiliated to University of Kerala), Thiruvananthapuram, Kerala, India

Bibekananda Nayak Materials Chemistry and Interfacial Engineering Department, CSIR-Institute of Minerals and Materials Technology, Bhubaneswar, India;
Academy of Scientific and Innovative Research (AcSIR) Ghaziabad, Ghaziabad, India

Ganesh Chandra Nayak Department of Chemistry and Chemical Biology, Indian Institute of Technology (ISM) Dhanbad, Dhanbad, Jharkhand, India

Eliezer Quadro Oreste Laboratório de Eletroespectro Analítica (LEEA), Escola de Química e Alimentos, Universidade Federal do Rio Grande, Rio Grande, RS, Brazil

D. Parajuli Research Center for Applied Science and Technology, Tribhuvan University, Kirtipur, Kathmandu, Nepal

Shubham Kumar Patial Academy of Scientific and Innovative Research (AcSIR), Ghaziabad, India;
Materials Science and Sensor Applications, CSIR-Central Scientific Instruments Organisation, Chandigarh, India

Siddappa A. Patil Centre for Nano and Material Sciences, Jain (Deemed-to-be University), Jain Global Campus, Kanakapura, Bangalore, Karnataka, India

Deepthi N. Rajendran Department of Physics, Government College for Women (Affiliated to University of Kerala), Thiruvananthapuram, Kerala, India

Saleja Sahoo Department of Chemistry, Veer Surendra Sai University of Technology, Siddhi Vihar, Burla, Sambalpur, Odisha, India

Satyanarayan Sahoo Department of Chemistry, Berhampur University, Brahmapur, Odisha, India

Roja Sahu School of Pharmacy and Life Sciences, Centurion University of Technology and Management, Balangir, Odisha, India

Anuj Sharma Materials Science and Sensor Applications, CSIR-Central Scientific Instruments Organisation, Chandigarh, India

Shavita Academy of Scientific and Innovative Research (AcSIR), Ghaziabad, India;
CSIR-Central Scientific Instruments Organisation, Chandigarh, India

Lipsa Shubhadarshinee Department of Chemistry, Veer Surendra Sai University of Technology, Siddhi Vihar, Burla, Sambalpur, Odisha, India

Sambhavi Shukla School of Computer Science, UPES Bidholi Campus, Dehradun, India

Salim Hassan Siddiki Department of Chemistry and Chemical Biology, IIT (ISM) Dhanbad, Dhanbad, India

Armeen Singh Department of Electronics and Communication Engineering, Punjab Engineering College, Chandigarh, Punjab, India

Suman Singh Academy of Scientific and Innovative Research (AcSIR), Ghaziabad, India;
Materials Science and Sensor Applications, CSIR-Central Scientific Instruments Organisation, Chandigarh, India

Susmi Anna Thomas Department of Physics, Government College for Women (Affiliated to University of Kerala), Thiruvananthapuram, Kerala, India

T. S. Xavier Center for Advanced Materials Research, Department of Physics, Government College for Women, Thiruvananthapuram, Kerala, India

Rohit Yadav Energy Materials and Photovoltaics Laboratory, Department of Physics, National Institute of Technology, Warangal, Telangana, India

Present Status and Prospects of MXene Research

Sambhavi Shukla and Pankaj Arora

Abstract The MXene family has turned the tables in the materials field since its first reported work in 2011. This is primarily because of the unique underlying physical and chemical properties of these 2D nanomaterial. High metallic conductivity and greater robustness provided by MXene sheets have already given them a leading edge over transition metal dichalcogenides and graphite. The MXene family has now entered into a vast platform of sensor applications, wherein they are acting as a quintessential element. Incorporating MXene flakes above the metal substrates provides stronger corrosion protection, a large surface area, and increased hydrophilicity. Due to its stability in aqueous solutions, it is an active co-catalyst that captures the photo-generated charges utilized for sustainable hydrogen production. The most commonly used MXene, i.e., $Ti_3C_2T_x$, is reported to exhibit intense plasmonic behavior in the near to mid-infrared region. Strongly tunable optical properties of $Ti_3C_2T_x$, Nb_2C, and Ta_2C have also emerged as promising SERS substrates. The excellent mechanical strength and flexibility exhibited by crumpled MXene hybrid nano-coatings strive to act as electrodes, which were acquired for designing bendable high-performance supercapacitors. MXene films provide non-toxicity and hydrophilicity towards miniscule organisms, making them biologically compatible. By tailoring the properties using different MXene geometrical configurations (films, nanostructures, or 3D network layered structure), a plethora of applications in different domains are coming up. These applications also depend on the type of terminated groups utilized; for example, the –OH terminated groups have the strongest binding energy strength compared to their counterparts. So far, with over 30 MXene stoichiometric varieties reported, the future holds promising prospects of incorporating the MXene family into myriad real-world applications. The challenge lies in the experimental validation/realization of more and more MXene composites and the possibility of tweaking their intrinsic properties for large-scale commercial production. Therefore, the chapter attempts to gather the current state-of-the-art

---

S. Shukla
School of Computer Science, UPES Bidholi Campus, Dehradun, India

P. Arora (✉)
Smart Materials & Sensing Devices (SMSD) Lab, Department of Electrical and Electronics Engineering, Birla Institute of Technology and Science Pilani, Pilani, Rajasthan, India
e-mail: pankaj.arora@pilani.bits-pilani.ac.in

© The Author(s), under exclusive license to Springer Nature Singapore Pte Ltd. 2025
S. Moharana and R. N. Mahaling (eds.), *Surface Engineering of MXenes*, Engineering Materials, https://doi.org/10.1007/978-981-96-3640-2_1

research carried out over the past few years to unravel the exquisite features of the MXene family and its composites. Consequently, the chapter will attempt to thoroughly analyze the present situation and future opportunities facing the MXene family in creating effective and long-lasting composites.

Keywords MXene · 2D nanomaterial · MXene fabrication · MXene characterization · MXene applications

1 Introduction

Among 2D nanomaterials, Graphene (Gr) already grasped the market firmly owing to its distinct properties and emerging prospects in real-time. At the time, Molybdenum disulfide (MoS_2) was also among the well-known transition metal dichalcogenides with a high absorption efficiency reaching almost 5%. By 2011, the 2D transition metal carbonitrides, carbides, and nitrides (MXene) family had emerged as an unconventional contender in the research and development domain. Most probably because MXene was, in general, providing a "complete package" for most of the properties combined in one. The MXene hydrophilicity showcased the advantage of being employed for sensing purposes in the biological environment since they have a comparable refractive index as water and the immobilization of the biomolecules over the sensor's surface [1]. In contrast to most of the transition metal dichalcogenides, which are predominantly semiconducting, the MXene family exhibited metallic conductivity and maintained a higher optical absorption and mechanical robustness in the entire wavelength range. A prime constraint faced by Metal–Organic Frameworks (MOFs) that limits their extensive utilization is low conductivity and poor stability. With the advent of MXene composites, excellent structural stability and strong metallic conductivity could be witnessed [2]. Last but not least, the ability of MXene to modify its intrinsic properties by varying different surface functional groups has turned the tables in the materials field. These terminations alter the surface energy, enhancing mechanical, optical, electrical, and electronic properties. Thus, in the quest to decode the exquisite properties of the MXene family and its composites, the chapter aims to compile the present state-of-the-art research conducted over recent years. By tailoring the properties using different MXene geometrical configurations (films, nanostructures, or 3D network layered structure), a plethora of applications in different domains are coming up. The synthesis of 2D MXenes involves two main approaches: bottom-up and top-down mechanisms. Through the top-down mechanism, crystal quantities in bulk are exfoliated into MXene single-layered sheets. On the other hand, the bottom-up arrangement focuses on the progress of MXenes from atoms/molecules [3].

Employing bottom-up methodologies for MXene synthesis produces large-scale processability, which gives this methodology an edge over many other 2D materials made in modest quantities. Additionally, when manufacturing $Ti_3C_2T_x$ at the 1 or 50 g batch scale, no discernible changes in the physical properties were noticed

[4]. This chapter's first section entails the fabrication techniques and methodologies followed over the years to synthesize pure MXene and its composites. This is followed by discussing the several properties (mechanical, structural, and optical) in Sect. 2 that vary with the type of fabrication technique implemented and can also be tuned following the surface terminations (X) in the $Ti_3C_2T_x$ family. The proceeding section discusses diverse surface terminations and MAX/non-MAX phases, resulting in different applications. Finally, in the last section, the existing challenges of the fabrication methods are discussed along with the future scope of this rising 2D nanomaterial in terms of numerous applications that it can potentially support.

2 Fundamentals of MXene with Historical Background

Figure 1 illustrates the various aspects of MXene and its composites that will be covered in this chapter subsequently. MXene represents the Transition Metal Dichalcogenide (TMD) family, which closely resembles the structural properties of Gr. The MXene is typically synthesized from the precursors to the MXene family, i.e., the MAX phase. MAX phases are transition metals interleaved with carbon/nitrogen atoms and elements from periodic table 13–16 to stabilize the structure. The switch from the MAX state to the MXene phase has increased the latter's power, as now it possesses mixed metal and ceramic characteristics.

Since its first development by Drexel University in 2011, different forms of single and multi-layered MXene have witnessed phenomenal evolution through many rigorous research works reported over the years. Figure 2 shows a brief timeline of MXene development in the recent years. From 2019 onwards, the MXene family has been exhaustively explored for diverse applications, including electromagnetic interference (EMI) shielding [5]. Subsequently, extensive research on surface functionalization of MXene nanosheets grabbed a lot of attention, thus expanding the output sensing performance. Different surface terminations can lead to fine-tuning of electrical and chemical reactions. After that, near varied terminal functionalization, stable nanocomposites of MXenes forming robust heterostructures started forming the basis for solving many real-time applications. On the sidelines, researchers studied MXene's distinct electrical structure and suitability for quantum dot integration to develop qubits. Simultaneously, MXene-based quantum computing [6] and MXene-based smart materials [7], among others, have started gaining traction for future potential applications. Figure 2b shows the pumping wavelengths of MXene and other TMDCs. Herein, MXene is representative of covering a wide range of the wavelength regime favorable for photonic applications [6].

With over 30 MXene stoichiometric varieties reported, the future holds promising prospects of incorporating the MXene family into myriad real-world applications. The challenge lies in the experimental validation/realization of more and more MXene composites and the possibility of tweaking their intrinsic properties for large-scale commercial production. Therefore, the chapter will aim for a comprehensive

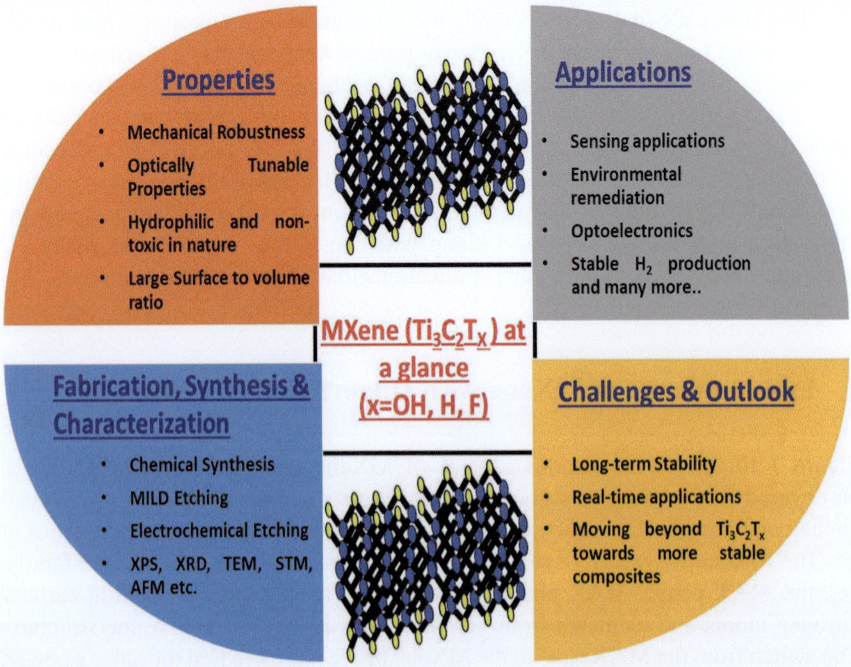

Fig. 1 An illustration showing various aspects of MXene sheets/nanostructures and composites

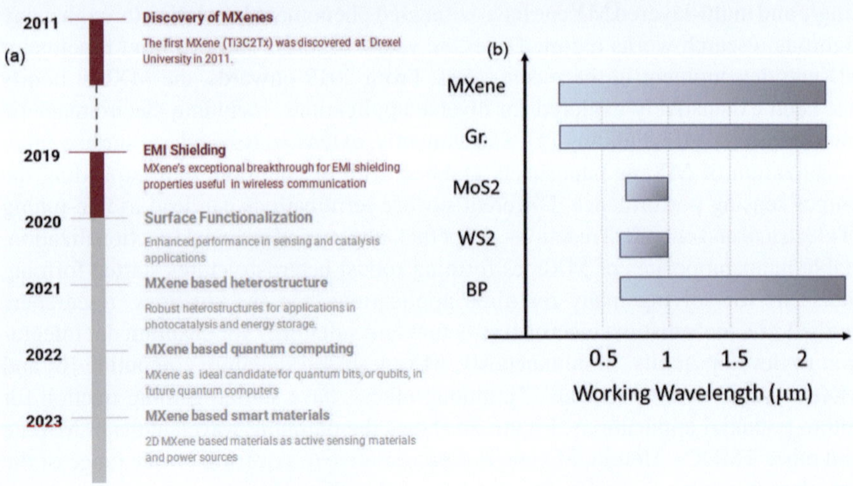

Fig. 2 **a** Brief timeline of MXene development since its first synthesis in 2011. **b** A working wavelength comparison between different 2D nanomaterials from Ref. [6]

review, highlighting the current status and prospects incurred for the MXene family in developing efficient and durable composites.

3 MXene Fabrication Techniques and Characterization

(a) Fabrication Techniques

Over the past ten years since the advent of MXene, there has been significant growth in the chemical and structural nature of synthesized MXenes. By immersing TiO_2 and AlC_2 fine powders in 50% hydrofluoric acid (HF) at room temperature for two hours, Naguib et al. produced the first MXene [7]. Typically, they are 2D metal carbides, nitrides, along with carbo-nitrides having the form $M_{n+1}X_nT_x$—where M is group 3–6 early transition metals and n + 1 layers (where n = 1–4), interspersed carbon plus nitrogen atoms ('X') of n layers; T is surface endings on the exposed M layers, for example: –F, –O, –OH; The atoms of the X element occupy M's interstitial positions in the hexagonal crystal sub lattice. This fascinating material is fabricated in the following three steps:

1. MXene precursors, i.e., MAX ($Mn_{+1}AXn$), from which the crystal structure is derived, are synthesized layered-wise. Herein, 'A' pertains to a group of 11–16 atoms such as Aluminum (Al), Silicon (Si), Gallium (Ga), or non-MAX-phase layered materials with more than one 'A' atomic layer ($M_{n+1}A_2X_n$) or with carbide layers of A elements ($M_nA_3X_{n+2}$). Of the ten distinct 'A' components comprising groups 13 and 14, only Al has been effectively etched from MAX phases to make MXenes. Over 70 phases of MAX have been documented thus far. Different phases of MAX are a result of varying etching conditions. Application-specific etching conditions are essential for achieving high yields and converting MAX phases into MXenes [8].
2. The atomic layers of A are removed through etching, which exfoliates the precursors and forms weakly bonded MXene multilayers. This etching process cleaves the weaker M–A bonds (compared to M–X bonds), resulting in under-coordinated M metallic surfaces, rapidly re-saturating by reacting with T_x species from the etchant.
3. Single to few-layer MXene sheets are yielded by delamination, which isolates the exfoliated multilayer MXene sheets.

Apart from the current groups, –O, –OH, and –F, the compositional control of surface terminated groups, significantly affecting the properties of MXenes, encompasses various halogens (–Cl, –Br, –I), chalcogens (–S, –Se, –Te), and imido (–NH) groups at the outer surfaces of the MXene layer. Although these developments have given new directions to synthesize MXenes, they depend heavily on wet acid etching employing hazardous HF solutions.

The synthesis of multi-layered MXene flakes is usually implemented using wet chemical etching (HF acid as etchant), which was present until 2013 when organic molecules were integrated, and the sheets were separated from one another to isolate single-layer MXene flakes [9]. The etching process and delamination circumstances influence the quality, total crystallinity, flaws, and MXene flakes' surface functionalization and delamination effectiveness. Larger MXene flakes with lower fault concentrations are often produced under milder etching and delamination conditions [10]. Additionally, crucial parameters are the etching temperature and duration. For instance, high etching temperature and longer etching duration can result in the complete transformation to MXene when synthesizing $Ti_3C_2T_x$ in Hydrochloric (HCl) acid and Lithium Fluoride (LiF). Still, the above parameters can also lower the delamination yield and flakes' quality because of a higher concentration of flaws [11]. Alkali treatment is also employed for MXene synthesis to avoid hazardous HF etchants due to its effective linkage with the Al element. However, the extracted elements subsequently oxidize, so precise monitoring is essential [12]. Another technique to produce MXene via a fluorine-free approach is electrochemical etching. In this technique, the isolation of the MXenes by selectively etching the stacked MAX phase and conducting electrolyte design research is required. For example, the anodic etching of Ti_3AlC_2, Ti_2AlC, as well as Ti_3SiC_2 MAX phases using aqueous solutions of Sodium Chloride (NaCl), HCl, and HF as electrolytes have facilitated the doing away of both Titanium (Ti) and (Al/Si) atoms, leading to the creation of amorphous carbon [13]. Over the years, additional intercalants such as isopropylamine and tetra-butylammonium hydroxide were employed to increase the delamination yield [14]. Sonication is crucial in blending, where the intercalant is applied between two-dimensional sheets. Longer and stronger sonication durations will result in smaller, more flawed flakes and may produce concentrations that differ from un-sonicated samples. The resulting flakes of the colloidal solution can be separated by centrifugation under specific environmental conditions. The Minimally Intensive Layer Delamination (MILD) method introduced in 2016 led to isolating larger MXene flakes, now among the most famous techniques [15].

In this context, Table 1 lists some essential examples of various fabrication techniques. HF etching is the most commonly used fabrication technique [7]. Additionally, etching via other methods like fluoride salts [11], molten salts [16], and electrochemical etching [17] are also reported. These are tabulated from top-down and/or bottom-up approaches.

The hydrothermal technique is another thriving approach for producing multi-patterned, highly-refined 2D MXene composites for utilisations on a large scale, preventing contact with highly toxic HF vapor. In a comparative analysis of the synthesis of Ti_3C_2 acquired with a conventional etchant (HF), the Ti_3C_2 produced via the hydrothermal method led to getting rid of the Al layer, increased lattice parameter as well as the highest multilayered width among two-dimensional MXenes [19]. Other methods, like combining a strong acid and a fluoride salt, are among the standard methods to synthesize MXenes [11]. Likewise, myriad combinations of acids (Sulphuric acid (H_2SO_4)) along with salts containing fluorine ions (Sodium Fluoride (NaF), ferric fluoride (FeF_3), Potassium Fluoride (KF), Cesium Fluoride (CsF), Calcium Fluoride (CaF_2)) have also been used to synthesize MXenes [11]. The first nitride-based MXene ($Ti_4N_3T_x$) was created by Urbankowski et al. by

Table 1 MXene fabrication methods, as reported over the years, based on top-down and bottom-up approaches

Method	Year [Ref.]	Preparation method	Functional groups	Structure	Advantages	Disadvantages	Applications
Top-down methods	[7]	HF etching	–F, –O, –OH	Accordion-like	Simple, many applications, room temperature	Many defects, dangerous, poor stability	Electronics (MXetronics), sensors (wearable and biosensors), heavy metal adsorption, gas–liquid separation membranes, tribology (adhesion, lubrication), drug delivery
	[11]	HCl–Fluoride salt etching (clay method)	–F, –O, –OH	Clay-like	High yield, large size, small defects, intercalant not required	Long reaction time, non-room temperature	Wearable and biomedical sensors, energy storage, heavy metal adsorption
		Bi-fluoride etching method: NH_4HF_2	–F	Accordion-like	Quick reaction, thermally stable, large layer spacing	Only a few MXenes can be created, non-room temperature	
	[16]	Molten salt etching	–O	Accordion-like	Wide application	Complex post-processing, poor hydrophilicity, high temperatures	Energy storage (batteries and supercapacitors), catalytic electrodes
	[17]	Alkali etching	–OH, –O	Compact layered	High yield	High temperature, highly alkaline—dangerous	Biomedical applications—antimicrobial, antibacterial, cancer theranostics, biosensors; energy storage applications

(continued)

Table 1 (continued)

Method	Year [Ref.]	Preparation method	Functional groups	Structure	Advantages	Disadvantages	Applications
		Electrochemical etching	–Cl, –OH	Compact layered	Low temperature	Low yield	Biomedical and energy storage, catalytic conversion applications
	[18]	Halogen etching	–OH, –X, –O	Layered	Environment stability, good structural integrity	Low yield, suitable for few MXenes, slow reaction	Catalytic electrodes (photo electrocatalysis), batteries, supercapacitors
Bottom-up methods	[19]	Chemical vapor deposition (CVD)	No functional groups	Film	Few defects, a large area	Low yield, high cost	Fundamental studies of MXene properties—2D transport, thickness-dependent properties (catalytic, optical, magnetic, electronic)

heating a mix of molten fluoride salts and $Ti_4N_3T_x$ powders to 550 °C in an argon environment [16]. Since the etching procedure takes several hours to several days to synthesize MXene from the MAX phase, new works discuss alternative fabrication process flow, such as exposing the etchant solution and MAX phase to acoustic waves. The acoustic force enables the delamination of the produced MXene, thus making a yield comparable to the conventional synthesis methods [20]. In another report, surface acoustic waves were utilized to create delaminated MXenes in the presence of LiF in water [20]. After synthesis, the substance might need to be stored before being used. Storage can be challenging, as the MXene flakes are prone to oxidation. Some well-known techniques for storing the resulting MXene solutions are refrigeration, filtering the colloidal solution, vacuum-sealing the resultant film, and storing it in organic solvents [21].

The slot die coating technique is widely used in industrial scale fabrication of MXene sheets, the fabrication procedure for which is depicted in Fig. 3a [22]. For the same, Fig. 3b illustrates the improved conductivity by the transferred $Ti_3C_2T_x$ sheet's preservation of the smooth surface network and near interlayer distance. On the other hand, after drying the condensed droplet on the surface, multilayer $Ti_3C_2T_x$ aggregates were produced [23]. Similarly, a schematic representation of the iodine-assisted synthesis process is shown in Fig. 3e, f. Herein, 2D iodine-MXene sheets could be separated and dispersed in water for additional research with sufficient efficiency using hand shaking in HCl solution [18]. Similarly, the synthesis of MXene/COF composite membranes, routinely used for dye-aqueous solutions, is demonstrated [24].

The development of stable MXene sheets is a prerequisite for real-time applications. This called for MXene composites to evolve rapidly. Thus, Fig. 4 presents some of the high-performance MXene nanocomposite fabrication process flow. A fascinating depiction of the fabrication process flow of MXene-based cellulose composite is shown in Fig. 4a. The synthesis shown in the schematic is a potential window for future applications in the field of supercapacitors [25]. Meanwhile, Fig. 4b presents the synthesis procedure for $Ti_3C_2T_x/Fe_3O_4$ composite, which results in better strength and broadband properties, as reported by [26]. Firstly, a solution of sodium acetate, ethylene glycol, and diethylene glycol is prepared to synthesize $Ti_3C_2T_x/Fe_3O_4$. Then, to create a homogenous mixture, MXene particles are added and mixed to get the desired output. By mixing $FeCl_3 \cdot 6\ H_2O$ to the solution and using the solvothermal technique, the consequent product is obtained. Figure 4c depicts the preparation of MnO_2/MXenes nano-composite when a few mg of MXenes were mixed with mM concentration of $MnSO_4$ solution and subjected to sonication. Post stirring, the mixture was gradually added to $KMnO_4$ solution, and it was stirred again for an additional 15 min at 60 °C. The resultant product had a 1:1 MXene to MnO_2 ratio after it had been cleaned, filtered, and vacuum-dried for ten hours at 80 °C [27]. Likewise, the $Ti_3C_2T_x$/Ag NPs combination is presented schematically in Fig. 4d. The solutions of $Ti_3C_2T_x$ NSs and Ag NPs were mixed, ultrasonicated, and filtered by vacuum-assisted to form a hybrid film. This is followed by digital photos of the Ag NPs, $Ti_3C_2T_x$ NSs, and the mixture dispersions [28].

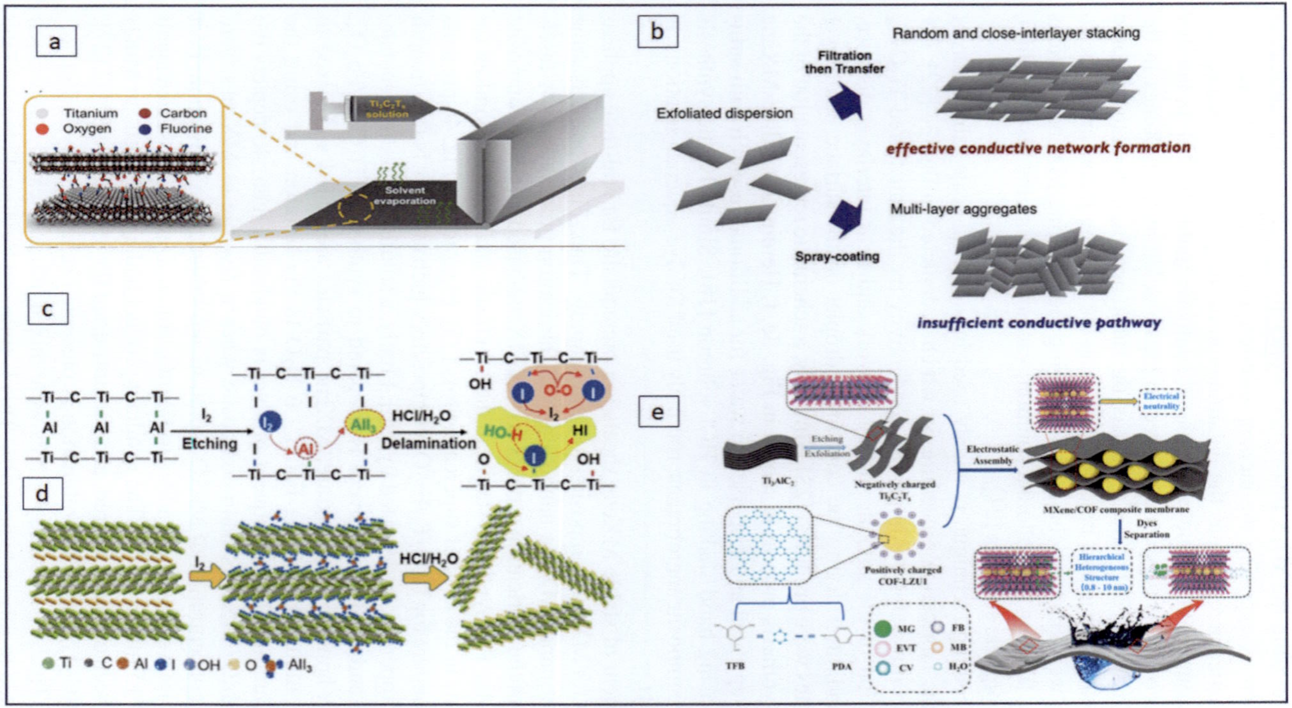

Fig. 3 Previously reported works on MXene fabrication techniques implemented for some of the real-time applications. **a** Illustration of coating the MXene surface procedure via slot-die technique; Reproduced with permission from Springer Nature, Copyright 2024 [22]. **b** Schematic of exhibiting the shape of the spray-coated and transferred $Ti_3C_2T_x$ MXene films; Reproduced from Ref. [23] [https://doi.org/10.35848/1882-0786/acbbb8]; under the terms of the CC BY 4.0 license [https://creativecommons.org/licenses/by/4.0/]. **c**, **d** schematic representation of the iodine-assisted synthesis process of MXene; Reproduced from Ref. [18] [https://doi.org/https://doi.org/10.1002/anie.202015627]; under the terms of the CC BY 4.0 license [https://creativecommons.org/licenses/by-nc-nd/4.0/], **e** schematic of MXene-COF based membrane composite; Reproduced with permission from Elsevier, Copyright 2022 [24]

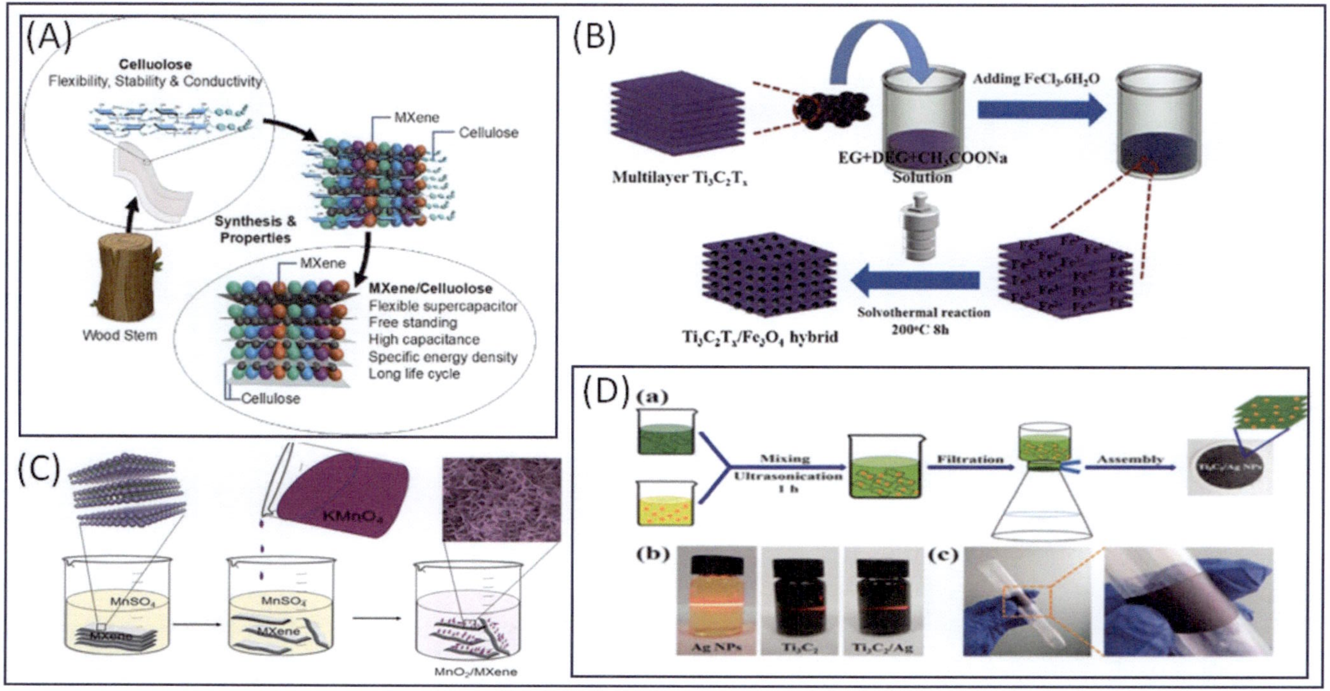

Fig. 4 Previously reported works on MXene-based composite fabrication techniques implemented for some of the real-time applications. **A** Properties and synthesis of MXene/cellulose composites; Reproduced with permission from Elsevier, Copyright 2024 [25]. **B** Fabrication steps shown for the $Ti_3C_2T_x/Fe_3O_4$ composite; Reproduced with permission from Elsevier, Copyright 2023 [26]. **C** MnO_2/MXenes nano-composite; Reproduced with permission from American Chemical Society, Copyright 2016 [27]. **D** Fabrication procedure presented for MXene along with Ag nanoparticles; Reproduced with permission from American Chemical Society, Copyright 2016 [28]

(b) **Characterization**

Establishing a characterization platform for MXene includes quantifying coordination interactions, fundamental composition, and termination structure and developing precise quantitative testing methods. This is crucial for advancing MXene research. Accurate characterization of MXene terminal surface groups is essential for studying regulation mechanisms and their impact on properties and performance. However, many terminal functional groups are unstable, posing challenges for characterization. Currently, techniques like X-ray Photoelectron Spectroscopy (XPS), Energy Dispersive X-ray Spectroscopy (EDX), X-ray Absorption Spectroscopy (XAS), and Electron Energy Loss Spectroscopy (EELS) are typically used to analyze MXene surface terminations both qualitatively and quantitatively. Through the statistical study of MXene crystal structure, X-ray diffraction (XRD), a fundamental method for characterizing crystal structures, indirectly shows the influence of MXene surface termination on crystal structure and interlayer environment. Analysis of the appearance of new peaks, variations in peak positions, and changes in half-peak width provide insights into atomic arrangement and evolution of induced variations in structural complexity [29]. MXene was investigated for cancer photothermal therapy. The resulting MXene sheets underwent X-ray Diffraction (XRD) characterization to elucidate their crystal structure. XRD analysis was conducted using settings of 45 kV and 20 mA, with data acquired within the range of $2\theta = 5°–100°$. Transmission electron microscopy (TEM) was reportedly used to verify the MXene sheets' morphology [30].

X-ray Photoelectron Spectroscopy (XPS) offers precise measurements of inner electron binding energy and chemical shifts, providing insights into molecular structure, valence states, element composition, content, chemical states, and molecular configuration. This robust technique significantly supports the exploration of diverse surface terminations in MXene surface termination research. Regarding Raman spectroscopy, the MXene spectrum typically comprises four distinctive regions: the formant, thin section, T_x area, and carbon area. Characterization of MXene involves analyzing changes in peak displacement, intensity, and width within these regions. Various surface terminations, intercalated and adsorbed species, can influence lattice vibrations, resulting in peak shifts and broadening. The etched $Ti_3C_2T_x$ spectra of LiF–HCl and HF–HCl show identical Alg (C) peak locations and strong –OH surface groups. Nonetheless, the HF-etched multilayer $Ti_3C_2T_x$ spectrum predominates –O, which shifts the A1g peak to 711 cm^{-1}. This shift primarily occurs due to the reduction in pH resulting from HCl addition, leading to surface protonation [29].

Researchers documented the process flow of Nitride-based MXene, yielding multilayered Ti_2NT_x (T representing terminals on the surface), which subsequently produced few-layered Ti_2NT_x via centrifugation. The MXene formation was confirmed, and comprehensive characterization was conducted using Raman spectroscopy, XRD, XPS, FESEM-EDS, TEM, STM, and AFM techniques. The synthesized Ti_2NT_x's Surface-enhanced Raman scattering (SERS) activity was explored by creating SERS substrates based on paper, silicon, and glass. A remarkable Raman enhancement factor of 10^{12} was achieved using rhodamine 6G as the model compound with a 532 nm incident wavelength. Moreover, explosives at trace level were detected using a SERS substrate (paper-based) with Ti_2N (MXene) as the active material. Investigation and analyses of Raman spectra and SERS activity for the MXenes synthesis were performed utilizing a WITec alpha 300R confocal Raman microscope equipped with an excitation laser of 532 nm. A 2 μL analyte was applied onto the SERS substrates, and 10 Raman spectra were collected over 5 s using a 50× air objective with a numerical aperture of 0.55 [31].

Li-intercalated 2D Ti_2CT_z materials were synthesized, and their resulting transport and optical properties underwent characterization. Spectroscopic ellipsometry measurements revealed optical properties consistent with metallic behavior. The optically determined resistivity parameter closely aligns with that determined via electrical measurements. In Ti_2CT_z, with all occupied terminal sites, the total count of O, OH, and F terminations should be 2. However, the fact that z is less than 2 suggests that most M atoms are not terminated. The precise cause of this phenomenon remains unclear. It is conjectured that some terminations may not get absorbed in the vacuum of the XPS. Nevertheless, from a thermodynamic perspective, it is pretty unlikely that these terminations will disappear when the films are exposed to air. Therefore, during air characterization, all M atoms are anticipated to terminate [32].

Synchrotron-based characterizations have demonstrated significant usage and precision in investigating the electronic architecture of the associated reaction mechanisms of MXenes. Synchrotron X-ray diffraction (SXRD) and in situ, SXRD are potent devices for detailing MXene structures with exceptional resolution, including interlayer spacing, crystal planes, and phase changes. Synchrotron X-ray photoelectron spectroscopy (SXPS) offers rapid measurements and superior detection accuracy due to intense synchrotron irradiation. With inherent surface sensitivity, SXPS is well-suited for identifying elements' chemical states under extreme vacuum conditions. Moreover, adjusting photon energies allows the testing depth to be finely tuned within a broad range of 2–50 nm, enabling non-destructive depth-resolved analysis of MXenes. In addition to other synchrotron radiation techniques, static and dynamic X-ray absorption fine structure (XAFS) spectroscopy offers notable advantages in discerning atomic interactions within solid, liquid, and gaseous states because of its sensitivity to short-range order and element-specific nature. Commonly applied in nanomaterials research, including crystalline and amorphous materials, XAFS spectroscopy facilitates the determination of local and electronic structures—a task challenging for conventional methods like TEM, XRD, and XPS. Therefore, the study suggests that with the advancement of synchrotron light sources and the evolving needs of MXenes' structural and mechanistic studies in various applications, synchrotron radiation-based techniques such as SXRD, SXPS, synchrotron radiation-based FTIR, and APXPS should be developed and effectively utilized to advance the understanding and utilization of MXenes [33]. However, due to accuracy limitations, establishing accurate quantitative assessments of MXene is still challenging. Moreover, various functionalization poses challenges for applications dealing with detection or resolution. Therefore, it is essential to develop new characterization approaches to advance the investigation of MXene surface terminations.

4 Properties

The fabrication technique imparts unique properties to MXene and its composites. Some of the properties are as follows:

(a) **Mechanical Properties of MXene**

MXenes must possess mechanical strength in order to be utilized in applications such as functional wearable fabrics. The characteristics of MXene fibers in both their pure and composite forms have been covered. As the number of MXene layers increases, Young's modulus falls; typically, nitride-based MXenes have greater values than carbide MXenes [34]. Although they have lower elastic properties than Gr, MXenes show maximum bending characteristics, making them suitable for composite reinforcement materials. Additionally, M components also impact MXene's mechanical properties. The good mechanical stability of M_2X is indicated by the estimated breaking strength, which is expected to fall between 92 and 161 N/m [35].

Generally speaking, the main factor maintaining MXene's elastic stiffness is its ability to couple, which is inversely related to the layer distance. The separation of two $Ti_3C_2OH_x$ layers (0.57 Å) is substantially less as compared to O as well as F terminations (~2.5 Å) because of the creation of H-bond between hydroxide groups, which results in a greater normal-to-the-plane elastic modulus [35]. Besides, enhancement in the mechanical parameters can further be instilled by polymers such as polyvinyl alcohol (PVA) and chitosan [36]. Ling et al. reported that 10 wt.% PVA in Ti_3C_2 paper roll could sustain ~15,000 fold its weight [36]. Moreover, by incrementing the displacement of Ti_3C_2 nanosheets, chitosan may also increase the tensile strength of Ti_3C_2 films (about 43.52 MPa with 14 weight percent chitosan) [37].

- For Pure MXene Fibers

The built-in characteristics of MXene fibers are affected by several variables, including the solidification bath, compactness, interfacial interactions, sheet diameter, and sheet orientation. Various approaches have been investigated to produce MXene with improved mechanical characteristics. For example, a higher spin rate can lead to more regular structures of MXene fibers. The subsequent solidification via ammonium ions leads to higher tensile strength, elongation, Young's modulus, and toughness. Since sheet disorder and pre-defect also affect the properties, defect control must be monitored simultaneously to maintain the abovementioned parameters. Ultra-dense MXene fibers fabricated using a rigorously monitored approach also aid in excellent tensile strength and ultra-high toughness.

- For MXene Composite Fibers

When it comes to composites of MXene, large-sized MXene sheets exhibited high tensile strength and Young's modulus compared to small-sized sheets. Compared to graphene, MXenes have superior interaction ability with polymeric matrices for composite effectiveness owing to the functionalities involved [38].

(b) **Electrical Properties of MXene**

MXene possesses impressive electrical conductivity with conductivity of over 11000 S cm^{-1} for monolayer. Due to variations in the etching process, the electrical conductivities of $Ti_3C_2T_x$ exhibited severe modification in areas like surface functionality groups, defect concentration, delamination yield, and lateral dimensions, among many more. Environmental moisture also impacts electrical conductivity, promoting sensor-based applications [34]. Surface variations can be further enhanced using thermal and alkaline treatments to improve electrical characteristics, thereby providing a two-order increase in conductivity's magnitude. This escalation is caused by the incorporation or elimination of intercalated ions/molecules and surface functionalities. Few experimental works have revealed strong electronic properties of MXenes that vary based on composition and surface terminations, with bare MXene being metallic and terminated MXenes exhibiting semiconductor or metallic behavior [8]. Nitride and carbonitride MXenes have far more excellent metallic characteristics than carbide MXenes because nitrogen has more electrons than other elements. The metallic property also varies upon the various 'X' atoms in $Ti_3C_2T_x$. For instance, as nitrogen is intertwined with more electrons than carbon, titanium carbide has less metallic qualities than titanium nitride [39]. It has been reported that pure MXenes are metallic and have a significant electron density close to the Fermi level. At the same time, certain experimental findings demonstrate that surface functionalization transforms MXenes into semiconductors [35]. Furthermore, the electrical characteristics of MXenes are influenced by the orientations of these terminating groups.

- For Pure MXene Fibers

Since the orientation of the crystal structure largely impacts the crystal lattice, which in turn manipulates the electrical properties, selecting large-size MXene usually enhances the same and reduces porosity. In the case of the chitosan solidification bath, large-size MXene nanosheets presented excellent electrical properties of 7748 S cm^{-1}, 2.2 times more than S–Ti_3C_2 fibers [38]. In a mixed coagulation bath of Ammonium chloride (NH_4Cl) and Ammonium hydroxide (NH_4OH), assembled MXene sheets achieved a high electrical conductivity of 7713 S cm^{-1}. Self-supporting MXene hydrogels from MXene colloidal nanosheets produced highly aligned fibers with high conductivity and increased orientation, increasing from 8260 to 12504 S cm^{-1} [34].

- For MXene Composite Fibers

Composite fiber having higher MXene content instills improved properties. For example, an ordered arrangement of MXene sheets between Graphene Oxide (GO) templates with an MXene content of up to 95 wt.% obtained an enhanced conductivity of 290 S cm^{-1} [38]. By incorporating minute carbon nanotube (CNT) concentrations (1 wt.%), long-range connections were established, again greatly enhancing conductivity to 1142.08 ± 40.04 S cm^{-1} [34].

(c) Optical Properties of MXene

Transparent, optically conducting electrode machines, photovoltaic, photocatalytic, and UV light spectrum absorption depend on it. $Ti_3C_2T_x$ films can potentially absorb UV-visible light with wavelengths ranging from 300 to 500 nm. 91% transmittance was measured at a layer thickness of 10 nm. Moreover, it may gather a broad and powerful band of absorption at 700–800 nm, depending on film thickness, indicating the creation of pale green films, which renders it extremely effective in photo-thermal therapy applications [40].

MXene and hybrid composite thin films are typically transparent. To be more precise, Ti_3C_2 MXene allows more than 97% of visible light to pass through per nanometer of thickness [39], but oxidation occurs in the presence of light or moisture, suggesting storage in a dark, oxygen-free environment. As a result, it is recommended that MXene colloids be stored in sans-oxygen, dark environments under refrigeration.

(d) Structural Properties of MXene (Both Thermal and Oxidation Stability)

Amongst the MXene family, "X atoms" inhabit an octahedral interstitial position, akin to the MAX phase predecessors, whereas "M" atoms are arranged in a densely packed arrangement. MXenes have a hexagonal close-packed crystal structure with a distinct arrangement of the 'M' atom, going from M_2X to M_3X_2 to M_4X_3. At the same time, the octahedral interstitial sites are occupied by X atoms [40]. Khazaei et al. used computer calculations to stimulate the crystal formations of $M_{n+1}X_n$ ($n = 1, 2$, or 3), in which two M layers encircle one X layer [35].

O-and/or OH-terminated MXene species were proposed to be steady following its exfoliation from the MAX phase among the F, OH, and O terminations, considering OH-groups will replace the –F surface terminations at the time of rinsing and/or storing in water [8]. One of the most popular techniques for making MXene materials more stable and less expensive overall is adding polymers, which broadens the range of possibilities for these materials [35].

Even though oxidation treatment can yield intriguing structures and characteristics, oxidation should generally be avoided, especially for MXene compounds with great conductivity and an appropriate 2D layered structure. This brings us to the conclusion that stability is a crucial parameter given the involvement of water/oxygen amidst MXenes synthesis. When MXene is transformed to titanium oxide on carbon sheets, a precipitate that ranges in color from black or green to bluish-white develops. Researchers have shown that at temperatures up to 500 K, anatase titanium oxide forms in TiO_2 nanosheets and is subsequently transformed into rutile titanium oxide at higher temperatures. However, MXene is stable up to 1300 °C in an argon atmosphere. In addition, the oxidation rate of MXene colloidal solution can be decreased through storage in a dark place, a refrigerator, or under a complete vacuum. Thus, in chemically etched MXenes stored in solutions, the –OH and –O groups are the main terminations as they are stable compared to the –F termination group. It must also be noted that these functionalities at the surface end play a vital role in tuning the properties of MXenes. However, relative to the MXene sheet's layered structure, MXene flakes are relatively stable in dry air [39]. Modeling is essential in comprehending MXene's structure; however, modeling the parameters for MXene is usually considered uniformly terminated, which does not suffice with practical implementation. More rigorous modeling is still required to investigate the adsorption of various surface groups within certain MXenes to realize the complex structure of the latter [8].

The environment and nature of MXenes impact their stability at high-temperature. As seen by their phase diagrams, higher-quality MXene flakes have better stability [15]. While the hexagonal crystal structure of bare Ti_3C_2 may be preserved even at 1200 °C under the same atmosphere, the $Ti_3C_2T_x$ (T = OH or F) is stable at 800 °C in argon (Ar) [41]. On the other hand, $Ti_3C_2T_x$ partially oxidizes to TiO_2 nanocrystals at 200 °C and completely changes over rutile TiO_2 in oxygen at a temperature of 1000 °C [42]. It has also been discovered that the oxidation products are regulated by heat-treatment methods and temperature [35].

Table 2 lists various mechanical and electrical characteristics that are particular to the geometrical configuration of MXene, along with the nanocomposites it forms. Conductivity is as high as 7700 S cm^{-1} [43], along with some impressive mechanical properties like excellent tensile strength of 500 MPa [44] and efficient Young's modulus and robustness in the structural design.

Figure 5 highlights a few of the properties of MXene nanosheets. Figure 5a discusses the thermal stability of MXene by showing SEM images of the material at two different temperatures, i.e., 200 °C and 1000 °C, respectively. At 200 °C, the quasi-structure was kept, while at 1000 °C, the crystal structure integrated to form a net shape [42]. For the same, Fig. 5c ascertains that the fracture strain is independent of the Ti_3C content. The mechanism for such improved mechanical characteristics is crack deflection, grain delamination, etc., as cited in [47]. Similarly, Fig. 5d shows the electrical characteristics of the memristors based on the Ag–V_2C MXene-Vanadium (w) configuration of intermediate layers [48].

5 Applications

Altering the surface functionality tunes the characteristics of MXenes, which also influences the real-time applications. Thus, aqueous HF and alkali-based etchings to achieve T_x and halogen-based etchings are best suited for compositional control. In contrast, Lewis acid molten salt etchings strive for newer T_x functionalities to improve the chemical selectivity. Another benefit of electrochemical etching of MAX electrodes is that MXene electrodes can be produced without needing extra transfer or electrode preparation operations. Producing large-area, ultra-thin MXene sheets using physical or CVD development of MXenes is a useful method for meeting the needs of applications requiring a greater degree of controlled thickness and defect density. Thus, evaluating the features with an application selective application is critical. Accordingly, the synthesis of MXene nanostructures can be tailored for enhanced performance and selectivity.

All techniques can be tailored for a particular use. Spray coating is useful for producing large-area nano-films for electromagnetic interference shielding [17], while vacuum-assisted filtration and rolling are better suited for creating battery

Table 2 Few of the previously reported mechanical and electrical properties of MXene

Year	Materials	MXene content (wt.%)	Preparation method	Tensile strength (MPa)	Young's modulus (GPa)	Toughness (MJ m^{-3})	Conductivity (S cm^{-1})	References
2020	$Ti_3C_2T_x$/ Nanoyarns	–	Electrospinning	40	–	0.27	1195	[45]
2020	$Ti_3C_2T_x$/rGO	60	Wet spinning	110	–	3.8	743.1	[35]
2020	$Ti_3C_2T_x$/PU	23.1	Wet spinning	14	1.15	–	22.6	[35]
2020	$Ti_3C_2T_x$	100	Wet spinning	63.9	29.6	0.09	7713	[43]
2021	$Ti_3C_2T_x$	100	Wet spinning	118	–	1.09	7200	[46]
2021	$Ti_3C_2T_x$ aerogel	100	Dynamic sol–gel spinning	1.1	–	–	100	[46]
2021	$Ti_3C_2T_x$ ribbon	100	Wet spinning	–	–	–	2458	[46]
2022	MXene @alginate	74.1	Coating	–	–	–	1083	[40]
2022	GO/$Ti_3C_2T_x$ @RC	70	Wet spinning	92.2	9	3.0	167.5	[40]
2022	$Ti_3C_2T_x$/CNT	91	Wet spinning	161	–	–	1715	[40]
2023	$Ti_3C_2T_x$	100	Wet spinning	205.5	–	–	3545	[34]
2023	$Ti_3C_2T_x$/ polyrotaxane	–	Wet spinning	500	–	60	1000	[34]
2023	$Ti_3C_2T_x$	100	Wet spinning	154	–	–	8558	[34]

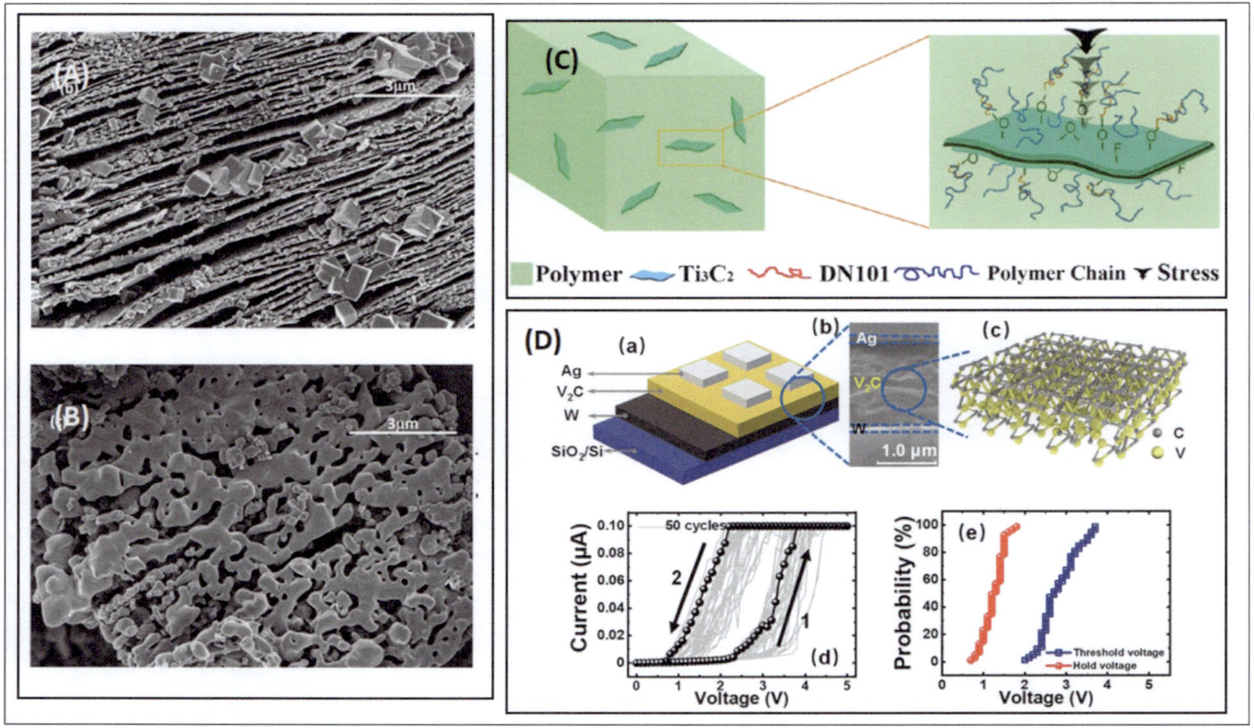

Fig. 5 Previously reported works highlighting properties of the MXene family. MXene SEM image at varying temperature conditions, i.e., **A** 200 °C, **B** 1000 °C in the exposure of O_2; Reproduced with permission from Elsevier, Copyright 2014 [42]. **C** Mechanical properties improvement mechanism; Reproduced with permission from Elsevier, Copyright 2015 [47] (**D**). **a** Schematic of Ag/V_2C/W based memristor. **b** SEM for the material composite, **c** atomic arrangement of the memristor. **d** I–V characteristics of the memristor, **e** extracted threshold voltage characteristics; Reproduced with permission from Ref. [48] [https://doi.org/https://doi.org/10.3390/nano11112860], under the terms of the CC BY 4.0 license [/voic0htps:/creativecommonsorg/licenses/by/4.0//vic0:ili.:i.li..tts://crtivcs.r/licss/y/./]

or electrochemical capacitor electrodes. It is also possible to use other methods like slurry rolling, inkjet printing, or screen printing [36]. On the other hand, spin coating can provide thin, even coatings for electronics or optics [49]. The MXene family demonstrates diverse properties beyond those mentioned, including the phenomenon of surface plasmon resonance, expanding their applications in fields such as sensors. With advanced optoelectronic properties, the optimized thickness of MXene layers enhances plasmonic behavior at the nano metal–dielectric interface whenever the latter is employed as a dielectric material between the intermediate layers [50]. Introducing MXene increases the sensor device sensitivity and detection accuracy, enhancing the resolution of nano-sensing devices. Hybrid nanostructures of MXene composites have opened new potential in bio and gas sensing applications, depending on their wavelength-specific deployment. There are reasonably reported works from the visible to near-infrared regions [51]. The net increment in sensitivity achieved was 3.45% with the inclusion of monolayer MXene. This is due to the biosamples' adequate binding strength on the MXene surface, which results from their greater surface area, chemical stability, and improved absorption characteristics [52]. Some of the applications are discussed below, forming the basis for the subsequent chapters.

Figure 6 presents some of the exciting works reported that have effectively utilized MXene's intrinsic properties to be extensively used in skin wound repair [54], industry applications [55], biosensing purposes [53], and photovoltaic applications [56], among many more. Detailed viewpoints on this matter shall be explained in the subsequent sections below.

(a) **MXene as Supercapacitor**

Regarding wearable/flexible electronics, MXene has been reportedly used as a supercapacitor. This is because they are lightweight and chemically stable, which enhances the prerequisites, i.e., durability and flexibility for a wearable device. However, the aggregation of the MXene layers is a bottleneck towards its extensive utilization. For this purpose, tailoring diffusion ion resistance is reported to be an effective strategy. Thus, it has been experimentally seen that the intrinsic properties of MXene layers are highly favorable for wearable supercapacitors. However, the stacking of MXene flakes can cause concern as it degrades the electrical conductivity [56]. This issue of self-stacking has been addressed by Yan et al., who introduced rGO sheets between MXene layers that act as spacers, resulting in ultra-high conductivity. This illustrates how interlayer spacing will promote the creation of flexible, highly integrated supercapacitors that will have superior volumetric and rate capabilities in the future [57]. The significance of interlayer spacing was discussed, wherein the carbon intercalated layers amidst the MXene interspacing also led to enhanced conductivity and specific capacitance. The choice of precursors and selective conditions of HF etching also enables the use of MXene supercapacitor electrodes in various electrolytes. Besides, the nano-lakes' size also affects the electrochemical activity of the supercapacitance. Smaller lateral size leads to a shorted diffusion ion path, thus exposing more electrolytes to the electrode surface. Likewise, the terminations introduced also manipulate the surface chemistry, impacting the electrochemistry of the supercapacitors. For example, O-based terminations have the best capacitance

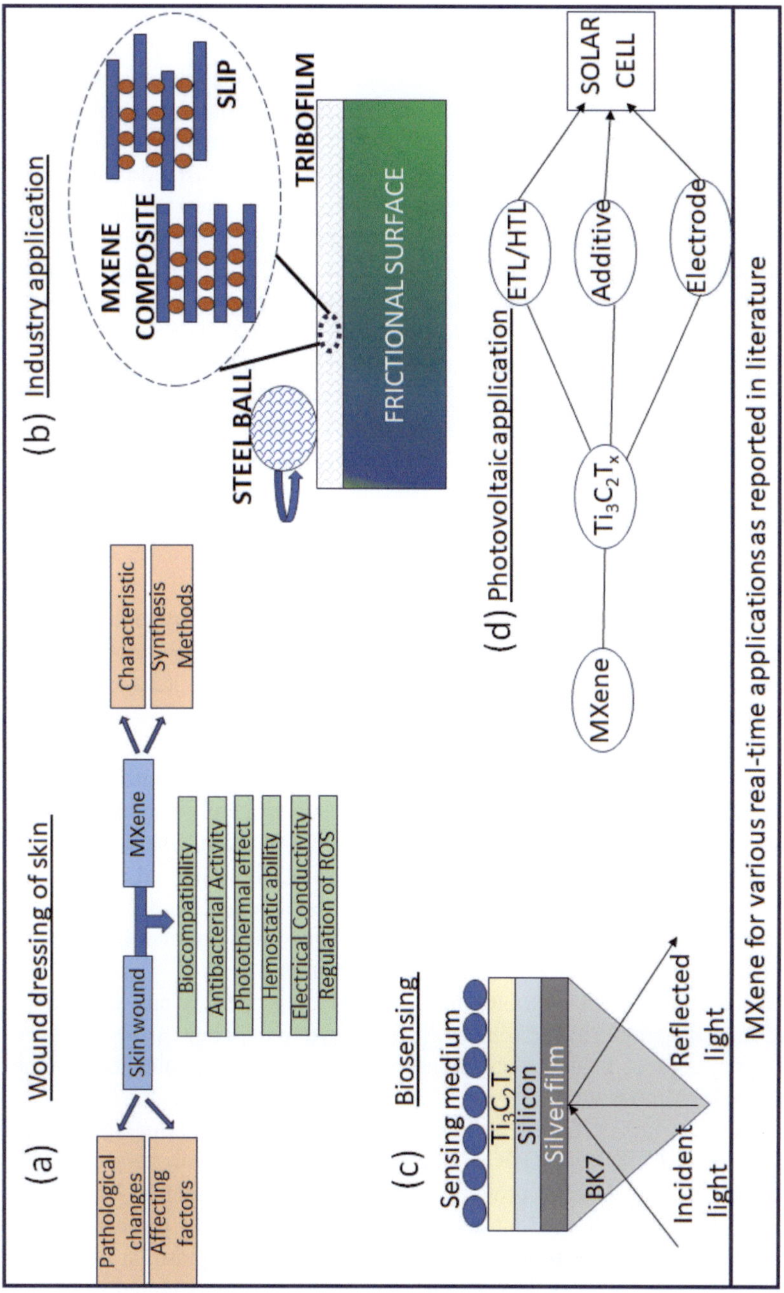

Fig. 6 Previously reported works on MXene employed for some of the real-time applications, **a** employing MXene nanosheets for wound dressing of skin [53], **b** industrial applications [54], **c** biosensing utilizing angle interrogation over silver as the plasmonic film [52], **d** photovoltaic application [55]

reported according to studies. The kind of electrolyte employed also affects parameters like gravimetric capacitance, aerial capacitance, volumetric capacitance, etc. When three-molar H_2SO_4 is utilized as the electrolyte, a high gravimetric capacitance of around 653F/g is noted [58]. Although Li-ion batteries are well-known candidates in terms of energy storage, the limited power density due to lesser ion transportation makes MXene-based electrodes a potential alternative. Thus, by establishing a trade-off between storage cycles and storage capacity, optimizing high performance with architectured sizes and shapes, and electrochemically ultra-fast redox reactions, high-performance MXene-based supercapacitance can be achieved [59].

The doping of heteroatoms is also one useful approach to increase the overall surface capacitance. In particular, MXene supercapacitors' energy storage capacity is significantly doped when nitrogen is added. The obtained N-doped MXene exhibits appreciable capacitance performance and stability when the mass ratio of chitosan (non-toxic precursor) and MXene is 1:1 at 800 °C. An N–Ti link is formed in MXene, wherein the N atom is added to replace the C and Ti atoms' electron binding. Additionally, N atoms in MXene combine with C atoms to create C–N=C, which increases the interlayer distance and enhances MXene's electrochemical capabilities. Subsequently, the N atom's lone pair electron and the enhanced interlayer distance, strong electrochemistry, and exceptional stability are observed [60].

(b) **MXene as Quantum Dots (QDs)**

MXene QDs are spherical, fluorescent, and have sufficient stability and photostability. They are a few nanometers in size. MXene QDs are highly promising for applicability in biomedical engineering, energy conversion and storage systems, biosensing, water purification, optoelectronics, and (electro)catalysis, among other applications. The QDs made from MXenes have unique optical and electrical characteristics in addition to conductivity, hydrophilicity, and biocompatibility that are comparable to those of MXenes. MXene QDs' superior light-harvesting qualities make them feasible for applications involving light conversion into heat. The MXene QDs exhibit promising properties in optics, photothermal conversion, photochemotherapy, fluorescence detection, and bioimaging. Some of them are Superior physicochemical, photoelectric activity, high water dispersibility, strong colloidal and photostability, and ease of functionalization.

Because of their excellent photoluminescent (PL) properties and excitation-dependent emission properties, MXene QDs are less cytotoxic. They can be employed to detect a plethora of analytes, including environmental contaminants, disease biomarkers, and biomolecules.

Due to surface imperfections and size effects, the hydrothermal procedure used to prepare Ti_3C_2 MXene QDs to develop sensors for investigating histidine concentration in human samples produced a brilliant blue fluorescence. The sensor provided a detection limit of 3.52 nm and operated with little detection time. Because of the enhanced electron transfer capacity and surface plasmon resonance impact of the Au nanoparticles, MXene QDs loaded with them exhibited increased electrochemiluminescence. This led to their use in biomedical applications.

MXene QDs rivets to enhance Nickel–Cobalt-based Layered Double Hydroxides (LDH) stability and electrochemical activity. These QDs were riveted over the interface of LDH using the hydrothermal method. On the surface of Ni–Co LDH@MXene QDs, it was shown that the electrons were redistributed around the metal atoms., enhancing the surface activity. MXene QD-coated LDH was found to have enhanced electron transport and structural stability. Furthermore, as the unpaired electrons increased, MXene QDs enhanced the electrolyte ion adsorption. The findings of the area and discharge time after riveting the LDH with MXene QD indicate an enhancement in the energy storage feature.

Due to the exceptional potential of the bioconjugates of QDs for the bio-targeting, bioimaging, and biolabeling applications that they may perform, researchers have recently highly suggested MXene QDs. MXene QDs' superior optical qualities, including their strong near-infrared (NIR) absorption capabilities and great biocompatibility, allow for their direct use in photothermal therapy and immunomodulation. Photothermal therapy increases the temperature of the solution dependent on the concentration of QDs, laser power, and photon absorption by utilizing the photon-to-heat conversion characteristic of MXene QDs.

A broad-spectrum water splitting using a solar simulation system was demonstrated using a plasmonic nanostructure of Ti_3C_2 nanosheets modified with Ti_3C_2 QDs and gold nanorods (Au NRs/TDTS). On the Ti_3C_2 nanosheets, Au seeds were distributed, providing locations for the nucleation and development of nanorods. The Au nanorods exhibited surface plasmon resonance within the NIR spectrum and facilitated interfacial charge transfer to the QD-decorated nanosheets [61].

When used for biomolecule detection, these QDs have a wider dynamic range, lower detection limits, better reproducibility, fast analytical response, high sensitivity, specificity, and accuracy. These qualities also make them useful for medicine, clinical diagnostics, water analysis, food, and the detection of biowarfare agents. The primary means of observing the detection effects are the restoration of fluorescence along with quenching in the sensor. These phenomena are predominantly due to the inner filter effects (IFE) or quantum confinement issues of the QDs. Better outcomes can be achieved by further enhancing the QDs' detection properties through surface termination modifications [62]. According to research, MXene QDs can be used as co-catalysts to meet energy-related needs. A composite of carbon nanotube augmented with Ti_3C_2 and MoS_2 QDs (MoS_2QDs@$Ti_3C_2T_x$QDs@MWCNTs) was used in the direct methanol fuel cell as an effective bifunctional catalyst for the oxygen reduction reaction (ORR) and the methanol oxidation process (MOR) [63].

(c) **MXene in Solar Cell Applications**

In general, reports about MXenes' use in solar cell applications have only been available since the latter quarter of 2018. Utilizing nanoscale $Ti_3C_2T_x$ quantum dots (TQDs), the perovskite absorber was reported to be designed that led to long-term ambient and light stability, as well as an excellent Power Conversion Efficiency (PCE) without hysteresis of 20.72% as opposed to 18.31% for the benchmark device. This is because of the increased crystallinity of the perovskite layer and the solar cell's matching alignment of energy levels. The authors have thoroughly studied and listed the MXene layers used in solar cells as additive layers, electrodes, or electron/hole transport layers. Open-circuit voltage (Voc), Power conversion efficiency (PCE), and short-circuit current density (Jsc) are used to evaluate the output performance [64].

The counter electrode (CE) of quantum dot-sensitized solar cells (QDSCs) was made of $Ti_3C_2T_x$ MXene. Chen et al. presented a composite CE in 2019 that included hydrothermally produced CuSe nanoparticles atop $Ti_3C_2T_x$ MXene nanosheets that were screen-printed over a graphite monolayer. With a greater specific surface area and improved electrical conductivity for electron transfer, this composite CE provides more productive sites to minimize polysulfide electrolytes [65].

Perovskite solar cells (PSCs) are gaining traction in photovoltaics as an effective alternative to modern energy harvesters. Adding MXene improves PSCs' operational stability and conductivity by lowering the recombination barrier and creating efficient charge transfer pathways. To improve the interfacial energy-level alignment, enhance the extraction of holes, alleviate recombination at the carbon/perovskite electrode interface, and help eliminate energy-level mismatch, MXene is employed as an interfacial layer between perovskites and electrodes. Furthermore, strong interactions between the MXene and the under-coordinated Pb atoms are made possible by the surface functional groups of the Ti_3C_2 MXene nanosheets.

MXene-based single-walled carbon nanotubes (SWCNT) composites were created using metallic SWCNTs to boost the photovoltaic output performance of PSCs. With the optimized composition as reported by Bali et al., the device configuration achieved a maximum PCE of 21.42%, a J_{sc} of 25.09 mA cm^{-2}, and a V_{oc} of 1.073 V. The reduced oxygen vacancies on the surface of tin dioxide (SnO_2) and the enhanced conductivity made possible by carbon nanotubes confirmed a low trap density and improved charge extraction and transport characteristics, as evidenced by the enhanced PL spectra and decreased charge transfer resistance. Future applications will see a rise in exploring the wide range of undiscovered MXenes materials and optimizing optoelectronic qualities to enhance PCE further. In addition to PSCs, various perovskite-based optical devices, such as photodiodes, photodetectors, and phototransistors, etc., have attracted the attention of the MXenes [66]. An integrated thin-film MXene-based Metamaterial (MM) absorber was designed to increase solar cell absorbance capacity. Contrasting the carbon-nano tube MM absorber operating at visible light spectrum, which is appropriate for enhancing the efficiency of solar cells with the suggested MM absorber design using $Ti_3C_2T_x$ MXene material. It has been demonstrated that the former could achieve the same absorption band and level with a thinner MXene material (roughly 50% reduction in thickness). With an absorption value of almost 97%, it was demonstrated that an impressive absorbance value throughout the EM regime (300–714 THz) was observed. The electromagnetic characteristics of the MXene material must first be assessed before implementing such an absorber. The impact of the suggested absorber's geometrical parameters on its absorption capability was then examined to increase its absorption capability. Thus, the two most important factors in optimizing the absorbance capabilities of the suggested absorber are reported to be the periodicity of MXene nanostructure and MXene thickness [67].

MXenes can be employed in myriad approaches in solar cells, such as a back, common, or flexible hybrid electrode. In contrast to carbon electrodes, MXene electrodes have the potential to improve power conversion efficiency by reducing series resistance, increasing open-circuit voltage, and facilitating hole injection rate. The interface between the materials of the solar cell channel enhances the hole/electron transport layer, electron potential barrier, electron mobility, synergistic effects, and hole transport/collection significantly because of the tunable work function. More importantly, the effectiveness of charge separation at a solar cell junction created between the MXene and Silicon can be substantially enhanced since MXenes have an adjustable work function. MXene-silicon-based heterojunction solar cells use MXenes as a transparent conducting electrode. MXenes are a transparent conducting layer that increases the inherent potential for electron–hole pair separation and promotes charge transport [68].

Defect passivation investigations by thermal admittance spectroscopy (TAS) are detected when examining the MXene and the halide perovskite layers, including the interface. This suggests that shallow defects are processed at the bulk of the thin films. Deep defects affecting the device efficiency are also passivated at the surface level. Consequently, MXene-based PSCs reported an efficiency of approximately 22%, while the control device's efficiency was 20.56%. Stability evaluations conducted in various conditions (including genuine outdoor analysis, continuous light irradiation, and darkness) show that the MXene layer is always used to extend the lifespan of photosynthesis capacitors. It was found that the device's lifespan increased under real operational outside situations (ISOS-O), where the solar cells displayed a T_{80} at about 600 h, which is particularly noteworthy. This is the first paper on the stability analysis of MXene-based PSCs performed outdoors (ISOS-O).

It can be concluded that interface defect passivation and ion migration suppression play a significant role in the MXene-modified devices' enhanced outdoor stability performance. MXenes paved the path to highly stable perovskite solar cells. The optimal stability condition under continuous light irradiation (ISOS-L) and in real outside conditions (ISOS-O) has been demonstrated via the insertion of the H3pp within the halide perovskite absorber and the MXene hole transport layer (HTL) [69].

Furthermore, due to the unique electrical, optical, and mechanical qualities of MXenes, additional advancements in optoelectronic devices, for example, photodetectors and light-emitting diodes, can be made in addition to solar cell applications. Nevertheless, the enhancement of solar cell absorbance capacity remains unexplored.

(d) **MXene for Energy Storage Applications**

Significantly, first-principle calculations have anticipated substituting –OH and –F groups paired with Phosphorus and Silicon will increase Li capacity to over 1200 and 1700 mAh g^{-1}, respectively. Presumably, Na and K ion batteries will follow suit with this trend, owing to potential enhancements in MXene's metallicity, reduction in ion diffusion barriers, and the facilitation of multilayer alkali ion adsorption through modification of surface [70].

To ascertain the stability of the S, N, and GO or rGO (SNMG-40) film for extended storage across diverse environments, the capacitance retention of the film following storage under ambient conditions or immersion in H_2SO_4 electrolyte for varying durations was examined. Surprisingly, after 100 days of storage, the SNMG-40 hybrid thin film showed about 100% capacitance retention under ambient conditions and roughly 95.7% in aqueous H_2SO_4 electrolyte, affirming its stability for prolonged storage across different conditions [71].

(e) **MXene for Sensor and Environmental Applications**

MXene-based sensors have been extensively explored across various categories, including mechanical, gas, electrochemical, optical, and multifunctional sensors. For instance, a strain sensor utilizing a hybrid network of $Ti_3C_2T_x$ nanoparticles (NPs) demonstrated a Gauge Factor (GF) exceeding 178.4, a 0–53% sensing range, and a 0.025% detection limit. Due to MXenes' 2D layered structure and modifiable surface terminations, they are regarded as excellent platforms for gas sensing. $Ti_3C_2T_x$ is a viable option for optical sensing applications because of its broad optical absorption from visible to short-wave infrared and good photothermal conversion efficiency. MXene has been utilized extensively in electrochemical sensors to detect small molecules, environmental pollutants, and biomarkers. The sensing performance was notably enhanced through the electrochemical hybridization of $Ti_3C_2T_x$ with rGO. The resulting $Ti_3C_2T_x$/rGO sensor achieved a remarkable detection limit of 0.67 nM. Furthermore, an electrochemical sensor employing $Ti_3C_2T_x$/Au–Pd as the sensing material was developed to detect organophosphorus pesticides, commonly used in agriculture to enhance crop productivity but associated with severe environmental concerns. Regarding biomarkers, the plenty of surface terminations (such as –OH, –F, and –O) on MXenes render them suitable platforms for immobilizing active materials. For instance, tetrahedral DNA nanostructures were effectively immobilized on the surface of $Ti_3C_2T_x$ nanosheets [72]. A research team has created a wearable, flexible epidermal sensor using MXene/polyampholytes hydrogel. This innovative material, formed by combining MXene with polyampholyte hydrogel, exhibits exceptional properties such as self-recovery, self-healing, self-adhesion, and stretchability, which enhance the sensor's sensing capabilities significantly. Using a sophisticated MXene/shear stiffening hybrid, Gong's team devised a wearable sensor that improves electromagnetic interference shielding performance. This smart wearable sensor can be comfortably attached to various body regions to track multiple motions. Apart from its sensing application, this material effectively shields humans from harmful electromagnetic radiation, demonstrating a shielding effect of up to 58 dB [73].

The sensor can also suffer from humid environmental conditions. Oxidation issues can compromise MXene's electrical characteristics in humid environments. As a result, a cotton fabric with a rough construction structure was used to offer the superhydrophobic Polypyrole (PPy)/MXene sensor. Consequently, the sensor performed well in real-time human movement monitoring and showed outstanding detection limit, sensitivity, and a wider linear range. For example, MXene-based sensors are suitable for water-resistant smart electronics because of their exceptional properties, such as their extreme hydrophobicity, self-cleaning ability, and alkali and acid corrosion resistance. The $Ti_3C_2T_x$ MXene termination-modified sensor created another wearable exhaled breath monitoring device. The development of the sensing probe involved simultaneously reducing its hydrophilicity and increasing $Ti_3C_2T_x$ MXene's sensitivity to volatile organic compounds by adding hydrocarbon terminations. The sensor demonstrated strong ethanol detection capabilities at ambient temperature and enhanced humidity stability [74].

(f) MXene as Composites for Battery Applications

Most of the MXene battery demonstrations have been carried out in coin cells, mainly to analyze the electrochemical characteristics of the active material. Yet, in the future, it will be necessary to put supercapacitors and pouch-type batteries together and connect them with flexible or wearable technology. MXene has played a crucial role in enabling batteries to achieve a massive capacity while enabling quick charge discharge thanks to its special chemistry and suitable interlayer spacing. These achievements stem from MXene's intrinsic properties, like high electrical conductivity, a readily identifiable layered structure, and customization capacity, which allow electrode tailoring for specific targets. Furthermore, by exploring topics such as micro-supercapacitors (MSCs), hybrid capacitors, and batteries beyond Li-ion, MXene has proven to be beneficial in enabling supercapacitors and batteries to overcome conventional boundaries. The use of MXenes as anodes in lithium-ion batteries (LIBs) is a promising alternative for electric vehicles, and Naguib et al. were the first to suggest this possibility. In contrast to graphene or MAX (Ti_2AlC), this newly found material shows a surface area increase of about ten times. Ti_2CT_x MXene can function as an anode in Na-ion hybrid capacitors, according to a publication by Wang et al. The intercalation/adsorption of Na ions onto the surface of MXene nanosheets enhances the interlayer distance between the MXene layers. The extensive family of MXenes, such as V_2CT_x, can also function as a cathode electrode material when utilized as a cathode. An asymmetric supercapacitor was constructed with a hard carbon anode and V_2CT_x as a cathode, delivering 50 mAh g^{-1} for a 3.5 V window. It was shown that the performance of intercalating Na and Li ions between the V_2CT_x layers was comparable. As a result, V_2CT_x MXene exhibits potential as a material with a wide working voltage window [75].

(g) MXene as Composites for Biomedical and Industrial Applications

A highly sensitive device was created to track any surging behavior in primary hippocampus neurons and conduct label-free dopamine detection. It was built on an MXene-FET-based biosensor. The potential for the active surface of multilayered Ti_3C_2 to come into touch with tiny biomolecules and change the conductivity signal is provided by the as-synthesized MXene stripes. For example, the Au/MXene nanocomposite platform was created to detect sensitive enzymatic glucose. With a comparatively good sensitivity of 4.2 μA mM^{-1} cm^{-2} and a detection limit of 5.9×10^{-6} M, a generalized trend for GOx/gold/MXene/Nafion/Glassy carbon electrode biosensors has been observed, exhibiting a linear amperometric responsiveness in the glucose concentration range of 0.1 to 18×10^{-3} M. Overall, the current is rising as the concentration of glucose rises [76].

The presence of a few-layer $Ti_3C_2T_x$ SPR biosensors with MXene on their surface can have increased sensitivity through its absorption properties. The benefit of this proposed structure lies in its heightened sensitivity. Specifically, improvements in the sensitivity are 16.8%, 28.4%, 46.3%, and 33.6% for the suggested surface plasmon resonance biosensors utilizing Au with four layers of $Ti_3C_2T_x$, Ag with seven layers of $Ti_3C_2T_x$, Al with twelve layers of $Ti_3C_2T_x$, and Cu with nine layers of $Ti_3C_2T_x$, respectively.

Zhao et al. developed a versatile MXene smart fabric by depositing $Ti_3C_2T_x$ onto a cellulose fiber non-woven fabric. MXene's metal-like electrical conductivity offers the fabric exceptional thermal characteristics, resulting in effective bacteria control around wounds during bacterial infection treatment. This innovation is poised to advance the field of wearable electronic healthcare devices because of its multifunctional, flexible fabric. Utilizing Ti_3C_2 MXene as the active material, the commercial cellulose fiber nonwoven fabrics (cellulose fabric) were transformed into multifunctional MXene-based smart fabrics (M-fabric) using a straightforward repeated dip-coating and drying process. As wearable electronic devices, these resilient and adaptable M-fabrics have a lot of potential uses in biomedical therapy and healthcare. Furthermore, after optimizing the MXene coating, the multipurpose textiles have good Joule-heating capabilities, acceptable breathability, and an accurate humidity response. Consequently, these M-fabrics are comfortable to wear on the body and show potential for use in breath monitoring and as a smart thermotherapy platform featuring temperature warning features. Furthermore, as shown by in vivo studies, the heat produced can be evenly distributed to skin afflicted with bacteria for bacterial ablation, hastening the healing of infectious wounds. These integrated multifunctional smart textiles have potential uses as wearable electronic device platforms in medical therapy and portable healthcare. The resultant multifunctional M-fabric shows promise for medical treatment and personal healthcare due to its exceptional bacterial ablation capacity, excellent Joule heating performance, sensitive humidity responsiveness, and satisfactory structural and performance reliability [77].

Hydrogels based on MXenes are excellent choices for biomedical uses. Recently, there has been a significant interest in using MXene-based hydrogels for drug delivery, biosensors, bioimaging, and bone regeneration. Hydrogel formulation of MXenes considerably increases its durability and allows for the advancement of lightweight, customizable MXene-based materials. MXene-based hydrogels have potential use in tissue engineering, drug delivery, wound healing, and biosensing. By combining the properties of hydrogel networks and MXene nanosheets, researchers are exploring new ways to address critical biological issues [78].

A two-dimensional (2D) MXene layer, silver metal, and a dielectric buffer layer of Cytop were used in a study on a long-range surface plasmon resonance (LRSPR) sensor. The objective was to improve penetration depth (PD), figure of merit (FoM), and sensitivity. Simulations were carried out for various incident angles on a CaF_2 prismatic configuration with a detection wavelength of 1550 nm. The Transfer Matrix Method (TMM) is an analytical approach that reported the highest FoM achieved as

571.99/RIU with a Cytop thickness of 400 nm and an Ag layer of 58 nm adjacent to the MXene layer. The maximum PD reached was 1447.95 nm, obtained with a Cytop thickness of 1000 nm and an Ag layer of 50 nm. These impressive PD and FoM values suggest the long-range SPR sensor's potential for precisely detecting large biomolecules such as cancer cells, oligonucleotides, Escherichia coli, and B.anthracis bacteria [79]. Yet another SPR-based sensor had a graphene-mediated heterostructure (MXene-graphene-MXene), demonstrating leptospirosis bacteria detection in rodent urine at a wavelength of 633 nm. Sensor design utilized the Kretschmann configuration, with the TMM technique employed to simulate reflectivity. The optimization of the sensor design aimed at achieving maximum sensitivity and minimizing reflectance (R_{min}). As a result, the sensor exhibited maximum sensitivities of 214.51°/RIU for polyuria and 211.71°/RIU for oliguria detection [80].

MXene nanosheets served as quenching agents to impede the fluorescent signals emitted by fluorescent sensing probes before detecting target molecules in fluorescent biosensors. Fluorescence resonance energy transfer (FRET)-based biosensing systems have emerged for exosome detection, leveraging Ti_3C_2 MXene nanosheets in conjunction with Cy3-labeled CD63 aptamer. CD63, a transmembrane protein found in exosomes, represents a compelling target molecule due to its encapsulation of vital biomolecules like DNA and RNA, enabling non-destructive cellular monitoring. MXenes have recently been combined with TMD nanomaterials, notably MoS_2, which shares a similar research interest with MXenes. For instance, a nanohybrid material comprising MoS_2 and MXene ($Ti_3C_2T_x$) was proposed for developing an electrochemical aptasensor to detect thyroxine [81]. This study introduced an electrochemical aptasensor that integrates magnetic separation with a nuclease-powered walking DNA nanomachine for detecting Mucin 1 (MUC1). The proposed electrochemical aptasensor demonstrated MUC1 detection capabilities ranging from 5 pg/mL to 50 ng/mL, achieving a detection limit of 0.72 pg/mL. Consequently, the devised nuclease-powered walking DNA nanomachine offered an effective approach for quantitatively analyzing proteins. Thus, this proposed method not only furnished a reliable aptasensing platform for MUC1 assessment in human serum but also holds promise for future clinical analysis and diagnosis by employing cancer cells as models [82].

(h) **MXene as Nanofillers for Polymer Applications**

Monomers have the potential to undergo polymerization over the MXene nanosurface, thereby augmenting the structural robustness of the resulting MXene. On the other hand, strong covalent bonds are formed between both polymers and MXene layers through in-situ polymerization. An ex-situ blending strategy is used for MX-CP composites due to its ability to precisely manipulate their composition and specified polymer structures. Throughout hybridization, the level of interaction between polymer chains and MXene flakes is governed by van der Waals forces, hydrogen bonding, and electrostatic assembly. Monomers are immediately polymerized on the MXene surface using the in-situ polymerization technique. Using in-situ methods to polymerize the MXene surface improves the distance between layers and guarantees that MXene is distributed evenly throughout the polymer matrix. These techniques produce composites with enhanced conductivity, ion exchange, and capacitance, which makes them desirable for energy storage. Integrating MXene with a polymer can mitigate capacity degradation by improving electrical conductivity, preventing the aggregation of MXene flakes, thereby imparting mechanical durability toward the composite.

Given the crucial role of conjugated polymers (CPs) in metal-ion batteries (MIBs), there is potential in the MXene and CPs integration for mitigating MXene aggregation, thereby enhancing structural integrity in polymers. MXene-CPs represent a potential avenue for enhancing endurance, expanding voltage range, and increasing mass loading in batteries. However, MXene-CP nanoarchitecture is probably in its early phases of development, and continuing studies ought to include the wider range of the MXene family, including V_2C, Mo_2C, Nb_2C, Ta_4C_3, bimetal transition carbides, and beyond [83].

Extensive research has been at play to integrate MXenes into polymers to boost the membrane system's functionality for water treatment. This method provides numerous benefits, such as better rejection of solute, enhanced water permeability, and increased resistance to fouling. MXenes have also been extensively studied as adsorbents in water treatment, particularly $Ti_3C_2T_x$, which shows promise for removing heavy metals like Pb, Hg, Cu, and Cd. Studies indicate MXenes possess excellent selectivity for ion sorption and capacity for reducing various pollutants found in wastewater. Combining MXene nanosheets with silver nanoparticles (AgNPs) has resulted in highly efficient membrane systems. Including AgNPs is thought to expand the confined spaces between MXene layers, creating additional pathways for rapid water transport. They demonstrate bactericidal properties, thereby enhancing water filtration systems. rGO was intercalated between MXene layers to establish highly permeable pathways. This innovative composite achieved 198.8 L m^{-2} h^{-1} bar^{-1} water flux and 100% dye rejection. The synergistic combination of two distinct 2D nanomaterials, GO and MXene, presents a remarkable approach to enhancing membrane properties, elevating water permeability, and improving rejection rates. This amalgamation can yield durable membranes with well-organized stacking and minimal flaws. A nanofiltration membrane was developed using $Ti_3C_2T_x$ MXene and polyamide, resulting in an increase in water permeability from 8.65 to 12.16 L m^{-2} h^{-1} bar^{-1}, along with excellent rejection rates of 96.85 and 93.35% for perfluorochemicals. Perfluorochemicals pose significant challenges in water treatment due to their bioaccumulative nature in humans and organisms [84].

Incorporating polymers with MXenes can yield synergistic benefits such as enhanced mechanical stability, thermal resistance, flexibility, and toughness. Consequently, numerous MXene aerogels incorporating polymers have been explored for diverse applications, spanning sensors, EMI shielding, microwave absorption, thermal regulation, energy storage and conversion, and water purification. Crosslinking MXenes with polymeric materials enhances electrical and thermal conductivity and bolsters the mechanical properties of polymers when investigated with other 2D nanosheet-based materials. The robust oxygen/fluorine-surface groups of MXene facilitate strong coordination with other composite derivatives through covalent bonding, electrostatic interaction, or van der Waals forces. Several polymer-crosslinked MXene aerogels have emerged for pressure sensors due to their mechanical resilience, substantial surface area, and high conductivity. However, further research is warranted to ascertain the durability of polymer-incorporated MXene aerogels under extreme temperatures, high humidity, and chemical exposure [85].

6 Challenges and Future Scope

The most prevalent challenge is the large-scale production of single-layer flakes with controllable surface chemistry [3]. The creation of MXenes primarily focuses on carbides, with minimal research on nitride preparation since the A layer in nitrides is much harder to peel off than carbides [86]. Although it is environment-friendly as the Ti_3AlC_2 MAX phase produced from TiO_2 precursor, recycled Al scrap, and waste-recovered carbon, inert atmospheres under high temperatures are one of the prerequisites for routine synthesis methods that limit the economic viability of the material to be utilized in sensing applications for mass scale production [87]. Halogen etching is gaining momentum since the latter shows stability in water for at least two weeks. Still, the low product yield limits the employability of halogen etching for MXene synthesis [88]. Three main issues that need to be resolved to influence the mechanical assessment of MXenes are the lack of surface control in MXenes, the occurrence of geometric vacancies, and contacts between weak composite materials. Thus, subsequent studies should focus on how the synthesis process develops to ascertain its defects and differentiate between different functional groups of pristine components. It may not yet be possible to estimate all mechanical characteristics and prominent functional groupings using theory and experimentation. Experimental assessment of MXene mechanical properties, especially in two-dimensional materials, remains challenging, often relying on nano-indentation methods.

References

1. Shukla, S., Arora, P.: Design and comparative analysis of aluminum-MoS2 based plasmonic devices with enhanced sensitivity and Figure of Merit for biosensing applications in the near-infrared region. Optik (Stuttg) **228**, 166196 (2021). https://doi.org/10.1016/j.ijleo.2020.166196
2. Yang, H., Zhang, G.-X., Zhou, H.-J., Sun, Y.-Y., Pang, H.: Metal-organic frameworks meet MXene: new opportunities for electrochemical application. Energy Mater. Adv. **4**, 33 (2023). https://doi.org/10.34133/energymatadv.0033
3. Zhan, X., Si, C., Zhou, J., Sun, Z.: MXene and MXene-based composites: synthesis, properties and environment-related applications. Nanoscale Horizons **5**(2), 235–258 (2020). https://doi.org/10.1039/C9NH00571D
4. Lim, K.R.G., Shekhirev, M., Wyatt, B.C., Anasori, B., Gogotsi, Y., Seh, Z.W.: Fundamentals of MXene synthesis. Nat. Synth. **1**(8), 601–614 (2022). https://doi.org/10.1038/s44160-022-00104-6

5. Han, M., et al.: Beyond Ti3C2Tx: MXenes for electromagnetic interference shielding. ACS Nano **14**(4), 5008–5016 (2020). https://doi.org/10.1021/acsnano.0c01312
6. Zuo, Y., et al.: Broadband multi-wavelength optical sensing based on photothermal effect of 2D MXene films. **9**(1), 123–131 (2020). https://doi.org/10.1515/nanoph-2019-0338
7. Naguib, M., et al.: Two-dimensional nanocrystals produced by exfoliation of Ti3AlC2. Adv. Mater. **23**(37), 4248–4253 (2011). https://doi.org/10.1002/adma.201102306
8. Lei, J.C., Zhang, X., Zhou, Z.: Recent advances in MXene: preparation, properties, and applications. Front. Phys. **10**(3), 276–286 (2015) (Higher Education Press). https://doi.org/10.1007/s11467-015-0493-x
9. Mashtalir, O., et al.: Intercalation and delamination of layered carbides and carbonitrides. Nat. Commun. **4**(1), 1716 (2013). https://doi.org/10.1038/ncomms2664
10. Halim, J., et al.: Synthesis and characterization of 2D molybdenum carbide (MXene). Adv. Funct. Mater. **26**(18), 3118–3127 (2016). https://doi.org/10.1002/adfm.201505328
11. Ghidiu, M., Lukatskaya, M.R., Zhao, M.-Q., Gogotsi, Y., Barsoum, M.W.: Conductive two-dimensional titanium carbide 'clay' with high volumetric capacitance. Nature **516**(7529), 78–81 (2014). https://doi.org/10.1038/nature13970
12. Zou, G., et al.: Hydrogenated Core-Shell MAX@K2Ti8O17 Pseudocapacitance with ultrafast sodium storage and long-term cycling. Adv. Energy Mater. **7**(18), 1700700 (2017). https://doi.org/10.1002/aenm.201700700
13. Lukatskaya, M.R., et al.: Room-temperature carbide-derived carbon synthesis by electrochemical etching of MAX phases. Angew. Chem. Int. Ed. Engl. **53**(19), 4877–4880 (2014). https://doi.org/10.1002/anie.201402513
14. Naguib, M., Unocic, R.R., Armstrong, B.L., Nanda, J.: Large-scale delamination of multi-layers transition metal carbides and carbonitrides 'MXenes.' Dalt. Trans. **44**(20), 9353–9358 (2015). https://doi.org/10.1039/C5DT01247C
15. Lipatov, A., Alhabeb, M., Lukatskaya, M., Boson, A., Gogotsi, Y., Sinitskii, A.: MXene materials: effect of synthesis on quality, electronic properties and environmental stability of individual monolayer Ti3C2 MXene flakes. Adv. Electron. Mater. **2** (2016). https://doi.org/10.1002/aelm.201670068
16. Urbankowski, P., et al.: Synthesis of two-dimensional titanium nitride Ti4N3 (MXene). Nanoscale **8**(22), 11385–11391 (2016). https://doi.org/10.1039/C6NR02253G
17. Yang, S., et al.: Fluoride-free synthesis of two-dimensional titanium carbide (MXene) using a binary aqueous system. Angew. Chem. Int. Ed. Engl. **57**(47), 15491–15495 (2018). https://doi.org/10.1002/anie.201809662
18. Shi, H., et al.: Ambient-stable two-dimensional titanium carbide (MXene) enabled by iodine etching. Angew. Chemie Int. Ed. **60**(16), 8689–8693 (2021). https://doi.org/10.1002/anie.202015627
19. Geng, D., et al.: Direct synthesis of large-area 2D Mo2C on in situ grown graphene. Adv. Mater. **29**(35), 1700072 (2017). https://doi.org/10.1002/adma.201700072
20. El Ghazaly, A., et al.: Ultrafast, one-step, salt-solution-based acoustic synthesis of Ti3C2 MXene. ACS Nano **15**(3), 4287–4293 (2021). https://doi.org/10.1021/acsnano.0c07242
21. Zhang, C.J., et al.: Oxidation stability of colloidal two-dimensional titanium carbides (MXenes). Chem. Mater. **29**(11), 4848–4856 (2017). https://doi.org/10.1021/acs.chemmater.7b00745
22. Mallakpour, S., Azadi, E.: Novel methodologies and materials for facile fabrication of nanofiltration membranes. Emergent Mater. **5**(5), 1263–1288 (2022). https://doi.org/10.1007/s42247-021-00278-3
23. Shibata, Y., Suizu, R., Awaga, K., Hirotani, J., Omachi, H.: Fabrication of MXene transparent conductive films via transfer process (2023)
24. Gong, X., Zhang, G., Dong, H., Wang, H., Nie, J., Ma, G.: Self-assembled hierarchical heterogeneous MXene/COF membranes for efficient dye separations. J. Memb. Sci. **657**, 120667 (2022). https://doi.org/10.1016/j.memsci.2022.120667
25. Shaheen, I., et al.: Fabrication of MXene/cellulose composite-based flexible supercapacitor: synthesis, properties, and future perspectives. J. Energy Storage **87**, 111513 (2024). https://doi.org/10.1016/j.est.2024.111513

26. Pershaanaa, M., et al.: MXenes and their transformation to composites for potential applications. Mater. Today Commun. **35**, 106143 (2023). https://doi.org/10.1016/j.mtcomm.2023.106143
27. Rakhi, R.B., Ahmed, B., Anjum, D., Alshareef, H.N.: Direct chemical synthesis of MnO2 nanowhiskers on transition-metal carbide surfaces for supercapacitor applications. ACS Appl. Mater. Interfaces **8**(29), 18806–18814 (2016). https://doi.org/10.1021/acsami.6b04481
28. Li, L., Zhang, N., Zhang, M., Wu, L., Zhang, X., Zhang, Z.: Ag-nanoparticle-decorated 2D titanium carbide (MXene) with superior electrochemical performance for supercapacitors. ACS Sustain. Chem. Eng. **6**(6), 7442–7450 (2018). https://doi.org/10.1021/acssuschemeng.8b00047
29. Tang, M., et al.: Surface terminations of MXene: synthesis, characterization, and properties. Symmetry **14**(11) (2022). https://doi.org/10.3390/sym14112232
30. Shurbaji, S., Manaph, N.P.A., Ltaief, S.M., Al-Shammari, A.R., Elzatahry, A., Yalcin, H.C.: Characterization of MXene as a cancer photothermal agent under physiological conditions. Front. Nanotechnol.**3** (2021). https://www.frontiersin.org/journals/nanotechnology/articles//10.3389/fnano.2021.689718
31. Soundiraraju, B., George, B.K.: Two-dimensional titanium nitride (Ti2N) MXene: synthesis, characterization, and potential application as surface-enhanced raman scattering substrate. ACS Nano **11**(9), 8892–8900 (2017). https://doi.org/10.1021/acsnano.7b03129
32. Halim, J., et al.: Electronic and optical characterization of 2D Ti2C and Nb2C (MXene) thin films. J. Phys. Condens. Matter **31**(16), 165301 (2019). https://doi.org/10.1088/1361-648X/ab00a2
33. Wang, C., Chen, S., Song, L.: Tuning 2D MXenes by surface controlling and interlayer engineering: methods, properties, and synchrotron radiation characterizations. Adv. Funct. Mater. **30**(47), 2000869 (2020). https://doi.org/10.1002/adfm.202000869
34. Yang, M., et al.: Recent advances in MXene-based fibers: fabrication, performance, and application. Small Methods **7**(10) (Wiley) (2023). https://doi.org/10.1002/smtd.202300518
35. Gao, L., et al.: MXene/polymer membranes: synthesis, properties, and emerging applications. Chem. Mater. **32**(5), 1703–1747 (2020). https://doi.org/10.1021/acs.chemmater.9b04408
36. Ling, Z., et al.: Flexible and conductive MXene films and nanocomposites with high capacitance. Proc. Natl. Acad. Sci. U. S. A. **111** (2014). https://doi.org/10.1073/pnas.1414215111
37. Hu, C., Shen, F., Zhu, D., Zhang, H., Xue, J., Han, X.: Characteristics of Ti3C2X-Chitosan films with enhanced mechanical properties. Front. Energy Res. **4**(JAN) (2017). https://doi.org/10.3389/fenrg.2016.00041
38. Meng, W., et al.: Advances and challenges in 2D MXenes: from structures to energy storage and conversions. Nano Today **40**, 101273 (2021). https://doi.org/10.1016/J.NANTOD.2021.101273
39. De, S., Acharya, S., Sahoo, S., Nayak, G.: Current trends in MXene research: properties and applications. Mater. Chem. Front. **5**(19), 7134–7169 (2021). https://doi.org/10.1039/D1QM00556A
40. Kumar, J.A., et al.: Methods of synthesis, characteristics, and environmental applications of MXene: a comprehensive review. Chemosphere **286**, 131607 (2022). https://doi.org/10.1016/J.CHEMOSPHERE.2021.131607
41. Wang, K., Zhou, Y., Xu, W., Huang, D., Wang, Z., Hong, M.: Fabrication and thermal stability of two-dimensional carbide Ti3C2 nanosheets. Ceram. Int. **42**(7), 8419–8424 (2016). https://doi.org/10.1016/J.CERAMINT.2016.02.059
42. Li, Z., et al.: Synthesis and thermal stability of two-dimensional carbide MXene Ti3C2. Mater. Sci. Eng. B **191**, 33–40 (2015). https://doi.org/10.1016/j.mseb.2014.10.009
43. Lee, S.H., et al.: Room-temperature, highly durable Ti3C2Tx MXene/graphene hybrid fibers for NH3 gas sensing. ACS Appl. Mater. Interfaces **12**(9), 10434–10442 (2020). https://doi.org/10.1021/acsami.9b21765
44. Yang, R., Chen, X., Ke, W., Wu, X.: Recent research progress in the structure, fabrication, and application of MXene-based heterostructures. Nanomaterials **12**(11) (MDPI) (2022). https://doi.org/10.3390/nano12111907

45. Seyedin, S., et al.: MXene composite and coaxial fibers with high stretchability and conductivity for wearable strain sensing textiles. Adv. Funct. Mater. **30** (2020). https://doi.org/10.1002/adfm.201910504
46. Montes García, V., et al.: Inside back cover—Advanced functional materials. vol. 31, no. 10 (2021)
47. Zhang, H., et al.: Preparation, mechanical and anti-friction performance of MXene/polymer composites. Mater. Des. **92**, 682–689 (2016). https://doi.org/10.1016/j.matdes.2015.12.084
48. Wang, Y., et al.: Artificial neurons based on Ag/V2C/W threshold switching memristors. Nanomaterials **11**(11) (2021). https://doi.org/10.3390/nano11112860
49. Mariano, M., et al.: Solution-processed Titanium Carbide MXene films examined as highly transparent conductors. Nanoscale **8** (2016). https://doi.org/10.1039/C6NR03682A
50. Shukla, S., Arora, P.: Investigation of 2D nanomaterials on MXene (Ti3C2Tx)-based aluminum plasmonic devices for biosensing in the near-infrared region. Appl. Phys. A **128**(9), 785 (2022). https://doi.org/10.1007/s00339-022-05947-6
51. Chao, C.-T.C., Kooh, M.R.R., Lim, C.M., Thotagamuge, R., Mahadi, A.H., Chau, Y.-F.C.: Visible-range multiple-channel metal-shell rod-shaped narrowband plasmonic metamaterial absorber for refractive index and temperature sensing. Micromachines **14**(2) (2023). https://doi.org/10.3390/mi14020340
52. Kumar, R., Pal, S., Prajapati, Y.K., Saini, J.P.: Sensitivity enhancement of MXene based SPR sensor using silicon: theoretical analysis. Silicon **13**(6), 1887–1894 (2021). https://doi.org/10.1007/s12633-020-00558-3
53. Zhang, Z., Qi, Z., Kong, W., Zhang, R., Yao, C.: Applications of MXene and its modified materials in skin wound repair. Front. Bioeng. Biotechnol. **11** (2023). https://www.frontiersin.org/journals/bioengineering-and-biotechnology/articles/https://doi.org/10.3389/fbioe.2023.1154301
54. Jaya Prakash, N., Kandasubramanian, B.: 18—Industrial applications of MXene nanocomposites. In: Mavinkere Rangappa, S., Parameswaranpillai, J., Yashas Gowda, T.G., Siengchin, S., Seydibeyoglu, M.O. (eds.) Nanoparticle-Based Polymer Composites, pp. 481–503. Woodhead Publishing (2022)
55. Akanbi, O., et al.: Research advances on 2D Mxenes for photovoltaic applications. J. Adv. Mech. Eng. Appl. **4**(1) (2023). https://doi.org/10.30880/jamea.2023.04.01.006
56. Iravani, S., Varma, R.S.: MXene-based wearable supercapacitors and their transformative impact on healthcare. Mater. Adv. **4**(19), 4317–4332 (2023). https://doi.org/10.1039/D3MA00365E
57. Yan, J., et al.: Flexible MXene/graphene films for ultrafast supercapacitors with outstanding volumetric capacitance. Adv. Funct. Mater. **27**(30), 1701264 (2017). https://doi.org/10.1002/adfm.201701264
58. Sohan, A., Banoth, P., Aleksandrova, M., Nirmala Grace, A., Kollu, P.: Review on MXene synthesis, properties, and recent research exploring electrode architecture for supercapacitor applications. Int. J. Energy Res.**45**(14), 19746–19771 (2021). https://doi.org/10.1002/er.7068
59. Panda, S., Deshmukh, K., Khadheer Pasha, S.K., Theerthagiri, J., Manickam, S., Choi, M.Y.: MXene based emerging materials for supercapacitor applications: recent advances, challenges, and future perspectives. Coord. Chem. Rev. **462**, 214518 (2022). https://doi.org/10.1016/j.ccr.2022.214518
60. Pu, L., et al.: N-doped MXene derived from chitosan for the highly effective electrochemical properties as supercapacitor. Adv. Compos. Hybrid Mater. **5**(1), 356–369 (2022). https://doi.org/10.1007/s42114-021-00371-5
61. Nie, Y., Liang, Z., Wang, P., Ma, Q., Su, X.: MXene-derived quantum dot@gold nanobones heterostructure-based electrochemiluminescence sensor for triple-negative breast cancer diagnosis. Anal. Chem. **93**(51), 17086–17093 (2021). https://doi.org/10.1021/acs.analchem.1c04184
62. Xu, G., et al.: Preparation of Ti3C2T MXene-derived quantum dots with white/blue-emitting photoluminescence and electrochemiluminescence. Adv. Opt. Mater. **6**(24), 1800951 (2018). https://doi.org/10.1002/adom.201800951

63. Yang, X., et al.: Multiwall carbon nanotubes loaded with MoS2 quantum dots and MXene quantum dots: Non–Pt bifunctional catalyst for the methanol oxidation and oxygen reduction reactions in alkaline solution. Appl. Surf. Sci. **464**, 78–87 (2019). https://doi.org/10.1016/j.apsusc.2018.09.069
64. Yin, L., et al.: MXenes for solar cells. Nano-Micro Lett. **13**(1), 78 (2021). https://doi.org/10.1007/s40820-021-00604-8
65. Chen, Y., Wang, D., Lin, Y., Zou, X., Xie, T.: In suit growth of CuSe nanoparticles on MXene (Ti3C2) nanosheets as an efficient counter electrode for quantum dot-sensitized solar cells. Electrochim. Acta **316**, 248–256 (2019). https://doi.org/10.1016/j.electacta.2019.05.132
66. Bati, A.S.R., et al.: 1D-2D synergistic MXene-nanotubes hybrids for efficient perovskite solar cells. Small **17**(32), 2101925 (2021). https://doi.org/10.1002/smll.202101925
67. Korkmaz, H., Hasar, U.C., Ramahi, O.M.: Thin-film MXene-based metamaterial absorber design for solar cell applications. Opt. Quantum Electron. **55**(6), 530 (2023). https://doi.org/10.1007/s11082-023-04810-z
68. Shi, Z., Khaledialidusti, R., Malaki, M., Zhang, H.: MXene-based materials for solar cell applications. Nanomaterials **11**(12) (2021). https://doi.org/10.3390/nano11123170
69. Karimipour, M., Paingott Parambil, A., Tabah Tanko, K., Zhang, T., Gao, F., Lira-Cantu, M.: Functionalized MXene/Halide perovskite heterojunctions for perovskite solar cells stable under real outdoor conditions. Adv Energy Mater. **13**(44), 2301959 (2023). https://doi.org/10.1002/aenm.202301959
70. Das, P., Wu, Z.-S.: MXene for energy storage: present status and future perspectives. J. Phys. Energy **2**(3), 032004 (2020). https://doi.org/10.1088/2515-7655/ab9b1d
71. Liao, L., Jiang, D., Zheng, K., Zhang, M., Liu, J.: Industry-scale and environmentally stable Ti3C2T MXene based film for flexible energy storage devices. Adv. Funct. Mater. **31**(35), 2103960 (2021). https://doi.org/10.1002/adfm.202103960
72. Pei, Y., et al.: Ti3C2TX MXene for sensing applications: recent progress, design principles, and future perspectives. ACS Nano **15**(3), 3996–4017 (2021). https://doi.org/10.1021/acsnano.1c00248
73. Aziz, A., Asif, M., Ashraf, G., Iftikhar, T., Hussain, W., Wang, S.: Environmental significance of wearable sensors based on MXene and graphene. Trends Environ. Anal. Chem. **36**, e00180 (2022). https://doi.org/10.1016/j.teac.2022.e00180
74. He, J., et al.: Wearable superhydrophobic PPy/MXene pressure sensor based on cotton fabric with superior sensitivity for human detection and information transmission. Colloids Surfaces A Physicochem. Eng. Asp. **642**, 128676 (2022). https://doi.org/10.1016/j.colsurfa.2022.128676
75. Garg, R., Agarwal, A., Agarwal, M.: A review on MXene for energy storage application: effect of interlayer distance. Mater. Res. Express **7**(2), 22001 (2020). https://doi.org/10.1088/2053-1591/ab750d
76. Lin, H., Chen, Y., Shi, J.: Insights into 2D MXenes for versatile biomedical applications: current advances and challenges ahead. Adv. Sci. **5**(10), 1800518 (2018). https://doi.org/10.1002/advs.201800518
77. Zhao, X., et al.: Smart Ti3C2Tx MXene Fabric with Fast Humidity response and joule heating for healthcare and medical therapy applications. ACS Nano **14**(7), 8793–8805 (2020). https://doi.org/10.1021/acsnano.0c03391
78. Selvaraj, S., Chauhan, A., Verma, R., Viswanathan, K., Subbarayan, R., Ghotekar, S.: Multifunctional biomedical applications of MXene-based hydrogels: a review. Inorg. Chem. Commun. **164**, 112457 (2024). https://doi.org/10.1016/j.inoche.2024.112457
79. Kumar, R., Pal, S., Prajapati, Y.K.: MXene-mediated long-range surface plasmon resonance (LRSPR) sensor at near-infrared wavelength. Plasmonics **18**(5), 1787–1798 (2023). https://doi.org/10.1007/s11468-023-01896-9
80. Kumar, R., Pal, S., Pal, N., Verma, A., Prajapati, Y.K.: MXene-graphene-MXene–mediated heterostructure-based surface plasmon resonance sensor for the detection of leptospirosis bacteria in rodent urine. Plasmonics (2024). https://doi.org/10.1007/s11468-024-02246-z
81. Yoon, J., Shin, M., Lim, J., Lee, J.-Y., Choi, J.-W.: Recent advances in MXene nanocomposite-based biosensors. Biosensors **10**(11) (2020). https://doi.org/10.3390/bios10110185

82. Li, Z.Y., Li, D.Y., Huang, L., Hu, R., Yang, T., Yang, Y.H.: An electrochemical aptasensor based on intelligent walking DNA nanomachine with cascade signal amplification powered by nuclease for Mucin 1 assay. Anal. Chim. Acta **1214**, 339964 (2022). https://doi.org/10.1016/j.aca.2022.339964
83. Vallem, S., Venkateswarlu, S., Song, S., Kim, J., Li, Y., Bae, J.: Flexible MXene-conjugated polymer nanoarchitectures: current developments and future frontiers in battery technology. Coord. Chem. Rev. **510**, 215778 (2024). https://doi.org/10.1016/j.ccr.2024.215778
84. Ajith, S., Almomani, F., Qiblawey, H.: Emerging 2D MXene-based polymeric membranes for water treatment and desalination. J. Environ. Chem. Eng. **12**(2), 112078 (2024). https://doi.org/10.1016/j.jece.2024.112078
85. Phadtare, V.D., Parale, V.G., Lee, K.-Y., Choi, H., Bae, Y.-S., Park, H.-H.: Polymer-incorporated MXene aerogels: synthesis, classification, and properties. Polym. Rev. **64**(3), 761–817 (2024). https://doi.org/10.1080/15583724.2024.2319585
86. Shuai, T.-Y., Zhan, Q.-N., Xu, H.-M., Huang, C.-J., Zhang, Z.-J., Li, G.-R.: Recent advances in the synthesis and electrocatalytic application of MXene materials. Chem. Commun. **59**(27), 3968–3999 (2023). https://doi.org/10.1039/D2CC06418A
87. Halim, J., et al.: X-ray photoelectron spectroscopy of select multi-layered transition metal carbides (MXenes). Appl. Surf. Sci. **362**, 406–417 (2016). https://doi.org/10.1016/j.apsusc.2015.11.089
88. Krämer, M., et al.: Near-atomic-scale perspective on the oxidation of Ti3C2T MXenes: insights from atom probe tomography. Adv. Mater. **36**(3), 2305183 (2024). https://doi.org/10.1002/adma.202305183

Synthesis and Fabrication Techniques of MXene

Shubham Kumar Patial, Anuj Sharma, and Suman Singh

Abstract Since their discovery in 2011, MXenes have garnered substantial attention from researchers due to their unique characteristics. These 2D transition metal carbides, nitrides, and carbonitrides exhibit versatile surface chemistry and adjustable structures. Possessing a combination of hydrophilicity and excellent metallic conductivity, these materials—available in over 40 compositions, offer distinctive properties such as metal-like electrical conductivity. This makes them valuable in applications ranging from energy storage and optoelectronics to biomedical, communications, and environmental technologies. The rich physical and chemical diversity of MXenes has prompted the exploration of various synthesis routes for producing nanosheets and derivatives. Successful methods should yield structures with high ionic transport properties, stability, and electrical conductivity. This chapter delves into the synthesis routes employed for the production and fabrication of MXenes. The discussion encompasses emerging etching methods, including HF etching, in situ HF-forming etching, electrochemical etching, alkali etching, and molten salt etching methods, as well as delamination strategies. The aim is to comprehensively review the synthesis procedures, theoretical yield, and properties of synthesized MXenes, with a focus on their application. Additionally, to enhance properties, the discussion extends to composites of MXene-based structures, aiming for high performance. The chapter concludes by addressing present challenges and outlining the future scope for large-scale and feasible production of MXenes and their derivatives.

Keywords Hydrophilicity · Optoelectronic · Environmental technologies · Electrochemical etching · Delamination

S. K. Patial · S. Singh (✉)
Academy of Scientific and Innovative Research (AcSIR), Ghaziabad, India
e-mail: ssingh@csio.res.in

S. K. Patial · A. Sharma · S. Singh
Materials Science and Sensor Applications, CSIR-Central Scientific Instruments Organisation, Chandigarh, India

© The Author(s), under exclusive license to Springer Nature Singapore Pte Ltd. 2025

S. Moharana and R. N. Mahaling (eds.), *Surface Engineering of MXenes*, Engineering Materials, https://doi.org/10.1007/978-981-96-3640-2_2

Abbreviations

2D	Two Dimensional
3D	There Dimensional
TMC	Transition Metal Chlorides
DOS	Density of States
EMI	Electromagnetic Interference
CVD	Chemical Vapor Deposition
PE	PLD Plasma-enhanced Pulsed Layer Deposition
CNT	Carbon Nanotubes
RPM	Rotation Per Minutes
AFM	Atomic Force Microscopy
FDM	Fused Deposition Modeling
LOM	Laminated Object Manufacturing
CAD	Computer-Aided Design
MOF	Metal-organic frameworks

Objective of the Chapter
The objectives of this chapter are to shed light on the fundamentals of MXene chemistry, different synthesis methods that have been reported for successful synthesis of 2D MXenes, and their fabrication process. Insights into various synthesis methods provide the foundation for the commercial production of MXene materials and are highlighted in this chapter.

1 Introduction

Since the discovery of Graphene (2004), 2-D materials have attracted interest from researchers around the globe [1]. Materials like hexagonal BN, MoS_2, hydroxides, and transition metal oxides have been harnessed for the early stages of 2D material research. Graphene got most of the attention among these materials due to its varied applications [2]. Due to the quantum confinement effect, these materials exhibit extraordinary physical and chemical properties. A class of 2D materials, MXenes, are structurally similar to graphene. These materials have seen tremendous research growth since Naguib et al.'s first development of Ti_3C_2 in 2011 [3]. These materials are derived from carbides, nitrides, and carbonitrides of transition metals and have a 2D sheet structure like graphene (hence MXene). MXenes are made by etching the MAX phase of these transition metals. The 2D nanostructure of MXene has enhanced the abilities of the MAX phase materials due to surface terminations and the presence of metal atoms in the intrinsic sites.

MXenes have a general formula of $M_{n+1}X_nT_x$, where M designates early transition metal, X denotes the inclusion of C or N atoms, and T represents the surface termination (–F, –O, –OH) on the outermost exposed layer of MXenes [4]. The number 'n' denotes the layers, which means n + 1 denotes the layers of the transition metal (M), which are interleaved by n layers of X. MXene synthesis requires an etching process. Specific atomic layers are etched from ternary carbide or nitride MAX phases. The MAX phase was reported by Jeitschko et al., who termed these precursors TMC, which was further changed to the MAX phase for a more prominent family of ternary carbides and nitrides [5]. MAX phase is the precursor phase of MXene, which has a typical formula of $M_{n+1}AX_n$ (n = 1, 2, or 3) with M being an early transition metal, the A is an element from the groups 13 and 14 in the periodic table, and X is usually a C or N (as shown in Fig. 1) [4, 6, 7]. They have a hexagonal layered structure with transition metal carbides and nitrides as units. MAX phase has [8]. Traditional photocatalysts with semiconductor bandgap, including but not limited to TiO_2, GCN ($g\text{-}C_3N_4$), and CdS, suffer from some drawbacks like the inability to absorb the visible-NIR light and the ability to recombine the photo-generated charge carriers [9, 10].

Some of the distinct properties of MXenes include (i) their surface morphology: atomically thin structure, and the electrical conductivity of the metal atom ensures quick and successful migration of charge carriers from bulk to surface; (ii) their surface terminations: these materials have functional groups as surface terminations. Although all MXenes are 2D nanomaterials, their properties are governed by the synthesis method used, electrode fabrication, and the method used for MXene assembly [4, 6, 12, 13].

2 Significance of MXene

MXenes have a unique set of properties that are enabled due to their structure. Being a 2D material, they have the structure of a 2D sheet with surface terminations. Unlike graphene, they are made of three or more elements and require an etching process to retain their structure from their MAX phase [14]. Since they have better electrochemical and optoelectronic properties than most of the other materials, MXenes have seen varied applications. Some of these applications include saturable absorbers, energy storage, non-linear photonics, field-effect transistors, conductive electrodes, and biomedicine, which have outshone conventional materials [9, 15]. Furthermore, the properties of MXenes can be tuned by utilising variations of ratios of M or X elements (in some cases, variation in surface terminations can also lead to enhanced properties) [16]. With these properties and advantages over traditional electrode materials, there are several drawbacks for MXenes that should be noted too. These drawbacks include easy restacking, poor stability in an oxygen-rich atmosphere, and low flexibility. Furthermore, the synthesis of MXenes requires acid etching, which utilises hazardous solutions for the procedure. This has also had an adverse effect on the commercial aspect of MXenes [3, 6, 8].

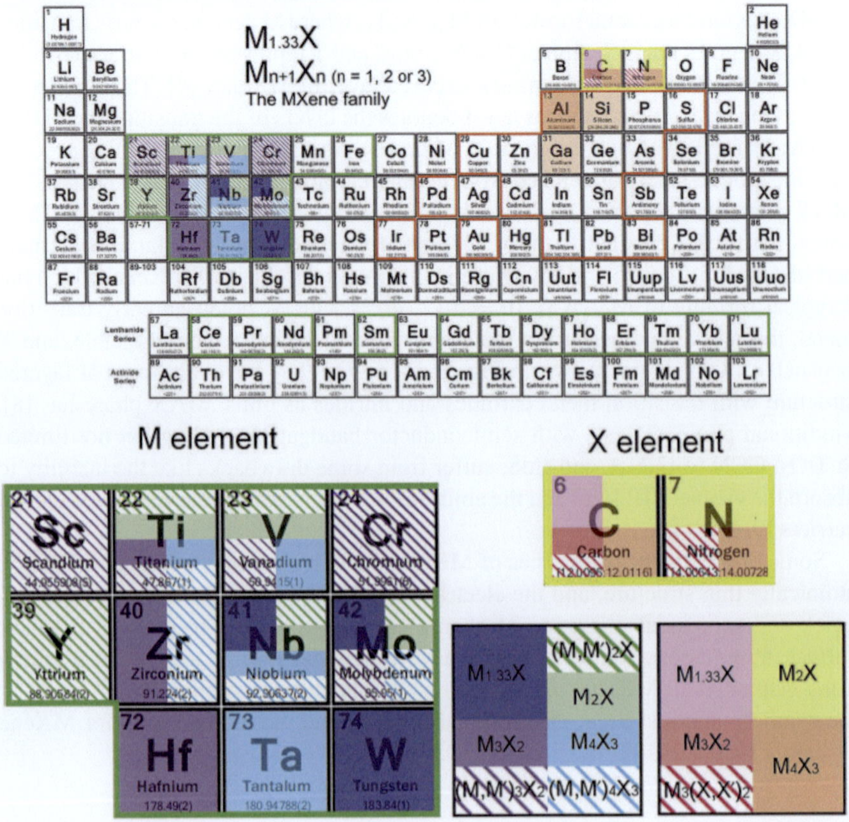

Fig. 1 Color-coded periodic table with elements that are used in the synthesis of MXenes experimentally [11]. Copyright 2019. Reproduced with permission from Elsvier

3 Fundamentals of MXenes

As stated earlier, MXenes are formed by etching the MAX phase of transition metal carbides, nitrides, and carbonitrides, specifically, the A group in their MAX phase precursor is removed to form a 2D sheet. Figure 2 inculcates some of the properties and potential applications of MXenes and MXene-composites. The structure in the MAX phase governs the structure of MXene. Extensive research in this area resulted in approximately 155 MAX phase precursors that can be formed from the combination of 14 M and 16 A elements. In the MAX phase, the bond between M and A atoms is chemically active and relatively easier to break than the M–X ionic bonds and the mixed covalent bond that forms the sheet. This is why the etching process selectively breaks M–A bonds to form MX layers (MXenes) [17, 18].

To fully establish the fundamentals of MXenes, their structure and atom placement should be properly explored. Based on the literature, the crystals of MXene have a

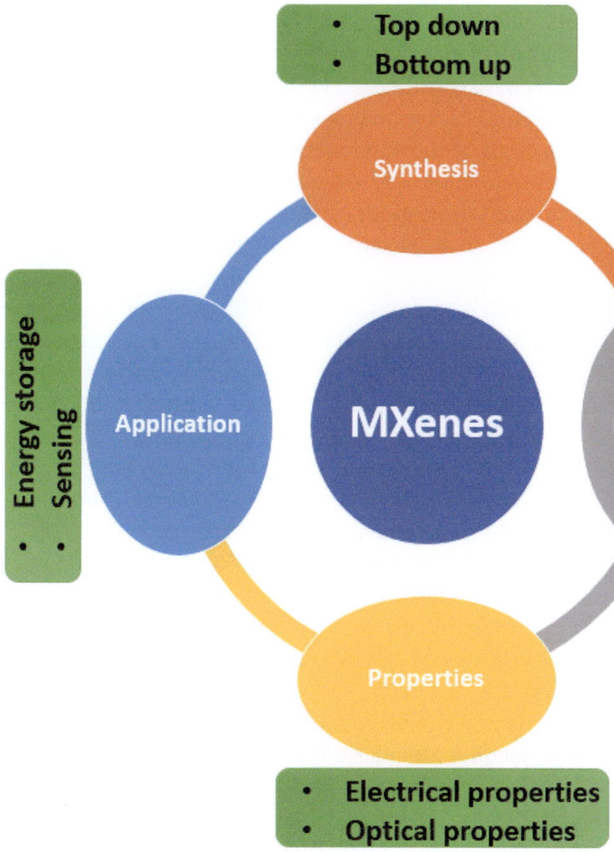

Fig. 2 MXene materials and their overview

hexagonal close-packed structure with differing M atoms. The conjugation of M atoms and X atoms can be of the order M_2X, M_3X_2 and M_4X_3 for structural stability purposes. Computer simulations help predict the properties and structure, which in turn, helps set the research orientation for synthesis [9, 17–19].

4 Properties and Applications

MXenes are metallic compounds similar to their precursor, the MAX phase. However, due to surface modifications, some tend to become semiconductors. Some properties of MXenes that make them an interesting choice of materials for future investigations include Dirac dispersion near the Fermi level, high electronic anisotropy, and low work function. The electronic properties of MXenes and their derivatives are best described in terms of density of states (DOS) and band gap energy (E_g) [20].

MXenes are generally studied under two classes which are based on their topographies (trivial and non-trivial). The functional groups present in the MXene structure further enhance the electronic behaviour of these 2D sheets. Studies show that due to the high electronegativity of surface terminations (the presence of oxygen and fluorine groups), the work functions of the structure increase, and the presence of hydroxyl groups imparts low work function due to the dipole effect. Hence, the presence of functional groups as surface terminations can impact the density of states on the fermi level and, consequently, affect the electrical conductivity and band structure [20–23].

When the MAX phase is etched, or the covalent bond breakage of M–A takes place, MXenes are formed, which can be stabilised to form free-standing films [22]. Stabilisation of these MXene layers often requires delamination intercalation or both. The composition and versatility of MXenes and their surface terminations (functional groups) can be extremely rewarding. Surface modification can enhance the properties and characteristics of these materials without destroying their original structure and morphology. This versatility bestows them with various compelling mechanical properties which greatly complement their electrical properties [20].

Research in MXene-polymer interaction suggests that MXenes can be incorporated with a variety of polymers by employing strategies of ex-situ blending or in-situ polymerisation. This is usually due to the presence of an abundant amount of active surface terminations. In the in-situ polymerisation process, strong covalent bonds are formed during the process, and only a few monomers are attached to the MXene surface [24, 25]. Therefore, ex-situ blending is preferred over in-situ due to the merits it provides. Ex-situ blending provides a well-defined polymer structure, and the composition of the MXene/polymer composite can be controlled. Because of the exceptional electronic properties of MXenes, MXene/polymer membranes have seen their use in electromagnetic interference (EMI) shielding and energy storage devices. In addition, they have also been explored in various other fields, such as filtration, sensor applications, and wearable electronics [23, 26–28].

5 Structural Overview

The structure of MXenes is similar to graphene. The sheets are derived from the MAX phase, and stability is achieved by a cross-network of metal atoms and the C/N (sometimes both are present in the structure), and the properties are enhanced by the surface terminations, without destroying the original structure and morphology. M atoms in MXenes are grouped in a close-packed configuration, akin to their MAX predecessors, whereas X atoms occupy the octahedral interstitial spaces [29]. There are three ways to pack: M_3X_2–M_3X_2, M_4X_3–M_4X_3, and M_2X–M_2X. The overall crystal structure of MXenes is hexagonally close-packed. But M atoms are now arranged differently—they are now M_3X_2 and M_4X_3 instead of M_2X. M atoms exhibit ABABAB ordering (hexagonal close-packed stacking) in M_2X, while ABCABC ordering (face-centred cubic stacking) is present in M_3C_2 and M_4C_3. For instance, while $Mo_3C_2T_x$ and Mo_4C_3Tx, where M would be in the ABCABC ordering, are unstable, Mo_2CT_x is stable. In order to create the ordered double transition metals $Mo_2TiC_2T_x$ and $Mo_2Ti_2C_3T_x$, the latter two have been stabilised by adding another M element (for instance, Ti) to the structure [30]. The surface chemistry of structures that are totally composed of surfaces governs the properties and their characteristics [22]. About 20 distinct compositions have been synthesised experimentally from the approximately 100 stoichiometric structures that theoretical calculations indicate may be obtained. Among the materials in this class that have been investigated the most are Ti_3C_2, Ti_2N, Nb_4C_3, Nb_2C, and V_4C_3. The parent Ti_3AlC_2 MAX phase's Al (A layer) was removed, forming voids between the Ti_3C_2 layers. It is evident that there are areas where the layers of nanosheets are adhered to one another by hydrogen bonds and van der Waals forces [29]. Experimentally synthesised MXenes have one of the structures as shown in Fig. 3.

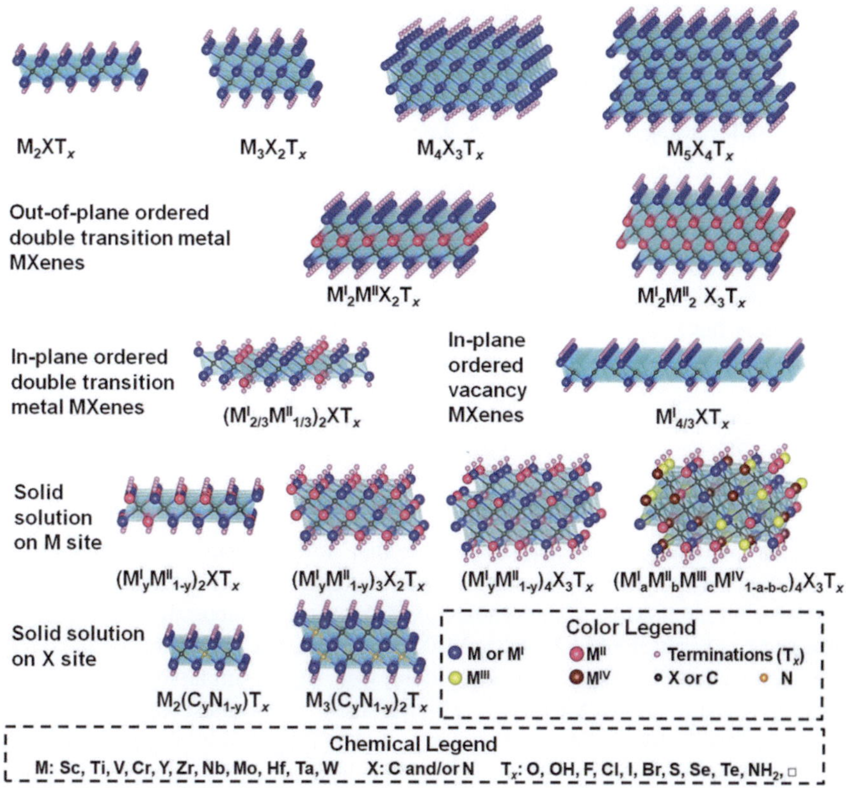

Fig. 3 Schematic representation of the structure of MXenes [31]. Copyright 2021. Reproduced with permission from Wiley

6 Synthesis Methods

Various synthesis methods have been reported in scientific articles. In general, the synthesis of MXenes requires selective splitting of MAX layers in MX layers by severing the M–A bonds. This occurs in three stages: (i) MAX phase precursor preparation, (ii) etching of the A layer that leaves a single layer of MX, and (iii) intercalation and exfoliation techniques for retrieval of these MX layers. This can be achieved via two approaches, namely, top-down and bottom-up approaches. A detailed discussion of the approaches and methods of synthesis will be discussed in the upcoming sections [20, 28, 32, 33].

6.1 Top-Down Approaches

The top–down approach, as the name suggests, requires the splitting of 3D structures into 2D nanosheets. For MXenes, this approach is described as breaking the M–A bond from the MAX phase (precursors for MXenes synthesis) to obtain a single-layered structure (MXenes) [23, 26, 34]. Synthesis of good quality MXenes is influenced by several factors: the selection of appropriate precursor material, etching agent, condition of synthesis, and the intercalation-delamination process. The top-down approach is the simplest and most conventional approach for MXene synthesis as it takes the 3D MAX phase and etches it to form 2D nanosheets. This approach poses drawbacks in the form of an etchant that is used for the etching process. Researchers have been trying to eradicate the use of acid etching and some have succeeded in this venture. Synthesis methods under this approach are as follows:

6.2 Precursors

Computational analysis has suggested more than 150 MXenes combinations from the MAX phase, but the experimental realisation of the majority of these structures is still under research. Most of the synthesised MXenes are structurally defined by the formula $M_{n+1}X_n$. In the MAX phase, M–X bonds are ionic bonds and are generally stronger than the M–A bonds (covalent nature) [35]. Bonds similar to these M–X and M–A also exist in non-MAX phases, which leads to the synthesis of MXenes from these non-MAX phases. One such example is Mo_2CT_x MXene from a non-MAX phase Mo_2Ga_2C. The energy difference between these bonds (M–A and M–X) allows the exfoliation of these precursors for MXene synthesis [36, 37]. This difference in the bond energy is the critical factor for successful exfoliation and production of stable MXene layers from their precursors. The exfoliation process requires the precursor phases to be placed in an etchant (usually acidic) or molten salt [38, 39]. The mechanical exfoliation process exploits the energy difference to isolate a single-layer crystal of these MAX and non-MAX phases by providing shear stress at M–A bonds. It is to be noted that upon practical application, other factors, such as the choice of etchant and the procedure also play a significant role in successful MXene production [40]. Tremendous research efforts have been put forth for the synthesis and commercialisation of MXene-based materials; a brief timeline for the evolution of synthesis methods is shown in Fig. 4, a timeline showing the development of MXene synthesis. By using HF etching, the first MXene ($Ti_3C_2T_x$) was produced in 2011. 2013 saw the successful delamination of the accordion-like multilayer MXenes to monolayer using DMSO. MXenes were prepared in 2014 using HCl + LiF etchants and bifluoride salts etching [41]. TBAOH has shown itself to be a versatile organic intercalation agent in 2015 [42]. The bottom-up CVD approach for MXene synthesis was disclosed in the same year. Single-layer MXene nanosheets were, however, produced by the mechanical stripping technique. 2016 saw the preparation of MXenes using the non-MAX phase. Melted salt etching with fluorine was used to successfully prepare the first nitride MXenes. Afterwards, a MILD approach was developed from the HCl + LiF etching technique. In 2017, MXenes were prepared using electrochemical etching. 2018 saw the successful synthesis of fluorine-free MXenes using concentrated alkali hydrothermal etching [43]. A paper on fluorine-free molten salt etching was published in 2019. The molten salt approach of universal Lewis acid was reported in 2020 [44].

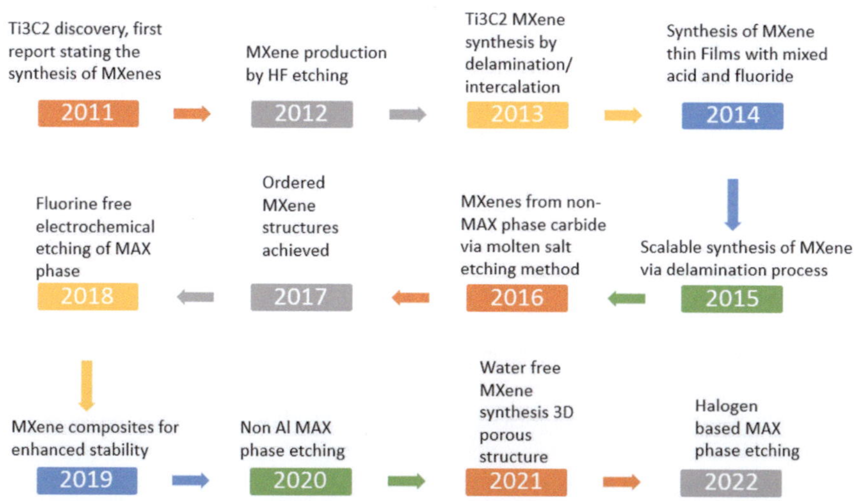

Fig. 4 Timeline of MXene synthesis and synthesis methods. The HF etching method was the first method used for MXene synthesis, and recently, halogen etching has been used successfully on a laboratory scale

6.3 Chemical Etching Process

The top-down approach for MXene synthesis requires the selective removal of the A atomic layer from the precursor, which leaves M–X layer intact and stable. Gogotsi et al. saw success in etching the $TiAlC_2$ phase into a 2D nanosheet structure of composition $Ti_3C_2T_x$ using hydrogen fluoride as an etchant [35]. The chemical reactions that lead to the synthesis of this MXene can be described as:

$$Ti_3AlC_2 + 3HF \rightarrow AlF_3 + 3/2\ H_2 + Ti_3C_2 \quad (1)$$

$$Ti_3C_2 + 2H_2O \rightarrow Ti_3C_2(OH)_2 + H_2 \quad (2)$$

$$Ti_3C_2 + 2HF \rightarrow Ti_3C_2F_2 + H_2 \quad (3)$$

This method is by far the most common method for producing MXenes and can be used on a variety of MAX phases [39]. Figure 5 shows the typical synthesis of MXene from the MAX phase using the HF etching method. The exfoliation of layers of the MAX phase into a 2D structure produces MXenes, which can be seen in Fig. 5a. The synthesized materials are shown in Fig. 5b, and the material characterization results are depicted in Fig. 5c, d. The hydrogen fluoride etchant can be replaced with some other chemically active etchant. A study done by Ghidiu et al. reported the use of LiF and HCl as an etchant to fabricate $Ti_3C_2T_x$ nanoflakes [45]. The use of lithium fluoride and hydrochloric acid over hydrogen fluoride significantly increased the yield of MXene. Similar to this study, a combination of KF, NaF, FeF_3, and NH_4F with HCl has also been reported [45].

6.4 Electrochemical Etching

Chemical etching may produce materials with improved corrosion resistance and electrical conductivity; the layers are seldom singular. This means that although the materials have enhanced properties of the MAX phase, they still need to be modified as only a few nm of depth from the phase surface got etched [47–49]. This limitation of chemical etching can be overcome through the process of electrochemical etching as shown in Fig. 6. The exfoliation at the molecular level can be seen in Fig. 6a, and the process used for exfoliation can be seen in Fig. 6b. The synthesized materials Fig. 6c, d are characterized for their crystallinity via XRD scans in Fig. 6e. This one-step procedure for producing energy storage devices with MXene was introduced in 2020. The production of MXene and the subsequent multistage assembly of devices are the two processes that the authors suggested combining [42]. The fact that neither acidic nor alkaline etchants are used in this procedure, nor does any reaction involve contact with the outside world, makes it unique as well. Feng et al. utilised this

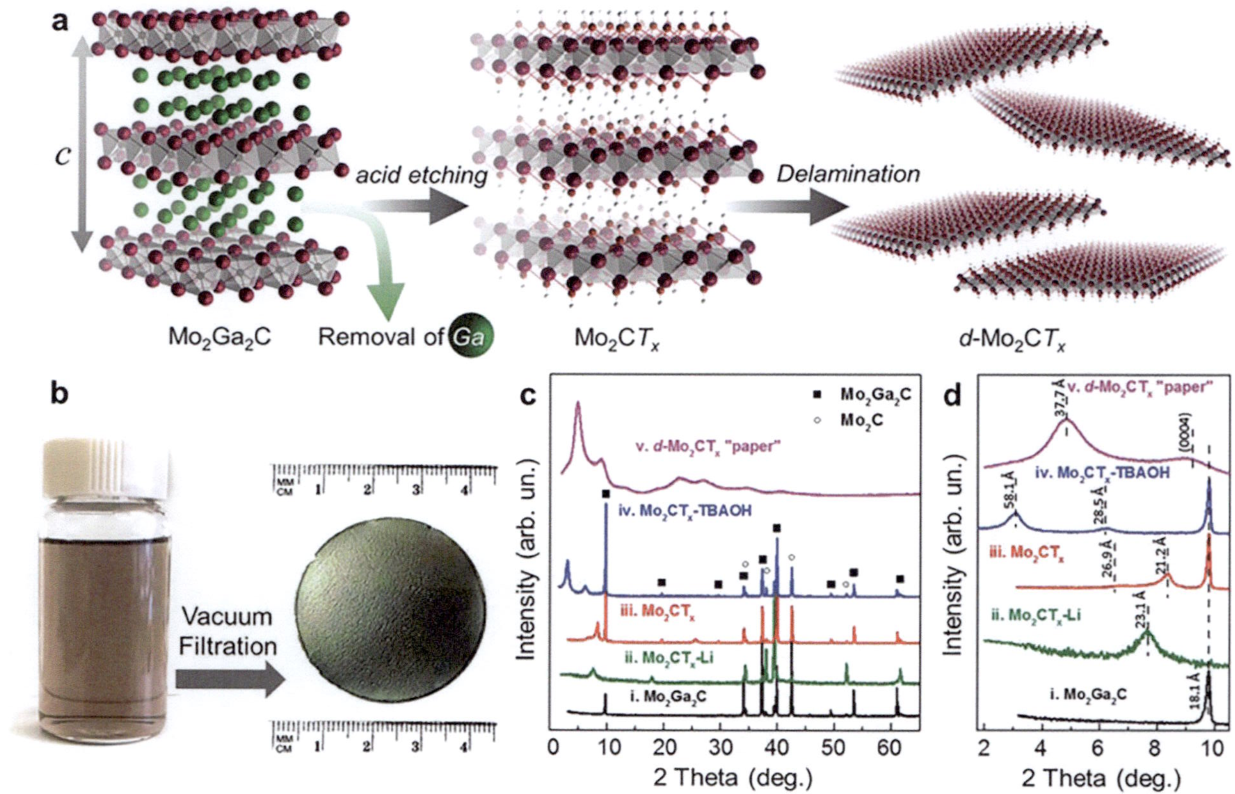

Fig. 5 **a** Schematic showing synthesis and delamination of Mo_2CT_X MXene. **b** Digital photos of delaminated and free-standing "paper" produced by filtration of delaminated solution. **c** XRD patterns of, i. Mo_2Ga_2C (black), ii. Mo_2CT_X–Li (green), iii. Mo_2CT_X (red), iv. Mo_2CT_X (blue) intercalated with TBAOH, and v. (purple) free-standing "paper" after TBAOH intercalation. **d** XRD for 2Θ range from 2 to 10.5°, showing the shift in the 0002 peak [46] Copyright 2016. Reproduced with permission from Wiley

method to produce $Ti_3C_2T_x$ MXene sheets from bulk Ti_3AlC_2 anode. The binary aqueous solution of NH_4Cl and TMAOH (tetramethylammonium hydroxide) was used as an electrolyte in a two-electrode system for the electrochemical etching process [50]. This approach of MXene synthesis has several advantages associated with it: (i) the bonds between M–A can be easily broken by rapid anodic etching induced by chloride ions, (ii) the exfoliated layers of MXenes have only –O and –OH terminations (the structure is devoid of –F terminations), and (iii) this method provides high yield for single- or bilayer MXenes (reported yields are as high as 90%) [40, 51]. Pang et al. demonstrated a thermal-assisted electrochemical etching method for the production of a variety of MXenes, in which, mild heat facilitates the etching process of the MAX phase [52].

6.5 Intercalation

Intercalation refers to the selective etching of the A-layer elements from MAX phases, followed by the insertion of various species between the MXene layers (Fig. 7). Intercalation of Ti_3C_2 with hydrazine monohydrate (HM) showed an increase in their c-lattice parameters (c-LP), which contributed to its higher electrical resistivity than other non-intercalated samples [54–56]. Various intercalating agents have been reported, including various metal ions like Li^+, Na^+, and K^+. Organic molecules like dimethyl sulfoxide (DMSO) and tetrabutylammonium hydroxide (TBAOH) can also act as proper intercalants [57–59]. These agents act as a barrier and expand the interlayer spacing of MXenes, which leads to weak interlayer interactions that can be broken easily to form nanosheets.

The organic base, TMAOH, reacts with the Al atomic layer in the gallery to promote key processes, including cleavage of the Ti–Al bonds through Al hydrolysis, modification of the TC unit surface with $Al(OH)_4^-$, and the intercalation of bulky TMA^+ into the gallery, in a single step. The cleavage of Ti–Al metallic bonds and bulky-ion intercalation facilitates the subsequent disassembly of the precursor-layered crystals into their elementary layers [60].

Several intercalation methods can be used, such as the hydrothermal method, introduction of a repulsion group, ion-assisted intercalation, etc. These methods inhibit the formation of stable layers of said MXenes once the MAX phase is etched, and the resultant species will be rich in functional groups such as surface terminations. Halim et al. demonstrated a large-scale synthesis method that was the first of its kind [46]. They prepared Mo_2CTx MXenes with a single-layer nanosheet structure via the intercalation method with TBAOH as the intercalating agent. Upon characterisation of the synthesised materials, the XRD pattern showed two peaks corresponding to c-LPs of 28.5 and 58.1 Å, respectively. Another report by Lipatov et al. demonstrated the synthesis of MXene flakes via two synthesis routes: chemical etching and intercalation [24]. The flakes derived from the intercalation process demonstrated superior

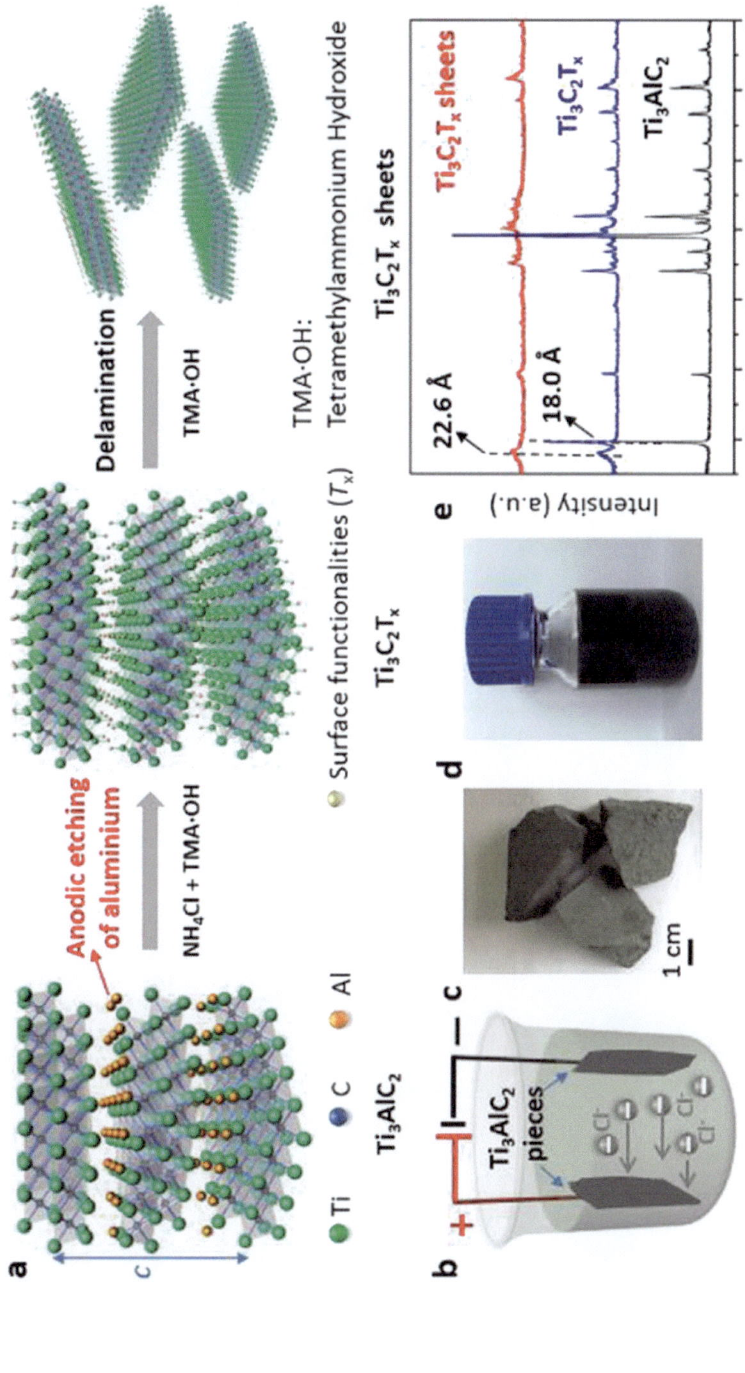

Fig. 6 Electrochemical etching of bulk Ti_3AlC_2 in a binary aqueous electrolyte. **a** Schematic of the process. **b** The configuration of an electrochemical cell. **c** Optical image of the as-received bulk Ti_3AlC_2. **d** The aqueous dispersion of delaminated $Ti_3C_2T_x$ (0.15 mg mL^{-1}). **e** X-ray diffraction patterns of Ti_3AlC_2, $Ti_3C_2T_x$, and $Ti_3C_2T_x$ film [53] Copyright 2018. Reproduced with permission from Wiley

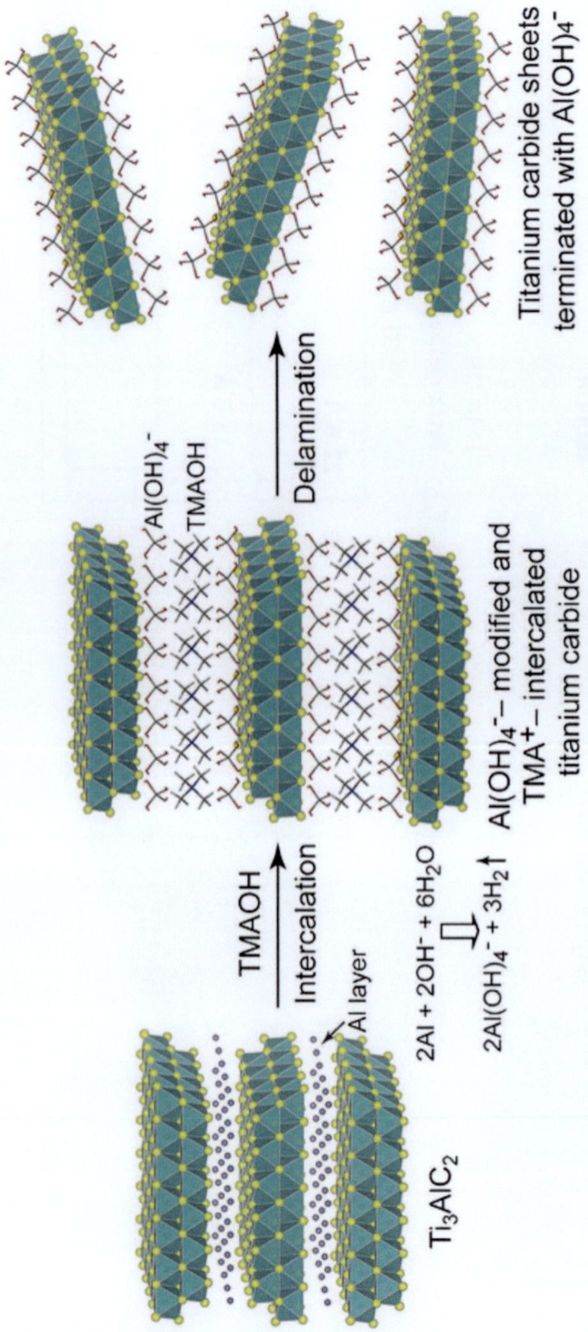

Fig. 7 Schematic illustration showing the intercalation and delamination process [60]. Copyright 2016. Reproduced with permission from Wiley

properties than the flakes derived from the chemical etching method. These flakes had a larger size, uniform morphology, and cleaner surface, which can be attributed to the intercalation method.

6.6 Delamination

Delamination refers to the process of exfoliating the layers of the MAX phase into MXenes to produce the desired nanosheets (Fig. 7 demonstrates both intercalation and delamination processes used in MXene synthesis). The process involves the removal of A layer from the MAX phase. This leaves behind thin layers of MX, also called MXenes. Delamination is a necessary procedure to split a layered material into stabilised individual layers so that probing for their characteristics can be done in their 2D state [61]. The necessity also arises due to the fact that a multilayered MXene has many times stronger interlayer interactions than graphite; thus, superficial mechanical exfoliation produces low-yield layers [62]. Intercalation is a prominent and effective way of delamination as it can be used to delaminate m-MXenes into d-MXenes by equipping appropriate intercalants.

6.7 Exfoliation

Simulations regarding the synthesis and structure of MXenes from MAX phases via chemical etching approaches with HF as etchant leads to the adsorption of H and F atoms on the surface of metal atoms that, in turn, weakens the bond between metal ions [63]. This creates mixed surface terminations for the MXene, which suggests that the Ax distribution is largely independent of the composition. As per reports, the M–A bond strength in carbide MAX phases is a detrimental factor for exfoliation energy calculation within the structure of the desired MXene. This energy also depends on the M–A composition for both the case of nitrides and carbides. The M–A composition with smaller atoms and higher valence electrons has the tendency to form stronger bonds than their counterparts. This, in turn, leads to diminishing the difference between the M–X and M–A bond energy [64–66]. A practical example to explain this case will be the effect of an M atom with higher valency (it should require lower exfoliation energy) for Ti_3AlC_2. Upon comparison with a theoretical MAX structure of Mo_3AlC_2, the relative ease of exfoliating Ti_3AlC_2 compared to the Mo_2AlC_2 reflects the importance of M–A composition.

7 Bottom-Up Approaches

Bottom-up approaches for the synthesis of MXenes involve starting from atomic or molecular precursors and building up the MXene structure layer by layer, typically through chemical reactions or physical processes. While using bulk precursors offers scalability advantages, researchers often prefer creating materials from their elemental composition due to the precise control it affords over material chemistry, allowing for tailored customisation of material designs. Notable bottom-up synthesis methods for MXenes are Molten salt-assisted synthesis, chemical vapour deposition (CVD), and plasma-enhanced pulsed layer deposition (PE-PLD). These methods enable scientists to fine-tune the properties of MXenes, such as thickness, surface chemistry, and electronic properties, making them highly desirable for a range of applications from energy storage to catalysis [67].

7.1 Precursors

Synthesis of MXenes using bottom-up approaches, requires precise control over precursor materials to achieve the desired properties and structures. Typically, MXenes are synthesised from their parent MAX phases, which are ternary compounds composed of transition metals (M), group A elements (A), and carbon or nitrogen (X). The selection of suitable precursors is a crucial aspect of the bottom-up synthesis process. Transition metals like titanium, vanadium, and niobium are often chosen because they form stable carbides or nitrides under appropriate conditions [68]. Carbon and nitrogen sources like graphite, graphene, or carbonitride precursors are used to extract the A layers from MAX phases, leaving behind MXene layers. The choice of etchants is also essential in the synthesis process. They are used to selectively remove the A layers to expose the MXene structure. Common etchants include hydrofluoric acid (HF) or its derivatives, which dissolve the A layers without affecting the MXene layers [69]. The successful bottom-up synthesis of MXenes with desired properties and functionalities requires careful consideration of precursor materials. Precursor selection is vital for tailoring the composition, morphology, and properties of synthesised MXenes, influencing their applications in various fields [70].

7.2 Molten Salt-Assisted Synthesis

Molten salt synthesis is a highly significant method for producing MXenes. Key milestones in the advancement of molten-salt-assisted synthesis methods for MXene production are depicted in Fig. 8. This process involves exploiting the distinctive cleavage energies within MAX (M: transition metal; A: group 13 or 14 element; X: carbon or nitrogen) phases to selectively etch layers and generate MXenes. Salt systems, which encompass both positively and negatively charged ions, offer versatile pathways for MXene synthesis due to their thermodynamic advantages at elevated temperatures [71]. By interplaying various salts, eutectic mixtures can be obtained, enabling lower melting temperatures and precise control over reaction conditions. Nitrate, chloride, and carbonate salts are commonly employed in these systems, facilitating the synthesis of MXenes with adjustable characteristics. Molten salts not only serve as reaction media but also as designer solvents, manipulating the physicochemical properties of reactants to enhance dissolution and modify surface properties. However, managing salt quantities is crucial to prevent excess salt from hindering reaction kinetics and complicating post-treatment processes, as observed in some systems [72]. Additionally, the solubility of precursors in molten salts can influence reaction rates, particle size distribution, and morphology [73].

Pure molten-salt etching represents a pioneering method demonstrated in the early synthesis of cubic MXenes. Notably, the use of molten salts has enabled the synthesis of nitride-based MXenes, which were previously challenging with aqueous acidic solutions [75]. Innovations in molten-salt-assisted synthesis include the noble-metal replacement reaction and element-exchanging methods, offering avenues for engineering MXenes with tailored properties and compositions. These techniques have broadened the scope of MXene synthesis, enabling the production of MXenes with high lateral dimensions and controlled morphologies [76]. Moreover, recent advancements in Lewis-acidic molten salt etching have provided avenues for selective and controlled removal of 'A' layers from MAX phases, offering insights into MXene synthesis mechanisms and paving the way for the development of new MXene materials with enhanced properties and functionalities [77].

7.3 Chemical Vapor Deposition (CVD)

The Chemical Vapor Deposition (CVD) method has emerged as a promising avenue due to its cost-effectiveness, scalability, and precise control over synthesis conditions. This technique has found widespread application in the production of MXenes, particularly in generating single-crystal samples. It is well-established that several key growth parameters, including temperature, pressure, gas flow rate, and catalyst type, dictate the outcome of the CVD process [78]. Through careful adjustment of these parameters, researchers can exert precise control over the size, morphology, layer number, and quality of the resulting MXenes materials [79]. Against this backdrop, there has been a notable increase in research efforts focused on fabricating high-quality MXenes using the CVD method in recent years. This trend underscores the growing recognition of CVD as a versatile and practical approach for synthesising MXenes with tailored properties and characteristics [61]. One notable

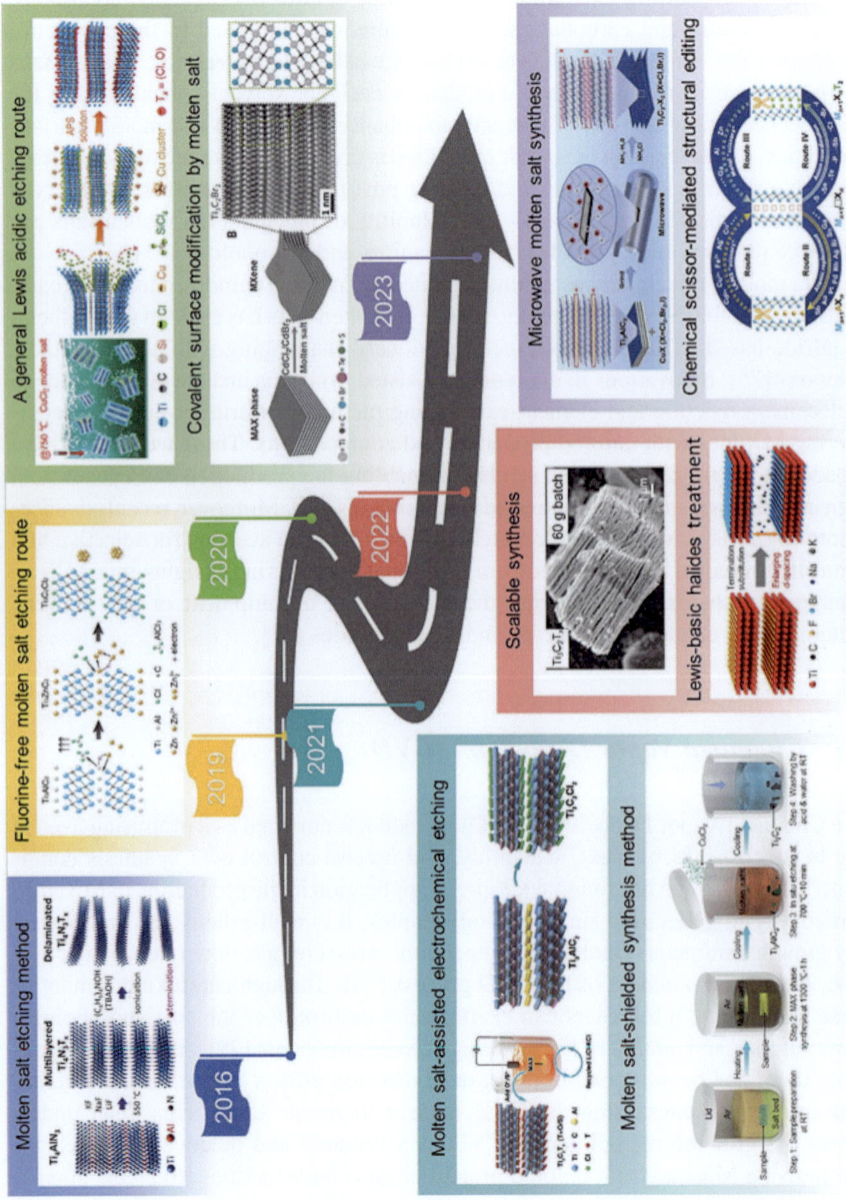

Fig. 8 Chronology of molten-salt-assisted synthesis techniques for MXenes [74]. Copyright 2023. Reproduced with permission from Elsevier

synthesis method involves the large-scale production of ultrathin α-Mo_2C crystals using methane as a carbon source and copper foil as a catalyst, as demonstrated by Cheng's group [80]. Through meticulous adjustment of growth parameters such as temperature and time, uniform α-Mo_2C crystals with nanoscale thicknesses were achieved. Another significant development in MXene synthesis via CVD is the utilisation of gold (Au) cluster/α-Mo_2C catalyst systems, as evidenced by recent research [81]. This approach explores the catalytic capabilities of Au clusters, offering potential advantages over traditional copper catalysts.

7.4 Plasma-Enhanced Pulsed Layer Deposition (PE-PLD)

The synthesis of MXenes by the Plasma-Enhanced Pulsed Layer Deposition (PE-PLD) method involves a sophisticated process that utilises plasma to create thin layers of MXene materials. PE-PLD combines aspects of both pulsed laser deposition (PLD) and plasma-enhanced chemical vapour deposition (PECVD) techniques. In this method, a target composed of precursor materials, typically MAX phase powders, is bombarded with high-energy laser pulses, causing ablation and ejection of atoms or molecules from the target surface. These ablated species then deposit onto a substrate in a controlled environment under the influence of a plasma. Zhang et al. utilised PE-PLD to synthesise titanium carbide ($Ti_3C_2T_x$) MXene thin films from Ti_3AlC_2 MAX phase precursors, demonstrating a facile route for large-scale production with tunable properties [82]. Similarly, Yang et al. reported the synthesis of vanadium carbide (V_2CT_x) MXene films via PE-PLD, showcasing the versatility of this method for synthesising diverse MXene compositions [83]. These studies highlight the potential of PE-PLD as an effective technique for the scalable production of MXene.

8 Fabrication Techniques

Nanomaterials have been proven to be extraordinary class of materials. Demands for commercialisation of these materials have seen a massive upsurge in the past few years. Commercialisation of any material requires the fabrication of components from the material. In other words, the material should be stable, cost-effective, readily available, and sustainable. Various nanomaterials have been studied and applied for practical application, but only a few have been successful in the commercialisation process. 2D nanosheets, like MXenes, have extraordinary responses individually, but synthesis and regulation of these sheets for a scaled-up process provide a new direction for research.

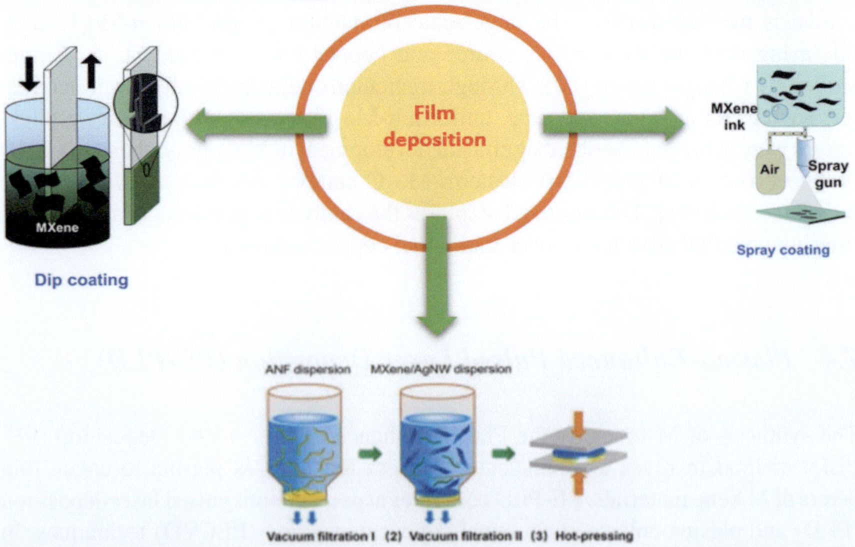

Fig. 9 Film deposition as a fabrication method for MXene-derived materials. Three commonly used methods for film deposition include vacuum filtration, spin coating and spray coating [84]. Copyright 2023. Reproduced with permission from Wiley

8.1 Film Deposition

Thin films from MXenes can be synthesised using various reported methods, including spray coating (Fig. 9), spin coating (Fig. 9), vacuum filtration (Fig. 9), interfacial assembly, electrochemical polymerisation, spin-spray coating, physical vapour deposition, etc. [85–87]. Some methods are under research for further efficiency/yield enhancement. Among these methods, the most common include spray coating, spin coating, and vacuum filtration, as shown in Fig. 9. These methods are discussed further:

8.2 Spray Coating

As the method indicates, the MXene/MXene-composite solution is pressurised and dispensed in the form of aerosols. When a compressed liquid that is stored in a volume-defined container is dispensed, evaporation takes place, and the solution is converted into aerosols as presented in Fig. 9. The dispersion of these aerosols is directional and controllable [88]. Deposition of these aerosols is also controllable by tuning the pressure and distance between the nozzle and substrate. Cai et al. reported the fabrication process of MXene-CNT nanostructures on latex as substrate using the spray coating technique [89]. The coating was provided layer-by-layer for a smooth surface and involved spraying of individual layers of both MXene and CNT solutions on the substrate, followed by drying using a nitrogen gun.

Other methods, like the spin coating method or vacuum filtration method of fabrication, have higher Rs values that can be attributed to inter-flake distancing and defects during fabrication. Spray coating is a solution-based method that has emerged for the fabrication of defect-free MXene structures into thin films for application [90, 91]. This method is reported to be the fastest fabrication method among others. The thickness of these coated films can be affected by various factors like spray flow, spray speed, substrate, and frequency of spraying. Gogotsi et al. reported the fabrication of MXene films with varying thickness from 5 to 70 nm using this method [92].

8.3 Spin Coating

The spin coating method is a standard method for the fabrication of thin films of different materials. Subsequently, it has been employed to fabricate MXene thin films. As seen in Fig. 9, this method utilises centripetal force as the governing principle for the fabrication process; a substrate is spun at high speed with a liquid solution on its surface. As the spinning goes on, centripetal force and surface tension act on the fluid, and an even layer of the liquid is distributed on the surface of the substrate. Removal of solvent by evaporation helps in attaining thin films with widths ranging from a few nanometers to a few micrometres. The governing factors of this method include the rotation speed and the concentration of MXene in the solution present. In simple terms, higher RPM leads to less area coverage by the film on the substrate surface; on the other hand, a higher concentration of MXene in the solution will cover the larger surface area on the substrate and increase the chances of deposition of a uniform film. These parameters govern the thickness and uniformity of the deposited film and can be used to manipulate the thickness and uniformity of the deposited film. Zhang et al. reported the fabrication of Ti_3C_2 thin films on a silica substrate by using this method [93]. The analytical characterisation of the deposited film by performing different microscopic techniques like electron microscopy and AFM provided a thickness of 2.2 μm. Another report by Huang et al. worked on the fabrication of Ti_3C_2 thin films using this method. In this work, the silica substrate was pre-treated with hydrophilic groups for uniformity and integrity of the deposited thin film [94]. They reported the deposition of Ti_3C_2 thin film on the silica substrate with a film thickness of 142 nm.

8.4 Vacuum Filtration

For any thin film deposition process, uniformity and integrity of the film are required for successful application. As seen in Fig. 9 the vacuum filtration method or vacuum-assisted filtration method is the simplest method for the fabrication of MXene thin films [95]. This method requires a dispersion solution of MXenes for film fabrication.

Briefly, the dispersion solution is fed to a suction filtration device where MXene sticks to the porous membrane's surface after the filtration process and forms a layer. The thickness of the fabricated film depends on the MXene concentration, and filtration provides the uniformity of the film on the filter membrane. This layer on the filter membrane is dried off and peeled for further application [96].

9 3D Printing

The field of advanced materials synthesis and fabrication has seen a remarkable transformation with the emergence of 3D printing. The precision and versatility of this technology have revolutionised the production of MXenes, enabling the fabrication of complex structures with unprecedented intricacy and efficiency. Additive manufacturing processes allow for the creation of intricate geometries and tailored architectures of MXenes-based components, offering precise control over their properties and functionalities [97]. The use of 3D printing to fabricate MXenes opens up new possibilities in fields such as energy storage, electronics, and biomedical applications. The ability to customise MXene structures on a microscale level holds the promise of enhanced performance and novel functionalities, driving the advancement of MXene-based materials toward real-world applications and technological breakthroughs [98]. As researchers explore the interactions between 3D printing and MXene fabrication, the possibilities for material design continue to expand, paving the way for a future of limitless potential [99]. Figure 10 outlines the primary printing methods utilising MXene and their associated applications.

9.1 Traditional Techniques for Printing of MXenes

The text outlines two prevalent methodologies in 3D printing, Direct Writing, and Inkjet Printing, highlighting their capability to utilise standard ink-like substances, which are essentially MXenes dispersed in solvents. However, it also notes that alternative 3D printing methods might necessitate specific material engineering or integration techniques for effective adaptation. Detailed explanations of each method are provided as follows.:

Inkjet Printing is identified as a method ideally suited for materials in liquid form, offering material savings. The technique employs inks, which are solutions or dispersions of a solute in a solvent. It operates by propelling a precise quantity of ink from a nozzle into a chamber by rapidly reducing the chamber volume in a quasi-adiabatic manner through piezoelectric action. An external voltage causes the liquid-filled chamber to contract, resulting in a shockwave that forces a droplet of liquid out of the nozzle [101]. Electrodes produced via IJP are characterised by their thin-film nature and increased porosity, which enhances rate capability. Notably,

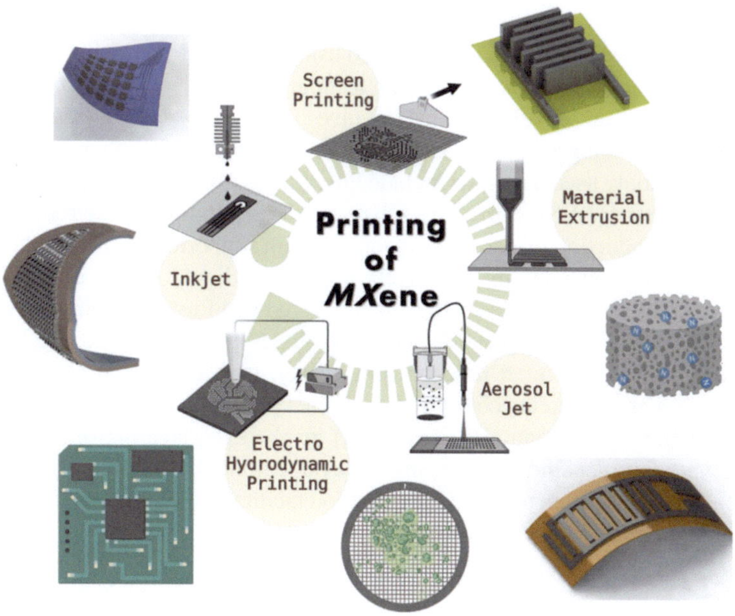

Fig. 10 Provide an overview of the advancements in MXene-based printing methods and their associated applications [100]. Copyright 2023. Reproduced with permission from Springer

this technique has been leveraged to make lithium-sulfur batteries with considerable capacities, exemplified by an 850 $mAhg^{-1}$ sulfur capacity [102].

Direct Ink Writing relies on the extrusion of ink that acts as a paste under conditions where it exhibits shear-thinning behaviour. The inks for DIW are typically printed material mixed with a solvent or carrier fluid and various additives. These inks are pressured out of a nozzle, creating a continuous ribbon-like structure that can be manipulated in terms of width, height, and form. The ink then solidifies through mechanisms like solvent evaporation, thermal curing, or UV radiation based on the specific ink formulation and its intended application [103].

Freeze Nano Printing is presented as a nanofabrication strategy that employs cryogenic temperatures to fabricate intricate nanostructures precisely. This technique involves depositing material onto a substrate and then rapidly freezing it to achieve a solid, crystalline form [104]. The solidified structure undergoes an etching process that selectively removes material to achieve the desired pattern or structure. By integrating Inkjet Printing and Direct Ink Writing with traditional freeze-casting methods, complex graphene aerogels with structured designs have been developed using FNP, suggesting the potential for superior specific power and energy [105].

Stereolithography uses a laser to incrementally solidify layers of a liquid photopolymer resin incrementally, thus constructing a 3D object. The method involves selectively exposing a vat of liquid photopolymer resin to a laser beam, causing the resin to solidify and bond to the layer below. This layering process continues until the object is fully formed [106]. Various ceramic/polymer dielectric capacitors have been successfully fabricated using a blend of traditional tape-casting ceramic production and SLA 3D printing techniques [107].

Fused Deposition Modeling (FDM) involves the layer-by-layer construction of a 3D object by extruding a thermoplastic filament, heated to its melting point, through a nozzle. Commonly used materials include Poly Lactic Acid (PLA) and Acrylonitrile Butadiene Styrene (ABS), with active substances incorporated for electrode printing, enabling the production of porous components with supporting substrates [108].

Binder Jetting operates by selectively applying a liquid binder onto a bed of powder to create a solid part, which is then sintered in a furnace to bind the powder particles together and eliminate any residual binder. This technique is capable of producing porous structured objects from plaster, with the option to introduce a secondary material to create composite structures [109].

Laminated Object Manufacturing (LOM) involves the layering of materials by cutting sheets into specific shapes using a laser or cutting tool, applying an adhesive, and then laminating them together. This method facilitates the creation of 3D models by bonding layers of different materials [110].

9.2 Printing Parameters

Efficient and cost-effective preparation of MXenes on a large scale requires strict adherence to benign reaction conditions, including ambient air, non-corrosive etchants, and room temperature. Machine learning models are necessary to decipher the complex chemistry and degradation processes intrinsic to MXene-based energy storage devices, ensuring precise prognoses. To ensure reliable chemical stability, aqueous suspensions of single- or few-layer Ti_3C_2 sheets derived from reconfigured Ti_3AlC_2 are crucial. However, extensive research efforts are necessary to achieve stringent control over MXene's chemical stability. MXene's surface area must be enhanced to facilitate electrolyte infiltration and shorten ion diffusion lengths, enabling improved rate capability in practical applications. Strategies to enhance interaction with interfaces in MXene-based nanocomposites are necessary to prevent resistance, pulverisation, exfoliation, and aggregation during cycling.

Transitioning from material characterisation to actual testing is vital for MXene-based complete cells. Generating full-cell electrochemical data at substantially higher current densities and more significant areal mass loadings is essential for advancing the development of grid energy storage devices. Additionally, exploring alternative MXenes, such as vanadium (V) and niobium (Nb)-based variants, opens up promising avenues for upgraded applications beyond traditional lithium-ion batteries. Research efforts encompassing diverse energy storage systems, including aluminium-ion, potassium-oxygen, calcium-ion, lithium-carbon dioxide, and magnesium-ion batteries, are indispensable to addressing contemporary environmental and energy challenges comprehensively. Optimising printing parameters for

MXenes requires a multifaceted approach involving large-scale production, chemical stability enhancement, surface area augmentation, interface optimisation, and comprehensive electrochemical testing. These endeavours are pivotal in unlocking the full potential of MXenes in diverse energy storage applications, contributing to sustainable energy solutions for the future.

9.3 Challenges in 3D Printing Processing

Challenges in 3D Printing Processing encompass a broad spectrum of issues ranging from optimisation of shape design to productivity constraints and health and safety concerns. The optimisation of shape design poses a significant challenge in achieving cheaper production costs by minimising raw material usage, manufacturing time, and energy requirements while maximising component performance factors like weight, strength, and surface area [111]. However, achieving an optimal design space is complex and often requires a delicate balance between various design parameters. However, pattern-making for additive manufacturing (AM) introduces complexities in incorporating hierarchical complexity within components and integrating multiple materials, necessitating the development of new design tools and reconsideration of conventional design methodologies [112]. Pre- and post-processing stages are crucial for refining the printed components, involving tasks such as support removal, surface finishing, and defect detection, each presenting unique challenges [113]. Various methods used for printing, including layered fabrication and alternative approaches, come with their own set of benefits and drawbacks, affecting the mechanical properties and quality of the printed objects [114]. Controlling errors throughout the printing process, including data preparation, processing, and material issues, remains a persistent challenge due to the need for standardised quality assurance protocols [115]. Printing with multiple materials introduces additional complexities in modelling, production, and material compatibility, necessitating advanced CAD systems and careful consideration of material science principles [116]. Maintenance and hardware issues, such as machine configuration and upkeep, contribute to production cycle delays and require continuous monitoring to maintain accuracy and efficiency [117]. Alignment, slicing problems, and speed constraints further complicate the printing process, impacting build quality, accuracy, and throughput [118]. Health and safety considerations, including the emission of harmful particulates and volatile organic compounds, highlight the importance of stringent pollution control measures in 3D printing facilities [119]. Moreover, cybercrime and copyright violations pose risks to intellectual property protection as 3D printing technology becomes more ubiquitous. Low productivity and scalability limitations hinder the widespread commercialisation of 3D printing, particularly for mass production applications, necessitating ongoing technological advancements and regulatory frameworks to address these challenges effectively [120].

10 Composite Formation

MXenes have several properties that boast the advantages of them over traditional materials. Some of them include their electrochemical and optoelectronic properties and their surface terminations that grant them adsorbing capabilities. Even with these properties, there are several drawbacks associated with the use of MXenes. Some of these drawbacks include easy restacking, poor stability in an oxygen atmosphere, low flexibility, etc. In order to commercialise and scale-up the production process, stabilisation of MXenes can be done by using MXene-based composites. Composites are conjugations of two or more materials in order to combine the benefits of each material and surpass the advantages of individual materials. Reports suggest that MXene materials can be integrated successfully with other materials for performance enhancement. These materials include metals, polymers, ceramic materials, nanotubes/wires, metal–organic-frameworks, and various other nanomaterials. Some of these composites will be discussed in this section.

10.1 Polymer-MXene Composites

As discussed earlier, the drawbacks of MXene structures hinder the scale-up process of their production. In order to tackle the situation and increase their stability, MXenes can be used in conjugation with other materials to make composites. One such candidate is the use of polymers, as they have the advantages of low production cost, easy modification processes, flexible structure, etc. MXenes have abundant surface termination that allows them to be incorporated with a variety of polymers via ex-situ blending or in-situ polymerisation.

However, due to insufficient polymerisation energy, the polymerisation of MXenes with monomers is restricted to a handful of monomers. These monomers can be polymerised on the MXene surface via physical agitation, UV irradiation or electrochemical polymerisation. The prediction of electron transfer was made by Chen et al., who reported two main types [121]. According to them, when they performed charge-transfer-induced polymerisation with the monomer 3,4-ethylenedioxythiophene (EDOT), the electrons could shift from high energy orbits (HOMO) of EDOT to low energy orbits (LUMO) of MXenes or vice-versa. For successful electron transfer, the distance between monomer and MXenes should be as low as possible. So far, many polymers like poly vinyl alcohol (PVA), poly ethylene oxide (PEO), polybenzimidazole (PBI), poly vinylidene fluoride (PVDF), etc., have been incorporated for the fabrication of MXene-polymer composite.

As most of the MXene structures have a negative charge on their surface terminations, they facilitate electrostatic solid interaction with positively charged polymers.

11 Metal-MXene Composites

Metal ions/metallic species can be incorporated with MXenes to produce a Metal-MXene composite that benefits from the conductivity of both metal ions and MXenes. These composite structures can be synthesised using the chemical reduction method. Briefly, the chemical reduction method produces metal nanoparticles, which can either react with the surface terminations of MXene or get trapped in the interstitial sites [122]. Rakhi et al. synthesised a novel nanocomposite of $Au/Ti_3C_2T_x$ by chemical reduction method and employed the material for sensing application. Liu et al. synthesised a bimetallic nanocomposite of RhNi/MXene via a facile wet chemical synthesis method [123]. The synthesised nanocomposite showed remarkable hydrogen production from hydrous hydrazine. Han et al. produced surface-modified MXenes by annealing $Ti_3C_2T_x$ nanosheets at a very high temperature (800 °C in Ar atmosphere) with TiO_2 nanoparticles and amorphous carbon. This annealing process produced a sandwiched nanostructure of the MXene with nanoparticles and carbon [124].

Nanostructure, exceptional mechanical properties, electromagnetic adsorption, enhanced electrical conductivity, etc., are some of the properties of these composites. These properties encourage the use of these composites in various conditions. Hu et al. predicted that if the concentration of MXene ($Ti_3C_2T_x$) were to be raised above three wt.%, then the mechanical properties of MXene-metal composite ($Ti_3C_2T_x/Al$) might enhance [125]. Due to the tough interface that exists between MXene and Metal, the applied stress should be successively conveyed to the MXene particles when metal is distorted. The one-step synthesis procedure of MXene-Metal composites has enabled the fabrication of metals with MXenes. Literature has suggested that MXene-metal composites have been demonstrated as better materials than their predecessors as they have enhanced properties of both metal and MXene materials [126].

12 Ceramic-MXene Composites

As discussed earlier, MXenes have exceptional mechanical properties that can be conjugated with ceramic materials in order to enhance electrical conductivity and high-temperature phase stability. MXenes have negative zeta potential, which can be attributed to the presence of –F, –Cl, –O and –(OH) as surface terminations. This results in the adsorption of MXene nanosheets on the positively charged materials (ceramic particles in this case) due to electrostatic interaction between MXenes and ceramic particles. The presence of stiff and electrically conductive MXenes improves fracture properties and electron transport in MXene-ceramic composites.

The preparation of MXene-ceramic composite materials is dependent on the ceramic material used. Fei et al. reported the synthesis of MXene-reinforced alumina ceramic composites by mixing multilayer particles of $Ti_3C_2T_x$ with alumina in ethanol via ball milling for 20 h [127]. They reported an increase in fracture toughness, bending strength, and hardness by approximately 300%, 150%, and 300%, respectively. Cygan et al. performed surface modification of 2D Ti_3C_2 MXene by sputtering the multilayer MXene sheet with Mo and Ti, and a composite of this structurally modified MXene and alumina was made [128]. The sputtering of transition metal ions was done to create a protective shield for the prevention of oxidation of MXene in the oxide matrix. The incorporation of MXenes in ceramic structures may result in the enhancement of ceramic properties. MXenes have the potential to improve the electrical, optical and magnetic properties of ceramic structures, which increases the scope of application ranging from EMI shielding to capacitive, dielectric, and electrically conductive additives or energy storage.

13 Metal–Organic Framework-MXene Composites

Metal–organic frameworks (MOF) are hierarchically porous materials that have similar applications as MXenes. MOFs are grown in situ, and they form a suspension solution in water. Recent trends suggest that MXene-MOF composites and their derivatives have received increased attention in energy-related fields. This is due to the fact that both these materials are electrically conductive and are used in energy storage applications. These materials benefit each other in various ways: (i) MXenes have negatively charged surface terminations that can act as valid substrate to support MOF growth, (ii) this growth of MOF on MXene creates highly conductive channels with shortened ion-diffusion pathway, (iii) MOF-MXene composite structures have strong interfacial interaction that offers superior structural integrity and stability, and (iv) the MOF-MXene composite structure is tunable which diversify their application.

Among the MOFs, Ni–MOF has several advantages over other MOFs for composite formation. Ni–MOF has a hierarchically porous structure that provides increased surface area for adsorption. Additionally, Ni–MOF offers high electrical conductivity and thermal stability. Zhang et al. included these advantages and further enhanced the properties by incorporating MXene flakes to produce MXene-MOF composite electrodes [129]. The study suggested the role of MXene as a structure-directing agent that forms a micro belt-like structure, preventing the accumulation of Ni–MOF nanosheets. Qu et al. also reported a MOF/MXene composite with Ni–MOF/$Ti_3C_2T_x$ hybrid nanosheet prepared through an ultrasonic method [130]. According to the study, they concluded that nickel-based structures offer better high-performance supercapacitors due to their phase transition between Ni (II) and Ni (III) complexes. They conducted various tests with the hybrid electrode and demonstrated high specific capacitance (867.3 Fg^{-1} at 1 Ag^{-1}) and stability over repeated charge/discharge cycles (87.1% after 5000 cycles at 5 Ag^{-1}).

14 Challenges and Future Directions

In this chapter, we discussed the structure and various synthesis techniques that have been reported for MXene synthesis. This chapter included the synthesis of MXenes from the MAX phase and the fabrication of MXene-based materials. There have been 30 reported MXene materials that have been experimentally synthesised as compared to 150+ theoretically predicted MAX phases [131]. The fabricated materials are further limited due to stability and compatibility issues [36]. This section deals with the challenges faced during the synthesis of MXenes from MAX phases, their future prospects, and broad recommendations for further research in this area.

15 Current Challenges in MXene Synthesis

Even though significant research progress has been achieved in the synthesis methods and fabrication techniques, some challenges still exist. One such challenge is associated with the synthesis of MXenes. The synthesis methods that are currently in use are labour-intensive and time-consuming. Application of MXene-based structures has been reported, but issues such as agglomeration, re-stacking, and oxidation of MXene nanosheets hinder their performance [132, 133]. The preparation of MXene-based materials is still in the early stages due to (i) the synthesis method, quality, size, and surface terminations/functional groups present, (ii) in the assembly process of MXene-based materials, the MXene nanosheets are prone to folding and loose stacking that can deform the heterostructure and produce low-quality materials, and (iii) scalable synthesis procedure to produce MXene in bulk quantities without significantly altering the structure and properties [132].

16 Emerging Trends and Future Prospects

The practical application of MXenes requires the fabrication of MXene-based materials. MXene thin films have widespread applications, including sensors, photodetectors, motors, energy storage devices, catalysis, and separation. These applications require controllability, reproducibility, and robustness of MXene-based thin films. Thus, the fabrication of MXene-based thin films requires further development. In recent years, 3D printing technology has been seen as an emerging trend used for the fabrication of MXene-based materials on a large scale. The technology promises a preparation process that is less time-consuming and more efficient.

17 Recommendations for Further Research

Various theoretically predicted MXene structures are yet to be experimentally confirmed for application. Since its discovery, HF-based etching has been the most commonly used synthesis method, which needs to be revised in its own essence. HF-free selective etching should be explored as (i) the highly corrosive nature of HF makes the method environmentally hazardous, (ii) MXene nanosheets synthesised from HF etching are prone to having defects in their structure, (iii) the synthesised structure exhibits poor oxidation stability, and (iv) these structures require additional post-synthesis process (intercalation or delamination) to enhance the quality of nanosheets.

Research growth in scale-up of the synthesis process and bulk production of MXene and MXene-based materials is necessary to produce affordable nanomaterials with enhanced properties over conventional materials. Although scalable production of MXenes is challenging as etchants used in the synthesis process are corrosive and hazardous materials, fluorine-free synthesis methods offer a safer and easier alternative. The use of alternative methods may change the surface composition and interfere with the properties, but it may present an opportunity to engineer the properties of the synthesised materials.

18 Conclusion

In this chapter, detailed insights into synthesis methods and fabrication techniques for MXenes have been provided. As discussed throughout the chapter, the majority of reports of MXene synthesis require selective etching of MAX phase precursors for the simple reason of breaking M–A bonds to get ultrathin 2D nanosheets. The etching process can be assisted by the use of a suitable etchant (like HF or fluorine salts), followed by intercalation or delamination. The quality and surface chemistry of synthesised MXenes depend on the etchant and can vary for each etchant. Some reports suggest the use of specific etchants due to their accompanying surface variation, which enhances the performance of MXene products. It can be said without a doubt that the decade of consistent research and experimental efforts has witnessed growth in the synthesis and fabrication of MXene-based products. With these advances, there are certain drawbacks to MXene synthesis that turn off the commercial use of MXenes. One such drawback is the use of HF etchant. Exercising HF-free etching is essential for the synthesis of MXenes because (i) the etchant is highly corrosive in nature and has subsequent environmental hazards, (ii) the synthesised MXenes are prone to structural defects and are generally unstable in a single nanosheet state, and (iii) additional steps like intercalation or delamination are required for removal of these defects and stabilisation of the structure. These reasons make the scale-up of the synthesis method virtually impossible, and other synthesis methods with fewer drawbacks should be explored.

The fabrication of MXene-based products requires a stable structure of these nanosheets. The synthesis methods that provide stability and sturdy structures can be used for MXene synthesis, and fabrication techniques (like film deposition, 3D printing, etc.) can be utilised in order to form MXene-based products. There are various reports of new synthesis methods with certain advantages, but the research in this area is still in its early stages. Upon scale-up implementation, many drawbacks withhold the use of these materials for commercial purposes. Despite the progress and growth toward the commercialisation process, challenges regarding implementation still exist. As discussed earlier, 3D printing technology is a promising lead for the successful implementation of these 2D structures. This technology offers the possibility of manufacturing large-scale MXene-based products.

References

1. Chaudhari, N.K., et al.: MXene: an emerging two-dimensional material for future energy conversion and storage applications. J. Mater. Chem. A **5**(47), 24564–24579 (2017)
2. Sohan, A., et al.: Review on MXene synthesis, properties, and recent research exploring electrode architecture for supercapacitor applications. Int. J. Energy Res. **45**(14), 19746–19771 (2021)
3. Naguib, M., et al.: Two-dimensional nanocrystals produced by exfoliation of Ti3AlC2. Adv. Mater. **23**(37), 4248–4253 (2011)
4. Zhu, J., et al.: Recent advance in MXenes: a promising 2D material for catalysis, sensor and chemical adsorption. Coord. Chem. Rev. **352**, 306–327 (2017)
5. Jeitschko, W., Nowotny, H., Benesovsky, F.: Carbides of formula T2MC. J. Less Common Metals **7**(2), 133–138 (1964)
6. Pang, J., et al.: Applications of 2D MXenes in energy conversion and storage systems. Chem. Soc. Rev. **48**(1), 72–133 (2019)
7. Sun, Y., Chen, D., Liang, Z.: Two-dimensional MXenes for energy storage and conversion applications. Mater. Today Energy **5**, 22–36 (2017)
8. Maleski, K., et al.: Size-dependent physical and electrochemical properties of two-dimensional MXene flakes. ACS Appl. Mater. Interfaces **10**(29), 24491–24498 (2018)
9. Hantanasirisakul, K., Gogotsi, Y.: Electronic and optical properties of 2D transition metal carbides and nitrides (MXenes). Adv. Mater. **30**(52), 1804779 (2018)
10. Song, Y., et al.: Nonlinear few-layer MXene-assisted all-optical wavelength conversion at telecommunication band. Adv. Opt. Mater. **7**(18), 1801777 (2019)
11. Verger, L., et al.: MXenes: an introduction of their synthesis, select properties, and applications. Trends Chem. **1**(7), 656–669 (2019)
12. Naguib, M., et al.: 25th Anniversary Article: MXenes: A New Family of Two-Dimensional Materials. Adv. Mater. **26**(7), 992–1005 (2014)
13. Zhang, Y., et al.: Adsorptive environmental applications of MXene nanomaterials: a review. RSC Adv. **8**(36), 19895–19905 (2018)
14. Khazaei, M., et al.: Electronic properties and applications of MXenes: a theoretical review. J. Mater. Chem. C **5**(10), 2488–2503 (2017)
15. Xu, B., et al.: Ultrathin MXene-micropattern-based field-effect transistor for probing neural activity. Adv. Mater. **28**(17), 3333–3339 (2016)
16. Lin, H., Chen, Y., Shi, J.: Insights into 2D MXenes for versatile biomedical applications: current advances and challenges ahead. Adv. Sci. **5**(10), 1800518 (2018)
17. Dahlqvist, M., et al.: Prediction and synthesis of a family of atomic laminate phases with Kagomé-like and in-plane chemical ordering. Sci. Adv. **3**(7), e1700642 (2017)

18. Frey, N.C., et al.: Prediction of synthesis of 2D metal carbides and nitrides (MXenes) and their precursors with positive and unlabeled machine learning. ACS Nano **13**(3), 3031–3041 (2019)
19. Gao, G., et al.: Monolayer MXenes: promising half-metals and spin gapless semiconductors. Nanoscale **8**(16), 8986–8994 (2016)
20. Huang, S., Mochalin, V.N.: Understanding chemistry of two-dimensional transition metal carbides and carbonitrides (MXenes) with gas analysis. ACS Nano **14**(8), 10251–10257 (2020)
21. An, X., et al.: The synergetic effects of Ti3C2 MXene and Pt as co-catalysts for highly efficient photocatalytic hydrogen evolution over g-C3N4. Phys. Chem. Chem. Phys. **20**(16), 11405–11411 (2018)
22. Barsoum, M.W.: The MN+1AXN phases: a new class of solids: Thermodynamically stable nanolaminates. Prog. Solid State Chem. **28**(1), 201–281 (2000)
23. Tang, H., et al.: MXene–2D layered electrode materials for energy storage. Progr. Nat. Sci.: Mater. Int. **28**(2), 133–147 (2018)
24. Lipatov, A., et al.: Effect of synthesis on quality, electronic properties and environmental stability of individual monolayer Ti3C2 MXene flakes. Adv. Electron. Mater. **2**(12), 1600255 (2016)
25. Liu, Y., et al.: Binder-free layered Ti3C2/CNTs nanocomposite anodes with enhanced capacity and long-cycle life for lithium-ion batteries. Dalton Trans. **44**(16), 7123–7126 (2015)
26. Simon, P., Gogotsi, Y.: Materials for electrochemical capacitors. Nat. Mater. **7**(11), 845–854 (2008)
27. Wang, Y., et al.: Stimulation of surface terminating group by carbon quantum dots for improving pseudocapacitance of Ti3C2Tx MXene based electrode. Carbon **180**, 118–126 (2021)
28. Wu, X., et al.: Polymer-Ti3C2Tx composite membranes to overcome the trade-off in solvent resistant nanofiltration for alcohol-based system. J. Membr. Sci. **515**, 175–188 (2016)
29. Pogorielov, M., et al.: MXenes—A new class of two-dimensional materials: structure, properties and potential applications. Nanomaterials **11**(12), 3412 (2021)
30. Anasori, B., Lukatskaya, M.R., Gogotsi, Y.: 2D metal carbides and nitrides (MXenes) for energy storage. Nat. Rev. Mater. **2**(2), 16098 (2017)
31. Naguib, M., Barsoum, M.W., Gogotsi, Y.: Ten years of progress in the synthesis and development of MXenes. Adv. Mater. **33**(39), 2103393 (2021)
32. Deysher, G., et al.: Synthesis of Mo4VAlC4 MAX phase and two-dimensional Mo4VC4 MXene with five atomic layers of transition metals. ACS Nano **14**(1), 204–217 (2020)
33. Zhang, C.J., et al.: Oxidation stability of colloidal two-dimensional titanium carbides (MXenes). Chem. Mater. **29**(11), 4848–4856 (2017)
34. Syamsai, R., Grace, A.N.: Ta4C3 MXene as supercapacitor electrodes. J. Alloy. Compd. **792**, 1230–1238 (2019)
35. Naguib, M., et al.: Two-dimensional transition metal carbides. ACS Nano **6**(2), 1322–1331 (2012)
36. Huang, W., et al.: Recent advances in functional 2D MXene-based nanostructures for next-generation devices. Adv. Func. Mater. **30**(49), 2005223 (2020)
37. Liu, F., et al.: Research progress on construction and energy storage performance of MXene heterostructures. J. Energy Chem. **62**, 220–242 (2021)
38. Cheng, L., et al.: Two-dimensional transition metal MXene-based photocatalysts for solar fuel generation. J. Phys. Chem. Lett. **10**(12), 3488–3494 (2019)
39. Tao, Q., et al.: Two-dimensional Mo1.33C MXene with divacancy ordering prepared from parent 3D laminate with in-plane chemical ordering. Nat. Commun. **8**(1), 14949 (2017)
40. Shahzad, F., et al.: 2D transition metal carbides (MXenes): applications as an electrically conducting material. Adv. Mater. **32**(51), 2002159 (2020)
41. Halim, J., et al.: Transparent conductive two-dimensional titanium carbide epitaxial thin films. Chem. Mater. **26**(7), 2374–2381 (2014)

42. Xu, C., et al.: Large-area high-quality 2D ultrathin Mo2C superconducting crystals. Nat. Mater. **14**(11), 1135–1141 (2015)
43. Li, T., et al.: Fluorine-free synthesis of high-purity Ti3C2Tx (T = OH, O) via Alkali treatment. Angew. Chem. Int. Ed. **57**(21), 6115–6119 (2018)
44. Huang, X., Wu, P.: A facile, high-yield, and freeze-and-thaw-assisted approach to fabricate MXene with plentiful wrinkles and its application in on-chip micro-supercapacitors. Adv. Func. Mater. **30**(12), 1910048 (2020)
45. Ghidiu, M., et al.: Conductive two-dimensional titanium carbide 'clay' with high volumetric capacitance. Nature **516**(7529), 78–81 (2014)
46. Halim, J., et al.: Synthesis and characterization of 2D molybdenum carbide (MXene). Adv. Func. Mater. **26**(18), 3118–3127 (2016)
47. Wu, X., et al.: Recent advances on tuning the interlayer coupling and properties in van der Waals heterostructures. Small **18**(15), 2105877 (2022)
48. Yang, Q., et al.: MXene/graphene hybrid fibers for high performance flexible supercapacitors. J. Mater. Chem. A **5**(42), 22113–22119 (2017)
49. Ye, C., et al.: CdS/Ti3C2 heterostructure–based photoelectrochemical platform for sensitive and selective detection of trace amount of Cu2+. Anal. Bioanal. Chem. **414**(12), 3571–3580 (2022)
50. Feng, Y., et al.: High dielectric and breakdown properties achieved in ternary BaTiO3/MXene/PVDF nanocomposites with low-concentration fillers from enhanced interface polarization. Ceram. Int. **45** (2019)
51. Sharma, S.K., et al.: MXenes based nano-heterojunctions and composites for advanced photocatalytic environmental detoxification and energy conversion: a review. Chemosphere **291**, 132923 (2022)
52. Pang, S.-Y., et al.: Universal strategy for HF-free facile and rapid synthesis of two-dimensional MXenes as multifunctional energy materials. J. Am. Chem. Soc. **141**(24), 9610–9616 (2019)
53. Yang, S., et al.: Fluoride-free synthesis of two-dimensional titanium carbide (MXene) using a binary aqueous system. Angew. Chem. Int. Ed. **57**(47), 15491–15495 (2018)
54. Mashtalir, O., et al.: Intercalation and delamination of layered carbides and carbonitrides. Nat. Commun. **4**(1), 1716 (2013)
55. Luo, J., et al.: Pillared structure design of MXene with ultralarge interlayer spacing for high-performance lithium-ion capacitors. ACS Nano **11**(3), 2459–2469 (2017)
56. Hantanasirisakul, K., et al.: Effects of synthesis and processing on optoelectronic properties of titanium carbonitride MXene. Chem. Mater. **31**(8), 2941–2951 (2019)
57. Naguib, M., et al.: Large-scale delamination of multi-layers transition metal carbides and carbonitrides "MXenes." Dalton Trans. **44**(20), 9353–9358 (2015)
58. Qian, A., et al.: Surface functional groups and electrochemical behavior in dimethyl sulfoxide-delaminated Ti3C2Tx MXene. Chemsuschem **11**(21), 3719–3723 (2018)
59. Hu, A., et al.: One-step synthesis for cations intercalation of two-dimensional carbide crystal Ti3C2 MXene. Appl. Surf. Sci. **505**, 144538 (2020)
60. Xuan, J., et al.: Organic-base-driven intercalation and delamination for the production of functionalized titanium carbide nanosheets with superior photothermal therapeutic performance. Angew. Chem. Int. Ed. **55**(47), 14569–14574 (2016)
61. Zeng, M., et al.: 2D WC single crystal embedded in graphene for enhancing hydrogen evolution reaction. Nano Energy **33**, 356–362 (2017)
62. Cao, S., et al.: 2D/2D heterojunction of ultrathin MXene/Bi2WO6 nanosheets for improved photocatalytic CO2 reduction. Adv. Func. Mater. **28**(21), 1800136 (2018)
63. Liu, F., et al.: Pursuit of a high-capacity and long-life Mg-storage cathode by tailoring sandwich-structured MXene@carbon nanosphere composites. J. Mater. Chem. A **7**(28), 16712–16719 (2019)
64. Zhuang, Y., Liu, Y., Meng, X.: Fabrication of TiO2 nanofibers/MXene Ti3C2 nanocomposites for photocatalytic H2 evolution by electrostatic self-assembly. Appl. Surf. Sci. **496**, 143647 (2019)

65. Cheng, L., et al.: Boosting the photocatalytic activity of CdLa2S4 for hydrogen production using Ti3C2 MXene as a co-catalyst. Appl. Catal. B **267**, 118379 (2020)
66. Zhang, Y., et al.: Flexible and freestanding heterostructures based on COF-derived N-doped porous carbon and two-dimensional MXene for all-solid-state lithium-sulfur batteries. Chem. Eng. J. **428**, 131040 (2022)
67. Oyehan, T.A., et al.: MXenes: synthesis, properties, and applications for sustainable energy and environment. Appl. Mater. Today **35**, 101993 (2023)
68. Zamhuri, A., et al.: MXene in the lens of biomedical engineering: synthesis, applications and future outlook. BioMed. Eng. OnLine **20** (2021)
69. Verger, L., et al.: Overview of the synthesis of MXenes and other ultrathin 2D transition metal carbides and nitrides. Curr. Opin. Solid State Mater. Sci. **23**(3), 149–163 (2019)
70. Jiang, L., et al.: 2D single-and few-layered MXenes: synthesis, applications and perspectives. J. Mater. Chem. A **10**(26), 13651–13672 (2022)
71. Song, Q., et al.: Carbonization of transition metals in molten salts. Phys. Chem. Chem. Phys. **21**(32), 17801–17810 (2019)
72. Naguib, M., et al.: On the topotactic transformation of Ti2AlC into a Ti–C–O–F cubic phase by heating in molten lithium fluoride in air. J. Am. Ceram. Soc. **94**(12), 4556–4561 (2011)
73. Fu, Y.-D., et al.: The role of group III, IV elements in Nb 4 AC 3 MAX phases (A = Al, Si, Ga, Ge) and the unusual anisotropic behavior of the electronic and optical properties. Phys. Chem. Chem. Phys. **19**(23), 15471–15483 (2017)
74. Wang, F., et al.: Advances in molten-salt-assisted synthesis of 2D MXenes and their applications in electrochemical energy storage and conversion. Chem. Eng. J. **470**, 144185 (2023)
75. Dubois, S., et al.: A new ternary nanolaminate carbide: Ti3SnC2. J. Am. Ceram. Soc. **90**(8), 2642–2644 (2007)
76. Dash, A., et al.: Molten salt shielded synthesis of oxidation prone materials in air. Nat. Mater. **18**(5), 465–470 (2019)
77. Roy, C., Banerjee, P., Bhattacharyya, S.: Molten salt shielded synthesis (MS3) of Ti2AlN and V2AlC MAX phase powders in open air. J. Eur. Ceram. Soc. **40**(3), 923–929 (2020)
78. Wu, T., et al.: Fast growth of inch-sized single-crystalline graphene from a controlled single nucleus on Cu–Ni alloys. Nat. Mater. **15**(1), 43–47 (2016)
79. Yuan, Q., Yakobson, B.I., Ding, F.: Edge-catalyst wetting and orientation control of graphene growth by chemical vapor deposition growth. J. Phys. Chem. Lett. **5**(18), 3093–3099 (2014)
80. Xu, C., et al.: Large-area high-quality 2D ultrathin Mo2C superconducting crystals. Nat. Mater. **14** (2015)
81. Öper, M., et al.: Controlled CVD growth of ultrathin Mo2C (MXene) flakes. J. Appl. Phys. **131**, 025304 (2022)
82. Xu, F., et al.: Synthesis and photocatalytic H2-production activity of plasma-treated Ti3C2Tx MXene modified graphitic carbon nitride. J. Am. Ceram. Soc. **103**(2), 849–858 (2020)
83. Yang, X., et al.: Heterogeneous Ni–MOF/V2CTx–MXene hierarchically-porous nanorods for robust and high energy density hybrid supercapacitors. J. Mater. Chem. A **10**(22), 12225–12234 (2022)
84. Syduzzaman, M., et al.: Nanotechnology for high-performance textiles: a promising frontier for innovation. ChemNanoMat **9**(9), e202300205 (2023)
85. Chernikova, V., Shekhah, O., Eddaoudi, M.: Advanced fabrication method for the preparation of MOF thin films: liquid-phase epitaxy approach meets spin coating method. ACS Appl. Mater. Interfaces **8**(31), 20459–20464 (2016)
86. Salles, P., et al.: Automated scalpel patterning of solution processed thin films for fabrication of transparent MXene microsupercapacitors. Small **14**(44), 1802864 (2018)
87. Kim, S.J., et al.: Interfacial assembly of ultrathin, functional MXene films. ACS Appl. Mater. Interfaces **11**(35), 32320–32327 (2019)
88. Zhang, M., et al.: Formation of new MXene film using spinning coating method with DMSO solution and its application in advanced memristive device. Ceram. Int. **45**(15), 19467–19472 (2019)

89. Cai, Y., et al.: Stretchable Ti3C2Tx MXene/carbon nanotube composite based strain sensor with ultrahigh sensitivity and tunable sensing range. ACS Nano **12**(1), 56–62 (2018)
90. Zhao, M.-Q., et al.: Scalable manufacturing of large and flexible sheets of MXene/graphene heterostructures. Adv. Mater. Technol. **4**(5), 1800639 (2019)
91. Lee, G.S., et al.: Mussel inspired highly aligned Ti3C2Tx MXene film with synergistic enhancement of mechanical strength and ambient stability. ACS Nano **14**(9), 11722–11732 (2020)
92. Hantanasirisakul, K., et al.: Fabrication of Ti3C2Tx MXene transparent thin films with tunable optoelectronic properties. Adv. Electron. Mater. **2**(6), 1600050 (2016)
93. Zhang, Z., et al.: Flexible MXene composed triboelectric nanogenerator via facile vacuum-assistant filtration method for self-powered biomechanical sensing. Nano Energy **88**, 106257 (2021)
94. Huang, H., et al.: Scalable, and low-cost treating-cutting-coating manufacture platform for MXene-based on-chip micro-supercapacitors. Nano Energy **69**, 104431 (2020)
95. Zou, Q., et al.: MXene-based ultra-thin film for terahertz radiation shielding. Nanotechnology **31**(50), 505710 (2020)
96. Huang, Y.-L., Bian, S.-W.: Vacuum-filtration assisted layer-by-layer strategy to design MXene/carbon nanotube@MnO2 all-in-one supercapacitors. J. Mater. Chem. A **9**(37), 21347–21356 (2021)
97. Gómez, I.J., et al.: Recent advances on 2D materials towards 3D printing. Chemistry **3**, 1314–1343 (2021). https://doi.org/10.3390/chemistry3040095
98. Gupta, N., Saha, P., Bhattacharya, P.: 3D printing of MXenes-based electrodes for energy storage applications. Recent Progr. Mater. **5**(2), 1–23 (2023)
99. Zhang, Y.Z., et al.: MXene printing and patterned coating for device applications. Adv. Mater. **32**(21), 1908486 (2020)
100. Li, T., et al.: Review of additive manufacturing with 2D MXene: techniques, applications, and future perspectives. Progr. Addit. Manuf. **8**(6), 1587–1617 (2023)
101. Huang, J., et al.: Electrochemical properties of LiCoO2 thin film electrode prepared by ink-jet printing technique. Thin Solid Films **516**, 3314–3319 (2008)
102. Zhang, Q., et al.: 3D printing of graphene aerogels. Small **12**(13), 1702–1708 (2016)
103. Chen, L.-F., et al.: Synthesis of nitrogen-doped porous carbon nanofibers as an efficient electrode material for supercapacitors. ACS Nano **6**(8), 7092–7102 (2012)
104. Wu, Z.-S., et al.: Three-dimensional nitrogen and boron co-doped graphene for high-performance all-solid-state supercapacitors. Adv. Mater. **24**(37), 5130–5135 (2012)
105. Peng, C., et al.: 3D printed electrochemical energy storage devices. J. Mater. Chem. A **7** (2019)
106. Eckel, Z.C., et al.: Additive manufacturing of polymer-derived ceramics. Science **351**(6268), 58–62 (2016)
107. Ning, H., et al.: Holographic patterning of high-performance on-chip 3D lithium-ion microbatteries. Proc. Natl. Acad. Sci. **112**(21), 6573–6578 (2015)
108. Zhang, D., et al.: Fabrication of highly conductive graphene flexible circuits by 3D printing. Synth. Met. **217**, 79–86 (2016)
109. Czyżewski, J., et al.: Rapid prototyping of electrically conductive components using 3D printing technology. J. Mater. Process. Technol. **209**(12), 5281–5285 (2009)
110. Xu, B., et al.: Laminated fabrication of 3D queue micro-electrode and its application in micro-EDM. Int. J. Adv. Manuf. Technol. **80**(9), 1701–1711 (2015)
111. Gardan, N., Schneider, A.: Topological optimization of internal patterns and support in additive manufacturing. J. Manuf. Syst. (2014)
112. Kim, S., Moon, S.K.: A part consolidation design method for additive manufacturing based on product disassembly complexity. Appl. Sci. **10**(3), 1100 (2020)
113. Navangul, G., Paul, R., Anand, S.: Error minimization in layered manufacturing parts by stereolithography file modification using a vertex translation algorithm. J. Manuf. Sci. Eng. **135**, 031006 (2013)

114. Brajlih, T., et al.: Speed and accuracy evaluation of additive manufacturing machines. Rapid Prototyp. J. **17**, 64–75 (2011)
115. Tong, K., Lehtihet, E.-A., Joshi, S.: Parametric error modeling and software error compensation for rapid prototyping. Rapid Prototyp. J. **9**, 301–313 (2003)
116. Kou, X.Y., Tan, S.T.: Heterogeneous object modeling: a review. Comput. Aided Des. **39**(4), 284–301 (2007)
117. Christ, S., et al.: Fiber reinforcement during 3D printing. Mater. Lett. **139**, 165–168 (2015)
118. Roberson, D.A., Espalin, D., Wicker, R.B.: 3D printer selection: a decision-making evaluation and ranking model. Virtual Phys. Prototyp. **8**(3), 201–212 (2013)
119. Stephens, B., et al.: Ultrafine particle emissions from desktop 3D printers. Atmos. Environ. **79**, 334–339 (2013)
120. khorram niaki, M., Nonino, F.: Impact of additive manufacturing on business competitiveness: a multiple case study. J. Manuf. Technol. Manag. **28** (2016)
121. Chen, C., et al.: Charge transfer induced polymerization of EDOT confined between 2D titanium carbide layers. J. Mater. Chem. A **5**(11), 5260–5265 (2017)
122. Khan, M.U., et al.: Preparations and applications of MXene-metal composites: a review. Coatings **12**(4), 516 (2022)
123. Rakhi, R.B., et al.: Novel amperometric glucose biosensor based on MXene nanocomposite. Sci. Rep. **6**(1), 36422 (2016)
124. Han, M., et al.: Ti3C2 MXenes with modified surface for high-performance electromagnetic absorption and shielding in the X-band. ACS Appl. Mater. Interfaces. **8**(32), 21011–21019 (2016)
125. Hu, J., et al.: Mechanical properties and frictional resistance of Al composites reinforced with Ti3C2Tx MXene. Chin. Chem. Lett. **31**(4), 996–999 (2020)
126. Wyatt, B.C., Nemani, S.K., Anasori, B.: 2D transition metal carbides (MXenes) in metal and ceramic matrix composites. Nano Convergence **8**(1), 16 (2021)
127. Fei, M., et al.: MXene-reinforced alumina ceramic composites. Ceram. Int. **43**(18), 17206–17210 (2017)
128. Cygan, T., et al.: Microstructure and mechanical properties of alumina composites with addition of structurally modified 2D Ti3C2 (MXene) phase. Materials **14**(4), 829 (2021)
129. Zhang, C., et al.: Achieving quick charge/discharge rate of 3.0 V s^{-1} by 2D titanium carbide (MXene) via N-doped carbon intercalation. Mater. Lett. **234**, 21–25 (2019)
130. Qu, C., et al.: Nickel-based pillared MOFs for high-performance supercapacitors: design, synthesis and stability study. Nano Energy **26**, 66–73 (2016)
131. Arróyave, R., et al.: Out-of-plane ordering in quaternary MAX alloys: an alloy theoretic perspective. Mater. Res. Lett. **6**(1), 1–12 (2018)
132. Bhat, A., et al.: Prospects challenges and stability of 2D MXenes for clean energy conversion and storage applications. NPJ 2D Mater. Appl. **5**(1), 61 (2021)
133. Yuan, K., et al.: A two-dimensional MXene/BN van der Waals heterostructure as an anode material for lithium-ion batteries. Phys. Chem. Chem. Phys. **24**(22), 13713–13719 (2022)

Surface Characterization Techniques of MXene Films and Powders

Susmi Anna Thomas, Jayesh Cherusseri, Drisya Mottammal, Asheesh Kumar, and Deepthi N. Rajendran

Abstract Ultrathin two-dimensional (2D) materials have profound interest from recent years due to its extraordinary physical and chemical properties. Within these different 2D materials, transition metal nitrides/carbides, generally known as MXenes have more preference in accordance to their structure and other properties. An analysis of their properties making them suitable for multifunctional applications is necessary to introduce. Thus, there are various characterization techniques used to distinguish their properties in order to meet these application requirements. Among different characterization techniques used for MXenes, surface characterization techniques have more preference, in order to achieve details on the presence of functional groups, surface active sites, structural tunability, etc. Here in the present chapter, we establish an outline on the basic and major surface characterization techniques used in MXene films and powders.

Keywords MXene · X-ray photoelectron spectroscopy · Transmission electron microscopy · X-ray photoemission electron microscopy · X-ray absorption spectroscopy

S. A. Thomas · D. Mottammal · D. N. Rajendran
Department of Physics, Government College for Women (Affiliated to University of Kerala), Thiruvananthapuram, Kerala, India

J. Cherusseri (✉)
Department of Chemistry (BK21 FOUR), Research Institute of Advanced Chemistry, Gyeongsang National University, Jinju 52828, Republic of Korea
e-mail: drjayeshpuli@gmail.com

A. Kumar
Department of Mechanical Engineering, Mahatma Gandhi Institute of Technology, Hyderabad, Telangana, India

© The Author(s), under exclusive license to Springer Nature Singapore Pte Ltd. 2025

S. Moharana and R. N. Mahaling (eds.), *Surface Engineering of MXenes*, Engineering Materials, https://doi.org/10.1007/978-981-96-3640-2_3

1 Introduction

Nanotechnology has prominent advancements in the current century, due to its applications in multifunctional fields [1–6]. Two-dimensional (2D) transition metal nitrides, carbides, and carbonitrides, generally named it as MXene, received profound attention in different fields from its first introduction in the year 2011 itself [2, 7–10]. The layered architecture and extraordinary physicochemical properties such as higher thermal and electronic conductivity, tunability in the electronic bandgap, hydrophilic surface character, and optical transparency make it as different among others [11–17]. This introduced new class of materials follows the chemical formula $M_{n+1}X_nA_x$ (MAX) and consists of $M_{n+1}X_n$ lamellae having 'M' is the transition metal such as Cr, Ti, V, Nb, and 'X' is the carbon or nitrogen and the parameter 'n' ranging from 1 to 4 and 'A' is the elements such as Al, Ga, Ge, Si, etc., from the groups 13–16 in periodic table [18–20]. The hexagonal MAX phase materials have layered architecture, where M layers are densely packed, and the octahedral sites are occupied with X atoms, while layers of M and X are bonded through A atoms [21, 22]. In a MAX phase, M-X bonds possess a mixed character of ionic/covalent/metallic characteristics, while the M-A bond mostly exhibits a metallic nature [2, 23–25]. Consequently, due to strong interlayer bonding present in MAX phase, mechanical exfoliation or shearing of these bonds is not possible unlike in the case of graphite and TMDs, here weak van der Waals force grip the layers together [26, 27]. MAX phase materials have unusual mixture of properties which contribute to the ceramics and metals [28]. Similar to ceramic materials, it possesses good stiffness, thermal stability, brittleness, and strength, while it is capable to conduct electricity and heat as like as metals [29, 30]. MXenes are prepared through the selective etching of A layer from MAX phase, which consists of three or higher number of atomic layer thick 2D materials which hold various features in comparison to three-dimensional (3D) parent precursor [31]. If the n values in MAX phase range from 1 to 3, the resultant MXene have three possible forms of lattice structures such as: M_2X, M_3X_2, and M_4X_3 (3, 5 or 7 atomic layers) [32]. The first MXene synthesized was $Ti_3C_2T_x$, and there is vast research was established to synthesize other MXene compounds. It is possible to obtain MXene with greater than one M element [33, 34]. The structure is obtained in two various forms such as: ordered phase and solid solution. In the case of the ordered phase, mono or bi-layer component of one transition metal is sandwiched between another component of transition metal layer (such as $(Mo_2Ti)\ C_2$ and $(Ti_2Ta_2)\ C_3)$) but in the latter case, there exists a random distribution of two different transition metals in M layers (for example $(Ti,V)_2C$ and $(Ti,Nb)_4C_3$) [35]. In addition to this, various etching procedures for MXene introduce the difference in layer thickness and surface functional groups, which leads to tuning the properties of MXenes [36, 37]. Within different structures consisting of MXenes, MXene-related films hold higher conductivity, prominent ultra-thin thickness, and higher flexibility introducing prominent advancements in different fields, most specifically in the case of electromagnetic shielding and supercapacitor [38–41]. After a few years of evolution of MXene [42, 43], there are different studies that are introduced for the development of MXene films to these applications. In the case of MXene films and powders, it is necessary to introduce a more detailed analysis on their surface characteristics to depict their application features. Here the present chapter gives a detailed analysis of surface characterization techniques utilized for MXene films and powders. It is sure that this proposed chapter will definitely provide a deep understanding of these surface characterization tools used in MXene films and powders.

2 Surface Characterization Using X-Ray Photoelectron Microscopy (XPS)

By using a low-cost approach of combining a self-propagation high temperature synthesis and grinding route to prepare Ti_3AlC_2 precursor and then it is utilized to synthesis $Ti_3C_2T_x$. Here MAX phase is synthesized from elemental powders consisting of titanium, aluminium, and carbon black with the stoichiometry of 3:2:1.5 corresponding to Ti:Al:C [44]. Here pre-ball mixture consisting of Ti, Al, and carbon black is introduced to an isolated graphite cup. After the introduction of combustion wave and difference in delay time, rotor is dropped to lower base of reactor for a duration of 20–30 s, to grind proposed mixture. The rotor holds a diameter of 25 mm having rotational speed of 240 rpm and an applied force value of 50 N. Here rotor introduces the equalization of concentration hold by reacting substances and it is involved in chemical synthesis of unreacted initial components. During this proposed time, the prepared materials are totally in the underwent phase and in transformations of structural character. The temperature during this self-propagation high temperature synthesis is 1700 ± 10 °C and the complete duration for this procedure is 30–35 s. The maximum value of synthesized powder per one of self-propagation high temperature synthesis procedures is 50 ± 5 g. Here this synthesized powder is further kept for grounding with an agate mortar to obtain a particle size less than 38 μm. Here MXene is synthesized through etching A-layer from the MAX phase precursor through a less aggressive route, called it as minimally intensive layer delamination. Here the authors of present work utilized 1 g amount of Ti_3AlC_2 fine powder and slowly introduced it into the mixture consisting of 1 g LiF and 20 ml volume of HCl for a time duration of 24 h at a temperature of 35 °C, through stirring with magnetic stirrer. The resultant solution obtained after etching is kept for repetitive washing using deionized water and centrifuged to introduce a neutral pH. The resultant mixture is kept for vacuum-assisted filtration process and dried in the vacuum chamber at a temperature of 60 °C for 24 h to prepare multilayer powder. After this process, 0.2 g of $Ti_3C_2T_x$ powder is redispersed in a 20 ml volume of deionized water in a glass jar and the obtained aqueous suspension is bath sonicated for a duration of 1 h in continuous argon bubbling. The obtained suspension is stable after performing a sonication for 1 h at 3500 rpm, it verifies the delamination of MXene to single 2D flakes. Through XPS analysis, they evaluated the surface states including oxidation state/valency of prepared MXene powder. The XPS spectra of Ti, C, O, and F in prepared MXene with peak fitting are shown in Fig. 1. The Gaussian–Lorentzian fitting of XPS peak corresponds to Ti $2p_{3/2}$ ($2p_{1/2}$) in the MXene confirming that titanium is present in this structure having five doublets (Ti2p $2p_{3/2}$–Ti $2p_{1/2}$) having fixed area ratio equal to 2:1. Here the components of Ti 2p are fitted as Ti–C, Ti(II), Ti(III) that corresponds to Ti bonded by carbon having binding energy (BE) of 455.5 (461.7) eV and Ti atoms are bonded by –O or –OH at a BE of 456.9 (462.9) eV and 456.0 (461.9) eV, respectively. In the proposed case, Ti–C, Ti(II), Ti(III) species introduced a major contribution of about 77% of XPS spectrum corresponding to Ti2p. Here Ti $2p_{3/2}$ component is centred at 458.3 (464.0) and 459.4 (465.2) eV corresponds to Ti ions having formal valency of +4 (TiO_2 and $TiO_{2\text{-}x}F_x$) which results from spontaneous surface oxidation. In the XPS spectrum, the O1s region of MXene is fitted through components related to C–Ti–O_x and C–Ti–$(OH)_x$ at a BE value of 630.9 and 532.1 eV, respectively. Peak relates to adsorbed water is present at a BE value of 533.7 eV. The carbon is present in three bonds. In carbon, the peak present at a BE value of 282.2 eV relates to carbide component of Ti–C in MXene (here approximately 37% of photoemission intensity appears at C1s region. The other two peaks are corresponding to graphitic C–C, introduced from selective Ti dissolution during chemical etching and surface contaminations of C–O is present at 285 and 286.5 eV, respectively. The prominent component in F1s region at a BE of 685 and 685.5 eV corresponds to C–Ti–F_x and $TiO_{2\text{-}x}F_x$, respectively. It also introduces the existence of a smaller quantity of AlF_x at BE = 686.6 eV and $Al(OF)_x$ at BE of 687.8 eV, which is more likely introduced from residue after etching reaction of Ti_3AlC_2.

In addition to the powdered form of MXene powder, XPS spectra found prominent interest in MXene films also. The synthesis and characterization of MXene films are more deeply needed to analyse because these film forms introduce a direct utilization of them in applications. There is a report based on preparation of multilayered, monolayered, and flakes of $Mo_2TiC_2T_z$ and $Mo_2Ti_2C_3T_z$ (MXenes) through cold pressing approach [45]. The delaminated thin films of these MXenes are synthesized through filtering colloidal suspensions, it is named as d-21 and d-22. Here later one is synthesized through mixing a quantity of 2 g of monolayered filtered powder having 20 ml of 54–56 wt.% of tetrabutylammonium hydroxide (TBAOH) and stirred it for 1 h at room temperature. The resultant mixture is centrifuged at 5000 rpm for a duration of 5 min and supernatant is decanted and collected. The obtained supernatant is kept for washing 3 times with deionized water and kept for centrifugation at 5000 rpm for a duration of 1 min per cycle, then finally it is decanted to remove any TBAOH leftover. Finally after performing synthesis procedures, obtained films are annealed at a temperature of 180 ºC for 18 h. Here they analysed the prepared MXene thin films using surface analysis system that employs monochromic Al-Kα X-ray beam to irradiate surface of sample. The photoelectrons are received at a take-off angle of 180º between the sample surface and path to analyser. The charge neutralization is established with dual beam charge neutralizer which irradiates ion beam and electrons having low-energy inorder to shift the recorded BE. Here the authors have taken survey spectra having pass energy value of 117.5 eV, having step size of 0.1 eV. The high-resolution spectra corresponding to Mo 3d, Ti 2p, C 1s, O

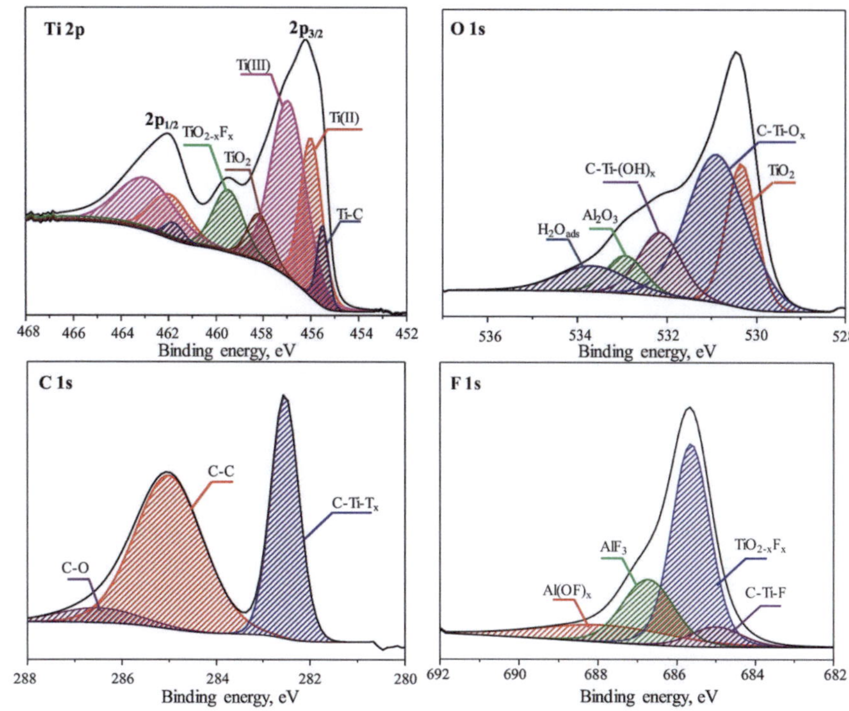

Fig. 1 XPS spectra of Ti 2p, O 1s, C 1s, and F 1s for $Ti_3C_2T_x$ MXene synthesized from self-propagation high temperature synthesis of ground MAX phase powder. Reproduced with permission from [44] Copyright (2019) Elsevier Inc.

1s, F 1s, and Al 2p at a pass energy value of 11.75 eV are taken, having step size of 0.05 eV. Here the BE scale of the entire XPS spectra was referred to Fermi-edge (E_f), fixed to a BE value of zero eV. In order to obtain XPS spectra, samples for analysis are mounted over a double-side tape and it is connected electrically to sample holder with copper wire. The analysis is done without using Ar ion sputtering. With software package CasaXPS version 2.3.16 RP 1.6, the authors of the present work performed quantification using intensity of core-level and peak fitting of core-level spectra. In advance to quantification and peak fitting procedure, subtraction of background contribution is done with Shirley function. Figure 2a represents high resolution XPS spectra for Mo 3d region in multilayered (ML)-21 (lower spectrum region) and d-21 (upper spectrum). In these prepared samples, the major part of almost greater than 79% of signal corresponds to the Mo atoms which reside in $Mo_2TiC_2T_z$ MXene and compared with Mo 3d in $Mo_2Ti_2C_3Tz$ (Fig. 2d). These proposed species relate to Mo atoms in MXene through Mo atoms bonded to C atoms in MX block and Tz or C–Mo–Tz surface terminations. For BE of 229.6 eV, which is obtained to be 1.5 eV greater than that of Mo atom present in parent MAX phase. This shift in BE value is typically observed in MXene and it happens when replacing Al layers with highly electronegative species. There is a minor component of Mo about 8–14% corresponds to Mo^{5+} and Mo^{6+}, these peaks are observed at BE of 231.2 and/or 232.6 V, respectively, it is due to the presence of surface oxides. It needs only two species for fitting d-$Mo_2Ti_2C_2$ spectrum. For peak present at 228.4 eV corresponds to binary carbide, Mo_2C, it appears as secondary phase (contains approximately 5.7 atomic percentage) in actual MAX phase. The small difference occurs in BE value of Mo^{6+} corresponding to reduced intensity of its peak and concomitant defect for fitting obtained result. The high-resolution spectra in ML-(lower spectra) correspond to Ti 2p region and d-21 (upper spectra) films are given in Fig. 2b. The related results for the 22 are given in Fig. 2e. In the entire procedures of fitting, minimum four components Ti–C, Ti (II), Ti (III), and Ti (IV)/TiO_2 are utilized. The corresponding BE values are approximately 455, 455.7, 457.2, and 459 eV, respectively. The first three contributions correspond to Ti atoms present in MXene films and fourth contribution is from TiO_2, having BE of 459 eV. In O1s region, ML-21 (lower) and d-21 (upper spectra) are shown in Fig. 2c and the corresponding spectra related to ML-(lower) and d-22 (upper) films fitted using four components are given in Fig. 2f. The fitted component possesses a BE of 530 eV corresponding to molybdenum oxide (MoO_x), TiO_2, and/or terminations of oxygen fixed on B or HCP sites (atom here labelled as O(I) in Fig. 2). This henceforth labelled in the final composition as $O(I)_x$. The fitted component with BE of 530 eV corresponding to surface terminations of O on A or FCC sites, the atom labelled here as O(II) in Fig. 2, and it is referred to final formulae C–Mo–O(II)x. The peak formed at BE of 532 and 533 eV corresponds to C–Mo–$(OH)_x$ and adsorbed water respectively.

The XPS spectra of C 1s region of ML-21 (lower spectra) and d-21 (upper spectra) films are given in Fig. 3a. Here C spectra correspond to ML-22 (lower) and d-22 (upper) films are shown in Fig. 3c, which is fit by six components. Here the first component is fixed to a BE value of approximately 282.5 eV, named it as C–Mo/Ti–T_x, which corresponds to bonding of C atoms to both the Mo and Ti atoms and it exists in both compositions. This obtained BE value is the same as one for C in parent MAX phases. The peak observed at BE of approximately 282 eV only exists in e $Mo_2Ti_2C_3T_z$ composition is corresponds to C atoms which are bounded only towards Ti atoms, which means that C atoms only exist in the centre of Mo–Ti–Ti–Mo block (Fig. 2). In ML-21 sample, highest contribution of about 57% is from graphitic carbon (C–C), but there exists some contribution by carbon contamination (CHx, C–O, and COO). The obtained graphitic carbon is a major byproduct during the synthesis of MXene. The contamination from carbon is deposited without any criterion from the surroundings. Then performing delamination, these same five species exist in it, albeit exist in various ratios. Here graphitic carbon content is reducing to a value of 33%, but the quantity of contamination by hydrocarbon increasing to 35%. The BE value of C–Mo/Ti–T_x component makes a shift through higher range of approximately 0.5 eV. Then after performing delamination process, which makes a correlation to improvement in total oxidation state. The highest component that exist in both ML and d-22 is again graphitic carbon having C–C about 46 and 57% respectively), but contamination of carbon by elements such as CH_x, C–O, and COO. More surprisingly, the quantity of graphitic carbon is increasing after delamination,

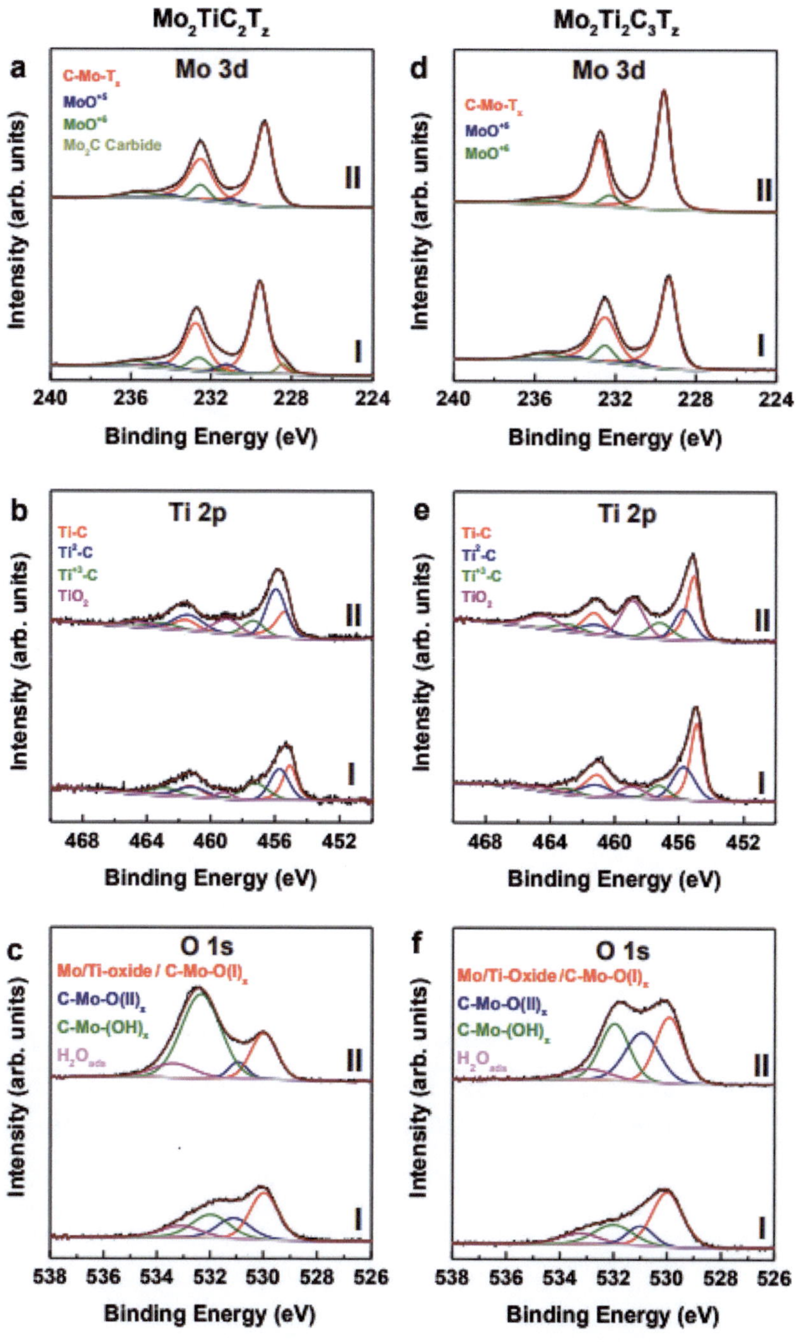

Fig. 2 Curve fitted XPS spectra for **a** Mo 3d, **b** Ti 2p, and **c** O 1s region in ML-(I) and delaminated (II) thin films for a composition of 21; **d** similar to **a** for 22, **e** similar to **b** for 22, **f** is same as **c** for the composition of 22. Peak represents different species which exist. Here peak colours and labels are coordinated. Reproduced with permission from [45] Copyright (2019) Elsevier Inc.

rather than reduced as in the case of $Mo_2TiC_2T_z$, but there exists a reduction in hydrocarbon contamination. The high-resolution XPS spectra of F 1s region in ML-21 (lower spectra) and d-21 (upper spectra) films are given in Fig. 3b and compared with F 1s in $Mo_2Ti_2C_3T_z$ (Fig. 3d). In fitting XPS spectra, when compared with all the fits, there is least ambiguous are those with F.

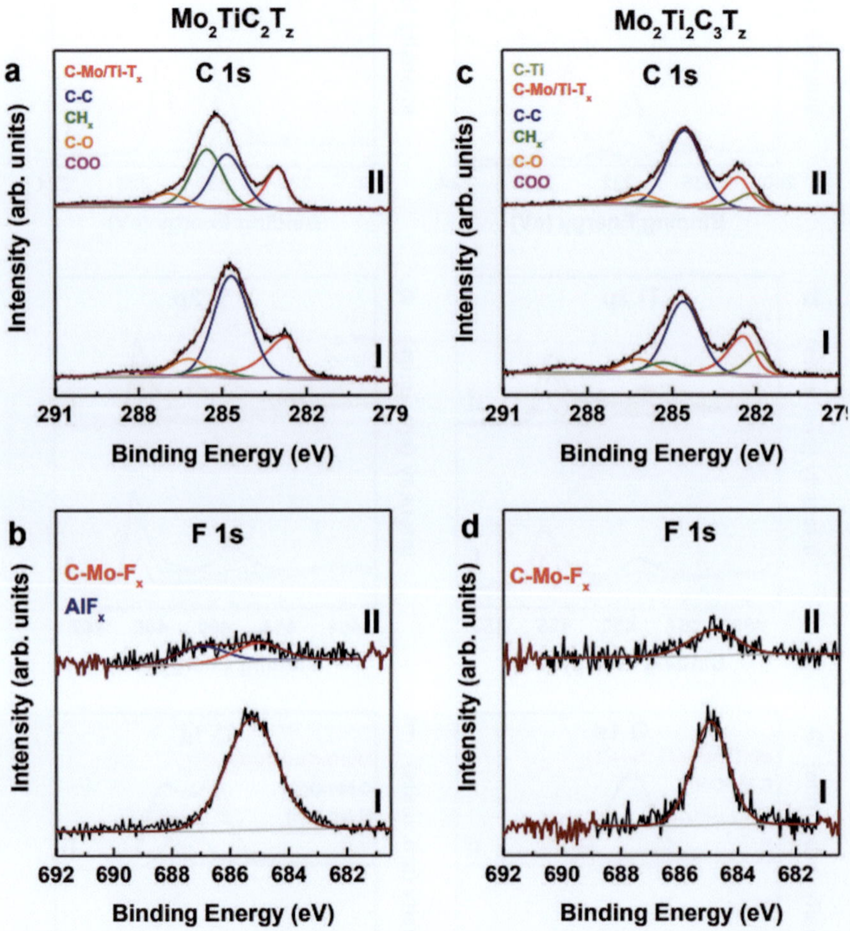

Fig. 3 Curve fitted XPS spectra for **a** C 1s, **b** F 1s region in ML-(I) and d-(II) thin film in 21, **c** similar to **a** but in 22, **d** same to **b** but in 22. Different peaks present here show different species that are assumed to exist. Peak colours and labels are coordinated. Reproduced with permission from [45] Copyright (2019) Elsevier Inc.

2.1 Transmission Electron Microscopy (TEM)/ High-Resolution TEM (HRTEM)

TEM imaging approach is a major surface characterization technique, which is more complex compared to other techniques such as scanning electron microscopy. In order to operate TEM, it requires wide operator training and extensive sample preparation routes. The major benefit by using TEM technique is that it facilitates imaging magnification in the range of five orders of magnitude and spatial resolution beyond its lattice structure, it is found to be major parameters in 2D material analysis. TEM images are obtained through transmission of highly coherent electron beam along a sample and by detecting the scattered electrons [46]. Here sample image is generated in image plane of electromagnetic objective lens and collected it by a CCD camera. Due to passing electron beam through sample, thickness parameter is a major criterion, because there exists a higher amount of material degradation which affects the quality of image. The coherent TEM imaging approach promotes more in depth information about characterization of microstructure, in compared with SEM, which facilitates the possibility to detect immediate contrast difference exist between crystal orientation, phase and conducting statistical analysis to flake size and dimension. In addition to this, lattice fringes observation in images having high-resolution provides interpretable ideas about structure which includes lattice spacing and relations orientations. The oxidation chemistry behind the 2D transition metal carbide MXene has new research interest with respect to their protection and application. But this oxidation character and mechanism of degradation that exist in MXenes, particularly with time in oxygen conditions at room temperature is not yet discovered. There are different class of experimental and theoretical routes which are introduced to study the early stage of oxidation phenomena behind HF-etched $Ti_3C_2T_x$ MXene in oxygen atmosphere at room temperature. The aberration-corrected environmental TEM coupled through reactive molecular dynamic simulation (MD) approach shows a crystal plane dependent oxidation state in $Ti_3C_2T_x$ MXene and expansion in oxide corresponds to difference occurs in coordination and charge in Ti superficial atoms and the presence of channel between nearby MXene layers on various crystal planes [46]. Here MXene is synthesized by HF etching route from its MAX phase using HF solution. Then a 50 mg of prepared MXene ground powder is quickly transferred from N2 glove box to ethanol solution of 10 mL and kept for ultrasonic dispersion. The solution consists of $Ti_3C_2T_x$ dropped on holey C film supported over copper grids and it is vacuum dried. The Spectra 300 TEM apparatus which works at 300 kV having Cs corrector for objective and condenser lens and is used to characterize chemical composition and atomic structure in $Ti_3C_2T_x$ MXene before and after performing oxidation. The simulations of HRTEM imaging were done on hypothesized atomic model which consists of α-TiO_2 and C disordered region with the aid of SimulaTEM software, it is a program related to Cowley and Moodie Multislice process. The parameters for selected image simulation hold an acceleration voltage of 300 kV, specimen thickness value of 2.1 nm which is close to actual thickness value of observing object with defocus of approximately 115 nm, and spherical aberration coefficient of objective lens is 0.1 mm. The interplanar distance measurements are done on HRTEM images using FFT in a Digital Micrograph software and the inverse FFT image is received through Digital micrograph software through the selection of desired diffraction spots present in FFT image. To receive the early-stage (0–65 min) oxidation character of MXene at room temperature, variation in pristine MXene microstructure is evaluated by in-situ aberration corrected environmental transmission electron microscopy (AC-ETEM) under O_2 gas atmosphere with a pressure of approximately 1×10^{-2} mbar and a complimentary in-situ Raman spectroscopy was used under air at room temperature, as given in Fig. 4. By considering benefit from open environmental chamber having specimen placed in high pressure ETEM column, as given by a schematic representation in Fig. 4a. The discontinuous evaluation of MXene oxidation in ETEM having minimal interference of electron beam (e^-) is fist considered to achieve intrinsic oxidation character. In consistent to HAADF evaluation, AC high resolution TEM (ACHRTEM) and related indexed fast Fourier transform (FFT) image represent that pristine MXene holds a stacked layer arrangement before oxidation (Fig. 4b(i)). With respect to the increase in oxidation time to 15 min in ETEM column, there exists a formation of amorphous oxide layer on nanosheet surface and extending to interior of structure (Fig. 4b(ii)). The analysis of oxide thickness measurement reveals that $Ti_3C_2T_x$ (010) plane have high oxidation rate rather than (001) plane. Similarly, high oxidation rate holds by (010) crystal plane also exists in some of the pristine MXene particles etched in HF.H_2O solution. This introduced the mechanism of oxidation observed in TEM measurement environment is practicable to discuss real-world oxidation in MXene. After finishing a period of 30 min, this completely oxidized sample introduces many amounts of fine TiO_2 anatase (tetragonal α phase having space group of I41/amd) nanocrystal having size in the range of 1–2 nm in diameter and there is an apparent lateral expansion in a direction perpendicular to (001) plane of nanosheet (Fig. 4b(iii)). With respect to the increase in exposure time of O_2 to 45 and 55 min, a brookite new TiO_2 phase (orthorhombic β phase having space group value of Pcab) begins to introduce in the prepared samples Fig. 4b(iv, v)). There exists a growth of fine-structured α and β-TiO_2, having coalescence of small number of crystals to large ones and there exists a new TiO_2 nucleation. This TiO_2 nanocrystals make a continuous growth to a diameter of approximately 10 nm, when oxidation time is obtained to be 65 nm (Fig. 4b(vi)). The inverse FFT image (Fig. 4c) introduced using characteristic diffraction spot by β{221}Copin TiO_2 (given in Fig. 4b(iv)) introduced the distribution and formation in β-TiO_2 phase in oxidized sample for a time duration of 65 min in addition to initially introduced α-TiO_2. To support the evidence obtained from AC-HRTEM analysis, the authors of the present work introduced in-situ Raman analysis for the sample and it shows the formation of β-TiO_2 in the sample oxidized at 1 h (Fig. 4d). Additionally, there exists a gradual reduction in peak intensity hold by α-TiO_2, disappearance of β-TiO_2 peak, and an increase in peak intensity in rutile TiO_2, with respect to the increase in oxidation time from 30 min to 20 days, also found in Raman spectra. Consequently, this obtained observation demonstrates sequential precipitation happened at early stage of oxidation in α-TiO_2 and β-TiO_2. In addition to this, transformation happened from α-TiO_2 (or β-TiO_2) to γ-TiO_2 and the generation of disordered C nanosheet during later oxidation. The simulations of HRTEM image (Fig. 4e) based upon hypothesized atomic model consist α-TiO_2 and disordered C region representing that regions possess bright contrast to Fig. 4b(vi) results from exposed C nanosheet having amorphous character among TiO_2 nanoislands. The higher oxidation rate within 30 min due to rapid transition of MXene to oxide. Full transition of MXene into oxide is finished at approximately 30 min. The reduced oxide growth present between 30 and 55 is through gradual expansion of oxide nanosheet in a direction perpendicular to (001) during oxidation (Fig. 4f). A pictorial representation of these observations obtained in evolution of MXene oxidation, consisting of this MXene amorphization, oxide expansion on (001) plane as well as TiO_2 nucleation, growth, and transition is given in Fig. 4g. In addition to this, $Ti_3C_2T_x$/oxide transition is finished within approximately 30 min, evaluation of precipitation, growth, and transition of TiO_2 after a time duration of 30 min is also prominent because the given processes have close relation to final TiO_2 nanoislands formation found in samples which are oxidized for 20 days.

To obtain earlier stage (0–70 s) features of MXene oxidation need a continuous in-situ AC-HRTEM imaging procedure, which inevitably introduces electron beam that affect during oxidation. In prior to continuous in-situ imaging, influence of electron beam on intrinsic oxidation reduction of MXene is initially evaluated. The obtained results show that there exists a negligible influence of electron beam on structural change of this MXene. Figure 5a represents different typical AC-HRTEM cross-sectional images obtained at various stage of MXene oxidation process at room temperature. The MXene possesses a complete layered architecture, with smaller amount of oxidation on surface at 0 s (Fig. 5a(i)). At shorter period of oxidation time at 10 s (Fig. 5a(ii)), 20 s (Fig. 5a(iii)), 30 s (Fig. 5a(iv)), 50 s (Fig. 5a(v), and 70 s (Fig. 5a(vi)), there exists the formation of amorphous oxide continuously produced starting from outside and proceeds to interior portion of particle, which results to an improvement in thickness on both (001) and (010) plane of nanosheet, in addition to this the oxidation rate of (001) plane have greater value. Additionally, an increase in oxide thickness towards interior portion of material, results to oxidation in thickness of oxide layer on (001) plane in outward direction, which represents oxide nanosheet expansion in perpendicular to (001) plane through oxidation. The in-situ analysis of basal plane (001) oxidation of MXene nanosheets is very crucial to confirm result obtained through cross-sectional analysis. It is observed that nanosheets basal plane proceed same amorphization from the edge to interior having inhomogeneous oxidation rate (Fig. 5b(i–vi)). Thus, oxidation character on basal plane of nanosheets generally consistent with observed cross-section. By consider different oxidation rate and expansion of oxide on various crystal planes (Fig. 5c, d), a pictorial representation of characteristic steps involved in MXene oxidation is shown in Fig. 5e. More specifically, disordering from outside to inside of nanosheet having high oxidation rate hold by (010) plane than (001) plane and simultaneous expansion of oxide on (001) plane is verified in earlier oxidation stage of the prepared MXene. The present study introduced the early stage in evolution of structure present in MXene with the appearance of oxygen and it provides further tailoring of MXene performance which employs an oxidation strategy.

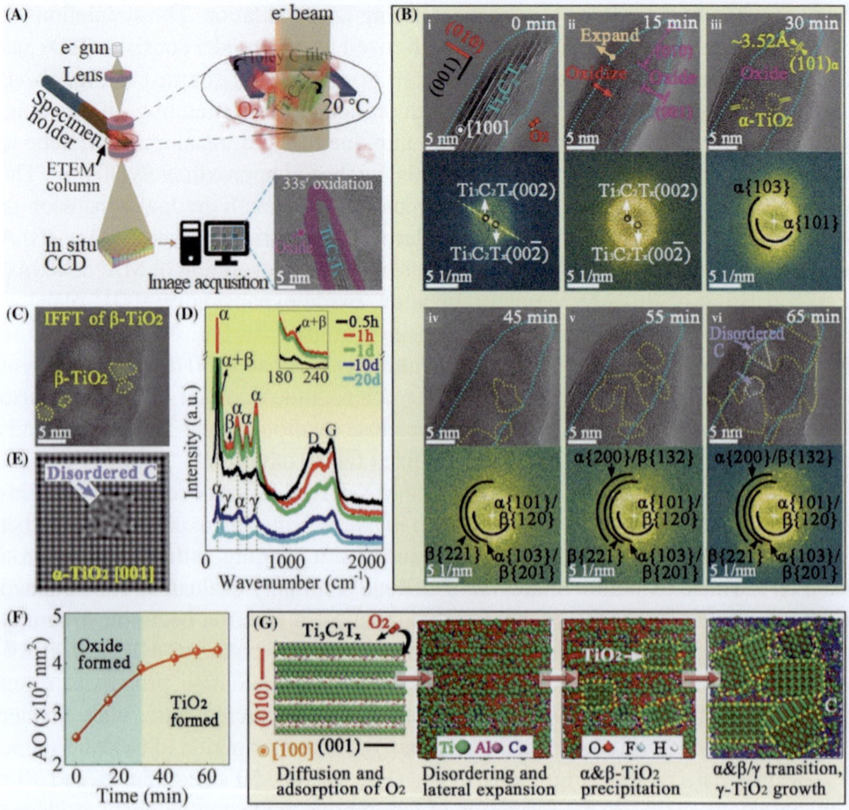

Fig. 4 Discontinuous in-situ AC-HRTEM analysis having minimum electron beam interference during $Ti_3C_2T_x$ exposure to O_2 continuous flow at room temperature and pO_2 approximately 1×10^{-2} mbar and in-situ Raman evaluation of MXene oxidation in air at room temperature which show microstructural evolution of this MXene oxidation; **a** Pictorial representation of open environmental chamber having specimen in high pressure TEM column, which shows the experiment of in-situ AC-HRTEM oxidation used in the present study; **b** Cross-sectional time resolved AC-HRTEM image for MXene oxidized for the durations of (i) 0 min, (ii) 15 min, (iii) 30 min, (iv) 45 min, (v) 55 min, and (vi) 65 min. The FFT index is given below each HRTEM image. The dotted lines with cyan colour represent the actual position of $Ti_3C_2T_x$ surface before performing oxidation. The dotted line having cyan transparent line in (ii) represents interface present between prepared MXene and oxide; **c** inverse FFT image which utilizes diffraction spots corresponding to β-TiO_2 in B(vi), shows the phase and distribution of β-TiO_2 nanocrystal formed in oxide; **d** in-situ Raman spectra of MXene samples having various oxidation times in air present at room temperature; **e** simulation of HRTEM image based upon atomic model; **f** growth rate hold by amorphous oxide which is measured through increment in oxide area with respect to time from **b**; **g** Pictorial representation of evolution of MXene upon room temperature oxidation (here AO is the oxide area). Reproduced with permission from [46] Copyright (2024) Wiley Verlag VCH Ltd.

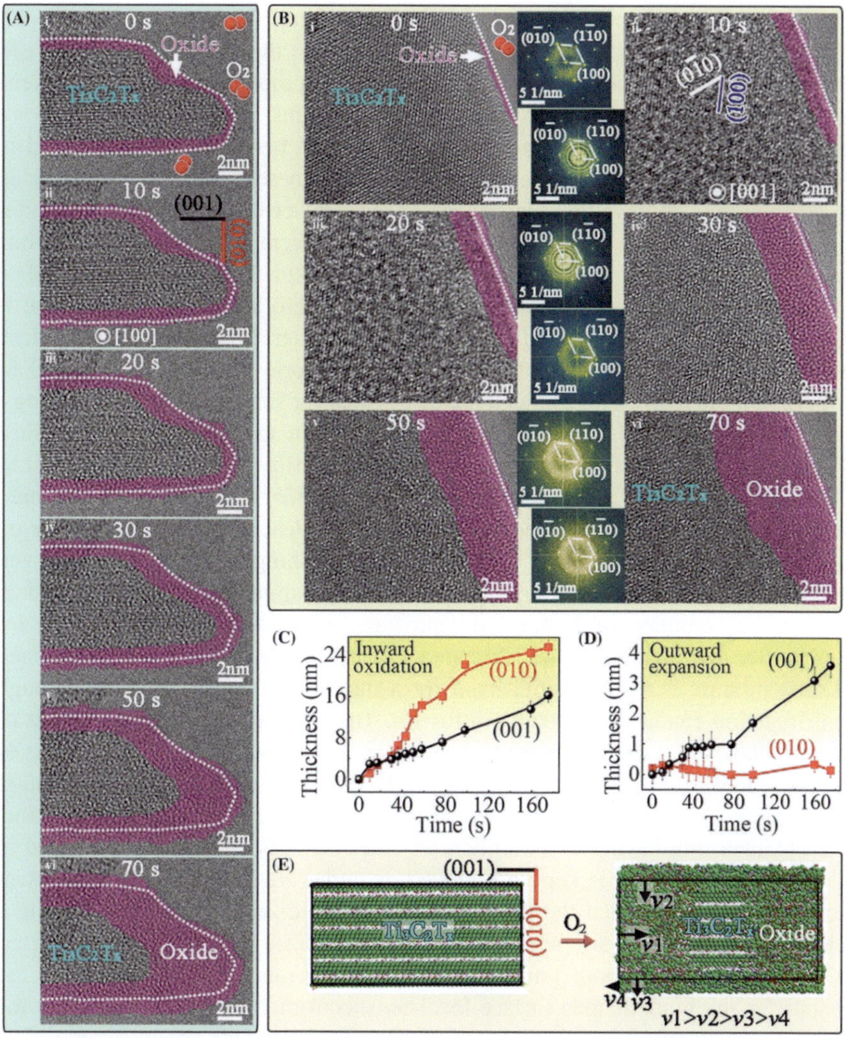

Fig. 5 In-situ AC-HRTEM analysis having continuous exposure of electron beam which shows MXene's earlier stage of oxidation process (before 70 s) with respect to exposure in continuous O_2 gas flow at room temperature and pO_2 approximately 1×10^{-2} mbar; **a**, **b** are time-resolved cross-sectional image obtained. The FFT images indexed in **b** are from its corresponding connected AC-HRTEM image, which represents microstructural degradation in MXene. The overlaid areas present in AC-HRTEM image represent the formation of amorphous oxide during oxidation; **c**, **d** thickness of oxide with respect to time, indicates inward and outward expansion of (010) and (001) plane in MXene. The error bar exists around each of these data points represent the standard error of mean value; **e** pictorial representation of characteristic features of earlier oxidation process of MXene sample with respect to in-situ observation. v in **e** represents increase in rate of oxidation thickness. Reproduced with permission from [46] Copyright (2024) Wiley Verlag VCH Ltd.

In addition to the powdered form of these MXene powders, there is a report that makes analysis on features hold by MXene films through TEM and HRTEM analysis. There is a report which studies the surface properties of $Ti_3C_2T_z$ MXene films through TEM and HRTEM analysis [47]. For this, the MAX phase Ti_3AlC_2 for MXene is prepared through solid–liquid reaction between Ti, TiC, and Al in the molar ratio of 1:2:1, respectively. They mixed these precursor powder using an agate mortar, followed by placing the mixed powder in alumina crucible and it is heated in horizontal tube under an Ar flow of 5 sccm at 5 ºC/min heating rate to 1450 ºC, for 120 min. The obtained sample is kept for cooling, then crushed with mortar and pestle and sieved by 450 mesh. Here a multilayered MXene is synthesized by slow immersion of 1 g of MAX powder in polytetrafluoroethylene (PTFE) bottle containing 20 mL of 12 M hydrochloric acid (HCl) and a 2.3 M lithium fluoride (LiF). The obtained mixture is kept for subsequent agitation using a magnetic stirrer coated with PTFE at 35 ºC for 24 h in an oil bath and it is washed with 40 ml of 1 ml HCl followed through three times further washing in 40 mL of 1 M HCl, after that washed in 40 mL of 1 M lithium chloride. The multilayer structured MXene is then washed thrice using 40 mL of N2 deaerated deionized water up to a dark supernatant obtained. For each of this washing stage, 40 mL deionized water was introduced to multilayer powder in centrifuging tube, then hand-shaken it for a time duration of 1 min and centrifuged at a condition of 4200 × g for 1 min, after this supernatant was decanted. After performing washing, the obtained MXene mixture is agitated with hand for a duration of 5 min followed through centrifugation condition of 2058 × g for 1 h. The obtained supernatant consists of $Ti_3C_2T_z$ flakes in water, it possesses a concentration of 1–4 mg/mL. The authors of this work fabricated a free-standing flexible $Ti_3C_2T_z$ film through vacuum filtration of concentrated colloidal solution of 15 mL by a nanopolypropylene membrane. The obtained film having 34 mm diameter and thickness of 5 μm is then dried in air for overnight and it is kept in a glove box under N2 flow. Here in the present work, the authors studied the effect of exposing oxidized MXene films having a thickness of micron or thin nanometer and it is oxidized to flakes while dispersed in solution. In brief (about 1 min) higher frequency range of about 10 MHz, the nanometer amplitude ranged surface-localized electromechanical vibration is in the form of surface reflected bulk waves (SRBW), which produce effective removal of its oxide layer, thereby allowing recovery of their electrical and electrochemical performance under ambient experimental conditions without using any chemical additives or reducing agent. Here the low-magnification and high-magnification TEM analysis of prepared MXene films are performed at operating voltage values of 100 and 200 kV, respectively. The samples for analysis is prepared by grinding MXene films using mortar and pestle, then deposit the suspension of this obtained flake in isopropanol on holey carbon-coated copper grid. The analysis of film using TEM is given in Fig. 6a, b, in prior to (Fig. 6a) and followed through (Fig. 6b) SRBW exposure at highest intensity value of 25 dBm indicates a prominent reduction in coverage by uniform layer of TiO_2, that initially coated over surface of oxidized film. The oxide layer removal with SRBW irradiation is further evaluated by contrast in HRTEM images taken between control (that is unexcited) and samples that are recovered SRBW, as shown in Fig. 6c, d. More clearly, the HRTEM image taken for control (oxidized film) given as Fig. 6d shows characteristic plane (0110) is prepared MXene, in addition to this disorder in lattice due to existence of surface oxide layer. In addition to this, the authors found a lattice spacing value of 3.5 and 3.3 Å corresponds to (101) and (110) plane of rhombic structured anatase and prismatic rutile phase in TiO_2, the obtained anatase (101) TiO_2 plane is further confirmed from selected area electron diffraction (SAED) pattern, in addition to hexagonal diffraction pattern features of $Ti_3C_2T_z$ MXene. On the other hand, HRTEM image of film subjected to SRBW having higher intensity value of 25 dBm (Fig. 6d) is found to be devoid of rutile and anatase TiO_2, while retaining their characteristic lattice spacing of $Ti_3C_2T_z$ and diffraction plane, thus confirms the removal of prominent proportion of oxide layer from film surface.

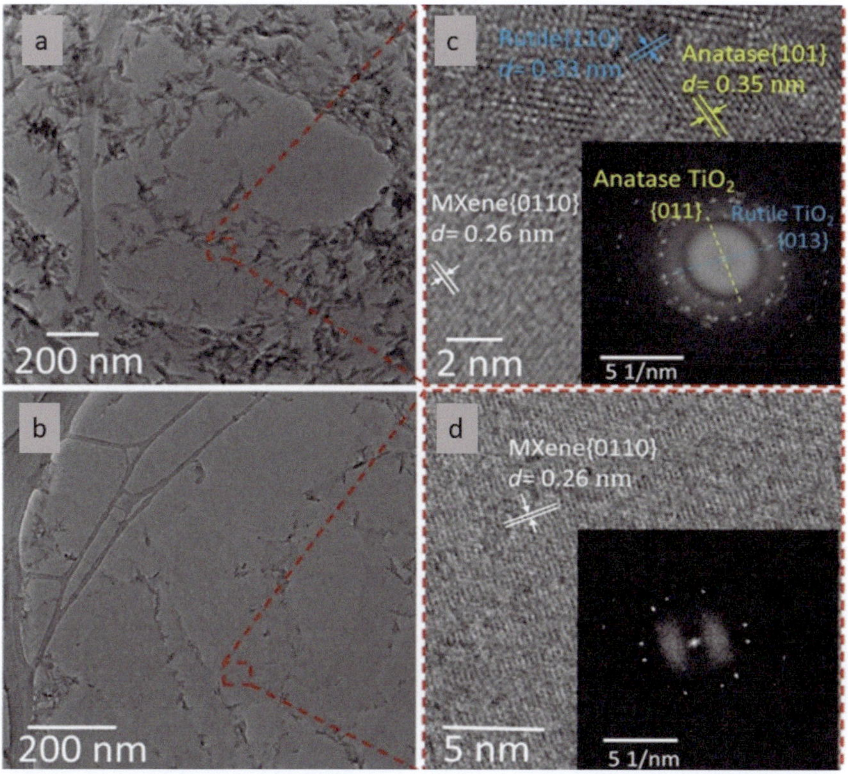

Fig. 6 **a**, **b** TEM and **c**, **d** HRTEM images of oxidized (**a**, **c**) and SRBW-recovered (**b**, **d**) $Ti_3C_2T_z$ MXene films (inset shows the corresponding SAED pattern). Reproduced with permission from [47] Copyright (2023) Springer Nature

2.2 Scanning Transmission Electron Microscopy (STEM)

MXenes usually possess an oxidization character with respect to time in ambient condition. Thus, it is necessary to introduce a basic understanding related to its material stability, particularly with respect to time and under ambient experimental conditions. Beyond the synthesis condition optimization, characterization of property and theoretical studies of structural characterizations on atomic level of MXene are scarce. STEM analysis was introduced to atomic level investigation on MXenes [48]. There is a report based upon STEM analysis of Nb_2C MXene with respect to the existence and coordination of surface adsorbed species, and its influence on structural stability with respect to time [48]. Here Nb_2C MXene is synthesized through selective etching of the Al layer from its corresponding MAX phase (Nb_2AlC). Here MAX phase is synthesized through pressureless sintering approach and it is further immersed in aqueous HCl (48%) to prepare multilayer MXene. To prepare the Nb_2C monolayer, MXene multilayer powder is intercalated with tetrabutylammonium hydroxide (TBAOH) in the water followed through delamination and sonication for a duration of 1 h, which yields smaller than 0.5 mg/mL suspension of MXene colloidal flakes in water. The STEM analysis samples are prepared through drop-casting of obtained supernatant to a lacey carbon film suspended with copper TEM grid and dry it. A low-magnification STEM image of a freshly prepared single layer Nb_2CT_x sheet along (0001) c-axis is given in Fig. 7a. The prepared sheet possesses a large lateral dimension value greater than 2 μm and homogeneous contrast indicates that the sheet is clean and it is uniformly thick. Figure 7b represents atomically resolved STEM image of Nb_2CT_x sheet, together with corresponding FFT. It is observed that this prepared MXene possesses a highly ordered and close-packed atomic structure, which possesses an extra complex of apparent holes (dark colour) and highly contrasting atomic column (bright) in an ordered mosaic. This atomic structure obtained is unprecedented and it does not resemble to honey-comb patterned atomic arrangement expect from Nb_2C MXene structure aligned along c-axis. The STEM image obtained from the same TEM sample after being stored for a time duration of 5 days in vacuum condition is shown in Fig. 7c.

Amazingly, STEM image obtained for Nb_2CT_x MXene exhibited an expected honey-comb-like pattern as verified through FFT. Locally, this obtained structure shows a disordered appearance. The authors introduced an additional exposure (6 days after performing synthesis) to ambient condition, which increased the disordered pattern, thereby this honey-comb architecture is retained only in the small regions, as shown in Fig. 7d. More specific to say that, at the present point 2D architecture held by macroscopic sheet is found to be intact and there exists an absence of observation corresponding to degradation at low magnification range. The authors observed that fresh sample is found to be not similar to $M_2X\ T_x$ structure and it is surprising that it tends to evolve to the regular appearance of M_2X with respect to time. To evaluate this character, it is very important to interpret atomic structure of fresh sample. Thus, STEM image shown in Fig. 8a is divided into basic building blocks. The evaluated M_2X model structure is projected along c-axis and

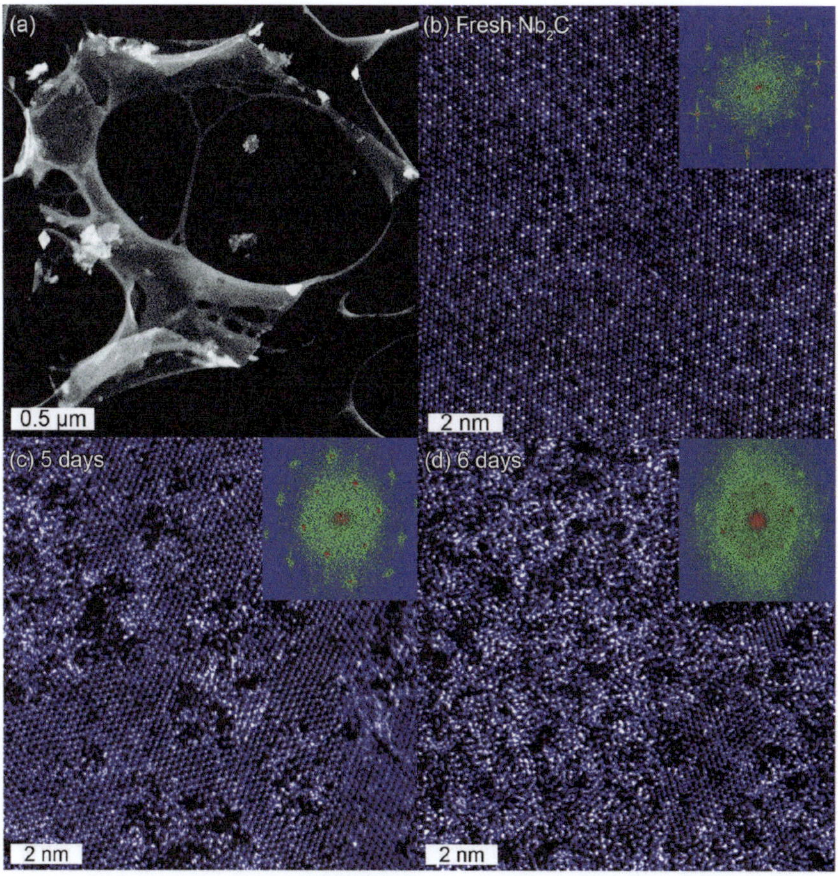

Fig. 7 **a** Complete view of monolayer fresh Nb_2CT_x MXene sheet. STEM images which are atomically resolved and corresponding FFT pattern as inset along c-axis to form **b** fresh, **c** 5, and **d** 6 days after performing synthesis. Reproduced with permission from [48] Copyright (2018) Royal Society of Chemistry

overlaid on STEM image (Fig. 8b) through positioning model honeycomb core above 'holes' in image. The proposed model matches with images well, which means all 'holes' make alignment on honeycomb core. In addition to this, highly intense atomic columns make decoration to cores and it doesn't coincide with M atomic column of model structure. Furthermore, some cores act as hosting atoms having the same intensity as Nb atoms which aligned with model structure. These various columns are colour coded to Fig. 8c, where green, blue, and pink correspond to model structure columns, highly intense columns, and columns having intensity equal to model structure columns, respectively. It is also found that high intense atomic column locally produces a large honeycomb close packed pattern structure (represented with red lines in Fig. 8c). Thereby, the obtained fresh sample structure corresponds to

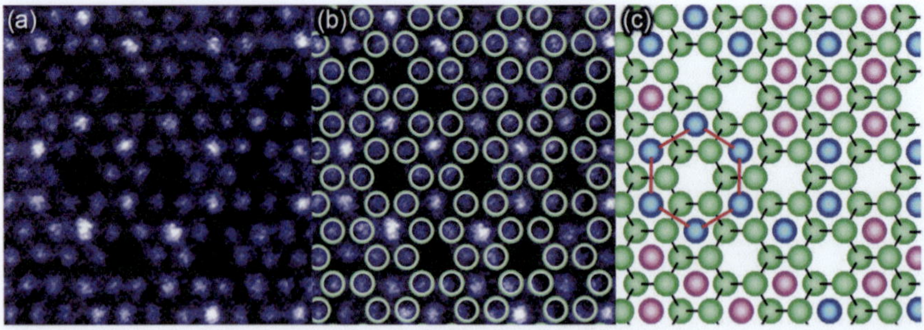

Fig. 8 **a** STEM images atomically resolved from fresh Nb_2CT_x MXene, it is **b** having M_2X model structure overlaid, and **c** the crystal structure sketch with honeycomb cores filled by low (coloured by pink) and highly intense atomic column (blue). High intense atomic column which form large honeycombs are represented by red line in **c**. Reproduced with permission from [48] Copyright (2018) Royal Society of Chemistry

M_2X model structure having extra atomic column of same or high intensity which decorates these honeycomb cores.

There is a reports based on selective etching of Al layer from Ti_2CT_z and Nb_2CT_z film from its MAX phase using LiF and HCl solution [49]. To etch Ti_2AlC film, the solution for etching is prepared by introducing 1.2 g LiF salt to 20 mL taken from 12 M HCl. The prepared solution is stirred at room temperature for 15 min to facilitate total dissolution of LiF salt in HCl. Then subsequently, the process of stirring was stopped and the MAX phase thin film (Ti_2AlC) was immersed in solution for a time duration of 0.3 h at room temperature. For Nb_2AlC thin film preparation, the solution contains LiF, and HCl is synthesized through adding 3 g LiF to a 40 mL of 12 M HCl and it is stirred completely. After completing stirring process, Nb_2AlC films are immersed in an etching solution for a time duration of 3 h at a temperature of 55 °C and it is rinsed with distilled water, followed through ethanol and finally dried in blowing Ar gas. But during this stage, there exists an intercalation of multilayers by Li, thus it is termed as Ti_2CT_z-yLi and Nb_2CT_z-yLi. Here STEM analysis of prepared films is performed using Linköping double-Cs-corrected Titan3 G2 60–300 (FEI, Eindhoven, Netherlands) operated at 300 kV. The STEM images obtained for Ti and Nb-based films having thickness in the range of 40 and 20 nm are shown in Fig. 9. Here Fig. 9a represents a typical overview image obtained for Ti_2CT_z-yLi on top of Ti_2AlC–TiC mixed incubation layer developed over Al_2O_3. The atomically resolved Ti_2CT_z-yLi layer is shown in Fig. 9b. A typical overview image obtained for Nb_2CT_z-yLi (Fig. 9c) and atomically resolved Nb_2CT_z-yLi film (Fig. 9d) is evaluated. At the range 11.4 ± 0.2 and 11.8 ± 0.3 Å, the values obtained for δ correspond to Ti- and Nb-based films respectively, calculated from TEM analysis is in agreement with XRD.

Fig. 9 STEM images obtained along [11¯0] zone axis of Al_2O_3; **a** total view obtained for Ti_2CT_z-yLi on Ti_2AlC–TiC top mixed incubation layer developed over Al_2O_3, **b** Ti_2CT_z-yLi layers atomically resolved, **c** total view of Nb_2CT_z-yLi over Al_2O_3 and **d** Nb_2CT_z-yLi layers atomically resolved. Reproduced with permission from [49] Copyright (2019) IOP Publishing Ltd.

2.3 Raman Spectroscopy

Raman spectroscopy is an analysis tool used to establish measurements related to bonding in MXene structure through phonons and lattice vibrations. Raman spectroscopy identifies molecular fingerprints and it is found to be sensitive to amorphous compounds and trace of transition metal oxide, which is found in MXene structure. There is a report based on multiwavelength Raman spectra analysis of $Ti_3C_2T_x$ to evaluate its synthesis condition, material state (single flake, colloidal solution, and stacked flake), and the intercalated species in Raman spectra [50]. Here MXene is synthesized by selective etching of the Al layer present in Ti_3AlC_2, the different synthesis approaches possible for MXene are shown in Fig. 10a. They utilized an inverted reflection mode Renishaw InVia instrument arranged with 20 × (NA = 0.4), 63 × (NA = 0.7), 55 × (NA = 0.45), and 100 × (NA = 1.3) objective and diffraction-based room-temperature spectrometer. In He–Ne (633 nm) and diode (785 nm) laser, they utilized a 1200 line/mm diffraction grating, and in the case of Ar^+ laser (488 and 514 nm emission), they used 1800 line/mm grating. Here laser power is obtained in the range of approximately 0.5–1 mW. The mapping is performed through raster scanning in StreamLine mode at different x-axis step and the inspectral range was limited by grating. For 1200 line/mm grating, number obtained for wavenumber per scan is in the range of 400 cm^{-1}, thus to scan 100–800 cm^{-1}, they performed a scan twice and manually stitched it. To measure the solution, the authors utilized a glass quartz tube capillary having inner diameter value of 0.5 mm. Before start to evaluate complex and broad spectra hold by $Ti_3C_2T_x$, it is necessary to introduce theoretical studies related to its vibrational features, thereby helping to navigate multiple Raman bands. Here proposed MXene is in D_{3d} point group and a unit cell, not belong to molecule, vibrations given in Mulliken symbols given by $4E_g + 2A_{1g} + 4E_u + 2A_{2u}$. Here E_g and A_{1g} represent Raman-active modes, but E_u and A_{2u} are IR-active. This is very important to indicate that E_g and E_u are double degenerate modes, thus it is appeared as one in the irreducible representation. This corresponding to four Raman active modes, E_g in-plane vibration between Ti and C atoms and A_{1g} is the out-of-plane vibration between Ti and C atoms. Two others represent in-plane and out-of-plane vibration by C atoms. If surface groups such as –O and –F are introduced, there exists an increase in number of atoms per unit cell, thus number of vibrations also increase. With respect to theoretical and experimental data, the authors represented a complete peak assignment to $Ti_3C_2T_x$. Raman spectrum of MXene have multiple features in 100–800 cm^{-1} range. The Raman spectrum represented by MXene have different regions, as shown in Fig. 10b. Firstly, there exists a resonant peak, obtained by the use of 785 nm laser, and coupled to a plasmonic peak. The next one is flake region which contain E_g(Ti, C, O) and A_{1g}(Ti, C, O) mode, corresponds to in-plane and out-of-plane vibration of titanium atom in outer layer as well as carbon and the surface groups respectively. This corresponds to vibration of flake as the group vibration by Ti, C and surface group. This introduced vibration is stiffest, consists of maximum number of atoms from MXene unit cell. In addition to this, 230–470 cm^{-1} region represents in-plane (E_g) vibration of surface group attached towards titanium atoms. Here proposed region is affected by the surface atoms; thus, it is potentially used to probe surface chemistry of MXene and their change, resulting from chemical/electrochemical reaction. Here region exists between 580 and 730 cm^{-1} corresponding mostly to carbon vibrations (it is both E_g and A_{1g}). Here later region is utilized to fingerprint surface group in different in-situ electrochemical analysis.

The different samples analysed using Raman microscopy are shown in Fig. 11a. The Raman spectra of multilayer powder, clay, delaminated film, single flake over Si/SiO_2 substrate, and colloidal solution, are shown in Fig. 11b. All products are obtained through etching of Ti_3AlC_2, having exception of clay (MXene slurry with liquid crystalline characteristics), here LiF-HCl etching is utilized. Figure 11b represents difference in Raman spectrum obtained from various samples, show two distinct sharp peaks of A_{1g} (Ti, O, C) in the range 200 cm^{-1} and A_{1g} (C) around the 720 cm^{-1} wavenumber regions. It is prominent that A_{1g} (C) possesses biggest shift in comparison to other peaks. Here, the lowest wavenumber holds by this peak (711 cm^{-1} in multilayer powder after HF etching and 719 cm^{-1} for HF-HCl etching) is obtained in the multilayer form, here restrictions to vibrations are by stacking and small interlayer spacing. After intercalant is given, higher amount of water is introduced between the layers, which expands interlayer spacing. Here appearance of water stiffens out-of-plane vibration and value of interlayer spacing is large, which shifts A_{1g} (C) peak to approximately 720 cm^{-1}. Prominently, similar shift in multilayer form is observed in filtered and dried desiccator film. It is possible to explain with appearance of less amount of water in dried film, stiffens out-of-plane vibration and preventing the flakes sliding with respect to each other. It is prominent to observe that even though the intercalant has appeared in filtered film, vibrations exhibit same frequency. In other words, for single flake deposited over Si/SiO_2 substrate, out-of-plane vibration is restricted with substrate on one side only, which leads to asymmetric peak, covering range from 720 to 736 cm^{-1}. It is prominent that, spectra are collected by different laser excitation. But there is no shift found when change the laser. The peak shift obtained to high wavenumber range of 726 cm^{-1} for MXene flake dispersed in water, here both sides of flakes are surrounded with H_2O molecules. It is due to

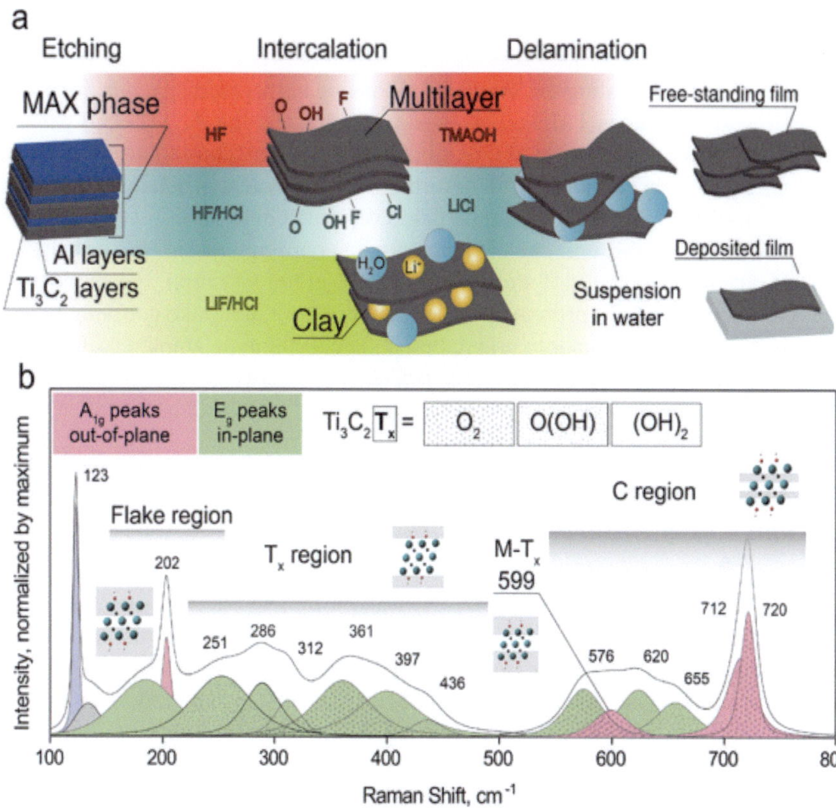

Fig. 10 Synthesis of $Ti_3C_2T_x$ MXene and assignment of Raman peak; **a** Pictorial representation of $Ti_3C_2T_x$ MXene synthesis and delamination procedure used in the study and $Ti_3C_2T_x$ schematics at various stage of processing, which includes etching, intercalation and delamination; **b** deconvoluted Raman spectrum received from $Ti_3C_2T_x$ film synthesized through HF-HCl etching and it is excited with 785 nm laser. Reproduced with permission from [50] Copyright (2020) American Chemical Society

the less restrictions in liquid media and the flakes make free movement. A same peak present at 725 cm^{-1} is found in the claylike MXene, it is same as solution, flakes are surrounded through H_2O molecules having motion is less restricted than dried restacked film. Thus, few monolayers in water permit unrestricted vibration, same as colloidal solution. This obtained observation indicates peak presence is not only related to surface group but to flake-to-flake interaction. It is possible to explain through $Ti_3C_2T_x$ flakes stacking in the same manner like parent MAX phase. After flakes become delaminated and restacked, the ratio increases to 1.2, which means coupling present between flakes are weakened through flakes random orientation with respect to each other. Here M and X layer ordering is reduced after performing etching, in comparison to flakes random orientation with respect to each other. Interestingly, the A_{1g} (C) possesses higher intensity in these conditions, which indicates vibration of carbon layer does not depend upon flake orientation. A reverse effect is found in single flake on Si/SiO_2 substrate. Here ratio between A_{1g}(C)/A_{1g}(Ti,C,O) is obtained to be 0.9, indicating that A_{1g}(Ti,C,O) having higher intensity, it is the vibration by a whole, same as TMDs. For aqueous dispersion, the A_{1g}(Ti,C,O) peak is vanished by random orientation of flakes in the space and by continuous movement. The surface groups in the range (230–470 cm^{-1}) are prominent in delaminated flakes restacked in filtered film, also for single flakes deposited over Si/SiO_2 substrate. For water, this can be established through reduction in coupling present between flakes, it affects out-of-plane peaks, it strengthens surface group vibration. For monolayer flake, the region is highly prominent, since it doesn't contain flake coupling which weakens A_{1g} peak. Here random orientation in flakes within delaminated film plane which broadens surface functional group peaks.

2.4 X-ray Photoemission Electron Microscopy (XPEM) and X-Ray Absorption Spectroscopy (XAS)

Soft x-ray absorption spectroscopy (XAS) was prominent element-selective route which facilitates investigation of transition metal fine electronic structure and highly sensitive to its surface chemistry. Here higher chemical sensitivity holds by XAS at Ti L-edge is prominently applied to differentiate between different Ti oxide species. It is possible to differentiate oxide due to its orbital difference in electronic coordination, here TiO, Ti_2O_3, and TiO_2 consists of two, one and no electron in it respectively. X-ray photoemission electron spectroscopy (XPEM) facilitates acquiring x-ray absorption spectra under ultra-high vacuum condition having 30 nm spatial resolution. There is a study based on characterization of pristine and urea-intercalated MXene ($Ti_3C_2T_x$) electrodes and related to Ti–O bonding nature in the vacuum and water with having XAS at the Ti L and O K edges [51]. The spatially resolved oxidation with MXene individual flakes are studied through XPEM. Additionally, they used in-situ XAS on MXene aqueous dispersion to observe influence of solvation effect on structure hold by MXene. The solid samples are spin coated under air on the conductive Si substrate for prepare multilayer isolated single flakes of the $Ti_3C_2T_x$ (pristine), urea-intercalated $Ti_3C_2T_x$ and thin TiO_2 nanoparticle film. In order to mimic the evaluated bulk sample, the authors evaluated oxidation state of Ti atoms having relatively thick flake which hold more or less same flake size. It possesses some difficulty in XPEM analysis with respect to their long-time duration to finish the experiment. Thus, they characterized samples through XAS using partial electron yield (PEY) mode having XPEM endstation at UE-49-PGM, it is a beamline of synchrotron BESSY II light source operated with HZB. The TiO_2 anatase particle has mean particle size of 4–8 nm and it is stabilized with 10 wt% nitric acid. Here drop casting films over Si conductive substrates for entire samples are dried firstly in the air and analysed in the Total Electron Yield Mode (TEY) in vacuum conditions at a beamline of BESSY II using LiXEdrom endstation. The water (pure), multilayer dispersed

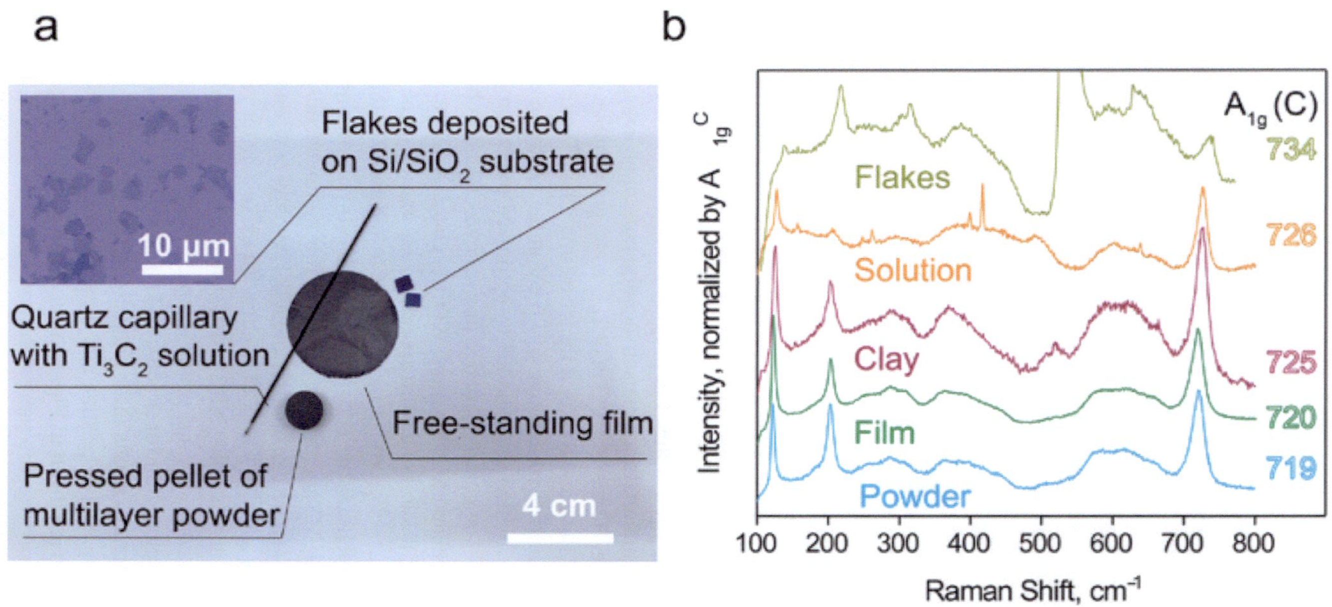

Fig. 11 Raman spectra obtained from various MXene states; **a** Raman spectroscopy versatility, all the samples given in photo have different $Ti_3C_2T_x$ states analysed by Raman microscope; **b** Raman spectra correspond to a multilayer powder, preparation of clay by LiF-HCl etching, single flake on Si/SiO_2 substrate, here laser with 785 nm is used. For the case of single flake, the authors used 633 nm. A prominent peak for Si present at 521 cm^{-1} was introduced from Si/SiO_2 substrate. Sharp peaks given in the liquid samples case are attributed to setup used in the study. Reproduced with permission from [50] Copyright (2020) American Chemical Society

$Ti_3C_2T_x$, urea-intercalated $Ti_3C_2T_x$, and urea in water is analysed with membrane-based flow cell which utilize Total Fluorescence Yield (TFY) recorded by XUV-100 silicon photodiode. They used a thick silicon nitride membrane having thickness of 50 nm with window size of 0.5 mm × 0.5 mm, for isolate liquid from a vacuum chamber fixed at a pressure range of 10^{-6} mbar, required to perform soft XAS analysis. After performing characterization for samples, flow cell flushed with deionized water and acquired the reference spectra at O K-edge for validating the absence of sample contamination left in cell. Here energy value of O Kedge is calibrated for water pre-edge at the value 535.0 eV. To establish a clarity, intensity of O K edge XAS is normalized to intensity before and after main edge of water. The Ti L-edge is normalized to dried MXene L_3 (t_{2g}) peak present at 458.1 eV. In order to distinguish the oxidation state of Ti atoms, XAS analysis was before and after the intercalation of urea in various conditions. Figure 12a represents XPEM image taken for vacuum at Ti L-edge of $Ti_3C_2T_x$ MXene flake before and after intercalation of urea. These obtained flakes are a few micrometers in diameter, having sharp edge. Here yellow and green coloured area in Fig. 12a act to represent specific regions given respectively as centre and edge., it is possible to extract XAS of Ti L-edge from these images. Here Ti L-edge the is corresponding to excitation of core levels Ti $2P_{3/2}$ (L_3-edge) and Ti $2P_{1/2}$ (L_2-edge) to the unoccupied Ti 3d state. In accordance to the ligand field introduced through binding by oxygen, each Ti L_3 and L_2 edges split into two different sub peaks, which corresponds to electronic state with t_{2g} and e_g symmetry. The e_g band have higher sensitivity to local environment due to direct pointing of Ti e_g orbital to 2p orbital of surrounding oxygen atoms. For proposed MXene flake, the splitting of L_3-edge to t_{2g} and e_g subbands are found at each location present on flake, although it is efficiently defined at edge. Here onset region of Ti L-edge, given in Fig. 12b, differs between edge and centre of flake and it is used for the estimation of Ti oxidation state. Here onset energy value shifting from range 454.6 eV at the centre in comparison with 455 eV at flake edge, which corresponds to optimized oxidation state found to be slight greater than Ti^{3+} at centre, but edge position was found to be relatively near to Ti^{4+}. In urea intercalated MXene flake, XPEM-related XAS introduces that L_3 split into two prominent sub-band corresponds to t_{2g} and e_g peak. The value of pre-edge onset energy at 455.8 eV is independent to location on these MXene and represents large contribution by Ti^{4+}. The Ti atoms stay in prominently high oxidation state relative with MXene pristine flake and it is homogeneously oxidized over complete flake. Here oxygen distribution and local bonding environment is also possible to characterize through XPEM at O K-edge. To make a better visualization for distribution of oxygen across flake, XPEM difference image was taken and it is shown in Fig. 12c. At high contrast value which is close to flake edge illustrates high oxygen content in this given region for pristine MXene, but the content of oxygen was found to be highly homogeneous after intercalation of urea. The comparison for O K edge XAS corresponds to pristine and urea introduced $Ti_3C_2T_x$ at centre and edge position of flakes are given in Fig. 12d. Here XAS of O K edge possesses two prominent peaks at energy region present between 529 and 535 eV, introduced from transition between O 1s to O 2p state and it is hybridized by Ti 4s and 4p state. The special character of MXene at 530.8 and 533.2 eV corresponds to t_{2g}

and e_g orbitals, respectively. Although, in the case of pristine $Ti_3C_2T_x$, the peak intensity of e_g is obtained to be comparatively low value at centre, which suggests low mean oxidation state relative to its edges. Further, make intercalation of urea, e_g peak present at 533.2 eV is well resolved, as given in Fig. 12d. The O K edge spectra thereby emphasize the same oxidation state across urea intercalated $Ti_3C_2T_x$ flake, which is in agreement with results obtained for Ti L edge. It is found that contribution by intercalated urea is possible to appear between 532.5 and 534.0 eV. These features found between 537 and 540 eV at urea introduced $Ti_3C_2T_x$ edge position which corresponds to SiO_2 substrate. The high oxidation state value of Ti at edges of pristine $Ti_3C_2T_x$ MXene is probably due to high exposure by water molecules and the oxygen from an ambient environment as part of aqueous dispersion and the drying procedure in air. In addition to this, oxidation executes faster at the edge positions of MXene flake relative to its basal plane, it is observed to be eventually contribute to total reduction for conductivity of these MXene.

3 Future Perspectives and Conclusion

From the present chapter, we found that MXene films and powders have a major role in various applications. The surface characterization techniques of MXene films and powders show that it is possible to establish their basic features through this analysis. The utilization of a possible and favourable characterization tool delivers how the features of MXene powder and flakes meet the application criterion. The major demerit underlying these characterization tools is the correlation of results obtained from each of these techniques. Sometimes, it gives contradictory results in each characterization technique, so a unified theoretical route is needed to correlate these results. Another drawback associated with these surface characterization techniques used in MXene films and powders is their influence on environmental and experimental conditions. In addition to this, the analysis has a demerit in the case of composite/heterostructure preparation with other layered materials such as graphitic carbon nitrides [52, 53], transition metal oxides [54–57], boron carbon nitride [58–61], transition metal dichalcogenides [62–66], etc. In this particular case, an in depth idea about intercalation/formation mechanism, etc., of these composites is not able to characterize through these techniques, it needs some advanced techniques for its evaluation. Thus, it is necessary to optimize these conditions for experiments before they start. Through the proper optimization of these parameters and by evaluation of its surface properties, we can establish MXene-based films and powders as prominent candidates in various applications in accordance to their demand.

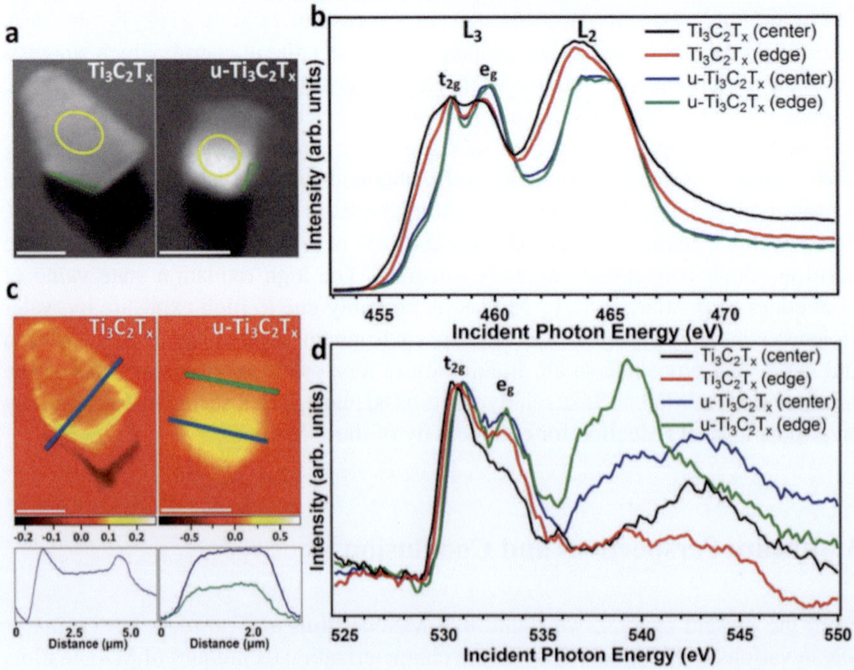

Fig. 12 **a** XPEM micrograph at Ti L edge taken at excitation energy value of 463.9 eV (black and white) shows pristine $Ti_3C_2T_x$ flake and urea intercalated MXene (here scale bar is 2 µm). The regions given in centre and edge portion in **b** and **d** are highlighted with yellow and green colour, respectively; **b** XPEM Ti L-edge XAS single flakes present in $Ti_3C_2T_x$ and urea intercalated $Ti_3C_2T_x$; **c** oxygen distribution over the MXene individual flakes obtained from the difference in mean XPEM micrograph at O K-edge in t_{2g} and e_g region (530–534.5 eV) which is relative to background (524.5–529 eV), corresponding line profile along flakes are given in underneath; **d** XPEM O K edge XAS of pristine and urea intercalated MXene flakes. Reproduced with permission from [51] Copyright (2020) American Chemical Society

References

1. Cherusseri, J., Sharma, R., Kar, K.K.: Nanotechnology advancements on carbon nanotube/polypyrrole composite electrodes for supercapacitors. In: Handbook of Polymer Nanocomposites. Processing, Performance and Application: Volume B: Carbon Nanotube Based Polymer Composites, pp. 479–510 (2015)
2. Cherusseri, J., et al.: Rapid synthesis of cobalt manganese phosphate by microwave-assisted hydrothermal method and application as positrode material in supercapatteries. Sci. Rep. **14**(1), 26550 (2024)
3. Kumar, D., et al.: Technology computer-aided design device simulation of nanowall-based CdS/CdTe solar cell for energy conversion. Energy Storage, e498 (2023)
4. Baby, A., et al.: Azopyridine as a linker molecule in polyaniline-grafted graphene oxide nanocomposite electrodes for asymmetric supercapacitors. ACS Appl. Energy Mater. **6**(20), 10442–10456 (2023)

5. Pallavolu, M.R., et al.: Scalable synthesis of binder-free hierarchical MnCo2O4 nanospikes/Ni (OH) 2 nanosheets composite electrodes for high-capacity supercapatteries. J. Energy Storage **73**, 108999 (2023)
6. Mishra, R., Cherusseri, J., Joseph, K.: Thermal and crystallization behavior of micro and nano fibrillar in-situ composites. In: Micro and Nano Fibrillar Composites (MFCs and NFCs) from Polymer Blends, pp. 213–231. Elsevier (2017)
7. Thomas, S.A., Cherusseri, J.: A review of Nb2CT x MXene as an emerging 2D material: synthesis, applications in rechargeable batteries and supercapacitors, progress, and outlook. Energy & Fuels (2023)
8. Thomas, S.A., Cherusseri, J., Rajendran, D.N.: Hierarchical two-dimensional layered nickel disulfide (NiS2)@ PEDOT: PSS nanocomposites as battery-type electrodes for battery-type supercapacitors with high energy density. Electrochem **5**(3), 298–313 (2024)
9. Kumar, D., et al.: Nanocarbon assisted green hydrogen production: development and recent trends. Int. J. Hydrogen Energy (2023)
10. Thomas, S.A., et al.: "Water-in-salt" electrolyte—toward high-voltage aqueous supercapacitors. In: Supercapacitors, pp. 289–315. Elsevier (2024)
11. Zaed, M., et al.: Utilization of recycled materials for low-cost MXene synthesis and fabrication of graphite/MXene composite for enhanced water desalination performance. Sep. Purif. Technol. 129055 (2024)
12. Thomas, S.A., Cherusseri, J., and Rajendran, D.N.: Strategically-designed hierarchical polypyrrole-modified manganese-doped tin disulfide (SnS2) nanocomposite electrodes for supercapatteries with high specific capacity. Electrochimica Acta 144910 (2024)
13. Zhang, J., et al.: Scalable manufacturing of free-standing, strong Ti3C2Tx MXene films with outstanding conductivity. In: MXenes, pp. 519–538. Jenny Stanford Publishing (2023)
14. Cherusseri, J., Kar, K.K.: Recent progress in nanocomposites based on carbon nanomaterials and electronically conducting polymers. Polym. Nanocompos. Inorg. Org. Nanomater. 229–256 (2015)
15. Nallapureddy, J., et al.: Morphology controlled synthesis of MnCo2S4 nanoflower structures for different metal precursors used for energy storage applications. J. Power Sour. **613**, 234883 (2024)
16. Cherusseri, J., Varma, S.J.: electrospun manganese oxide-based composites as anodes for lithium-ion batteries. Electrospinning Adv. Energy Storage Appl. 345–371 (2021)
17. Banerjee, S., et al.: Applications of supercapacitors. In: Handbook of Nanocomposite Supercapacitor Materials I: Characteristics, pp. 341–350 (2020)
18. Thomas, S.A., Cherusseri, J.: Recent advances in synthesis and properties of zirconium-based MXenes for application in rechargeable batteries. Energy Storage, e475
19. Qi, J.-Q., et al.: MXene-wrapped ZnCo2S4 core–shell nanospheres via electrostatic self-assembly as positive electrode materials for asymmetric supercapacitors. Rare Met. **41**(8), 2633–2644 (2022)
20. Zaed, M., et al.: Synthesis and characterization of hierarchical Ti3C2Tx MXene/graphitic-carbon nitride/activated carbon@ luffa sponge composite for enhanced water desalination. Open Ceram. **19**, 100645 (2024)
21. Bärmann, P., et al.: Scalable synthesis of MAX phase precursors toward titanium-based MXenes for lithium-ion batteries. ACS Appl. Mater. Interfaces. **13**(22), 26074–26083 (2021)
22. Cherusseri, J., et al.: Polymer-based composite materials: characterizations. Compos. Mater. Process. Appl. Charact. 37–77 (2017)
23. Dahlqvist, M., Barsoum, M.W., Rosen, J.: MAX phases–Past, present, and future. Mater. Today (2023)
24. Wang, J., et al.: A weldable MXene film assisted by water. Matter **5**(3), 1042–1055 (2022)
25. Richard, B., et al.: Minireview on fluid manipulation techniques for the synthesis and energy applications of two-dimensional MXenes: advances, challenges, and perspectives. Energy & Fuels (2023)
26. Thomas, S.A., Cherusseri, J., Rajendran, D.N.: 2D nickel sulfide electrodes with superior electrochemical thermal stability along with long cyclic stability for supercapatteries. Energy Technol. 2301641 (2024)

27. Thomas, S.A., Cherusseri, J., Rajendran. D.N.: Rapid synthesis of hierarchical cobalt disulfide nanostructures by microwave-assisted hydrothermal method for high performance supercapatteries. ACS Appl. Electron. Mater. (2024)
28. Gao, H., et al.: Solid-solution MAX phase TiVAlC assisted with impurity for enhancing hydrogen storage performance of magnesium hydride. J. Colloid Interface Sci. **652**, 979–988 (2023)
29. Cherusseri, J.: Polymer-ionic liquid gel electrolytes for lithium-ion batteries. In: Polymer Electrolytes for Energy Storage Devices, pp. 235–253. CRC Press (2021)
30. Banerjee, S., et al.: Capacitor to supercapacitor. In: Handbook of Nanocomposite Supercapacitor Materials I: Characteristics, pp. 53–89 (2020)
31. An, Y., et al.: MXenes for advanced separator in rechargeable batteries. Mater. Today (2022)
32. Gao, G., O'Mullane, A.P., Du, A.: 2D MXenes: a new family of promising catalysts for the hydrogen evolution reaction. ACS Catal. **7**(1), 494–500 (2017)
33. Hu, M., et al.: Emerging 2D MXenes for supercapacitors: status, challenges and prospects **49**(18), 6666–6693 (2020)
34. Huang, Y., et al.: Flexible MXene films for batteries and beyond. Carbon Energy **4**(4), 598–620 (2022)
35. Salim, O., et al.: Introduction to MXenes: synthesis and characteristics. Mater. Today Chem. **14**, 100191 (2019)
36. Li, P.-X., et al.: Bidirectionally aligned MXene hybrid aerogels assembled with MXene nanosheets and microgels for supercapacitors. Rare Met. **42**(4), 1249–1260 (2023)
37. Ying, G., et al.: Conductive transparent V2CTx (MXene) films. FlatChem **8**, 25–30 (2018)
38. Li, X., et al.: Advances in MXene films: synthesis, assembly, and applications. Trans. Tianjin Univ. **27**, 217–247 (2021)
39. Baby, A., et al.: Azopyridine as a linker molecule in polyaniline-grafted graphene oxide nanocomposite electrodes for asymmetric supercapacitors. ACS Appl. Energy Mater. (2023)
40. Zaed, M., et al.: Synthesis of titanium carbide (TiC) using recycled precursor for potential scale-up of low-cost MXene. Open Ceram. **19**, 100637 (2024)
41. Zaed, M., et al.: Low-cost synthesis of Ti3C2Tx MXene-based sponge for solar steam generation and clean water production. Ceramics Int. (2024)
42. Zaed, M., et al.: Synthesis of Ti3C2Tx MXene@ Carbon-Enhanced cellulose fiber composite-based photothermal absorber for sustainable water desalination. Mater. Today Sustain. **28**, 100971 (2024)
43. Zaed, M., et al.: Hierarchical Ti3C2Tx MXene@ Honeycomb nanocomposite with high energy efficiency for solar water desalination. Chemosphere **366**, 143459 (2024)
44. Pazniak, A., et al.: Ti3C2Tx MXene characterization produced from SHS-ground Ti3AlC2. Mater. Des. **183**, 108143 (2019)
45. Halim, J., et al.: XPS of cold pressed multilayered and freestanding delaminated 2D thin films of Mo2TiC2Tz and Mo2Ti2C3Tz (MXenes). Appl. Surf. Sci. **494**, 1138–1147 (2019)
46. Liu, Y., et al.: The mechanism of room-temperature oxidation of a HF-etched Ti3C2T x MXene determined via environmental transmission electron microscopy and molecular dynamics. InfoMat, e12536 (2024)
47. Ahmed, H., et al.: Recovery of oxidized two-dimensional MXenes through high frequency nanoscale electromechanical vibration. Nat. Commun. **14**(1), 3 (2023)
48. Palisaitis, J., et al.: On the structural stability of MXene and the role of transition metal adatoms. Nanoscale **10**(23), 10850–10855 (2018)
49. Halim, J., et al.: Electronic and optical characterization of 2D Ti2C and Nb2C (MXene) thin films. J. Phys. Condens. Matter **31**(16), 165301 (2019)
50. Sarycheva, A., Gogotsi, Y.: Raman spectroscopy analysis of the structure and surface chemistry of Ti3C2Tx MXene. In: MXenes, pp. 333–355. Jenny Stanford Publishing (2023)
51. Al-Temimy, A., et al.: Enhancement of Ti3C2 MXene pseudocapacitance after urea intercalation studied by soft X-ray absorption spectroscopy. J. Phys. Chem. C **124**(9), 5079–5086 (2020)

52. Thomas, S.A., et al.: Graphitic carbon nitride (g-C3N4): Futuristic material for rechargeable batteries. J. Energy Storage **68**, 107673 (2023)
53. Chen, L., et al.: Ultrathin 2D graphitic carbon nitride on metal films: underpotential sodium deposition in adlayers for sodium-ion batteries. Angew. Chem. Int. Ed. **59**(23), 9067–9073 (2020)
54. Thomas, S.A., Cherusseri, J., Rajendran, D.N.: Recent progresses in the synthesis and strategic designs of sustainable carbon-based fibrous electrodes for flexible batteries. RSC Sustain. (2025)
55. Thomas, S.A., et al.: Polyoxometalates and redox-active molecular clusters for supercapacitors. In: Supercapacitors, pp. 221–243. Elsevier (2024)
56. Thomas, S.A.: Layered two-dimensional black phosphorous-based hybrid electrodes for rechargeable batteries. J. Energy Storage **73**, 109068 (2023)
57. Banerjee, S., et al.: Handbook of Nanocomposite Supercapacitor Materials I. Springer Nature Switzerland AG (2020)
58. Thomas, S.A., et al.: Boron Carbon Nitride (BCN): an emerging two-dimensional material for rechargeable batteries. Energy & Fuels (2024)
59. Thomas, S.A., Cherusseri, J.: Boron Carbon Nitride (BCN): emerging two-dimensional nanomaterial for supercapacitors. J. Mater. Chem. A (2023)
60. Giusto, P., et al.: Boron carbon nitride thin films: from disordered to ordered conjugated ternary materials. J. Am. Chem. Soc. **142**(49), 20883–20891 (2020)
61. Thomas, S.A., et al.: Graphitic carbon nitride and their derivatives. In: Handbook of Functionalized Carbon Nanostructures: From Synthesis Methods to Applications, pp. 1–38. Springer (2024)
62. Thomas, S.A., Cherusseri, J.: Strategically designing layered two-dimensional SnS2-based hybrid electrodes: a futuristic option for low-cost supercapacitors. J. Energy Chem. (2023)
63. Manzeli, S., et al.: 2D transition metal dichalcogenides. Nat. Rev. Mater. **2**(8), 1–15 (2017)
64. Thomas, S.A., Cherusseri, J., Rajendran, D.N.: Functionalized carbon nanostructures for hydrogen storage. In: Handbook of Functionalized Carbon Nanostructures: From Synthesis Methods to Applications, pp. 1–39. Springer (2024)
65. Thomas, S.A., Cherusseri, J., Rajendran, D.N.: Functionalized carbon nanostructures for hydrogen storage. In: Handbook of Functionalized Carbon Nanostructures: From Synthesis Methods to Applications, pp. 1471–1509. Springer (2024)
66. Thomas, S.A., et al.: Designing PEDOT-based hybrid electrodes for supercapacitors by electrospinning strategy. Discov. Electrochem. **1**(1), 6 (2024)

Surface Functionalization Strategies of MXene: A Critical Overview of Its Improved Physical, Chemical, and Electrochemical Properties

M. A. Anu and T. S. Xavier

Abstract The research field has recently focused on 2D structured materials because of their excellent electronic, mechanical, chemical, and optical properties. MXenes have been studied extensively in the last few years since they are a group of 2D structured materials with high electrical conductivity and mechanical strength. They are obtained by removing element 'A' from the parent (MAX) phase; '$M_{n+1}AX_n$' to '$M_{n+1}X_nT_x$' where M is the early transition metal (Ti, Nb, Hf, V, etc.) X is carbides or nitrides, and T is the surface terminating functional group (e.g., O, F, OH, Cl). The distinct hydrophilicity, high conductivity, and adjustable surface terminations of MXenes have drawn much interest in various sectors. Hence, appropriately modifying the MXene is a better opening for multiple applications. This chapter describes the surface functionalization strategies of MXene and its improved physical, chemical, and electrochemical properties. The physicochemical properties of MXenes can be controlled by their surface chemistry, which also solves issues with MXene oxidation and dispersion stability. The environment, heat treatment temperature, and synthesis techniques all impact the surface functionalization groups. The surface chemistry of MXene can be modified by post-surface functionalization techniques or by the chemical etching procedure. Over the previous three years, porous MXenes have undergone significant design and development. Thus far, several appropriate topologies for porous MXenes have been developed using artificial procedures and applied in numerous applications, all demonstrating enhanced performance. The functional group modification by heat treatment, heteroatom doping, hybridization with polymer, and organic molecule intercalation prevent re-stacking problems and concurrently improve the through-plane ionic enhance the inherent physicochemical properties of MXene. The observed significant improvement when switching from MXene to modified MXene was attributed to the development of the electroactive area and the ion transport capabilities, supported by the morphological opening with

M. A. Anu · T. S. Xavier (✉)
Center for Advanced Materials Research, Department of Physics, Government College for Women, Thiruvananthapuram, Kerala, India
e-mail: xavierkattukulam@gmail.com

M. A. Anu
e-mail: anu232@gmail.com

© The Author(s), under exclusive license to Springer Nature Singapore Pte Ltd. 2025

S. Moharana and R. N. Mahaling (eds.), *Surface Engineering of MXenes*, Engineering Materials, https://doi.org/10.1007/978-981-96-3640-2_4

bigger interlayer space and pores through the MXene layers. So, modified MXene is a promising candidate for improved electrochemical performance.

Keywords MXene · Terminal groups · Surface chemistry · Physiochemical properties · Electrochemical behaviour

1 Introduction

MXenes are an advanced type of 2D materials and have been the subject of much research due to their numerous surface terminal groups, which are used to increase physiochemical properties, making them hopeful for many applications [1]. They have fascinating features including rich surface functional groups, excellent solubility in solvents, and metal-like high conductivity for facilitating electron mobility, electrical conductivity, and hydrophobicity. Its high dispersion and chemical interaction with other materials are facilitated by its abundance of active sites [2]. MXenes can be synthesized by the elimination of the 'A' layers, primarily consisting of IIIA or IVA elements, often Al or Si, from the precursor MAX phase. The bond between M and A produces a stronger metallic feature than transition metal oxides and, graphene that contain van der Waals bonds between the layers. It can be produced by chemically etching layers designated as 'A' since the M–A bond is somewhat weaker than the M–X bond. A hexagonal close-packed structure with a $P6_3/mmc$ space group symmetry is therefore produced, with functional groups forming at the surface where n + 1 'M' layers encase n 'X' layers [3]. The typical formula for MXenes after etching is $M_{n+1}X_nT_x$, where Tx stands for the surface functional groups such as Oxygen (O), Chlorine (Cl), Fluorine (F), hydroxyl (OH) [4]. They usually undergo complete termination during wet chemical etching, incorporating surface atoms such as –O, –F, and –OH, etc. Depending on the particular etchant used, the occurrence of surface functional groups varies during the etching process. They are available in several formulations, such as M_2X, M_3X_2, and M_4X_3 based on their structure [5]. MXenes can be tuned to doping or surface functionalization based on end-user specifications. Surface functionalization requires a complex preparatory procedure [6]. The outermost layers of Ti atoms are connected to the functional groups of 'T', which facilitates the comprehension of the two-dimensional structures of MXenes [7].

The arrangement of 'M' atoms and 'Tx' allows MXenes to be categorized into six groups, (1) S-MXene: solid-solution MXenes with arbitrary 'M' elements and hybrid 'Tx' (e.g., $Ti_{2-y}V_yCT_x$); (2) MXenes with a single type of 'Tx' and an identical 'M' element are known as t-MXenes; (3) MXenes with a single type of 'M' element and a hybrid 'Tx' element are known as n-MXenes; (4) o-MXene: MXenes with hybrid 'Tx' and out-of-plane ordered 'M' components (e.g., $Mo_2Ti_2C_3T_x$) (5) i-MXene: MXenes with hybrid 'Tx' and in-plane ordered 'M' (e.g., $Mo_{2/3}Y_{1/3})_2CT_x$) (6) h-MXene: high-entropy MXenes such as $TiVNbMoC_3T_x$ and $TiVCrMoC_3T_x$ that have a variety of 'M' components (e.g., $TiVNbMoC_3T_x$ and $TiVCrMoC_3T_x$) [3]. Surface terminations can occur at three distinct locations: (i) right on the transition metal atoms, (ii) within a hollow site situated between the top metal atoms, and (iii) within a hollow site located between the subsequent stacking layer of X atoms [8, 9]. On MXenes' surface, the functional groups terminated arbitrarily. The –F and –O do not have precise spatial connections. However, these functional groups are likely to distribute preferentially in the vacancies between the titanium atoms on the MXenes' surface. Furthermore, the determination of the bonding conditions and elemental composition of various surface terminations on the MXenes necessitates the utilization of a variety of analytical techniques [10]. The lamellar stacks that constitute MXene membranes have regular, aligned sub-nanometre channels that utilize an abundance of surface functional groups [11].

During the synthesis process, surface termination groups such as fluorine, oxygen, and hydroxyl are added, which causes the MXene surface to be hydrophilic [1]. Hydrophilic electronegative surface functional groups like –OH, –O, and –F are typically inherited by MXenes produced by aqueous fluoride-containing acid etching. It was confirmed that different MXenes bonded with H_2O molecules are reasonably strong. Significant bonding energy can also be used to anchor other anionic terminations to the MXene surface [8]. There exists a connection between the variables –OH and –F. Furthermore, it should be noted that there is no neighbouring –OH group present, and water molecules form hydrogen bonds with the –OH groups [12]. The functional groups that have been adsorbed (−O, –OH, and –F) are dispersed arbitrarily across the surface of Ti_3C_2Tx and exhibit a preference for remaining on the highest sites of the Ti(C) atoms within the monolayer of Ti_3C_2 [10]. Even though F is more important than O, the O atom preferentially occupies the same site whereas the F terminations exclusively occupy A-sites, which are on top of the core Ti atoms. Hence, the O atoms live on a different site in addition to the favoured site occupied by the F atoms, which likely results in the apparent disarray seen before heating [13]. The way that MXenes interact with other compounds is largely determined by these surface functional groups [7].

Typically, the MXene's surface is randomly distributed with mixed surface terminal groups. The Fourier Transform Infrared Spectroscopy (FT-IR) method is used to confirm the functional groups in MXene. The direct bonding of –F and –OH on the MXene's surface was confirmed using qualitative NMR studies. The neutron scattering on Ti_3C_2Tx suggests that the van der Waals interaction was responsible for the binding between O and F, while the hydrogen bonding is between O and F and the nearby OH groups. Furthermore, there appeared to be a positive hydrogen bonding between the intercalated molecules of water and surface O and OH groups [14].

Considerable work has been done to take advantage of new synthesis techniques for various MXenes. The two types of synthesis methodologies are the bottom-up approach (from components to MXenes thin film) and the top-down approach (from MAX to MXenes). The characteristics, surface functional groups, consistency, and morphologies of MXenes are significantly influenced by the synthesis procedures [15]. When $Ti_3C_2T_x$ is etched with HF, more fluorine (–F) groups are present than oxygen (–O) and hydroxyl (–OH) groups. On the other hand, functional groups containing oxygen are predominantly obtained when $Ti_3C_2T_x$ is etched by a mix of

HCl and LiF. Furthermore, $Ti_3C_2T_x$ etched with HF contains four times as many –F groups as $Ti_3C_2T_x$ generated by HCl–LiF etching. The alkaline solution is often regarded as a practical etching agent to create MXene by accomplishing the transition from –F to –OH and substituting –OH groups for the Al layer. Since the functional groups are fluoride-free, the resulting $Ti_3C_2(OH)_2$ nanosheets perform better in terms of energy storage than those made using the HF etching approach. In addition to being less harmful to the environment, fluorine-free etching techniques also don't result in the production of functional groups that include fluorine, which enhances their ability to store energy [12]. A fluoride-free etching technique that is effective entails the anodic corrosion of titanium aluminium carbide (Ti_3AlC_2) using a binary aqueous electrolyte. Ammonium hydroxide undergoes in-situ intercalation as a result of the corrosion that dissolves the aluminium. Chloride ions facilitate quick anodic etching of aluminium and weaken the Ti–Al bonds. To open the edges of the etched anode and promote the etching of the underlying surfaces, ammonium hydroxide (NH_4OH) intercalation is a crucial step [16]. Halide ions can exchange places with functional groups and other atoms in a novel type of surface reaction that can be efficiently participated in by MXenes having terminals –Cl and Br. It provides previously unattainable control over the properties, chemistry, and structure of the surfaces of MXene materials [17]. So far, no MXene without surface terminal groups has been found. MXenes without surface terminal groups would exhibit distinct features and could potentially possess less stability. The inclusion of surface terminal groups serves to solidify the MXene structure and boost its performance in diverse applications. Figure 1 represents the effects of surface modification.

Fig. 1 Schematic representation of the effects of surface modification

2 Methods of Surface Functionalization

It is discovered that MXene's intrinsic shortcomings, including restacking, easy oxidation, low flexibility, and worse electrochemical performance, make it unable to meet the requirements of a variety of applications. Many researchers have worked extremely hard to create functional MXene by an organic molecule or cation intercalation, heteroatom doping, polymer hybridization, and other methods to address these shortcomings [12]. The applicability of MXenes is limited due to their limited solubility in weakly polar non-polar polymers. In order to enhance its practicality, it might be essential to efficiently functionalize MXenes by employing ion, organic, and inorganic intercalation changes to control their morphology. The active sites, surface functional groups, pore structure, and electron transfer ability of MXenes are greatly influenced by their structure and morphology. The features of MXenes can be enhanced through organic modification, which involves the introduction of certain functional groups to improve their physical and chemical properties. For this modification, coupling agents and polymers are frequently employed. These compounds inhibit the occurrence of spontaneous oxidation, promote the stability of the application, improve the capacity to disperse and link with other substrates, and increase compatibility and reactivity with the polymer matrix [18]. The electronic structures, stabilities, and physical and electrochemical properties of MXene were shown to be significantly impacted by surface functional groups proven in the majority of computational studies conducted on the MXene [14]. Surface functionalization of MXenes is accomplished using a variety of ways to tune their characteristics to specific uses. Chemical approaches introduce functional groups by aqueous reactions or organic molecule attachment, whereas physical methods such as plasma and thermal treatments alter surface chemistry. Electrochemical procedures like electrodeposition and ion intercalation improve conductivity and energy storage. These techniques allow MXenes to be tailored for better performance in energy storage, sensing, catalysis, and medicinal applications. MXenes can be modified by surface functionalization through the following methods. Figure 2 displays the schematic representation of the methods of surface functionalization.

3 Small Molecules Intercalation

Surface modification of MXenes with a range of small molecules offers a straightforward and affordable way to progress the performance attributes of MXenes, like mechanical stability, electrical properties, and solution stability [19]. The interlayer spacing of MXene-based electrodes is a crucial variable in controlling their electrochemical properties. A sufficient interlayer separation will let electrons and ions diffuse more easily, providing channels for the transfer of electrolyte electrons and ions. The most efficient way to increase interlayer space is through the intercalation of molecules [12]. Cation intercalation is a way to modify MXenes [20].

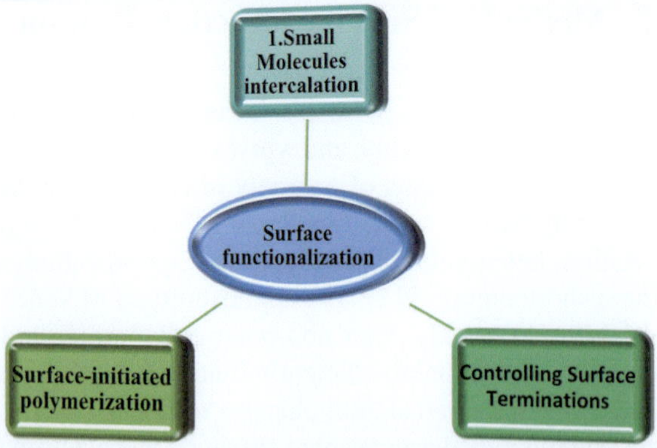

Fig. 2 Schematic representation of methods of surface functionalization

The etching process plays a vital role in defining the surface chemistry of MXene, including the type and proportion of functional groups that are present. Using NH_4HF_2 as the etchant, the interlayers of $Ti_3C_2T_x$ were expanded by the intercalation of $–NH_3$ and $–NH_4^+$ functional groups. This resulted in a noteworthy rise in the interlayer spaces of the final products [21]. The $Mo_2TiC_2T_x$ compound is synthesized through a process including the etching of Mo_2TiAlC_2 using hydrofluoric acid (HF) and subsequent delamination facilitated by the intercalation of tetrabutylammonium hydroxide (TBAOH). This technique decreases the concentration of –F terminations, leading to the intercalation of tetrabutylammonium (TBA^+) and H_2O. This intercalation process increases the space between flakes and consequently raises the electrical resistance [22]. Ti_3C_2Tx samples with surface terminations consisting of fluorine (–F) were subjected to treatment in a salt solution containing LiCl, KCl, and K_2CO_3. This treatment led to an expansion of the spacing between the layers of the material and an increase in capacitance. These changes were ascribed to the intercalation of potassium-oxygenated ions and the replacement of fluorine terminations with oxygen (–O) terminations [23]. The occurrence of electronegative termination in MXenes outcomes in a decrease in the density of states (DOS) at the Fermi surface. This is achieved by the creation of a new energy band positioned below the Fermi surface. Both –OH and –F groups have comparable effects, however, an O-group can take two electrons [8]. The application of oxygen plasma treatment leads to a notable enhancement in the peak of $TiO_{2-x}F_x$ and the terminations containing O. While the sample cannot be fully restored to its original state, it effectively decreases the presence of oxygen-containing terminations and partial oxides in the $Ti_3C_2T_x$ film. The application of plasma treatment can regulate the terminations of MXenes that include oxygen, hence enabling a reversible alteration in their conductivity [24].

The surface termination of MXenes was modified using a hydrothermal method, which involved replacing –OH and –F terminal groups with –O groups. As a consequence, there was a change in the chemical potential that caused a shift towards the edges of the band, which resulted in a rise in the bandgap and an enhancement of the Seebeck coefficient. The Seebeck coefficient of the improved MXene films exhibited a negative correlation with the lowering pH level of the solvents. This indicates that an alkaline atmosphere is advantageous for the alteration of the surface terminal groups. With temperature rise, the Seebeck coefficient observed an increase, whereas the electrical conductivity displayed a declining pattern, possibly attributed to the intensified surface functionalization modification impact, which is visible in Fig. 3a–c. At elevated temperatures, a greater amount of –F and –OH functional groups were successfully eliminated, whereas the number of –O functional groups increased and the impacts are shown in figure d-f. With the intercalation of cation, there is a high electrical conductivity, leading to a remarkable power factor of about 44.98 mW m^{-1} K^{-2} at ambient temperature and excellent elasticity [25].

Heteroatom doping is a useful technique to introduce the pseudo capacitance effect by increasing the spacing between the interlayers and hence improving the electrochemical properties. The most popular doped strategies N, S, or P are effective and feasible to improve the electrochemical properties of MXenes. The addition of the heteroatoms enhances MXene's active sites, expanding its electrochemically active region and enhancing the material's capacity for energy storage and catalysis [12]. N-doping can enhance electrochemical active sites and improve electrical conductivity, which opens up more possibilities for energy conversion and storage. The N elements are distributed uniformly in the N-doped MXene, which was produced via solvothermal and hydrothermal processes [26]. The ability to store electrochemical energy is decreased by reducing the gap between layers and taking care of restacking issues, which impede the flow of ions and electrons. Sulphur doping is thought to be a very successful strategy for overcoming these obstacles [27]. P can efficiently lengthen MXene's electron transport pathway and boost the density of its active sites, increasing the electrocatalytic application of MXene in hydrogen evolution reaction (HER) [28]. The study examined the CO_2RR and the HER activity of the pure Nb_2CO_2, doping of nonmetal, and modification of metal Nb_2CO_2 systems through first-principles calculations. It has been discovered that Nb_2CO_2 is the most stable functional group. However, pure Nb_2CO_2 is unfit for HER and CO_2RR catalysts [29]. The incorporation of Cu_{2+} into $Ti_3C_2T_x$ causes the Ti– and ions surface functional groups of MXene to undergo modifications. This results in the disruption of the electrostatic effect of $Ti_3C_2T_x$ and the loss of the barrier effect that the functional groups possess. Because of this, the lattice spacing is increased, and the electrochemical performance is enhanced [30]. Ag nanoparticles were inserted between $Ti_3C_2T_x$ interlayers by an in-situ reduction process, which uncovered active areas and improved the specific capacitance. The a-Ag/$Ti_3C_2T_x$ electrode exhibits an augmented specific capacitance and a twofold increase in intercalation pseudocapacitance. The increased distance between layers enables efficient and consistent movement of ions, leading to an outstanding rate performance with 88.3% retention of capacitance and 100% stability even after 20,000 cycles. This is a feasible approach to improve the $Ti_3C_2T_x$ negative electrode's electrochemical energy storage through the intercalation and termination processes [31].

The surface functionalization of MXene with heterogeneous atoms can enhance its capacity for charge storage. The technique of fast in-situ phosphorus doping at low temperatures is employed to fabricate functionalized $Ti_3C_2T_x$ MXene ($Ti_3C_2T_x$-P),

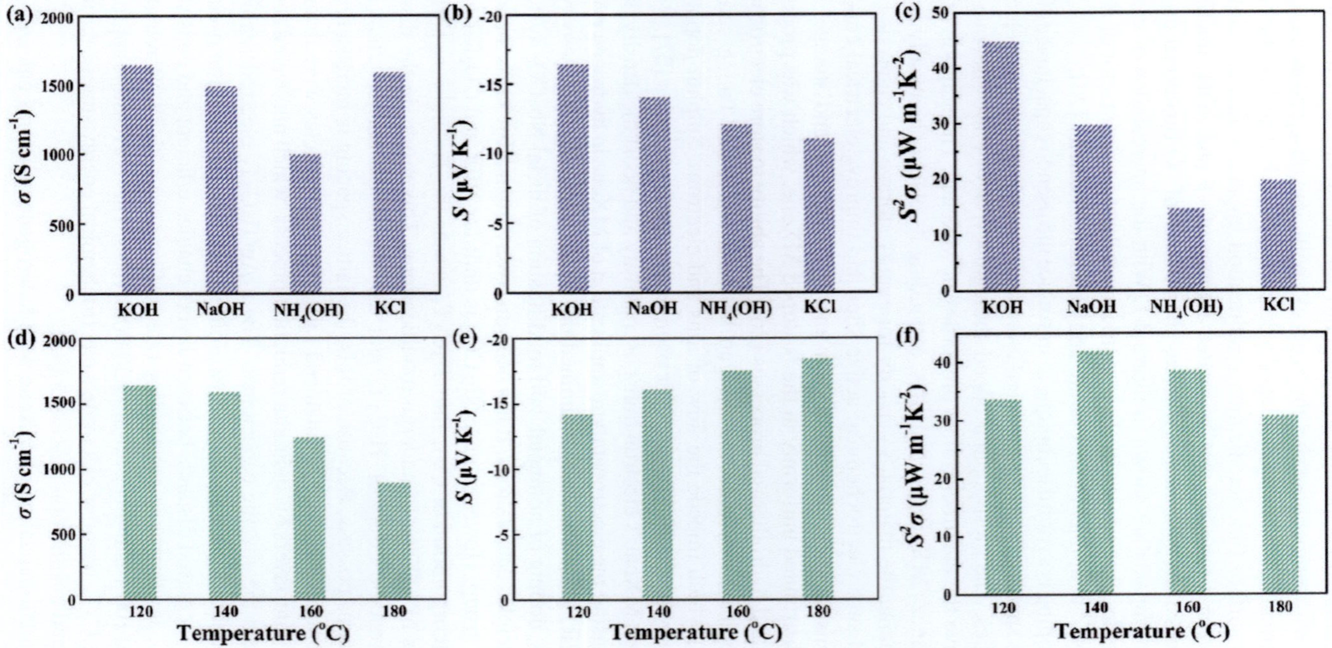

Fig. 3 **a** Electronical conductivity, **b** Seebeck coefficient, and **c** the power factor of K–$Ti_3C_2T_x$ with hydrothermal treatment by different kinds of alkali. **d** Electrical conductivity, **e** The Seebeck coefficient, and **f** the power factor of K–$Ti_3C_2T_x$ with hydrothermal treatment at different temperatures. Reproduced from Ref. [25]. Copyright 2020, Elsevier

which results in an expansion of the layer gap and the formation of P–C and P–O bonds. This technique enhances the composition, structure, and electrochemical properties of $Ti_3C_2T_x$ while also offering a means to modify and add functionality to MXene materials [32].

A method combining surface modification and in-situ intercalation is used to enhance the electrochemical performance of $Ti_3C_2T_x$. A novel Ti_3C_2 nanocomposite incorporating Fe_3O_4 particles is synthesized by an in-situ growing process and suggests a surface modification, such as carbon coating to advance the electrochemical properties of the nanocomposite. Carbon coatings provide protection for active components, reduce electrolyte breakdown, and enhance conductivity, resulting in the advance of high-capacity MXene-based anode structures [33].

Solvothermal treatment can adjust the surface functionalization of MXenes, with deionized water and ethanol being used as the low-toxic solvents. This process allows easy changes in layer spacing and termination, replacing the –F termination with –OH and increasing the proportion of O-containing functional groups on the surface of MXenes, narrowing the separation between conduction and valence bands [34]. UV-ozone treatment can also induce the Ti–O bond of MXenes. The additional O-containing terminations that occupy several active sites originally associated with the –F termination are responsible for the shortening of the MXene bandgap [35]. The incorporation of two silver (Ag) particles and even the distribution of multi-walled carbon nanotubes (MWCNTs) on the surface can enhance the layered structure of V_2CT_x and increase the available surface area for the electrolyte ions. The introduction of Ag particles and MWCNTs exhibited a bridging phenomenon, whereby the functional groups present on the surface of V_2CT_x collaborated to establish a conductive channel that facilitated the flow of electric current to the surrounding medium. The incorporation of MWCNTs resulted in the consumption of termination groups on the surface of the composites and a reduction in the number of functional groups on the surface of M–V. Consequently, this controlled an important growth in the fraction of V^{4+} in M–V composites [36]. A hybrid approach is proposed for the implementation of a robust MXene-based memristor, which integrates both ionic mobility and electrical conduction. Surface modification employing ionic celestine blue and a 2D heterojunction design stimulates electronic transition and enhances the repeatability of the device. This method presents a hopeful strategy for the development of 2D MXenes in ionotropic engineering, which can contribute to the advancement of in-memory computing technology [37]. Through KOH treatment, K+ ions are intercalated into the interlayer of Nb_2CT_x MXene, which prevents self-restacking and eliminates unfavourable –F surface terminations. Through this process, the material's capacitance and electrochemical performance are improved by the introduction of additional active sites. Additionally, annealing treatment removes most of the –F termination groups on Nb_2CT_x MXene and replaces a significant number of –OH functional groups with –O terminations. This leads to the formation of O-terminated Nb_2CT_x (400-KOH–Nb_2CT_x), which demonstrates enhanced adsorption capability and lower diffusion barriers. As a result, it achieves exceptional ion storage capacity and facilitates fast ion diffusion and transport [38].

Small molecule intercalation in MXenes entails introducing ions or small organic molecules between MXene layers by methods such as solution-based or electrochemical intercalation. Metal ions, such as Li^+ and Na^+, and tiny organic molecules are common intercalants. This method improves electrical conductivity, mechanical characteristics, and surface chemistry. It can also enhance thermal stability and encourage the creation of new phases or compounds. Improved energy storage in batteries and supercapacitors, greater catalytic activity, increased sensor sensitivity, and potential usage in water purification and environmental remediation due to changed adsorption characteristics are all possible applications. For successful use in a variety of technologies, intercalating small molecules onto MXenes poses several obstacles that must be addressed. Among these difficulties are maintaining chemical compatibility between MXene surfaces and intercalants to avoid deterioration, precisely controlling interlayer spacing to maximize material attributes, and distributing intercalants uniformly for reliable performance. Understanding structural changes and validating effective intercalation depend on the accurate characterization of intercalated MXenes. Furthermore, reliable performance in applications like energy storage and catalysis depends on guaranteeing long-term stability and durability under various environmental circumstances.

4 Surface-Initiated Polymerization

Several surface terminal groups such as –O, –OH, and –Cl, functionalize the surface of MXene based on the needs of the chemistry of the polymer matrix [39]. MXene can be combined with polymer to address the limitation of convenient repositioning of MXene. Numerous polymers are utilized in the manufacture of nano-composites incorporating MXenes, owing to their cost-effectiveness and intriguing flexibility [12]. The optimization of interfacial compatibility and interaction between polymer matrix and MXenes has been confirmed using the efficient method of surface modification [40]. The contact between the polymer matrix and MXenes is boosted by the covalent functionalization of MXene. To reduce interfacial defects, the selection of a surface modification agent is essential as it functions as a bridge between the nanofillers and the polymer. Silane coupling agents are frequently utilized due to their plentiful, adaptability, and efficiency [40]. Owing to their high modulus, high aspect ratio, and functionalized surfaces, MXenes are also added to the polymeric matrix to prepare functional polymers [20]. MXenes possess a multitude of active terminal groups on their surface, enabling the in-situ polymerization or ex-situ blending of polymers. The production of MXene/polymer composites often involves the utilization of an ex-situ blending technique with non-covalent bond interaction MXene and polymers such as Poly dimethyl diallyl ammonium chloride (PDDA) and polyvinyl alcohol (PVA), which offers several advantages including the ability to achieve well-defined tunable compositions and polymer structures. Therefore, the hydrogen bonding and electrostatic interactions can effectively enhance the interaction between

MXenes and polymers [19]. By using hydrogen bonds, van der Waals forces, or electrostatic interactions, the ex-situ blending method enables precise control over the structure and composition of the MXene/polymer nanocomposite [12]. The integration of MXene with polymers can be achieved through in-situ polymerization with the covalent bonds between the MXenes and polymers, such as polyaniline (PANI), polypyrrole (PPy), and poly-3,4-ethylenedioxythiophene (PEDOT) [12]. Boota et al. proposed an easy method for in-situ polymerization of pyrrole restricted between the $Ti_3C_2T_x$ layers. After the fluorine is added to the polypyrrole (chain propagation) to terminate the MXene layers, the process proceeds to create larger chains of polypyrrole, which eventually turn conductive. Strong bonding between the PPy and $Ti_3C_2T_x$ surfaces can be induced by the polymerization of pyrrole by the surface termination groups of the $Ti_3C_2T_x$ layers leading to improved charge transfer [41]. By using the carbonization process and in-situ surface modification, the zwitterionic polymer with nitrogen-doped MXene nanosheets (PEIS@N-MXene) was effectively created. Without the need for a surfactant, the obtained PEIS@N-MXene shows enhanced diffusion stability in water for ten days. When added to aqueous lubricants, the PEIS@N-MXene shows exceptional lubrication behaviour when compared to pure water. At a concentration of 2.5 weight percent, the average coefficient of friction (COF) decreases to 0.109 from 0.633 and the wear volume decreases by 42.3%. A hydrated layer surrounds the nanoparticles on the MXenes surface due to the presence of many, strongly polar oxygen-containing functional groups that can establish hydrogen bonds with nearby molecules of water via dipole forces. However, MXene forms a hydration layer that is 1–2 molecular layers thick. Since the zwitterionic polymer PEIS is easily stretched in water and has a greater polarity and the binding force with the water molecules, the hydration layer is thicker. Also due to mutual repulsion, the hydration layer is elastic. Figure 4a demonstrates that the friction coefficient (COF) of MXene, N-MXene, and PEIS@N-MXene is much lower than that of pure water, with PEIS@N-MXene having the lowest COF of 0.109. Figure 4b shows that PEIS@N-MXene lowers the depth and width of the wear scar, exhibiting improved wear resistance over pure water. Figure 4c demonstrates that the addition of 2.5 wt.% PEIS@N-MXene provides the best anti-friction and anti-wear performance. Lower concentrations do not completely cover the friction contact area, resulting in insufficient lubrication, whereas concentrations of more than 2.5 wt.% produce stable tribological qualities [42].

Stearic acid (SA) was added to $Ti_3C_2T_x$ MXene to improve its surface compatibility with hydrophobic PLA chains [20]. $Ti_3C_2T_x$ MXene modified with n-octyltriethoxysilane added to a PLA matrix improves the membrane's biological characteristics and ultimate tensile strength, suggesting that it may be used as a guided bone regeneration membrane. This indicates that one useful method for creating multifunctional polymer composites is to use inorganic MXenes as fillers in polymer matrices throughout the composite production process [43]. Poly(diallyl dimethyl ammonium chloride) (PDDA) solution is used in a straightforward and scalable manner to exfoliate and functionalize Ti_3C_2 (MXene) during ball milling. During wet ball milling, positively charged PDDA clings to negatively charged MXene,

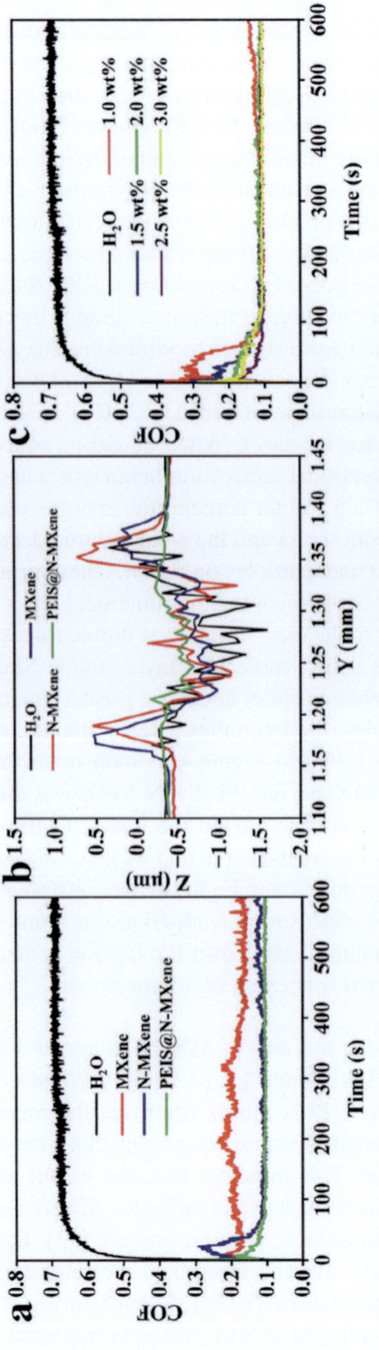

Fig. 4 The tribological performance of lubricants with 2.5 wt.% MXene, N-MXene and PEIS@N-MXene. **a** COF curves; **b** cross-sectional profiles of wear scars in the y-axis direction, **c** COF curves of PEIS@N-MXene with 0, 1.0, 1.5, 2.0, 2.5, and 3.0 wt.%. Reproduced from Ref. [42]. Copyright 2023, Elsevier

preventing MXene sheets from stacking together and enhancing exfoliation. Moreover, the addition of MXene nanosheets, which effectively transfer the load to neighbouring polymer chains, improves the mechanical characteristics and improves the compatibility between the MXene-PDDA nanosheets [44]. Amino-functionalization plays a crucial role in improving the characteristics of MXene. The surface of MXene nanosheets was functionalized using L-Cysteine. The potentiodynamic polarization tests and electrochemical impedance spectroscopy were used to study the anticorrosion capabilities of a composite waterborne epoxy (WEP) coating that contains functionalized MXene (fMX). The utilization of green-modified MXene presents a viable approach for enhancing the corrosion resistance of the coating [45]. The amino-functionalization of few-layered $Ti_3C_2T_x$ sheets was achieved through treatment with 3-aminopropyl triethoxysilane and was used as reinforced additives into the aqueous epoxy coating. The tribological and anti-corrosion capabilities of composite coatings incorporating amino-functionalized MXene are attributed to the effective dispersion of fillers within the polymer matrix and strong adhesion between the coating and the metal substrate [46]. A method was developed to modify MXene nanosheets by crosslinking them with polyethyleneimine using dopamine. This modified MXene was then used to strengthen an aqueous epoxy coating by a simple spraying approach. The outstanding anti-corrosion capacity of WEP coating is attributed to the effective dispersion of functional filler, increased physical barrier, and interface compatibility [47].

Hydrophilic MXene with the –OH functional group can be readily dissolved in water without the incorporation of surfactants [39]. Hexaethylene glycol mono dodecyl ether ($C_{12}E_6$), a non-ionic surfactant, was added to MXene to enhance the molecular interactions between the nanosheets. $C_{12}E_6$ molecules form strong hydrogen bonds with the –F or –O groups on the surface of the MXene. A new fan-like texture is confirmed through the incorporation of $C_{12}E_6$ between MXene nanosheets [48]. The improved properties of the composite coatings primarily rely on three key factors: (a) the effective dispersion and compatibility of amino-functionalized $Ti_3C_2T_x$ within the organic matrix, (b) the strong adhesion between the coating and the metal substrate, and (c) inherent characteristics of the $Ti_3C_2T_x$ sheets. The 3-methacryloxy propyl trimethoxysilane (γ-MPS@MXene) was employed to functionalize MXene nanosheets, which were then added to vinyl ester resin (VER) using a solution blending technique. This improved the thermal, mechanical, dynamic thermomechanical, anti-corrosion, and UV ageing resistance properties of VER nanocomposites [40]. In biological lubrication systems, polymer chains are typically employed because of their numerous functional groups, which enable them to encapsulate water molecules and generate a thick hydration layer that improves lubricity. Enhancing the hydration capacity of water-based lubricant additives through surface functionalization using charged and hydrophilic polymers is a successful strategy [49].

Surface-initiated polymerization of MXenes involves generating polymer chains from their surfaces to improve characteristics and extend applications. This process can use methods such as grafting-from, in which initiators are fixed to the MXene surface, and grafting-to, in which prepared polymers are attached. Atom Transfer

Radical Polymerization (ATRP), Reversible Addition-Fragmentation Chain Transfer (RAFT), and Ring-Opening Polymerization (ROP) are examples of common polymerization processes. It enhances the dispersibility and stability of MXenes while also allowing for customized surface characteristics. Applications include better energy storage in batteries and supercapacitors, higher sensor sensitivity, improved catalytic activity, and possible composite applications, resulting in flexible MXene materials suited for a wide range of sophisticated technologies. Even though, it has some challenges that must be overcome before it can be used successfully in advanced technologies. These problems include achieving homogeneous initiator distribution and regulating polymer chain length to ensure consistent characteristics. Ensuring adequate surface functionalization is critical for maintaining active groups without passivation. The reactions are sensitive to environmental conditions, necessitating efficient, repeatable techniques to prevent deterioration and maintain high yields. Scalability concerns require managing industrial production costs and time while ensuring consistent performance across batches. Accurate surface and polymer characterization are critical for validating successful functionalization and understanding material behaviour. The long-term stability and performance reliability of polymerized MXenes are also important for their practical use in a variety of applications.

5 Controlling Surface Terminations

All surface modification techniques rely on chemical reaction approaches, which necessitate complex procedures and precise control over reaction conditions. Additionally, surface functional groups are typically consumed during the chemical reaction, along with the MXene skeleton's degradation [14]. The introduction and regulation of mixed terminations (–F, –OH, and –O) during synthesis affects the characteristics of several MXenes. MXenes exhibit unusual hydrophobicity, increased environmental sensitivity, and a positive surface potential due to the addition of external terminal groups [24]. Based on DFT calculation results, it was found that the removal of the –OH in mixed functional groups should be prioritized during heat treatment. However, the presence of water molecules covering this –OH group prevented thermogravimetry mass spectrometry (TG-MS) from detecting its signal. In addition, the removal of fluorine atoms with high electronegativity can alter the MXenes band structure and enhance their electrical conductivity [24]. The incorporation of external terminal groups has the potential to alter the hydrophobic-hydrophilic nature of MXenes or render them responsive to eco-friendly factors. The hydrophilic group underwent a reaction with the hydroxyl (–OH) terminal present on the surface of Ti_3C_2, resulting in the formation of many stable ether bonds. Upon alteration, the Ti_3C_2 films exhibited a hydrophobic nature due to the occurrence of $-CF_3$ groups [50].

Surface functional group evolutions of MXenes throughout the heat treatment process are primarily influenced by temperature and atmospheric conditions. The conductivity and environmental stability of MXene are enhanced as the treatment temperature rises, resulting in a reduction in the distribution of terminations on the surface. Specifically, the terminations were reduced in the series of –OH, –F, and –O. Hence, this particular procedure is frequently utilized to enhance the conductivity of transparent conductive films based on MXenes [24]. The –O termination content was raised during heat treatment by adding alkaline additives. The intercalation of $Ti_3C_2T_x$ with the dimethyl sulfoxide (DMSO) in the KOH solution, substituting –OH terminations for –F terminations. MXenes conductivity is improved by heat treatment at 450 °C in an argon environment, and hydrophilicity is enhanced by an abundance of –OH terminations [51]. The fraction of O-containing functional groups (–O and –OH) on the surface of MXenes is mostly impacted by the use of alkaline additives and reductive atmosphere during the treatment of heat. Termination-free surfaces are obtained by heating $Ti_3C_2T_x$ in a vacuum at 650 °C to remove –F terminations and adding a hydrogen environment at 700 °C to react with –O terminations. According to DFT calculations, termination-free $Ti_3C_2T_x$ should have a greater metallicity; nevertheless, because of its instability, it must be passivated by other terminations using carbon dioxide adsorption [52]. The application of UV-ozone can stimulate the formation of surface Ti–O bonds in MXenes. The presence of a doublet in the Ti 2p spectra, which corresponds to Ti–O bonds, originated from Ti_3C_2Tx entirely. The magnitude of the doublet exhibited an upward trend as the duration of UV-ozone treatment extended, suggesting an elevated concentration of terminations containing oxygen. The observed rise in O-containing terminations resulted in a reduction in the bandgap of MXene, as these terminations occupied active sites that were formerly associated with the –F termination [53].

Heat treatment is a popular method for modifying the surface termination of MXenes. Various atmospheres and additives can result in an extensive variety of surface functionalizations. For optimal preservation of the atomic and chemical structure of MXenes, it is essential to carry out heat treatment under an inert or in a vacuum or reductive atmosphere. MXene sample that underwent annealing in an N_2/H_2 atmosphere showed the most impressive electrochemical performance. This can be ascribed to its high carbon quantity and low fluorine quantity on the sample's surface after annealing [54]. After annealing at 380 °C in a vacuum, oxygen takes up at least two separate adsorption sites, while fluorine fills the face-centered cubic adsorption site. The strong diffusion of more than 1 eV in the electrical structure is ascribed to Ti–O bonds. Ti_3C_2Tx's work function changes with surface composition, ranging from 3.9 to 4.8 eV due to the desorption of the –OH groups. When fluorine desorption occurred at higher temperatures (500–750 °C), the work function dropped to 4.1 eV. Work function values of uniformly terminated Ti_3C_2 surfaces are largely dependent on the interaction between the surface moieties and the local dipoles [55]. Temperatures between 300 and 775 °C were used to anneal $Ti_3C_2T_x$, which led to a notable decrease in –F termination types and a rise in sample conductivity. After the electron irradiation and 500 °C annealing, huge voids arose, and –O terminations were lost. Figure 5a shows the behaviour of Ti_3CNTx during the de-intercalation process, with a focus on the release of H_2O. The mass spectroscopy (MS) data reveals a large peak for H_2O at a mass-to-charge ratio (m/e) of 18, centered at 150 °C. This suggests that H_2O starts to de-intercalate from the MXene structure at low temperatures. Figure 5b presents in situ Electron Energy Loss Spectroscopy (EELS) data for Ti_3CNT_x after various annealing steps. Figure 5c depicts ex situ X-ray Diffraction (XRD) data for Ti_3CNT_x, which indicate that the c-lattice parameter contracts after annealing at 150 °C. This shrinkage indicates a decrease in interlayer space, confirming the partial de-intercalation of water. The smaller inter-layer spacing lends credence to the claim that de-intercalation increases electronic conductivity by lowering inter-flake resistance [22].

The process of producing Ti–C vacancies through H_2 annealing is responsible for the saturation magnetization of Ti_3C_2. This results in consistent amounts of –O concentration across the annealing temperature range of 100–500 °C. C–Ti–OH functional groups were converted into C–Ti–O and O–Ti–O functional groups, directing the development of Ti–C vacancies in addition to C and Ti vacancies. The DFT calculations and magnetization enhancement demonstrated a notable enhancement in saturation magnetization caused by Ti–C vacancies. This differentiation allowed for a distinction between nonmagnetic completely terminated Ti_3C_2 MXenes and ferromagnetic pristine Ti_3C_2 [19]. $Ti_3C_2T_x$ MXenes with hydroxyl terminal groups have a work function ranging from 1.6 to 2.8 eV, whereas $Ti_3C_2T_x$ MXenes terminated with oxygen exhibit a higher work function between 5.75 and 6.25 eV. Hydroxyl (−OH) groups have a greater influence on the functioning of MXene compared to oxygen (−O) and fluorine (−F) functional groups [56].

For MXenes to perform optimally in a variety of applications, surface termination control is essential. To modify surface chemistry and improve functions like wettability and electrochemical behaviour, chemically altering surface groups like hydroxyl, fluorine, or oxygen are required. Especially in corrosive settings, maintaining persistent surface passivation enhances material stability and inhibits undesirable reactions. Conductivity and catalytic activity are affected by the electronic and crystallographic orientations and surface terminations that are controlled during synthesis. Surface termination effectiveness and uniformity are evaluated using quantitative analysis techniques such as FTIR and XPS. By optimizing these terminations, MXenes can achieve superior performance in energy storage, sensors, and catalysis, as well as ion storage capacity and biocompatibility. The effective implementation of MXenes in many technologies requires the resolution of notable issues related to controlling their surface terminations. Two major challenges are preserving homogeneity throughout MXene layers to maintain constant material characteristics and guaranteeing the long-term stability of added functional groups under changing environmental conditions. Controlling surface reactivity is essential to prevent unintentional chemical interactions that can change the functionality of MXene. Accurately characterizing surface terminations in the context of MXene's intricate structure and interlayer interactions becomes more challenging. Optimizing MXene features like conductivity and catalytic activity requires an understanding of how controlled terminations affect these attributes. Concerns about scalability emerge when procedures are scaled up to regulate surface terminations for industrial applications consistently and economically. By resolving these issues, MXenes' full promise in areas like medicinal applications, catalysis, and energy storage will be realized.

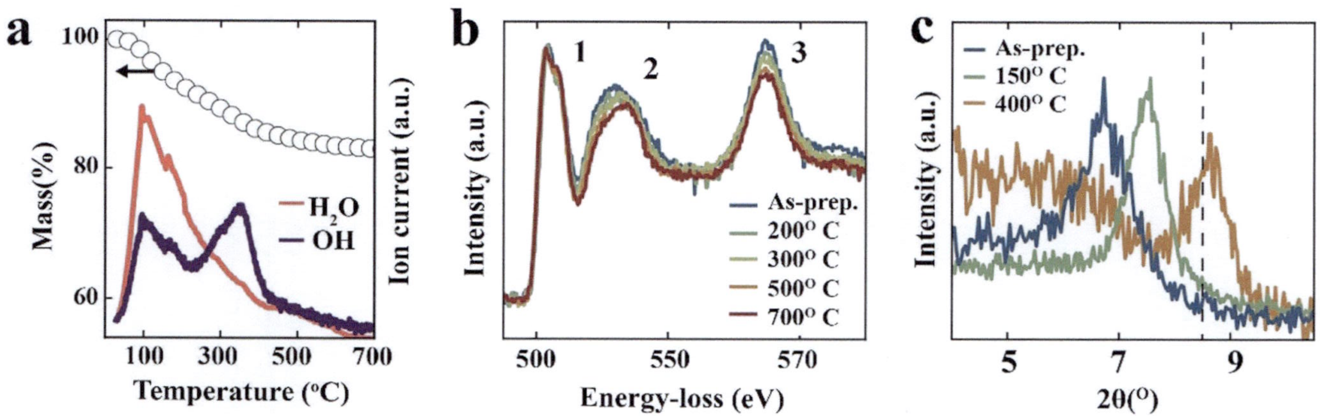

Fig. 5 Influence of intercalated species on the MXene resistance and dR/dT. **a** TGA-MS of Ti_3CNT_x showing the loss of H2O intercalants and –OH termination species. **b** In situ EELS of the normalized O K-edge of Ti_3CNT_x indicating the loss of H_2O intercalants with annealing. **c** Ex situ XRD of the Ti_3CNT_x (002) peak for an as-prepared thick sample, i.e., vacuum filtered and free-standing, as well as thick samples annealed at 150 °C (in vacuum) and 400 °C (in Ar). The peak shift to higher 2θ values represents a decrease in the inter-layer spacing. The dashed vertical line represents the approximate position of the $Ti_3C_2T_x$ (002) peak after annealing at 150 °C. Reproduced from Ref. [22]. Copyright 2019, Nature communications

6 Effect of Surface Functionalization on MXene Properties

The surface functional groups of MXene have a significant influence on its properties. It is possible to enhance the properties of the materials by using technical methods to increase the content of particular functional groups. The strong connection between the surface functional groups is responsible for these characteristics of MXenes. The presence of bonds between M, X, and functional groups exhibit a blend of metallic, covalent, and ionic types, along with a rich surface chemistry. Therefore, MXenes demonstrate distinctive physiochemical and electrochemical properties [3]. The synthesis processes, morphologies, topologies, compositions, and the quantity and type of surface terminal groups all have a noteworthy effect on the properties of MXene and their possible uses [57]. Functional groups change the electronic structure of MXene, influencing its bandgap, conductivity, and optical features including emission and absorption. MXene's longevity is increased by its improved chemical stability against oxidation and corrosion, and its improved surface wettability enhances its compatibility with matrices and solvents in composite materials. Functional groups act as active sites for reactions, which further increases the catalytic activity of MXene. Furthermore, functionalization improves the interfacial bonding in composite materials, which raises their toughness and tensile strength. These modifications maximize the use of MXene materials in a variety of applications, such as catalysis, energy storage, and sensing.

6.1 Physical Properties

MXenes are well-known for their excellent physical characteristics, which arise from their metal conductivity, active sites, and the hydrophilicity of their terminal groups. The electrical characteristics including conductivity and electron transport capacities are influenced by the –O and –F terminations. Conductivity is lower in MXenes with –O terminations than in those with –OH terminations. The conductivity fluctuates according to the –O to –OH ratio. MXenes with F-termination have lower electrical conductivity in comparison to their counterparts with O-termination. The NH_2-terminated MXenes also demonstrate electrical conductivity [58].

The electrical and structural characteristics of $Ti_{n+1}N_n$ and $Ti_{n+1}C_n$ (n = 1–9) MXene monolayers with the surfaces terminated by –OH, –O, –H, and –F groups are metallic, except for Ti_2CO_2, which is semiconducting. Functional groups are likely to form an electron-donor connection with the Ti atoms. MXenes are magnetic in the absence of terminations, mainly because of Ti atoms on the surface. The magnetism is eliminated upon functionalization. Except for semiconducting Ti_2CO_2, all functionalized MXenes are metallic because of substantial orbital hybridization between the Ti 3d and the O 2p orbitals close to the Fermi level. Thicker MXenes (n ~ 5) have distinct electrical characteristics than the thinnest and different properties will probably be displayed by the thickest and thinnest MXenes [59]. Oxygen-functionalized MXenes typically show thinner layer thicknesses and reduced lattice parameters compared to those terminated with hydroxyl and fluorine. Most oxygen-functionalized MXenes exhibit superior mechanical strength compared to those functionalized with fluorine or hydroxyl groups [60]. Cr_2N becomes semi-metallic when it develops –F, –O, or –OH surfaces because it has more valence electrons than Cr_2C [61].

The doping of MXenes with non-metallic materials with weak electronegative properties helps convert the semiconductor material into a metallic state [24]. Surface functional groups on certain transition metals can even form bands. The Fermi level density of states (DOS) can be affected, which can alter the band structure and electrical conductivity [62]. MXenes without surface terminations behave metallically, mainly because of the abundance of transition metal d electrons, which drive a high density of states (DOS) nearby to the Fermi level. The addition of surface functional groups causes a significant change in the band structure of MXene, giving MXenes a semiconducting nature. Through the hybridization of the p orbital of the termination group and the d orbital of the transition metal, the peak density of states is reduced, leading to the creation of a band gap. By raising the d orbital above the Fermi level, the properties of MXene undergo a shift, transitioning from a metal to a semiconductor. As the atomic number of the transition metal increases, the band gap also increases. Therefore, with the introduction of termination groups, both MAX phases and MXenes exhibit semiconductor properties [63].

MXenes are renowned for their exceptional mechanical strength due to their strong covalent bonds within 2D layers and weak van der Waals interactions. They are used as reinforcing additives in composite materials to enhance tensile strength and toughness. Their incorporation into polymers, metals, and ceramics results in composites with improved performance, making them valuable for structural and engineering applications. Carbides provide enhanced structural stability in comparison to nitrides, wherein the mechanical robustness of $M_4X_3O_2$ MXene is predominantly impacted by the strength of M–O bonding [64]. Through surface modification, Stearic acid an efficient modifier, was able to effectively improve the mechanical properties of Ti_3C_2Tx MXene [20]. Temperature changes during annealing modify the electronic properties of $Ti_3C_2T_x$ for various surface terminations. While oxygen is present in numerous locations, fluorine is only found in the face-centered cubic adsorption site. The electrical structure exhibits considerable dispersion, which arises from the Ti–O bonds. Work function, ranging from 3.9 to 4.8 electron volts (eV), is influenced by surface composition. This variation is mostly dependent on the interaction between local dipoles and surface moieties [55].

Surface functionals that establish bonds with outer layer transition metal atoms can significantly alter electronic band structures. MXenes ended with F atoms exhibit the highest electronic transmission, but the electronic transmission is significantly reduced when the surface is functionalized with O atoms [8]. Ti_2CT_x compounds demonstrate minimal adsorption and reflectivity, but Ti_2CO_2 possesses an optical gap as a result of its semiconducting properties. The optical characteristics of the MXenes are closely connected to their band structure and greatly influenced by surface terminations. The electronegative groups containing oxygen (–O) can increase the size of the bandgap, which improves the absorption of light in low-frequency optical ranges [8].

According to theoretical predictions, the majority of non-terminated MXenes have magnetic properties, because surface Ti atoms have 3d electrons on their surface [59]. Membranes with improved thermal and mechanical stability are produced by $Ti_3C_2T_x$ MXene's covalent functionalization and high affinity for the polymer matrix [65]. These magnetic properties can be significantly altered by adding surface terminations. After surface termination, initially magnetic $Ti_3C_2T_x$ and $Ti_4C_3T_x$ may become non-magnetic. When –OH and –F terminations are used, Cr_2CT_x retains its ferromagnetism; when –O terminations are used, it loses it [66]. Regardless of surface termination, Mn_2NTx is ferromagnetic. The energy gaps of the antiferromagnetic insulating state were influenced by the type of functional group present. When V_2C is terminated with a –F or –OH, it becomes an antiferromagnetic semiconductor with a tiny bandgap instead of a metal. An antiferromagnetic semiconductor can be created from the ferromagnetic semi-metal Cr_2C by adding –F, –Cl, –OH, or –H end groups [61]. The elastic constant, also known as Young's modulus, is expected to increase when terminal –O functionalities are present after etching as opposed to –F, –OH, or bare MXene. This is a result of the –O functional groups' strong coordination with the surface transition metal, which lowers surface imperfections. MXene exhibits bending rigidity in addition to its inherent stiffness, which enables it to maintain its flexibility as 2D sheets and adhere effectively to substrates [67].

Hydrophilic functional groups are highly prevalent on the surfaces of MXene, providing a two-dimensional semiconductor manufacturing substrate. This particular attribute promotes strong contact with other photocatalytic materials, resulting in increased durability of charges and higher efficiency in the transfer of charges [68]. MXenes are inherently endowed with a negative charge by surface terminal groups, such as –OH, –O, and –F. Therefore, guest cations and the negatively charged MXenes may become attracted to one other by multivalence cations intercalation as a result of electrostatic attraction. It is therefore able to limit the growth of the interlayer distance of the layers of MXenes that results from the pillar effect [69]. The presence of the –OH groups can increase hydrophilicity, which makes it very appealing for use in water-based systems. Modifying surface termination groups of $Ti_3C_2T_x$ MXenes can affect their functioning purpose. Charge transfer between the functional groups explains the existence of induced dipole moments in $Ti_3C_2T_x$ MXenes. Moreover, alterations in the collective surface dipole moments resulting from surface relaxation have a role in this phenomenon [68]. The hydrophilic feature of the MXene is strengthened by the formation of surface terminal groups on its surfaces, including –OH, –O, and –F. This property aids in the retention of water molecules for enhanced detection sensitivity. The abundance of many active sites on the surface of MXene facilitates the adsorption efficiency of metal ions. In this scenario, the Pb^{2+} ions have the ability to adhere to the surface of MXene due to chemical and electrostatic interactions. The connection between the MXene layers and the surface functional groups is improved by the intercalation of Pb^{2+} cations, which encourages the absorption process by extending the distance between the MXene interlayers [70]. The detection of Pb^{2+} ions by an Au-coated arc-shaped fiber SPR sensor was compared with that of a Ti_2C MXene-Au-coated SPR sensor. The particular benefits of Ti_2C MXene in detecting Pb^{2+} ions allow for higher performance with good repeatability and stability for Au-coated arc-shaped fiber deposited with Ti_2C MXene. MXene's huge surface area increases surface contact, and the development of the functional groups (–F, –O, and –OH) on its surfaces encourages hydrophilic qualities, which make it easier for water molecules to be trapped. This opens up the possibility of investigating further MXene materials in combination with optical fiber SPR sensing devices through experimentation. The incorporation of surface terminal groups increases the surface wettability of MXenes [7].

A novel strategy is put forward to keep MXene from restacking and improve MOFs' electrical conductivity. The re-stacking problem between the layers of MXene during the preparation phase can be successfully resolved with the hard template method. A conductive network for potential electrodes is created by coupling 2D-modified MXene with a highly redox-active NiCo-MOF. MgO nanoparticles physically alter MXene layers by acting as solid spacers. These layers are easily removed with mild acidity. This modification facilitates the movement of electrolytic ions between the enlarged MXene layers, enhancing their efficiency during the process of charging and discharging. The CV curve, charging-discharging cycles, and Nyquist plot of the modified samples are shown in Fig. 6a–c. During MXene etching, hydroxyl groups are surface exposed which makes it possible for them to weakly attract metal ions. Surface defects are produced during MOF creation by this mechanism, which also creates oxygen vacancies to make up for MOF deficiencies [71].

In a variety of applications, surface functionalization dramatically modifies the physical characteristics of MXenes. By strengthening the interfacial contacts in composites, it increases mechanical strength and toughness. The electronic structure of MXene is altered by the introduction of functional groups, which affects conductivity and electronic transport characteristics. MXene's optical characteristics, which are essential for sensors and optoelectronics, are changed by surface chemical changes. Additionally, functionalization modifies surface wettability, enhancing compatibility with various applications and liquid interactions. Additionally, it improves MXene's chemical and thermal stability, which is essential for robustness under challenging conditions. These modifications show how surface functionalization can be tailored to MXene materials for a wide range of technological applications. Even though surface functionalization of MXenes is advantageous, some issues affect their physical characteristics. Controlling surface coverage, guaranteeing the long-term stability of functional groups, and achieving uniform and consistent functionalization are challenging issues. Complexity is added by functional groups' compatibility with application needs while maintaining MXene's intrinsic features. The method is further complicated by the need for scalability and quality maintenance during industrial-scale functionalization, accurate surface characterization, and high prices. To fully realize the potential of surface-functionalized MXenes in cutting-edge applications, several issues must be resolved.

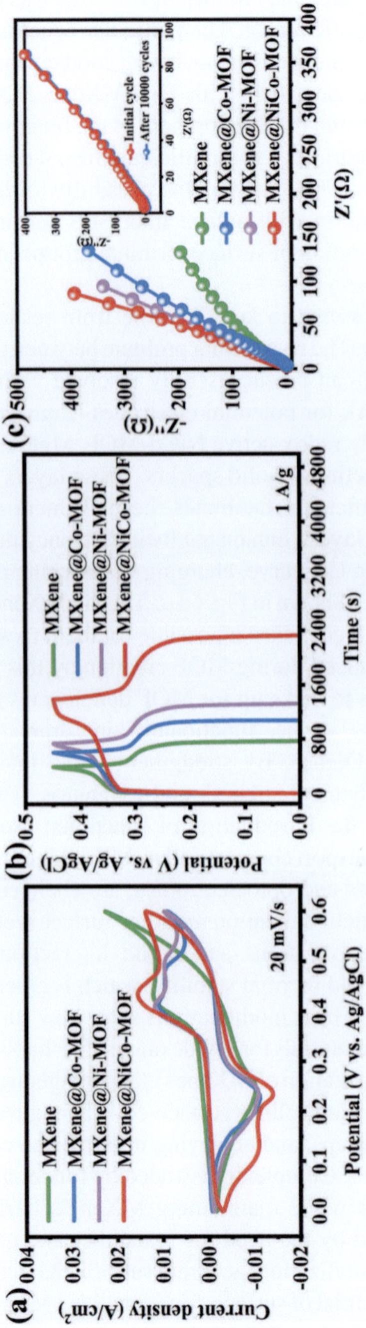

Fig. 6 **a** CV curves at a scan rate of 20 mV s^{-1}; **b** GCD curves at a current density of 1 A g^{-1}; **c** EIS curves. Reproduced from Ref. [71]. Copyright 2024 Elsevier

6.2 Chemical Properties

MXenes are unique and well-known for their remarkable chemical characteristics, which result in distinct chemical reactivity and interaction properties by their terminal groups' (which can be either O or –OH). MXene surfaces that are fully functionalized are more stable than those that are only partially functionalized [5]. MXenes' distinct surface chemistry and electrical structure give rise to their catalytic qualities, which render them adaptable options for a range of catalytic uses [64]. The surface of MXene exhibits a multitude of active sites, including vacancies, defects, and functional groups, which facilitate the processes of gas adsorption and charge transfer. Consequently, MXene materials possess exceptional suitability as the sensing materials for the gas sensors with high sensitivity. The functional groups on the MXenes' surface can interact with gas molecules, causing variations in the electronic properties of the material [67].

Crystal structure, 'Tx' terminations, and chemical composition all have a prominent influence on the electrical properties of MXenes. To be more precise, in the two-dimensional sandwich-like structure of MXenes, transport of electrons to the 'X' layers in the MXene lattice can be facilitated by the replaceable 'M' core layers through M–X bonds [3]. MXenes exhibit an adjustable interlayer distance, which enables rapid intercalation and deintercalation of solvent ions of varying radii. This characteristic makes MXenes highly attractive for use in both mono- and multi-valent ion batteries. The expansion of interlayer space in MXenes is commonly achieved through the introduction of the ions, the molecules, and the polymers, as observed in the delamination techniques, moderate interlayer gap including cations has a significant effect on the conductivity of the MXenes. However, excessively wide interlayer spacing can reduce both the structural stability and conductivity of MXenes [72]. The O atoms act as the active sites that have strong H radical interaction capabilities, making O^-/OH-terminated MXenes the favoured choice for hydrogen evolution reaction (HER). Because of their mild O–H interaction, O-terminated MXenes have a moderate attraction for H radicals, which makes hydrogen desorption easier [73].

MXenes are rich in terminal groups that include oxygen, which can be utilized to modify their surface chemically or covalently. This makes it possible for MXenes to add particular functional groups to acquire new capabilities, enhancing their capacity to extract contaminants from wastewater. When it comes to removing contaminants caused by pharmaceuticals, using MXene and the MXene-functionalized composites offers several benefits through the adsorption and advanced oxidation processes (AOPs). By modifying the surface chemistry of MXenes, their effectiveness in removing pollutants can be enhanced. This can be achieved through alterations or functionalization that boost their adsorption affinity for specific contaminants. Various materials, including Ti_3C_2–MoS_2, Ti_3C_2@Fe_3O_4@β-CD, $BiFeO_3$/MXene, Sodium alginate/MXene/$CoFe_2O_4$, and others, have shown excellent absorption capabilities for antibiotics [74].

Surface functional groups in MXene are essential for collecting charges and developing conductivity routes. These groups have an interaction with the transition metal, which has an impact on its valence electrons. They also have the ability to accept electrons, which helps to balance the electron density in MXenes. The band structure of MXene is impacted by the strength of its connection with transition metal. This interaction demonstrates the direct influence of surface functional groups on the electronic properties of MXene, which in turn affects its charge storage and electrical conductivity [62]. Proton adsorption kinetics can be improved by fluorine termination, although conductivity is decreased. MXenes etched using standard HF techniques inevitably result in the replacement of –O groups with –F inert groups, which lowers the conductivity and the number of active sites [64]. The surface area of MXene offers lots of reactive sites for forming composites, and its surface functional groups are highly electronegativity and reactive. This facilitates the in-situ nucleation of MOFs by encouraging the deprotonation of organic ligands, which in turn act as potent electron donors. Moreover, the strong lamellar structure and substantial specific surface area of MXene are preserved by the covering of metal–organic frameworks (MOFs) which inhibits metallic oxidation and nanosheet stacking [75]. MXene has low dispersion stability in hydrophobic base oils because of its intrinsic surface terminations with hydrophilic groups. The hydrophobicity of MXenes can be customized and the interlayer gap increased by using the right chemical functionalization techniques, such as grafting polymers or tiny molecules. When MXene nanosheets are functionalized with the ionic liquids added to base oil, the lubricating oil's low friction and wear-resistant properties are enhanced synergistically [76].

The functional groups on the surface of MXenes are also necessary for stability. MXene film stability can be effectively increased by hydrogen annealing to reduce the number of –OH terminations. The resultant film has poor dispersibility in the aqueous solutions, which makes it tough to reprocess the material, which is a disadvantage of the treatment [61]. Ti_3C_2 MXene with F/OH termination shows stability in an Ar environment up to 800 °C, but MXene with Cl termination stays stable up to 750 °C [77]. When compared to hydroxyl- and fluorine-terminated MXenes, the oxygen-terminated MXenes (such as $Ti_3C_2O_2$) show higher thermodynamic stability. The reason for this increased stability is that the oxygen groups prevent the Ti–C structure from deforming, while other terminating species cause the bond instability modes [78].

Positively charged materials can produce nanocomposites by electrostatic contact with MXene due to the occurrence of negative terminal groups on its surface [79]. Schottky junctions at a semiconductor–metal interface can be readily formed via hydrogen bonding and electrostatic interaction, which can enhance the conductive properties of the MXene composite [80]. In $Ti_3C_2T_x$ MXenes, functional groups are often dispersed randomly over nanosheet surfaces. Moreover, etching and delamination circumstances have a significant influence on the surface termination profiles. For instance, there will be significantly more –O termination than –F group if the HF solutions at very low concentrations are utilized for the exfoliation. Even more, water molecules bound with –O/–OH terminations were present in $Ti_3C_2T_x$ exfoliated with an HCl–LiF etching solution [21].

MXenes' chemical characteristics are greatly improved by surface functionalization, which also improves their stability, reactivity, and interactions with other compounds. Particular functional groups can be added to change the hydrophilicity or hydrophobicity, enhance oxidation and corrosion resistance, and adjust chemical reactivity. These modifications improve the ion exchange capacity and catalytic activity of MXenes, increasing their suitability for use in arid conditions, water purification, and energy storage. Thus, functionalization enables the precise modification of MXenes' chemical characteristics to satisfy particular application needs. MXene surface functionalization comes with several difficulties that affect its chemical characteristics. It is challenging to achieve a uniform distribution of functional groups, which results in uneven properties. Because these groups tend to disintegrate or desorb, it is difficult to ensure their long-term stability under various environmental circumstances. It can be challenging to regulate a functional group's reactivity to prevent undesirable side effects, especially in reactive settings. Requiring sophisticated and expensive analytical methods, functional groups must be accurately characterized. Furthermore, it can be challenging and expensive for industrial applications to scale up the functionalization process while preserving control and reproducibility. To maximize the chemical characteristics of MXenes for practical application, these obstacles must be overcome.

6.3 Electrochemical Properties

The charge storage performance of the MXenes is directly impacted by various surface terminations. Different theoretical capacities result from the varied molecular mass of the –F, –O, –OH, or –Cl terminations, even when they share the same intercalation mechanism. Additionally, the functional groups on MXenes affect the kinetics of the process, leading to a variety of electrochemical performances [81]. Poor storage performance may arise from the material's reduced active surface areas due to the spontaneous accumulation of the surface functional groups [1]. Tunable surface chemistry with diverse terminal groups can enhance the electrochemical properties by increasing the number of active sites and facilitating ion adsorption [7]. Tunability in surface terminations is a major benefit of the electrochemical approach. The composition of the electrolyte used will determine which specific anionic species can couple with the exposed MXene surface following A-element etching. The ability to modify the MXene surface to a desired degree enables the formation of different functional groups and improves its properties for a wide range of applications [12].

The supercapacitive properties of MXenes are influenced by the surface chemistry and the specific surface area. Experiments have confirmed that certain surface functionalizations of MXenes can reduce the capacitance of the supercapacitors [3]. MXenes can be synthesized by electrochemical etching method. The outcome of this method is that Ti_2CT_x shows a functional group of –O, –Cl, and –OH instead of –F [15]. The nature of the surface functional groups significantly influences electrochemical parameters, including ionic/electronic conductivity, hydrophily, and also energy storage capacity. One of the essential elements controlling the electrochemical energy storage process is electrical conductivity [6]. For example, ion transport can be impeded by surface groups such as –F and –OH, which lowers electrochemical capacity [82]. Mesoporous $Ti_3C_2T_x$ nanosheets were synthesized at room temperature by a wet-chemical technique. First, an etching agent (HCl/LiF) was used to create $Ti_3C_2T_x$ nanosheets, which were then suspended in water. $Ti_3C_2T_x$ nanoflakes were oxidized by dissolved O_2 molecules when this suspension was combined with $CuSO_4$ solution. After being treated with HF solution, TiO_2 nanoparticles were eliminated, leaving behind highly porous $Ti_3C_2T_x$ nanoflakes that had a lower number of –F, more –OH, and –O functional groups and more defects. When mixed with MWCNTs, these nanoflakes resulted in electrodes with enhanced electrical conductivity and electrochemical characteristics [83]. The synthesis of the O-free, and Cl-terminated MXenes (referred to $Ti_3C_2Cl_x$) was achieved by the molten salts ($FeCl_2$) etching procedure. Magnet cleaning was used in place of the oxidative chemical washing phase in the APS process. Subsequently, the surface termination of $Ti_3C_2Cl_x$ in Li_3N was modified from Cl_4 to N_4 (with a small amount of O) through a subsequent thermal treatment, resulting in the formation of N-containing $Ti_3C_2T_x$. The N-containing $Ti_3C_2T_x$, which has been changed on its surface, displays a distinct electrochemical pattern in the sulfuric acid aqueous electrolytes at extremely high rates (303 F/g at 2 V/s). This is a redox reaction that resembles a capacitance and happens over the whole voltage range of 1.0 V [84].

The double-layer structure of $Ti_3C_2T_x$ MXene electrodes at the atomic scale is influenced by the termination of their halogen elements. The electric double-layer capacitance was enhanced by terminating with the halogen atoms of higher atomic numbers, specifically I > Br > Cl > F. The observed rise in capacitance can be attributed to the reduced valence electron of the terminating atoms, which possess lower electronegativity. This characteristic enables the electrostatic accumulation of electrons on the surface of the electrode. This consistent pattern serves as a foundation for contemplating the design of MXene surfaces with increased capacitance [85]. MXenes possess dual discharge platforms as a result of the reversible redox behaviours exhibited by the multi-halogen groups (–ClBr, –ClI, –BrI, and –ClBrI) [86]. The incorporation of halogen (–Cl, –Br, and –I) and amino-rich functional groups into MXene nanosheets would result in an increase in interlayer spacing, leading to a notable improvement in the accessibility of metal ions during the electrochemical process. Steric chloride termination in Ti_2CT_x (LiF/HCl) electrodes increases interlayer space, hence enhancing Li-ion accessibility. This results in superior capacitive performance. When utilized in a Li-ion hybrid capacitor, the electrode achieves high energy and power density. The study provides a thorough knowledge of the Li-ion storage mechanism in MXene, highlighting the critical function of surface termination groups [87]. An advancement in cathode materials for the aqueous zinc ion batteries is represented by halogenated $Ti_3C_2T_x$ MXenes, which are produced via a Lewis-acidic-melt etching technique. These new halogenated versions (containing Br and I or their combinations) work exceptionally well in contrast to previously identified MXenes with components like Cl, F, O, and OH, which had restricted capabilities and lacked a discharge platform. They are extremely competitive in battery applications due to their steady discharge platform, high energy density, large capacity, and great rate performance. Researchers created MXenes with binary

(ClBr or ClI), ternary (ClBrI), and single (Br or I) halogen terminals by adjusting the surface chemistry; each displayed unique electrochemical property. MXenes that ended with Br or I exhibited discharge platforms at about 1.6 V and 1.1 V, respectively, with the specific capacities of 135 and 97.6 mAh g^{-1}. Dual discharge platforms and specific capacities of 117.2 and 107.4 mAh g^{-1} were observed those with multiple halogens, such as BrI or ClBrI. Conversely, MXenes that ended in Cl or that were made the traditional way with HF etching ($Ti_3C_2(OF)$) showed reduced capacities and no obvious discharge platform. The enhanced electrochemical properties of the halogenated MXenes is ascribed to the reversible redox reactions, underscoring the possibility of customised surface chemistry to augment MXene performance across an array of applications [86]. Figure 7a–c illustrates the relationship between electrochemical behavior and terminal compositions in halogenated MXene using cyclic voltammetry (CV) curves. Figure 7d shows galvanostatic charge–discharge profiles (GCD), which confirm the distinct electrochemical behavior of MXenes. Figure 7f shows the good rate performance of $Ti_3C_2Br_2$ and $Ti_3C_2I_2$. Figure 7e depicts the total capabilities of $Ti_3C_2Br_2$ and $Ti_3C_2I_2$, which are comprised of two contributions.

Thermal annealing was used to analyse the electrochemical behaviour of the Ti_3C_2Tx-MXene to comprehend the behaviour of the surface terminal groups and the role that structural water plays in regulating MXene's ability to store redox charges. When surface functional groups are combined, different electrochemistry is produced than when HF-based techniques are used. It is accepted that good electrochemical performance requires keeping MXene's desirable surface functional groups intact [88]. Controlling the surface electrical structure of Ti_3C_2Tx MXene through bio-thermal treatment with ascorbic acid solution is a simple method. Consequently, supercapacitors utilizing controlled electrodes of Ti_3C_2TxMXene (2 h treated of Ti_3C_2Tx MXene, 2-MX) demonstrate a notable reduction in self-discharge. When compared to p-MX with more –OH termination, the self-discharge rate of 2-MX with fewer –OH groups decreased by almost 20% [89].

Optimizing the performance of the MXenes electrodes heavily depends on fine-tuning the content and the distribution of the surface functional groups. After K^+ intercalation and the removal of functional groups (OH^-/F^-), the intercalation pseudo capacitance increases threefold compared to the pristine MXene. The enhanced electrochemical properties can be ascribed to the numerous empty spaces between layers in Ti_3C_2 and the low amount of surface groups present. In contrast to –F terminations, –O groups offer a larger number of pseudo-capacitance active sites [90]. The poor electrochemical properties of MXene are caused by the negative effects of –OH and –F groups on the surface, which hinder the movement of electrons and electrolyte ions. Conversely, –O groups have advantageous electrochemical characteristics [91]. An approach involves modifying the surface termination, expanding the interlayer spacing, and constructing a three-dimensional structure to produce a high-performance $Ti_3C_2T_x$ MXene. The negative electrode demonstrates remarkable specific capacitance, outstanding cycling constancy, and notable rate capability. This solution tackles the constraints of energy storage caused by restacking problems and the occurrence of -F surface termination [92].

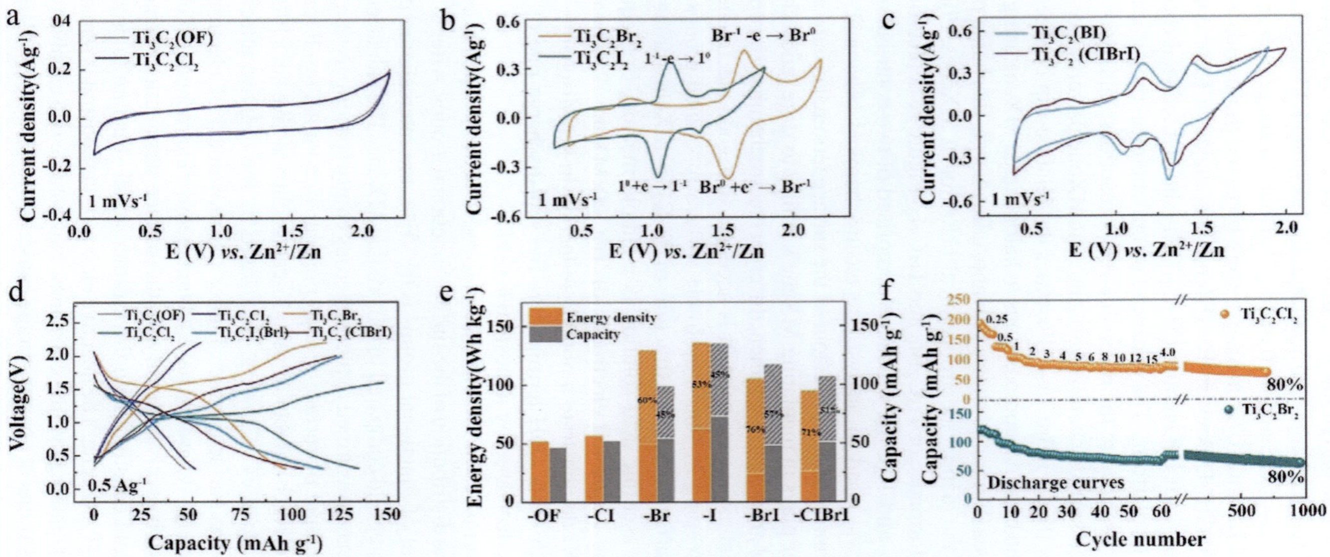

Fig. 7 Electrochemical performance of halogenated Ti_3C_2 MXenes. **a–c** Cyclic voltammogram of $Ti_3C_2(OF)$, $Ti_3C_2Cl_2$, $Ti_3C_2Br_2$, $Ti_3C_2I_2$, $Ti_3C_2(BrI)$, and $Ti_3C_2(ClBrI)$ at a scan rate of 1 mV s^{-1}. **d** Typical GCD curves of Ti_3C_2 MXenes with different terminals at the current density of 0.5 A g^{-1}. **e** Energy density and capacity of Ti_3C_2 MXenes with different terminals. The dashed area presents the contribution from the discharge platform region. **f** Rate capability and long-term cycling performance of $Ti_3C_2Br_2$ and $Ti_3C_2I_2$. Reproduced from Ref. [86]. Copyright 2021, American Chemical Society

MXenes operate well in energy storage and conversion devices when their electrochemical characteristics are surface functionalized. Functional groups can change the electrode potential, add more active sites to boost charge storage capacity, and change electronic pathways to alter electrical conductivity. By preventing structural deterioration during charge–discharge cycles, functionalization enhances cycling stability in batteries and supercapacitors. It can also help or impede ion diffusion, which can impact the rate performance of devices. More active sites for electrochemical reactions can be provided by it by increasing the effective surface area. MXenes are customized for cutting-edge uses in batteries, supercapacitors, and other energy devices using these changes. MXene surface functionalization presents difficulties that affect their electrochemical characteristics. It is challenging to achieve consistent functionalization across MXene surfaces, which results in inconsistent electrochemical performance. It is difficult to guarantee the long-term stability of functional groups under electrochemical cycling circumstances since they can deteriorate or separate over time, reducing the lifespan of the device. It is difficult and demands exact control to manage the kind and density of functional groups to maximize electrochemical properties without sacrificing other qualities. To further complicate matters, accurate identification of functional groups and their impact on electrochemical behaviour necessitates sophisticated analytical techniques. For industrial applications, it is essential to scale up the functionalization process while keeping costs under control and repeatability intact. To fully realize the potential of surface-functionalized MXenes in the advancement of electrochemical devices such as supercapacitors and batteries, these problems must be addressed.

In conclusion, there is a close relationship between the surface terminations and the mechanical, electrical, and electrochemical properties. Numerous applications are made possible by the surface terminations' versatility. To maximize the benefits of MXene, a great deal of work has gone into optimizing the kinds and numbers of surface functional groups.

7 Applications

The application greatly benefits from the surface terminal functional groups of MXenes. By employing different post-processing techniques, the properties of MXenes can be tailored to enhance their electrical and mechanical characteristics as well as their environmental durability. This opens up new possibilities for their utilization in catalysis, sensing, electromagnetic shielding, energy storage and conversion, and various other domains. To manufacture surface-modified MXene materials for various applications, an MXene with the appropriate concentrations of termination groups and interlayer spacing is required. MXenes are employed in a variety of applications due to their excellent conductivity and environmental stability. These characteristics are improved by atomic doping and material compounding, for example, nitrogen-doped Ti3C2Tx electrodes and oxygen-resistant nano plating in lithium storage and hydrogen evolution devices [93]. MXenes provide a diverse array of functional groups that facilitate electron transport, accommodate large active sites, exhibit exceptional mechanical strength, and demonstrate hydrophilic properties. Consequently, MXenes find extensive utilization in the field of supercapacitors [3]. The absorbance coefficients of F and OH-terminated Ti_3C_2 surfaces are reduced, making them well-suited for transparent electrode applications [8]. MXene's behaviour in detecting gas is significantly influenced by the ratio of functional groups on its surface [94].

At a temperature of 220 °C, the subsequent treatment greatly improved the exfoliation and functionalization of Ta_4C_3Tx nanosheets with the addition of –OH and –O groups. The functional groups enhance ion diffusion to enable Faradaic reactions. The composite achieved a capacitance of 447 mF cm^{-3} at a rate of 1 mVs^{-1}, maintaining 90% of its capacitance after 10,000 cycles. Thus, the TTO hybrid structure has the potential to serve as a bioelectrode material with exceptional energy storage capabilities, making it suitable for applications that require precise sizing [95]. During annealing, the $Ti_3C_2T_x$ material underwent a removal of the –F functional groups. The presence of –O and –OH groups on the $Ti_3C_2T_x$ surface promoted the polymerization of aniline monomers. Through a clever intercalation technique, the restacking of $Ti_3C_2T_x$ was prevented and the interlayer spacing was expanded, resulting in enhanced charge transport. The combination of surface modification, increased interlayer spacing, and a three-dimensional structure in A–$Ti_3C_2T_x$/PANI led to impressive results. The specific capacitance reached 652.3 at 1 Ag^{-1}, with an outstanding 99% capacitance retention after 10,000 cycles. Additionally, the volume capacitance achieved 15.9 Fcm^{-3} at 5 mVs^{-1}, establishing it as a highly promising micro supercapacitor (MSC) [92]. By incorporating a $Ti_3C_2T_x$ MXene layer on the LDH, a protective coating is created on the Al alloy, which proves to be an exceptionally efficient anti-corrosion technique. From a scientific perspective, the $Ti_3C_2T_x$ MXene layer on the upper surface exhibits a remarkable ability to act as a barrier against oxygen, due to its impermeability. The surface terminations of the $Ti_3C_2T_x$ can restrict the movement of water and –Cl due to their Lewis acid and –OH properties [96].

The modification of the surface functional groups and crystal structure of $Ti_3C_2T_x$ with Cu^{2+} results in the development of a composite material $Ti_3C_2T_x$@Cu. According to the results of physicochemical characterizations, Cu^{2+} is responsible for introducing alterations in the surface and crystal structure features of MXene. The exposed Ti-ions on the surface of $Ti_3C_2T_x$ undergo a transition from a low valence to a high valence, which causes the material to become more stable in aluminium batteries [30]. The elimination of –F functional groups and the conversion of –OH to –O in the annealed $Ti_3C_2T_x$ sample led to a nearly two-times improvement in specific capacity than to the unannealed samples when employed as an electrode for lithium-ion batteries [12]. The hydrophilic nature of MXenes makes them promising for biomedical applications, by their nanoscale size and planar topology. Nevertheless, they tend to be unstable and accumulate in the physiological environment. To enhance systemic circulation, biocompatibility, and targetability, it is necessary to perform surface functionalization on two-dimensional MXenes. Existing techniques generate end functional groups, but, surface modification can enhance the stability of dispersion, alter surface activity, and increase degrading qualities in living organisms with less toxicity [97].

The specific features of surface-functionalized MXenes make them indispensable for a wide range of electrochemical applications. They improve lithium-ion batteries' energy density and cycling stability by acting as electrode materials that improve conductivity and stability. In supercapacitors, functionalized MXenes also maximize surface area and charge storage capacity, enabling quick charge–discharge cycles for high-power applications. Because of their changed surfaces, MXenes are better able to catalyze activities that are essential for energy conversion and environmental cleanup, such as hydrogen evolution reactions (HER) and oxygen reduction reactions (ORR). By improving the sensitivity and selectivity of electrochemical sensors, functionalized MXenes can detect gases, biomolecules, and contaminants, leading to the advancement of monitoring and diagnostic technology. By adsorbing and breaking down pollutants, they also play a crucial part in electrochemical water treatment systems, which promote environmental sustainability. Their adaptability also includes electrochemical capacitors and flexible electronics, where they improve conductivity and capacitance for uses in wearable electronics and energy storage. The aforementioned applications highlight the noteworthy influence of surface-functionalized MXenes in diverse electrochemical domains, propelling advancements in energy storage, environmental monitoring, and flexible electronics.

8 Conclusion

The rational design of MXene could be improved by surface functionalization, which could increase its application possibilities beyond selective etching. Functional MXenes perform better because they can give more electrochemically active sites and better electronic structures. There are several ways to achieve these functional MXenes, including adjusting terminations, doping heteroatoms, enlarging interlayer space, and hybridizing with polymers. Their perfect surface properties, superior mechanical qualities, abundance of active sites, and vast diffusion space, make them useful in energy storage, catalysis, sensors, EMI shielding, and mechanical actuators. MXenes' varied surface functionalization enables customization of their properties to fit various application scenarios. Their processability as inks and thin profiles also make it possible to fabricate innovative gadgets. MXenes are a prospective possibility for a widespread variety of applications at the junction of engineering and materials science because of their reconfigurability and the broad range of properties, despite being a relatively new field of study. Efforts are still being made to maximize the technological potential of MXenes and build upon the initial discoveries.

References

1. Alli, Y.A., Bamisaye, A., Nancy, P., Zachariah, S.M., Oladoye, P.O., Bankole, O.M., Akamo, D.O., Chkirida, S., Anuar, H., Thomas, S.: MXene composites: properties, synthesis and its emerging application in rechargeable batteries. J. Energy Storage **77**, 109954 (2024). https://doi.org/10.1016/j.est.2023.109954
2. Wang, T., Ma, X., Gong, B., Zhu, C., Xue, P., Guo, L., Tian, X., Shen, X., Min, Y., Xu, Q., Cao, H.: Bio-inspired Ti3C2Tx MXene composite coating for enhancing corrosion resistance of aluminum alloy in acidic environments. J. Colloid Interface Sci. **658**, 865–878 (2024). https://doi.org/10.1016/j.jcis.2023.12.143
3. Bi, W., Gao, G., Li, C., Wu, G., Cao, G.: Synthesis, properties, and applications of MXenes and their composites for electrical energy storage. Prog. Mater. Sci. **142**, 101227 (2024). https://doi.org/10.1016/j.pmatsci.2023.101227
4. Wang, L., Han, M., Shuck, C.E., Wang, X., Gogotsi, Y.: Adjustable electrochemical properties of solid-solution MXenes. Nano Energy **88**, 106308 (2021). https://doi.org/10.1016/j.nanoen.2021.106308
5. Kitchamsetti, N., Cho, J.S.: A roadmap of recent advances in MXene@MOF hybrids, its derived composites: Synthesis, properties, and their utilization as an electrode for supercapacitors, rechargeable batteries and electrocatalysis. J. Energy Storage **80**, 110293 (2024). https://doi.org/10.1016/j.est.2023.110293
6. Syamsai, R.: MXenes: synthesis, properties, and electrochemical performance of titanium, vanadium, and tantalum carbide MXenes as supercapacitor electrodes. Elsevier Ltd. (2022). https://doi.org/10.1016/B978-0-323-89952-9.00003-8
7. Bilibana, M.P.: Electrochemical properties of MXenes and applications. Adv. Sens. Energy Mater. **2**, 100080 (2023). https://doi.org/10.1016/j.asems.2023.100080
8. Jiang, X., Kuklin, A.V., Baev, A., Ge, Y., Ågren, H., Zhang, H., Prasad, P.N.: Two-dimensional MXenes: from morphological to optical, electric, and magnetic properties and applications. Phys. Rep. **848**, 1–58 (2020). https://doi.org/10.1016/j.physrep.2019.12.006
9. Enyashin, A.N., Ivanovskii, A.L.: Two-dimensional titanium carbonitrides and their hydroxylated derivatives: structural, electronic properties and stability of MXenes Ti3C2−xNx(OH)2 from DFTB calculations. J. Solid State Chem. **207**, 42–48 (2013). https://doi.org/10.1016/j.jssc.2013.09.010
10. Wang, X., Shen, X., Gao, Y., Wang, Z., Yu, R., Chen, L.: Atomic-scale recognition of surface structure and intercalation mechanism of Ti3C2X. J. Am. Chem. Soc. **137**, 2715–2721 (2015). https://doi.org/10.1021/ja512820k
11. Aldhaher, A., Rabiee, N., Iravani, S.: Exploring the synergistic potential of MXene-MOF hybrid composites: a perspective on synthesis, properties, and applications. Hybrid Adv. **5**, 100131 (2024). https://doi.org/10.1016/j.hybadv.2023.100131
12. Li, G., Lian, S., Wang, J., Xie, G., Zhang, N., Xie, X.: Surface chemistry engineering and the applications of MXenes. J. Mater. **9**, 1160–1184 (2023). https://doi.org/10.1016/j.jmat.2023.08.003
13. Persson, I., Näslund, L.Å., Halim, J., Barsoum, M.W., Darakchieva, V., Palisaitis, J., Rosen, J., Persson, P.O.Å.: On the organization and thermal behavior of functional groups on Ti3C2 MXene surfaces in vacuum. 2D Mater. **5** (2018). https://doi.org/10.1088/2053-1583/aa89cd
14. Yu, H., Wang, Y., Jing, Y., Ma, J., Du, C.F., Yan, Q.: Surface modified MXene-based nanocomposites for electrochemical energy conversion and storage. Small **15**, 1–20 (2019). https://doi.org/10.1002/smll.201901503
15. Baig, M.M., Gul, I.H., Baig, S.M., Shahzad, F.: 2D MXenes: synthesis, properties, and electrochemical energy storage for supercapacitors—A review. J. Electroanal. Chem. **904**, 115920 (2022). https://doi.org/10.1016/j.jelechem.2021.115920
16. Yang, S., Zhang, P., Wang, F., Ricciardulli, A.G., Lohe, M.R., Blom, P.W.M., Feng, X.: Fluoride-free synthesis of two-dimensional titanium carbide (MXene) using a binary aqueous system. Angew. Chemie Int. Ed. **57**, 15491–15495 (2018). https://doi.org/10.1002/anie.201809662

17. Kamysbayev, V., Filatov, A.S., Hu, H., Rui, X., Lagunas, F., Wang, D., Klie, R.F., Talapin, D.V. Covalent surface modifications and superconductivity of two-dimensional metal carbide MXenes. Science **369**, 979–983 (2020). https://doi.org/10.1126/science.aba8311
18. Li, N., Huo, J., Zhang, Y., Ye, B., Chen, X., Li, X., Xu, S., He, J., Chen, X., Tang, Y., Zhu, Y., Ling, K., Zhu, R.: Transition metal carbides/nitrides (MXenes): properties, synthesis, functional modification and photocatalytic application. Sep. Purif. Technol. **330**, 125325 (2024). https://doi.org/10.1016/j.seppur.2023.125325
19. Mozafari, M., Soroush, M.: Surface functionalization of MXenes. Mater. Adv. **2**, 7277–7307 (2021). https://doi.org/10.1039/d1ma00625h
20. Yi, Z., Yang, J., Liu, X., Mao, L., Cui, L., Liu, Y.: Enhanced mechanical properties of poly(lactic acid) composites with ultrathin nanosheets of MXene modified by stearic acid. J. Appl. Polym. Sci. **137**, 1–8 (2020). https://doi.org/10.1002/app.48621
21. Xiao, Z., Ruan, S., Kong, L.B., Que, W., Zhou, K., Liu, Y., Zhang, T.: Synthesis and properties of MXenes (2020). https://doi.org/10.1007/978-3-030-59373-5_2
22. Hart, J.L., Hantanasirisakul, K., Lang, A.C., Anasori, B., Pinto, D., Pivak, Y., van Omme, J.T., May, S.J., Gogotsi, Y., Taheri, M.L.: Control of MXenes' electronic properties through termination and intercalation, Nat. Commun. **10** (2019). https://doi.org/10.1038/s41467-018-08169-8
23. Guo, L., Jiang, W.Y., Shen, M., Xu, C., Ding, C.X., Zhao, S.F., Yuan, T.T., Wang, C.Y., Zhang, X.Q., Wang, J.Q.: High capacitance of MXene (Ti3C2Tx) through intercalation and surface modification in molten salt. Electrochim. Acta **401**, 139476 (2022). https://doi.org/10.1016/j.electacta.2021.139476
24. Tang, H., Wang, R., Shi, L., Sheremet, E., Rodriguez, R.D., Sun, J.: Post-processing strategies for improving the electrical and mechanical properties of MXenes. Chem. Eng. J. **425**, 131472 (2021). https://doi.org/10.1016/j.cej.2021.131472
25. Liu, P., Ding, W., Liu, J., Shen, L., Jiang, F., Liu, P., Zhu, Z., Zhang, G., Liu, C., Xu, J.: Surface termination modification on high-conductivity MXene film for energy conversion. J. Alloys Compd. **829** (2020). https://doi.org/10.1016/j.jallcom.2020.154634
26. Wen, Y., Rufford, T.E., Chen, X., Li, N., Lyu, M., Dai, L., Wang, L.: Nitrogen-doped Ti3C2Tx MXene electrodes for high-performance supercapacitors. Nano Energy **38**, 368–376 (2017). https://doi.org/10.1016/j.nanoen.2017.06.009
27. Sun, S., Xie, Z., Yan, Y., Wu, S.: Hybrid energy storage mechanisms for sulfur-decorated Ti3C2 MXene anode material for high-rate and long-life sodium-ion batteries. Chem. Eng. J. **366**, 460–467 (2019). https://doi.org/10.1016/j.cej.2019.01.185
28. Yoon, Y., Tiwari, A.P., Choi, M., Novak, T.G., Song, W., Chang, H., Zyung, T., Lee, S.S., Jeon, S., An, K.S.: Precious-metal-free Electrocatalysts for activation of hydrogen evolution with nonmetallic electron donor: chemical composition controllable phosphorous doped vanadium carbide MXene. Adv. Funct. Mater. **29**, 1–12 (2019). https://doi.org/10.1002/adfm.201903443
29. Cheng, Y., Xu, X., Li, Y., Zhang, Y., Song, Y.: CO2 reduction mechanism on the Nb2CO2 MXene surface: effect of nonmetal and metal modification. Comput. Mater. Sci. **202**, 110971 (2022). https://doi.org/10.1016/j.commatsci.2021.110971
30. Ni, J., Li, W., Li, H., Huo, X., Xiao, S.: Simple Cu2+ modification significantly improves the electrochemical performance of MXene (Ti3C2Tx) in aluminum batteries. J. Electroanal. Chem. **953**, 118000 (2024). https://doi.org/10.1016/j.jelechem.2023.118000
31. Li, Z., Jiang, M., Wu, F., Wu, L., Zhang, X., Li, L.: Synergistic in-situ intercalation and surface modification strategy for Ti3C2Tx MXene-based supercapacitors with enhanced electrochemical energy storage. J. Energy Storage **84**, 1–9 (2024). https://doi.org/10.1016/j.est.2024.110772
32. Wei, X., Cai, M., Yuan, F., Lu, D., Li, C., Huang, H., Xu, S., Liang, X., Zhou, W., Guo, J.: The surface functional modification of Ti3C2Tx MXene by phosphorus doping and its application in quasi-solid state flexible supercapacitor. Appl. Surf. Sci. **606**, 154817 (2022). https://doi.org/10.1016/j.apsusc.2022.154817
33. Kong, F., He, X., Liu, Q., Qi, X., Sun, D., Zheng, Y., Wang, R., Bai, Y.: Further surface modification by carbon coating for in-situ growth of Fe3O4 nanoparticles on MXene Ti3C2

multilayers for advanced Li-ion storage. Electrochim. Acta **289**, 228–237 (2018). https://doi.org/10.1016/j.electacta.2018.09.007
34. Jin, Z., Fang, Y., Wang, X., Xu, G., liu, M., Wei, S., Zhou, C., Zhang, Y., Xu, Y.: Ultra-efficient electromagnetic wave absorption with ethanol-thermally treated two-dimensional Nb2CTx nanosheets. J. Colloid Interface Sci. **537**, 306–315 (2019). https://doi.org/10.1016/j.jcis.2018.11.034
35. Yang, L., Dall'Agnese, C., Dall'Agnese, Y., Chen, G., Gao, Y., Sanehira, Y., Jena, A.K., Wang, X., Gogotsi, Y., Miyasaka, T.: Surface-modified metallic Ti3C2Tx MXene as electron transport layer for planar heterojunction perovskite solar cells. Adv. Funct. Mater. **29**, 1905694 (2019)
36. Zhang, Y., Lu, W., Zhou, J., Sun, D., Li, H.: The multiple synthesis of layered V2CTx-MXene composites with enhanced electrochemical properties, J. Alloys Compd. **929**, 167276 (2022). https://doi.org/10.1016/j.jallcom.2022.167276
37. Ling, S., Lin, S., Wu, Y., Li, Y.: Toward highly-robust MXene hybrid memristor by synergetic ionotronic modification and two-dimensional heterojunction. Chem. Eng. J. **486**, 150100 (2024). https://doi.org/10.1016/j.cej.2024.150100
38. Medhin Ashebo, M., Liu, N., Yu, F., Ma, J.: Surface functional modification of Nb2CTx MXene for high performance capacitive deionization. Sep. Purif. Technol. 127125 (2024). https://doi.org/10.1016/j.seppur.2024.127125
39. Faruk, O., Adak, B.: Recent advances in PEDOT:PSS integrated graphene and MXene-based composites for electrochemical supercapacitor applications. Synth. Met. **297**, 117384 (2023). https://doi.org/10.1016/j.synthmet.2023.117384
40. Chen, W., Wu, Z., He, X., Su, Y., Zheng, G., Oh, S.K., Mei, C.: Achieving superior anti-corrosion properties of vinyl ester resin coatings via compositing with 3-methacryloxy propyl trimethoxysilane functionalized MXene nanosheets. Polym. Test. **127**, 108203 (2023). https://doi.org/10.1016/j.polymertesting.2023.108203
41. Boota, M., Anasori, B., Voigt, C., Zhao, M.Q., Barsoum, M.W., Gogotsi, Y.: Pseudocapacitive electrodes produced by oxidant-free polymerization of pyrrole between the layers of 2D titanium carbide (MXene). Adv. Mater. **28**, 1517–1522 (2016). https://doi.org/10.1002/adma.201504705
42. Chen, X., Xue, S., Yan, Y., Bai, W., feng Du, S. Liu, Q. Ye, F. Zhou, Zwitterionic polymer-functionalized nitrogen-doped MXene nanosheets as aqueous lubricant additive. Tribol. Int. **186**, 108625 (2023). https://doi.org/10.1016/j.triboint.2023.108625
43. Chen, K., Chen, Y., Deng, Q., Jeong, S.H., Jang, T.S., Du, S., Kim, H.E., Huang, Q., Han, C.M.: Strong and biocompatible poly(lactic acid) membrane enhanced by Ti3C2Tz (MXene) nanosheets for Guided bone regeneration. Mater. Lett. **229**, 114–117 (2018). https://doi.org/10.1016/j.matlet.2018.06.063
44. He, L., Wang, J., Wang, B., Wang, X., Zhou, X., Cai, W., Mu, X., Hou, Y., Hu, Y., Song, L.: Large-scale production of simultaneously exfoliated and functionalized MXenes as promising flame retardant for polyurethane. Compos. Part B Eng. **179**, 107486 (2019). https://doi.org/10.1016/j.compositesb.2019.107486
45. Li, S., Huang, H., Chen, F., He, X., Ma, Y., Zhang, L., Sheng, X., Chen, Y., Shchukina, E., Shchukin, D.: Reinforced anticorrosion performance of waterborne epoxy coating with eco-friendly L-cysteine modified Ti3C2Tx MXene nanosheets. Prog. Org. Coat. **161**, 106478 (2021). https://doi.org/10.1016/j.porgcoat.2021.106478
46. Yan, H., Cai, M., Li, W., Fan, X., Zhu, M.: Amino-functionalized Ti3C2Tx with anti-corrosive/wear function for waterborne epoxy coating. J. Mater. Sci. Technol. **54**, 144–159 (2020). https://doi.org/10.1016/j.jmst.2020.05.002
47. Chen, Y., Zhan, Y., Dong, H., Li, Y., Yang, X., Sun, A., Chen, X., Zhu, F., Jia, H.: Two-dimensional lamellar MXene nanosheets/waterborne epoxy composite coating: Dopamine triggered surface modification and long-term anticorrosion performance. Colloids Surf. A Physicochem. Eng. Asp. **674**, 131865 (2023). https://doi.org/10.1016/j.colsurfa.2023.131865
48. Xia, Y., Mathis, T.S., Zhao, M.Q., Anasori, B., Dang, A., Zhou, Z., Cho, H., Gogotsi, Y., Yang, S.: Thickness-independent capacitance of vertically aligned liquid-crystalline MXenes. Nature **557**, 409–412 (2018). https://doi.org/10.1038/s41586-018-0109-z

49. Lu, Q., Zhang, T., He, B., Xu, F., Liu, S., Ye, Q., Zhou, F.: Enhanced lubricity and anti-wear performance of zwitterionic polymer-modified N-enriched porous carbon nanosheets as water-based lubricant additive. Tribol. Int. **167** (2022). https://doi.org/10.1016/j.triboint.2021.107421
50. Zhao, J., Yang, Y., Yang, C., Tian, Y., Han, Y., Liu, J., Yin, X., Que, W.: A hydrophobic surface enabled salt-blocking 2D Ti3C2 MXene membrane for efficient and stable solar desalination. J. Mater. Chem. A. **6**, 16196–16204 (2018). https://doi.org/10.1039/c8ta05569f
51. Jiang, Y., Sun, T., Xie, X., Jiang, W., Li, J., Tian, B., Su, C.: Oxygen-functionalized ultrathin Ti3C2Tx MXene for enhanced electrocatalytic hydrogen evolution. Chemsuschem **12**, 1368–1373 (2019). https://doi.org/10.1002/cssc.201803032
52. Persson, I., Halim, J., Lind, H., Hansen, T.W., Wagner, J.B., Näslund, L.Å., Darakchieva, V., Palisaitis, J., Rosen, J., Persson, P.O.Å.: 2D transition metal carbides (MXenes) for carbon capture. Adv. Mater. **31**, 1–5 (2019). https://doi.org/10.1002/adma.201805472
53. Yang, L., Dall'Agnese, C., Dall'Agnese, Y., Chen, G., Gao, Y., Sanehira, Y., Jena, A.K., Wang, X.F., Gogotsi, Y., Miyasaka, T.: Surface-modified metallic Ti3C2Tx MXene as electron transport layer for planar heterojunction perovskite solar cells. Adv. Funct. Mater. **29**, 1–8 (2019). https://doi.org/10.1002/adfm.201905694
54. Rakhi, R.B., Ahmed, B., Hedhili, M.N., Anjum, D.H., Alshareef, H.N.: Effect of Postetch Annealing gas composition on the structural and electrochemical properties of Ti2CTx MXene electrodes for supercapacitor applications. Chem. Mater. **27**, 5314–5323 (2015). https://doi.org/10.1021/acs.chemmater.5b01623
55. Schultz, T., Frey, N.C., Hantanasirisakul, K., Park, S., May, S.J., Shenoy, V.B., Gogotsi, Y., Koch, N.: Surface termination dependent work function and electronic properties of Ti3C2Tx MXene. Chem. Mater. **31**, 6590–6597 (2019). https://doi.org/10.1021/acs.chemmater.9b00414
56. Singh, S.K., Tiwari, A.K., Paliwal, H.K.: A holistic review of MXenes for solar device applications: synthesis, characterization, properties and stability. FlatChem **39**, 100493 (2023). https://doi.org/10.1016/j.flatc.2023.100493
57. Mahmood, M., Hussain, I., Mahmood, S., Shahzad, F.: 2D MXenes: synthesis, properties, and electrochemical energy storage for supercapacitors—A review. J. Electroanal. Chem. **904**, 115920 (2022). https://doi.org/10.1016/j.jelechem.2021.115920
58. Peng, M., Dong, M., Wei, W., Xu, H., Liu, C., Shen, C.: The introduction of amino termination on Ti3C2 MXene surface for its flexible film with excellent property. Carbon N. Y. **179**, 400–407 (2021). https://doi.org/10.1016/j.carbon.2021.04.049
59. Xie, Y., Kent, P.R.C.: Hybrid density functional study of structural and electronic properties of functionalized $Ti_{n+1}Xn$ (X = C, N) monolayers. Phys. Rev. B Condens. Matter Mater. Phys. **87**, 1–10 (2013). https://doi.org/10.1103/PhysRevB.87.235441
60. Zha, X.H., Luo, K., Li, Q., Huang, Q., He, J., Wen, X., Du, S.: Role of the surface effect on the structural, electronic and mechanical properties of the carbide MXenes. Europhys. Lett. **111** (2015). https://doi.org/10.1209/0295-5075/111/26007
61. Khazaei, M., Arai, M., Sasaki, T., Ranjbar, A., Liang, Y., Yunoki, S.: OH terminated two-dimensional transition metal carbides and nitrides (MXenes) as ultralow work function materials. Adv. Funct. Mater. **23**, 2185–2192 (2013)
62. Kim, H., Alshareef, H.N.: MXetronics: MXene-enabled electronic and photonic devices. ACS Mater. Lett. **2**, 55–70 (2020). https://doi.org/10.1021/acsmaterialslett.9b00419
63. Khazaei, M., Arai, M., Sasaki, T., Chung, C.Y., Venkataramanan, N.S., Estili, M., Sakka, Y., Kawazoe, Y.: Novel electronic and magnetic properties of two-dimensional transition metal carbides and nitrides. Adv. Funct. Mater. **23**, 2185–2192 (2013). https://doi.org/10.1002/adfm.201202502
64. Kumar, S., Kumari, N., Seo, Y.: MXenes: versatile 2D materials with tailored surface chemistry and diverse applications. J. Energy Chem. **90**, 253–293 (2024). https://doi.org/10.1016/j.jechem.2023.11.031
65. An, W., Xue, R., Yuan, P., Gao, L., Xu, Q., Hu, D., Ma, J.: Design strategies of MXene-based coatings towards multifunctional application. Surf. Interfaces **46**, 104083 (2024). https://doi.org/10.1016/j.surfin.2024.104083

66. Zhao, S., Kang, W., Xue, J.: Manipulation of electronic and magnetic properties of M2C (M = Hf, Nb, Sc, Ta, Ti, V, Zr) monolayer by applying mechanical strains. Appl. Phys. Lett. **104** (2014). https://doi.org/10.1063/1.4870515
67. Reddy, M.S.B., Aich, S.: Recent progress in surface and heterointerface engineering of 2D MXenes for gas sensing applications. Coord. Chem. Rev. **500**, 215542 (2024). https://doi.org/10.1016/j.ccr.2023.215542
68. Ashraf, T., Ali, B., Ashraf, S., Imran, M., Fazal, M.T., Iqbal, J.: Recent advancement in MXene based heterojunctions toward CO2 photo-reduction and H2 production applications: a review. FlatChem. **44**, 100620 (2024). https://doi.org/10.1016/j.flatc.2024.100620
69. Li, J., Lu, M., Zheng, W., Zhang, W.: Ion-intercalation architecture for robust functionalization of two-dimensional MXenes. Energy Storage Mater. **64**, 103068 (2024). https://doi.org/10.1016/j.ensm.2023.103068
70. Ahmad, H., Makhfuz, M.J.M., Yusoff, N., Zakaria, R.: The effect of Ti2C MXene on the performance of optical fiber-based surface plasmon resonance sensor towards lead detection. Mater. Sci. Eng. B. **302**, 117232 (2024). https://doi.org/10.1016/j.mseb.2024.117232
71. Wei, J., Hu, F., Pan, Y., Lv, C., Bian, L., Ouyang, Q.: Design strategy for metal–organic framework assembled on modifications of MXene layers for advanced supercapacitor electrodes. Chem. Eng. J. **481**, 148793 (2024). https://doi.org/10.1016/j.cej.2024.148793
72. Ghidiu, M., Lukatskaya, M.R., Zhao, M.Q., Gogotsi, Y., Barsoum, M.W.: Conductive two-dimensional titanium carbide "clay" with high volumetric capacitance. Nature **516**, 78–81 (2015). https://doi.org/10.1038/nature13970
73. Gao, G., O'Mullane, A.P., Du, A.: 2D MXenes: a new family of promising catalysts for the hydrogen evolution reaction. ACS Catal. **7**, 494–500 (2017). https://doi.org/10.1021/acscatal.6b02754
74. Sanga, P., Al-mashriqi, H.S., Chen, J., Qiu, H.: A mechanistic view of removing pharmaceutical-induced pollutants by MXene and MXene-functionalized composites via adsorption and advanced oxidation process. J. Environ. Chem. Eng. **12**, 111685 (2024). https://doi.org/10.1016/j.jece.2023.111685
75. Zhang, H., Li, Z., Hou, Z., Mei, H., Feng, Y., Xu, B., Sun, D.: Self-assembly of MOF on MXene nanosheets and in-situ conversion into superior nickel phosphates/MXene battery-type electrode. Chem. Eng. J. **425**, 130602 (2021). https://doi.org/10.1016/j.cej.2021.130602
76. Guo, R., Liu, Z., Chen, X., Dou, X., Ye, Q., Liu, X.: J Tribol. Int. 109526 (2024). https://doi.org/10.1016/j.triboint.2024.109526
77. Lu, J., Persson, I., Lind, H., Palisaitis, J., Li, M., Li, Y., Chen, K., Zhou, J., Du, S., Chai, Z., Huang, Z., Hultman, L., Eklund, P., Rosen, J., Huang, Q., Persson, P.O.Å.: Ti_{n+1}Cn MXenes with fully saturated and thermally stable Cl terminations electronic supplementary information (ESI). Nanoscale Adv. **1**, 3680–3685 (2019). https://doi.org/10.1039/c9na00324j
78. Zheng, Z., Guo, C., Wang, E., He, Z., Liang, T., Yang, T., Hou, X.: The oxidation and thermal stability of two-dimensional transition metal carbides and/or carbonitrides (MXenes) and the improvement based on their surface state. Inorg. Chem. Front. **8**, 2164–2182 (2021). https://doi.org/10.1039/d1qi00041a
79. Kshetri, T., Tran, D.T., Le, H.T., Nguyen, D.C., Van Hoa, H., Kim, N.H., Lee, J.H.: Recent advances in MXene-based nanocomposites for electrochemical energy storage applications. Prog. Mater. Sci. **117**, 100733 (2021). https://doi.org/10.1016/j.pmatsci.2020.100733
80. Liang, X., Yan, Y., Liu, G., Hou, S., Hou, S.: Constructing magnetically functionalized sandwich-like Eu-MOF/PDA@MXene hybrids for extremely sensitive molecular imprinted sensor toward trace detecting trimethylamine oxide. Chem. Eng. J. **483**, 149301 (2024). https://doi.org/10.1016/j.cej.2024.149301
81. Huang, Z., Farahmandjou, M., Marlton, F., Guo, X., Gao, H., Sun, B., Wang, G.: Surface and structure engineering of MXenes for rechargeable batteries beyond lithium. J. Mater. **10**, 253–268 (2024). https://doi.org/10.1016/j.jmat.2023.10.001
82. Li, J., Yuan, X., Lin, C., Yang, Y., Xu, L., Du, X., Xie, J., Lin, J., Sun, J.: Achieving high pseudocapacitance of 2D titanium carbide (MXene) by cation intercalation and surface modification. Adv. Energy Mater. **7** (2017). https://doi.org/10.1002/aenm.201602725

83. Ren, C.E., Zhao, M.Q., Makaryan, T., Halim, J., Boota, M., Kota, S., Anasori, B., Barsoum, M.W., Gogotsi, Y.: Porous two-dimensional transition metal carbide (MXene) flakes for high-performance Li-Ion storage. ChemElectroChem **3**, 689–693 (2016). https://doi.org/10.1002/celc.201600059
84. Liu, L., Zschiesche, H., Antonietti, M., Daffos, B., Tarakina, N.V., Gibilaro, M., Chamelot, P., Massot, L., Duployer, B., Taberna, P.L., Simon, P.: Tuning the surface chemistry of MXene to improve energy storage: example of nitrification by salt melt. Adv. Energy Mater. **13** (2023). https://doi.org/10.1002/aenm.202202709
85. Shimada, T., Takenaka, N., Ando, Y., Otani, M., Okubo, M., Yamada, A.: Relationship between electric double-layer structure of MXene electrode and its surface functional groups. Chem. Mater. **34**, 2069–2075 (2022). https://doi.org/10.1021/acs.chemmater.1c03328
86. Li, M., Li, X., Qin, G., Luo, K., Lu, J., Li, Y., Liang, G., Huang, Z., Zhou, J., Hultman, L., Eklund, P., Persson, P.O.Å., Du, S., Chai, Z., Zhi, C., Huang, Q.: Halogenated Ti3C2MXenes with electrochemically active terminals for high-performance zinc ion batteries. ACS Nano **15**, 1077–1085 (2021). https://doi.org/10.1021/acsnano.0c07972
87. Kajiyama, S., Szabova, L., Iinuma, H., Sugahara, A., Gotoh, K., Sodeyama, K., Tateyama, Y., Okubo, M., Yamada, A.: Enhanced Li-ion accessibility in mxene titanium carbide by steric chloride termination. Adv. Energy Mater. **7**, 1–8 (2017). https://doi.org/10.1002/aenm.201601873
88. Gopalakrishnan, M., Kao-ian, W., Aupama, V., Etesami, M., Ganesan, S., Theerthagiri, J., Choi, M.Y., Mohamad, A.A., Kheawhom, S.: Metal-organic framework/mxene heterostructure and its derivatives as electrode materials for rechargeable Zn-based batteries: Design strategies and perspectives. Chem. Eng. J. **484**, 149624 (2024). https://doi.org/10.1016/j.cej.2024.149624
89. Pu, S., Wang, Z., Xie, Y., Fan, J., Xu, Z., Wang, Y., He, H., Zhang, X., Yang, W., Zhang, H.: Origin and regulation of self-discharge in MXene supercapacitors. Adv. Funct. Mater. **33**, 1–10 (2023). https://doi.org/10.1002/adfm.202208715
90. Hui, X., Ge, X., Zhao, R., Li, Z., Yin, L.: Interface chemistry on MXene-based materials for enhanced energy storage and conversion performance. Adv. Funct. Mater. **30**, 1–37 (2020). https://doi.org/10.1002/adfm.202005190
91. Xie, Y., Naguib, M., Mochalin, V.N., Barsoum, M.W., Gogotsi, Y., Yu, X., Nam, K., Yang, X., Kolesnikov, A.I., Kent, P.R.C.: Role of surface structure on Li-ion energy storage capacity of two-dimensional transition-metal carbides (2014)
92. Liu, W., Zheng, Y., Zhang, Z., Zhang, Y., Wu, Y., Gao, H., Su, J., Gao, Y.: Ultrahigh gravimetric and volumetric capacitance in Ti3C2Tx MXene negative electrode enabled by surface modification and in-situ intercalation. J. Power Sour. **521**, 230965 (2022). https://doi.org/10.1016/j.jpowsour.2021.230965
93. Wu, X., Wang, Z., Yu, M., Xiu, L., Qiu, J.: Stabilizing the MXenes by carbon nanoplating for developing hierarchical nanohybrids with efficient lithium storage and hydrogen evolution capability. Adv. Mater. **29**, 1607017 (2017). https://doi.org/10.1002/adma.201607017
94. Hajian, S., Khakbaz, P., Moshayedi, M., Maddipatla, D., Narakathu, B.B., Turkani, V.S., Bazuin, B.J., Pourfath, M., Atashbar, M.Z.: Impact of different ratios of fluorine, oxygen, and hydroxyl surface terminations on Ti3C2Tx MXene as Ammonia sensor: a first-principles study. Proc. IEEE Sens. 1–4 (2018). https://doi.org/10.1109/ICSENS.2018.8589699
95. Rafieerad, A., Amiri, A., Sequiera, G.L., Yan, W., Chen, Y., Polycarpou, A.A., Dhingra, S.: Development of fluorine-free tantalum carbide MXene hybrid structure as a biocompatible material for supercapacitor electrodes **2100015**, 1–12 (2021). https://doi.org/10.1002/adfm.202100015
96. Jia, W., Cao, H., Wang, T., Min, Y., Xu, Q.: Electrochemical behavior and anti-corrosion property of Ti3C2Tx MXene/LDH heterostructured coating on aluminum alloy. Surf. Coat. Technol. **463**, 129551 (2023). https://doi.org/10.1016/j.surfcoat.2023.129551
97. Soman, S., Kulkarni, S., Pandey, A., Dhas, N., Shirur, K.S., Gude, R.S., Vidya, S.M., Nayak, S., George, S.D., Mutalik, S.: Unlocking the power of MXenes—Crafting a 2D nanoplatform for tomorrow: synthesis, functionalization, stability, and biomedical applications. Mater. Today Commun. **38**, 107711 (2024). https://doi.org/10.1016/j.mtcomm.2023.107711

MXene-Based Advanced Materials: Mechanical and Catalytic Properties and Their Key Applications

Drisya Mottammal, Susmi Anna Thomas, Jayesh Cherusseri, Asheesh Kumar, and Deepthi N. Rajendran

Abstract Environment, water, and energy are mandatory for living things in our Earth today and tomorrow. Conservation of all the natural resources for future generation is the duty of every individual living today. Discovery of different types of materials and technologies all over the world made the life of human being easier than in the past. Day by day increasing environmental pollution damages our nature very hard in many ways. Carbon emission during burning of fossil fuels is highly dangerous and should be controlled for better nature around us. Hydrogen energy is a good option for reducing the emission of carbon to atmosphere. Varieties of newly emerged materials are used as catalysts for efficient hydrogen production through photocatalytic and electrocatalytic water splitting. Still now, large scale production of hydrogen via the green route is in the infant stages. The use of two-dimensional layered MXenes as catalysts for green hydrogen production is one of the effective methods and new ideas are emerging in this field every day. Peculiar properties of MXenes like high surface area to volume ratio, electrical conductivity, mechanical properties, and catalytic properties are favourable for photocatalysis and electrocatalysis. Photocatalytic dye degradation is another use of MXenes-based advanced materials for waste water treatment. Thus, the properties of MXenes-based advanced materials along with different catalytic applications are included in this chapter in detail.

Keywords MAX phase · MXene · Catalyst · Photocatalysis · Electrocatalysis

---

D. Mottammal · S. A. Thomas · D. N. Rajendran (✉)
Department of Physics, Government College for Women (Affiliated to University of Kerala), Thiruvananthapuram, Kerala, India
e-mail: deepthinrphysics@gmail.com

J. Cherusseri
Department of Chemistry (BK21 FOUR), Research Institute of Advanced Chemistry, Gyeongsang National University, Jinju 52828, Republic of Korea

A. Kumar
Department of Mechanical Engineering, Mahatma Gandhi Institute of Technology, Hyderabad, Telangana, India

© The Author(s), under exclusive license to Springer Nature Singapore Pte Ltd. 2025

S. Moharana and R. N. Mahaling (eds.), *Surface Engineering of MXenes*, Engineering Materials, https://doi.org/10.1007/978-981-96-3640-2_5

1 Introduction

Environmental pollution and energy scarcity are the two fundamental issues faced today. Clean environment and energy sources are a must for human beings to sustain life on Earth and is duty of every individual to conserve nature and energy resources for the next generation. Both clean environment and green energy resources are hot topic among researchers and they are always trying to solve difficulties for associated in this field [1–5]. Nowadays, most of the energy demand is satisfied from non-renewable energy resources like coal, natural gas, crude oil, and nuclear energy [6–11]. As we all know, these resources are not readily replaceable and search for new source of energy is necessary to satisfy energy demands [12–16]. Hydrogen is such an ideal fuel currently using in diverse fields since it is environmentally friendly when it is produced from water [17–21]. Using photocatalytic and electrocatalytic water splitting, hydrogen fuel can be produced with zero carbon emission [22–26]. When we consider environmental pollution, carbon dioxide emitted during the burning of carbonaceous fuel is dangerously increasing day by day. To limit this carbon emission and to protect the environment from natural calamities as a result of global warming, hydrogen fuels are the best option [27, 28]. Along with energy pure water is also important for every living thing on Earth. Water scarcity is a big problem today itself. Overuse of water resources, fertilizers, and pesticides, lack of waste water treatment, etc., are some of the reasons for water scarcity. Among them, waste water treatment has prime importance because it will damage water bodies, living organisms in water bodies, and finally affect human being also. Thus, waste water should treat properly before entering rivers. Photocatalysis and electrocatalysis are the two possible options to remove the chemical contaminants from the water. Advancements in the field of nanotechnology enable the synthesis of novel nanomaterials for a variety of applications [29–38]. These include carbon nanomaterials [39–41], transition metal dichalcogenides [42–48], polymers [49, 50], carbon nitride [51, 52], etc.

MXenes are 2-dimensional (2D) layered transition metal carbides, nitride or carbonitrides that emerged in material science world in 2011 [53–55]. $Ti_3C_2T_x$ was the first member of this ever-growing MXene family discovered at Drexel University [56, 57]. MXenes are derived from their precursor MAX phase generally via wet-chemical exfoliation methods [58]. MAX phase has the general formula $M_{n+1}AX_n$ ($n = 1, 2$ or 3), where M is the early transition metal (Ti, V, Cr, Nb, Mn, Ta, Zr, etc.), A is the element from groups 13, 14, and 15 (Al, In, Si, Ge, Pb, Sn, As, etc.) and X represents either C or N [59]. MAX phases have hexagonal layered structure with space group $D_{6h}^4-P6_3/mmc$ and where layer of A is sandwiched between M_{n+1} and X_n slabs. Bonding between M and A in MAX phase is metallic in nature and is very strong whereas bonding between M and X can be covalent or ionic or metallic in nature [60, 61]. Strength of metallic bonding between M–X is comparatively weaker than the bonding between M–X is favourable for the etching of layer A to generate MXenes. MXenes have the general formula, $M_{n+1}X_nT_x$. Where T represents the surface termination groups like –OH, –F, –Cl, –O etc. and x is the number of terminations [62, 63]. Hydrophilicity nature exhibited by surface of MXenes have resulted from these surface termination groups. Stability of MXene depends on value 'n', if n is high then MXene will be more stable. $Ti_3C_2T_x$ is the first MXene that originated from its parent MAX phase Ti_3AlC_2 by taking hydrofluoric acid as an etchant [64]. There are more than 60 MAX phases currently known to exist and the utilization of new synthesis approaches invented more than 200 stable MXenes such as $Ti_3C_2T_x$, Ti_2CT_x, TiN_3T_x, $Zr_3C_2T_x$, Nb_2CT_x, Mo_2CT_x, $Ta_4C_3T_x$, etc. [63, 65].

Superior physical and chemical properties including maximum electrical conductivity, larger surface area to volume ratio, diverse surface mechanisms, excellent electrochemical features, mechanical stability, optical properties, and hydrophilicity of MXenes grabbed the attention of researchers from 2011 onwards [66, 67]. MXenes find enormous applications in diverse filed like energy storage, catalysis, supercapacitor, sensors, electromagnetic interference shielding, water purification energy conversion, etc. [68–73]. There is a report in which Ti_3C_2 MXene is employed as co-catalyst along with $CdLa_2S_4$ for photocatalytic hydrogen generation [74]. $CdLa_2S_4$/Ti_3C_2 nanocomposite is prepared by growing $CdLa_2S_4$ on MXene surfaces. BY varying the wt% of MXene in the composite, different samples are prepared and tested for hydrogen evolution. $CdLa_2S_4$/Ti_3C_2 nanocomposite with 1.0 wt.% of Ti_3C_2 has the maximum rate of hydrogen production (11182.4 μmol g^{-1} h^{-1}). This rate is almost 13.4 times more than that of the hydrogen production rate of pure $CdLa_2S_4$. $CdLa_2S_4$/Ti_3C_2 photocatalysts also exhibit excellent stability even after six cycles of continuous use. Superior electrical performances of the Ti_3C_2 MXenes result into the faster charge separation and enhance the rate of hydrogen generation. Meng et al. fabricated barium titanate/MXene composite via facile route and used as piezo-photocatalyst for photocatalytic dye degradation [75]. Synergistic action of tetragonal barium titanate and two dimensional MXene nanosheets in barium titanate/MXene composite results into higher electrical conductivity and energy conversion rate during oxidation process. The composite can degrade 90% of Rhodamine B (RhB) in waste water within 15 min. Increased number of reactive sites on the surface of the barium titanate/MXene composite adsorb and degrades the dye molecules easily. The contact between barium titanate and MXene helps for efficient electron transfer and the photo response capacity and piezoelectric properties are responsible for the improved catalytic action of the composites. Jun and his team used $Ti_3C_2T_x$ MXene as sonocatalyst for dye degradation during ultrasonic treatment [76]. Ultrasonic treatment of wate water is oxidative process using hydroxyl radicals produced via decomposition of water molecules. They have tried $Ti_3C_2T_x$ MXene for the first time to degrade organic dyes (methylene blue (MB) and acid blue 80) present in waste water. The synergistic indices for removal of organic dyes are more than 1 in all the situations. The degradation kinetics follow pseudo-first order and the MXene based sonocatalyst have excellent reusability.

In this chapter, the first section discusses different MXene-based advanced materials and their characteristics and features. The peculiar features including both catalytic and mechanical properties of MXenes and its composites are described in the second section. And the application section contains diverse field in which

MXenes are employed as catalyst. Catalytic applications of different MXenes and MXene-based advanced materials are included in the last section.

2 MXene-Based Advanced Materials

Removal of 'A' element selectively from the parent MAX phase produces MXenes with structural formula $M_{n+1}X_nT_x$. Elements in the periodic table used for the construction of the MAX phases are shown in Fig. 1. Max phases exist in three different ways and have hexagonal closed pack structure. In MAX phase, A element is sandwiched between closely packed M layers, and the X atoms are occupied at the octahedral sites. MAX phases can exist in three ways depending upon the value of n (n = 1, 2 and 3). M_2AX (n = 1), M_3AX_2 (n = 2), and M_4AX_3 (n = 3) are the three MAX phases and are included in Fig. 2. Out of these MAX phases, M_2AX is the most investigated one. MXenes are derived from these MAX phases via removal of A layer and separation of MX-slabs resulting into M_2X, M_3X_2, and M_4X_3 phases. The structural transformation of MAX into MX phase is shown in Fig. 2. For M_2X phase, the layers of transition metal atoms are arranged in the pattern ABABAB with hexagonal close packing structure, and in M_3X_2 and M_4X_3, layers are ordered by following the ABCABC pattern and have face-centred cubic structure [77, 78].

The top view of monolayer of M_3X_2 is presented in Fig. 3a. MXenes can exist in $[MX]_nM$ arrangement also with two M elements. In this layered structure, X layers are sandwiched between layers of M. Such double MXenes with two M elements exist in two different ways. One is solid solution phase and the other one is ordered phase. In solid solution MXenes, M layer is formed by the random arrangement of two different transition metals while, in ordered phase, a sandwiched structure is formed by intercalating single or double layer of one transition metal with other layer of transition metal. $(Ti, Nb)_2C$, $(Nb, Zr)_4N_3$, and $(Ti, V)_3CN$ are some of the examples for solid solution phase of MXenes. (Mo_2Nb) CN, (Mo_2V_2) C_3, and (Cr_2Ta) N_2 are some of the examples for ordered phase MXenes [78]. Stability is more in ordered phase MXenes compared to solid solution MXenes according to

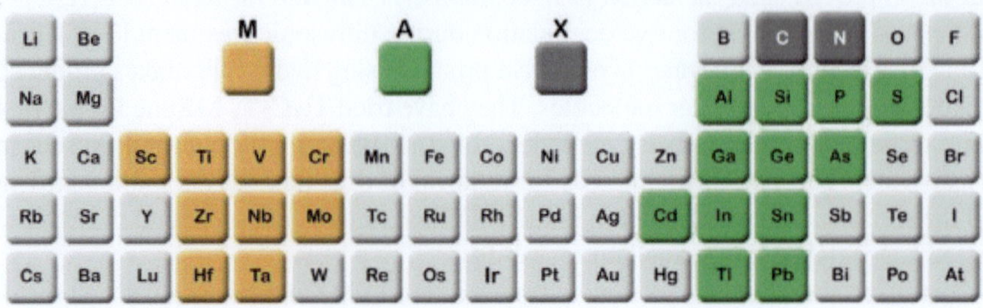

Fig. 1 Elements in the periodic table utilized for the preparation of MAX phase. Reproduced with permission from [79]. Copyright (2020) John Wiley & Sons Australia

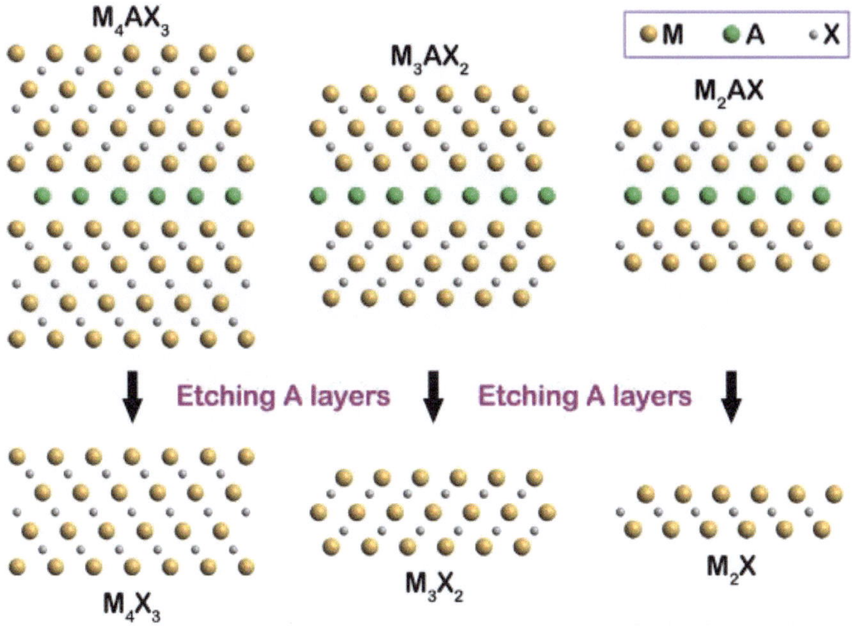

Fig. 2 Structural representation of three MAX phases and the corresponding MXene phases. Reproduced with permission from [79]. Copyright (2020) John Wiley & Sons Australia

density function theory [80]. Molecular structure model of a single layer $M_3X_2T_z$ is depicted in Fig. 3b.

Number of MXenes have been experimentally synthesized and studied from 2014 onwards and in which Ti_3C_2, Ti_2N, Nb_4C_3, Nb_2C, Mo_2C, and V_4C_3 are some of the most investigated MXenes. There is a report in which, Halim and team fabricated transparent conducting 2D-Titanium Carbide epitaxial thin film via selective etching of aluminium from Ti_3AlC_2 film [81]. Hydrofluoric acid (HF) and ammonium bifluoride (NH_4HF_2) are selected as etchants for the preparation of Ti_3C_2 MXene from

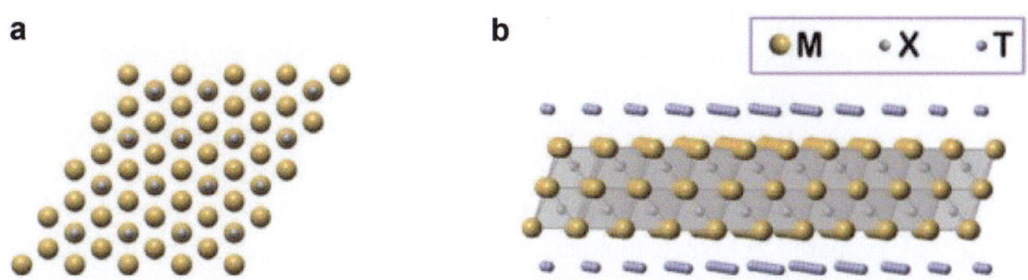

Fig. 3 **a** Top view of the M_3X_2 monolayer; **b** molecular structure model of single layer $M_3X_2T_z$. Reproduced with permission from [79]. Copyright (2020) John Wiley & Sons Australia

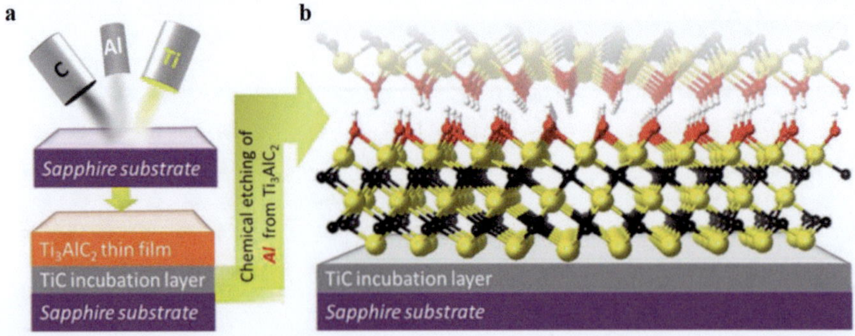

Fig. 4 **a** Formation of nanometre sized TiC incubation layer on sapphire substrate (0001) and deposition of Ti_3AlC_2 over incubation layer; **b** schematic representation of OH-terminated Ti_3C_2 MXene after etching of Al. Reproduced with permission from [81]. Copyright (2014) American Chemical Society

Ti_3AlC_2 film. Preparation of epitaxial Ti_3C_2 MXene film involved different steps and are presented in Fig. 4a, b.

Preparation of Ti_3C_2 MXene starts from the sputter-deposition of Ti_3AlC_2 followed by etching of the layer Al and results into the 2D-$Ti_3C_2T_x$, where x-coming from the aqueous HF etchant and it can be any of the surface termination group –O, –OH or –F. HF is the only etchant tried for the preparation of MXenes before this report but here they used NH_4HF_2 as etchant for the MXene synthesis. NH_4HF_2 is less hazardous and mild etchant compared to HF is attracting factor. X-Ray diffraction (XRD) pattern of the Ti_3AlC_2 confirmed the epitaxial growth of it and is further confirmed from transmission electron microscopy (TEM) analysis. In $Ti_3C_2T_x$, there is an increase in lattice parameter and arises from the shift in 2θ value from 18.6 (Ti_3AlC_2) to 19.8 Å ($Ti_3C_2T_x$). X-ray photoelectron spectroscopy (XPS) revealed the change in the bonding nature of Ti and C atoms in Ti_3AlC_2 and $Ti_3C_2T_x$. In MXene, the valance electrons of Ti and C atoms are removed by the effect of surface functional group and the surface intercalation effect. The elimination of Al from the MAX phase is confirmed from the high-resolution spectrum. Utilization of NH_4HF_2 instead of HF led to the formation of $(NH_4)_3AlF_6$ and the intercalation of NH_3 and NH_4^{+1} with MXene layers is investigated from the XPS spectra of N1s. Absence of surface termination groups like O, H, and F again confirmed the 2D structure formation of MXene. Thickness of the formed MXene film is identified from TEM images and found that atomic layers in $Ti_3C_2T_x$-IC (etched by NH_4HF_2) are uniformly spaced compared to the layers of $Ti_3C_2T_x$. This reflects the mild nature of the NH_4HF_2 etchant. Transparency of $Ti_3C_2T_x$ and $Ti_3C_2T_x$-IC is higher than that of the Ti_3AlC_2.

Wu et al. prepared 2D-Nb_2C nanosheets through a modified chemical exfoliation method by selecting HF as the etchant along with tetramethylammonium hydroxide as intercalating agent [82]. A schematic representation of the synthesis procedure is shown in Fig. 5.

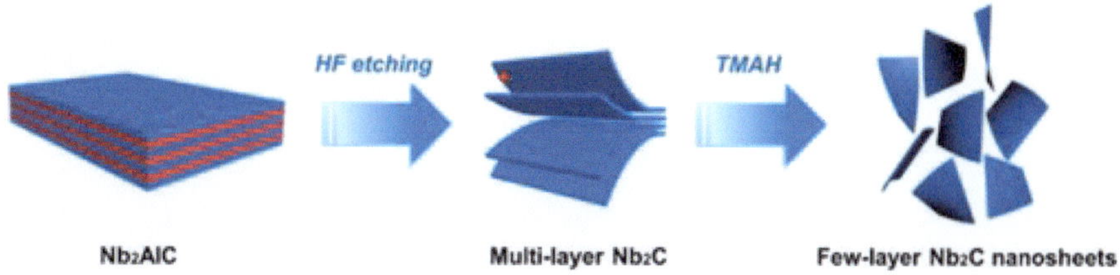

Fig. 5 Schematic representation of the synthesis of layered Nb_2C MXene. Reproduced with permission from [82]. Copyright (2021) American Chemical Society

XRD pattern of MAX phase Nb_2AlC, Nb_2C after HF treatment and intercalation are depicted in Fig. 6a. The absence of peak corresponding to MAX phase in Nb_2C after HF treatment indicated the formation of MXene. A strong diffraction peak at 7.9^0 in HF treated Nb_2C sample is attributed to crystalline Nb_2C. The addition of intercalating agent further shifts the peak (002) and confirms the formation of few-layered nanosheets of Nb_2C with considerable interlayer distances. From atomic force microscopy (AFM), thickness of the Nb_2C nanosheet is evaluated and is 4.58 nm. TEM analysis (Fig. 6b) is utilized for confirming the formation of thin Nb_2C nanosheets and the lateral size of the sheets is found to be 400 nm to 1.5 μm. Homogeneous distribution of Nb and C in the Nb_2C nanosheets is verified from energy dispersive X-ray spectroscopy (EDS) measurements.

Ahn et al. synthesized single-layered 2D-Ti_3C_2 MXene film via spin-casting method along with low temperature vacuum-annealing procedure for the fabrication of transparent conducting electrode [83]. They obtained higher electrical conductivity (11668 Scm^{-1}), optical transmittance (85%), and work function (5.1 eV) for the prepared electrode. A schematic representation of the synthesis procedure of Ti_3C_2 MXene is depicted in Fig. 7a.

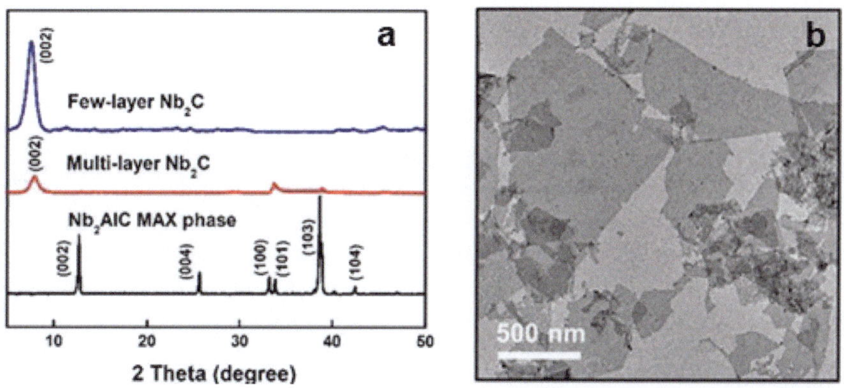

Fig. 6 **a** XRD pattern of Nb_2AlC, multi-layered Nb_2C, and few layered Nb_2C; **b** TEM image of Nb_2C nanosheets. Reproduced with permission from [82]. Copyright (2021) American Chemical Society

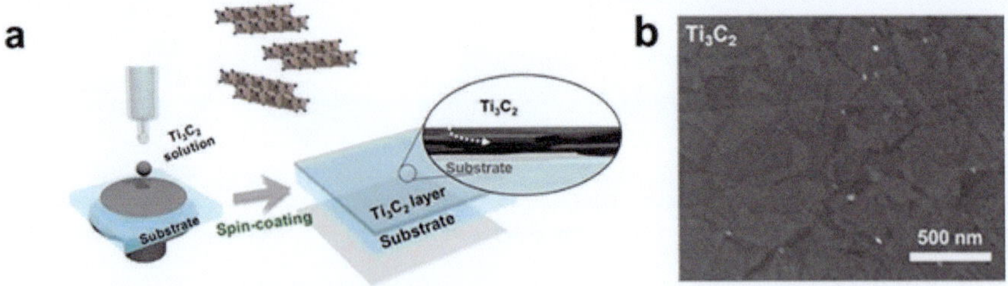

Fig. 7 **a** Schematic representation of the synthesis procedure of single-layered 2D-Ti_3C_2 MXene film; **b** TEM image of the Ti_3C_2 MXene. Reproduced with permission from [83]. Copyright (2020) John Wiley & Sons

As synthesized MXene film has a thickness of around 1 nm and is near to the theoretically calculated thickness of the monolayer of Ti_3C_2. TEM image of the Ti_3C_2 MXene nanosheet is presented in Fig. 7b and the layers are overlapped to form the nanosheets. By altering the thickness of the film from 29 to 7.9 nm, sheet resistance is tuned from 98 to 115 Ω sq^{-1}. They also compared the surface work function of synthesized Ti_3C_2 MXene and indium tin oxide electrode film by analysing the spatial surface potential. MXene film has a slightly higher work function (5.0 eV) compared to indium tin oxide (4.8 eV). The solution preparation technique eliminates the presence of surface functional group such as –OH, –Cl, –F, and –O on MXenes, Density functional theory predicated that the change in work function is associated with the presence of surface functional group. Work function for Ti_3C_2 MXene will range from 1.6 eV with –OH termination group up to 6.25 eV with –O surface termination group. Vacuum annealing is employed for the further removal of surface functional group in order to improve the work function of the MXene film. XPS is utilized for finding the presence of functional group after vacuum annealing. Increase in annealing temperature from 100 to 200 °C slightly increases the work function of the MXene to 5.1 eV. But further increases in annealing temperature (250 °C) decreases the work function of the MXene film due to the high-temperature induced –F termination. Selecting this solution processed MXene film as transparent conducting electrode, this team fabricated phosphorescent organic LEDs with maximum current density 102 Cd A^{-1} and maximum power efficiency 103.7 Im W^{-1}. The external quantum efficiency is calculated to be 28.5% for this fabricated organic LED.

Liang and team synthesized new 2D-titanium carbonitride ($Ti_2C_{0.5}N_{0.5}T_x$) MXene, which is the second carbonitride MXene after Ti_3CNT_x [84]. They synthesized $Ti_2C_{0.5}N_{0.5}T_x$ MAX phase, powder of multilayered $Ti_2C_{0.5}N_{0.5}T_x$, and monolayered nanosheets of $Ti_2C_{0.5}N_{0.5}T_x$ via in situ HF etching procedure. Multilayered $Ti_2C_{0.5}N_{0.5}T_x$ has a specific capacity of 182 $mAhg^{-1}$ at 20 mAg^{-1}, which is the highest capacity achieved among the reported multilayered MXenes. Structural and morphological features of the MXenes are identified from XRD, AFM, TEM, and SEM procedures. XRD pattern of the Ti_2AlCN MAX phase, multilayered $Ti_2C_{0.5}N_{0.5}T_x$, and $Ti_2C_{0.5}N_{0.5}T_x$ paper are depicted in Fig. 8a. Aqueous HF etching is initially employed for the preparation of MXenes but due to the weight loss at lower concentration, they switched etching procedure into in situ HF etching. For that HCl/KF mixture is used as etchant for the preparation of carbonitride MXene. Broadening and lower angle shift in (002) peak of MXene phase compared to MAX phase indicated the successful elimination of Al to form layered $Ti_2C_{0.5}N_{0.5}T_x$ MXene. In MAX phase, (002) peak is located at 13.13° while it is observed at 7.42° for MXene. Reasonable shift in this peak position revealed the increase in interlayer distance in $Ti_2C_{0.5}N_{0.5}T_x$ MXene. Intercalating effect of water, cations, and other etching products is the reason for improved d-spacing in MXene in comparison with MAX phase. EDS is used for the final confirmation of formation of MXene from the MAX phase along with the removal of surface termination elements. Accordion-like morphology is missing in the SEM images of the $Ti_2C_{0.5}N_{0.5}T_x$ MXene prepared via HF etching. Large-scale delamination is achieved in MXene with the help of intercalating agents. Tetramethylammonium hydroxide is the intercalating agent selected here to get maximum layer expansion. Thickness of the MXene nanoflake is around 2.5 nm for 2–3 layers from the AFM analysis (Fig. 8b) and which confirmed the formation of few layered MXene. TEM images (Fig. 9a) of the $Ti_2C_{0.5}N_{0.5}T_x$ MXene nanosheets are recorded and confirmed the formation of few layered MXene and a hexagonal symmetry of the planes of MXene is verified from the selected area diffraction (SAED) pattern (provided in the inset image).

Figure 9b shows the cross-sectional SEM images of the $Ti_2C_{0.5}N_{0.5}T_x$ MXene paper after filtration of the colloidal solution with an inset image shows the photograph of MXene paper. Electrical conductivity of the $Ti_2C_{0.5}N_{0.5}T_x$ MXene with thickness 2.5 μm is around 435 Scm^{-1}. Thus, selection of HCl/KF mixture as etchant along with delamination is an interesting procedure to obtain MXenes with better electrical conductivity, electroactive surface area, and faster ion transport for high specific capacity. Here, the authors of this work used this MXene as electrode material in sodium-ion battery and achieved a specific capacity of 182 $mAhg^{-1}$. These electrodes also exhibit long-term stability for 500 cycles.

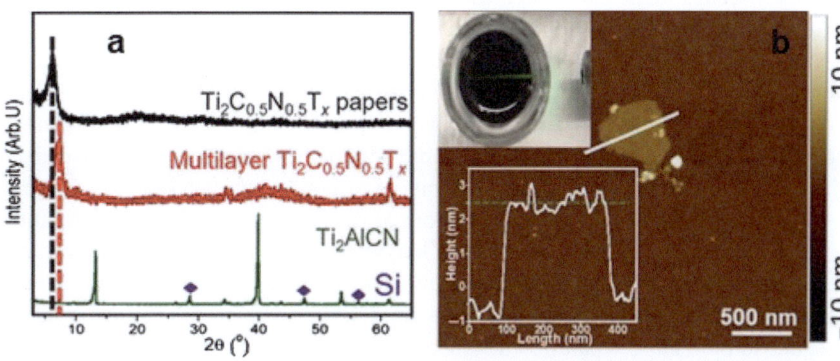

Fig. 8 **a** XRD pattern of the Ti_2AlCN MAX phase, multilayered $Ti_2C_{0.5}N_{0.5}T_x$ and $Ti_2C_{0.5}N_{0.5}T_x$ paper; **b** AFM image of the $Ti_2C_{0.5}N_{0.5}T_x$ MXene coated over a silicon wafer. Reproduced with permission from [84]. Copyright (2021) UESTC and John Wiley & Sons Australia, Ltd.

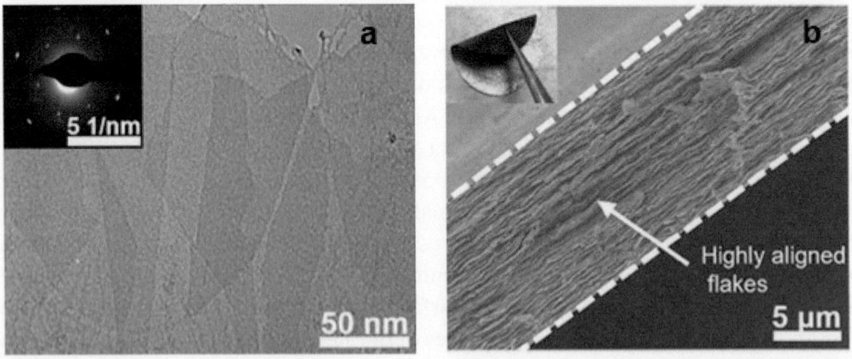

Fig. 9 **a** TEM images of the $Ti_2C_{0.5}N_{0.5}T_x$ MXene nanosheets (inset image shows the SAED pattern); **b** cross-sectional SEM images of the $Ti_2C_{0.5}N_{0.5}T_x$ MXene paper (inset image shows the photograph of MXene paper). Reproduced with permission from [84]. Copyright (2021) UESTC and John Wiley & Sons Australia, Ltd.

3 Properties of MXenes

3.1 Mechanical Properties of MXenes

Mechanical properties of MXenes are crucial in finding different field of applications even though, there are lot of things to be explored in mechanical behaviour of MXene. Combination of MXene (M–C or M–N or MCN), their structure, and surface termination groups attached are the fundamental factors in determining the mechanical properties of the MXenes. Density function theory is one of the tools used by researchers to identify the mechanical behaviour of MXenes. Number of literatures reports available in this area is very less. Thus, to understood the mechanical features MXenes ideas from three-dimensional crystalline binary transition metal carbides and nitrides are used along with the density function theory. In both MAX and MXene phases, C and N occupy their octahedral interstitial sites. Their (00l) plane in hexagonal closed pack structure is almost same to that of the (111) plane of M–X-rock-salt structure. The nature of bonding existing among transition metal and carbon or nitrogen can be covalent, ionic or metallic depending upon the d and s orbitals of transition metal and the p-orbitals of carbon or nitrogen [85, 86]. There is an ionic bonding between the electrons from M to X and resulted into the formation of covalent bonding in M–X structure. Geometry of the transition metal carbide/nitride/carbonitride crystal structure largely depend on the σ and π bond among the transition metal carbide and nitride. The bonding between two X atoms is negligible due to their large separation distances and resulted into weak interactions. The metallic bonding existing among transition metals through d-orbitals does not exist between carbon or nitrogen. On comparing the bond strength between M–X and M–M, former contributes more to the lattice bond energy [87]. Chakraborty and his team analysed the change in mechanical properties of Ti_2C MXene during alloying with boron (B) and vanadium (V) at Ti and C sites of Ti_2C respectively [88]. Theoretical calculations are performed using density function theory with the help Vienna Ab initio Simulation Package. Before analysing the mechanical property analysis of Ti_2C MXene, they investigated the electronic and magnetic structure of pure Ti_2C and $Ti_2(C_{0.5}, B_{0.5})$ and (Ti, V) C. They observed that intralayer and interlayer Ti–Ti coupling is respectively ferromagnetic and antiferromagnetic in nature. Analysis of total number of density states revealed the half metallic nature of Ti_2C MXene in ferromagnetic structure and it is exhibiting semiconducting behaviour with band gap around 0.2 eV in antiferromagnetic structure. Mechanical property variations in both pure and doped MXenes are identified from the stiffness, Young's modulus and critical strain calculated using the first principal techniques. From stress–strain curves for pure Ti_2C MXene and non-magnetic Ti_2C MXene, Ti_2C possess multiplanar hexagonal structure and belongs to quasi-2D. There is a general classification for 2D materials, in which materials belong to either of four classes named class A, B, C and D. Ti_2C MXene-based compounds belong to class D while graphene-based compounds are belonging to either class A or class B. To specify the elastic properties of qusi-2D materials belonging to class D, two Youngs modulus (E_1 and E_2) and two Poisson's ratio (v_{12} and v_{21}) are required. Zig-zag and armchair directions are respectively represented using 1 and 2. Density function theory analysis shows that Ti_2C MXene has greater E_1 compared to E_2 and E_1 and E_2 are differ from each other by 3–4%, thus they are nearly isotropic in nature. Poisson's ratio of Ti_2C system is also investigated and there is 3–5% difference between v_{12} and v_{21}. The difference in stress–strain curve for pure and B/V-doped Ti_2C is investigated and found that there exists a significant variation in elastic properties of both the materials. Young's modulus and the in-plane stiffness of $Ti_2(C0.5, B0.5)$ is decreased by 25–27% compared to Ti_2C. This decrease in stiffness in B-doped MXene is due to the weak Ti-B bonding compared to Ti–C bonding. Stress–strain curve for Ti_2C MXene is linear in small strain region and is non-linear in high-strain region. This will impart an anisotropic behaviour to mechanical properties and different critical strain (ε_{c1}) values in biaxial and uniaxial tensions along X and Y directions. Critical strain value is the maximum strain attained and after that the stress gets reduced and reached at instability. Biaxial critical strain for Ti_2C, TVC, and TBC are respectively 8, 11, and 6. Whereas uniaxial strain along X axis is 13, 18, and 14 for Ti_2C, TVC and TBC are respectively. It is 14, 18, and 15 along uniaxial Y axis for Ti_2C, TVC, and TBC, respectively.

Lee et al. studied the mechanical properties of pristine and functionalized $Ti_{n+1}C_nO_2$ and $Ti_{n+1}N_nO_2$ (n = 1, 2) MXenes using classical molecular dynamics simulations [89]. Molecular dynamic simulation is the widely used investigation method for understanding mechanical properties of materials. Here, this team used LAMMPS package for the analysis of mechanical properties of different MXenes. The selected eight MXenes from both titanium carbide and nitride family for mechanical property investigation. Ti_2N, Ti_2NO_2, Ti_3N_2, $Ti_3N_2O_2$, Ti_2C, Ti_2CO_2, Ti_3C_2, and $Ti_3C_2O_2$ are the selected MXenes for dynamic study. A schematic representation of the atomic structure of titanium nitride-based MXenes is shown in Fig. 10. The structure exhibits isotropic nature along zigzag and armchair directions.

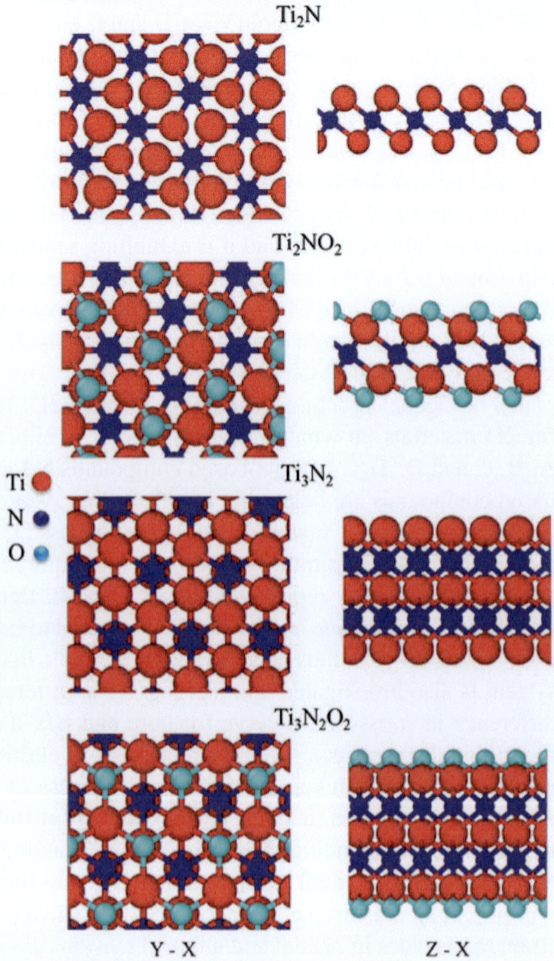

Fig. 10 Schematic representation of titanium nitride MXenes (Ti_2N, Ti_2NO_2, Ti_3N_2, and $Ti_3N_2O_2$). Y-X and Z-X plane are represented in left and right column, respectively. Reproduced with permission from [89]. Copyright (2021) Elsevier Inc.

Stress–strain curves for titanium nitride and carbide MXenes are depicted in Figs. 11 and 12, respectively. In titanium nitride MXenes, the addition of oxygen into Ti_2N and Ti_3N_2 produces 35 and 27% reduction in fracture stress and 36 and 24% decline in Toung's modulus respectively. There is no observed change in fracture strain. While in titanium carbide MXenes, addition of oxygen into Ti_2C and Ti_3C_2 increases Young's modulus by 20 and 42% and decreases fracture strain by 33 and 62% respectively. Stress–strain curve is also utilized to extract the mechanical parameters of different MXene structures. Highest Young's modulus (520 GPa) is obtained for Ti_2N in armchair direction and lowest is obtained for Ti_3C_2 (135 GPa) along armchair direction.

Comparison between Young's modulus of different MXenes revealed that titanium nitride-based MXenes are superior than that of titanium carbide-based MXenes. Analysis of fracture stress values revealed that nitrogen-based MXenes are exhibiting

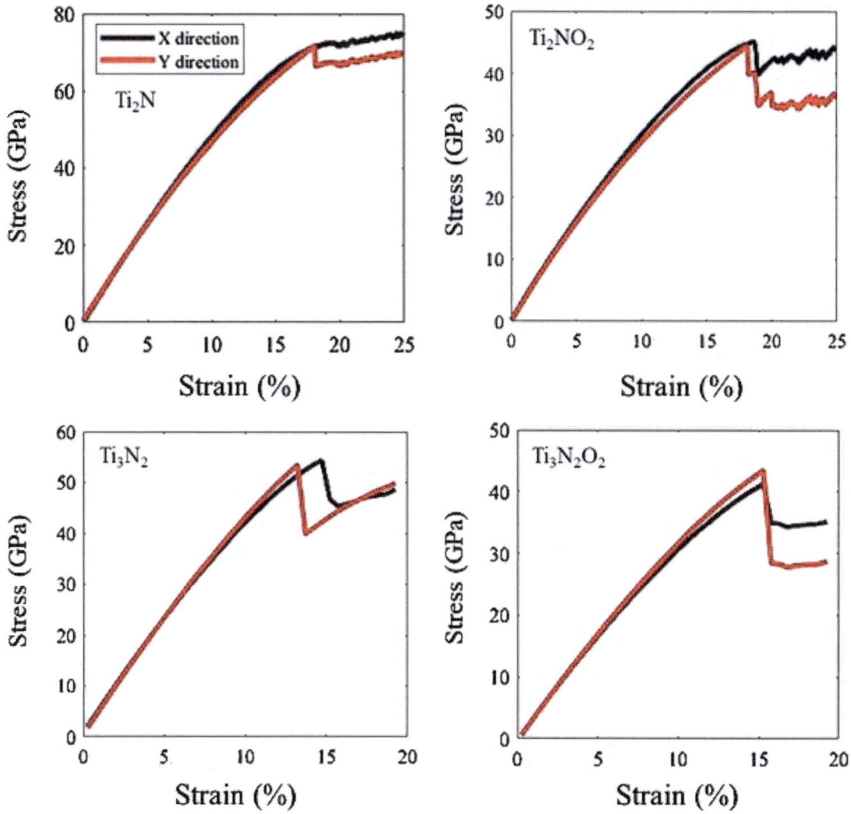

Fig. 11 Stress–strain curve for titanium nitride MXenes (Ti_2N, Ti_2NO_2, Ti_3N_2, $Ti_3N_2O_2$) along armchair and zigzag directions. Reproduced with permission from [89]. Copyright (2021) Elsevier Inc.

superior behaviour compared to carbon-based MXenes. Another important factor noticed in this analysis is the effect of oxidation on both carbon and nitrogen MXenes. Oxidation increases the Young's modulus in carbon-based MXenes whereas, it decreases the Young's modulus in nitrogen-based MXenes. Defect analysis is very important in the case of two-dimensional materials. To identify the effect of defect in Young's modulus of MXenes, different percentages of vacancies are applied along both the armchair and zigzag directions of the MXene systems. Calculated Young's modulus of different systems is analysed and reached a conclusion that there is a linear downward trend shows the interconnection between titanium and nitrogen vacancies with Young's modulus. Also, the stretching on defective system creates failure and rupture of at the original vacant place. Existence of 10% defect in MXene system decreases the Young's modulus by 28%. Vacancy defect in titanium atom affect Young's modulus more than nitrogen atom due to the presence of titanium atom in the middle layer of structure of the MXene.

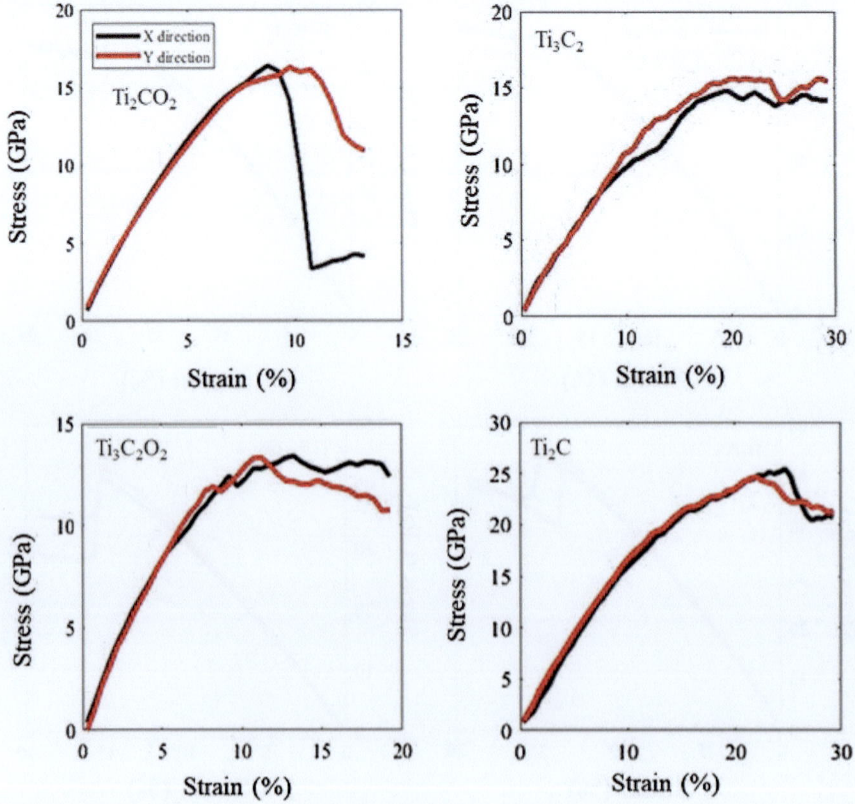

Fig. 12 Stress–strain curve for titanium carbide MXenes (Ti_2C, Ti_2CO_2, Ti_3C_2, and $Ti_3C_2O_2$) along armchair and zigzag directions. Reproduced with permission from [89]. Copyright (2021) Elsevier Inc.

There is another report in which Yan and his team fabricated $Ti_3C_2T_x$/poly (vinyl alcohol) (PVA) nanocomposite to study the thermal, mechanical and flame-retardant properties of it. MXene/PVA nanocomposites are prepared by the solution blending method [90]. Poly (vinyl phosphonic acid) (PVPA) and polyethylene polyamine (PEPA) are used as the flame retardants to enhance properties of the MXene nanosheets. By varying the wt% of PVA, PVPA, PEPA, and $Ti_3C_2T_x$, six different samples are prepared and property variations are investigated. SEM and TEM images (Fig. 13a, b) are utilized for the morphology analysis. Figure 13a and b are the images of multilayered stacked MXene after selective etching of Al from the MAX phase using LiF and HCl. Figure 13c is the single-layered MXene with hexagonal symmetry obtained after sonicating the multilayered MXene solution. TEM images of the single layered MXene nanosheet are shown in Fig. 13d and are an indication of the successful exfoliation of the nanosheets.

Fig. 13 SEM images of the **a**, **b** etched MXenes; **c** delaminated MXene; **d** TEM images of the delaminated MXene. Reproduced with permission from [90]. Copyright (2022) MDPI

The homogeneous dispersion of MXene, PEPA, and PVPA in PVA matrix is confirmed from the SEM images. Both PEPA and PVPA are also uniformly dispersed in PVA matrix. The absence of agglomeration in MXene/PVA nanocomposite is also evidence for the uniform dispersion. The mechanical properties of the PVA and the MXene/PVA nanocomposites are identified from the stress–strain curve and by finding the tensile strength, Young's modulus and elongation at break of the samples. Figure 14 shows the stress–strain curves obtained for the PVA and the MXene/PVA nanocomposites.

Addition of MXene into the polymer matrix enhances the tensile strength and Young's modulus of the PVA matrix. Tensile strength increases from 20 to 62.9% during the incorporation of MXene into the PVA matrix. Addition of MXene have the ability to distribute load on the PVA chain and helps to improve mechanical properties. In addition to this presence of hydrogen bonding delays the movement of polymer chain and enhances the tensile strength. Elongation at break of MXene / PVA nanocomposite is improved compared to PVA. Addition of 1.0 wt.% of MXene into PVA matrix improves the elongation at break by 318.8% and is about 16.3% more than that of the pure PVA. The presence of –OH, –F, and O functional groups in the system improves the integration among PVA and MXene materials and form a rigid network in the nanocomposite system. These factors enhance the mechanical behaviour of the composites.

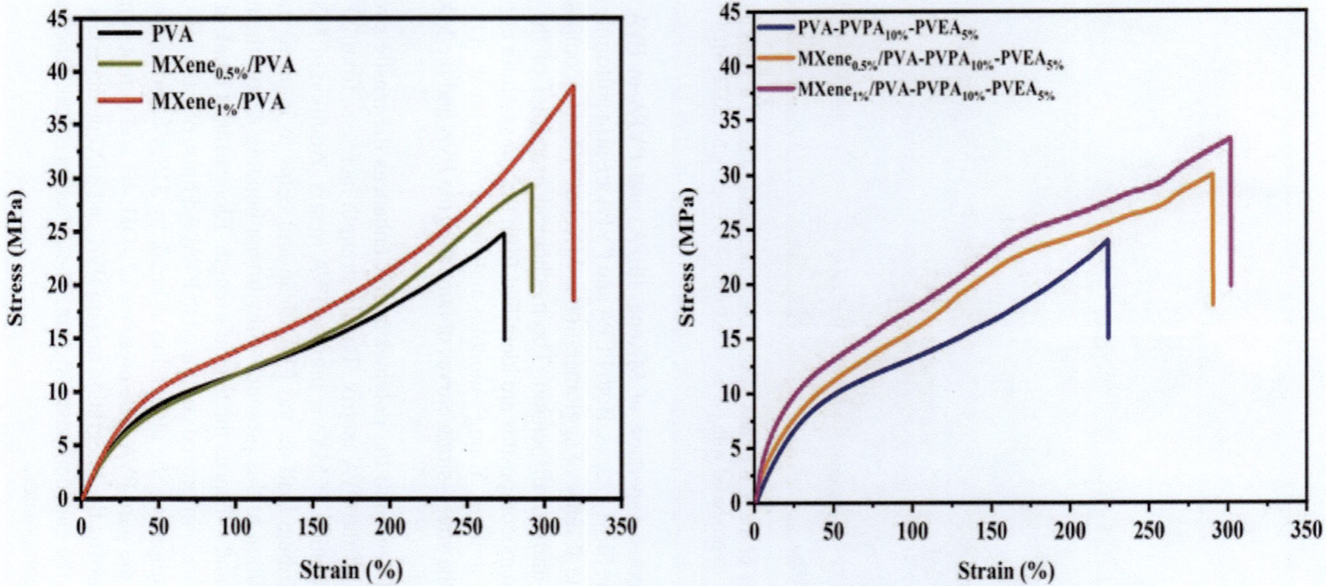

Fig. 14 Stress–strain curves for PVA and PVA/MXene nanocomposites. Reproduced with permission from [90]. Copyright (2022) MDPI

3.2 Catalytic Properties of MXenes

Two-dimensional MXenes have good electrical conductivity, high surface area to volume ratio, hydrophilicity, and easy functionalization. All these features largely enhance the catalytic properties of the MXenes. In earlier days, semiconducting materials such as ZnO, NiO, TiO_2, ZnS, CdS etc. [91] are the photocatalysts generally used for different purposes. Their low price, abundant sources, physical and chemical properties, oxidation capacity, and photocatalytic activity are the attractive features behind choosing them as catalysts. Even though, the large bandgap, higher recombination rate, and chance for forming agglomeration will reduces the catalytic activity. To overcome these difficulties, incorporation of semiconductor catalytic materials with two-dimensional material (MXenes, graphene, carbon nanotubes, and transition metal dichalcogenide) is one of the best ways. Thereby, we can reduce the charge recombination rate and can improve the speed of charge separation and will improve the catalytic properties of the systems.

Othman and his team fabricated silver (Ag) nanoparticles (NP) (AgNP)/TiO_2/$Ti_3C_2T_x$ MXene composites through hydrothermal synthesis route [92]. Ag as noble metal has good catalytic activity, electrical conductivity, and low cost made it suitable for incorporating with TiO_2 to improve the catalytic features. Palladium (Pd) is another noble metal element with remarkable ability to interact with surface of semiconducting catalytic materials. Thus, the incorporation of Ag and Pd along with TiO_2 and MXene produces drastic catalytic ability enhancement for the composite systems. For the preparation of AgNP/TiO_2/$Ti_3C_2T_x$ and PdNP/TiO_2/$Ti_3C_2T_x$ nanocomposite is prepared via hydrothermal method. Silver nitrate and palladium (II)acetate are noble metal precursors respectively for Ag and Pd. Both SEM and TEM images are used for the morphology analysis of the synthesized AgNP/TiO_2/$Ti_3C_2T_x$ and PdNP/TiO_2/$Ti_3C_2T_x$ nanocomposites. SEM image of the MXene have accordion-like multilayer structure which confirms the formation of MXene. In-situ ultrasonication route is used for the preparation of single-layered MXene sheet from the multilayered structure. Addition of TiO_2 into the MXene sheets enhances the surface roughness of the materials. For detailed analysis, TEM images of the TiO_2/$Ti_3C_2T_x$, AgNP/TiO_2/$Ti_3C_2T_x$, and PdNP/TiO_2/$Ti_3C_2T_x$ are depicted in Fig. 15a–c respectively, and the high-magnification images are provided as an inset. In Fig. 15a, the formation of TiO_2 nanocrystals on MXene surface is clear and it start from the edges and then covering the whole surfaces. TEM images of PdNP/TiO_2/$Ti_3C_2T_x$ composites consist of small amount of nanoparticle agglomeration on the surface whereas that of AgNP/TiO_2/$Ti_3C_2T_x$ contains different sized and shaped particles instead of agglomerative structure (Fig. 15b, c). For testing the photocatalytic ability of the as prepared catalysts, MB, and RhB dyes are selected and photocatalytic dye degradation is conducted under UV light irradiation for 15 min (Fig. 16a, b). Initially, they analysed the effect of catalyst to dye ratio on photocatalytic dye degradation capacity.

The materials exhibit 85% MB dye degradation at a low catalyst to dye ratio (1:1). The degradation percentage increses with increase in catalysts concentration and reaches a maximum at a ratio 10:1 in 15 min duration for MB. Whereas,

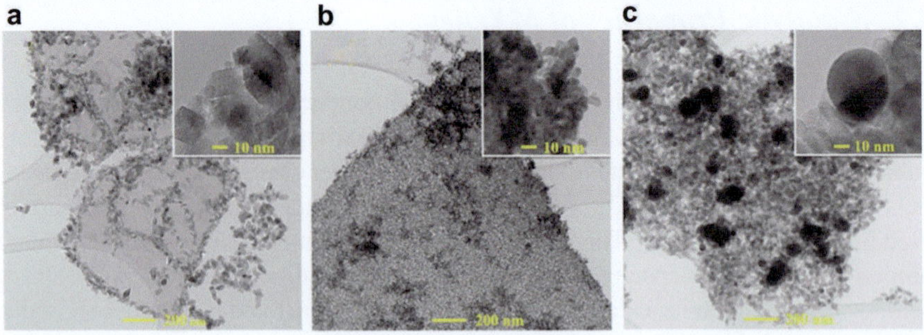

Fig. 15 TEM images (inset shows high-magnification image) of the **a** $TiO_2/Ti_3C_2T_x$; **b** AgNP/$TiO_2/Ti_3C_2T_x$ and **c** PdNP/$TiO_2/Ti_3C_2T_x$. Reproduced with permission from [92]. Copyright (2021) American Chemical Society

for AgNP/$TiO_2/Ti_3C_2T_x$, dye degradation (RhB) percentage for all the dosage is almost equal when compared to the $TiO_2/Ti_3C_2T_x$, PdNP/$TiO_2/Ti_3C_2T_x$ composites. The formation of anatase TiO_2 on the MXene surface improves the photocatalytic dye degradation efficiency of the composites. The surface plasmon resonance effect exhibited by noble metals is also a factor behind enhanced catalytic property of the composite system. The comparison between catalytic efficiency of AgNP/$TiO_2/Ti_3C_2T_x$ and PdNP/$TiO_2/Ti_3C_2T_x$ composites shows that former have the faster kinetics resulted from the size and distribution of the nanoparticles. The semi-spherical shape of the silver nanoparticle contributes more to the catalytic efficiency while the chance for nano agglomeration observed in palladium nanoparticle limit the catalytic performance of the PdNP/$TiO_2/Ti_3C_2T_x$ composite.

The authors of this work examined the effect of catalysts to dye ratio on catalytic performances of $Ti_3C_2T_x$, $TiO_2/Ti_3C_2T_x$, AgNP/$TiO_2/Ti_3C_2T_x$, and PdNP/$TiO_2/Ti_3C_2T_x$ composites for degradation of MB and RhB under solar light irradiation for 120 min (Fig. 17a and b). The results showed that AgNP/$TiO_2/Ti_3C_2T_x$ composite exhibits better performances compared to all the other composites. The efficiency of dye removal is more than 90% for AgNP/$TiO_2/Ti_3C_2T_x$ composite. In case of MB dye, highest degradation efficiency is obtained at a catalyst to dye ratio of 20:1 while, it is 30:1 for the RhB dye. As shown in UV irradiation results, here also the surface plasmon resonance effect of palladium and silver nanoparticles played an important role in enhancing the catalytic behaviour of the composite systems. For most of the MXene composites, the dye degradation efficiency is largely increased above a particular value of catalyst to dye ratio in both UV and solar light irradiation. The ratio is 2.5–5:1 for UV light and 10–20:1 for solar light illumination. Above these ratios, the catalytic dye removal performance of the composites is not improved due to the shielding effect of these two-dimensional materials.

This team also investigated the effect of solar irradiation time and photocatalytic kinetics on dye removal efficiency of $Ti_3C_2T_x$, $TiO_2/Ti_3C_2T_x$, AgNP/$TiO_2/Ti_3C_2T_x$ and PdNP/$TiO_2/Ti_3C_2T_x$ composites. For this, they selected the photocatalysts with catalyst to dye ratio 2.5:1 and 20:1 for UV light and solar light irradiation

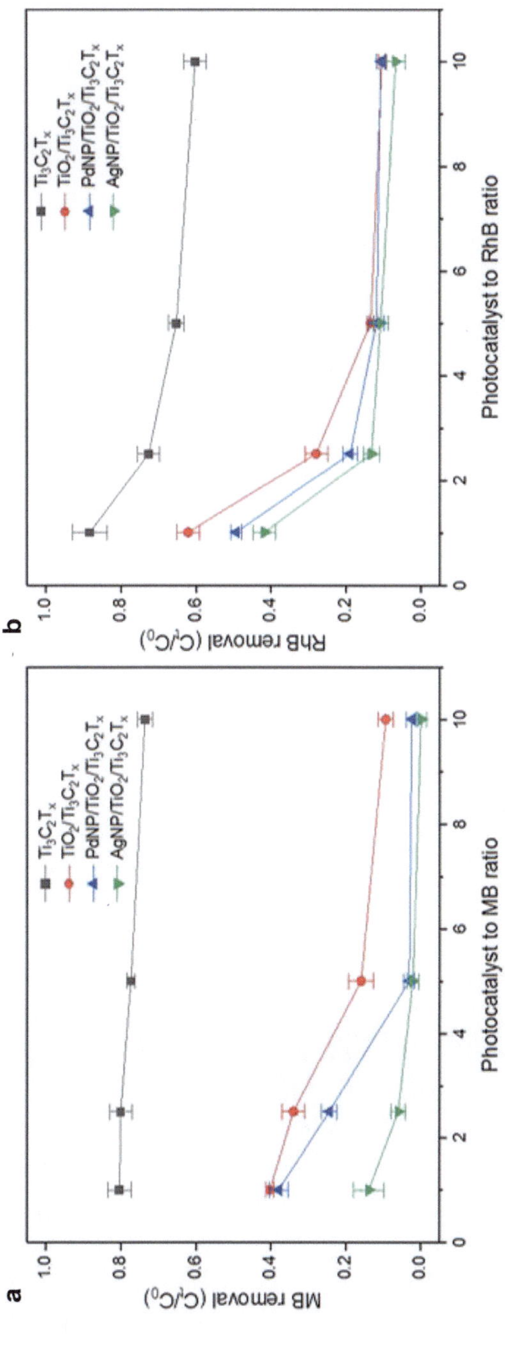

Fig. 16 Photocatalytic dye degradation capability of $Ti_3C_2T_x$, $TiO_2/Ti_3C_2T_x$, $AgNP/TiO_2/Ti_3C_2T_x$ and $PdNP/TiO_2/Ti_3C_2T_x$ under UV light irradiation by varying the amount of catalysts from 1 to 10 mg for **(a)** MB removal and **(b)** RhB removal. Reproduced with permission from [92]. Copyright (2021) American Chemical Society

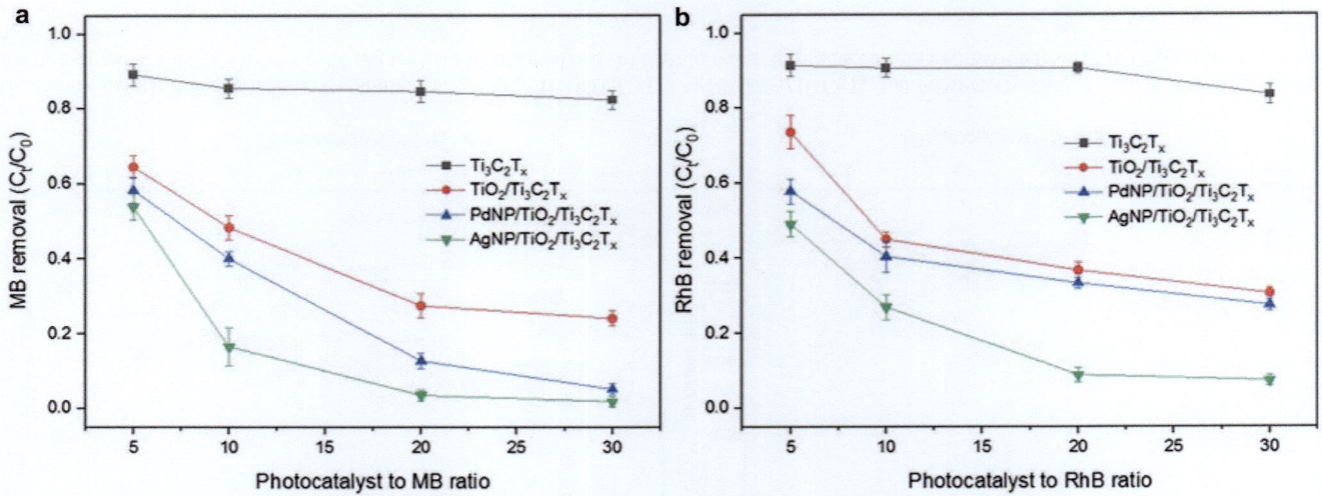

Fig. 17 Effect of catalyst to dye ratio on dye, **a** methylene blue (MB); and **b** Rhodamine B (RhB) degradation performance using $Ti_3C_2T_x$, $TiO_2/Ti_3C_2T_x$, $AgNP/TiO_2/Ti_3C_2T_x$, and $PdNP/TiO_2/Ti_3C_2T_x$ composites under solar light irradiation for 120 min. Reproduced with permission from [92]. Copyright (2021) American Chemical Society

respectively. Degradation time taken by the catalyst under sunlight irradiation is always more compared to UV light. This may be due to the delay in harvesting solar light. As UV light is provided by an experimental set up all the light can be easily captured by the system. The investigation of degradation kinetics of the catalysts under both UV and sunlight irradiation is conducted and found that AgNP/TiO_2/$Ti_3C_2T_x$ composite exhibits faster kinetics and degradation efficiency against MB and RhB dye. It is also noted that the degradation efficiency of AgNP/TiO_2/$Ti_3C_2T_x$ improved rapidly within 15 min and 60 min for UV and solar light irradiation, respectively. A similar response is obtained for other composites such as, $Ti_3C_2T_x$, TiO_2/$Ti_3C_2T_x$, and PdNP/TiO_2/$Ti_3C_2T_x$ also. As time proceeds, the concentration of dye decreases and will reduce the interaction between catalysts and dye also. The quantification of photocatalytic activity of AgNP/TiO_2/$Ti_3C_2T_x$ composite is identified using the Langmuir-Hinshelwood kinetics model and observed that the catalytic performance of AgNP/TiO_2/$Ti_3C_2T_x$ composite under UV and sunlight well matches with the pseudo-first-order kinetics. The degradation kinetics for MB is faster compared to RhB with AgNP/TiO_2/$Ti_3C_2T_x$ composites (Fig. 18a–c). This is due to the presence of nitrogen and sulfur atoms of the thiazine core in MB and can interact with metal atoms speedily. In case of RhB, which contains amino and carboxylic groups and are absorbed on metal surfaces and slow down the degradation kinetics. The higher negative charge density of MXene composites in an aqueous solution adsorb the cationic MB dye molecule easily. Whereas, the presence of positive and negative charges in structure of RhB reduces the electrostatic interactions. During the degradation of MB, the absorption peak at 664 nm reduces sooner than that of the absorption peak of RhB at 555 nm. Thus, the photodegradation follows a pseudo-first-order kinetics.

There is another report presented by Zhang et al. in which they incorporated TiO_2 into Ti_3C_2 MXene and degraded MB from waste water [93]. Hydrothermal synthesis technique is used for the preparation of TiO_2/MXene composite. Morphology of as synthesized MXene, TiO_2, and TiO_2/MXene composite are investigated using the SEM images as shown in Fig. 19a–d.

SEM images of MXene exhibit self-assembled book-like morphology and TiO_2 nanoparticle are spherical in shape with a diameter around 25 nm. In TiO_2/MXene composite, TiO_2 nanoparticles are homogeneously distributed on the surface of the MXene sheets. TEM images are also used for the verification of morphology of the TiO_2/MXene composite. TiO_2/MXene composite has exact crystal structure and has a lattice spacing of 0.35 nm and 1 nm corresponding to (101) plane of TiO_2 and (002) plane of MXene respectively. These observations confirmed the successful formation of TiO_2/MXene composites. Catalytic efficiency of the TiO_2/MXene composite in degrading MB is tested under three different conditions including dark, visible, and UV. Under dark condition, increases in mass ratio between MXene to MB from 0:10 to 1:3 increase the removal rate of MB from waste water from 2.26 to 14.37%. For visible and UV atmosphere, degradation percentage increases from 40.29 to 96.44% during the increases in mass ratio of MXene to TiO_2 to 1:4. These degradation percentage is around 9.26 and 1.44 times more than that of the degradation percentage of pristine TiO_2 nanoparticles. Reduction in band gap energy of TiO_2/MXene composite

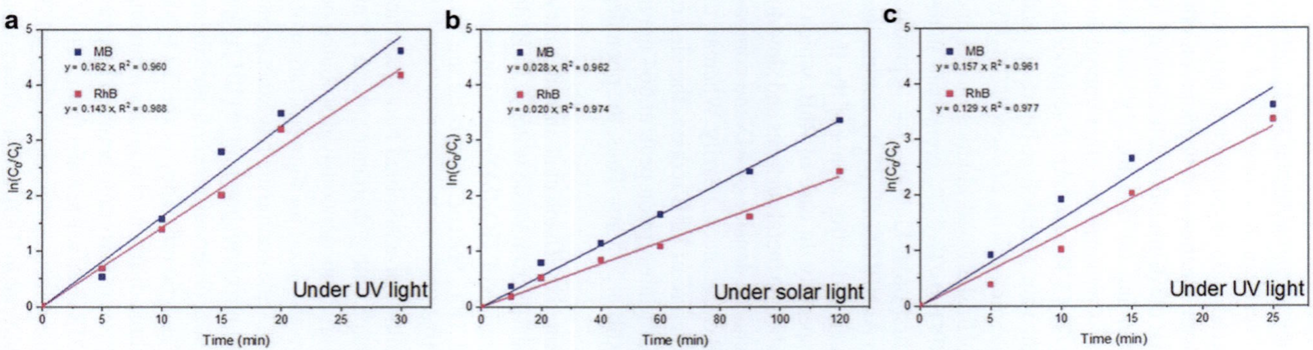

Fig. 18 Photodegradation kinetics of methylene blue (MB) and Rhodamine B (RhB) by AgNP/TiO_2/$Ti_3C_2T_x$ composite under. **a** UV light; **b** solar light; **c** UV light. Reproduced with permission from [92]. Copyright (2021) American Chemical Society

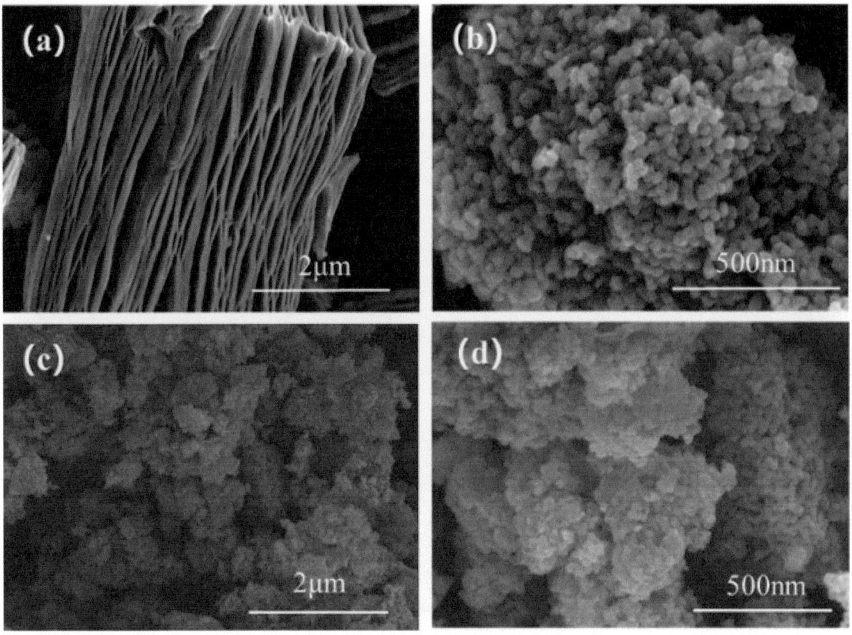

Fig. 19 SEM images of the **a** MXene; **b** TiO_2; TiO_2/MXene composite at different magnifications with scale bar, **c** 2 μm and **d** 500 nm. Reproduced with permission from [93]. Copyright (2023) Elsevier Inc.

improves the absorption of visible and UV light and cause improvement catalytic performances. In addition to that increased surface area to volume ratio of TiO_2/MXene composite (60.47 m^3 g^{-1}) reduces the chances for agglomeration of TiO_2 nanoparticles on MXene surface and exposes more TiO_2 active sites for reaction to take place. Even though, increase in MXene content in TiO_2/MXene composite above a threshold level will reduce the catalytic activity due to the light shielding effect exhibited by MXenes. They also analysed the effect of degradation time on the degradation of MB dye using TiO_2 and TiO_2/MXene composite under UV and visible conditions. Increase in reaction time decreases the concentration of MB in wastewater using TiO_2 and TiO_2/MXene composite. Where the degradation using MXene composite is faster than pristine TiO_2 nanoparticle in both UV and visible conditions as shown in Fig. 20a, b.

Degradation rate of MB using TiO_2 and TiO_2/MXene composite is 36.1% and 83.55% in 10 min respectively. After 60 min this degradation percentage becomes 65.54% and 96.44% respectively for TiO_2 and TiO_2/MXene composite under UV light. In visible light condition the degradation percentage is 3.21% and 30.69% respectively for TiO_2 and TiO_2/MXene composite in 10 min. But these values become 60.11% and 76.36% at equilibrium condition (60 min) respectively for TiO_2 and TiO_2/MXene composite. The degradation velocity of MB is calculated and is increased from 27.03% to 56.16% in UV and visible condition respectively within

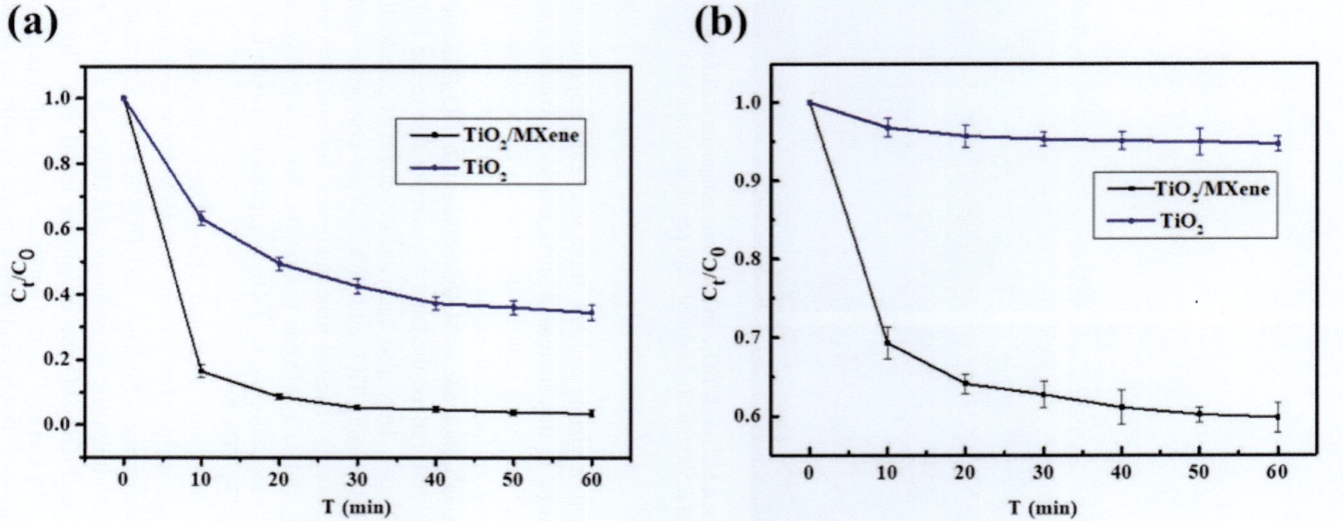

Fig. 20 Effect of degradation time in TiO_2 and TiO_2/MXene composite under. **a** UV; **b** visible conditions. Reproduced with permission from [93]. Copyright (2023) Elsevier Inc.

10 min. The stability and recycling efficiency are two important parameters for a photocatalyst during dye degradation. This is also examined by this and revealed that even after 5 cycles, the degradation rate for MB is 83.84%, which is a slight decrease. This shows the stability of the TiO_2/MXene composite for MB degradation from waste water.

4 Applications of MXenes

Unique properties exhibited by MXenes and its composites find promising applications in energy storage and catalysis. As photocatalysts, MXenes have the following applications including photocatalytic hydrogen evolution, photocatalytic carbon dioxide reduction, photocatalytic pollutant degradation, and nitrogen photo fixation and so many other photocatalytic processes [94]. MXenes are also used as electrocatalysts for hydrogen evolution reaction, oxygen evolution reaction, and oxygen reduction reaction [95]. Saini and colleagues prepared multilayered two-dimensional ZnO–Ti_3C_2 nanohybrid via one pot hydrothermal method and evaluated the performance of the hybrid material as both photo and electrocatalyst for hydrogen evolution reaction [96]. For the preparation of MXene, Al layer is removed from the Ti_3AlC_2 MAX phase followed by HF etching and then placed in a hydrothermal reactor at 180 °C for 12 h to get ZnO–Ti_3C_2 nanocomposites (Fig. 21). By varying the wt.% of ZnO (5, 10, and 15%) three composites are prepared and indicated as MXZnO-5, MXZnO-10, and MXZnO-15 respectively. Structural properties of Ti_3AlC_2, Ti_3C_2, pristine ZnO, and ZnO–Ti_3C_2 nanocomposites are identified from XRD pattern. Formation of Ti_3C_2 from Ti_3AlC_2 MAX phase is confirmed from XRD results, where (002) peak of Ti_3C_2 is shifted towards lower angle side. Selective etching of the Al layer from Ti_3AlC_2 MAX phase is confirmed from the increase in c parameter after HF treatment. Diffraction peaks of Ti_3AlC_2 MAX phase is well matches with the standard JCPDS card number 52-0875. Absence of high intense peak at 38.8° in Ti_3C_2 indicated the elimination of Al layer from the MAX phase to form MXene. Even though, there are impurity peaks present in Ti_3C_2 MXene and correspond to TiC and TiO_2 formed during etching process. By changing the wt.% of ZnO, three different composites are synthesized, and in which MXZnO-15 contains the highest intensity diffraction peaks. During ZnO addition, microstrain increases and reaches a maximum value from 24.67 to 37.05 and shows the incorporation of ZnO with the MXene. Thermogravimetric analysis of the synthesized composites is carried out and results revealed that there is a decrease in weight of Ti_3C_2 during rise in temperature from 85 to 170 °C. For ZnO–Ti_3C_2 composites, weight reduction is observed during room temperature to 200 °C resulted from the removal of functional groups like –F, –O, water, and –OH. Out of three composites sample MXZnO-15 is less thermal stability while MXZnO-5 exhibits highest thermal stability. UV-visible spectroscopy analysis shows that spectra of MAX phase are a straight line without absorption edge. But the presence of absorption edge at 520 nm in MXene revealed the presence of TiO_2 formed from MXene phase. The UV-Visible spectra of the ZnO–Ti_3C_2 composites have one absorption edge at 375 nm and result from the addition of ZnO nanoparticles. The band gap energy of the prepared composites is also evaluated from the Tauc plot and are 2.23, 2.69, 3.05, and 3.14 eV respectively for MXZnO-5, MXZnO-10, MXZnO-15, and pristine ZnO. Comparing to pristine ZnO, the band gap energy of the composite samples is decreased and is due to the presence of Ti_3C_2. In addition to that high degree of Ti_3C_2 doping creates more oxygen vacancies and reduces the band gap of the composite samples. Out of three composite samples MXZnO-10 has the lowest band gap and is expected to perform the catalytic activity more. Surface morphology of the composites is again investigated using the TEM images. Pristine ZnO has rod-like morphology over the MXene sheets. The pore size distribution analysis gives the mesopore voids ranges between 10 and 50 Å. Specific surface area of Ti_3C_2, MXZnO-5, MXZnO-10, MXZnO-15 are evaluated from Brunauer-Emmett-Teller (BET) analysis and are 6.55, 13.06, 25.58, and 31.79 $m^2\ g^{-1}$ respectively. All the MXene composites systems have higher specific surface area compared to pure MXene Ti_3C_2. Increased surface area is resulted from the addition of ZnO nanoparticles on the MXene layers. The MXene composite sample MXZnO-15 with high surface area is expected to perform both electrocatalytic and photocatalytic activity more. Optical properties of pure ZnO and the MXene composites are studied using the photoluminescence spectra. The sample with lowest photoluminescent intensity is related to the lower electron-hole recombination rate and will improve the photocatalytic activity. Both pure ZnO and MXene composites have green emission line around 500 nm and is resulting from the singly ionized oxygen vacancy. The presence of Schottky junction in all the MXene composites reduces the photoluminescent intensity compared to the pure ZnO. MXZnO-10 has the lowest photoluminescent intensity compared to other composite samples and is attributed to the reduced rate of electron-hole recombination and higher life time of charge carriers. Catalytic oxygen vacancies in MXene composites are identified from the electron paramagnetic resonance (EPR) spectra. Pristine ZnO have lowest EPR signal compared to the ZnO–Ti_3C_2 composites. Both pure ZnO and the ZnO-MXene composites show an EPR signal with G factor of 2.008 and is due to the electron captured on oxygen and carbon vacancies. Photocatalytic dye degradation efficiency of the pristine ZnO, Ti_3C_2, and ZnO–Ti_3C_2 composites are investigated using MB dye and the absorbance spectra obtained are shown in Fig. 22a–e. The absorbance spectra were recorded over a time range from 0 to 60 min using different ZnO–Ti_3C_2 composites. Analysis of the degradation efficiency of the composites shows that MXZnO-10 has the best degradation efficiency of 76.4% in 10 min and 99.2% efficiency in 60 min. While for pure ZnO, there is only 66.2% degradation in 60 min for MB dye. The UV-visible absorption spectra of the samples contain absorption bands at 665, 292, and 614 nm (shoulder peak) and undergo intensity variation within 10 min after catalytic reaction.

Figure 22f shows the plot between C/C_0 versus time for pure ZnO and Ti_3C_2 MXene, MXZnO-5, MXZnO-10, MXZnO-15 and MB. Where C and C_0 are the MB concentration at a time t and $t = 0$ respectively. Compared to ZnO-MXene composites, pure Ti_3C_2 MXene has the lowest photodegradation efficiency (7.4%)

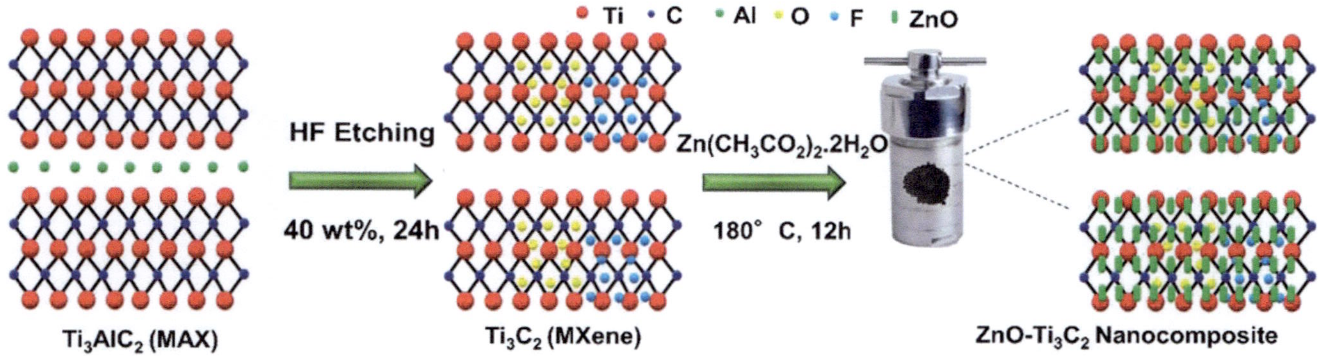

Fig. 21 A schematic representation of the synthesis procedure of ZnO–Ti_3C_2 nanocomposites. Reproduced with permission from [96]. Copyright (2022) American Chemical Society

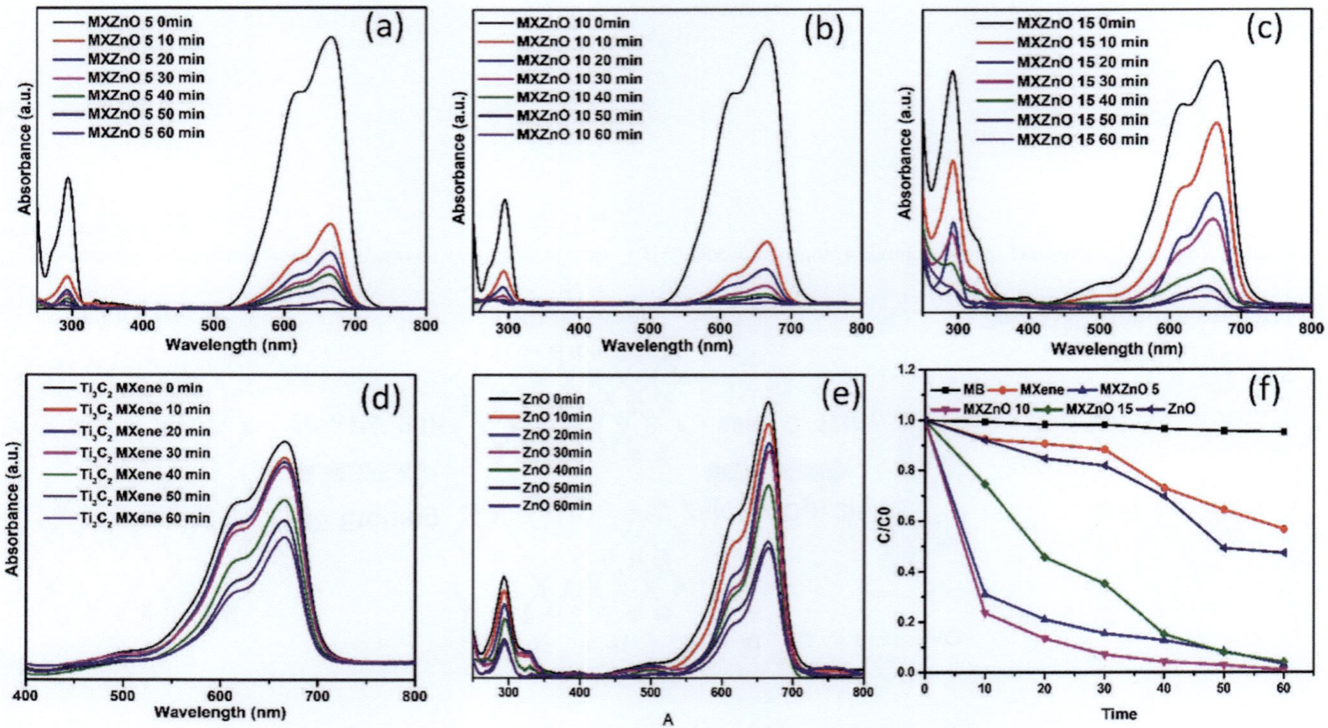

Fig. 22 Absorption spectra for **a** MXZnO-5; **b** MXZnO-10; **c** MXZnO-15; **d** pure Ti_3C_2 MXene; **e** pure ZnO. **f** The photocatalytic degradation efficiency for MB dye. Reproduced with permission from [96]. Copyright (2022) American Chemical Society

in 10 min. The addition of ZnO nanoparticles into MXene matrix lower the band gap of the composites and speed up the charge transfer towards water molecule. As Ti_3C_2 is an excellent electron acceptor, it can suppress charge carrier recombination to enhance catalytic reaction. If the absorption rate is higher, then the charge generation will be more and recombination will be less in MXZnO-10 and perform well during photocatalytic reduction. Stability of this MXZnO-10 catalyst is verified through performing catalytic reaction over six cycles under optimum conditions. Photodegradation efficiency is reduced to 92% after sixth cycle and is 7.3% less than that of the degradation efficiency at first cycle.

They also used the synthesized ZnO-MXene composites for green hydrogen production via electrocatalytic hydrogen evolution reaction. Figure 23a depicts the polarization curves for the catalysts ZnO, Ti_3C_2 MXene, MXZnO-5, and MXZnO-15 at a scan rate of 10 mV/s. MXZnO-5 and MXZnO-10 have an onset potential of 422 and 588 mV respectively. For these two electrocatalysts, overpotential is respectively 495 and 573 mV for MXZnO-5 and MXZnO-10. This overpotential very much higher than that of the standard overpotential of Pt/C (38 mV at a scan rate of 10 mV/s. Even though, MXZnO-5 has 40% reduction in onset overpotential and is around 18% in MXZnO-10 due to the incorporation of ZnO into MXene. Reduction in overpotential revealed the combined action of ZnO and MXene and the increment in a number of OH functional group on the surface of the ZnO-MXene composites.

Figure 23b shows a plot between capacitive current and scan rate, and from which the double layer capacitance is measured. MXZnO-5, MXZnO-10, and Ti_3C_2 MXenes have a double layer capacitance of 0.68, 0.53, and 0.59 mF/cm^2 respectively. Reduced double layer capacitance in MXZnO-10 results from the presence of ZnO nanoparticles, which can hinder the active sites in the nonfaradaic region. Nyquist plot for ZnO, Ti_3C_2 MXene, MXZnO-5, and MXZnO-10 are shown in Figure 23c. Due to the absence of Warburg response in the low frequency region, it is difficult to measure the diffusive layer resistance at the electrode–electrolyte interface. Thus, the charge transfer resistance of pure Ti_3C_2, MXZnO-5, and MXZnO-10 are evaluated and are respectively 127, 109, and 110 Ω. Figure 23d shows the Tafel slope for the catalysts ZnO, Ti_3C_2 MXene, MXZnO-5, and MXZnO-10. Tafel slope is highest for pure Ti_3C_2 MXene and is 139 mV/dec. All these results confirmed the extraordinary performances of the ZnO-MXene composites both as photo and electrocatalyst.

Ma et al. utilized $Ti_3C_2T_x$ MXenes as support for cobalt nanoparticle catalyst by altering the surface chemical properties of MXene [97]. MXene-supported cobalt nanoparticles are highly reactive for C–O decomposition and C–C coupling. HF etching procedure is utilized for the production of MXene from Ti_3AlC_2 MAX phase without affecting the Ti–C bonding. During the action of etchant, aluminium layers are removed and surface functional groups like hydroxyl, oxygen, and fluorine are bonded with titanium to form the MXene layers. Both SEM and TEM images of the samples are used to identify the morphology of the MXenes. As expected, the Ti_3AlC_2 MAX phase has stacked morphology. Etching imparts layered structure to the MXenes with a layer spacing of 1 nm. From selected area diffraction pattern, monocrystalline structure of the $Ti_3C_2T_x$ MXene is identified and number of titanium vacancies is observed. $Ti_3C_2T_x$ MXene has hexagonal close-packed structure.

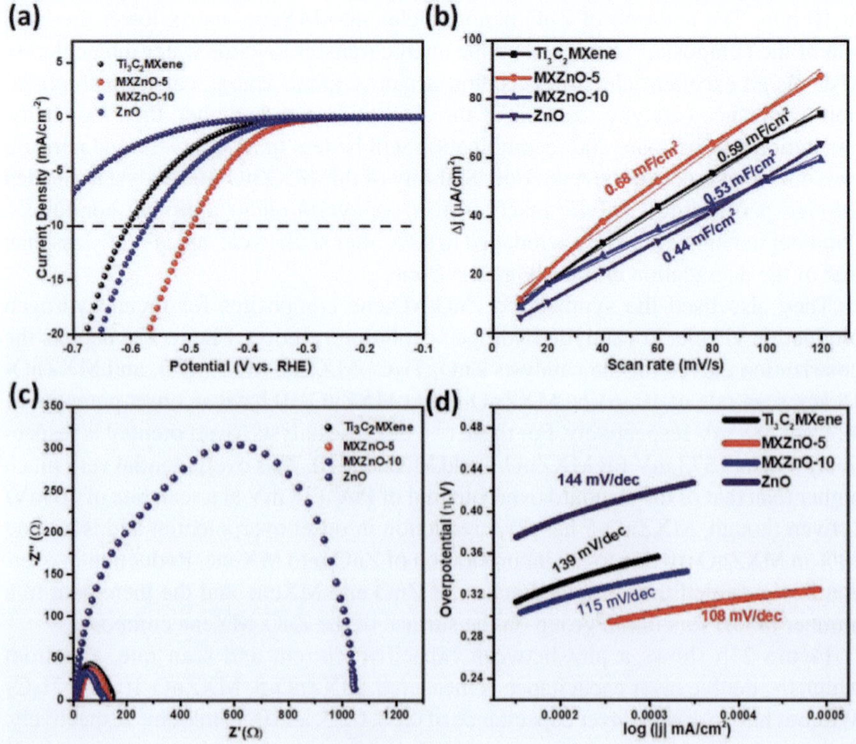

Fig. 23 **a** Hydrogen evolution reaction polarization curve for ZnO, Ti_3C_2 MXene, MXZnO-5 and MXZnO-10 at a scan rate of 10 mV/s; **b** plot of scan rate versus current density; **c** impedance spectra of ZnO, Ti_3C_2 MXene, MXZnO-5 and MXZnO-10; **d** Tafel slope for the catalyst ZnO, Ti_3C_2 MXene, MXZnO-5 and MXZnO-10. Reproduced with permission from [96]. Copyright (2022) American Chemical Society

Structural confirmation of the MXene done with XRD pattern and the shift in (002) peak in MAX phase from 9.5° to 8.9° confirmed the formation of $Ti_3C_2T_x$ MXene form corresponding MAX phase. Annealing with ammonia further shift the (002) peak towards 9.1° and which decreases the interlayer spacing to 0.97 nm. Another peak (004) also emerged at an angle 18.5° and which confirmed the nanosheet-like structure of the annealed MXene sheets. Effect of nitrogen doping on Co/MXene catalyst is identified from the carbon dioxide hydrogenation. The catalytic performance of the carbon dioxide hydrogenation as a function of temperature is recorded and the results showed that increases in temperature from 200 to 400 °C increase the carbon dioxide conversion rate. For the Co/MXene-NH_3 catalysts have a higher rate of carbon dioxide conversion compared to pure Co/MXene catalysts. The reaction rate reaches a maximum value of 10.6 and 30.4 μmol g^{-1} s^{-1} respectively for Co/MXene and Co/MXene-NH_3. At high temperature the reaction product is mostly carbon monoxide for Co/MXene catalyst whereas methane is the reaction product when catalyst changes to Co/MXene-NH_3. These results revealed the effect of nitrogen vacancies on carbon dioxide reduction using different MXene composites. Apparent activation energy for both the catalyst is evaluated and found that it is 34 and 59.1 kJ mol^{-1}respectively for Co/MXene and Co/MXene-NH_3 and this indicated the lower reducibility of Co/MXene compared to Co/MXene-NH_3.

5 Challenges and Future Perspectives

Here in this chapter, we discussed the characteristic features of the newly originated two-dimensional layered materials, MXenes and its composites. The mechanical and catalytic properties of MXenes and its composites are discussed in detail. Incorporation of metal nanoparticles, metal oxides, polymers, and other two-dimensional materials into MXene layers drastically changes the catalytic and mechanical properties of the MXene composites. Some of the photocatalytic and electrocatalytic applications of MXenes and their composites are discussed in the last section. Incorporation of metal oxides like TiO_2, ZnO into the MXene layer enhances the optical behaviour and thereby improving the catalytic efficiency. But the optimization of concentration of all these additional agents into the MXene matrix is a key factor in determining the features and properties. Discussion in the application part also revealed that MXene composites with metal oxides are rarely investigated for electrocatalytic hydrogen evolution reaction for green hydrogen production. Thus, it is necessary to reduce the overpotential of such MXene composites to find more practical applications in electrocatalytic water splitting for hydrogen production. Like this, there are difficulties in exploring practical applications with MXenes and its composites. Which includes:

- MXenes are excellent two-dimensional materials with outstanding mechanical properties. Incorporation of polymers, metal oxides, and newly emerged two-dimensional materials like transition metal dichalcogenides, graphene, carbon nitride, etc., will improve the mechanical features of the MXene composites. Moreover, selection of these material matching with MXenes and optimization of concentration of these additional components are important to obtain desired features.
- Addition of carbon nanotubes, single-walled carbon nanotubes, double-walled carbon nanotubes, and multi-walled carbon nanotubes into MXene layers will improve the specific surface area and enhance it features.
- Nowadays traditional synthesizing strategies are used by most of researchers to prepare MXenes for different applications. Inventing new strategies for the preparation of MXenes will be beneficial for reducing the use of toxic chemicals and to obtain more desired structures with improved properties and stability.
- New strategies for preparation of MXenes and MXenes-based compounds should be commercially suitable and the yield should be more compared to the conventional methods.

- For catalytic applications of MXenes and MXenes-based compounds strongly depend upon the position of surface functional group. Proper investigation is necessary regarding the role of surface functional group in determining the catalytic activity of MXenes.
- More investigation is needed in the field of MXenes and MXene-based advanced materials for electrocatalytic applications. This will help the world to find new energy sources to satisfy human needs.

6 Conclusions

This chapter discussed the basic features of MXenes and MXenes-based advanced composites in terms of their mechanical and catalytic properties, and applications. As a new generation material, there are a number of properties and features of MXenes and MXenes-based composites that are suitable for a wide variety of applications. There are a large number MXenes that theoretically predicted but not practically invented in the literature. Out of the practically synthesized MXenes and MXenes-based compounds, excellent properties are exhibited by most of the MXenes. Their mechanical properties are outstanding compared to other two-dimensional materials. In addition to that, incorporation of additional materials will uplift the mechanical and catalytic properties to the next level. The main drawbacks faced in MXenes-based system is the lack of simple, low-toxic, and cheap preparation methods. This is a challenge faced by the scientific community during the preparation of MXenes and MXenes-based compounds. Incorporation of additional materials is one of the best options to tune the properties of MXenes. This will enhance the properties of MXene-based materials for practical applications in the future.

References

1. Thomas, S.A.: Layered two-dimensional black phosphorous-based hybrid electrodes for rechargeable batteries. J. Energy Storage **73**, 109068 (2023)
2. Thomas, S.A., et al.: Polyoxometalates and redox-active molecular clusters for supercapacitors. In: Supercapacitors, pp. 221–243. Elsevier (2024)
3. Thomas, S.A., Cherusseri, J.: Recent advances in synthesis and properties of zirconium-based MXenes for application in rechargeable batteries. Energy Storage e475
4. Thomas, S.A., et al.: Designing PEDOT-based hybrid electrodes for supercapacitors by electrospinning strategy. Discov. Electrochem. **1**(1), 6 (2024)
5. Thomas, S.A., Cherusseri, J., Khalid, M.: A comprehensive review on tungsten oxide nanostructures-based electrochromic supercapacitors and machine learning models for design and process parameter optimization. Energy Storage e499
6. Thomas, S.A., Cherusseri, J., Rajendran, D.N.: Recent advancements on carbon fibers-based sustainable electrodes for flexible and wearable supercapacitors. RSC Sustain (2024)
7. Thomas, S.A., Cherusseri, J., Rajendran, D.N.: Functionalized carbon nanostructures for hydrogen storage. In: Handbook of Functionalized Carbon Nanostructures: from Synthesis Methods to Applications, pp. 1–39. Springer (2024)

8. Thomas, S.A., et al.: Translation of supercapacitor technology from laboratory scale to commercialization. In: Supercapacitors, pp. 371–395. Elsevier (2024)
9. Thomas, S.A., et al.: "Water-in-salt" electrolyte—Toward high-voltage aqueous supercapacitors. In: Supercapacitors, pp. 289–315. Elsevier (2024)
10. Kumar, D., et al.: Nanocarbon assisted green hydrogen production: development and recent trends. Int. J. Hydrogen Energy **50**, 118–141 (2024)
11. Baby, A., et al.: Azopyridine as a linker molecule in polyaniline-grafted graphene oxide nanocomposite electrodes for asymmetric supercapacitors. ACS Appl. Energy Mater. **6**(20), 10442–10456 (2023)
12. Banerjee, S., et al.: Applications of supercapacitors. In: Handbook of Nanocomposite Supercapacitor Materials I: characteristics, pp. 341–350 (2020)
13. Cherusseri, J., et al.: Rapid synthesis of cobalt manganese phosphate by microwave-assisted hydrothermal method and application as positrode material in supercapatteries. Sci. Rep. **14**(1), 26550 (2024)
14. Zaed, M., et al.: Utilization of recycled materials for low-cost MXene synthesis and fabrication of graphite/MXene composite for enhanced water desalination performance. Sep. Purif. Technol. **354**, 129055 (2025)
15. Tyagi, A., et al.: Characteristics of transition metal oxides. In: Handbook of Nanocomposite Supercapacitor Materials I: characteristics, pp. 91–123 (2020)
16. Kumar, D., et al.: Technology computer-aided design device simulation of nanowall-based CdS/CdTe solar cell for energy conversion. Energy Storage **6**(1), e498 (2024)
17. Zaed, M., et al.: Low-cost synthesis of Ti3C2Tx MXene-based sponge for solar steam generation and clean water production. Ceram. Int. (2024)
18. Cherusseri, J., Kar, K.K.: Recent progress in nanocomposites based on carbon nanomaterials and electronically conducting polymers. In: Polymer Nanocomposites Based on Inorganic and Organic Nanomaterials, pp. 229–256 (2015)
19. Thomas, S.A., Cherusseri, J., Rajendran, D.N.: Recent progresses in the synthesis and strategic designs of sustainable carbon-based fibrous electrodes for flexible batteries. RSC Sustain. (2025)
20. Ball, M., Wietschel, M.: The future of hydrogen–opportunities and challenges. Int. J. Hydrog. Energy **34**(2), 615–627 (2009)
21. Santhanam, K.S., et al.: Introduction to Hydrogen Technology. Wiley (2017)
22. Ahmed, A., et al.: Hydrogen fuel and transport system: a sustainable and environmental future. Int. J. Hydrog. Energy **41**(3), 1369–1380 (2016)
23. Singla, M.K., Nijhawan, P., Oberoi, A.S.: Hydrogen fuel and fuel cell technology for cleaner future: a review. Environ. Sci. Pollut. Res. **28**(13), 15607–15626 (2021)
24. Manoharan, Y., et al.: Hydrogen fuel cell vehicles; current status and future prospect. Appl. Sci. **9**(11), 2296 (2019)
25. Ismael, M.: Latest progress on the key operating parameters affecting the photocatalytic activity of TiO2-based photocatalysts for hydrogen fuel production: a comprehensive review. Fuel **303**, 121207 (2021)
26. Jain, I.P.: Hydrogen the fuel for 21st century. Int. J. Hydrog. Energy **34**(17), 7368–7378 (2009)
27. Crabtree, G.W., Dresselhaus, M.S.: The hydrogen fuel alternative. MRS Bull. **33**(4), 421–428 (2008)
28. Barton, J., Gammon, R.: The production of hydrogen fuel from renewable sources and its role in grid operations. J. Power Sour. **195**(24), 8222–8235 (2010)
29. Thomas, S.A., Cherusseri, J.: Strategically designing layered two-dimensional SnS2-based hybrid electrodes: a futuristic option for low-cost supercapacitors. J. Energy Chem. (2023)
30. Pallavolu, M.R., et al.: Scalable synthesis of binder-free hierarchical MnCo2O4 nanospikes/Ni(OH)2 nanosheets composite electrodes for high-capacity supercapatteries. J. Energy Storage **73**, 108999 (2023)
31. Nallapureddy, J., et al.: Morphology controlled synthesis of MnCo2S4 nanoflower structures for different metal precursors used for energy storage applications. J. Power Sour. **613**, 234883 (2024)

32. Cherusseri, J.: Polymer-ionic liquid gel electrolytes for lithium-ion batteries. In: Polymer Electrolytes for Energy Storage Devices, pp. 235–253. CRC Press (2021)
33. Cherusseri, J., Kar, K.K.: Hierarchical carbon nanopetal/polypyrrole nanocomposite electrodes with brush-like architecture for supercapacitors. Phys. Chem. Chem. Phys. **18**(12), 8587–8597 (2016)
34. Cherusseri, J., et al.: Novel mesoporous electrode materials for symmetric, asymmetric and hybrid supercapacitors. Nanotechnology **30**(20), 202001 (2019)
35. Cherusseri, J., et al.: Polymer-based composite materials: characterizations. In: Composite Materials: processing, Applications, Characterizations, pp. 37–77 (2017)
36. Cherusseri, J., Sharma, R., Kar, K.K.: Nanotechnology advancements on carbon nanotube/polypyrrole composite electrodes for supercapacitors. In: Handbook of Polymer Nanocomposites. Processing, Performance and Application: volume B: carbon Nanotube Based Polymer Composites, pp. 479–510 (2015)
37. Cherusseri, J., Sharma, R., Kar, K.K.J.C.: Helically coiled carbon nanotube electrodes for flexible supercapacitors. **105**, 113–125 (2016)
38. Cherusseri, J., Varma, S.J.: Electrospun manganese oxide-based composites as anodes for lithium-ion batteries. In: Electrospinning for Advanced Energy Storage Applications, pp. 345–371 (2021)
39. Banerjee, S., et al.: Capacitor to supercapacitor. In: Handbook of Nanocomposite Supercapacitor Materials I: characteristics, pp. 53–89 (2020)
40. Mishra, R., Cherusseri, J., Joseph, K.: Thermal and crystallization behavior of micro and nano fibrillar in-situ composites. In: Micro and Nano Fibrillar Composites (MFCs and NFCs) from Polymer Blends, pp. 213–231. Elsevier (2017)
41. Mishra, R.K., et al.: Heterogeneous integration of 2D materials with silicon complementary metal oxide semiconductor (Si-CMOS) devices. In: Beyond Si-Based CMOS Devices: materials to Architecture, pp. 149–179. Springer (2024)
42. Thomas, S.A., Cherusseri, J.: A review of Nb2CTx MXene as an emerging 2D material: synthesis, applications in rechargeable batteries and supercapacitors, progress, and outlook. Energy Fuels (2023)
43. Thomas, S.A., Cherusseri, J.: Boron carbon nitride (BCN): emerging two-dimensional nanomaterial for supercapacitors. J. Mater. Chem. A (2023)
44. Thomas, S.A., Cherusseri, J., Rajendran, D.N.: 2D nickel sulfide electrodes with superior electrochemical thermal stability along with long cyclic stability for supercapatteries. Energy Technol. 2301641 (2024)
45. Thomas, S.A., Cherusseri, J., Rajendran, D.N.: Rapid synthesis of hierarchical tin disulfide (SnS2) nanostructures by a microwave-assisted hydrothermal method for high-performance supercapatteries. ACS Appl. Electron. Mater. (2024)
46. Thomas, S.A., Cherusseri, J., Rajendran, D.N.: Nickel disulfide (NiS2): a sustainable low-cost electrode material for high-performance supercapacitors. Energ. Technol. **12**(7), 2400138 (2024)
47. Thomas, S.A., Cherusseri, J., Rajendran, D.N.: Hierarchical two-dimensional layered nickel disulfide (NiS2)@ PEDOT: PSS nanocomposites as battery-type electrodes for battery-type supercapacitors with high energy density. Electrochem **5**(3), 298–313 (2024)
48. Thomas, S.A., Cherusseri, J., Rajendran, D.N.: Rapid synthesis of hierarchical cobalt disulfide nanostructures by microwave-assisted hydrothermal method for high performance supercapatteries. ACS Appl. Electron. Mater. (2024)
49. Cherusseri, J., Kar, K.K.: Polypyrrole-decorated 2D carbon nanosheet electrodes for supercapacitors with high areal capacitance. RSC Adv. **6**(65), 60454–60466 (2016)
50. Cherusseri, J., Kar, K.K.: Ultra-flexible fibrous supercapacitors with carbon nanotube/polypyrrole brush-like electrodes. J. Mater. Chem. A **4**(25), 9910–9922 (2016)
51. Thomas, S.A., et al.: Boron carbon nitride (BCN): an emerging two-dimensional material for rechargeable batteries. Energy Fuels (2024)
52. Thomas, S.A., et al.: Graphitic carbon nitride (g-C3N4): futuristic material for rechargeable batteries. J. Energy Storage **68**, 107673 (2023)

53. Zaed, M., et al.: Synthesis of titanium carbide (TiC) using recycled precursor for potential scale-up of low-cost MXene. Open Ceram. **19**, 100637 (2024)
54. Zaed, M., et al.: Synthesis and characterization of hierarchical Ti3C2Tx MXene/graphitic-carbon nitride/activated carbon@ luffa sponge composite for enhanced water desalination. Open Ceram. **19**, 100645 (2024)
55. Zaed, M., et al.: Synthesis of Ti3C2Tx MXene@ Carbon-Enhanced cellulose fiber composite-based photothermal absorber for sustainable water desalination. Mater. Today Sustain. **28**, 100971 (2024)
56. Zaed, M., et al.: Hierarchical Ti3C2Tx MXene@ Honeycomb nanocomposite with high energy efficiency for solar water desalination. Chemosphere **366**, 143459 (2024)
57. Zaed, M., et al.: Unlocking desalination's potential: harnessing MXene composite for sustainable desalination. Chem. Eng. J. 156910 (2024)
58. Ahmad, K., Raza, W., Khan, R.A.: Ti3AlC2 MAX phase modified screen-printed electrode for the fabrication of hydrazine sensor. Micromachines **15**(5), 633 (2024)
59. Deysher, G., et al.: Synthesis of Mo4VAlC4 MAX phase and two-dimensional Mo4VC4 MXene with five atomic layers of transition metals. ACS Nano **14**(1), 204–217 (2019)
60. Li, P.-X., et al.: Bidirectionally aligned MXene hybrid aerogels assembled with MXene nanosheets and microgels for supercapacitors. Rare Met. **42**(4), 1249–1260 (2023)
61. Li, R., et al.: Flexible and high-performance electrochromic devices enabled by self-assembled 2D TiO2/MXene heterostructures. Nat. Commun. **12**(1), 1587 (2021)
62. Schultz, T., et al.: Surface termination dependent work function and electronic properties of Ti3C2Tx MXene. Chem. Mater. **31**(17), 6590–6597 (2019)
63. Tang, M., et al.: Surface terminations of MXene: synthesis, characterization, and properties. Symmetry **14**(11), 2232 (2022)
64. Gogotsi, Y., Anasori, B.: The rise of MXenes. In: MXenes, pp. 3–11. Jenny Stanford Publishing (2023)
65. Naguib, M., et al.: Two-dimensional nanocrystals produced by exfoliation of Ti3AlC2. In: MXenes. Jenny Stanford Publishing (2011)
66. Kumar, J.A., et al.: Methods of synthesis, characteristics, and environmental applications of MXene: a comprehensive review. Chemosphere **286**, 131607 (2022)
67. Jiang, X., et al.: Two-dimensional MXenes: from morphological to optical, electric, and magnetic properties and applications. Phys. Rep. **848**, 1–58 (2020)
68. Xiong, D., et al.: Recent advances in layered Ti3C2Tx MXene for electrochemical energy storage. Small **14**(17), 1703419 (2018)
69. Sfeir, A., et al.: Unlocking the potential of mxene in catalysis: decorated Mo2CTx catalyst for ammonia synthesis under mild conditions. J. Am. Chem. Soc. **146**(29), 20033–20044 (2024)
70. Wang, Z., et al.: Unraveling and regulating self-discharge behavior of Ti3C2Tx MXene-based supercapacitors. ACS Nano **14**(4), 4916–4924 (2020)
71. Pei, Y., et al.: Ti3C2TX MXene for sensing applications: recent progress, design principles, and future perspectives. ACS Nano **15**(3), 3996–4017 (2021)
72. Han, M., et al.: Beyond Ti3C2Tx: MXenes for electromagnetic interference shielding. ACS Nano **14**(4), 5008–5016 (2020)
73. Lim, G.P., et al.: Recent progress and new perspective of MXene-based membranes for water purification: a review. Ceram. Int. **48**(12), 16477–16491 (2022)
74. Cheng, L., et al.: Boosting the photocatalytic activity of CdLa2S4 for hydrogen production using Ti3C2 MXene as a co-catalyst. Appl. Catal. B **267**, 118379 (2020)
75. Meng, L., et al.: Synergistic barium titanate/MXene composite as a high-performance piezo-photocatalyst for efficient dye degradation. J. Colloid Interface Sci. **674**, 972–981 (2024)
76. Jun, B.-M., et al.: Ultrasonic degradation of selected dyes using Ti3C2Tx MXene as a sonocatalyst. Ultrason. Sonochem. **64**, 104993 (2020)
77. Nan, J., et al.: Nanoengineering of 2D MXene-based materials for energy storage applications. Small **17**(9), 1902085 (2021)
78. Anasori, B., Lukatskaya, M.R., Gogotsi, Y.: 2D metal carbides and nitrides (MXenes) for energy storage. In: MXenes, pp. 677–722. Jenny Stanford Publishing (2023)

79. Luo, J., et al.: Interfacial structure design of MXene-based nanomaterials for electrochemical energy storage and conversion. InfoMat **2**(6), 1057–1076 (2020)
80. Anasori, B., et al.: Two-dimensional, ordered, double transition metals carbides (MXenes). ACS Nano **9**(10), 9507–9516 (2015)
81. Halim, J., et al.: Transparent conductive two-dimensional titanium carbide epitaxial thin films. Chem. Mater. **26**(7), 2374–2381 (2014)
82. Wu, Z., et al.: Niobium and titanium carbides (MXenes) as superior photothermal supports for CO2 photocatalysis. ACS Nano **15**(3), 5696–5705 (2021)
83. Ahn, S., et al.: A 2D titanium carbide MXene flexible electrode for high-efficiency light-emitting diodes. Adv. Mater. **32**(23), 2000919 (2020)
84. Liang, K., et al.: Synthesis of new two-dimensional titanium carbonitride Ti2C0. 5N0. 5Tx MXene and its performance as an electrode material for sodium-ion battery. InfoMat **3**(12), 1422–1430 (2021)
85. Shi, C., et al.: Structure of nanocrystalline Ti3C2 MXene using atomic pair distribution function. Phys. Rev. Lett. **112**(12), 125501 (2014)
86. Wyatt, B.C., Rosenkranz, A., Anasori, B.: 2D MXenes: tunable mechanical and tribological properties. Adv. Mater. **33**(17), 2007973 (2021)
87. Zhang, Y., et al.: A theoretical study on the chemical bonding of 3d-transition-metal carbides. Solid State Commun. **121**(8), 411–416 (2002)
88. Chakraborty, P., et al.: Manipulating the mechanical properties of Ti2C MXene: effect of substitutional doping. Phys. Rev. B **95**(18), 184106 (2017)
89. Hatam-Lee, S.M., Esfandiar, A., Rajabpour, A.: Mechanical behaviors of titanium nitride and carbide MXenes: a molecular dynamics study. Appl. Surf. Sci. **566**, 150633 (2021)
90. Yan, X., et al.: Flame retardancy, thermal and mechanical properties of novel intumescent flame retardant/MXene/poly (vinyl alcohol) nanocomposites. Nanomaterials **12**(3), 477 (2022)
91. Luque-Morales, P.A., et al.: ZnO semiconductor nanoparticles and their application in photocatalytic degradation of various organic dyes. Materials **14**(24), 7537 (2021)
92. Othman, Z., et al.: Efficient photocatalytic degradation of organic dyes by AgNPs/TiO2/ Ti3C2Tx MXene composites under UV and solar light. ACS Omega **6**(49), 33325–33338 (2021)
93. Zhang, S., et al.: Photocatalytic degradation of TiO2 via incorporating Ti3C2 MXene for methylene blue removal from water. Catal. Commun. **174**, 106594 (2023)
94. Chirica, I.M., et al.: Applications of MAX phases and MXenes as catalysts. J. Mater. Chem. A **9**(35), 19589–19612 (2021)
95. Bai, X., Guan, J.: MXenes for electrocatalysis applications: modification and hybridization. Chin. J. Catal. **43**(8), 2057–2090 (2022)
96. Saini, B., et al.: Role of ZnO in ZnO nanoflake/Ti3C2 MXene composites in photocatalytic and electrocatalytic hydrogen evolution. ACS Appl. Nano Mater. **5**(7), 9319–9333 (2022)
97. Ma, J., et al.: Tuning the chemical properties of Co–Ti3C2Tx MXene materials for catalytic CO2 reduction. Small **17**(26), 2007509 (2021)

MXene: Antiviral and Antibacterial Properties

Arnet Maria Antony, Maruboina Hemanth Kumar, and Siddappa A. Patil

Abstract The rise in drug resistance in microbes towards conventional antibiotics and other medicines has made effective treatment of infections difficult and time-consuming, drastically affecting the quality of healthcare. Additionally, the exponential growth in population, accompanied by diminished hygiene due to the polluted environment and poor lifestyle, has also resulted in the increase of unanticipated and unusual ailments. Thus, there is a pressing need for innovation in efficient medicinal agents to address these challenging issues. Interestingly, the tunable properties of 2-dimensional materials, particularly MXene, have characteristically influenced their exploitation in the medical scenario in addition to their diverse applicability in a range of applications. This 2-dimensional material and its multivariate composites, have revealed promising antibacterial and antiviral activities owing to its potential in tailoring the surface functionalities, thermal stability, hydrophilicity, biocompatibility, and catalytic activity to one's desire. Despite the optimistic perspective, the bulk synthesis and functionalization of MXene are still a concern. Herein, we discuss the synthetic protocols for the preparation of MXenes and their composites, addressing their pros and cons. Following, the possible mechanistic interaction of these MXene composites with microbes to bring about antibacterial and antiviral properties. Finally, the opportunities MXenes and their composites hold in the field of medicine for the future.

Keywords MXene · MXene-organic/inorganic composites · Antibacterial · Antiviral

A. M. Antony (✉) · M. H. Kumar · S. A. Patil (✉)
Centre for Nano and Material Sciences, Jain (Deemed-to-be University), Jain Global Campus, Kanakapura, Bangalore, Karnataka, India
e-mail: a.maria@jainuniversity.ac.in

S. A. Patil
e-mail: p.siddappa@jainuniversity.ac.in

© The Author(s), under exclusive license to Springer Nature Singapore Pte Ltd. 2025

S. Moharana and R. N. Mahaling (eds.), *Surface Engineering of MXenes*, Engineering Materials, https://doi.org/10.1007/978-981-96-3640-2_6

1 Introduction

The upsurge in the rate of mortality due to microbial infections can be attributed to the rapid mutation in the microbes leading to strong resistance towards conventional drug molecules [1]. This results in the longstanding usage of high concentrations of conventional antimicrobial drugs for effective treatment of these infections [2]. Additionally, the poor hygiene of the surroundings, exponential population growth, poor lifestyle, and the ability of bacteria to persist on surfaces are major factors contributing to the widespread infections. Plentiful antimicrobial drugs are being produced on a daily basis and released into the market to overcome microbial resistance to pharmaceutical drugs and improve their efficacy [3, 4]. However, the new medications also lose their effectiveness once the microbes develop resistance to them. These issues have influenced the research community to identify innovative antimicrobial agents with low susceptibility toward drug resistance while simultaneously active against a range of microbial strains [5]. The bonding of the microbes to the material surface through various interactions, such as van der Waals force, hydrogen bonding, Coulombic force, plays a critical role in the biocidal activity [6].

In line with this, a wide range of systems like organic, inorganic, and organometallic compounds, nanomaterials, 2D materials, and their composites have been extensively explored for their antimicrobial activity. One such material that grabbed the attention of the scientific community is MXene, a 2D transition metal carbide, nitride, or carbonitride-based nanomaterial [7–10]. The name 'MXene' is derived from graphene for its similarity to the latter in its properties [11]. The synthesis of the first MXene Ti_3C_2 was in 2011 by the selective exfoliation of the 3D MAX phase (layered and hexagonal) Ti_3AlC_2 by hydrofluoric acid (HF) (Fig. 1a) [12]. The scanning electron microscopic (SEM) image of the MAX phase Ti_3AlC_2 prior to the treatment with HF shows the typical nature of a 3D bulk material (Fig. 1b). On HF treatment, the selective etching of the layers involving the A-group element bound to the transition metal layer by weak metallic bonds (in Ti_3AlC_2, it is the aluminium layers) occurs. This results in the exfoliation, which can be evidently seen in the SEM images, indicating the formation of multi-layered vermiculite clay-like morphology in the obtained MXene precursors (Fig. 1c–g). Sonication of these exfoliated MXene precursors results in the final 2D MXene, which can be obtained as nanosheets, nanotubes, or scrolls.

In the MXene, the carbide/nitride/carbonitride layer (denoted as the X layer) is packed between the layers of the transition metal (denoted as the M layers). By varying the ratio of the X and M layers, a series of unique MXenes have been designed and synthesized with tunable physical, chemical, and biological properties [13]. Predictions through theoretical calculations have indicated the fabrication of over 100 MXene compositions possible by varying the M, X, and T sites, tuning their material properties [14]. Of the various MXenes, the hexagonal MAX phased MXene has the general formula $M_{n+1}X_nT_x$ (M = early transition metal, n = 1 to 4, X = carbon/nitrogen/oxygen, T = surface terminations, and x = number of surface terminals per unit). The T group are mainly high surface energy moieties such as O,

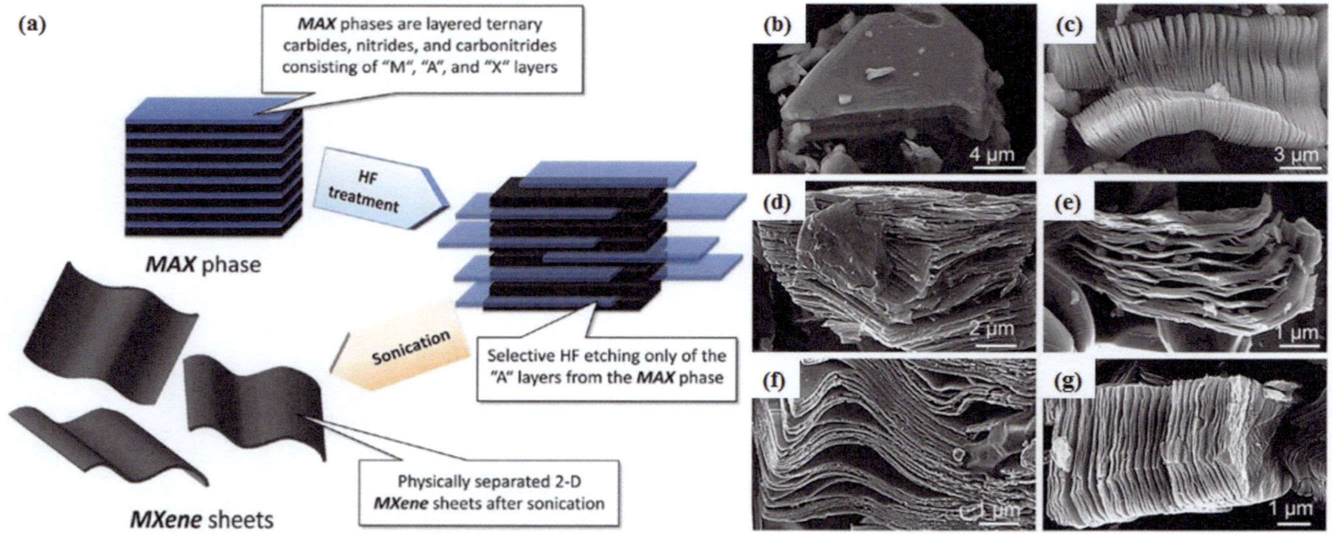

Fig. 1 **a** Illustration of the exfoliation of 3D MAX phase to MXenes; SEM images of **b** untreated 3D MAX phase and the exfoliation observed in **c** Ti3AlC_2, **d** Ti_2AlC, **e** Ta_4AlC_3, **f** TiNbAlC, and **g** Ti_3AlCN after treatment with HF. Reproduced from Ref. [11] (Open Access)

F, Cl, or OH groups, which greatly influence the electronic and magnetic nature [15, 16]. The layered structure of MAX phase MXene consists of XM_6 distorted octahedra with shared edges interweaved with individual layers of A-group elements, ascribing to their outstanding physical, chemical, and electronic properties [17, 18]. These intrinsic properties can be desirably modified to enhance their superiority in photocatalytic, photothermal, sensing, and medicinal applications. The most common MXenes are of the formulae M_2XT_x, $M_3X_2T_x$, $M_4X_3T_x$, and $M_5X_4T_x$ (Fig. 2) [19].

Owing to their higher biocompatibility and lower cytotoxicity, abundant surface area and functionalities, thermal and metallic conductivity, electrical and photothermal properties, MXenes have paved their way into widespread applications in the fields of biomedical, sensors, electronics, communications, energy storage, and environmental [20]. This chapter provides an overview of the synthesis of MXene-based systems and their prospective applications as antibacterial and antiviral agents in the treatment of bacterial and viral infections.

2 MXenes

2.1 Synthesis

The synthetic approach towards MXene preparation immensely affects the predicted properties of the final product. MXene precursors are usually synthesized by mixing or sintering, while the fundamental steps involved in MXene synthesis are the etching or exfoliation of the 3D MAX phased precursors, followed by the intercalation or delamination [21]. Both the top-down and bottom-up approaches have been employed for MXene synthesis [22]. The top-down approach involves mechanical, thermal, or chemical exfoliators, whereas the chemical vapour deposition (CVD) [23] and wet-chemical synthesis are the general protocols for the bottom-up approach. The complexity of the layer arrangement in MXenes is still posing challenges in the advancement of the bottom-up approach. The mechanical method for etching the A layer in the top-down approach is restricted by the metallic bond between the M and A layers. While in the case of employing the high temperature method, recrystallization of the formed layered MXene back to a bulk 3D carbide or nitride is induced, decreasing the development of MXene sheets [24]. Thus, the chemical method is preferred, where the liquid phase exfoliation is mostly used. Preliminarily, as depicted in Fig. 1a, the MXenes were synthesized by the etching of the transition metal layers of the MAX phase employing HF as the most common etching agent. Over the years, the hazardous HF has been replaced by various other etching agents such as alkalis and acids, molten salts, as well as electrochemical etching [25]. Through the etching process, in addition to exfoliation, the surface of the MAX phase layers is functionalized with T groups and they hold the layers together by secondary bonding such as hydrogen bonding and van der Waals interaction [26].

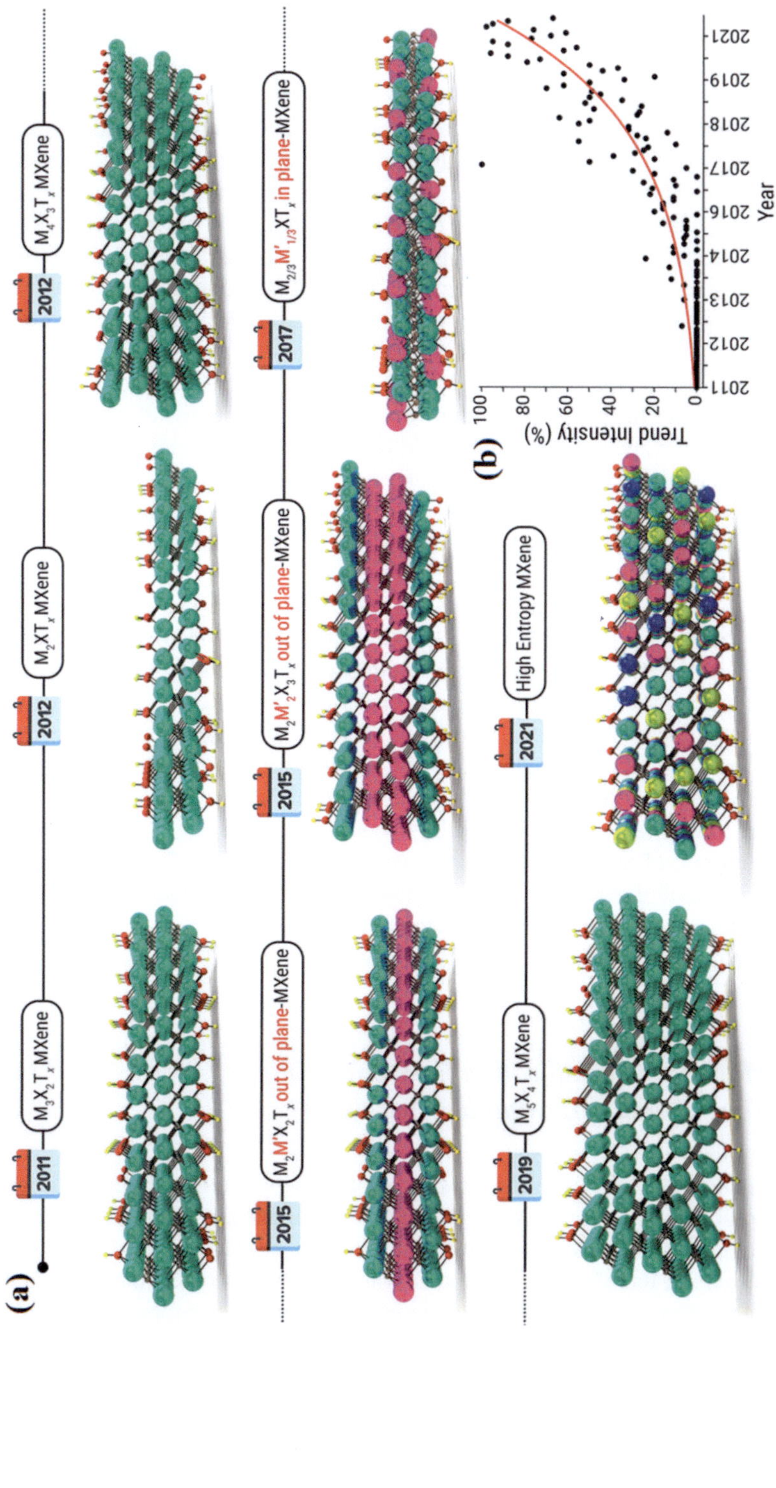

Fig. 2 **a** Typical structure and compositions of MXene over the years; **b** the monthly Google Trends data with the word "MXene" collected over the decade mined using the "pytrends" Python library. Reproduced from Ref. [19] (Open Access)

These secondary bonds further provide a platform for the intercalation between the layers. The delamination step is achieved by sonication or intercalation with suitable intercalants to increase the interlayer distance, so as to yield the MXenes sheets. The intercalation is enhanced by the surface negative charge of MXene making it easier for the organic and inorganic cations to intercalate. Some of the intercalants introduced are dimethyl sulfoxide, tetrabutyl/methyl ammonium hydroxides, urea, isoprolyamine, hydrazine monohydrate, diazonium salts, as well as inorganic cations (Al^{3+}, Mg^{2+}, Li^+, Na^+, K^+, NH_4^+) [27, 28]. Later, exfoliation by supercritical fluids was developed, enhancing the quality of the MXene [25, 29]. The hydrophilicity of MXenes, as a result of the T_x functionalities and the surface negative charge, assists in the easy dispersion of MXene in water even without the aid of a surfactant. This has greatly simplified the MXene synthesis [11, 30]. In 2014, Barsoum et al. reported the synthesis of Ti_3C_2 by exfoliation and delamination of MAX phase Ti_3AlC_2 in a solution of LiF and concentrated HCl [31]. The LiF–HCl mixture generates the HF in situ, bringing forth the exfoliation, while the simultaneous intercalation of the Li^+ ions into the layers prompts the hydration causing delamination. This work introduced a new path for the practice of safer and shortened processes for MXene synthesis. Over the years, various synthetic protocols have been devised for obtaining single or multi-layered MXenes of 0D (quantum dots), 1D (nanoribbons, nanobelts, nanotubes), 2D, 3D, and their composites (Fig. 3) [27, 32, 33].

A major downside of the conventional synthetic strategies is the use of harmful substances, drastic reaction conditions, lengthy procedures, and careful handling which significantly affect the commercial production of MXenes. Conversely, the traditional liquid-phase exfoliation produces multi-layered MXenes, which, due to their large lateral size, have poor biomedical efficiency. As an alternative, green synthetic approaches are targeted to provide viable strategies yielding safer and sustainable MXenes. Some of the green pathways for MXene synthesis are molten salt method, microwave hybrid methods, magnetron sputtering with or without coupling with pulsed laser deposition, and sol–gel method [34–39]. Additionally, the replacement of the toxic HF for exfoliation with environmentally friendly solvents (water, ethanol, isopropanol), reagents such as NaOH, HCl, molten $ZnCl_2$, or electrochemical/hydrothermal has made the process greener [40–43].

2.2 Surface Functionalization for Generating MXene-Organic/Inorganic Composites

The introduction of MXenes in the field of biosensors by Xu and group [44], for label-free probing of neurons, paved the path towards surface immobilization of MXenes for developing biosensors. Although water-based synthesis of MXenes has greatly improvised, their inadequacy and unavailability in biological environments are attributed to their instability and aggregation drawing the need for surface modifications to tackle the said shortcomings. These modifications can be brought about

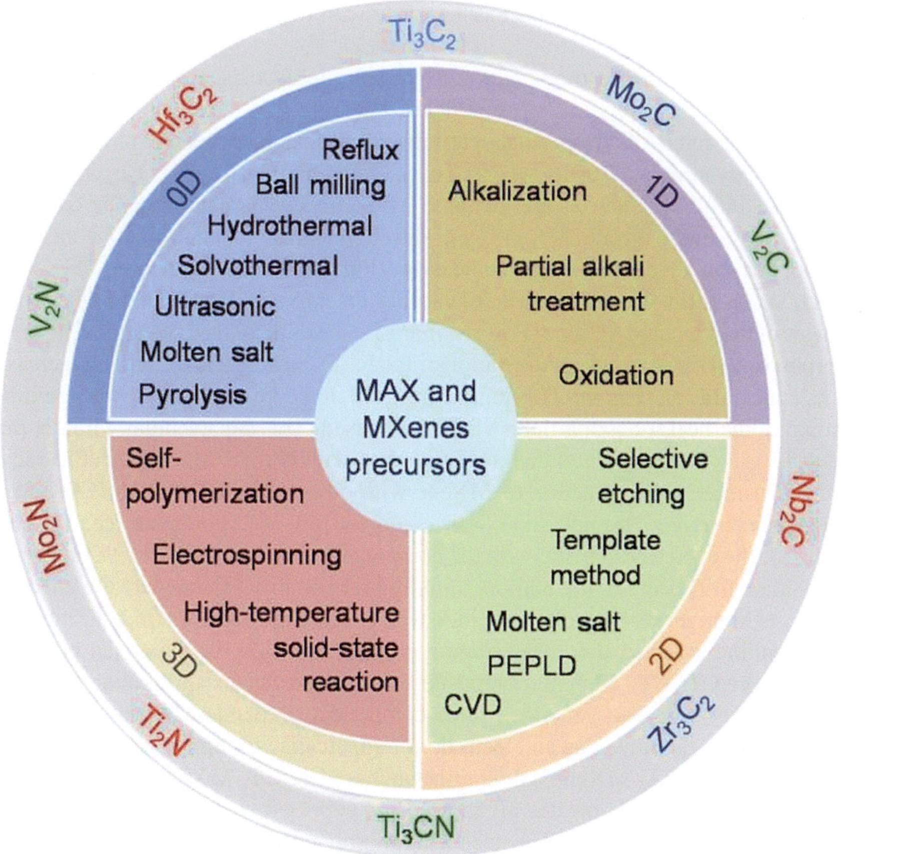

Fig. 3 Fabrication methods of 0D, 1D, 2D, and 3D forms of MXenes and their derivatives. Reproduced from Ref. [27] (Open Access)

by physical adsorption, hydrogen bonding, non-covalent interactions, or chemical bonding, resulting in alterations in the properties of the MXenes. The abundance of modifiable groups (F, Cl, O, and OH) on the surface of the MXene makes them fitting for surface functionalization with a range of moieties, both organic and inorganic systems. The moieties can range from biological molecules like enzymes, proteins, antibodies, biomarkers, to pharmaceutical drugs, chitosan, and polymers (polyvinyl pyrrolidone, polyethylene glycol) for improving material dispersion and hydrophilicity through polymerization or chemical linkers [45–48]. Wu et al. fabricated MXene-polypyrrole composite with imprinted sites on the surface as a modified electrode for the selective electrochemical sensing of *Salmonella* [45]. The sensor was synthesized by the electrochemical polymerization of pyrrole with MXene and *Salmonella* onto a glassy carbon electrode, followed by the removal of the bacterial template (Fig. 4a). The presence of MXene improved the electrical conductivity and interacted better with the bacterial surface with high specificity and selectivity towards *Salmonella.*

In addition to surface modification by organic molecules, the surface can also be decorated using inorganic nanoparticles, boosting the overall properties of the composite. A range of MXene-nanoparticle composites have been explored, where nanoparticles such as Au(0), Ag(0), TiO_2, CeO_2, Fe_3O_4, Co_3O_4, Cu_2O, etc are employed [49–58]. Chen et al. anchored magnetic Fe_3O_4 nanoparticles onto the surface of the multi-layered MXene by in situ chemical approach in alkaline medium for application in photothermal conversion (Fig. 4b) [50]. A series of MXene–Fe_3O_4 composites was developed by varying the MXene–Fe_3O_4 ratio. The SEM images of the prepared MXene–Fe_3O_4 composite series displayed the typical layered morphology with obvious manifestation of the Fe_3O_4 nanoparticles with an increase in its concentration (Fig. 4c–f). With the increase in the Fe_3O_4 content, the agglomeration of the Fe_3O_4 nanoparticles visibly increased. This can eventually hinder the light absorption capacity and thereby lower the photothermal conversion efficiency.

Furthermore, the interaction of MXene with other materials (1D and 2D materials, metal–organic frameworks) was explored [59–66]. For example, Fu and group synthesized a composite with the components graphene, Ti_2CT_x MXene, and polyaniline [59]. Herein, the authors initially functionalized the negatively charged Ti_2CT_x MXene surface by polymerization with polyaniline obtained from the radical-cation aniline. Further, the encapsulation of graphene onto the generated polyaniline anchored Ti_2CT_x MXene composite was carried out in an alkaline medium of pH 9 (Fig. 5a). The generated three-component graphene-MXene-polyaniline composite was employed as an electrode for supercapacitor application. The SEM images of the Ti_2CT_x MXene, polyaniline anchored Ti_2CT_x MXene, and graphene-MXene-polyaniline composites all showed the expected stacked sheets characteristic of MXenes (Fig. 5b–d). The anchoring of polyaniline onto the Ti_2CT_x MXene led to the granulation on the surface of the MXene, clearly apparent in the SEM image (Fig. 5c). In the final graphene-MXene-polyaniline composite, the graphene wraps onto the Ti_2CT_x MXene-polyaniline as seen in Fig. 5d. Thus, we can see many possible modifications of MXenes in appraising and improving the characteristic properties of MXenes. Besides doping of MXene with one other material, the incorporation of combinations of the above dopants as well is widely explored [67–70].

2.3 Medicinal Properties of MXenes

MXene and its composites, similar to other 2D materials, have displayed the promising potential for biomedical applications in healthcare, bioimaging, tissue engineering (neural, cardiac, bone), (bio)sensors (electrochemical, optical), pH-, electrical- and multi-responsive drug delivery systems, photothermal therapy, wound dressing, etc. (Fig. 6) [13, 71, 72]. The tunability of MXenes in blending material science with a range of other systems to attain tailored unique features is the key factor

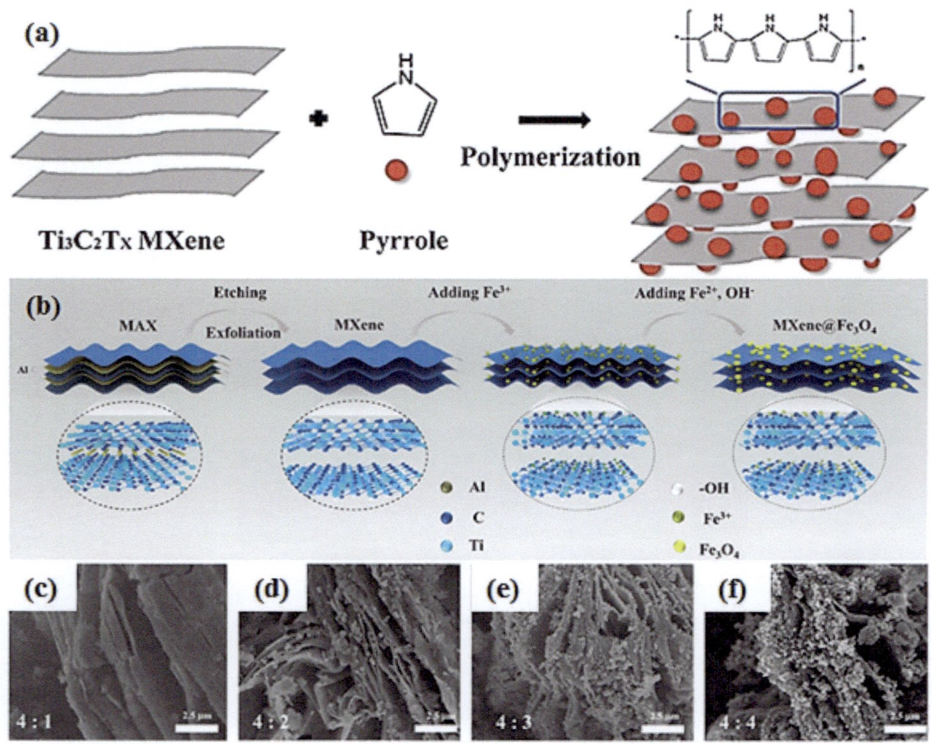

Fig. 4 Schematic representation of **a** surface modification of MXenes by polymerization with pyrrole. Reproduced from Ref. [45] with permission from Elsevier, Copyright 2024 **b** anchoring of Fe_3O_4 nanoparticles onto MXene, SEM images of MXene–Fe_3O_4 composites with MXene:Fe_3O_4 ratio of **c** 4:1, **d** 4:2, **e** 4:3, and **f** 4:4. Reproduced from Ref. [50] (Open Access)

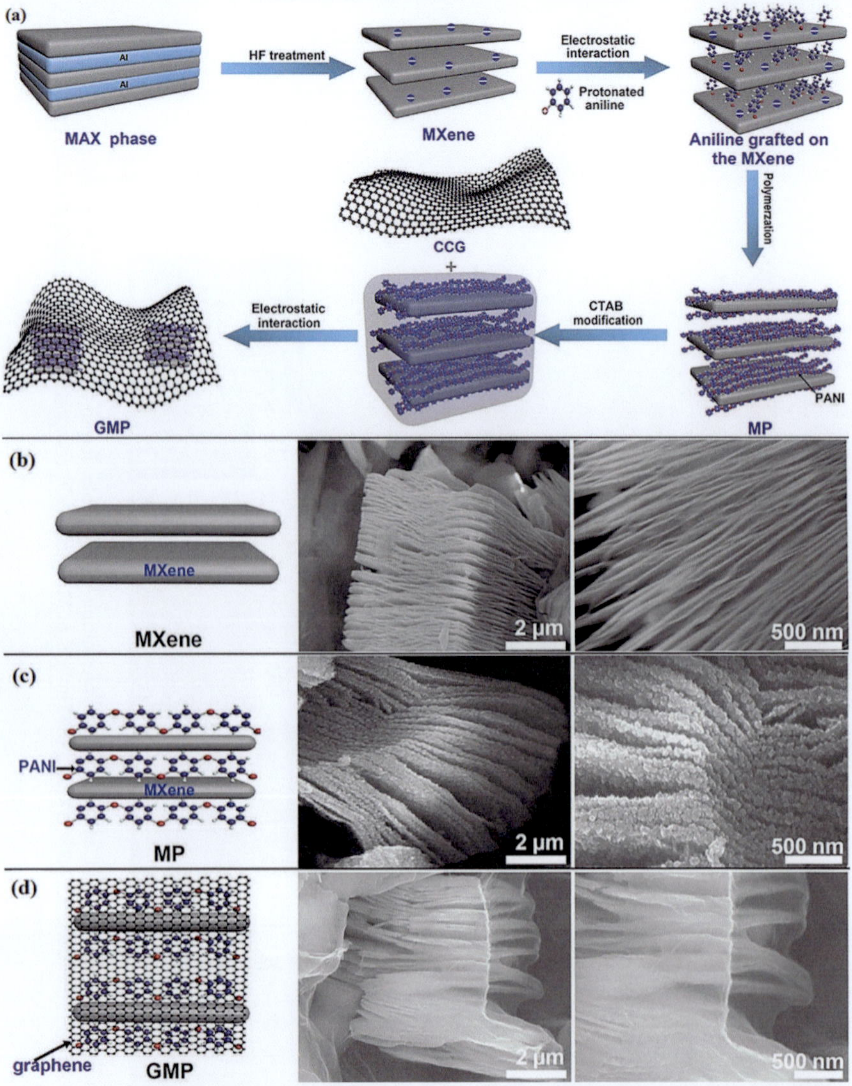

Fig. 5 **a** Schematic illustration for the synthesis of graphene-MXene-polyaniline composite, SEM images of **b** Ti_2CT_x MXene, **c** Ti_2CT_x MXene-polyaniline, and **d** graphene-MXene-polyaniline composite. Reproduced from Ref. [59] with permission from the American Chemical Society, Copyright 2024

for its versatility. This feat is achieved by controlling the nature of the transition metal and X element, surface functionalities, thickness, and hydrophilicity. This provides the researchers with a suitable platform to shape the MXenes and their composites with desired physical, chemical, thermal, mechanical, and electrical properties [73, 74]. For example, substances with high electrical conductivity can be appropriately

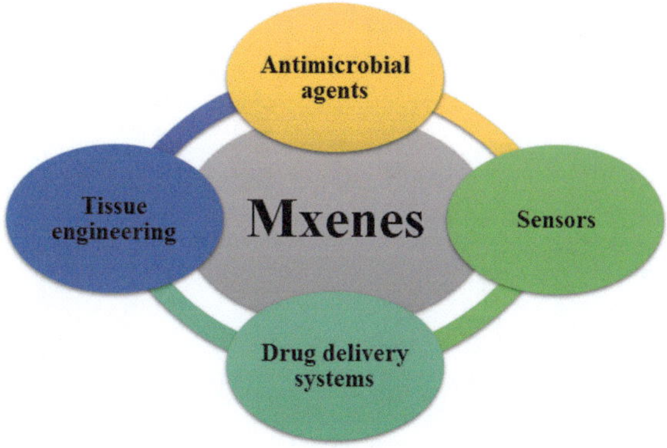

Fig. 6 Various biomedical applications of MXenes

exploited as sensors, as the electron mobility could improve sensitivity and selectivity of the sensor [75, 76]. Moreover, this electrical energy and biocompatibility are exploited in the field of tissue engineering [13]. Alternatively, for employment in the drug delivery system, the surface of the MXene plays an important role in the loading of the therapeutic agents and the stimuli-responsive release of the agent [77].

3 MXenes: Antibacterial Agents

The urgent need for alternative strategies like MXenes is highlighted by the present obstacles facing standard antibacterial and antiviral medicines, including host toxicity, biofilm development, virus resistance, antibiotic resistance, and environmental issues. The distinct characteristics and encouraging antimicrobial efficacy of MXenes render them appealing contenders for the creation of innovative antimicrobial therapies to tackle the escalating peril of antibiotic resistance [78]. For instance, Cao, M. et al. explored the synthesis of Ti_3C_2/γ-glycidoxypropyltrimethoxysilane (GPS) films by immersing aluminum alloy specimens with the Ti_3C_2 nanosheets/γ-GPS solution. After that, they assessed how effective they were at preventing bacterial growth. Using the plate-counting approach, the growth suppression was investigated against *Escherichia coli* (*E. coli*) at 37 °C for 24 h and achieved 90% of the cell death. This antibacterial activity of MXene sheets lies in their ability to disrupt bacterial membranes, leading to leakage of cytoplasmic materials [79]. Mangalaraja, R. V. et al. synthesized Cu–MXene nanocomposite through an etching process by utilizing $CuSO_4$ as a Cu source to investigate the antibacterial activity against both gram-negative *E. coli* and gram-positive *Staphylococcus aureus* (*S. aureus*) by using a zone of inhibition test. At 20 μg/mL concentration, the synthesized Cu–MXene nanocomposite had a strong biocidal activity against *E. Coli* (Fig. 7a) and *S. aureus* (Fig. 7b), with a 1.3 and 1.4 mm zone of inhibition. Additionally, it demonstrated

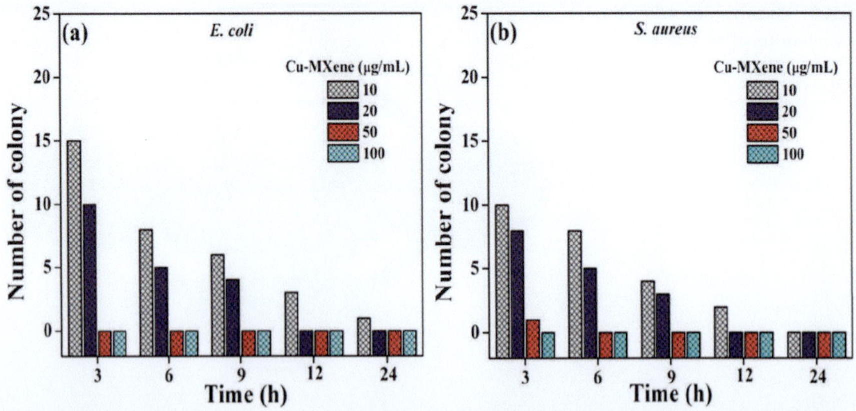

Fig. 7 Antibacterial performance of Cu–MXene nanocomposite synthesized by Mangalaraja et al. against **a** *E. coli* and **b** *S. aureus*. Reproduced from Ref. [80] with permission from Elsevier, Copyright 2024

outstanding antibacterial performance by successfully inhibiting (~99.99%) bacterial growth. The reason for improving its antibacterial action might be due to the addition of Cu to MXene's interlayer spacing [80].

Schauer, C. L. et al. prepared delaminated $Ti_3C_2T_z$ flakes within chitosan nanofibers ($Ti_3C_2T_z$/CS) using an electrospinning technique. First, they prepared the $Ti_3C_2T_z$ flakes using Ti_3AlC_2, later, a chitosan (CS) solution was added to the prepared $Ti_3C_2T_z$ MXene to obtain $Ti_3C_2T_z$/CS composite. An impressive 95% cell reduction rate against *E. coli* and 62% against *S. aureus* bacteria was achieved by direct mechanical destruction of cell membranes with 0.75 wt% $Ti_3C_2T_z$/CS for 4 h. Enhanced cell reduction against *E. coli* might be a result of MXene and CS's synergistic actions. Still, the decreased activity against *S. aureus* might be due to the stronger peptidoglycan cell wall of the Gram-positive *S. aureus* bacteria [81]. Yuen, A. C. Y. et al. reported the synthesis of Ti_3C_2/CS/Cu and Ag composites coated with polyurethane foam (PUF) through a layer-by-layer assembly method with $CuSO_4$ and $AgNO_3$ as metal precursors for imparting antimicrobial functionality. Among them, Ti_3C_2/CS/Ag-coated PUF showed higher activity by reducing 88.9% of *S. aureus* and 99.97% of *Pseudomonas aeruginosa* (*P. aeruginosa*) whereas Ti_3C_2/CS/Cu-coated PUF results in decreased cell reduction against both *S. aureus* and *P. aeruginosa* by 72.4 and 58.7%. The improved activity of Ti_3C_2/CS/Ag-coated PUF compared to Ti_3C_2/CS/Cu-coated PUF might be due to the enhanced thermal stability of Ag over Cu composite [67]. Chen, S. et al. synthesized cuprous oxide (Cu_2O) nanoparticles supported on MXene nanosheets (Cu_2O/MXene) and assessed their antibacterial activity against *P. aeruginosa* and *S. aureus* by the plate colony counting method. The Cu_2O/MXene composite was prepared by using $CuSO_4$ and $Ti_3C_2T_x$ as starting materials at room temperature, in which Cu_2O nanospheres were uniformly decorated on MXene by the electronic transition between Cu_2O and MXene nanosheets. The obtained nanocomposite exhibits excellent antibacterial activity, with bacteriostasis efficiencies of up to 97.04 and 95.59% against *P. aeruginosa* and *S. aureus*, respectively, by following the photoelectron transportation mechanism. Furthermore, Cu_2O nanoparticles' antibacterial properties are enhanced by the MXene nanosheets' large surface area and electrical conductivity, which promote effective electron transfer processes for enhanced antibacterial activity [82]. Zhang, Q. et al. designed the synthesis of a hydrogel scaffold based on MXene@CeO_2 nanocomposite for the antibacterial activity. In this study, the synthesis began with the fabrication of a hydrogel by combining MXene@CeO_2 nanocomposite with a polyethyleneimine (MEI)-grafted Pluronic F127 and oxidized sodium alginate. Later on, extensive in vitro studies were used to assess the scaffold's antibacterial efficacy for *E. coli*, *S. aureus*, and *methicillin-resistant S. aureus* (*MRSA*) and attained complete cell inhibition. The reason for the complete cell death was due to the cationic PEI attracting the negatively charged bacterial cell membrane and destroying its transmembrane potential, as well as the MXene nanocomposite's nano-knife-like structures could disrupt the integrity of cell membrane structures [83]. Xie, J. et al. developed ultra-small gold nanoclusters (AuNCs) by conjugation on MXene nanosheets ($Ti_3C_2T_x$–AuNCs). The selection of MXene nanosheets was based on their ability to physically destroy the bacterial cell membrane, and AuNCs were selected because they can generate reactive oxygen species (ROS) for bacterial DNA and cell membrane oxidation, which eventually leads to cell death. The MXene-AuNC conjugates' antibacterial effectiveness is rigorously assessed against both *S. aureus* and *E. coli* and achieved >98 and >96% of bacterial death [84]. Deng, Y. et al. described the synthesis of MXene/cobalt nanowire heterojunctions (Ti_3C_2/CoNWs/SPEEK) and also evaluated their antimicrobial activity against both *E. coli* and *S. aureus* by the spread plate method in both light and dark environments. Firstly, Ti_3C_2/CoNWs/SPEEK nanocomposite was designed by mixing both Ti_3C_2 MXene and CoNWs in a 2:1 ratio, then, through the spin coating technique, the obtained composite was coated on the SPEEK. After that, the obtained material showed a significant increase in antibacterial efficacy (92.74 and 80.1% against *E. coli* and *S. aureus*, respectively) under light irradiation compared to dark conditions. This improvement is ascribed to the heterojunctions' facilitation of the photo-activated production of ROS [85]. Hong, W. et al. synthesized 3 different types of MXenes (Ti_3C_2, V_2C, and Nb_2C) and investigated their antibacterial activity in vitro against 15 various types of bacteria at different wavelengths of light (532, 808, and 980 nm). Finally, Ti_3C_2 at 808 nm of light showed greater mortality rates of 94% only for 20 min. Of these, the Ti_3C_2 demonstrated noteworthy antibacterial efficacy, resulting in 95 and 99% cells killing for *E. Coli* and *S. aureus*, respectively. The greater efficacy of Ti_3C_2 was attributed to its ability to kill bacteria with light by both physical and through photothermal effect [86]. To capitalize on their potential to fight against microbial infections, Lin, D. et al. created graphitic carbon-supported TiO_2 nanoparticles generated from MXene (B-$Ti_3C_2T_x$ for bulk B-$Ti_3C_2T_x$, I for isopropylamine delaminated B-$Ti_3C_2T_x$, and W for water delaminated B-$Ti_3C_2T_x$) at temperatures ranging from 200–1000 °C for

1–12 h. Additionally, the created hybrid nanostructures (B-$Ti_3C_2T_x$, I-350 °C-1 h, I-350 °C-3 h, and I-350 °C-6 h) demonstrated outstanding antibacterial activities with about 52.63, 97.38, 70.12, and 65.68% of bacteriostasis (Fig. 8a) by the colony count method against *E. coli*. From the dose-dependent bactericidal curves of I-$Ti_3C_2T_x$-350 °C-1 h and B-$Ti_3C_3T_x$, the former displayed excellent bacteriostasis of 97.38% with 50 μg/mL concentration, while the latter showed about 42% bacteriostasis at the same concentration (Fig. 8b). After that, I-350 °C-1 h produced more oxidative stress by releasing ROS species than the control and B-$Ti_3C_3T_x$, which proved the greater interaction between bacterial cells and I-350 °C-1 h (Fig. 8c). Later on, to determine the influence of the light-induced activity, the bactericidal properties of I-$Ti_3C_2T_x$-350 °C-1 h, B-$Ti_3C_3T_x$, and pure TiO_2 nanoparticles were explored under dark and light environments. Among them, I-350 °C-1 h showed excellent antibacterial activity against *E. coli* in the presence of light (Fig. 8d) [87]. The above-discussed systems and their antibacterial performances are listed in Table 1.

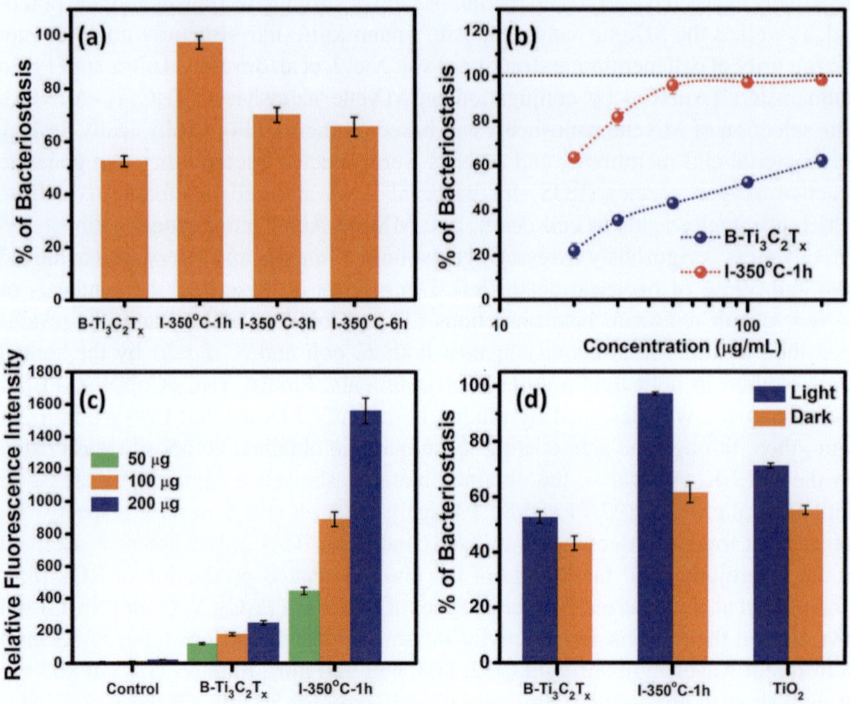

Fig. 8 **a** Bactericidal activities, **b** dose-dependent bactericidal curves, **c** relative production of ROS within cells, and **d** bactericidal properties of B-$Ti_3C_2T_x$, I-350 °C-1 h, and TiO_2 nanoparticles in both light and dark environments. Reproduced from Ref. [87] with permission from Elsevier, Copyright 2024

Table 1 Antibacterial activity of different types of MXenes

Entry	MXene composites	No. of microbes screened (n)	Highest activity against	MXene loading	Microbe concentration	Exposure (h)	% Inhibition	References
1	Ti_3C_2/γ-GPS	1	*E. coli*	0.02 wt.%	10^5 CFU mL^{-1}	24	90	[79]
2	Cu–MXene	2	*S. aureus*	10 μg mL^{-1}	10 μg mL^{-1}	–	~99.99	[80]
3	$Ti_3C_2T_z$/CS	2	*E. coli*	0.75 wt.%	~3 × 10^8 CFU mL^{-1}	4	95	[81]
4	Ti_3C_2/Chitosan/Ag	2	*P. aeruginosa*	–	100 μL	–	99.97	[67]
5	Cu_2O/MXene	2	*S. aureus*	40 μg mL^{-1}	1 μLmL^{-1}	3	97.04	[82]
6	MXene@CeO_2	3	*E. coli*, *S. aureus*, and *MRSA*	–	10^8 CFU mL^{-1}	2	100	[83]
7	$Ti_3C_2T_x$–AuNCs	2	*S. aureus*	11.7 μg mL^{-1}	-	2	>98	[84]
8	Ti_3C_2/CoNWs/SPEEK	2	*E. coli*	50 μg mL^{-1}	10^6 CFU mL^{-1}	0.3	92.74	[85]
9	Ti_3C_2	15	*S. aureus*	100 μg mL^{-1}	10^6-10^7 CFU mL^{-1}	20 min	99	[86]
10	I-350 °C-6 h	1	*E. coli*	100 μg mL^{-1}	10^6-10^7 CFU mL^{-1}	3	97.38	[87]

4 MXenes: Antiviral Agents

Among the many difficulties faced by conventional antiviral treatments are drug resistance, toxicity, a narrow range of efficacy, and high cost. MXenes offer a variety of modes of action, low toxicity, broad-spectrum efficacy, and the possibility of combination therapy, making them a promising choice to address these issues [88]. For instance, Yilmazer A. et al. explored the synthesis of four different types of MXenes–$Ti_3C_2T_x$, $Ta_4C_3T_x$, $Mo_2Ti_2C_3T_x$, and $Nb_4C_3T_x$ as novel and potential antiviral agents against the COVID-19-causing four various SARS-CoV-2 viruses. After that, they measured the number of viral copies and the vitality of Vero E6 cells to evaluate the viral inhibition, following which they proposed a proteomic analysis-based mechanism of viral suppression. Finally, $Ti_3C_2T_x$ achieved greater than 99% inhibition at 0.32 μg/mL, $Mo_2Ti_2C_3T_x$ achieved greater than 95% inhibition at 100 μg/mL, and there was no discernible shift in the viral copy number for $Ta_4C_3T_x$ and $Nb_4C_3T_x$. Here, $Ti_3C_2T_x$ uses other signalling pathways in addition to the cell surface to carry out its viral inhibitory function, and the host proteins have been demonstrated to interact with viral proteins to be key regulators of $Ti_3C_2T_x$-dependent antiviral action (Fig. 9) [89].

5 Mechanism for Antimicrobial Behavior of MXenes and Their Composites

Research is still ongoing to completely understand the process of MXenes' microbial inactivation. Nevertheless, several explanations have been postulated in light of experimental data encompassing photocatalytic mechanisms, photothermal mechanisms, and physical contact destruction [82, 86, 90].

5.1 Photocatalytic Mechanism

Light-induced microbial disinfection, or photocatalytic antimicrobial efficacy, is largely dependent on the generation of highly active ROS via photogenerated electron–hole (e^--h^+) pairs [91]. There are three main phases to the photocatalytic disinfection mechanism starting with the attachment of the microbial strain to the MXene nanosheet surface, the efficient separation of photogenerated e^--h^+ pairs, and ending with the interaction of the separated e^--h^+ pairs with oxygen (O_2) and water (H_2O) molecules to produce a variety of ROS [92]. Chen, S. et al. demonstrated the greater antibacterial activity of Cu_2O/MXene nanocomposite and enhanced photocatalytic antibacterial efficacy against *P. aeruginosa* and *S. aureus*. They clarified

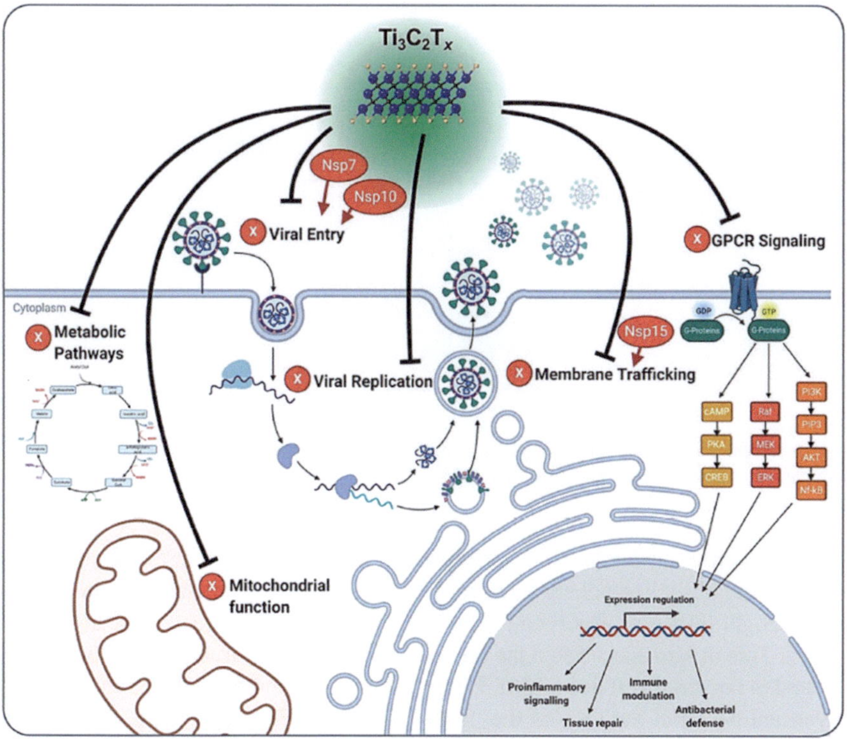

Fig. 9 The mechanism of viral inhibition based on MXene ($Ti_3C_2T_x$). Reproduced from Ref. [89] with permission from Elsevier, Copyright 2024

that when the synthesized nanocomposite was irrafiated with sunlight, the electrons selectively moved from Cu_2O to MXenes, successfully preventing the e^--h^+ pair recombination. Additionally, MXene takes up electrons from Cu_2O, producing a large number of electrons on its surface that improve charge transfer between Cu_2O/MXene nanosheets and bacteria. Finally, the accumulation of intracellular ROS, such as hydrogen peroxide (H_2O_2), superoxide anions ($\cdot O_2^-$), or hydroxyl radicals ($\cdot OH$), damages the cell membrane and ultimately results in cell death (Fig. 10) [82].

5.2 *Photothermal Mechanism*

The photothermal mechanism, an essential biomedical technique, uses photothermal agents to efficiently produce heat under specific power intensity light irradiation in order to kill affected cells [93]. Hong, W. et al. reported the antibacterial activity of Ti_3C_2 MXenes against 15 different types of bacteria under near-infrared (NIR) laser irradiation at 808 nm. Furthermore, they assessed the mechanism to prove the effect

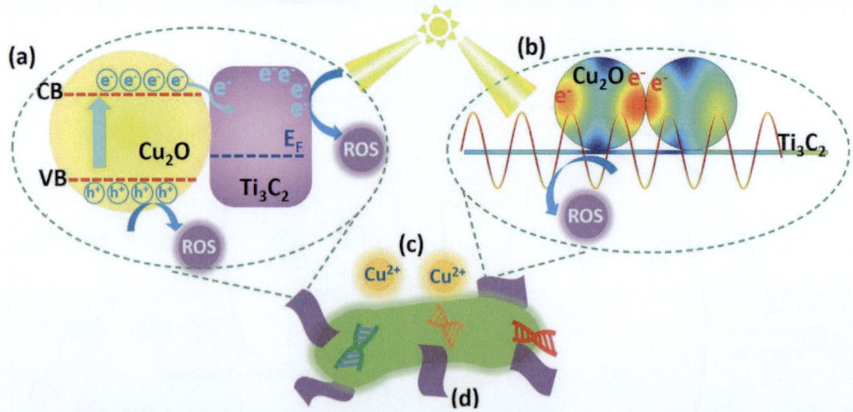

Fig. 10 Proposed antibacterial mechanism of action for Cu_2O/MXene. Reproduced from Ref. [82] with permission from Elsevier, Copyright 2024

of light by different methods. They explained that the sharp edges of Ti_3C_2 flakes provided the ability to adhere to or insert into bacterial cells, and the intensity of the NIR light (808 nm) absorbed by the Ti_3C_2 significantly increased the heat of the system. This in turn accelerated the breakdown of bacterial structures and ultimately resulted in bacterial cell death [86]. Deng, Y. et al. demonstrated the photothermally driven antibacterial activity of the Ti_3C_2/CoNWs/SPEEK against *E. Coli* and *S. aureus*. The enhanced activity of the material is due to photothermal-induced hyperthermia increasing the bacterial membrane permeability upon NIR light irradiation, facilitating the easy passage of NIR-activated ROS across the bacterial membrane into the bacteria to cause oxidative stress and significant protein leakage. In this case, normal cells are not harmed because, once the bacteria are eliminated, the Ti_3C_2/CoNWs/SPEEK will cease to produce heat and ROS with the removal of NIR light (Fig. 11) [85].

5.3 Physical Contact Destruction

MXenes can physically interact with the membranes of microorganisms because of their sharp edges and high surface roughness. This interaction has the potential to damage the cell membrane, jeopardizing its integrity and allowing internal organelles to seep out, ultimately resulting in cell death [79]. In this regard, MnO_2 and MoS_2/Ti_3C_2 nanosheets performed exceptionally well against *B. subtilis* and *E. coli*, as demonstrated by Strongin, D. R. et al. The MnO_2 and MoS_2 nanosheets arranged vertically over the MXene have sharp edges (as nano-knives), and thus the integrity of the membrane may be lowered as a result of this interaction, leading to cell death (Fig. 12) [90].

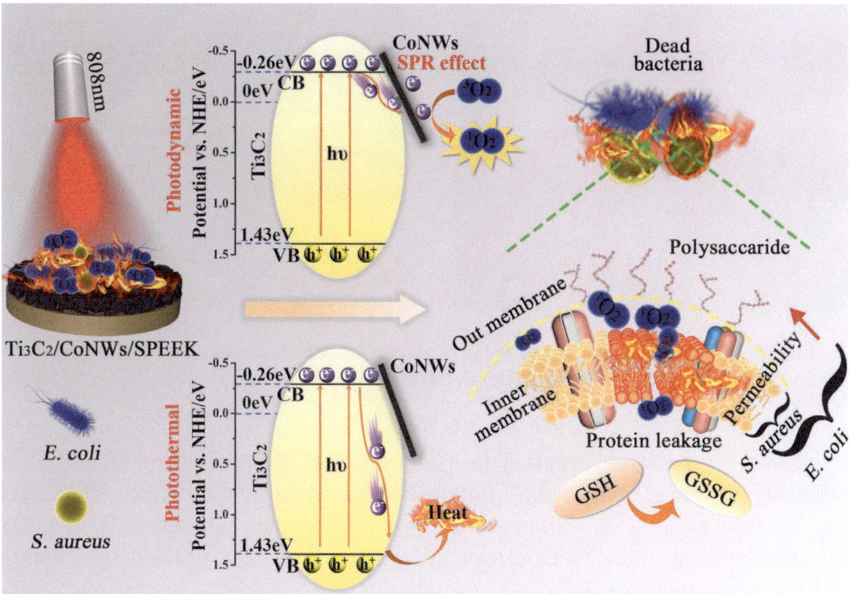

Fig. 11 Diagram representation of the Ti_3C_2/CoNWs/SPEEK synergistic antibacterial mechanism activated by NIR. Reproduced from Ref. [85] with permission from Elsevier, Copyright 2024

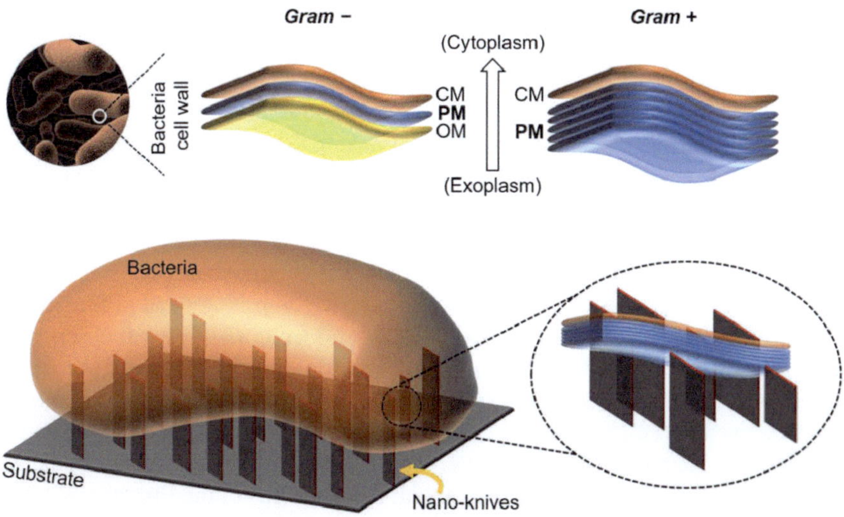

Fig. 12 **a** The cell membrane structure of gram-negative and gram-positive bacteria and **b** diagrammatic illustration of the direct physical contact between a substrate and the sharp edges of vertically oriented nanosheets on the surface of the bacteria. Reproduced from Ref. [90] with permission from the American Chemical Society, Copyright 2024

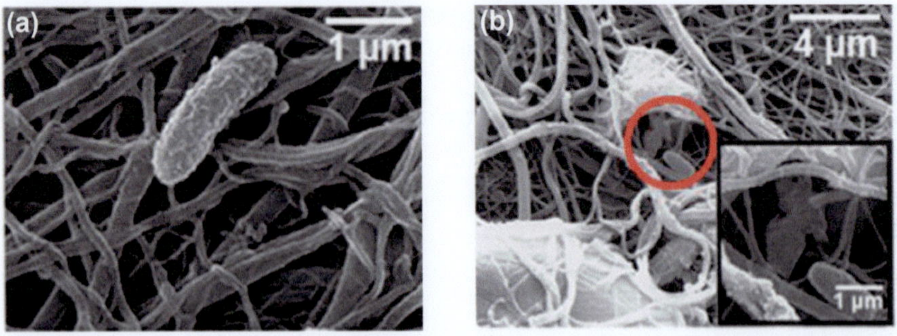

Fig. 13 SEM images of **a** intact and **b** killed *E. coli* bacterium on the 0.75 wt% $Ti_3C_2T_z$/CS nanofiber. Reproduced from Ref. [81] (Open Access)

Schauer, C. L. et al. reported the synthesis and the activity of $Ti_3C_2T_z$/CS composite against *E. Coli* and *S. aureus*. Finally, SEM images showed the intact (Fig. 13a) and killed (Fig. 13b) appearances of *E. coli* when exposed to $Ti_3C_2T_z$/CS mats, which reveales the direct mechanical degradation caused by MXene penetrating the bacterial membranes [81].

6 Biocompatibility and Toxicity

Regardless of the distinct properties of MXene and its composites, their biocompatibility and cytotoxicity need to be addressed to evaluate the potential risk to health and the environment. Though studies have claimed MXenes to be biocompatible, the extent of cytotoxicity is a crucial parameter for effective usage in biomedical applications. From the cytotoxicity studies explored for Ti_3C_2 by Jastrzebska and group [94], the MXene was found to display higher toxicity towards cancer than normal cells, suggesting the production of ROS, which is higher in cancer cells. Likewise, when Ti_3C_2 nanosheet was assessed for its influence on primary and differentiated cells derived from neural stem cells, no noticeable effect on viability was observed at 12.5 μg/mL, while a significant rise in apoptotic neural stem cells occurred at concentrations of 25 μg/mL and higher (Fig. 14a) [95]. Moreover, they presented a cellular compatibility with cell viability of over 70%. Similar in vivo studies with zebrafish embryos and mouse models with $Ti_3C_2T_x$ MXenes demonstrated high compatibility with no significant toxicity in line with the standards of the Fish and Wildlife Service [96, 97]. Not just MXenes, their composites too exhibited trivial toxicity. Dai et al. performed the in vivo study with MnO_x/Ti_3C_2-SP composite on healthy Kunming mice to understand its biocompatibility [49]. The composite was administered intravenously with varying doses of 5, 10, and 20 mg kg^{-1}. The histological characterization of the vital organs showed insignificant tissue damage (Fig. 14b). Although the in vitro and in vivo studies indicate biocompatibility and low toxicity, the in vivo distribution, accumulation, mechanistic pathway, and toxicity of MXenes still remain vague and need further venturing to invariably establish the safe usage of MXene and its composites.

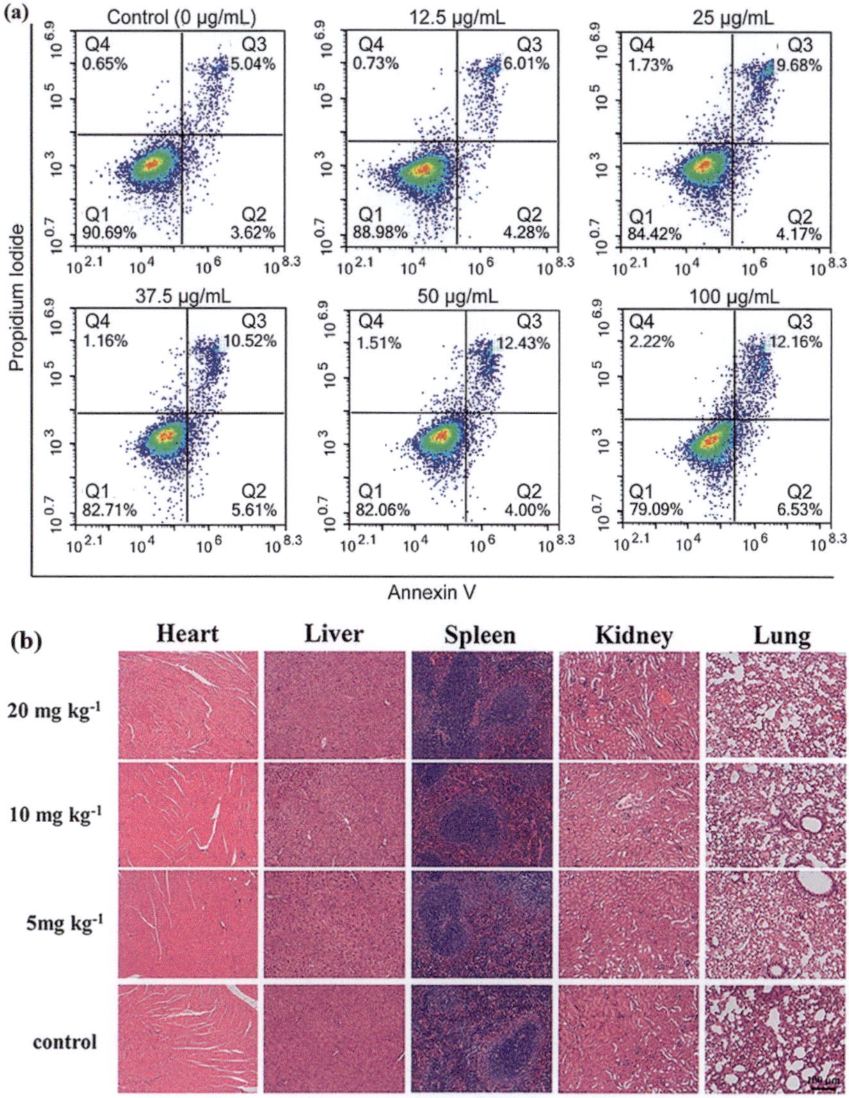

Fig. 14 **a** Representative flow cytometry results to analyze cytotoxicity of Ti_3C_2 nanosheets with Ti_3C_2-treated neural stem cells stained by annexin V and propidium iodide. Reproduced from Ref. [95] with permission from the American Chemical Society, Copyright 2024. **b** The histological characterization (H&E-stained tissue sections) of the heart, liver, spleen, lung, and kidney of healthy Kunming mice with various dosages of MnO_x/Ti_3C_2-SP composite. Reproduced from Ref. [49] with permission from the American Chemical Society, Copyright 2024

7 Conclusion and Future Perspective

To summarize, MXenes, although known for over a decade, have been poorly explored for their antimicrobial potential in the past few years. The demanding synthetic procedure of MXene are being improvised to greener and facile protocols, aiming at a greater expansion in the material synthesis and applications. These MXenes are assuring materials for a range of applications, especially in the biomedical field of biosensing, bioimaging, theranostic, therapeutic, tissue engineering, etc., attributed to their fascinating and tunable physiochemical, photo, thermal and electrical properties. Herein, we have briefly discussed the synthetic approaches, all possible surface modification, and biomedical applications of MXenes, with weightage given to their antibacterial and antiviral activities. Although the antibacterial property of MXenes and their composites has been widely explored, their evaluation as antivirals is still in its infancy. Studies have depicted the adequate biocompatibility of MXenes, promising a better future for these systems; however, cytotoxicity concerns still provide room for improvement. On the other hand, the complete mechanistic interaction between the MXene and microbe is unclear, needing further venturing to understand and thereby harness and develop proficient antimicrobial agents. Better understanding of the microbe-composite interaction, biocompatibility, and non-toxicity can pave the use of these MXenes and their composite in real-time products and therapies. There is still a long way to go for the application of MXene and its composites from laboratory scale to commercialization, exploiting their antimicrobial activity to wearable masks, aprons, and medical tools.

References

1. Skwarczynski, M., Bashiri, S., Yuan, Y., Ziora, Z.M., Nabil, O., Masuda, K., Khongkow, M., Rimsueb, N., Cabral, H., Ruktanonchai, U.: Antimicrobial activity enhancers: towards smart delivery of antimicrobial agents. Antibiotics **11**, 412 (2022). https://doi.org/10.3390/antibiotics11030412
2. Ramirez, J., Dartois, N., Gandjini, H., Yan, J.L., Korth-Bradley, J., McGovern, P.C.: Randomized phase 2 trial to evaluate the clinical efficacy of two high-dosage tigecycline regimens versus imipenem-cilastatin for treatment of hospital-acquired pneumonia. Antimicrob. Agents Chemother. **57**, 1756–1762 (2013). https://doi.org/10.1128/aac.01232-12
3. Balasubramaniam, B., Prateek, Ranjan, S., Saraf, M., Kar, P., Singh, S.P., Thakur, V.K., Singh, A., Gupta, R.K.: Antibacterial and antiviral functional materials: chemistry and biological activity toward tackling COVID-19-like pandemics. ACS Pharmacol. Transl. Sci. **4**, 8–54 (2020). https://doi.org/10.1021/acsptsci.0c00174
4. Ramezani Farani, M., Farsadrooh, M., Zare, I., Gholami, A., Akhavan, O.: Green synthesis of magnesium oxide nanoparticles and nanocomposites for photocatalytic antimicrobial, antibiofilm and antifungal applications. Catalysts **13**, 642 (2023). https://doi.org/10.3390/catal13040642
5. Luo, H., Yin, X.-Q., Tan, P.-F., Gu, Z.-P., Liu, Z.-M., Tan, L.: Polymeric antibacterial materials: design, platforms and applications. J. Mater. Chem. B **9**, 2802–2815 (2021). https://doi.org/10.1039/D1TB00109D

6. Li, W., Thian, E.S., Wang, M., Wang, Z., Ren, L.: Surface design for antibacterial materials: from fundamentals to advanced strategies. Adv. Sci. **8**, 2100368 (2021). https://doi.org/10.1002/advs.202100368
7. Zamhuri, A., Lim, G.P., Ma, N.L., Tee, K.S., Soon, C.F.: MXene in the lens of biomedical engineering: synthesis, applications and future outlook. Biomed. Eng. Online **20**, 1–24 (2021). https://doi.org/10.1186/s12938-021-00873-9
8. Iravani, P., Iravani, S., Varma, R.S.: MXene-chitosan composites and their biomedical potentials. Micromachines **13**, 1383 (2022). https://doi.org/10.3390/mi13091383
9. Iravani, S.: MXenes and MXene-based (nano) structures: a perspective on greener synthesis and biomedical prospects. Ceram. Int. **48**, 24144–24156 (2022). https://doi.org/10.1016/j.ceramint.2022.05.137
10. Iravani, S., Varma, R.S.: MXenes for cancer therapy and diagnosis: recent advances and current challenges. ACS Biomater. Sci. Eng. **7**, 1900–1913 (2021). https://doi.org/10.1021/acsbiomaterials.0c01763
11. Naguib, M., Mashtalir, O., Carle, J., Presser, V., Lu, J., Hultman, L., Gogotsi, Y., Barsoum, M.W.: Two-dimensional transition metal carbides. ACS Nano **6**, 1322–1331 (2012). https://doi.org/10.1021/nn204153h
12. Naguib, M., Kurtoglu, M., Presser, V., Lu, J., Niu, J., Heon, M., Hultman, L., Gogotsi, Y., Barsoum, M.W.: Two-dimensional nanocrystals produced by exfoliation of Ti_3AlC_2. In: MXenes. Jenny Stanford Publishing, pp. 15–29 (2011)
13. Lee, I.-C., Li, Y.-C.E., Thomas, J.L., Lee, M.-H., Lin, H.-Y.: Recent advances using MXenes in biomedical applications. Mater. Horiz. **11**, 876–902 (2024). https://doi.org/10.1039/D3MH01588B
14. Frey, N.C., Wang, J., Vega Bellido, G.I., Anasori, B., Gogotsi, Y., Shenoy, V.B.: Prediction of synthesis of 2D metal carbides and nitrides (MXenes) and their precursors with positive and unlabeled machine learning. ACS Nano **13**, 3031–3041 (2019). https://doi.org/10.1021/acsnano.8b08014
15. Bae, S., Kang, Y.G., Khazaei, M., Ohno, K., Kim, Y.H., Han, M.J., Chang, K.J., Raebiger, H.: Electronic and magnetic properties of carbide MXenes-the role of electron correlations. Mater. Today Adv. **9**, 100118 (2021). https://doi.org/10.1016/j.mtadv.2020.100118
16. Khazaei, M., Mishra, A., Venkataramanan, N.S., Singh, A.K., Yunoki, S.: Recent advances in MXenes: from fundamentals to applications. Curr. Opin. Solid State Mater. Sci. **23**, 164–178 (2019). https://doi.org/10.1016/j.cossms.2019.01.002
17. Michałowski, P.P., Anayee, M., Mathis, T.S., Kozdra, S., Wójcik, A., Hantanasirisakul, K., Jóźwik, I., Piątkowska, A., Możdżonek, M., Malinowska, A.: Oxycarbide MXenes and MAX phases identification using monoatomic layer-by-layer analysis with ultralow-energy secondary-ion mass spectrometry. Nat. Nanotechnol. **17**, 1192–1197 (2022). https://doi.org/10.1038/s41565-022-01214-0
18. Gogotsi, Y., Anasori, B.: The rise of MXenes. ACS Nano **13**, 8491–8494 (2019). https://doi.org/10.1021/acsnano.9b06394
19. Firouzjaei, M.D., Karimiziarani, M., Moradkhani, H., Elliott, M., Anasori, B.: MXenes: the two-dimensional influencers. Mater. Today Adv. **13**, 100202 (2022). https://doi.org/10.1016/j.mtadv.2021.100202
20. Chen, J., Fu, W., Jiang, F.-L., Liu, Y., Jiang, P.: Recent advances in 2D metal carbides and nitrides (MXenes): synthesis and biological application. J. Mater. Chem. B **11**, 702–715 (2023). https://doi.org/10.1039/D2TB01503J
21. Lim, K.R.G., Shekhirev, M., Wyatt, B.C., Anasori, B., Gogotsi, Y., Seh, Z.W.: Fundamentals of MXene synthesis. Nat. Synth. **1**, 601–614 (2022). https://doi.org/10.1038/s44160-022-00104-6
22. Ding, L., Wei, Y., Wang, Y., Chen, H., Caro, J., Wang, H.: A two-dimensional lamellar membrane: MXene nanosheet stacks. Angew. Chem., Int. Ed. Engl. **56**, 1825–1829 (2017). https://doi.org/10.1002/anie.201609306
23. Xu, C., Wang, L., Liu, Z., Chen, L., Guo, J., Kang, N., Ma, X.-L., Cheng, H.-M., Ren, W.: Large-area high-quality 2D ultrathin Mo_2C superconducting crystals. Nat. Mater. **14**, 1135–1141 (2015). https://doi.org/10.1038/nmat4374

24. Barsoum, M.W., El-Raghy, T., Farber, L., Amer, M., Christini, R., Adams, A.: The topotactic transformation of Ti3SiC2 into a partially ordered cubic Ti (C 0.67 Si0. 06) phase by the diffusion of si into molten cryolite. J. Electrochem. Soc. **146**, 3919 (1999). https://doi.org/10.1149/1.1392573
25. Naguib, M., Barsoum, M.W., Gogotsi, Y.: Ten years of progress in the synthesis and development of MXenes. Adv. Mater. **33**, 2103393 (2021). https://doi.org/10.1002/adma.202103393
26. Wang, H.-W., Naguib, M., Page, K., Wesolowski, D.J., Gogotsi, Y.: Resolving the structure of $Ti_3C_2T_x$ MXenes through multilevel structural modeling of the atomic pair distribution function. Chem. Mater. **28**, 349–359 (2016). https://doi.org/10.1021/acs.chemmater.5b04250
27. Amrillah, T., Abdullah, C.A.C., Hermawan, A., Sari, F.N.I., Alviani, V.N.: Towards greener and more sustainable synthesis of MXenes: a review. Nanomaterials **12**, 4280 (2022). https://doi.org/10.3390/nano12234280
28. Zhan, X., Si, C., Zhou, J., Sun, Z.: MXene and MXene-based composites: synthesis, properties and environment-related applications. Nanoscale Horiz. **5**, 235–258 (2020). https://doi.org/10.1039/C9NH00571D
29. Sun, Z., Fan, Q., Zhang, M., Liu, S., Tao, H., Texter, J.: Supercritical fluid-facilitated exfoliation and processing of 2D materials. Adv. Sci. **6**, 1901084 (2019). https://doi.org/10.1002/advs.201901084
30. Naguib, M., Unocic, R.R., Armstrong, B.L., Nanda, J.: Large-scale delamination of multi-layers transition metal carbides and carbonitrides "MXenes." Dalton Trans. **44**, 9353–9358 (2015). https://doi.org/10.1039/C5DT01247C
31. Ghidiu, M., Lukatskaya, M.R., Zhao, M.-Q., Gogotsi, Y., Barsoum, M.W.: Conductive two-dimensional titanium carbide 'clay'with high volumetric capacitance. Nature **516**, 78–81 (2014). https://doi.org/10.1038/nature13970
32. Wan, M., Jimu, A., Yang, H., Zhou, J., Dai, X., Zheng, Y., Ou, J., Yang, Y., Liu, J., Wang, L.: MXene quantum dots enhanced 3D-printed electrochemical sensor for the highly sensitive detection of dopamine. Microchem. J. **184**, 108180 (2023). https://doi.org/10.1016/j.microc.2022.108180
33. Xue, Q., Zhang, H., Zhu, M., Pei, Z., Li, H., Wang, Z., Huang, Y., Huang, Y., Deng, Q., Zhou, J.: Photoluminescent Ti_3C_2 MXene quantum dots for multicolor cellular imaging. Adv. Mater. **29** (2017) https://doi.org/10.1002/adma.201604847
34. Liu, X., Fechler, N., Antonietti, M.: Salt melt synthesis of ceramics, semiconductors and carbon nanostructures. Chem. Soc. Rev. **42**, 8237–8265 (2013). https://doi.org/10.1039/C3CS60159E
35. Gonzalez-Julian, J.: Processing of MAX phases: from synthesis to applications. J. Am. Ceram. Soc. **104**, 659–690 (2021). https://doi.org/10.1111/jace.17544
36. Tan, L., Guan, C., Tian, Y., Dang, P., Wang, S., Li, J., Li, W., Zhao, Z.: Synthesis and tribological properties of ultrafine Cr_2AlC MAX phase. J. Ceram. Soc. Jpn. **127**, 754–760 (2019). https://doi.org/10.2109/jcersj2.18184
37. Hu, J., Bultman, J., Patton, S., Zabinski, J.: Pulsed laser deposition and properties of $M_{n+1}AX_n$ phase formulated Ti_3SiC_2 thin films. Tribol. Lett. **16**, 113–122 (2004). https://doi.org/10.1023/B:TRIL.0000009721.87012.45
38. Siebert, J.P., Bischoff, L., Lepple, M., Zintler, A., Molina-Luna, L., Wiedwald, U., Birkel, C.S.: Sol-gel based synthesis and enhanced processability of MAX phase Cr_2GaC. J. Mater. Chem. C **7**, 6034–6040 (2019). https://doi.org/10.1039/C9TC01416K
39. Siebert, J.P., Patarakun, K., Birkel, C.S.: Mechanistic insights into the nonconventional Sol–Gel synthesis of MAX phase M_2GeC (M = V, Cr). Inorg. Chem. **61**, 1603–1610 (2022). https://doi.org/10.1021/acs.inorgchem.1c03415
40. Luo, S., Dong, S., Lu, C., Yu, C., Ou, Y., Luo, L., Sun, J., Sun, J.: Rational and green synthesis of novel two-dimensional WS_2/MoS_2 heterojunction via direct exfoliation in ethanol-water targeting advanced visible-light-responsive photocatalytic performance. J. Colloid Interface Sci. **513**, 389–399 (2018). https://doi.org/10.1016/j.jcis.2017.11.044
41. Li, T., Yao, L., Liu, Q., Gu, J., Luo, R., Li, J., Yan, X., Wang, W., Liu, P., Chen, B.: Fluorine-free synthesis of high-purity $Ti_3C_2T_x$ (T = OH, O) via alkali treatment. Angew. Chem. Int. Ed. **57**, 6115–6119 (2018). https://doi.org/10.1002/anie.201800887

42. Sun, W., Shah, S., Chen, Y., Tan, Z., Gao, H., Habib, T., Radovic, M., Green, M.: Electrochemical etching of Ti_2AlC to Ti_2CT_x (MXene) in low-concentration hydrochloric acid solution. J. Mater. Chem. A **5**, 21663–21668 (2017). https://doi.org/10.1039/C7TA05574A
43. Yang, S., Zhang, P., Wang, F., Ricciardulli, A.G., Lohe, M.R., Blom, P.W., Feng, X.: Fluoride-free synthesis of two-dimensional titanium carbide (MXene) using a binary aqueous system. Angew. Chem. **130**, 15717–15721 (2018). https://doi.org/10.1002/ange.201809662
44. Ferrari, M.: Cancer nanotechnology: opportunities and challenges. Nat. Rev. Cancer **5**, 161–171 (2005). https://doi.org/10.1038/nrc1566
45. Wang, O., Jia, X., Liu, J., Sun, M., Wu, J.: Rapid and simple preparation of an MXene/polypyrrole-based bacteria imprinted sensor for ultrasensitive *Salmonella* detection. J. Electroanal. Chem. **918**, 116513 (2022). https://doi.org/10.1016/j.jelechem.2022.116513
46. Lin, H., Chen, Y., Shi, J.: Insights into 2D MXenes for versatile biomedical applications: current advances and challenges ahead. Adv. Sci. **5**, 1800518 (2018). https://doi.org/10.1002/advs.201800518
47. Lin, H., Gao, S., Dai, C., Chen, Y., Shi, J.: A two-dimensional biodegradable niobium carbide (MXene) for photothermal tumor eradication in NIR-I and NIR-II biowindows. J. Am. Chem. Soc. **139**, 16235–16247 (2017). https://doi.org/10.1021/jacs.7b07818
48. Hroncekova, S., Lorencova, L., Bertok, T., Hires, M., Jane, E., Bučko, M., Kasak, P., Tkac, J.: Amperometric miniaturised portable enzymatic nanobiosensor for the ultrasensitive analysis of a prostate cancer biomarker. J. Funct. Biomater. **14**, 161 (2023). https://doi.org/10.3390/jfb14030161
49. Dai, C., Lin, H., Xu, G., Liu, Z., Wu, R., Chen, Y.: Biocompatible 2D titanium carbide (MXenes) composite nanosheets for pH-responsive MRI-guided tumor hyperthermia. Chem. Mater. **29**, 8637–8652 (2017). https://doi.org/10.1021/acs.chemmater.7b02441
50. Gao, Y., Tang, Z., Chen, X., Yan, J., Jiang, Y., Xu, J., Tao, Z., Wang, L., Liu, Z., Wang, G.: Magnetically accelerated thermal energy storage within Fe_3O_4-anchored MXene-based phase change materials. Aggregate **4**, e248 (2023). https://doi.org/10.1002/agt2.248
51. Liu, Z., Lin, H., Zhao, M., Dai, C., Zhang, S., Peng, W., Chen, Y.: 2D superparamagnetic tantalum carbide composite MXenes for efficient breast-cancer theranostics. Theranostics **8**, 1648 (2018). https://doi.org/10.7150/2Fthno.23369
52. Li, Z., Zhang, H., Han, J., Chen, Y., Lin, H., Yang, T.: Surface nanopore engineering of 2D MXenes for targeted and synergistic multitherapies of hepatocellular carcinoma. Adv. Mater. **30**, 1706981 (2018). https://doi.org/10.1002/adma.201706981
53. Singh, S., Numan, A., Khalid, M., Bello, I., Panza, E., Cinti, S.: Facile and affordable design of MXene-Co_3O_4-based nanocomposites for detection of hydrogen peroxide in cancer cells: toward portable tool for cancer management. Small **19**, 2208209 (2023). https://doi.org/10.1002/smll.202208209
54. Ding, R., Jiang, W., Ma, Y., Yang, Q., Han, X., Hou, X.: A highly sensitive MXene/AuPt/AChE-based electrochemical platform for the detection of chlorpyrifos. Microchem. J. **187**, 108425 (2023). https://doi.org/10.1016/j.microc.2023.108425
55. Jiang, S., Zhang, H., Li, Z., Chen, L., Yin, L., Liu, X.: Sponge-inspired MXene@CeO_2 detector for ultra-sensitive detection of glucose. Mater. Today Chem. **32**, 101638 (2023). https://doi.org/10.1016/j.mtchem.2023.101638
56. Lee, S.J., Lee, D.H., Lee, W.-Y.: An electrochemical sensor for capsaicin based on two-dimensional titanium carbide (MXene)-doped titania-Nafion composite film. Microchem. J. **185**, 108216 (2023). https://doi.org/10.1016/j.microc.2022.108216
57. Kim, J.H., Shin, J.H., Park, B., Cho, C.H., Huh, Y.S., Choi, C.-H., Park, J.P.: Harnessing protein sensing ability of electrochemical biosensors via a controlled peptide receptor–electrode interface. J. Nanobiotechnol. **21**, 100 (2023). https://doi.org/10.1186/s12951-023-01843-0
58. Cai, Z., Tu, B., Zhou, R., Xia, D., He, H., Gao, N., Chang, G., He, Y.: Ultrathin $Ti_3C_2T_x$ MXene-based electrochemical transistor for highly sensitive determination of nitrite. J. Electroanal. Chem. **928**, 117012 (2023). https://doi.org/10.1016/j.jelechem.2022.117012

59. Fu, J., Yun, J., Wu, S., Li, L., Yu, L., Kim, K.H.: Architecturally robust graphene-encapsulated MXene Ti_2CT_x@Polyaniline composite for high-performance pouch-type asymmetric supercapacitor. ACS Appl. Mater. Interfaces **10**, 34212–34221 (2018). https://doi.org/10.1021/acsami.8b10195
60. Lu, C., Xu, P., Wang, D., Fu, D.: Integrating MWCNTs-doped MXene with multi-spiral-channel architecture enables field effect transistor biosensor capable of ultrasensitive determination of methotrexate. Arab. J. Chem. **16**, 104943 (2023). https://doi.org/10.1016/j.arabjc.2023.104943
61. Lu, Z., Wei, K., Ma, H., Duan, R., Sun, M., Zou, P., Yin, J., Wang, X., Wang, Y., Wu, C.: Bimetallic MOF synergy molecularly imprinted ratiometric electrochemical sensor based on MXene decorated with polythionine for ultra-sensitive sensing of catechol. Anal. Chim. Acta **1251**, 340983 (2023). https://doi.org/10.1016/j.aca.2023.340983
62. Tian, L., Jiang, M., Su, M., Cao, X., Jiang, Q., Liu, Q., Yu, C.: Sweat cortisol determination utilizing MXene and multi-walled carbon nanotube nanocomposite functionalized immunosensor. Microchem. J. **185**, 108172 (2023). https://doi.org/10.1016/j.microc.2022.108172
63. Jinsong, T., Jin, H., Weizhang, L., Junyou, C., Jian, L., Xiangming, C., Hua, Z.: Integrating MXene/MWCNTs into aptasensor capable of ultrasensitive quantification of cTnI towards the diagnosis of acute myocardial infarction. Chem. Pap. **78**, 3205–3212 (2024). https://doi.org/10.1007/s11696-024-03304-w
64. Li, Y., Zhang, L., Wu, M., Ma, G., Motlak, M., Mahdi, A.: Novel electrochemical strategy for determination of anticancer drug flutamide based on MXene/MOF composite. Inorg. Chem. Commun. **155**, 111061 (2023). https://doi.org/10.1016/j.inoche.2023.111061
65. Zhang, Y., Wang, Z., Liu, X., Liu, Y., Cheng, Y., Cui, D., Chen, F., Cao, W.: A MXene/MoS_2 heterostructure based biosensor for accurate sweat ascorbic acid detection. FlatChem **39**, 100503 (2023). https://doi.org/10.1016/j.flatc.2023.100503
66. Wu, P., You, T., Ren, Q., Xi, H., Liu, Q., Qin, F., Gu, H., Wang, Y., Yan, W., Gao, Y.: Interface electronic engineering of molybdenum sulfide/MXene hybrids for highly efficient biomimetic sensors. Nano Res. **16**, 1158–1164 (2023). https://doi.org/10.1007/s12274-022-5038-3
67. Lin, B., Yuen, A.C.Y., Oliver, S., Liu, J., Yu, B., Yang, W., Wu, S., Yeoh, G.H., Wang, C.H.: Dual functionalisation of polyurethane foam for unprecedented flame retardancy and antibacterial properties using layer-by-layer assembly of MXene chitosan with antibacterial metal particles. Compos. B Eng. **244**, 110147 (2022). https://doi.org/10.1016/j.compositesb.2022.110147
68. Shen, J., Liu, J., Yang, S., Yao, X., Fa, H., Hou, C., Yang, M.: Novel electrochemical sensor based on PDA/MXene/MWCNTs/$NiCo_2O_4$ nanocomposites for rapid, sensitive, and selective detection of aflatoxin B1. Food Anal. Methods **16**, 1055–1068 (2023). https://doi.org/10.1007/s12161-023-02464-x
69. Feng, L., Qin, W., Wang, Y., Gu, C., Li, X., Chen, J., Chen, J., Qiao, H., Yang, M., Tian, Z.: $Ti_3C_2T_x$ MXene/Graphene/AuNPs 3D porous composites for high sensitivity and fast response glucose biosensing. Microchem. J. **184**, 108142 (2023). https://doi.org/10.1016/j.microc.2022.108142
70. Zhang, M., Liu, H., Wang, X.: Cholesterol oxidase-immobilized MXene/sodium alginate/silica@n-docosane hierarchical microcapsules for ultrasensitive electrochemical biosensing detection of cholesterol. J. Mater. Chem. B **11**, 1506–1522 (2023). https://doi.org/10.1039/D2TB02367A
71. Liu, H., Xing, X., Tan, Y., Dong, H.: Two-dimensional transition metal carbides and nitrides (MXenes) based biosensing and molecular imaging. Nanophotonics **11**, 4977–4993 (2022). https://doi.org/10.1515/nanoph-2022-0550
72. Park, H., Kwon, N., Park, G., Jang, M., Kwon, Y., Yoon, Y., An, J., Min, J., Lee, T.: Fast-response electrochemical biosensor based on a truncated aptamer and MXene heterolayer for West Nile virus detection in human serum. Bioelectrochemistry **154**, 108540 (2023). https://doi.org/10.1016/j.bioelechem.2023.108540
73. Lim, G.P., Soon, C.F., Ma, N.L., Morsin, M., Nayan, N., Ahmad, M.K., Tee, K.S.: Cytotoxicity of MXene-based nanomaterials for biomedical applications: a mini review. Environ. Res. **201**, 111592 (2021). https://doi.org/10.1016/j.envres.2021.111592

74. Mostafavi, E., Iravani, S.: MXene-graphene composites: A perspective on biomedical potentials. Nano-Micro Lett. **14**, 130 (2022). https://doi.org/10.1007/s40820-022-00880-y
75. Sreenilayam, S.P., Ahad, I.U., Nicolosi, V., Brabazon, D.: MXene materials based printed flexible devices for healthcare, biomedical and energy storage applications. Mater. Today **43**, 99–131 (2021). https://doi.org/10.1016/j.mattod.2020.10.025
76. Maleki, A., Ghomi, M., Nikfarjam, N., Akbari, M., Sharifi, E., Shahbazi, M.A., Kermanian, M., Seyedhamzeh, M., Nazarzadeh Zare, E., Mehrali, M.: Biomedical applications of MXene-integrated composites: regenerative medicine, infection therapy, cancer treatment, and biosensing. Adv. Funct. Mater. **32**, 2203430 (2022). https://doi.org/10.1002/adfm.202203430
77. Farani, M.R., Khiarak, B.N., Tao, R., Wang, Z., Ahmadi, S., Hassanpour, M., Rabiee, M., Saeb, M.R., Lima, E.C., Rabiee, N.: 2D MXene nanocomposites: Electrochemical and biomedical applications. Environ. Sci. Nano **9**, 4038–4068 (2022). https://doi.org/10.1039/D2EN00527A
78. Lam, S.-M., Jaffari, Z.H., Yong, Z.-J., Sin, J.-C., Zeng, H., Lin, H., Li, H., Mohamed, A.R.: Chapter 16—MXenes and their composites for potential antimicrobial applications. In: Sadasivuni, K.K., Deshmukh, K., Pasha, S.K.K., Kovářík, T. (eds.) Mxenes and Their Composites, pp. 525–551. Elsevier (2022)
79. Nie, Y., Huang, J., Ma, S., Li, Z., Shi, Y., Yang, X., Fang, X., Zeng, J., Bi, P., Qi, J.: MXene-hybridized silane films for metal anticorrosion and antibacterial applications. Appl. Surf. Sci. **527**, 146915 (2020). https://doi.org/10.1016/j.apsusc.2020.146915
80. Talreja, N., Ashfaq, M., Chauhan, D., Mangalaraja, R.: Cu-MXene: a potential biocide for the next-generation biomedical application. Mater. Chem. Phys. **294**, 127029 (2023). https://doi.org/10.1016/j.matchemphys.2022.127029
81. Mayerberger, E.A., Street, R.M., McDaniel, R.M., Barsoum, M.W., Schauer, C.L.: Antibacterial properties of electrospun $Ti_3C_2T_z$ (MXene)/chitosan nanofibers. RSC Adv. **8**, 35386–35394 (2018). https://doi.org/10.1039/c8ra06274a
82. Wang, W., Feng, H., Liu, J., Zhang, M., Liu, S., Feng, C., Chen, S.: A photo catalyst of cuprous oxide anchored MXene nanosheet for dramatic enhancement of synergistic antibacterial ability. J. Chem. Eng. **386**, 124116 (2020). https://doi.org/10.1016/j.cej.2020.124116
83. Zheng, H., Wang, S., Cheng, F., He, X., Liu, Z., Wang, W., Zhou, L., Zhang, Q.: Bioactive anti-inflammatory, antibacterial, conductive multifunctional scaffold based on MXene@CeO_2 nanocomposites for infection-impaired skin multimodal therapy. J. Chem. Eng. **424**, 130148 (2021). https://doi.org/10.1016/j.cej.2021.130148
84. Zheng, K., Li, S., Jing, L., Chen, P.Y., Xie, J.: Synergistic antimicrobial titanium carbide (MXene) conjugated with gold nanoclusters. Adv. Healthc. Mater. **9**, 2001007 (2020). https://doi.org/10.1002/adhm.202001007
85. Liu, Y., Tian, Y., Han, Q., Yin, J., Zhang, J., Yu, Y., Yang, W., Deng, Y.: Synergism of 2D/1D MXene/cobalt nanowire heterojunctions for boosted photo-activated antibacterial application. J. Chem. Eng. **410**, 128209 (2021). https://doi.org/10.1016/j.cej.2020.128209
86. Wu, F., Zheng, H., Wang, W., Wu, Q., Zhang, Q., Guo, J., Pu, B., Shi, X., Li, J., Chen, X.: Rapid eradication of antibiotic-resistant bacteria and biofilms by MXene and near-infrared light through photothermal ablation. Sci. China Mater. **64**, 748–758 (2021). https://doi.org/10.1007/s40843-020-1451-7
87. Rajavel, K., Shen, S., Ke, T., Lin, D.: Photocatalytic and bactericidal properties of MXene-derived graphitic carbon-supported TiO_2 nanoparticles. Appl. Surf. Sci. **538**, 148083 (2021). https://doi.org/10.1016/j.apsusc.2020.148083
88. Panda, S., Deshmukh, K., Mustansar Hussain, C., Khadheer Pasha, S.K.: 2D MXenes for combatting COVID-19 pandemic: a perspective on latest developments and innovations. FlatChem **33**, 100377 (2022). https://doi.org/10.1016/j.flatc.2022.100377
89. Unal, M.A., Bayrakdar, F., Fusco, L., Besbinar, O., Shuck, C.E., Yalcin, S., Erken, M.T., Ozkul, A., Gurcan, C., Panatli, O., Summak, G.Y., Gokce, C., Orecchioni, M., Gazzi, A., Vitale, F., Somers, J., Demir, E., Yildiz, S.S., Nazir, H., Grivel, J.-C., Bedognetti, D., Crisanti, A., Akcali, K.C., Gogotsi, Y., Delogu, L.G., Yilmazer, A.: 2D MXenes with antiviral and immunomodulatory properties: a pilot study against SARS-CoV-2. Nano Today **38**, 101136 (2021). https://doi.org/10.1016/j.nantod.2021.101136

90. Alimohammadi, F., Sharifian, M., Attanayake, N.H., Thenuwara, A.C., Gogotsi, Y., Anasori, B., Strongin, D.R.: Antimicrobial properties of 2D MnO_2 and MoS_2 nanomaterials vertically aligned on graphene materials and Ti_3C_2 MXene. Langmuir **34**, 7192–7200 (2018). https://doi.org/10.1021/acs.langmuir.8b00262
91. Jaffari, Z.H., Lam, S.M., Sin, J.C., Mohamed, A.R.: Constructing magnetic Pt-loaded $BiFeO_3$ nanocomposite for boosted visible light photocatalytic and antibacterial activities. Environ. Sci. Pollut. R **26**, 10204–10218 (2019). https://doi.org/10.1007/s11356-019-04503-9
92. Jaffari, Z.H., Lam, S.-M., Sin, J.-C., Zeng, H.: Boosting visible light photocatalytic and antibacterial performance by decoration of silver on magnetic spindle-like bismuth ferrite. Mater. Sci. Semicond. Process. **101**, 103–115 (2019). https://doi.org/10.1016/j.mssp.2019.05.036
93. Sun, W., Wu, F.G.: Two-dimensional materials for antimicrobial applications: graphene materials and beyond. Chem. Asian J. **13**, 3378–3410 (2018). https://doi.org/10.1002/asia.201800851
94. Jastrzębska, A., Szuplewska, A., Wojciechowski, T., Chudy, M., Ziemkowska, W., Chlubny, L., Rozmysłowska, A., Olszyna, A.: In vitro studies on cytotoxicity of delaminated Ti_3C_2 MXene. J. Hazard. Mater. **339**, 1–8 (2017). https://doi.org/10.1016/j.jhazmat.2017.06.004
95. Wu, W., Ge, H., Zhang, L., Lei, X., Yang, Y., Fu, Y., Feng, H.: Evaluating the cytotoxicity of Ti_3C_2 MXene to neural stem cells. Chem. Res. Toxicol. **33**, 2953–2962 (2020). https://doi.org/10.1021/acs.chemrestox.0c00232
96. Nasrallah, G.K., Al-Asmakh, M., Rasool, K., Mahmoud, K.A.: Ecotoxicological assessment of $Ti_3C_2T_x$ (MXene) using a zebrafish embryo model. Environ. Sci. Nano **5**, 1002–1011 (2018). https://doi.org/10.1039/C7EN01239J
97. Han, X., Huang, J., Lin, H., Wang, Z., Li, P., Chen, Y.: 2D ultrathin MXene-based drug-delivery nanoplatform for synergistic photothermal ablation and chemotherapy of cancer. Adv. Healthc. Mater. **7**, 1701394 (2018). https://doi.org/10.1002/adhm.201701394

MXene and Its Derivatives for Energy Storage Applications

Shavita, Jyoti Jangra, and Suman Singh

Abstract MXenes (2-D transition metal carbides, nitrides, and carbonitrides), a novel class of two-dimensional materials, have cumulative significant attention, particularly for their remarkable properties in energy-related domains. Their diverse structures, compositions, conductivity, and surface properties make them one of the most versatile families of 2D materials. Since the last decade, various MXenes and their derivatives have been developed. MXene-derived materials include carbon-based materials, metal oxides, and transition metal sulfides, and these materials have different compositions and can be prepared by oxidation, chlorination, nitridation, or sulfurization. Hence, this chapter discusses the recent enhancement of MXene materials and their derivatives in energy storage application batteries and supercapacitors, and the relationship between the structure and electrochemical performance of the MXene derivatives is discussed. This chapter will first discuss the current synthesis methods for MXene derivatives, emphasizing oxidation reactions and exploring the correlation between the structure of derivatives and as electrode material for different metal ion batteries and supercapacitor applications. A focal point of MXene derivatives is the interrelationship between their structure, properties, and applications. Finally, the limitation of MXene derivatives and their upcoming potential for energy storage applications. This overview is a valuable guide for researchers interested in the unfolding evolution of MXene-based materials for energy storage applications.

Keywords 2D materials · MXene · Supercapacitor · Batteries · Nanomaterials

Shavita and Jyoti Jangra—Equal Contribution for first authorship.

Shavita · J. Jangra · S. Singh (✉)
Academy of Scientific and Innovative Research (AcSIR), Ghaziabad, India
e-mail: ssingh@csio.res.in

CSIR-Central Scientific Instruments Organisation, Chandigarh, India

© The Author(s), under exclusive license to Springer Nature Singapore Pte Ltd. 2025

S. Moharana and R. N. Mahaling (eds.), *Surface Engineering of MXenes*, Engineering Materials, https://doi.org/10.1007/978-981-96-3640-2_7

1 Introduction

Two-dimensional (2D) structured materials gained attention after the discovery of graphene. Among various 2D materials, known as MXene, which are transition metal carbides, nitrides, or carbonitrides, have been widely used in the form of potential electrode materials because of their higher volumetric capacitance, energy density, and power density [1, 2]. MXene belongs to the 2D material family of nitrides or metal carbides with the general formula of $M_{n+1}X_nT_x$ and is synthesized from its MAX phase, where M represents the early transition metal, A denotes group III A or IVA elements, and X could be carbon, nitrogen element by etching the A layer from MAX phase. Tx is the termination group, and n can vary from 2 to 4. After their synthesis of MXene in 2011, MXene has become a wonder material that has received considerable attention in various fields and has been used in multiple applications like catalysts, sensors, and HER/OER [3].

MXene is synthesized from the MAX phase with structural arrangements, $M_{n+1}AX_n$, where M is a transition metal, and A is Al or Si, etched from the MAX phase to form MXene, X is carbon, nitrogen, or both. The M-X bond is more stable than the M-A bond; because of this, the etching of the A layer from MXene and $M_{n+1}X_n$ can be easily separated. In the etching process, the MXene surface gets covered by oxygen, hydroxyl, fluorine, or chlorine, after which $M_{n+1}AX_n$ changes to $M_{n+1}XT_x$, where T_x is the termination group. With the recent development in research, various MAX phases have been discovered, which has also increased attention to developing new MXenes. MXenes, because of their surface functional groups, similar structure, reduction ability, and many derivatives, have also been created via different synthesis methods, including surface functionalization, single atom doping, or by modifying the interlayer of MXene sheets. Transition metal bonds with halogen atoms are relatively weak, which enables surface modification to customize terminations. MXenes, due to surface defects, serve as platforms for single-atom doping, and their interlayer gaps and positively charged surfaces facilitate the intercalation of ions between their sheets. Also, by replacing M–C bonds with M–N or M–O bonds, nitride and oxide heterostructures are formed, which enhance the physical and chemical properties of these MXene derivatives.

This chapter introduces and explores the MXenes as supercapacitors, pseudocapacitors, and batteries. Special attention is given to MXenes-derived materials for energy storage applications, with a comprehensive discussion on various synthesis methods of MXene derivatives and previously reported MXenes for energy-related applications, supercapacitors, and metal ion batteries. It addresses the difficulties and future development outlooks for MXene derivatives.

2 MXene and Its Derivatives

MXene is a two-dimensional material with great potential for energy storage applications. It belongs to the family of two-dimensional inorganic compounds consisting of atomically thin layers of transition metal carbides, nitrides, or carbonitrides. MXene and its derivatives have been explored in various energy storage devices, such as sodium-ion batteries (SIBs), lithium-ion batteries (LIBs), water splitting, and electrochemical charge storage devices. Different physical and electrochemical capabilities of MXenes have made them a promising choice for metal-ion, metal-air, and metal-sulfur batteries. Despite its potential, MXene has some limitations regarding its stability in aqueous medium, for which researchers are working to overcome the limitation. The research and development of MXene and its derivatives for energy storage applications are ongoing, focusing on addressing the material's shortcomings and realizing its full potential.

MXene can be used as a precursor for a derived material, opening a new category of controllable nanomaterials. Various carbon-based materials, metal oxides, and even MOF can be prepared from MXenes and have been used in multiple applications. MXene and its derivatives have shown promising applications in energy storage owing to their high conductivity and adjustable surface functional groups. They offer diverse functions in batteries and supercapacitors, including double-layer and redox-type ion storage, ion transfer regulation, steric hindrance, ion redistribution, electrocatalysts, and electrodeposition substrates. The material's worth as a reliable electrode for electrochemical energy storage devices has been demonstrated, and it is expected to play a significant role in industrial applications.

The preparation strategies of various MXene derivatives, including oxidation, chlorination, nitridation, fluorination, and sulfurization, as well as their products for energy storage, have been widely studied. MXene derivatives are mainly prepared through the oxidation reaction, and the relationship between the structure and electrochemical performance of the MXene derivatives is being investigated. Various types of structures are possible as shown in Fig. 1.

3 Surface-Functionalized MXenes

The surface or physical properties of MXene can be modified or improved by changing the termination group during the synthesis process [5]. –O or –F termination groups can be achieved by conventional chemical etching methods, such as HF etching or fluorine-containing salts. The concentration of these termination groups can also be varied by changing the time and concentration of the etchant used and by various post-treatment methods [6, 7]. Schultz and coworkers studied the effect of various termination groups on the physical properties of MXenes [8]. Moreover, Density Functional Theory (DFT) has also been used to evaluate the effects of surface

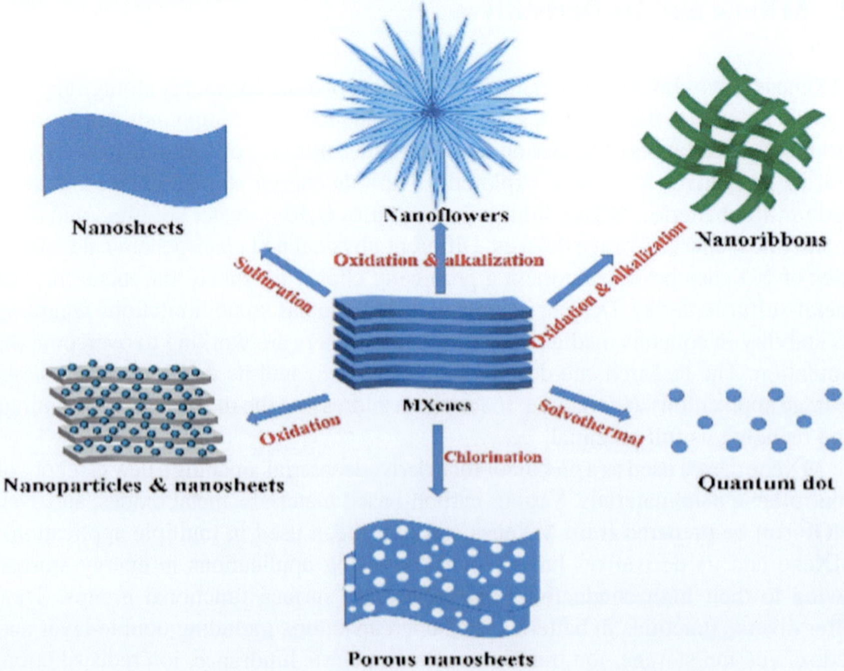

Fig. 1 Structures of MXene-derived materials. Reproduced with permission from Ref. [4]. Copyright 2020 Elsevier

termination on the electronic conductivity of MXene sheets. It was found that surface termination affects the Fermi-level density of states of MXene [9].

Surface termination in MXene can be manipulated to tune its physical and chemical properties. Hence, efforts have been directed toward investigating MXene with specific termination. Oxygen-enriched terminated MXene was synthesized through an alkalization process by Bao and coworkers [10]. First, the MAX phase Ti_3AlC_2 was selectively etched to form layered fluorine-enriched MXene $Ti_3Cl_2F_x$ via the HF etching method, and after the etching process, the synthesized MXene was allowed to react with KOH to form $Ti_3C_2O_x$. Oxygen-terminated MXene was achieved because Ti–F was thermodynamically unstable in an alkaline medium; under this condition, –F terminations were substituted by –OH terminations, generating a water-soluble fluoride compound. Lia and coworkers functionalized the $Ti_3C_2T_x$ MXene nanosheets with functional groups via fluorine groups through a surface treatment process as depicted in Fig. 2a [11].

Another approach was used by Xie et al. [9] to achieve fluorine-enriched MXene Ti_2CF_x via selective removal of the oxygen-terminated group from Ti_2CT_x MXene nanosheets. The method adopted to synthesize fluorine-terminated MXene was annealed at a lower temperature of 200 °C to remove –OH and –O surface terminations while maintaining the –F surface termination almost the same, resulting in fluorine-enriched MXene.

Talapin and coworkers used a covalent surface modification strategy with the help of a molten salt system to attach or remove the surface termination groups. Initially, $CdBr_2$ salt was used to form Br-terminated MXene, and as the M–Br bond is weak,

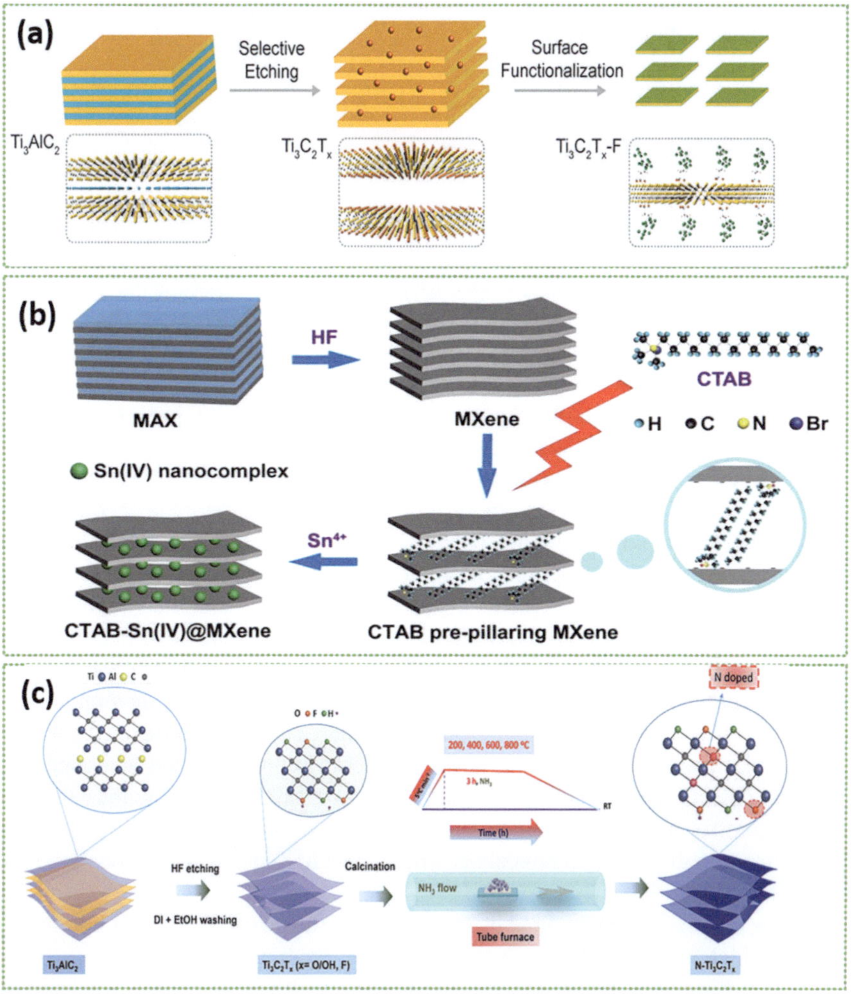

Fig. 2 **a** Surface functionalization of MXene flakes. Reproduced from Ref. [11] with permission from American Chemical Society, Copyright 2020; **b** Schematic illustration of preparation of CTAB–Sn(IV)@Ti_3C_2 by HF etching, CTAB prepillaring, and Sn^{4+} pillaring methods [35] with permission from American Chemical Society, Copyright 2016; **c** Schematic diagram of Synthesis of N-Doped MXene from the Ti_3AlC_2 MAX Phase Reproduced from Ref. [21] with permission from the American Chemical Society, Copyright 2019

making it suitable for covalent surface modification CsBr/KBr/LiBr eutectic system can also be used with Li_2S to replace the M–Br bond with the M–S bond. Then, the S-terminated MXene dispersed in n-butyl lithium intercalation and N-methyl formamide (NMF) solvent for delamination, resulting in O-, NH-, Cl-, Se-, Br-, Te-terminated MXenes, as well as bare MXenes without surface terminations. Another method used by Xiao et al. [12] for functionalized surface termination of MXene by Lewis basic halides $AlBr_3$/NaBr/KBr eutectic molten salt with different molar ratios was utilized to treat $Ti_3C_2T_x$, which facilitates the substitution of -F termination groups via nucleophilic reactions.

4 Single-Atom Doped MXenes

Introducing new atoms into the materials, which is known as doping, can effectively improve their physical or chemical properties. Theoretical and experimental methods have been explored for designing doped MXenes. Doping of MXene can be done in various ways, such as M-doped, X-doped, or T-substituted. Heteroatoms such as N, P, or O can be incorporated at the position of M, X, or T. As there are various possibilities for positioning the atoms, with such diverse possibilities, they possess remarkable opportunities in multiple applications. In-situ and ex-situ doping methods an active areas of interest nowadays [13, 14]. Metal ions can also be used for the doping of MXene, which offers the opportunity to tune their intrinsic properties. Metal ions are introduced into the host in various applications, such as energy storage or conversion. Electrodes in batteries exhibit redox behavior due to doped metal ions, and increased electrical conductivity is observed. There are two different processes: intercalation and doping. The intercalation process involves changes in stoichiometry and induces reversible changes in the host material, whereas doping involves irreversible structural changes that might change the material's intrinsic properties. The use of MXenes in electrochemical applications is still challenging because of their unfavorable electrochemical properties, which can be tackled by increasing the interlayer spacing of MXenes via single heteroatoms [15]. MXene, being hydrophilic, has an abundance of surface defects, a large surface area, and high electrical conductivity, and can be an ideal support for implanting single atoms [16, 17]. Various reports show significant improvement in the electrochemical performance of MXenes when doped with P, S, and N atoms [18, 19].

Doping of MXene can be done by two methods: the in-situ method or bottom-up method, which is doping the MAX phase with the atom, which is then followed by selective etching and exfoliation, and the second method is the ex-situ method or top-down method, which involves post-synthesis modification which allows dopant to be substituted at either at X or T sites. A recent method involves etching the MAX phase with molten salts to substitute the dopants on the termination group. Density functional theory (DFT) calculations were also performed to design, analyze, or predict materials' properties using first principles (ab initio) methods. DFT modeling of transition metal doped MXene shows that alternative d orbitals can impact the intrinsic properties of MXene. DFT analysis on X-substituted (B, N, Si, S, for C) MXenes shows that with the changes in M−X interatomic distance after doping, the layer thickness and electronic properties of the materials were also observed [20].

Non-metal atoms such as N or S can also be incorporated on the edges or basal planes of MXene via covalent bonds, forming single-atom doped MXene. Nitrogen is a well-known dopant widely used to incorporate carbon-based materials because of its similar atomic radius. N doping may also create oxygen vacancies that alter the electronic and optical properties of MXene. Lee et al. synthesized nitrogen-doped Ti3C2Tx by calcinating the MXene at different temperatures under an ammonia atmosphere, as shown in Fig. 2b [21]. The first N-doped MXene was synthesized by Naquib et al. in 2012 via an in-situ synthesis method, which involved mixing Ti, AlN, and graphite in a 3:1:1 ratio and then heating it at 1500 °C for 2 h to form Ti_3AlCN, followed by selective HF etching to obtain the N-doped MXene, (Ti_3CNT_x) [22]. Li et al. doped Nb_2C MXene with nitrogen atoms by using a hydrothermal method, and no changes in the morphology of MXene were observed [23]. Wang et al. [24] incorporated an N atom into the lattice structure of $Ti_3C_2T_x$ MXene to form nitrogen-doped $Ti_3C_2T_x$ MXene (N-$Ti_3C_2T_x$) using ammonia annealing treatment.

Similarly, V_2C MXene was doped by P atom via triphenylphosphine (TPP) using a heat treatment method by Yoon et al. [25], an observed increase in interplanar spacing in the case of P-doped V_2C (1.03 nm). In comparison, pristine V_2C had an interlayer spacing of 0.97 nm [26, 27]. Chen et al. used the ex-situ method to synthesize N-doped $Ti_3C_2T_x$ MXene with the help of the NH3/Ar plasma process, in which the nitrogen-doped content can be optimized by adjusting the ratio of ammonia and argon. First, the MAX phase was etched to synthesize MXene, which was then intercalated with NH_4HCO_3 ions by stirring the synthesized MXene overnight with an ammonium bicarbonate solution. The $NH_4HCO_3/Ti_3C_2T_x$ mixture was then freeze-dried and treated with a plasma NH_3/Ar gas to obtain a few-layered delaminated N-doped $Ti_3C_2T_x$MXene [28].

MXene can also be doped by other transition metals as Wang et al. [44] utilized the Mo vacancies of $Mo_2TiC_2T_x$ MXene, which were created by the electrochemical exfoliation to embed Pt single atoms and fabricated $Mo_2TiC_2T_x$-PtSA. Mo vacancies are created because of electrochemical exfoliation, resulting in the absorption of two H^+ ions on the surface of MXene, creating a Mo–OH_2 intermediate with a stronger Mo–O bond, causing the Mo–C bond to break and forming the Mo vacancies. These vacancies help absorb Pt ions from the electrolyte. The molten salt etching method involves metal byproducts that can occupy the vacancies in MXenes and can form single-atom-doped MXene. The molten salt etching method involves a reaction in which the metal by-product fills the vacancies in MXene and helps in the formation of single-atom doped MXene. In the molten salt etching, a metal byproduct tends to occupy vacancies in MXenes and form single-atom implanted MXenes [29]. Yang et al. used $ZnCl_2$ molten salt for the synthesis of MXene, in which zinc metal ions form the molten salt and get anchored on Ti vacancies of $Ti_3C_2Cl_x$ MXene layers. MAX and $ZnCl_2$ powder were heated at 550 °C for 10 h, which helped increase the Ti vacancies in MXene, and hence, more Zn ions can be doped in the basal planes of $Ti_3C_2Cl_x$ MXene. It was observed that after heating the mixture, the Zn content in Zn–MXene had reached about 0.87%, and hence, Zn atoms were precisely anchored on the Ti sites. Similarly, the Copper atom was also doped by employing the same molten salt etching method, and single-atom copper-immobilized MXene was fabricated by selectively etching $Ti_3(Al_{1-x}Cu_x)C_2$ quaternary MAX phases, with Al and Cu equally distributed in A layers. The selective Al atom removal was possible because of the saturated vapour pressure difference between Al- and Cu-containing compounds, which helped in the formation of volatile $AlCl_3$, while Cu atoms were preserved and anchored on the surface of $Ti_3C_2Cl_x$ MXene via Cu–O bonds [30].

5 Interlayer-Modified MXenes

MXenes, due to their termination group, possess negative charges on the surface and hence can attract the positively charged ions to intercalate between the sheets via electrostatic interaction [31, 32]. Monovalent cations involving Li^+, Na^+, or K^+ can be inserted into the layers, increasing the spacing between the sheets [33].

Song et al. [34] intercalated V_2C MXene with Co^{2+} ions by ion exchange method and reported that 80% of ions were inserted between the layers, whereas 20% of ions were attached to the surface of MXene. The spacing between the MXene sheets can also allow small molecules such as CTAB, urea, or hydrazine to be inserted. The strong interaction between cationic ions and negatively charged MXene, large-sized molecules can also be of interest in MXene. Tao et al. [35] fabricated Ti_3C_2 MXene (CTABSn(IV)@Ti_3C_2) by intercalating CTAB and Sn^{4+} ions as shown in Fig. 2c and developed a CTAB prepillaring and $Sn^{4+,}$ and observed an increase in interplanar spacing from 2.2 to 2.7 nm, which allows Sn^{4+} ions to be further intercalated.

6 Energy Storage Applications

In recent years, energy demand has increased a lot due to technological and industrial advancements, a growing population, high-end infrastructure, etc., which has created a sort of chaos and crisis in the energy sector. Fossil fuels have been used earlier to fulfil energy demands, but they pose challenges due to their finite availability and non-renewability. Renewable energy sources such as wind, solar, or tidal can reduce the dependence on fossil fuels. However, such energy sources are uncertain as the wind does not blow the whole day, and the sun does not shine constantly. Consequently, transitioning from fossil fuels and conventional energy sources has become imperative in modern world dynamics and technological expansion. These energy storage devices can play an inevitable role in the balancing act and facilitate the development of more sustainable infrastructure [36–38].

Energy storage devices can be of various types, such as mechanical energy storage devices, solar fuel cells, hydroelectric pumping storage, chemical energy storage devices, and electrochemical energy storage devices.

7 Energy Storage Devices and Technology

Energy storage devices can be of various categories, but here in this chapter, we are focusing on electrochemical energy storage devices (EESD) [39]. (EESD) include supercapacitors, which can be electrochemical capacitors, ultra-capacitors, or electric double-layer capacitors, and batteries. Both batteries and supercapacitors depend on electrochemical processes but have different charge-storing mechanisms, which helps in determining their relative energy and power density [40, 41]. Batteries can be non-rechargeable, which are known as primary batteries, and can be rechargeable, which are known as secondary batteries, whereas capacitors can be non-electrolytic, electrolytic, or electrochemical, which are known as supercapacitors. These devices are different in terms of their charge-storing mechanism, which can be faradic, which involves electrons in their reactions, or non-faradic, without electron transfer [42, 43]. However, these devices are similar in terms of their construction, consisting of electrodes separated by electrolyte and separator. There are various known batteries, among which the Li-ion batteries have gained more attention and are the most successful commercially available technology [44]. Consequently, nowadays, lithium-ion batteries are also predominant, especially in hybrid electric and electronic gadgets [45]. In LIBs, the chemical reaction occurs via two different mechanisms, which are either by displacement or by the intercalation of Li^+ ions. During the charging cycle, the Li^+ ion is deintercalated from the cathode and inserted into the anode, whereas, in the discharging cycle, a reverse process occurs in which the positive ions transport from the cathode to the anode, and reactions involve a faradic process in which can be identified from the CV profile which exhibits well-separated oxidation and reduction peaks. Recently, other alkali metals, alkaline metals, or transition metals have also been explored for battery applications as an alternative to the lithium metal ion [46]. Batteries often carry high energy density but low power density. As a result, research now focuses on developing new materials that can provide high energy and power densities [47]. However, the primary challenge for supercapacitors lies in reducing costs and enhancing energy density to levels comparable to batteries, ideally below 10 Wh/kg. Increasing capacitance and cell voltage are strategies to achieve this goal.

A capacitor is an energy-storing device that stores energy based on an electrostatic field. It generally consists of two closely spaced metal plates separated by a dielectric medium; when an external voltage is applied, one of the plates accumulates a positive charge, whereas the other accumulates a negative charge, resulting in charge separation and creating an electric field. Capacitors are known for their rapid charge and discharge capabilities but have very low energy density [39]. Electrolytic capacitors are the latest evolution of capacitor technology in which the electrolyte replaces the dielectric in a capacitor; it is known as an ultracapacitor or supercapacitor, in which ions are movable and can provide higher energy density. They are also known as electro-double-layer capacitors. Supercapacitors offer higher energy density than traditional capacitors and store charges/discharges faster than conventional capacitors; hence, they are known as supercapacitors. However, they have lower energy density and face high operational costs and voltage variations during operation. Supercapacitors have seen remarkable progress after introducing activated carbon/carbon composite electrodes. Hybrid supercapacitors, which emerged in 2001, combine a polymer-based material as the positive electrode with activated carbon as the negative electrode, resulting in an increase in the specific power density and cyclic stability. Supercapacitors with electrode material consisting of nanocomposite hybrid materials and exhibited good specific capacitance and cycle stability because of the integration of nanomaterials with the organic and inorganic components. Most of the devices assembled have aqueous electrolytes or organic electrolytes, which have a chance of leakage. Hence, all solid-state polymer supercapacitors were developed in 2010 and showed high specific capacitance, cycle stability, and flexibility. With the development of micro-supercapacitors and flexible supercapacitor devices, their application has also been extended to various applications such as sensor applications and wearables. Further innovations included fibre-like supercapacitors, multi-walled nanotubes, and nano-porous cellulose paper for enhanced electrode performance. These advancements collectively represent the evolution of supercapacitor technology towards higher performance, flexibility, and applicability in diverse fields, and the development has also been made towards novel electrode materials.

Electrical charges can be stored by a double-layer charging mechanism (on the surface) or a faradic mechanism (in the bulk of the material). For double layer mechanism (EDLC) as shown in Fig. 3a, electrode material must have a higher surface area and charge storage in such material is due to the accumulation ions at the interface of the electrode and electrolyte, and no chemical reaction takes place at the interface or inside the bulk of the electrode. The material should allow the easy insertion or deinsertion of the ions to be accumulated. In the case of pseudocapacitive behaviours, as shown in Fig. 3b, the Faradic mechanism, a reaction takes place between the electrode and electrolyte on the surface or near the surface of the electrode without the diffusion of ions within the bulk of the electrode. Hybrid energy storage devices are the hybrid of batteries and supercapacitors whose behaviour lies between that of EDLC and faradic process as shown in Fig. 3c, [39, 48].

Recently, research in energy storage technology has faced challenges in the performance and durability of such devices. Hence, there is a need to develop functional materials to enhance the performance of devices and promote the development of science and technology. At today's stage, the trend of developing new materials involves assembling the materials at the atomic or molecular level and controlling the size or morphology of materials to enhance the overall performance of the functional novel materials [49, 50].

Generally, carbon-based materials such as porous carbon, which also includes biomass-derived activated carbon [51], carbon nanotubes, or graphene-based materials, were used as electrode materials because of their high surface area [52]. Such materials show EDLC behaviour, and in spite of their high surface area and electrical conductivity, their delivered energy density is very low. Other materials such as metal oxides, transition metal chalcogenides, and metal–organic frameworks (MOFs) store charges by a pseudocapacitive mechanism, which is mainly focused

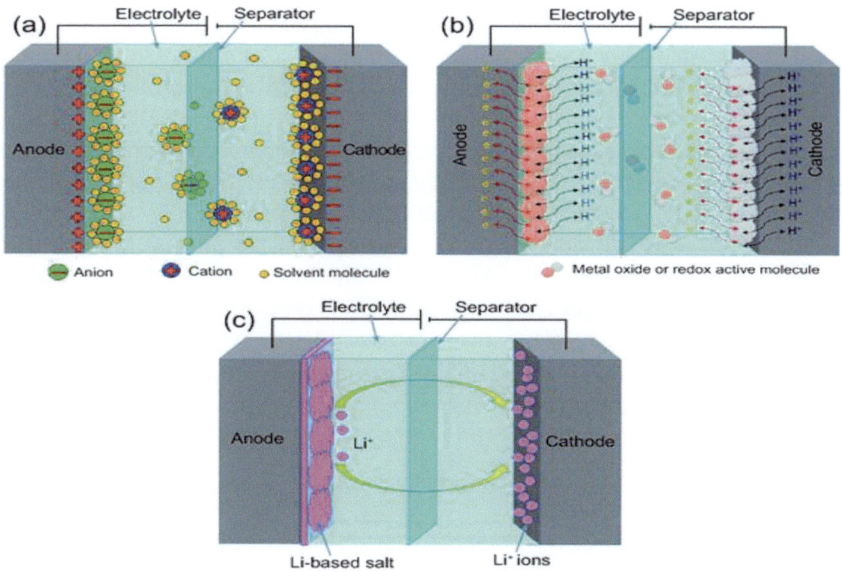

Fig. 3 Schematic representation of **a** electrical double-layer capacitor (EDLC), **b** pseudocapacitor (PC), and **c** hybrid supercapacitor (HSC). Reproduced from [54] with permission from American Chemical Society, Copyright 2016

on a reversible aggregation of electrolyte ions between electrode and electrolyte with faradaic processes [53]. However, metal oxides or MOFs exhibit low electrical conductivity, which might cause reduced power density and stability. Recently, after the introduction of graphene, 2D materials have been encouraging candidates for energy storage applications because of their electrical conductivity and high surface area.

8 MXenes as Supercapacitor Material

As researchers explore advanced materials for batteries and supercapacitors, the emphasis remains on achieving high energy density, high volume density, high power density, and prolonged cycle life. The development of supercapacitors alongside traditional battery technologies can be a promising approach for developing hybrid energy storage systems that can be used to address the increasing demand for energy storage solutions in the future. Energy storage is among the applications in which MXene is also explored. Metal ion batteries such as lithium-ion batteries, sodium-ion batteries, zinc-ion batteries, and supercapacitors have significant possibilities as an electrode material because of their layered structure (Fig. 4). Continuous ongoing research has explored new ways to enhance overall performance. Surface functionalized MXene, due to their modification on the surface chemistry, can dramatically affect the volumetric or gravimetric capacitance of the MXene derivative.

Sun and co-workers synthesized termination-modified $Ti_3C_2T_x$ MXene by cation intercalation and surface modification to improve more pseudocapacitive active sites and used them to study their energy storage performances. They reported that after intercalation of K^+ ion and surface modification, MXene exhibits a three times increase in the value of gravimetric capacitance than the pristine MXene. This increased value of gravimetric capacitance is due to the large interlayer spacing and lower surface groups present on the termination sites. KOH-treated MXene and removal of termination groups by calcinating at 673 K yielded higher specific capacitance as compared to the pristine Ti_3C_2 MXene electrode. After K^+ intercalation, a specific capacitance of 500 F/g was experienced with excellent cyclic stability of 99%

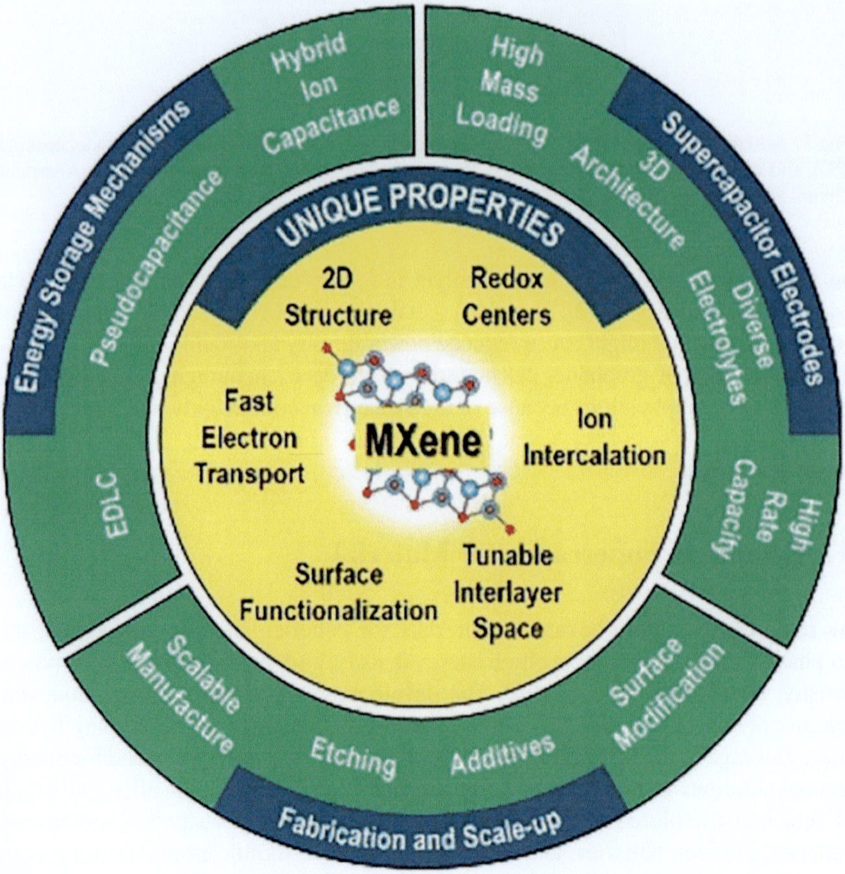

Fig. 4 Schematic representation of MXene properties and application for energy storage. Reproduced from [55] with permission from American Chemical Society, Copyright 2022

even after 10,000 cycles in 1M H_2SO_4 in three-electrode assemblies. The assembled symmetric device exhibited an energy density of 27.4 Wh/Kg in the electrolyte of 1M H_2SO_4 and 1M Li_2SO_4 at a current density of 1A/g, and after calcination with 400 °C KOH–Ti_3C_2, the volumetric capacitance of 19.5 Wh/L was achieved, which is even higher than the highest carbon-based supercapacitor devices. This improvement in capacitance can be due to the significant increase in space between the MXene layers and lower concentration of terminal groups, which increases the number of active sites exposed to the electrolytes and redox reactions [56]. Simon et al. [57] studied the effect of surface chemistry on the electrochemical performance in an acidic medium, as it is well known that MXene exhibits excellent electrochemical performance in an acidic solution. The presence of termination groups –F and –OH/–O but –F groups may be detrimental to charge storage and might not participate in the energy storage process. Hence, the chemical modification of MXene was done by treating the MXene with a fundamental solution to remove –F groups from the surface with –OH groups and delaminating with the organic solvents, and this modification on the surface was confirmed by XPS analysis. This modification on the MXene surface can lead to an increase of fourfold in the electrochemical performance of MXene and deliver 520 F/cm^3 at 2 mV/s, which is the higher than for previously reported MXene at that time. Also, if further modification is made, the electrochemical performance can be increased to a large extent. Wong et al. [58] had developed a time-saving electrochemical etching method to synthesize F-free and Cl-containing MXene by using LiOH and LiCl solution, which provided 92.2% efficiency. Delamination was done by simply sonicating the MXene solution without using a hazardous organic solvent. and studied the electrochemical performance of the synthesized material for supercapacitors. The MXene, when used as electrode materials, exhibited outstanding specific capacitance of 323.7 F/g or 1160 F/cm^3. The film prepared by using MXene was then also used in fabricating the quasi-solid state device Zinc ion hybrid capacitor, which delivered an energy density of 20.8 mWh/cm^3 or 249.9 μWh/cm^3 [58]. To investigate the effect of surface chemistry on the electrochemical process of MXene, the modification was done by intercalating the delamination process. As prepared, MXene exhibits a very low specific surface area and has termination groups of –F, –OH/O functional group, in which –F group might deteriorate the charge storage performance; hence Patrice and co-workers tried to modify the termination group chemically by two methods which are intercalation in which sheets are intercalated with polar molecules or cations by treating with the basic solution to replace the –F termination groups with –OH groups as Ti–F bond become weak in basic medium or by delamination with organic compounds which also increases the surface area. The termination of –F with –OH was confirmed from XPS analysis, which showed that the intensity of Ti–F peaks decreased, and a shift in the binding energy was observed towards a higher value as F has been removed from Ti as a salt. This change in surface chemistry led to enhanced volumetric capacitance of 520 cm^{-3} and gravimetric capacitance of 325 F/g at 2 m/V and electrochemical results [30]. It was reported that heteroatom doping could improve the conductivity of the sample and surface wettability and can also enhance the performance of supercapacitors [59]. In the case of N, O codoped materials are expected to enhance performance due to the interaction with nitrogen and carbon. Hence, N, O codoped carbon and MXene composite were synthesized by Ji and co-workers by in situ carbonization of ammonium citrate on MXene substrate. This heteroatom doping can prevent the re-stacking of MXene layers and can also lead to an enhancement in interplanar spacing. In addition to that, N and O elements can provide extra charges to the surface and act as extra pseudocapacitive sites, and also the synergic effect can make it an excellent electrochemical performance. CV was performed at different scan rates to study the kinetic behaviour. N, O co-doped C@Ti_3C_2 material with a nitrogen concentration of 9.5% delivered gravimetric capacitance of 250.6 F/g at a current density of 1 A/g and enhanced cyclic stability with capacitive retention of 94% even after 5000 cycles. Further, to evaluate their practical application, a symmetrical supercapacitor device was assembled, which delivered an energy density of 10.8 Wh/Kg at the power density of 600 W/kg and good cyclic stability with the retention up to 90% even after 5000 cycles and hence demonstrated the material N, O co-doped C@Ti_3C_2 as a promising material for the development of supercapacitor [60]. MXene-based heterostructures have recently been explored as active materials for energy-related applications because of their individual chemical and physical properties and also the additional synergistic effect between the two components [61]. Pang and co-workers reported the Ti_3C_2–Cu/Co hybrids by using the molten salt etching method, which provided an understanding of the molten salt synthesis method and new methods for synthesizing the active material based on MXene metal hybrid materials [62]. Cu/Co atoms and their interaction with MXene via the O atom were demonstrated, and due to the avoidance of the –F termination group, MXene intercalated with Cu atom exhibited high electrochemical performance. CV curve of CTAB–Sn(IV)@Ti_3C_2 shown in Fig. 5a, the current response can be attributed because of the electrochemical reaction of Li^+ and Sn(IV) nanocomplex or Ti_3C_2 matrix. As shown in Fig. 5b, even after 100 cycles electrode delivers a reversible capacity of 765 mAh/g. Figure 5c shows a stable capacity at various current density of CTAB–Sn(IV)@Ti_3C_2 and GCD curve is shown in Fig. 5d. At the current density of 0.1 A/g, capacity of 765.6 mAh/g still can be restored even after the 50 cycles and cyclic stability and stability up to 250 cycles is shown in Fig. 5e [35].

9 MXenes as Battery Material

A battery is a device that directly transforms chemical energy found in its active components into electric energy through the process of electrochemical oxidation-reduction (redox) [63] and is generally classified into two categories. The first type of batteries is those which are once discharged and then discarded [64]. The second one is discharged and recharged by reversing the current flow through the cell. Electrons are transferred from one material to another during this kind of reaction through an electric circuit. The actual electrochemical unit used to produce or store electric

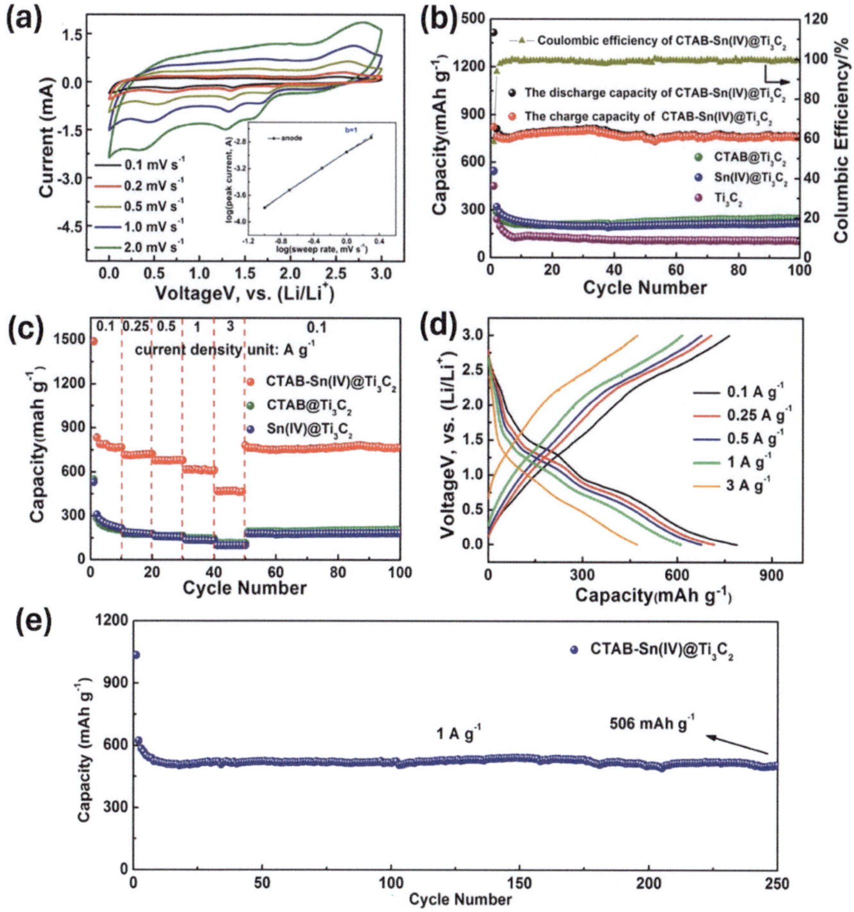

Fig. 5 **a** Scan rate variation of cyclic voltammetric curves of CTAB–Sn(IV)@Ti_3C_2. **b** Cycling performance of samples Ti_3C_2, CTAB@Ti_3C_2, Sn(IV)@Ti_3C_2, CTAB–Sn(IV)@Ti_3C_2 at current density of 0.1 A/g. **c** Rate performance of Ti_3C_2, CTAB@Ti_3C_2, Sn(IV)@Ti_3C_2, CTAB–Sn(IV)@Ti_3C_2. **d** GCD curves at different current densities of CTAB–Sn(IV)@Ti_3C_2 electrodes material. **e** Cyclic Stability of CTAB–Sn(IV)@Ti_3C_2 electrode material at the current density of 1 A/g. Reproduced from Ref. [35] with permission from the American Chemical Society, Copyright 2019

energy is called a cell, even though the term "battery" is frequently used. Structurally, a battery consists of three main components, which are an anode (the negative electrode), a cathode (the positive electrode), and an electrolyte (a solvent that permits the exchange of ions between the electrodes) [65]. A chemical reaction begins when it is connected to an external circuit. When a battery is discharging, the anode sends out electrons into the circuit, while the cathode takes them in [66]. To balance the charge, electrolytes allow the ions to move between the electrodes. When electrons move through a conductor, they create a flow of electric charge, which is known as an electric current. This current can be used to power various electronic devices [67]. When the battery is being charged, the process is reversed. An external voltage source induces electron flow from the cathode to the anode, storing energy in the battery for later use [68]. The type of material used in the construction of a battery determines the voltage. These materials with different chemistries provide different voltages and energy densities. Various factors like temperature, current, and the number of charge cycles it undergoes affect the capacity and efficiency of the batteries. Continuous research is currently being conducted to develop batteries that have higher energy density, faster charging times, and longer lifespans to meet the increasing demands of portable electronics, renewable energy storage, and electric transportation. MXenes, as discussed in the above sections, are becoming the material of choice for energy storage devices like batteries also. They belong to a larger family of transition metal carbides, nitrides, and carbonitrides [69]. In this context, "M" represents the transition metal, and "X" represents carbon and/or nitrogen. The layered structure of MXenes consists of transition metal atoms sandwiched between carbon or nitrogen layers, forming a deck of cards like stack. All these properties, like high electrical conductivity, mechanical strength, and chemical stability, make MXenes particularly exciting [12]. This promising candidate has a wide range of applications, such as energy storage devices like batteries, supercapacitors, and flexible electronics. They are also useful in sensors. Derivatives of MXenes refer to modified versions of these materials. These elements or functional groups are introduced to tailor their properties for specific applications [70]. This customization can greatly improve the performance of MXenes in various fields. For example, researchers can modify the surface chemistry of MXenes by introducing different surface terminations or functional groups. This modification can enhance the interaction of MXenes with other materials or make them more compatible with certain environments, thus improving their overall performance [71]. Their electrochemical performance and their stability in aqueous solutions can also be enhanced. It also enables their use as catalysts for important reactions. A good example of MXene derivative use in an application is in energy storage [40]. By adding MXenes to the electrodes of batteries and supercapacitors, their excellent conductivity can enhance the charge and discharge rates, thereby improving the overall efficiency of these energy storage devices.

With reference to the application of MXenes in batteries, recent research on electrochemical applications with various doped MXenes results in improved electrochemical performance for different metal ion batteries such as Lithium-ion, sodium-ion, lithium–sulfur (Li–S), and sodium-sulfur (Na–S) batteries, as well as supercapacitors and their hybrid systems [72]. Doping of different atoms can increase the value of the interplanar spacing, which further increases the c-lattice parameter and enlarges the surface area and moieties. The addition of defects to the lattice structure of the MXene materials increases electronic conductivity. It also facilitates the transfer of electrolyte ions into the MXene layers and creates a porous nature. Figure 6 The history and application of MXenes and their derivatives for aqueous rechargeable batteries are presented, followed by proposed future design tactics for various aqueous rechargeable batteries.

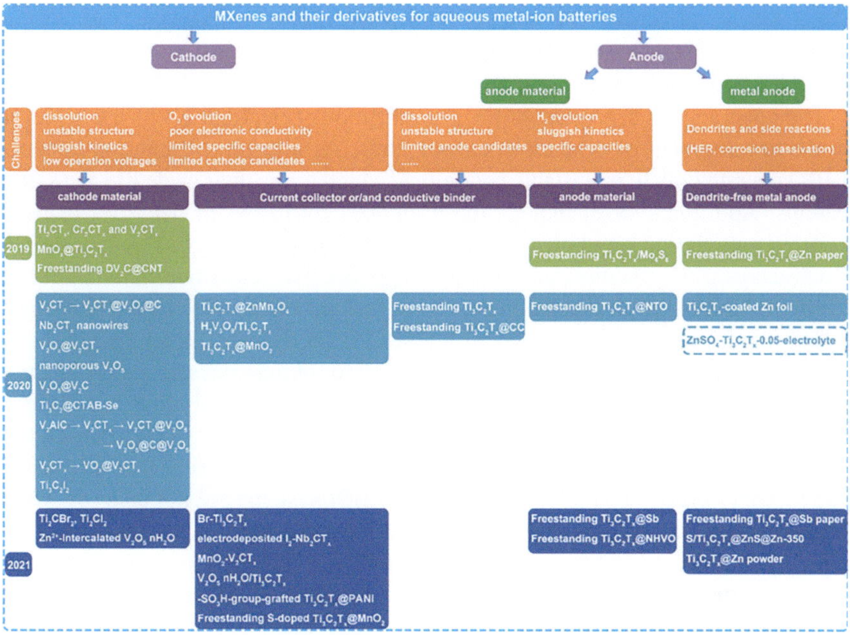

Fig. 6 Development and challenges of MXenes and their derivatives for water-based rechargeable batteries from 2019 to 2021. Reproduced from [73] with permission from Elsevier, Copyright 2022

9.1 Lithium-Ion Batteries

In the 1970s, Li-ion batteries became a commercial reality, and after that, an explosive increase in research interest in portable electronics revolutionized the industry [74]. Li-ion batteries have been the powerhouse of the personal digital electronic revolution since it began over two decades ago, around the time when Li-ion batteries were introduced into the commercial market. The constant advances in electronic items require more efficient Li-ion batteries to meet the increasing demand for better functionality [75]. Contemporary Li-ion batteries (LIBs) have an anode, a cathode, and a separator [76]. The separator is soaked in a non-aqueous electrolyte solution containing dissociated lithium salt, such as $LiPF_6$, and solvents like dimethyl carbonate (DMC), ethylene carbonate, EMC, and DEC. An anode consists of graphite with a stacked structure, providing space for the intercalation of Li-ions between its layers, while the cathode comprises a layered transition metal oxide, such as $LiCoO_2$. The anode consists of graphite layers, providing space for the intercalation of lithium ions between its different layers. However, the cathode is composed of different transition metal oxides. A separator acts as a barrier, preventing short-circuiting and allowing only lithium ions to exchange between the anode and cathode in the cell. In the course of charging, lithium ions are removed from the cathode, travel through a non-aqueous medium, i.e., electrolyte, and then inserted into graphite, which acts as an anode. The process is reversed during discharging. Lithium-ion batteries face a significant challenge of thermodynamic instability when in contact with the non-aqueous solution used in the LIB.

Their operation is dependent on the neutralization of electrodes, which is achieved by the development of protective surface films. As a result, lithium-ion batteries (LIBs) operate only in metastable conditions, which limits their cycle life and safety features. Additional cases include the hazardous formation of dendrites that can be dangerous during rapid charging because of lithium metal coating on graphite, which acts as anodes, significant volume changes in certain electrode materials during reversible lithium-ion intercalation processes, and inadequate electrical conductivity in the cathode material. As research progressed in LIB, different materials were used, which led to an emphasis on different subdisciplines of material science. In the early stages of research and development on Li-ion batteries, efforts are being made to develop rechargeable lithium-ion batteries with Li-metal anodes [77]. Tadiran Inc. developed a commercial AA battery in the mid-1990s for powering cellular phones. Due to safety problems, the Li–MoS_2 and Li–TiS_2 systems did not take off. Many failed because the time for charging Li metal-based batteries is very high, and we require prolonged cycle life, i.e., several hours of cycles. The Li–Li_xMnO_2 system, which was thought to be harmless (because of their internal safety mechanisms), failed due to the large time required for charging Li metal-based batteries to obtain high cycle life (several hours per cycle) [78]. For over three decades, the electrochemistry community has been interested in Li-sulphur batteries for their potential as rechargeable Li batteries.

Given the distinct structure and transport properties of MXene species, there is a motivation to consider them as alternative anode materials to graphite. This is especially true for fast-charging applications. Composite formation using MXene increases the electronic conductivity of LIB batteries (both anodes and cathodes). MXenes can store Li ions in their sheets due to their interlayer spacing of 0.7–1.1 nm, similar to graphite. A N-doped 2D Nb_2CTx MXene was synthesized using a hydrothermal route at a temperature of 423.15 K for 8 hours. The MXene contained 4.5 atomic percent of nitrogen, and the c-lattice parameter from 22.32 to 34.78 Å was increased. The voltage profile of N-Nb_2CTx MXene reflects the progress of the C-rate, which subsequently reduces the specific capacity of cells due to greater polarization and increased internal resistance, thereby limiting Li-ion transport kinetics. At a 0.2 C rate, the reversible capacity of N-Nb_2CTx is 360 mAh/g, which is nearly the same as commercial graphite's (372 mAh/g). the specific capacity of N-Nb_2CTx is 300, 170, and 120, at 0.5, 5, and 10 C, respectively. The EIS of Nb_2CTx and N-Nb_2CTx electrodes before cycling shows that N-Nb_2CTx has a lower resistance (Rs) of 2.2 Ω, indicating high electrode conduction and, similarly, lower electrode/ electrolyte interfacial resistance, compared to Nb_2CTx, which has an equivalent series resistance of 6.1 Ω. Figure 7a shows the Galvanostatic charge-discharge profile of N-Nb_2CT_x. The N-Nb_2CTx electrode has a lower charge transfer resistance than Nb_2CTx, which makes it possible to achieve long-term cycling with 92% capacity retention for up to 1500 cycles, as shown in Fig. 7b. and the Nyquist plot is shown in Fig. 7c. The improved composite performance is attributed to the presence of an N atom in the MXene lattice, which enhances the electrical conduction and also helps in modifying the electronic structure of the materials, which further increases the interlayer spacing [23].

9.2 Sodium-Ion Batteries

Na-ion batteries are also known as NIBs, SIBs, or Na-ion batteries. There are many kinds of Na^+-ion rechargeable batteries that utilize Na^+ ions as a charge carrier [79]. Nowadays, these batteries are considered highly economical and sustainable substitutes to Li-ion batteries for energy storage devices because of their widespread availability, high energy density, natural abundance, and comparatively lower sodium cost. In some cases, its working mechanism is similar to the variant of lithium-ion batteries, with continuous insertion and extraction of Li^+ and Na^+. It substitutes the intercalating agent lithium with sodium ions. Even the development of sodium-ion batteries started in the 1970s and early 1980, but in the 1990s, lithium-ion batteries showed greater commercial potential, leading to a decline in interest in sodium-ion batteries [79, 80]. Additionally, throughout those years, the general standard of materials and electrolytes was shortened, and managing sodium was difficult, hindering the ability to assess electrode performance accurately [81]. Recently, cathode materials for SIBs have seen the introduction of sodiated layer transition metal oxides and organic compounds phosphates. Technological advancements for anode materials have been made by utilizing specific carbonaceous materials, metal oxides or sulfides, and organic compounds in SIBs. Therefore, the essential aspect in the practical application of SIB depends on the anodic electrode material that demonstrates a better quality, such as high capacity, structure, and cycle stability. Even though both the batteries possess a similar working mechanism, i.e., rocking chair, the anionic material of LIB is unsuitable for SIB due to the larger diameter of Na (0.98 Å) than lithium (0.69 Å) [82]. Researchers have recently investigated different types of material, i.e., carbon-based material, phosphorus-based material, sodium alloys, metal oxide, sulfide, and other organic compounds for anodes in sodium-ion batteries. However, most material typically shows low initial Coulombic efficiency, limited specific capacity, and inadequate cycling stability. After the discovery of graphene in 2004, 2D materials have gained a lot of interest as anodic materials for SIB due to their 2D layered structure, planar dimensions, and numerous active sites, making them a promising material for replacing traditional anode materials in SIBs [83]. 2D materials provide a shorter path for ion movement and a larger surface area for sodium ion adsorption. Most of the 2D materials have been developed using the exfoliation method and direct synthesis. 2D materials are held together by weak van der Waals forces, which allow good mechanical flexibility. This property enables them to accommodate large sodium ions and substantial volume fluctuations during the sedation/dissociation processes. These materials also possess good electronic conductivity, providing better structural stability and larger capacity for SIBs.

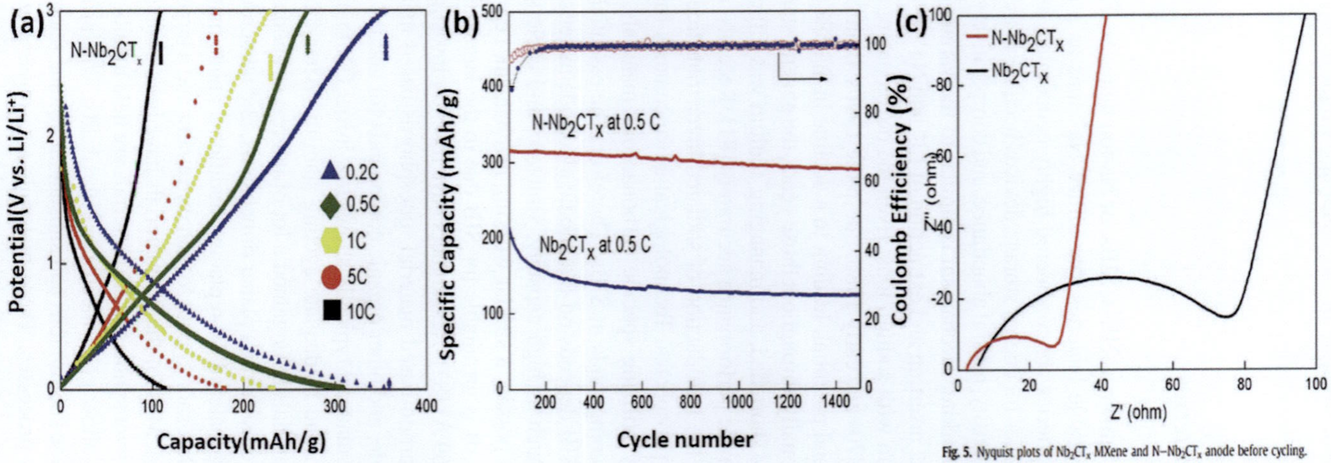

Fig. 7 **a** Galvanostatic charge–discharge profile of N-Nb_2CT_x **b** Specific capacity of material after no. of the cycles. **c** Impedance spectra of the material. Reproduced from [23] with permission from Elsevier, Copyright 2019

For the past decade, a novel class of 2D carbides, nitrides, and carbonitrides known as MXenes has been discovered through the stripping method, extracting them from layered precursors. This is typically expressed as $M_{n+1} X_nT_x$(where n = 1–3), M represents early transition metals (such as Ti, V, Ta, Cr, Sc, Zr, Mo, and Hf), and X represents carbon and nitrogen. Tx is denoted as a surface terminal, which includes hydroxyl, oxygen, and fluorine [84]. These materials possess excellent interlayer spacing, which allows for the easy intercalation of Na ions. MXene is also environmentally friendly and has high electronic conductivity, good mechanical properties, a large surface area, low working potential, and good chemical durability. Xiong et al. investigated the electrochemical behavior of $Ti_3C_2T_x$, which could demonstrate an initial discharge capacity of 270 mAh g^{-1}. Over subsequent initial cycles, $Ti_3C_2T_x$ demonstrated better cycle stability and maintained a specific capacity of approximately 100 mAh g^{-1} after 100 cycles [85]. Zhang Xiaogang et al. achieve a specific capacity of 267 mAh g^{-1} at a current density of 0.1 A g^{-1}, and it is maintained up to 76 mAh g^{-1} at a current density of 1 A g^{-1} after 1500 cycles [84]. Recently, Linlin Zhang et al. developed a three-dimensional porous structure of MXene doping with sulfur, which resulted in increased interlayer spacing. Sulfur-doped MXene (SPM) illustrates excellent performance as electrode materials for sodium ion battery anodes, with a capacity of 335.2 mA h g^{-1} after 490 cycles at 2 A g^{-1} and a long-term cycling performance of 256.1 mA h g^{-1} even after 2480 cycles at 5 A g^{-1} [86].

Similarly, MXene-based composite is also used as an anodic material for SIBs. Zhang et al. successfully achieved composite materials of CoS/MXene using the properties of both CoS and MXene. The MXene layers serve as a stable conductive material, enhancing the electronic conductivity of the composite. Its open 2D structure also exposes many active sites, facilitating rapid electrochemical reactions. From the composite, they achieve a reversible capacity of up to 267 mAh g^{-1} at the current density of 2 A g^{-1}, but after 1700 cycles, the capacity decline rate per cycle is nearly 0.00072%. Additionally, after 1000 cycles, from electron microscopy, it is confirmed that the structural integrity of the CoS/MXene composite remained intact, and even at high current density conditions, the capacity was 80 mAh g^{-1} after 1600 cycles in the CoS/MXene electrode [87]. The CoS/MXene composite's electrochemical performances are shown in Fig. 8. The CoS/MXene cyclic voltammetry (CV) curves in the first three cycles are shown in Fig. 8a. The data were collected at a scan rate of 0.1 mVs^{-1} within a voltage window of 0.4 and 2.9 V. Figure 8b demonstrates the CoS/MXene electrode's GCD curves at 2 A g^{-1}. Figure 8c–d illustrates the rate performance and the associated charge-discharge curves of the CoS/MXene electrode. Figure 8e presents the long-term cycling performances of both the CoS/MXene and MXene electrodes at a current density of 2 A g^{-1}.

Additionally, MXene and carbon-based composites are widely explored for sodium-ion batteries. Out of many carbonaceous materials, carbon nanotubes are mainly studied because of their 1D structure, high current density, low resistance, thin mechanical strength, and high surface area. Xie et al. fabricated a Ti_3C_2 MXene/CNT-based electrode for Na-ion batteries. In a composite, CNT offers good electrostatic bonding at the nano level. which hinders the stacking of 2D MXene nanochip. They achieve a volume capacity of 421 mAh cm^{-3} at 20 mA g^{-1}, and even after

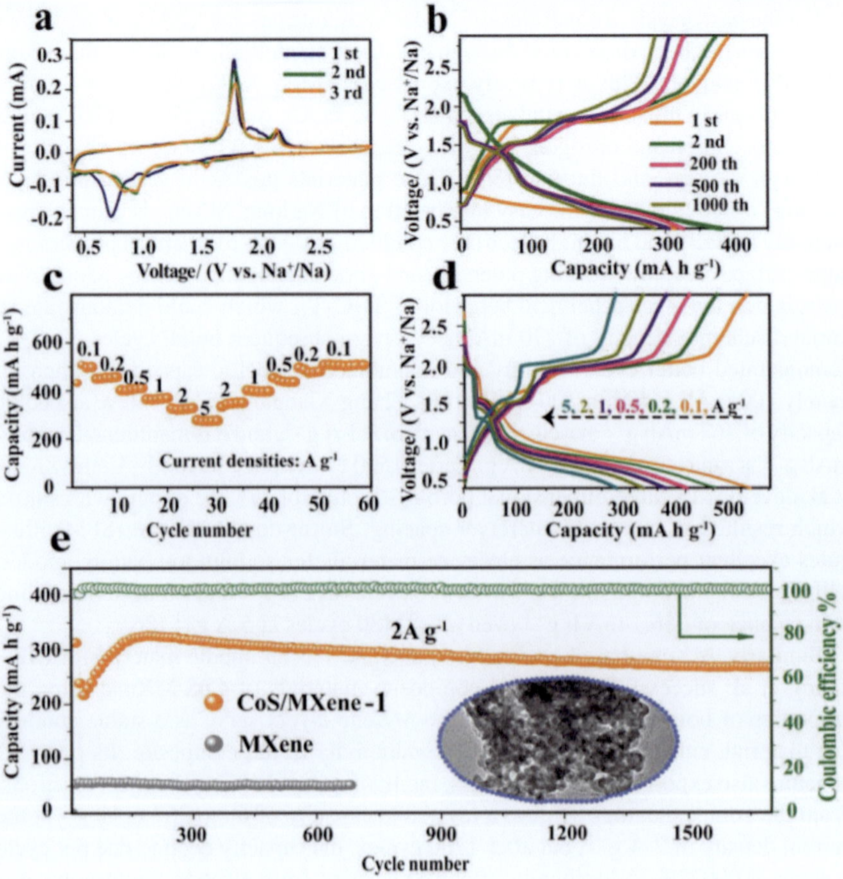

Fig. 8 Electrochemical density evaluation of CoS/MXene. **a** Cyclic voltammetry curves at a scan rate of 0.1 mVs^{-1}. **b** Charge–discharge curves at 2 A g^{-1}. **c** Rate performance and **d** Charge–discharge curves at different current densities. **e** The CoS/MXene composite's cycling performance at 2 A/g. Reproduced from [87] with permission from Elsevier, Copyright 2019

1500 cycles, the composite shows good cycle stability up to 189 mAh cm^{-3} at a current density of 5000 mA g^{-1} [88]. Other MXene composite-based electrodes were also fabricated for SIBs. Ren et al. fabricated a PANI-Ti_3C_2-based electrode for SIB and achieved a high specific capacitance of up to 164 F g^{-1} at 2 mV s^{-1}, which is 31 F g^{-1} higher than that of Ti_3C_2. It also shows good cycle stability of up to 96% after 300 cycles. Recently, MXene nanodots have also been utilized for Na ion storage. Zhang et al. synthesized micro-sized nanodots $Ti_3C_2T_x$ with red phosphorus using the ball milling method for SIBs applications. It shows good capacitance and cycle stability [84]. Although MXene and its composites have successfully addressed numerous limitations in SIBs, further research is required to apply SIBs in real-world scenarios.

9.3 Zinc Ion Batteries

Li-ion batteries (LIBs) are well-received in portable electronics due to their lightweight nature and stable performance. They also possess high energy density. However, safety and environmental concerns limit their scalability [89]. Aqueous Zn-ion batteries (ZIBs) are becoming increasingly popular for large-scale energy storage due to their safety, low cost, and non-toxicity. ZIBs have a double electron transport mechanism that involves transferring through by inserting or removing an anion into the cathode and then combining it with an anode, a specific chemical reaction can occur resulting in high volumetric energy density [90]. Despite these properties of ZIBs, they are encountering various challenges, like a lack of dependable host materials that can fast reversible Zn^{2+} storage electrochemically [91].

A composite material with a 3D nanobelt structure was made through pre-intercalation and alkalization, avoiding restacking and optimizing storage efficiency. The composite alkalized, an alkalized MXene-Zn, can replace traditional zinc foil to improve the anode of ZIS. The AMX-ZIS displays excellent energy storage characteristics when assembled with an AMX-Zn anode and PC cathode. AMX-ZIS is a material that boasts a high specific capacity of 75.2 mAh g^{-1} and an energy density of 60.2 $Whkg^{-1}$ (based on the mass of AMX-Zn anode and PC cathode). Even at a high-power density of 7.04 $kWkg^{-1}$, the energy density of the device remains at 21.08 $Whkg^{-1}$. Moreover, the asymmetric device exhibits an initial specific capacity of 92.5% after 10000 GCD cycles at 3.3 A g^{-1}. This work presents an effective approach for designing the next-generation high-performance energy storage system and optimizing the anode electrode of aqueous zinc-ion energy storage devices [92].

By precisely regulating the kinds of ions intercalated and altering the phases during the hydrated vanadate transition, V_2CT_x–MXene derivatives containing pre-intercalated ions can be produced. The Zn^+ ions diffusion kinetics can be accelerated by hydrated vanadate, as it possesses the intrinsic conductivity of V_2CT_x. Framework, i.e., internal conductive and effect of pre-intercalated Mn^{2+}, both these properties can effectively increase the rate performance of Zn^{2+}. The cycling stability of hydrated vanadate was also achieved by these properties. The MVO@VC cathode demonstrates a remarkably stable capacity of 289.6 mAh g^{-1}, retaining 92.9% of this capacity over 25,000 cycles at a high current density of 10 A g^{-1}. This performance outperforms the VOH cathode and all other V-based cathodes reported [93].

Zinc metal free anode $(NH_4)_2V_{10}O_{25}\cdot 8H_2O@Ti_3C_2T_x$ was prepared for the preparation of "rocking chair" Zinc ion batteries. The addition of $Ti_3C_2T_x$ not only improved the rate performance of $(NH_4)_2V_{10}O_{25}\cdot 8H_2O$ but also stabilized its nanobelt structure by providing conductive networks. The composite $(NH_4)_2V_{10}O_{25}\cdot 8H_2O@Ti_3C_2T_x$ electrode at 0.1 A g^{-1} in half-cell test showed an increase in specific capacity (514.7 mAh g^{-1}) with low potential (0.59 V vs Zn^{2+}/Zn). This was achieved due to the crystal structure of $(NH_4)_2V_{10}O_{25}\cdot 8H_2O$ and the assistance of $Ti_3C_2T_x$ nanosheets. After 6000 cycles, it retains 84.2% of its capacity at 5.0 A g^{-1}, exhibiting excellent cycling performance [94]. It is evident from the above analysis that MXenes play a crucial role in composite materials. Electronic conductivity and structural stability were increased by the MXene matrix. It boosts the electron and ion transfer capability, and the composite's dissolution and expansion/shrinkage during the charging and discharging process are prevented by the inhibitor. The combination of active materials and MXenes amalgamate their benefits, which will result in an overall improved performance of the composite.

The functionalized separator method efficaciously addresses cathode dissolution, shuttle effect, anode zinc dendrites, and side reactions. This enhances energy and power density and extends ZIB's lifespan. The separator was modified by assembling nanoporous NiO in MXenes suspension and coating it on the separator, known as the NiO/MXenes modified separator, which successfully achieved the purpose of improving the zinc anode. At 10 mAcm^{-2}, reversible cycle of 500 h was obtained. The NiO/MXenes layer has a synergistic effect that helps to evenly distribute Zn^{2+} and the electric field. Zinc deposition behaviour improved, and side reactions were suppressed by improving ion diffusion kinetics. At different current densities and areas, the Coulombic efficiency of Zn||Ti cells using pristine and MXene@NiO modified separators is shown in Fig. 9a–f. As seen in Fig. 9g, after 680 h, the cell with a clean separator exhibits an increase in voltage (2 mA cm^{-2} to 2 mAh cm^{-2}). On the other hand, stable polarization and a stable cycle performance up to 2400 hours are obtained with a modified separator under 1 mA cm^{-2} to 1 mAh cm^{-2} and 2 mA cm^{-2} to 2 mAh cm^{-2} circumstances. Zn||Zn cells with modified separators are tested using a spectrum of current densities with various cycle capacities (Fig. 9h) [95].

9.4 *Potassium-Ion Batteries*

Li-ion batteries are widely used in portable electronics and grid storage applications. However, limited lithium resources, rising costs, and safety concerns could hamper their use [96]. In response to these issues, potassium-ion batteries have promising results that can replace lithium-ion batteries for different energy storage applications. Sodium metal is abundant in nature, has a low cost, almost similar redox potential to LIB, and has high theoretical capacities. Potassium ion batteries give different electrochemical behaviour from lithium-ion batteries as the ionic radii of K^+ (1.38 Å) are large in comparison to Li^+(0.76 Å) [66]. Through theoretical and experimental studies, it has been shown that the storage capacities of pristine MXene were low for potassium, and their capacities decay fast. MXenes in their pure form are generally not suitable for use as anode materials alone; therefore, MXenes typically require modifications to enhance their electrochemical performance. In this way, by utilizing the exceptional electronic conductivity and ion transportation properties of MXenes, we can achieve better results.

Ti-based material has been widely studied for sodium ion and potassium ion batteries, especially sodium titanate (NTO) and potassium titanate (KTO). These materials are used due to their low operating potential, enormous chemical stability, eco-friendliness, and appropriate layer spacing [61]. The primary challenges that hinder its implementation are the slow diffusion of ions and the expansion of

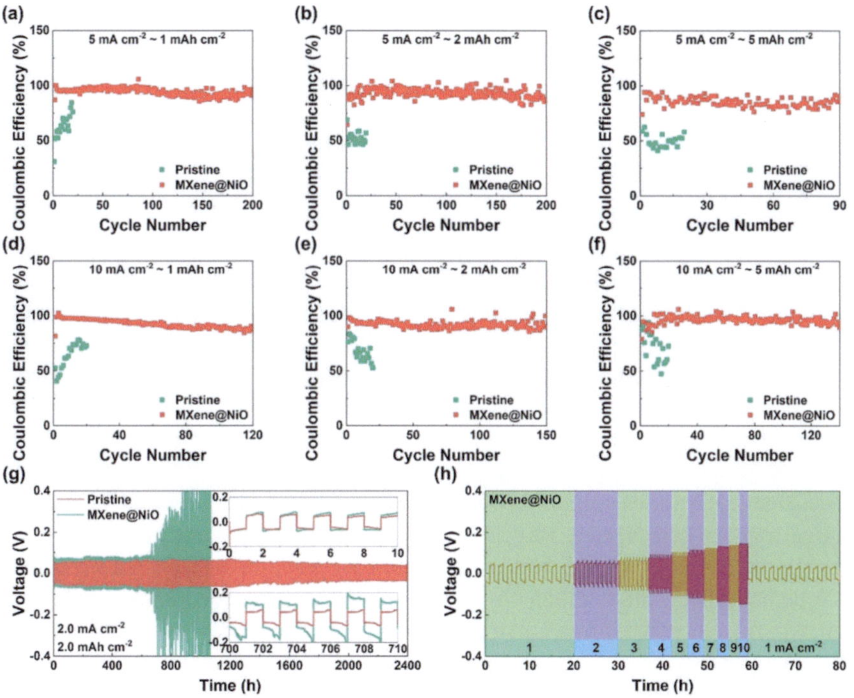

Fig. 9 **a–f** Coulombic efficiency of MXene@NiO modified and pristine Zn||Ti cells. **g** Zn||Zn cell cycling performance using pure and modified MXene@NiO separators. **h** Zn||Zn cells' rate capability with a modified MXene@NiO. Reproduced from [95] with permission from Elsevier, Copyright 2022

the lattice volume. Two-dimensional MXene and reduced graphene oxide (rGO) nanoflakes are used as precursors to prepare self-supporting membrane electrodes. This material consists of alternating layers of graphene and ultrathin nanoflakes of either sodium titanate or potassium titanate arranged in a sandwich-like structure. The NTO/rGO and KTO/rGO membrane electrodes demonstrated exceptional rate performance and long-cycle stability due to their flexibility. The outstanding electrochemical performance is attributed to the distinctive sandwich structure. Due to these sandwich structures, the diffusion distance between ions shortens, and electronic conductivity increases. The binding energy also increases between these layered nano chips sufficiently. KTO/rGO-10% acts as anode material for PIB, which exhibits a high initial discharge capacity of 10000mAh g^{-1}. After 700 cycles at a current density of 2 A g^{-1}, the capacity is maintained at 75 mAh g^{-1} [97].

Combining MXene with well-designed porous carbon material produces anodes with excellent rate capacity and ultrahigh structural stability and conductivity. Using poly (diallyldimethylammonium chloride) (PDDA)- N-rich porous carbon nanosheets (NPCNs)/Ti_3C_2 hybrids has several advantages; first, the hybrid body's favourable electrical conductivity is due to the 3D interconnected conductive network of N-rich porous carbon nanosheets and MXene. Secondly, the use of MXene reduces the adsorption energy of K^+, and the stacked structure of PDDA-NPCN/Ti_3C_2 expands the layer spacing, accelerating the ion transportation dynamics. Lastly, by intercalating K^+ ions, the sturdy structure of the hybrids could tolerate the volume change. All these properties enabled PDDA-NPCN/Ti_3C_2 hybrids to exhibit excellent electrochemical performances. The PDDA-NPCN/Ti_3C_2 anode achieved a high reversible capacity of 358.4 mAh g^{-1} at a current density of 0.1 A g^{-1}. Even after 2000 cycles, it maintained a reversible capacity of 252.2 mAh g^{-1} at a current density of 1.0 A g^{-1}, which is equivalent to a degradation of only 0.03% per cycle [98].

Organic compounds amalgamation with MXene proves to be a promising technology for potassium ion batteries as a cathode material that possesses high capacity, cost-effectiveness, and environmental friendliness [99]. Phenazine's structural, electronic characteristics, and pie-conjugated electronic structure, along with nitrogen heteroatoms that have lone pairs on them, render electrochemical activity; all these properties help phenazine to stand out among different compounds. Phenazine has various limitations, including its limited capacity and inadequate charge transfer capabilities. Researchers synthesized cyano-substituted diquinoxalinophenazine (6CN-DQPZ) by using a Schiff base reaction to resolve this problem. MXene nanosheets are combined with (6CN-DQPZ) using vacuum-assisted filtration, which leads to the formation of 6CN-DQPZ@MXene composites. The addition of cyano groups increases the electron affinity and provides additional redox sites, which enhances the electrochemical performance [100, 101]. Figure 10 shows the schematic synthesis of the composite formation, and Fig. 10b, c shows the SEM images of the MXene and 6CN-DQPZ. The HOMO and LUMO energy levels difference of DQPZ and 6CN-DQPZ are represented in the Fig. 10d. High flexibility and the cross-sectional SEM image of the composite are shown in Fig. 10e, f, respectively. The TEM images of the composite are shown in Fig. 10h. For the elemental detection, the EDAX was recorded, as shown in Fig. 10g. The 6CN-DQPZ@MXene composite shows exceptional electrochemical properties that maintain a reversible capacity of 238.8 mAh g^{-1} over 2000 cycles and have a capacity retention of 98.9%.

10 Challenges

The potential applications of MXene-based materials in electrocatalysis need further exploration, as there is still much to uncover regarding the catalytically active sites and reaction mechanisms. Defect engineering, such as vacancy defects and heteroatom doping, holds promise for enhancing the catalytic activity of MXenes in the future. The family of MXene must be expanded to create more $M_{n+1}X_nT_x$ with different metals (M) and variations of T and x. Managing and altering MXenes' surfaces will facilitate the production of MXenes with desired properties. Synthesizing MXenes without functional groups will result in excellent electrochemical performance. Additionally, preparing large-scale single-layered MXenes remains challenging. The practical application of MXene is quite difficult for energy storage

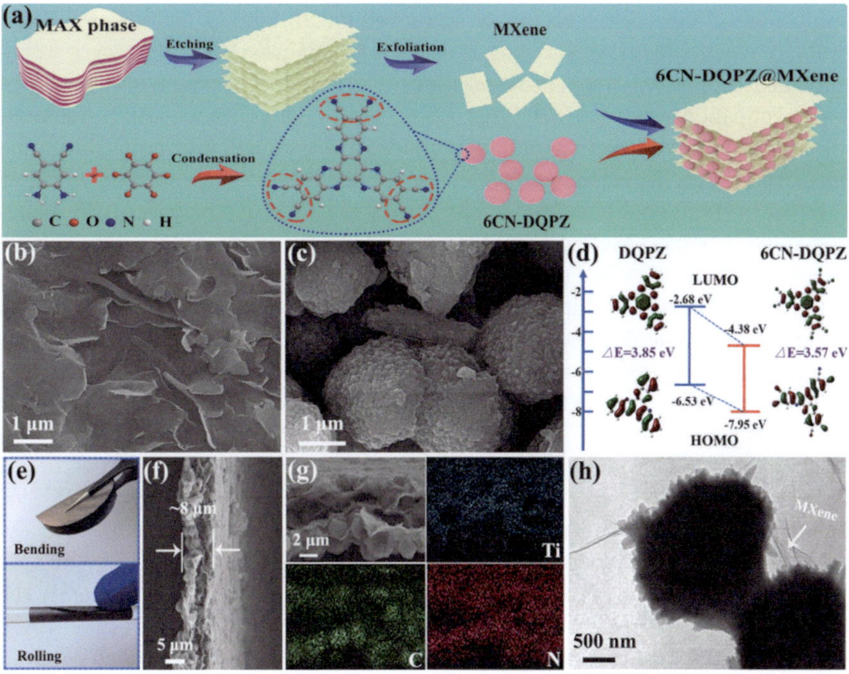

Fig. 10 **a** 6CN-DQPZ@MXene synthetic route. SEM images of **b** MXene and **c** 6CN-DQPZ. **d** Lowest occupied molecular orbital (LUMO)/highest unoccupied molecular orbital (HOMO) energy levels of DQPZ and 6CN-DQPZ. **e** High flexibility images of the 6CN-DQPZ@MXene composite. **f** Cross-sectional SEM image, **g** EDAX images, and **h** 6CN-DQPZ@MXene TEM image of a composite film. Reproduced from [101] with permission from Elsevier, Copyright 2022

systems. Current research on MXene-based ESDs is mainly limited to the laboratory scale.

11 Summary and Outlook

MXenes, due to their electronic, chemical, and morphological properties, can be used in multiple applications. MXenes edge sites, as well as basal plane, also show activities, whereas, in other 2D materials, catalytic activity is hindered at edge sites [102]. Besides all these properties, MXene's performance is distant from its perfect behavior in many applications. Material properties can be improved for optimal conduct for different applications by utilizing advancements in materials. This chapter offers a complete discussion of various advancements made by doping in MXenes for different energy storage applications. Various doping methodologies used in the process of synthesis have also been inspected and discussed. Catalytic site density was enhanced by increasing the metallic and conductive characteristics, resulting in a reduction of dependence on metals, improved metal stability, and lowered Fermi levels. In electrochemical reactions, it accelerates charge transfer, improves wettability, and suppresses competitive reactions. Onset potential was also lowered for some applications such as HER, OER, ORR, and CO_2RR [21, 103–107]. Additionally, it has improved cycling stability, reduced the production of polysulfide, and enhanced the specific capacity of alkali metal sulfur batteries up to ten times higher than that of traditional metal oxide-based cathodes in Li-ion batteries [23]. Moreover, it has increased cycling stability in Na–S batteries, restricted the development of polysulfide, and boosted redox activity.

For mobile devices, lithium batteries stand to be the most crucial power source. For small electronic devices or larger electric vehicles, these lithium-ion batteries also prove to be a very important energy source because of their longer cycling life, better energy storage capacity, and more environmental friendliness as compared to other alternatives.

Due to their more widely adjustable structural and chemical characteristics, doped MXenes will be very competitive when compared to previous 2D materials. MXene doping with different elements involves modifying the surfaces and bulk chemistry of MXenes to enhance properties for energy storage devices. Doping MXene with various elements can offer several advantages, including the ability to modify the surface and bulk chemistry of MXenes, resulting in unique features [108]. These features can be beneficial for a variety of energy storage systems. Doped MXenes can serve a variety of purposes in supercapacitors and batteries, such as regulating ion transfer, ion redistribution, steric hindrance, double-layer and redox-type ion storage, electrocatalysts, electrodeposition substrates, and more. $M_{n+1}X_nT_x$ has the capacity to hold ions of different sizes between its 2D layers and broadens the range of uses for MXenes, including lithium-ion batteries (LIBs) and other situations where electrode materials may be limited. The implementation of doped MXene-based devices is still in its early stages for future applications. Multifunctional solutions are required for greener energy. More advanced properties of these materials should be explored for more reliable electronic device formation. We would like to draw attention to some of the technological solutions that this exceptional material can offer. MXenes have shown exponential development in attention over the last ten years, as seen by over 11,358 articles [109]. After the discovery of MXene, chemists explored them through modification in structures by making variations in surface chemistry and numerous M-site chemistries so that there would be an increase in the applicability and usefulness of MXenes.

References

1. Ronchi, R.M., Arantes, J.T., Santos, S.F.: Synthesis, structure, properties and applications of MXenes: current status and perspectives. Ceram. Int. **45**(15), 18167–18188 (2019)
2. Lin, Z., et al.: MXenes as high-rate electrodes for energy storage. Trends Chem. **2**(7), 654–664 (2020)

3. Shen, X., et al.: Free-standing 3D alkalized Ti3C2Tx/Ti3C2Tx nanosheet membrane electrode for highly efficient and stable desalination in hybrid capacitive deionization. J. Mater. Chem. A **8**(37), 19309–19318 (2020)
4. Yu, L., et al.: Recent progress on MXene-Derived material and its' application in energy and environment. J. Power Sour. **490**, 229250 (2021)
5. Wang, C., Chen, S., Song, L.: Tuning 2D MXenes by surface controlling and interlayer engineering: methods, properties, and synchrotron radiation characterizations. Adv. Func. Mater. **30**(47), 2000869 (2020)
6. Thakur, R., et al.: Insights into the thermal and chemical stability of multilayered V2CTx MXene. Nanoscale **11**(22), 10716–10726 (2019)
7. Mashtalir, O., et al.: Kinetics of aluminum extraction from Ti3AlC2 in hydrofluoric acid. Mater. Chem. Phys. **139**(1), 147–152 (2013)
8. Schultz, T., et al.: Surface termination dependent work function and electronic properties of Ti3C2T x MXene. Chem. Mater. **31**(17), 6590–6597 (2019)
9. Hart, J.L., et al.: Control of MXenes' electronic properties through termination and intercalation. Nat. Commun. **10**(1), 522 (2019)
10. Lian, P., et al.: Alkalized Ti3C2 MXene nanoribbons with expanded interlayer spacing for high-capacity sodium and potassium ion batteries. Nano Energy **40**, 1–8 (2017)
11. Chen, W.Y., et al.: Surface functionalization of Ti3C2T x MXene with highly reliable superhydrophobic protection for volatile organic compounds sensing. ACS Nano **14**(9), 11490–11501 (2020)
12. Zhang, T., et al.: Simultaneously tuning interlayer spacing and termination of MXenes by Lewis-basic halides. Nat. Commun. **13**(1), 6731 (2022)
13. Kim, H., et al.: Thermoelectric properties of two-dimensional molybdenum-based MXenes. Chem. Mater. **29**(15), 6472–6479 (2017)
14. Gogotsi, Y., Anasori, B.: The Rise of MXenes, pp. 8491–8494. ACS Publications (2019)
15. Ge, H., et al.: Critical challenges and solutions: quasi-solid-state electrolytes for zinc-based batteries. Energy Environ. Sci. (2024)
16. Cui, C., et al.: Ultrastable MXene@ Pt/SWCNTs' nanocatalysts for hydrogen evolution reaction. Adv. Func. Mater. **30**(47), 2000693 (2020)
17. Lin, W., et al.: Atomic bridging modulation of Ir–N, S co-doped MXene for accelerating hydrogen evolution. J. Mater. Chem. A **10**(18), 9878–9885 (2022)
18. Mashtalir, O., et al.: The effect of hydrazine intercalation on the structure and capacitance of 2D titanium carbide (MXene). Nanoscale **8**(17), 9128–9133 (2016)
19. Lien, C.-W., et al.: Optimization of acetonitrile/water content in hybrid deep eutectic solvent for graphene/MoS2 hydrogel-based supercapacitors. Chem. Eng. J. **405**, 126706 (2021)
20. Khazaei, M., et al.: Novel electronic and magnetic properties of two-dimensional transition metal carbides and nitrides. Adv. Func. Mater. **23**(17), 2185–2192 (2013)
21. Le, T.A., et al.: Synergistic effects of nitrogen doping on MXene for enhancement of hydrogen evolution reaction. ACS Sustain. Chem. Eng. **7**(19), 16879–16888 (2019)
22. Naguib, M., et al.: Two-dimensional transition metal carbides. ACS Nano **6**(2), 1322–1331 (2012)
23. Liu, R., et al.: Nitrogen-doped Nb2CTx MXene as anode materials for lithium ion batteries. J. Alloy. Compd. **793**, 505–511 (2019)
24. Wen, Y., et al.: Nitrogen-doped Ti3C2Tx MXene electrodes for high-performance supercapacitors. Nano Energy **38**, 368–376 (2017)
25. Kan, D., et al.: Rational design of bifunctional ORR/OER catalysts based on Pt/Pd-doped Nb2CT2 MXene by first-principles calculations. J. Mater. Chem. A **8**(6), 3097–3108 (2020)
26. Yoon, Y., et al.: Precious-metal-free electrocatalysts for activation of hydrogen evolution with nonmetallic electron donor: chemical composition controllable phosphorous doped vanadium carbide MXene. Adv. Func. Mater. **29**(30), 1903443 (2019)
27. Mozafari, M., Soroush, M.: Surface functionalization of MXenes. Mater. Adv. **2**(22), 7277–7307 (2021)

28. Chen, X., et al.: Tunable nitrogen-doped delaminated 2D MXene obtained by NH3/Ar plasma treatment as highly efficient hydrogen and oxygen evolution reaction electrocatalyst. Chem. Eng. J. **420**, 129832 (2021)
29. Zhao, Q., et al.: Selective etching quaternary MAX phase toward single atom copper immobilized MXene (Ti3C2Cl x) for efficient CO2 electroreduction to methanol. ACS Nano **15**(3), 4927–4936 (2021)
30. Gu, J., et al.: Single zinc atoms immobilized on MXene (Ti3C2Cl x) layers toward dendrite-free lithium metal anodes. ACS Nano **14**(1), 891–898 (2020)
31. Simon, P., Gogotsi, Y.: Perspectives for electrochemical capacitors and related devices. Nat. Mater. **19**(11), 1151–1163 (2020)
32. VahidMohammadi, A., Rosen, J., Gogotsi, Y.: Review outlines the development and potential applications of a wide range of 2D metal carbides and nitrides. Science **372**, 6547 (2021)
33. Ghidiu, M., et al.: Ion-exchange and cation solvation reactions in Ti3C2 MXene. Chem. Mater. **28**(10), 3507–3514 (2016)
34. Wang, C., et al.: Atomic cobalt covalently engineered interlayers for superior lithium-ion storage. Adv. Mater. **30**(32), 1802525 (2018)
35. Luo, J., et al.: Pillared structure design of MXene with ultralarge interlayer spacing for high-performance lithium-ion capacitors. ACS Nano **11**(3), 2459–2469 (2017)
36. Nirogi, A., et al.: MXenes and their composites for supercapacitors and hybrid capacitors. In: Mxenes and Their Composites, pp. 371–396. Elsevier (2022)
37. Amir, M., Srivastava, S.K.: Analysis of harmonic distortion in PV–wind-battery based hybrid renewable Energy system for microgrid development. In: Applications of Computing, Automation and Wireless Systems in Electrical Engineering: Proceedings of MARC 2018. Springer (2019)
38. Lai, C.S., Locatelli, G.: Are energy policies for supporting low-carbon power generation killing energy storage? J. Clean. Prod. **280**, 124626 (2021)
39. Noori, A., et al.: Towards establishing standard performance metrics for batteries, supercapacitors and beyond. Chem. Soc. Rev. **48**(5), 1272–1341 (2019)
40. Rafik, F., et al.: Frequency, thermal and voltage supercapacitor characterization and modeling. J. Power Sour. **165**(2), 928–934 (2007)
41. Huang, S., et al.: Challenges and opportunities for supercapacitors. Apl Mater. **7**(10) (2019)
42. Gao, D., et al.: A survey of hybrid energy devices based on supercapacitors. Green Energy Environ. (2022)
43. Mathis, T.S., et al.: Energy storage data reporting in perspective—guidelines for interpreting the performance of electrochemical energy storage systems. Adv. Energy Mater. **9**(39), 1902007 (2019)
44. Dutta, A., et al.: A comprehensive review on batteries and supercapacitors: development and challenges since their inception. Energy Storage **5**(1), e339 (2023)
45. Peng, H.-J., Huang, J.-Q., Zhang, Q.: A review of flexible lithium–sulfur and analogous alkali metal–chalcogen rechargeable batteries. Chem. Soc. Rev. **46**(17), 5237–5288 (2017)
46. Kiai, M.S., Eroglu, O., Aslfattahi, N.: Metal-ion batteries: achievements, challenges, and prospects. Crystals **13**(7), 1002 (2023)
47. Ponnamma, D., et al.: Green synthesized materials for sensor, actuator, energy storage and energy generation: a review. Polym. Plast. Technol. Mater. **59**(1), 1–62 (2020)
48. Guan, L., Yu, L., Chen, G.Z.: Capacitive and non-capacitive faradaic charge storage. Electrochim. Acta **206**, 464–478 (2016)
49. Yu, F., et al.: Adsorption mechanism and effect of moisture contents on ciprofloxacin removal by three-dimensional porous graphene hydrogel. J. Hazard. Mater. **374**, 195–202 (2019)
50. Sun, Y., et al.: Nano-/micro-confined water in graphene hydrogel as superadsorbents for water purification. Nano-Micro Lett. **12**, 1–14 (2020)
51. Vijayakumar, M., et al.: Activated carbon fibres as high performance supercapacitor electrodes with commercial level mass loading. Carbon **140**, 465–476 (2018)
52. Chen, Z., et al.: Holey graphene-based nanocomposites for efficient electrochemical energy storage. Nano Energy **73**, 104762 (2020)

53. Wang, K., et al.: 2D metal–organic frameworks (MOFs) for high-performance BatCap hybrid devices. Small **16**(30), 2001987 (2020)
54. Yang, Y., et al.: Application of the supercapacitor for energy storage in China: role and strategy. Appl. Sci. **12**(1), 354 (2021)
55. Chen, Y., et al.: MXene-based electrodes for supercapacitor energy storage. Energy Fuels **36**(5), 2390–2406 (2022)
56. Li, J., et al.: Achieving high pseudocapacitance of 2D titanium carbide (MXene) by cation intercalation and surface modification. Adv. Energy Mater. **7**(15), 1602725 (2017)
57. Dall'Agnese, Y., et al.: High capacitance of surface-modified 2D titanium carbide in acidic electrolyte. Electrochem. Commun. **48**, 118–122 (2014)
58. Chen, J., et al.: Simplified synthesis of fluoride-free Ti3C2T x via electrochemical etching toward high-performance electrochemical capacitors. ACS Nano **16**(2), 2461–2470 (2022)
59. Tan, J., et al.: In situ nitrogen-doped mesoporous carbon nanofibers as flexible freestanding electrodes for high-performance supercapacitors. J. Mater. Chem. A **5**(45), 23620–23627 (2017)
60. Pan, Z., Ji, X.: Facile synthesis of nitrogen and oxygen co-doped C@ Ti3C2 MXene for high performance symmetric supercapacitors. J. Power Sour. **439**, 227068 (2019)
61. Wang, X., et al.: 2D/2D 1T-MoS2/Ti3C2 MXene heterostructure with excellent supercapacitor performance. Adv. Func. Mater. **30**(15), 0190302 (2020)
62. Bai, Y., et al.: MXene-copper/cobalt hybrids via Lewis acidic molten salts etching for high performance symmetric supercapacitors. Angew. Chem. **133**(48), 25522–25526 (2021)
63. Rubenbauer, H., Henninger, S.: Definitions and reference values for battery systems in electrical power grids. J. Energy Storage **12**, 87–107 (2017)
64. Schipper, F., Aurbach, D.: A brief review: past, present and future of lithium ion batteries. Russ. J. Electrochem. **52**, 1095–1121 (2016)
65. Dell, R., Rand, D.A.J.: Understanding Batteries, vol. 28. Royal Society of Chemistry (2001)
66. Yang, J.-L., et al.: Advanced aqueous proton batteries: working mechanism, key materials, challenges and prospects. EnergyChem **4**(6), 100092 (2022)
67. Crompton, T.R.: Battery Reference Book. Newnes (2000)
68. Dumont, O., De paepe, M., Lemort, V.: Carnot battery technology: a state-of-the-art review. J. Energy Storage **32**(10.1016) (2020)
69. Salim, O., et al.: Introduction to MXenes: synthesis and characteristics. Mater. Today Chem. **14**, 100191 (2019)
70. Wei, Y., et al.: Advances in the synthesis of 2D MXenes. Adv. Mater. **33**(39), 2103148 (2021)
71. Naguib, M., et al.: 25th anniversary article: MXenes: a new family of two-dimensional materials. Adv. Mater. **26**(7), 992–1005 (2014)
72. Dong, Y., Shi, H., Wu, Z.S.: Recent advances and promise of MXene-based nanostructures for high-performance metal ion batteries. Adv. Func. Mater. **30**(47), 2000706 (2020)
73. Tian, Y., et al.: MXenes and their derivatives for advanced aqueous rechargeable batteries. Mater. Today **52**, 225–249 (2022)
74. Yoshino, A.: The birth of the lithium-ion battery. Angew. Chem. Int. Ed. **51**(24), 5798–5800 (2012)
75. Deng, D.: Li-ion batteries: basics, progress, and challenges. Energy Sci. Eng. **3**(5), 385–418 (2015)
76. Huang, Y.: The discovery of cathode materials for lithium-ion batteries from the view of interdisciplinarity. Interdisc. Mater. **1**(3), 323–329 (2022)
77. Etacheri, V., et al.: Challenges in the development of advanced Li-ion batteries: a review. Energy Environ. Sci. **4**(9), 3243–3262 (2011)
78. Aurbach, D., et al.: On the surface chemical aspects of very high energy density, rechargeable Li–sulfur batteries. J. Electrochem. Soc. **156**(8), A694 (2009)
79. Aslam, M.K., et al.: 2D MXene materials for sodium ion batteries: a review on energy storage. J. Energy Storage **37**, 102478 (2021)
80. Yabuuchi, N., et al.: Research development on sodium-ion batteries. Chem. Rev. **114**(23), 11636–11682 (2014)

81. Hwang, J.-Y., Myung, S.-T., Sun, Y.-K.: Sodium-ion batteries: present and future. Chem. Soc. Rev. **46**(12), 3529–3614 (2017)
82. Wu, Y., Yu, Y.: 2D material as anode for sodium ion batteries: recent progress and perspectives. Energy Storage Mater. **16**, 323–343 (2019)
83. Zhong, W., et al.: MXene-derivative pompon-like Na2Ti3O7@ C anode material for advanced sodium ion batteries. Chem. Eng. J. **378**, 122209 (2019)
84. Aslam, M.K., Xu, M.: A mini-review: MXene composites for sodium/potassium-ion batteries. Nanoscale **12**(30), 15993–16007 (2020)
85. Xiong, D., et al.: Recent advances in layered Ti3C2Tx MXene for electrochemical energy storage. Small **14**(17), 1703419 (2018)
86. Zhang, L., et al.: Sulfur atom doped three-dimensional porous MXene assisted efficient sodium ion batteries. Dalton Trans. (2024)
87. Zhang, Y., et al.: Circuit board-like CoS/MXene composite with superior performance for sodium storage. Chem. Eng. J. **357**, 220–225 (2019)
88. Xie, X., et al.: Porous Ti3C2T x MXene for ultrahigh-rate sodium-ion storage with long cycle life. ACS Appl. Nano Mater. **1**(2), 505–511 (2018)
89. Fan, Z., et al.: Modified MXene/holey graphene films for advanced supercapacitor electrodes with superior energy storage. Adv. Sci. **5**(10), 1800750 (2018)
90. Tang, B., et al.: Issues and opportunities facing aqueous zinc-ion batteries. Energy Environ. Sci. **12**(11), 3288–3304 (2019)
91. Dunn, B., Kamath, H., Tarascon, J.-M.: Electrical energy storage for the grid: a battery of choices. Science **334**(6058), 928–935 (2011)
92. Li, Z., et al.: Exploration of Metal/Ti3C2 MXene-derived composites as anode for high-performance zinc-ion supercapacitor. J. Power Sour. **506**, 230197 (2021)
93. Zhu, X., et al.: Ion-intercalation regulation of MXene-derived hydrated vanadates for high-rate and long-life Zn-Ion batteries. Energy Storage Mater. **45**, 568–577 (2022)
94. Wang, X., et al.: Tailoring ultrahigh energy density and stable dendrite-free flexible anode with Ti3C2Tx MXene nanosheets and hydrated ammonium vanadate nanobelts for aqueous rocking-chair zinc ion batteries. Adv. Func. Mater. **31**(35), 2103210 (2021)
95. An, Y., et al.: Highly reversible Zn metal anodes enabled by freestanding, lightweight, and zincophilic MXene/nanoporous oxide heterostructure engineered separator for flexible Zn-MnO2 batteries. ACS Nano **16**(4), 6755–6770 (2022)
96. Liu, C., et al.: In situ growth of three-dimensional MXene/Metal–organic framework composites for high-performance supercapacitors. Angew. Chem. **134**(11), e202116282 (2022)
97. Zeng, C., et al.: Ultrathin titanate nanosheets/graphene films derived from confined transformation for excellent Na/K ion storage. Angew. Chem. **130**(28), 8676–8680 (2018)
98. Zhao, R., et al.: Self-assembled Ti3C2 MXene and N-rich porous carbon hybrids as superior anodes for high-performance potassium-ion batteries. Energy Environ. Sci. **13**(1), 246–257 (2020)
99. Li, L., et al.: Phenazine anodes for ultralongcycle-life aqueous rechargeable batteries. J. Mater. Chem. A **8**(48), 26013–26022 (2020)
100. Jeong, G.S., et al.: Electrochemical characteristics of cyanoquinones as organic cathodes for high-potential sodium-ion batteries. ACS Sustain. Chem. Eng. **8**(30), 11328–11336 (2020)
101. Shi, M., et al.: Multiple redox-active cyano-substituted organic compound integrated with MXene for high-performance flexible aqueous K-ion battery. Chem. Eng. J. **450**, 138238 (2022)
102. Pan, H.: Ultra-high electrochemical catalytic activity of MXenes. Sci. Rep. **6**(1), 32531 (2016)
103. Yoon, Y., et al.: Enhanced electrocatalytic activity by chemical nitridation of two-dimensional titanium carbide MXene for hydrogen evolution. J. Mater. Chem. A **6**(42), 20869–20877 (2018)
104. Ling, C., et al.: Transition metal-promoted V2CO2 (MXenes): a new and highly active catalyst for hydrogen evolution reaction. Adv. Sci. **3**(11), 1600180 (2016)

105. Li, P., et al.: High-throughput theoretical optimization of the hydrogen evolution reaction on MXenes by transition metal modification. J. Mater. Chem. A **6**(10), 4271–4278 (2018)
106. Dey, A., et al.: Cu2O/CuO heterojunction catalysts through atmospheric pressure plasma induced defect passivation. Appl. Surf. Sci. **541**, 148571 (2021)
107. Peng, Q., et al.: Cu single atoms on Ti2CO2 as a highly efficient oxygen reduction catalyst in a proton exchange membrane fuel cell. J. Mater. Chem. A **7**(45), 26062–26070 (2019)
108. Wang, R., et al.: Element-doped Mxenes: mechanism, synthesis, and applications. Small **18**(25), 2201740 (2022)
109. Naguib, M., et al.: Two-dimensional nanocrystals produced by exfoliation of Ti3AlC2. In: MXenes, pp. 15–29. Jenny Stanford Publishing (2011)

Recent Progress in MXene Research for Sensor and Environmental Applications

Lenon Medina Ferreira, Ana Paula de Oliveira Lopes Antunes, Isaac dos Santos Nunes, Ananda Fagundes Guarda, Eliezer Quadro Oreste, and Daiane Dias

Abstract MXenes are two-dimensional nanomaterials with several admirable properties that have attracted interest in sensing systems and environmental applications. These materials present high surface area, regular layered structure, adjustable elemental composition, high chemical activity, excellent electrical conductivity, and unique mechanical properties. The synthesis of MXenes can be approached from two perspectives. One is the top-down method, which involves processes like chemical etching or photo-Fenton, and the other is the bottom-up approach, exemplified by techniques such as chemical vapor deposition. Because of its broad applicability, studies have explored various possibilities for modifying MXenes, including organic and inorganic modifications, resulting in different chemical groups on the surface. Incorporating these groups into MXenes can facilitate the development of sensors for detecting various chemical compounds with wide-ranging applications. Recent advances have broadened the scope of MXenes-based materials, opening new horizons for environmental monitoring, photocatalysis, adsorption, and wastewater treatment applications. This chapter aims to provide an overview of the recent innovations in applying MXenes to develop sensing platforms for monitoring organic and inorganic contaminants in environmental samples. Furthermore, the main challenges and opportunities to be explored in new applications are also addressed.

Keyword MXene · Modification · Sensor · Environmental contaminants

L. M. Ferreira · A. P. de Oliveira Lopes Antunes · I. dos Santos Nunes · A. F. Guarda · E. Q. Oreste · D. Dias (✉)
Laboratório de Eletroespectro Analítica (LEEA), Escola de Química e Alimentos, Universidade Federal do Rio Grande, Rio Grande, RS, Brazil
e-mail: daianedias@furg.br

I. dos Santos Nunes
Federal Institute of Education, Science and Technology of São Paulo, Suzano Campus, Mogi das Cruzes Avenue 1501, 08673-010 Suzano, SP, Brazil

© The Author(s), under exclusive license to Springer Nature Singapore Pte Ltd. 2025
S. Moharana and R. N. Mahaling (eds.), *Surface Engineering of MXenes*, Engineering Materials, https://doi.org/10.1007/978-981-96-3640-2_8

1 Introduction

The MAX phases are three-dimensional (3D) materials with the formula $M_{n+1}AX_n$ (where "n" can be 1, 2, 3, or 4). In this formula, "M" represents an early transition metal, such as Sc, Ti, V, Cr, Zr, Nb, Mo, Hf, Ta, and W; "A" is an element from groups 13 to 16 of the periodic table, such as Al, Si, P, S, Ga, Ge, As, In, Sn, and Pb; "X" represents carbon and/or nitrogen [75]. These functional groups are derived from the etching process and impart semiconductor properties to MXenes, enabling their semiconductor behavior and various applications in environmental analysis [1]. Many compounds have been synthesized from different combinations of the MAX phase, exhibiting properties such as high structural rigidity and good electrical and thermal conductivity [53].

MXenes are compounds derived from the MAX phase, obtained by removing the "A" phase through chemical etching. This process disrupts the three-dimensional structures of the MAX phases, generating thin two-dimensional (2D) nanosheets of a new material called MXene. MXene originates from its similarity in hexagonal crystallinity to other two-dimensional materials, such as graphene [75].

The evaluation of the effective transformation of MAX into MXene can be made by X-ray diffraction (XRD) characterization, which presents differences in peak evaluation. About the properties, terminal groups can be changed to obtain new properties. MXenes are usually negatively charged [69], but this modification can make the material positively charged [80], increasing its application for some pollutant removal. Zeta potential can evaluate the surface charge. Another property that influences MXene's application is the surface area, which can be assessed using the BET method. Superficial characterization can focus on the chemical composition, presented by Fourier-Transform Infrared Spectroscopy (FTIR) or X-ray photoelectron spectroscopy (XPS), as an example, or on the structure, evaluated by Scanning Electron Microscopy (SEM) or Transmission Electron Microscopy (TEM) that can confirm the format of particles and size. These properties can indicate some applications of MXenes as sensors for environmental analysis, for example.

Sensors play an essential role in modern society, serving as tools that convert various forms of physical data, such as temperature, pressure, humidity, motion, and chemical composition, into signals that can be measured, analyzed, and acted upon [95]. In this context, MXenes, a two-dimensional (2D) material similar to graphene, has gathered significant attention due to their potential use in developing and modifying sensors. Their high surface area, modifiable surface, and high electrical conductivity allow for the creation of sensors capable of delivering high sensitivity, selectivity, and rapid response times [107]. These characteristics make MXenes ideal candidates for the development of new sensors.

With the growing concern for tracking pollutant levels in the environment, MXene-based sensors emerge as a compelling choice for detecting inorganic contaminants. Nowadays, most MXene-based sensors applied for sensing inorganic contaminants follow two distinct mechanisms: chemiresistive and electrochemical. Chemiresistive sensors usually detect toxic gases such as ammonia and H_2S [16], while electrochemical sensors detect heavy metals [26].

Among the possibilities for applying MXene-based sensors, an area of study of these devices is focused on the detection and monitoring of pollutants of an organic nature since these are capable of bioaccumulating, are toxic, mutagenic, and carcinogenic [79]. In the literature, most of the sensors studied and developed based on MXene are electrochemical, chemiresistive, and those that work through adsorption–desorption mechanisms [48, 94, 101]. In this scenario, such sensors are widely used to detect dyes, pesticides, phenolic, and pharmaceutical compounds in samples, mainly water, food, and atmospheric air [101].

2 Synthesis of MXenes

2.1 Precursors

The new materials called MAX phases have the important characteristic of simultaneously presenting desirable metal properties, such as resistance, thermal and electrical conductivity, and ceramic characteristics, such as high melting point and hardness [105]. Additionally, the materials are fatigue-resistant and, in many cases, suitable for cryogenic or refractory environments [23].

MAX phases are composed of polycrystalline laminated structures of carbides and nitrides. Its general formula is described by $M_{n+1}AX_n$, where "n" can be 1, 2, 3, or 4. In this representation, the letter "M" represents a transition element, and the letter "A" represents a chemical element from group A of the periodic table, mainly elements belonging to groups 13–16 [52]. The letter "X" represents carbon and/or nitrogen elements. The combination of these elements gives rise to different MAX phases, described as M_2AX, M_3AX_2, M_4AX_3, and M_5AX_4 [75].

MAX phases comprise unit cells with a metal M on the edges and X atoms in the center, forming a tetrahedral geometry with bonds classified as intense between the atoms [14]. When analyzing the three-dimensional structure, the layers formed by the unit cells are connected and form an octahedral geometry, interspersing the layers with phase A in pure form [27]. The bond formed between phase A and the tetrahedra is classified as weaker compared to the bonds between the atoms of the tetrahedron due to the metallic bond between elements M and A [150]. Furthermore, the difference between the types of MAX phase is in the number of layers of metal M. Initially, the value of n reported in the literature ranged from 1 to 3. However, with advances in the area, n up to 6 values are currently reported [113].

MAX phases can be classified according to their respective "n" values. The MAX phase of type M_2AX has two M layers between the pure A phase, receiving a classification of 211 [40]. The MAX phase of type M_3AX_2 has three M layers between the pure A phase, receiving a classification of 312 [151]. The MAX phase of type

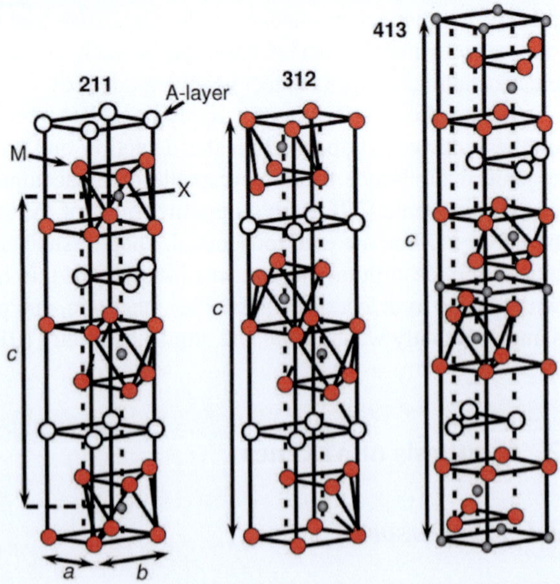

Fig. 1 Crystal structure diagrams illustrating MAX phases with varying values of n. Reprinted with permission from [151]. Copyright 2024, Elsevier

M_4AX_3 has four M layers between the pure A phase, receiving a classification of 413 [40]. More structural information can be seen in Fig. 1 [150].

The first generation of MAX phases, which corresponds to more than 50 described molecules [27], was classified as 211 [8], 312 [102], and 413 [24], according to the number of n synthesized and available at the time. Currently, new MAX phases of higher orders are described in the literature, such as M_5AX_4 (514) [121], M_6AX_5 (615) [70], and M_7AX_6 (716) [35]. It is essential to highlight that the MAX phases of higher orders, such as 615 and 716, are found only as secondary components or in an interconnected way in other MAX phases of lower orders, such as M_2AX and M_3AX_2 [113]. Due to the development of MAX phases with higher orders, there has been a significant increase in the number of compounds in the literature, with over 150 molecules described thus far [11]. Table 1 presents examples of the MAX phase, categorized according to the proportion of atoms in its structure, here called series (211, 312, 413, 514, 615, and 716), in addition to the metal used in the M phase [113].

The MAX phase synthesis is based on the mixture of substances that will give rise to the material. Each MAX phase is composed of a metal M (transition metal), an element A (group A element), and an element X (carbon and/or nitrogen). In this way, it is possible to conduct the synthesis by mixing the components in an elementary form or mixing binary compounds containing the elements that will give rise to the MAX phase [111].

One of the first synthetic routes listed in the literature was the synthesis through the mixture of powders of the constituent elements of the MAX phase. In a more simplified way, this occurred at high temperatures, initially forming M_xA_y compounds

Table 1 Examples of MAX phases according to their series and the metal used in the M phase

Serie	Phase MAX									
	Sc	Ti	Ta	V	Cr	Zr	Nb	Mo	Hf	Mn
211	Sc_2InC, Sc_2SnC	Ti_2CdC, Ti_2AuN, Ti_2GaC, Ti_2AlC, Ti_2InC, Ti_2TlC, Ti_2AlN, Ti_2InN, Ti_2SnC, Ti_2GaN, Ti_2SC, Ti_2ZnC, Ti_2ZnN, Ti_2GeC, Ti_2PbC	Ta_2GaC, Ta_2AlC	V_2AlC, V_2GaC, V_2GaN, V_2GeC, V_2ZnC, V_2SnC, V_2PC, V_2AsC	Cr_2GaC, Cr_2GaN, Cr_2GeC, Cr_2AlC	Zr_2InC, Zr_2TlC, Zr_2TlN, Zr_2PbC, Zr_2SnC, Zr_2SC, Zr_2AlC, Zr_2SB, Zr_2InN, $(Zr,Nb)_2(Al,Sn)C$	Nb_2AlC, Nb_2InC, Nb_2SnC, Nb_2AsC, Nb_2PC, Nb_2SC, Nb_2CuC, Nb_2SB, Nb_2GaC	Mo_2GaC, Mo_2AuC	Hf_2InC, Hf_2TlC, Hf_2SnC, Hf_2SnN, Hf_2PbC, Hf_2SC, Hf_2Sb	Mn_2GaC
312		Ti_3AlC_2, Ti_3InC_2, Ti_3GaC_2, Ti_3GeC_2, Ti_3SiC_2, Ti_3SnC_2, Ti_3ZnC_2	Ta_3AlC_2	V_3AlC_2		Zr_3AlC_2				
413		Ti_4AlN_3, Ti_4GaC_3, Ti_4SiC_3, $Ti4GeC_3$	Ta_4AlC_3	V_4AlC_3			Nb_4AlC_3	$(Mo,V)_4AlC_3$		
514		$(Ti_{0.5}Nb_{0.5})_5AlC_4$, $Ti_5Al_2C_3$						Mo_4VAlC_4		
615			Ta_6AlC_5							
716		Ti_7SnC_6								

Adapted with permission from [113]. Copyright 2024, Elsevier followed by the dissolution of the M_xX_y phase into compounds, resulting in the synthesized MAX phase [150]. An example of this synthesis is the formation of the MAX phase through the mixture of powders of the elements in elemental form, in the proportion 2Ti:1Al:1C (Ti_2AlC), using the self-propagating synthesis method at high temperature [31]. During this synthesis, after the ignition step, an exothermic reaction (heat release) occurs between the titanium element and carbon, resulting in the initial formation of TiC. The heat released from this occurrence leads to aluminum fusion, forming a liquid phase formed by Al–Ti. Then, the TiC crystals dissolve in this liquid phase, forming the MAX Ti_2AlC phase [116].

As previously described, diatomic molecules M_xX_y and M_xA_y are formed during the synthesis of the MAX phase. Suppose these intermediate compounds are employed as starting materials. In that case, the energy barrier for the synthesis of the MAX phase can be considerably reduced, allowing synthesis with greater purity and/or a decrease in the temperature applied in the reaction [151]. Due to the weaker M-A bonds between the layers of phase A than the M-X bonds within the layers, it is possible to synthesize the MAX phase through the intercalation of the A layers into M_xX_y compounds with x vacant sites [104].

In the literature, one of the earliest records of the synthesis of Ti_3AlC_2 involved the combination of Al with TiC, carried out at a temperature of 1000 °C for one hour [104]. This temperature is approximately 500 °C lower than the temperature used during the in-situ synthesis with spark plasma sintering (SPS) using elemental powders of the constituents of the MAX phase.

2.2 *Preparation Techniques*

The technique applied to MXenes preparation is crucial as it directly influences the resulting material's physical and chemical properties. The main preparation techniques are based on top-down and bottom-up mechanisms. However, regardless of the technique used, the material obtained after synthesis may still undergo surface modification to become compatible with the desired application [146].

2.2.1 Top-Down Approach

The top-down approach can be considered a classical preparation technique, in which MXene is synthesized from the MAX phase by selective removal of the A element followed by exfoliation of layers [58, 59, 146]. This approach is often carried out through wet chemical procedures and recent alternative methods such as selective dry or microwave-assisted extraction [90, 91].

Wet chemical methods involve using strong reagents, electrochemical exfoliation, and mechanical/hydrothermal processes to selectively remove the A element from the intermediate layer of the MAX phase [91]. Corrosive reagents commonly used include those containing fluoride ions, such as ammonium fluoride (NH_4F), hydrofluoric acid (HF), and a mixture of lithium fluoride (LiF) and hydrochloric acid (HCl) [9, 91]. Below is an example of a reaction for MXene synthesis occurring between the MAX phase and HF as the corrosive medium [58, 59].

1. Mn + 1AXn(s) + 3HF(aq) → Mn + 1Xn(s) + AF_3(aq) + 1.5H_2(g)
2. Mn + 1Xn(s) + 2HF(aq) → Mn + 1XnF_2(s) + H_2(g)
3. Mn + 1Xn(s) + 2H_2O(l) → Mn + 1Xn$(OH)_2$s + H_2(g).

In the first step of the reaction, the A element of the MAX phase reacts with fluoride ions to form AF and release hydrogen gas [58, 59, 131, 146]. Thus, the exfoliation of MXene adjoined layers occurs, which subsequently spontaneously reacts with water or HF to facilitate surface terminations. In this method, an essential factor is the concentration of HF, which can directly influence the synthesis time, which is generally faster when concentrated HF is used. Although diluted acid solutions can ensure excellent safety and lower environmental impacts, the synthesis becomes slower and requires increased temperature [91].

An alternative is the in situ synthesis of HF, which occurs with fluoride salts (such as NaF, LiF, or $NaBF_4$) in the presence of HCl. This method is known as minimally intensive layer delamination (MILD) and can be considered relatively safer because there is no direct exposure to concentrated HF [91, 122]. However, there is a growing demand for efficient fluoride-free synthesis methods with potential applications for the formed MXene [91, 133]. One possibility is alkaline hydrothermal synthesis, in which the MAX phase reacts directly with a concentrated alkaline solution (such as NaOH), forming MXene containing –OH or –O in its structure, as [65] reported. In addition to this, other recent methods are mentioned in the literature, such as MXene synthesis by HCl electrochemical etching, Lewis's acid molten salt etching, and halogen-based etching [133].

Selective dry extraction procedures stand out as an alternative to wet chemical methods. They are a recent approach to MXene production and are very similar to chemical vapor deposition (CVD). For this, a flask containing a specific amount of the MAX phase is placed inside a tubular furnace, where a vapor of the corrosive agent (such as I2, for example) flows at high temperatures. The temperature employed in this step must be optimized, as it can directly influence the reaction between the iodine vapor and the A element of the MAX phase [91, 103]. Figure 2 shows a scheme of this procedure, where uniformly distributed halogen-terminated MXene and gaseous halides of the A element are formed.

Another alternative for MXene synthesis is the microwave-assisted hydrothermal method, as [90] described. This method's main advantage is the synthesis time, which is much shorter than the conventional method of obtaining MXene [90]. Figure 3 presents a schematic synthesis of MXene using microwaves.

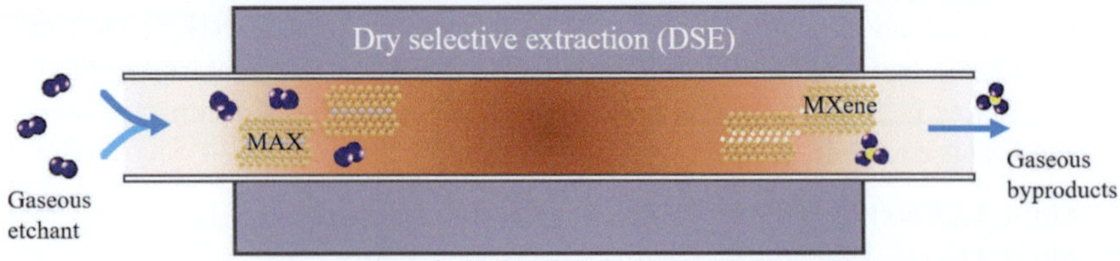

Fig. 2 Scheme of the selective dry extraction method for MXene synthesis. Reprinted with permission from [103]. Copyright 2024, Elsevier

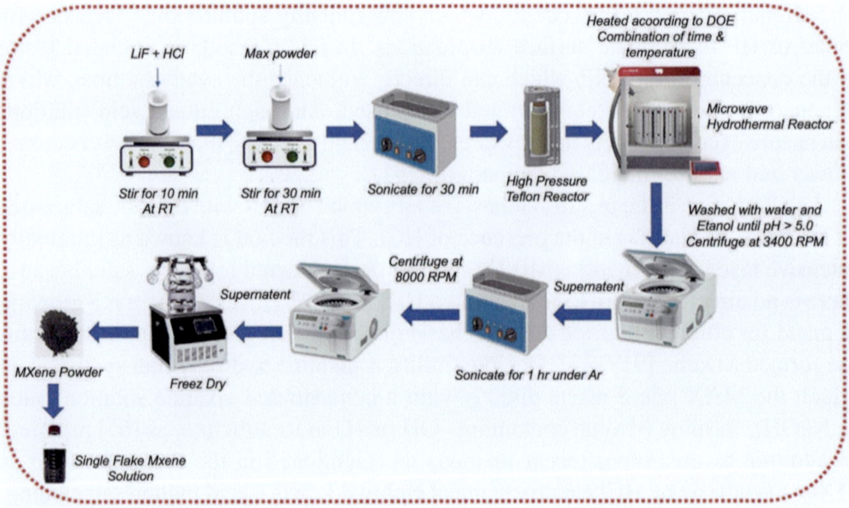

Fig. 3 Diagram of the microwave-assisted method for MXene synthesis. Reprinted with permission from [90]. Copyright 2024, Elsevier

2.2.2 Bottom-Up Approach

Many studies use the top-down method for MXene synthesis; however, the formation of small-sized crystals and irregular thickness are considered disadvantages of this method for specific applications [92]. To overcome these issues, bottom-up approaches for MXene synthesis enable material design customization [91]. Additionally, this method allows the synthesis of MXene with different stoichiometries and their heterostructures, which are hardly obtained by the top-down method [92].

A widely described synthesis is CVD, where a MXene film is deposited onto a metallic surface when gaseous precursors flow over a heated substrate [91, 122]. During this process, volatile, unstable byproducts are also formed, which are eliminated by the flowing gas inside the reaction chamber [109]. This synthesis provides MXenes with new morphologies containing more easily accessible surfaces and exposed catalytically active edges [128]. Another recent synthesis reported in the

Table 2 Comparison of top-down and bottom-up approaches for MXene synthesis

	Top-down approach	Bottom-up approach
Precursor	Initiate from the MAX phase of the 3D structure	Initiate from atom to crystal growth of MXene film
Method	It involves chemical etchants such as HF acid	Apply chemical vapor deposition or, salt-templating method, or plasma-enhanced deposition
Synthesis condition	Pressure-less synthesis	Control gas flow as a carbon source
Temperature	Require room or low temperature for synthesis	Require high temperature of ~1000 °C for synthesis
Morphology	Large irregular MXene sheets could be produced with a lateral size of a few hundred nanometers and thin sheets with a thickness between 10 and 200 nm	Produce defect-free and highly crystalline thin film of multilayers with a lateral size of between 10 and 100 μm
Surface properties	Functionalized –OH and –O after synthesis	Lack of functional groups after synthesis

Reprinted with permission from [146]. Copyright 2024, Elsevier

literature is Plasma-Enhanced Pulsed-Layer Deposition (PEPLD) [91]. Pulsed-laser deposition (PLD), which is a traditional thin-film growth procedure, is combined with Plasma-Enhanced CVD to produce variously sized monocrystalline films, which the number of laser pulses can control [131, 148]. Compared to conventional CVD, the plasma-enhanced method synthesizes higher-quality material at relatively lower temperatures [91, 92, 131].

In addition to the gas-phase procedures, direct solid-state synthesis is another bottom-up method that can be applied. This synthesis combines all constituents in their original states through solid-state reactions at high temperatures. This synthesis can obtain products with higher yields compared to CVD-based methods. Furthermore, the possibility of precisely controlling the termination group based on the choice of the precursor is an essential advantage of this method [91, 92].

When both described methods (top-down and bottom-up) are compared, some differences lead to products with distinct properties. Table 2 lists these main differences regarding the precursor, method, synthesis conditions, temperature, morphology, and surface properties.

2.3 *Modifications of MXenes*

MXenes possess numerous advantages and potential applications, yet they still lack the necessary properties for better application in specific fields. Thus, surface modification is an alternative to substantially improve the performance and stability of

MXenes [82]. This modification involves surface functionalization with various functional groups, leading to the obtainment of new materials with unique properties [112].

Functionalized MXenes can be obtained through the surface modification of original MXenes by covalent or non-covalent bonds [82]. Non-covalent modification can be achieved through van der Waals forces interactions, hydrogen bonds, and electrostatic attraction [82]. The characteristic of this modification is the possibility to preserve much of the material's intrinsic properties, besides being mostly a simple and economical procedure [66, 82, 117]. As for covalent surface modification, it can occur through small molecules (such as alkali metal hydroxides, acid halides, acid anhydrides, epoxy compounds, and organic amines), heteroatoms, or surface-initiated polymerization by macromolecules [82]. A characteristic of covalent modification is its irreversibility, permanently altering the MXene surface [93].

Some properties of MXenes can be enhanced through their modification with various small molecules, such as mechanical stability, electrical property, and stability in solution [82]. Recently, modifications with nanoparticles and inorganic nanostructures have gained prominence, mainly to improve the photoelectrochemical properties of MXenes. Among the developed materials, noble metal nanoparticles, metal oxides, metal hydroxides, metal sulfides, and others are used [112].

Regarding obtaining functionalized MXenes with macromolecules, modification can occur through in-situ polymerization with strong covalent bonds [33, 82]. However, the modification most frequently performed is through the mixture between MXene and ex situ polymers, which has the advantage of using well-defined polymeric structures and more adjustable compositions than the in-situ method [82]. This surface modification with polar or nonpolar polymers is a promising strategy to improve MXene properties, such as stability and conductivity, enabling various applications [61, 82, 112]. Another possibility of modification is with heteroatoms, which are required to improve the electrochemical performance of MXenes, applied, for example, in Na-ion batteries and Li-ion batteries, which occurs due to the increase in the interlayer spacing of MXenes caused by single heteroatoms [82].

Overall, MXene modification directly influences mechanical properties (such as stiffness), electronic properties (such as semi-conductivity, metallicity, and topological insulative), magnetic properties, optical properties (such as refractive index, reflectivity, absorption, and transmittance), hydrophilicity/hydrophobicity, and others [82]. Regarding application, specifically functionalized MXenes generally exhibit better performance compared to original MXenes, yet there should be a study on the structural and specific properties of the new material to understand the actual effect of functionalization [82, 93]. Figure 4 shows examples of MXene modification, obtaining functionalized surface materials.

As shown in Fig. 4, the study carried out by [29] led to the formation of a superficially modified MXene with TiO_2 nanoparticles. This process occurred by annealing treatment, and according to the authors, the Ti layer on the top and bottom was oxidized by the oxygen groups on the surface. The authors also conclude that the modified material has improved dielectric and conductive properties.

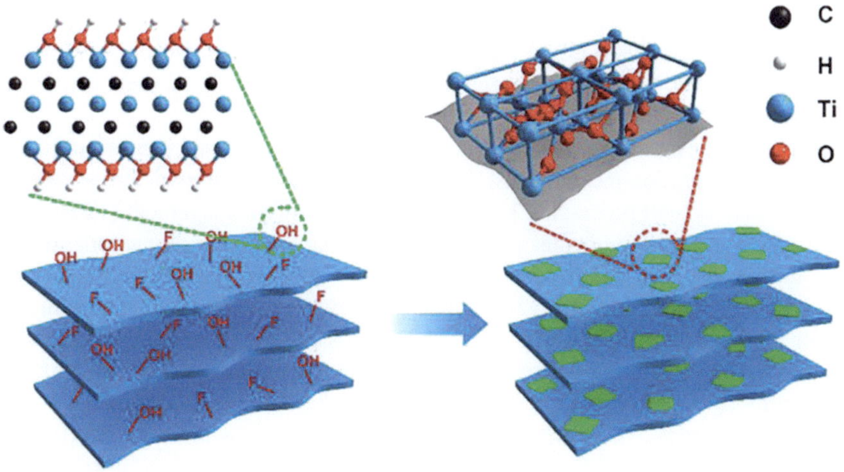

Fig. 4 Surface modification process of $Ti_3C_2T_x$ MXenes after annealing. Reprinted with permission from [29]. Copyright 2024, American Chemical Society

3 Properties and Characterization Applied for MXenes as a Sensor

3.1 Structure of MXenes-Based Sensor

MXenes, as a material, have captured the scientific community's attention due to their properties, including a high surface area, electrical conductivity, and a modifiable surface [135]. This versatility allows for the development and application of MXenes in a wide range of sensors used in various analytical techniques. Consequently, the structure of MXene-based sensors can vary depending on the method and type of sensor. However, most MXene-based sensors operate using three distinct mechanisms: chemiresistivity, electrochemistry, and adsorption–desorption [94, 101].

Chemiresistive MXene-based sensors are commonly used for gas detection. Usually, the substrates fabricating gas sensors can be divided into semiconductors and conductors [54]. Semiconductor sensors, such as Metal oxide semiconductors (MOS), require high temperatures for optimal signal-to-noise ratio [86]. This happens because this type of sensor needs high activation energy, making it challenging to use in portable devices [54]. On the other hand, conductive sensors produce low noise but lack sufficient gas adsorption sites for strong signals. In this situation, Kin and collaborators have suggested using Menes to solve the problem of the high-temperature needs of MOS-based sensors [54]. Their sensor comprises two Au electrodes printed 100 μm apart in a SiO_2 wafer. Between the two electrodes, a film of Ti $Ti_3C_2T_x$ (fully terminated by hydroxyl (−OH) and oxygen (−O) groups) was deposited. The gas sensing mechanism in MXenes is based on the physical or chemical adsorption of gas molecules on their surface, which leads to variations in resistance due to the absorption of gases [71]. Using this sensor, the authors could determine acetone, ethanol, ammonia, propanal, nitrogen dioxide, sulfur dioxide, and carbon dioxide with an empirical LOD of 50 ppb at room temperature.

Humidity sensors serve as a prime example of MXene-based sensors utilizing adsorption–desorption mechanisms. These types of sensors are used across various sectors, such as food processing [145], respiratory health management [125], agriculture [56], and environmental surveillance [17]. An ideal humidity sensor should have high sensitivity, rapid response time, and be of easy miniaturization, facilitating large-scale production [129]. MXenes have great potential in the humidity sensor application since the material's surface has many gas adsorption sites and functional groups that facilitate the adsorption of water molecules from the air [30].

A humidity sensor modified with MXenes can be built, as suggested by Wang and collaborators [128], by coating an interdigital electrode with Cu/Ni/Au (Fig. 5) and depositing a $CuO/Ti_3C_2T_x$ in the superficies of the electrode by drop casting. This sensor exhibits an n-type response, where its resistance decreases as relative humidity (RH) increases. At low humidity, water molecules chemisorb on the CuO and $Ti_3C_2T_x$ surfaces through a double hydrogen bond, and as RH increases, more water vapor physically adsorbs onto this layer. This process leads to the ionization of water molecules under an electrostatic field, generating H_3O^+ ions and causing protons to be transported via ionic conductivity. A water layer forms and penetrates the sensor material at high humidity levels, changing the sensor's electrical properties.

In this work, the authors have shown that the sensor modified with a $CuO/Ti_3C_2T_x$ had a high sensitivity sensor of 451 KΩ/% RH, approximately 282000 times the response of a Ti_3C_2/polyelectrolyte (1.6 Ω/% RH), has a short response time (0.5 s), recovery time (1 s), and a low hysteresis value.

Electrochemical MXene-based sensors usually incorporate a glassy carbon electrode (GCE) as their foundation. A glass carbon electrode is built within a body crafted from an insulating material, usually a polymer, with a glass carbon disk positioned at one end of the electrode [108]. This glass carbon disk will directly contact

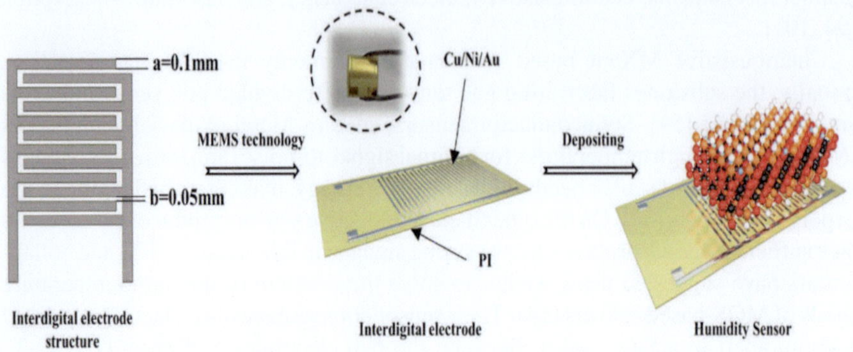

Fig. 5 Structure of a humidity sensor. Reprinted with permission from [129]. Open-source material, RSC Publications

the analytes during the experiment. The conductive material within the electrode's body is usually crafted using silver, copper, or platinum wires. This conductor will connect the glass carbon disk with the electrochemical equipment. This electrode type can be modified by drop-casting a solution with MXene within. Lee and collaborators proposed a similar modification, suggesting the modification of a GCE by drop-casting an MXene-titania-Nafion composite film [62]. This electrode was used to determine the concentration of capsaicin in pepper samples, and it presented a significant increase in the oxidation current towards capsaicin compared to that at a GCE and the titania-Nafion composite-modified GC electrode, having a detection limit of 2.5×10^{-8} M.

3.2 *Properties of MXenes-Based Sensor*

MXenes are a class of 2D nanomaterials that have gained considerable attention for environmental protection. Their key characteristics include high surface area, abundant terminal groups, and distinctive 2D layered structures, like graphene [69]. MXenes present active sites, easy functionalization, biocompatibility, high porosity, hydrophilicity, great interlayer spacing, chemical stability, and greater adsorption-reduction capacity [4].

The determination of BET surface area is an essential characterization of environmental applications of MXenes in adsorption, catalytic, and sensors, for example. MXenes with larger surface areas exhibit higher adsorption capacity [144]. Better Pb(II) adsorption was justified by enlarging surface area and significant presence of functional groups [149] employing adsorbents with surface area varying from 6.4 to 129.2 $m^2\ g^{-1}$. However, Wu et al. [136] present that for non-agglomerated MXenes, the surface area could be higher than 1.800 $m^2\ g^{-1}$. The authors classified a produced material as a superabsorbent with an adsorption capacity of 2.750 mg g^{-1} for crystal violet removal.

Pristine MXenes are negatively charged when in dispersion, such as distilled water, tap water, and saline, but these charges are affected by the material structure and surrounding environment [69]. Evaluation of surface charges can be made using zeta potential. Besides, MXenes are generally negatively charged; modifications on the surface can make them positively charged, enhancing the adsorption capacity for specific pollutants [80]. Zhang et al. [149] functionalized MXene with amino groups. After that, the zeta potential evaluation was higher than that of pristine MXene. The adsorbent was used to adsorb Pb(II), and the best results were attributed to the BET surface area after amino-functionalization and stronger forces between $–NH_2$ and Pb(II).

The regeneration of adsorbents is a crucial point for application in industrial-scale processes. The ideal adsorbent is a material that presents high adsorption capacity and retains this capacity across successive adsorption–desorption cycles after regeneration [85]. Regeneration depends on the stability of solids and can reduce operation costs. MXenes present excellent chemical stability, which makes these materials attract attention to environmental uses. Wang and co-workers presented good adsorption capacity from alkaline-treated MXene in the removal of Congo red (539.7 mg g^{-1}), tetracycline (1,053.3 mg g^{-1}), and Malachite green (7,111.3 mg g^{-1}) [130]. The authors reported that the solid was separated from the liquid with centrifugation and washed with distilled water to recover. After five cycles, the regeneration was above 85%, proving the material's superior reuse and regeneration capacity. Ghani and co-workers reported using an electrochemical process to regenerate $Ti_3C_2T_x$ MXene in removing ciprofloxacin from wastewater [21]. The material was regenerated entirely (up to 99.7% after 5 cycles) within 5 min, and the material was characterized, indicating the stability of the adsorbent. In another work, Ghani and collaborators used MXene nanosheets to remove tetracycline and found that the crystallinity was slightly reduced after 5 cycles of electrochemical regeneration [22]. The authors reported that adding new functional groups ensures the adsorptive capacity is retained. These findings support the possibility of functionalization of MXenes and their versatility.

Besides adsorption capacity, MXenes also possess a unique property to reduce pollutants while adsorbing them [4]. This is an advantage over other adsorbents and nanomaterials, and it is essential to know the interaction between the solid material and the pollutant [115]. These electrochemical properties are related to the treatment conducted in the MAX phase, replacing functional groups. MXenes with terminal groups have higher conductivity than pristine MXene [55]. For example, altering surface chemistry provides new chemical choices for designing high-rate redox capacitive materials for high-power storage [67, 68]. For electrochemical sensing, the highly conductive nature of MXene facilitates the mobility of electron transfer from the electrode to the electrolyte [81].

The thickness of MXenes is an essential factor for electrochemical applications. Conductivity decreases when thickness increases, revealing an inverse relationship between the thickness and conductivity of the MXene composite [28]. For Hadi and co-workers, the most desirable conductivity can be achieved with very thin MXene nanosheets [28]. The authors also found that larger MXene nanosheets exhibit higher conductivity by reducing the spacing between particles, increasing the potential for network formation.

The chemical stability of MXenes can be affected by water and dissolved oxygen, and light can promote the oxidation of colloidal MXene solutions, resulting in oxide formation [81]. Aqueous media is a critical challenge to MXene's extensive application. Materials like $Ti_3C_2T_x$ have moderate oxidation resistance and can be thermally degraded in carbon, TiO_2, and CO_2 at temperatures above 200 °C. To solve these problems, the functionalization of MXenes can enhance thermal and chemical stability [41]. Li and co-workers found that Ti_3C_2 with F/OH termination was thermally stable up to 800 °C in an Argon atmosphere and 200 °C in an oxygen environment [64]. According to Mohanapriya and co-workers, MXene's stability can be increased by preserving it in a dark, oxygen-free, low-temperature refrigeration environment [81].

MXenes have many intriguing properties that can lead to their application in various areas. Here, we present the most significant properties related to the environmental uses reported in the literature. However, there are numerous possibilities for assessing MXenes and discovering properties that may emerge in the coming years.

3.3 Characterization of MXenes-Based Sensors

Visual examination is the initial stage in characterizing MXenes because even in layered form, there is an evident color change when MAX is converted to a MXene. MAX is usually gray, and MXenes present distinct colors related to their optical properties, depending on their composition and structure [110].

X-ray diffraction (XRD) patterns are an easy way to evaluate the formation of MXenes. Shekhirev and co-workers present that more information can be obtained when the analysis is done by comparing patterns of MAX phase and from MXenes in multilayer form, employing wet and dried samples, and from delaminated samples [110]. The authors present that examining the peak for the dried multilayer powder gives information on the thickness of the MXene sheet. At the same time, a hydrated sample will provide the number of layers of water or intercalants between the layers. Comparing the XRD pattern is also essential to synthesizing MXenes by etching. The optimum duration of the procedure is determined experimentally when the peak corresponding to the MAX phases disappears [5, 42, 88]. Mashtalir and co-workers evaluated the influence of time and temperature by XRD patterns to produce Ti_3AlC_2 etching by hydrofluoric acid (HF) [78]. The evaluation was made by the absence of a peak of ~39°, as presented in Fig. 6a for 50 °C and 15 h.

Fourier-Transform Infrared Spectroscopy (FTIR) can be used to evaluate the chemical bonding involved. However, the presence of water between the layers of MXenes affects IR spectra [110] and can be confused with the –OH termination group on the surface of the sheets. Mustafa, Khurshid, and Isa prepared MXene composites to remove polycyclic aromatic hydrocarbons in wastewater and employed FTIR to characterize samples [84]. The authors also used the technique to evaluate the material after the removal. Results helped suggest the interaction mechanism of solids with analytes, maybe by reduction, oxidation, and electron donor–acceptor interactions.

X-ray photoelectron spectroscopy (XPS) is a surface-sensitive technique with a low penetration depth and can give information about chemical composition and oxidation states [2]. XPS is a very attractive analytical technique with high sensitivity. However, its spatial resolution is limited to the square micrometer range [6]. Some works have reported the correlation of results for FTIR and XPS spectra to infer hydrogen and covalent bonds [142], bidding mechanism [67, 68], and confirmed modifications [144]. XPS was already employed to verify the integration of MXenes in polymer matrix composite [49, 67, 68]. XPS and FTIR were employed by Jun and co-workers to explore the adsorption of Pb(II) onto MXene, confirming that the ion was well-adsorbed [43]. Regeneration of MXene was also confirmed by comparing spectra from MXene, MXene + Pb(II), and MXene after regeneration

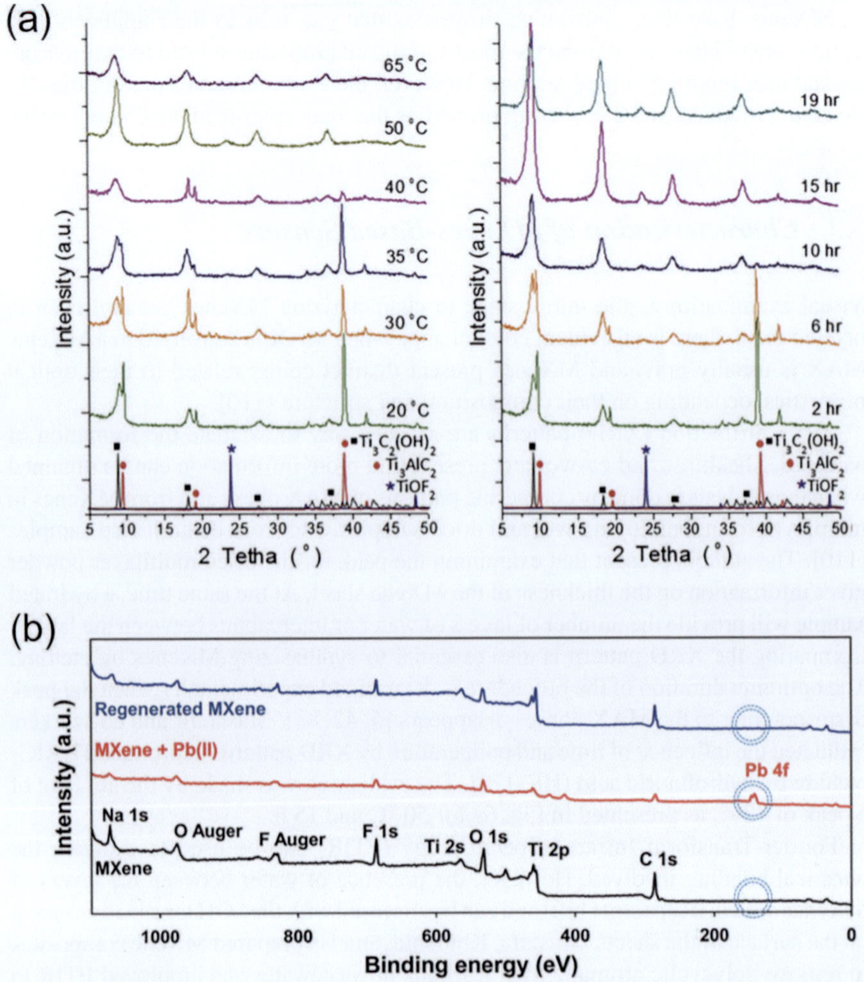

Fig. 6 **a** Diffractograms after etching of 50% HF as function of temperature and time at room temperature. Reprinted with permission from [78]. Copyright 2024, Elsevier, **b** Wide-scan XPS spectra related to Pb(II) adsorption by MXene. Reprinted with permission from [43]. Copyright 2024, RSC Publications

which reports the absence of Pb 4f peak, as presented in Fig. 6b. The findings indicate that these techniques cannot only be characterized but can also be used to evaluate the effectiveness of MXenes in environmental applications.

Raman spectroscopy is more applicable than FTIR; it is used to examine the structure of MXenes and detect oxidation. Like XRD, Raman spectroscopy provided information about the successful formation of MXene phases [42]. Raman also confirmed that MXenes were present in composites [44]. The sensitivity of Raman allows assessment of oxidized structure, presenting a reduction in the relative intensity of MXene peaks and the presence of a signal characteristic of the vibration of the oxygen bond with metal in the material structure [114].

To visualize structure and indicate the formation of MXenes, it is possible to use Scanning Electron Microscopy (SEM). Shekhirev and co-workers present that the appearance of "accordion-like" (Fig. 7a) most of the time is considered indicative of the success of MXenes formation [110]. The accordion-like structure was associated with it because the first syntheses were made in the immersion of highly concentrated HF. However, this does not describe multilayer MXene [110]. Accordion-like reactions come from escaped gases from the MAX structure because of the exothermic nature of HF reactions [5]. MXenes were already formed with a low concentration of HF (5%), resulting in a non-accordion-like structure (Fig. 7b). SEM can clearly show structure evolution by comparing the MAX phase and different conditions of synthesis. Liu and co-workers employed SEM images to show the formation of numerous tiny pores between layers after the foaming process, resulting in a well-defined porous MXene foam [72].

Transmission Electron Microscopy (TEM) is also used to probe the structure and morphology of MXenes. TEM images are obtained by transmitting a highly coherent electron beam through the sample, and the scattered electrons are detected. The thickness is too important because the electron beam passes through the sample. For MXenes as 2D materials, it's not a problem in most cases [6]. For a low-magnification image, TEM can reveal particles' average size and format (Fig. 7c). At high magnification, images can present the thickness and surface characteristics [10] (Fig. 7d). Some authors present results of High-Resolution Transmission Electron Microscopy (HRTEM). Alnoor and co-workers report that the dotted appearance of HRTEM does not directly relate to the atomic structure [6]. This is more related to the periodicity of crystalline structure and can provide information on crystallographic relations and determine crystalline spacing. Figure 7e presents the HRTEM image of MXene with the identification of crystalline spacing, and Fig. 7f presents spacing with the determination of crystalline planes, confirmed with XRD data.

As a 2D material, MXenes are generally dispersed in liquid media for environmental applications. In liquid media, the intensity of the attractive or repulsive forces between particles in a suspension is determined by the zeta potential [45]. The liquid media where the particles are dispersed consists of a stern and a diffusion layer. The free charges at the stern layer are strongly attracted to opposite charges on the MXene surface. In the diffusion layer, the ions are less connected. The zeta potential is the potential difference between the base liquid and the stern layer in contact with dispersed particles [76]. Zeta potential can be employed to evaluate changes in surface charges at different pH values [19, 80].

The pH strongly affects charges on the MXenes surface, so it is necessary to know the best environmental conditions to lead a process. For adsorption, pH at the zero-charge point (pH_{PZC}) refers to the pH value at which the adsorbent bears zero charges [12]. When the environment pH is lower than pH_{PZC}, surface groups will be protonated, resulting in a positively charged surface. These conditions are favorable for removing anionic pollutants. Additionally, when pH is higher than pH_{PZC}, surface groups will be deprotonated, showing a negatively charged surface on the

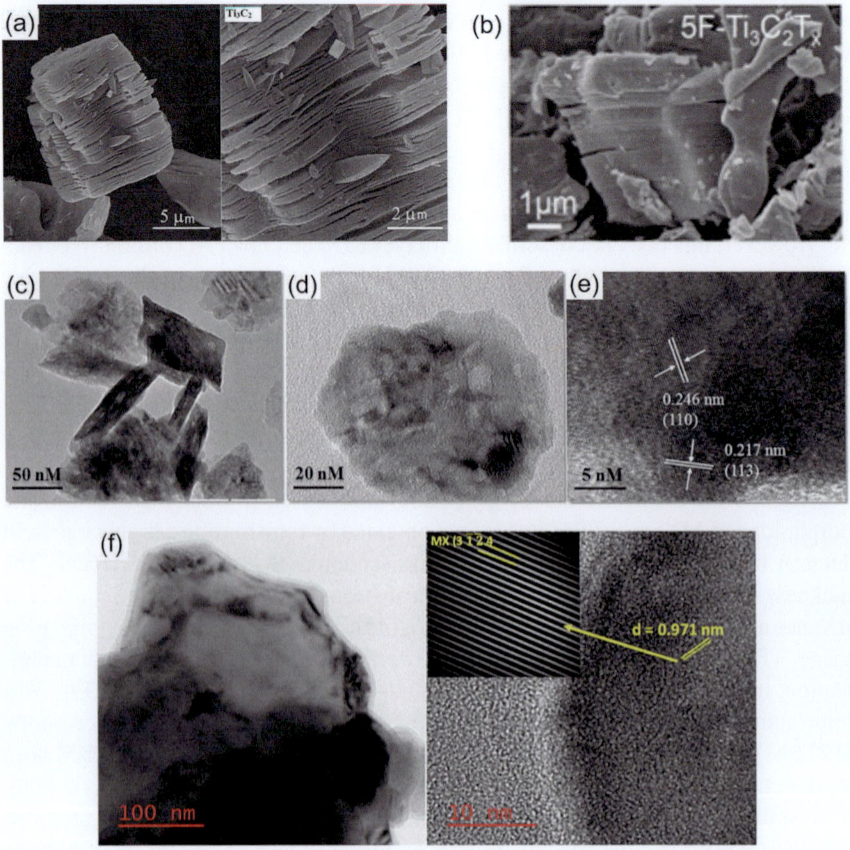

Fig. 7 **a** SEM images of Ti_3C_2 samples with "accordion-like" structure (Reprinted with permission from [77]. Copyright 2024, Elsevier, **b** SEM of $Ti_3C_2T_x$ powder produced with 5 wt.% HF (Reprinted with permission from [5]. Copyright 2024, American Chemical Society), TEM images at **c** low and **d** high magnification, **e** HRTEM with the determination of spacing planes (Reprinted with permission from [10]. Copyright 2024, Elsevier) and (f) HRTEM of MXene with spacing planes determination related to plane identification. Reprinted with permission from [46]. Copyright 2024, Elsevier

MXene. Cationic pollutant removal is favorable [12, 69]. The pH_{PZC} could be determined by zeta potential analysis [20]. Zeta potential was also used to explain the high adsorption of Pb(II) at pH > pH_{PZC} [25]. The technique was also used to indicate the existence of electrostatic adsorption of Hg(II) onto negatively charged MXene [36]. Wu and co-workers evaluated the influence of variables such as pH, temperature, salt, and humic acid, and zeta potential in the adsorption of four pharmaceutical compounds [138]. The authors' findings showed that pH and zeta potential represented a highly significant effect, and the other factors were almost negligible. These findings reinforce the importance of considering pH and zeta potential in MXenes applications in liquid media.

Characterization and evaluation of MXenes can provide valuable perceptions of their potential applications. Characterization techniques are employed to confirm synthesis results, verify the formation of MXenes, and corroborate property determinations.

4 MXenes for Environmental Sensing Applications

4.1 Organics Contaminants

Sensibility is essential for monitoring and controlling environmental pollutants [58, 59]. The major environmental concerns regarding organic pollution generally refer to contamination in water and food, but also in atmospheric air since the main classes of organic contaminants tend to bioaccumulate, presenting toxicity, mutagenicity, and carcinogenicity [79, 111, 139].

In this scenario, as can be seen in Table 3, the development of MXene-based materials has considerably expanded the field of sensing dyes [48, 73], pesticides [39, 119, 140, 153], pharmaceuticals [37, 98], phenolic compounds [38, 100, 124, 123, 134], and others.

Among the types and sensing mechanisms currently employed, electrochemical sensors are among the most investigated [101], as can also be seen in Table 3. This choice can be based on the advantageous characteristics of this type of sensor, such as minimum response and recovery time, high sensitivity and selectivity enabling detection limits at fento, nano, and pico levels, stability, and often portability [1, 83, 101]. Thus, in their work, Tumrani and collaborators reported anchoring Pd nanocubes (PdNCs) in partially oxidized MXene, obtained by stirring in a nitrogen atmosphere for 30 min [120]. The authors defend their thesis based on the synergism between the redox activity of PdNCs and the high conductivity of Ti_3C_2Tx, which enabled the detection by DPV of nitrofurantoin in hospital effluent, reaching LOD in the range of 0.01 nmol L^{-1}. On the other hand, Wu and collaborators developed MXene with a 2D structure with a –OH-terminated surface through an exfoliation process like graphene [134]. The material described as bioactive and stable, being a microenvironment favorable to the anchoring of enzymes such as tyrosinase, was used for the fast and ultrasensitive electrochemical study, by amperometry and CV, of phenol in water. The biosensor responded with good repeatability, reproducibility, and stability, achieving a linear range of 0.05–15.5 μmol L^{-1}, a LOD of 12 nmol L^{-1}, and a sensitivity of 414.4 mAM^{-1}.

However, sensors based on adsorption–desorption mechanisms and chemiresistive sensors are also technologies constantly associated with MXenes, given the material's properties, already been extensively described previously. Pazniak and collaborators demonstrated in their work a specific increase in the attention of MXenes due to partial oxidation at 350 °C, which maintains its electrical conductivity, but specifically improves its chemiresistive response to organic analytes in the air [94]. In the study,

Table 3 MXene for organic environmental contaminants sensing

Sensor	Mechanism	Detection technique	Analyte	Sample	LOD	References
Tyr–Ti_3C_2–CS/GCE	Electrochemical	CV and Amperometry	Phenol	Water	12 nmol L^{-1}	[134]
AchE-CS-$Ti_3C_2T_x$-GCE	Electrochemical	DPV	Malathion	Water	0.3 × 10^{-5} nmol L^{-1}	[153]
Au–In_2O_3–Ti_3	Adsorption–desorption	N/A	Formaldehyde	Atmospheric air	N/A	[74]
UiO-66–$Ti_3C_2T_x$	Adsorption–desorption	N/A	Methylene blue, Rhodamine B, Tetracycline	Wastewater	312, 285 and 476 mg g^{-1}*	[48]
Nb_2CT_x–Zn–C–NC/GCE	Electrochemical	DPV	4-nitrophenols	River water, tap water, and industrial wastewater	70 nmol L^{-1}	[38]
$Ti_3C_2T_x$	Chemiresistive	N/A	Ethanol	Synthetic air	N/A	[118]
$Fe_2(MoO_4)_3$–$Ti_3C_2T_x$	Chemiresistive	N/A	n-butanol and xylene	Atmospheric air	N/A	[156]
$MnMoO_4$–Ti_3C_2F/GCE	Electrochemical	DPV and CV	Hydroquinone and catechol	Water	0.26 and 0.30 nmol L^{-1}	[98]
$Ti_3C_2T_x$ nanosheets	Electromagnetic and chemical	SERS	Crystal violet, Rhodamine 6G, and Malachite Green	Ethyl alcohol solution	10, 100, 10 nmol L^{-1}	[73]
$Ti_3C_2T_x$–CNHs-β-CD-MOFs/GCE	Electrochemical	DPV	Carbendazim	Tomato juice	1.0 nmol L^{-1}	[119]

(continued)

Table 3 (continued)

Sensor	Mechanism	Detection technique	Analyte	Sample	LOD	References
$Ti_3C_2T_x$–rGO–GCE	Electrochemical	DPV and CV	Carbendazim	Cucumber and orange juice	0.67 nmol L^{-1}	[139]
Pd–MoS_2–$Ti_3C_2T_x$/SPCE	Electrochemical	SWV	Bisphenol A	Lake water	0.12 nmol L^{-1}	[124, 123]
Pt–$Ti_3C_2T_x$ nanocomposites/GCE	Electrochemical	DPV and CV	Bisphenol A	Water and fresh milk	32 nmol L^{-1}	[100]
AuNps–PPy–$Ti_3C_2T_x$/GCE	Electrochemical	Amperometry	Nitrofurantoin	Honey, pond water, and hospital wastewater	0.26 nmol L^{-1}	[124, 123]
$Ti_3C_2T_x$–BP–LIPG/GCE	Electrochemical	LSV	α-naphthalene acetic acid	Soil, water, tea, rice, wheat, corn	1.6 nmol L^{-1}	[155]
Pt–$Ti_3C_2T_x$–Si/SiO_2	Adsorption–desorption	N/A	Ethanol, methanol, isopropanol, and acetone	Synthetic air	N/A	[94]
$Ti_3C_2T_x$–MWCNTs	Electrochemical	DPV and CV	Catechol and hydroquinone	Industrial wastewater	6.6 and 3.9 nmol L^{-1}	[37]
$Ti_3C_2T_x$–Au–Pd/SPE	Electrochemical	DPV	Paraoxon	Pear and cucumber	1.75 ng L^{-1}	[152]
Pd–$Ti_3C_2T_x$–P	Electrochemical	DPV	Nitrofurantoin	Hospital waste effluent	0.01 nmol L^{-1}	[120]
$Ti_3C_2T_x$-PEDOT:PSS	Chemiresistive	N/A	Methanol	Synthetic air	N/A	[126]

N/A: Not available; Tyr = Tyrosinase; CS = Chitosan; GCE = Glassy carbon electrode; CV = Cyclic voltammetry; NPs = Nanoparticles; SWV = Square wave voltammetry; AchE = Acetycholinesterase; DPV = Differential pulse voltammetry; NC = N-doped carbon; SERS = Surface-enhanced Raman scattering; CNHs = Carbon nanohorns; CD = Cyclodextrin; MOFs = Metal organic frameworks; rGO = Reduced graphene oxide; SPCE = Screen-printed carbon electrode; PPy = Polypyrrole; BP = Phosphorene; LIPG = Laser-induced porous graphene; LSV = Linear sweep voltammetry; MWCNTs = Multi-walled carbon nanotube; SPE = Screen-printed electrode; PEDTO: PSS = Poly(3,4-ethylenedioxythiophene) doped with poly(styrene sulfonic acid) a comparison was made between pure MXene and those decorated with TiO_2, where the latter showed a considerable increase in response speed compared to the pure material, allowing selective recognition of the target species from other species with a similar chemical composition. Looking for an environment that resembles possible practical detection environments, specifically in the healthcare area, the material developed was applied to detect ethanol, methanol, isopropanol, and acetone.

In turn, among the various methods that can be used in the treatment of wastewater, adsorption–desorption studies can prove to be a very viable method for removing organic dyes, considering their economic advantages, high productivity, ease of operation, low cost, and environmentally friendly, which is further established given a green synthetic route for the adsorbent material [48, 87, 96, 106]. Thus, Kashif and collaborators aimed to synthesize the UiO-66-MXene composite through a solvothermal method, which consists of a technique applied to the synthesis of nanomaterials using a solvent under high pressure and temperature [48]. The set of parameters facilitates the interaction of precursors, and their manipulation makes it possible to shape the characteristics of the material as needed. The synthesized material was characterized using XRD, FTIR, and SEM techniques, which characterized the material based on its pore size and structure. To apply it to the adsorption study of the methylene blue and rhodamine B dyes and the antibiotic tetracycline, the adsorption process used was carried out by varying different parameters of the medium, such as adsorbent-adsorbate contact time, pH, initial concentration of the adsorbate, the proportion of the adsorbent and others, which can be seen in Figs. 8 and 9.

Results showed excellent adsorption performance, with high removal rates (98% for methylene blue, 97% for rhodamine B, and 99% for tetracycline). The kinetic process, in turn, was described by a pseudo-second-order kinetic model. The Langmuir isotherm well describes the adsorption of the analytes, and the maximum adsorption capacity achieved is defined at up to 312, 285, and 476 mg g^{-1}, respectively.

Finally, Table 3 also confirms that the class of pesticides [39, 119, 140, 153] and phenolic compounds [38, 100, 123, 124, 134] have received attention during the development of MXene for sensing.

Due to the ban on organochlorine pesticides, as an alternative, organophosphate pesticides have been widely used in agriculture to prevent the proliferation of pests [153]. Such products play an essential role in the field, simultaneously, they represent a problem for human health associated with several chronic issues, such as liver and respiratory diseases, cancer, and others [119, 153]. For this and other reasons, developing accurate methods for rapidly monitoring these pesticides, especially in water and food, becomes essential.

Carbendazim, or methyl 2-benzimidazole carbamate, is an example of a pesticide or fungicide used to prevent pests in fruits and vegetables, banned in some countries due to strong evidence of carcinogenicity and the capability of harming human reproduction capacity [119]. In this context, Tu and collaborators explored an electrostatically self-assembled composite based on MXene, carbon nanohorns (CNHs), and β-metal–organic cyclodextrin (β-CD-MOFs) as an electrochemical detection platform

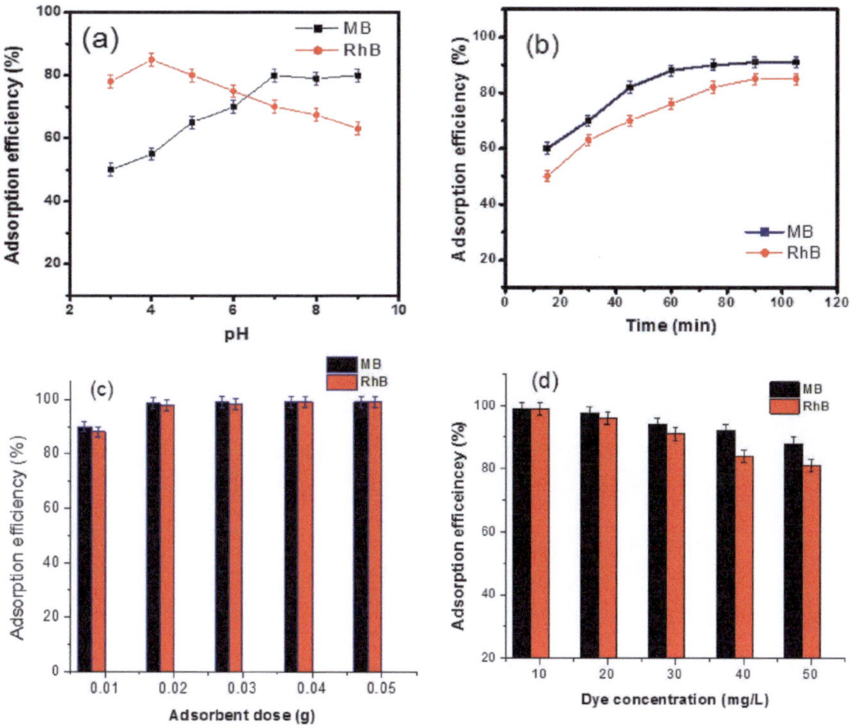

Fig. 8 Study of the variation of parameters in the adsorption of methylene blue and rhodiamine B. Reprinted with permission from [48]. Copyright 2024, Elsevier Publisher

for monitoring the pesticide carbendazim [119]. The authors state that the synergism of the host–guest recognition properties of β-cyclodextrin, the pore volume, and the high porosity of MOFs, combined with the high conductivity, high surface area, and availability of active sites of MXene-CNHs, enabled an increase in transport channels and electron transfer capacity in the analysis, by DPV, of the pesticide in tomato juice. Consequently, as a result, a sensitive, selective, reproducible, and stable electrochemical sensor was obtained, with a wide linear range from 3.0 nmol L^{-1} to 10.0 μmol L^{-1} and a LOD of 1.0 nmol L^{-1}.

On the other hand, Zhou and collaborators developed an acetylcholinesterase (AchE) biosensor for voltammetric detection, by DPV and CV, of malathion in water [153]. Based on chitosan nanosheets, the electrochemical behavior of the sensor was investigated by CV and EIS, and parameters such as pH, MXene concentration, enzymatic load, and others were optimized. Thus, given the excellent film formation capacity, the non-toxicity of chitosan, the high surface area of MXene, and its high electrical conductivity, the developed biosensor showed satisfactory recovery, reproducibility, and stability, with LOD of 0.3×10^{-5} nmol L^{-1} in a linear range of 1×10^{-14} to 1×10^{-18} mol L^{-1}. In Fig. 10, it is possible to see a schematic

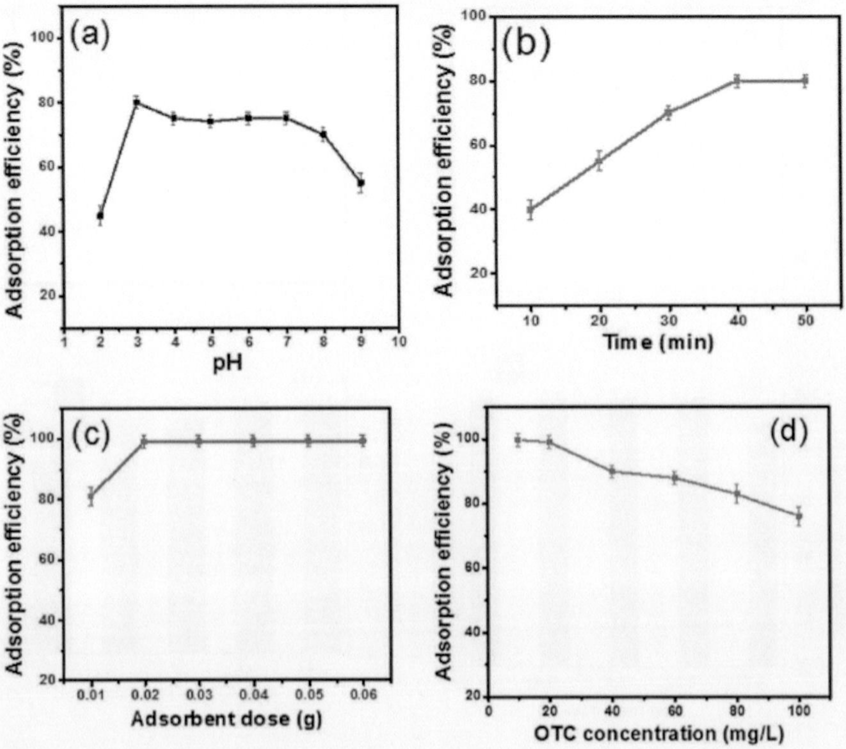

Fig. 9 Study of the variation of parameters in the adsorption of tetracycline. Reprinted with permission from [48]. Copyright 2024, Elsevier Publisher

of the deposition of the AchE-CS-$Ti_3C_2T_x$ film on GCE and the representation of malathion detection.

In turn, phenolic compounds, such as nitroaromatics, are compounds widely used on an industrial scale, whether as pesticides, plasticizers, and others [10, 38]. However, like the reported pesticides, such compounds can have severe environmental and human health consequences, making detection and monitoring platforms necessary. 4-nitrophenol is an essential example of a phenolic compound listed as a priority toxic pollutant in some environmental protection agencies [10]. In this way, Huang and collaborators sought to develop an electrochemical detection platform based on a self-assembled heterostructure composed of Nb_2CT_x and Zn–Co–NC [37]. Characterized by XPS, EDS, and SEM, the material showed excellent conductivity, high surface area, and catalytic activity, enabling the detection of 4-nitrophenol in a linear range from 1 to 500 μmol L^{-1}, with a LOD of 70 nmol L^{-1} and sensitivity of 4.65 $\mu A\ \mu M^{-1}\ cm^{-2}$.

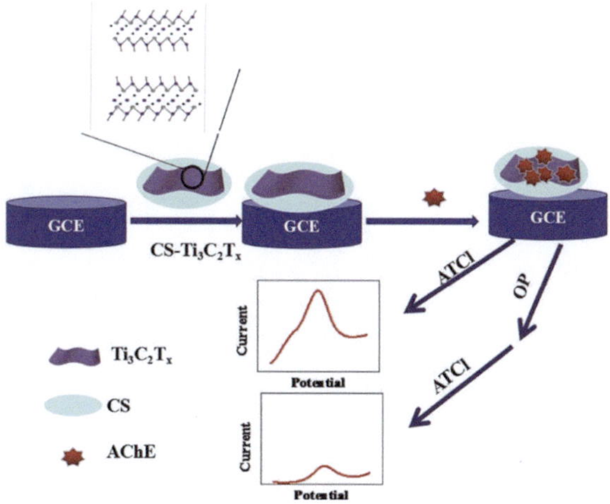

Fig. 10 Construction scheme of the AchE-CS-$Ti_3C_2T_x$-GCE sensor and malathion detection. Reprinted with permission from [153]. Copyright 2024, Elsevier Publisher

Therefore, considering the works presented here, the applicability of MXene-based sensors, whether electrochemical, chemiresistive, or based on adsorption–desorption mechanisms, is quite varied. These sensors are widely used to detect pesticides, dyes, phenolic compounds, and other analytes in various matrices, thus being essential in monitoring and controlling compounds that may harm the environment.

4.2 *Inorganic Contaminants*

The concentration of inorganic pollutants in the environment has increased with anthropogenic activities such as mining, industrial processes, and agriculture [16]. This rise in pollution represents a significant risk to humans and animals, as some of these pollutants can accumulate in water sources and food supplies [89]. MXene-based solutions are promising for detecting these hazardous substances in this scenario, as shown in Table 4.

As depicted in Table 4, the electrochemical sensors are the predominant technique for determining inorganic compounds. These sensors are commonly utilized for analyzing metals in water samples, as demonstrated by Zhu and collaborators [154]. After cleaning and polishing the surface of a GCE, they pipetted a 4 μL aliquot of a suspension containing 4 mg of Alk-Ti_3C_2 dissolved in 4 mL of anhydrous ethanol onto the surface. Afterward, the solvent was evaporated, and the SWASV was applied to determine Pb^{2+}, Cd^{2+}, Cu^{2+}, and Hg^{2+} in water samples. This combination achieved a LOD of 0.098, 0.041, 0.032, and 0.0130 ppm for Pb^{2+}, Cd^{2+}, Cu^{2+}, and Hg^{2+}, respectively.

The importance of monitoring metals comes from their nature. At the same time, some can play essential roles in several biological processes in the human body; their presence in excess poses significant risks to both human health and the environment

Table 4 MXene for inorganic environmental contaminants sensing

Sensor	Mechanism	Detection technique	Analyte	Sample	LOD	References
$Ti_3C_2T_x$	Chemiresistive	N/A	NH_3	Synthetic air	0.13 ppb	[54]
$Ti_3C_2T_x$/ZnTi LDHs	Chemiresistive	N/A	NH_3	Synthetic air	100 ppb	[97]
$Ti_3C_2T_x$/ZnO spheres	Chemiresistive	N/A	NO_2	Synthetic air	N/A	[143]
V_2CT_x/NaOH	Chemiresistive	N/A	NO_2	Synthetic air	N/A	[150]
V_2CT_x	Chemiresistive	N/A	H_2S	Synthetic air	3.504 ppm	[60]
$Ti_3C_2T_x$/Ag nanoparticles	Chemiresistive	N/A	H_2S	Synthetic air	35 ppb	[141]
$Ti_3C_2T_x$	Chemiresistive	N/A	CO_2	Air	N/A	[57]
$Ti_3C_2T_x$/BiNPs	Electrochemical	ASV	Pb^{2+}, Cd^{2+}	Tap and lake water	10.0, 12.4 ppb	[32]
PANI-Ti_3C_2/GCE	Electrochemical	ASV	Hg^{2+}	Tap and lake water	0.017 ppm	[15]
Alk-Ti_3C_2/GCE	Electrochemical	SWASW	Pb^{2+}, Cd^{2+}, Cu^{2+}, Hg^{2+}	Water	0.098, 0.04, 0.032, 0.0130 ppm	[154]
MXA-CuO/CC	Electrochemical	DPASV	Cd^{2+}, Pb^{2+}	Grain, water	0.3, 0.2 ppm	[132]
Melamine-doped rGO/MXene	Electrochemical	DPASV	Zn^{2+}, Cd^{2+} Pb^{2+}	Wheat, sorghum, corn, rice, water	N/A	[13]
TiO_2–$Ti_3C_2T_x$-CTAB-Chi-GCE	Electrochemical	DPV	NO_2^-	Milk	0.85 ppm	[127]
Nafion-$Ti_3C_2T_x$-GCE	Electrochemical	Chronoamperometry	BrO_3^-	Tap water	0.041 ppm	[99]
$Ti_3C_2T_x$/polyelectrolyte	Adsorption–desorption	N/A	Water (*humidity*)	Air	N/A	[7]

(continued)

Table 4 (continued)

Sensor	Mechanism	Detection technique	Analyte	Sample	LOD	References
$Ti_3C_2T_x/K_2Ti_4O_9$	Adsorption–desorption	N/A	Water (*humidity*)	Air	N/A	[137]
Ti_3C_2	Optical	Fluorescence	Fe^{3+}	$Fe(NO_3)_3$	1.4 ppm	[18]

N/A: Not avaliable; ASV: Anodic stripping voltammetry; SWASW: Square-wave anodic-stripping voltammetry; DPASV: Differential pulse anodic-stripping voltammetry; DPV: Differential pulse voltammetry

[26]. Notably, certain metals, like Cd, As, Cr, Se, Cu, Pb, Hg, and Zn, have been identified to cause harm to multiple biological systems, including the respiratory and neurological systems, the reproductive organs, and the digestive system, further underscoring the importance of their detection and analysis [51].

Although most of the papers in the literature make the analyses of metals in water samples, Chen and collaborators have proven that a MXene-based electrochemical sensor can also determine Zn^{2+}, Cd^{2+}, and Pb^{2+} in wheat, sorghum, corn, and rice [13]. They used a screen-printed electrode modified with a melamine-doped rGO/MXene film to determine these metals using differential pulse anodic stripping voltammetry. Meanwhile, Rasheed and collaborators have proven that electrochemical sensors can detect other compounds instead of metals [99]. They proposed a screen-printed electrode modified with a Nafion-$Ti_3C_2T_x$-GCE film capable of detecting bromate ions in tap water. Using this sensor, they achieved a LOD of 0,041 ppm for BrO_3^- in tap water.

As seen in Table 4, chemiresistive sensors are also essential in sensing inorganic contaminants in the environment. This is the second most used type of MXene-based sensor for detecting inorganic compounds. In the literature, papers are using this type of sensor for the detection of highly toxic inorganic gases, such as ammonia (NH_3), hydrogen sulfide (H_2S), nitrogen oxides (NO_2^-), sulfur oxides (SO_2^-), carbon monoxide (CO), which after entering the lung can cause itching or coughing, and when it finally reaches the bloodstream, it results in numerous diseases [3, 50]. Among these gases, ammonia needs special attention since this gas is excessively used in the food, textile, and chemical industries [16]. When small amounts of ammonia are inhaled, it can cause severe irritation and affect the respiratory tract in human beings, and exposure to NH_3 at higher concentrations can lead to death [47]. Qin and colleagues created a $Ti_3C_2T_x$-supported ZnTi–LDHs nanocomposite sensor for monitoring NH_3 in the air, achieving a LOD of 0.13 ppb [97].

Although uncommon, papers in the literature are developing optical MXene-based sensors for detecting inorganic contaminants. One example is the sensor developed by Gao and collaborators [18]. For the construction of the sensor, the authors added 20 mL of hydrofluoric acid into a Teflon container containing 2 g of bulk Ti_3AlC_2 powder, maintaining this mixture under constant stirring for 48 h at room temperature. In the process, the aluminum layers were etched, and the primary products were collected by centrifugation and washed until neutral pH. The solid obtained during this process was then dispersed into 50 mL of DMF and sonicated for another 48 h. After centrifugation, the supernatant containing the sensors was stored. The detection of the Fe^{3+} was achieved by fluorescence spectrometry by mixing 300 μL of the solution containing the sensors with a $Fe(NO_3)_3$ solution, achieving an LOD of 1.4 ppm.

5 Challenges and Perspectives

The past decade has achieved several advances in developing MXenes with environmental applications. The material's promising characteristics, such as high chemical and thermal stability, hydrophilic properties, and high absorption capacity for gases, organic molecules, and metals, make it suitable for environmental remediation. Despite these advantageous and promising characteristics, there are still limitations to overcome for its widespread and effective use in environmental issues [111].

One of the main challenges lies in the limited number of synthesized MXenes. Additionally, the number of compounds synthesized is still tiny compared to their theoretical potential, making it necessary to develop new synthesis routes [63]. Although there is potential for using MXenes as an antimicrobial material, its application is still considered to be at an early stage. Challenges include clarifying the mechanisms of action of MXenes as an antimicrobial material and their efficacy and safety in different environments [34].

The synthesis of new materials must elucidate the environmental impact of the new substance, from its production to final disposal, in a suitable location. The production of new MXenes suitable for environmental applications, with greater pollutant adsorption capacity and resistance to the environment, is dependent on research investment [111].

Different applications are possible as new MXenes are synthesized from different MAX phases. This includes new pollutant detection and removal devices. Furthermore, the area of materials allows for interdisciplinarity. Various areas, such as chemistry, biology, and environmental sciences, can contribute to elucidating and addressing challenges in solving critical environmental problems using MXenes [147].

The application of MXenes in the environmental area has excellent potential as it simultaneously presents desirable characteristics of a metallic and ceramic material. However, further advances are needed to address the limitations presented so that the material can be more widely used.

This chapter demonstrates that MXenes have great potential for modifying sensors due to their properties, mainly in the electrochemical area. With increased interest and knowledge about this material, its use will likely increase and become more known to the scientific community.

It is noticeable that significant advances in the use of MXene and its nanocomposites as a platform for detection and environmental monitoring have been achieved only in the last decade. However, given the significant advantages of using this material due to its characteristics already well explained in this chapter, MXenes have enormous potential for environmental applications, with diverse possibilities in organic pollutants such as dyes, pesticides, phenolic compounds, and others. To this end, the significant branches of detection, such as electrochemical, chemiresistive, and adsorptive, have become increasingly selective, reproducible, and sensitive, considering the use of MXene and mainly, through the synergism of properties of these materials when combined with others.

References

1. Ahmad, N., et al.: 2D MXenes and their composites; design, synthesis, and environmental sensing applications. Chemosphere 141–280 (2024)
2. Ahmaruzzaman, M.: MXenes and MXene-supported nanocomposites: a novel material for aqueous environmental remediation. R. Soc. Chem. Adv. **12**(53), 34766–34789 (2022)
3. Alarie, Y.: Toxicity of fire smoke. Crit. Rev. Toxicol. **32**, 259–289 (2002)
4. Algaradah, M.M.: MXene-based adsorbent materials for pollutants removal from water: current challenges and future prospects. Inorgan. Chem. Commun. (Elsevier) **161**, 112113 (2024)
5. Alhabeb, M., et al.: Guidelines for synthesis and processing of two-dimensional titanium carbide ($Ti_3C_2T_x$ MXene). Chem. Mater. **29**(18), 7633–7644 (2017)
6. Alnoor, H., et al.: Exploring MXenes and their MAX phase precursors by electron microscopy. Mater. Today Adv. 9 (2021)
7. An, H., et al.: Water sorption in MXene/polyelectrolyte multilayers for ultrafast humidity sensing. Am. Chem. Soc. Appl. Nano Mater. **2**(2), 948–955 (2019)
8. Aryal, S., et al.: A genomic approach to the stability, elastic, and electronic properties of the MAX phases. Phys. Status Solidi **251**(8), 1480–1497 (2014)
9. Aslam, M., et al.: MXenes as theranostics: diagnosis and therapy including in vitro and in vivo applications. Appl. Mater. Today **35** (2023)
10. Balasubramanian, P., et al.: Simplistic synthesis of ultrafine $CoMnO_3$ nanosheets: an excellent electrocatalyst for highly sensitive detection of toxic 4-nitrophenol in environmental water samples. J. Hazard. Mater. **361**, 123–133 (2019)
11. Benchakar, M., et al.: One MAX phase, different MXenes: a guideline to understand the crucial role of etching conditions on $Ti_3C_2T_x$ surface chemistry. Appl. Surf. Sci. **530**, 147209 (2020)
12. Bilal, M., et al.: MXenes: the emerging adsorbents for the removal of dyes from water. J. Mol. Liquids **385** (2023)
13. Chen, Y., et al.: A sensitive electrochemical sensor based on 3D porous melamine-doped rGO/MXene composite aerogel for the detection of heavy metal ions in the environment. Talanta **256**, 124294 (2023)
14. Chen, X., Bei, G.: Toughening mechanisms in nanolayered MAX phase ceramics-A review. Materials **10**(4), 366 (2017)
15. Cheng, H., Yang, J.: Preparation of Ti_3C_2-PANI composite as sensor for electrochemical determination of mercury ions in water. Int. J. Electrochem. Sci. **15**(3), 2295–2306 (2020)
16. Devaraj, M., et al.: A review on MXene and its nanocomposites for the detection of toxic inorganic gases. Chemosphere **302**, 134933 (2022)
17. Fan, Y., et al.: Evolution of rust layers on carbon steel and weathering steel in high humidity and heat marine atmospheric corrosion. J. Mater. Sci. Technol. **39**, 190–199 (2020)
18. Gao, X., et al.: N,N-Dimethyl Formamide regulating fluorescence of MXene quantum dots for the sensitive determination of F^{e3}+. Nanoscale Res. Lett. **16**(1), 160 (2021)
19. Gao, L., et al.: Preparation of $Ti_3C_2T_x$-based MXene composite aerogel and its adsorption performance for dibutyl phthalate. Colloids aSurf. A: Physicochem. Eng. Aspects Elsevier **686**, 133452 (2024)
20. Geng, L., et al.: Polyamidoamine dendrimers modified MXene and its high-performance adsorption behavior for Cr(VI). J. Mol. Liquids **399** (2024)
21. Ghani, A.A., et al.: Adsorption and electrochemical regeneration of intercalated $Ti_3C_2T_x$ MXene for the removal of ciprofloxacin from wastewater. Chem. Eng. J. **421** (2021)
22. Ghani, A.A., et al.: Adsorption and electrochemical regeneration of 2D magnetic MXene nanosheets loaded with tetracycline. Chem. Eng. J. **467** (2023)
23. Ghosh, N.C., Harimkar, S.P.: Consolidation and synthesis of MAX phases by Spark Plasma Sintering (SPS): a review. In: Advances in Science and Technology of $M_{n+1}AX_n$ Phases: 47–80 (2012)

24. Griseri, M., et al.: Ta-based 413 and 211 MAX phase solid solutions with Hf and Nb. J. Eur. Ceram. Soc. **40**(5), 1829–1838 (2020)
25. Gu, P., et al.: Experimental and theoretical calculation investigation on efficient Pb(ii) adsorption on etched Ti_3AlC_2 nanofibers and nanosheets. Environ. Sci. Nano **5**(4), 946–955 (2018)
26. Gumpu, M.B., et al.: A review on detection of heavy metal ions in water—An electrochemical approach. Sens. Actuators B Chem. **213**, 515–533 (2015)
27. Gupta, S., Barsoum, M.W.: On the tribology of the MAX phases and their composites during dry sliding: a review. Wear **271**(9–10), 1878–1894 (2011)
28. Hadi, Z., et al.: Predicting of electrical conductivity for polymer-MXene nanocomposites. J. Market. Res. **28**, 4229–4238 (2024)
29. Han, M., et al.: Ti_3C_2 MXenes with modified surface for high-performance electromagnetic absorption and shielding in the X-band. Am. Chem. Soc. Appl. Mater. Interfaces **8**(32), 21011–21019 (2016)
30. Han, M., et al.: Polydopamine-modified MXene/cellulose nanofibers composite film for self-powered humidity sensing and humidity actuating. Nano Energy **123**, 109445 (2024)
31. Hashimoto, S., et al.: Formation mechanism of Ti_2AlC under the self-propagating high-temperature synthesis (SHS) mode. Mater. Res. Bull. **47**(5), 1164–1168 (2012)
32. He, Y., et al.: Preparation and application of Bismuth/MXene nano-composite as electrochemical sensor for heavy metal ions detection. Nanomaterials **10**(5), 866 (2020)
33. He, S., et al.: Preparation strategies and applications of MXene-polymer composites: a review. Macromol. Rapid Commun. **42**(19), 2100324 (2021)
34. He, X. et al.: Emerging 2D MXenes for antibacterial applications: current status, challenges, and prospects. Coord. Chem. Rev. **492**, 215275 (2023)
35. Hu, C., et al.: New phases discovery in MAX family. Int. J. Refract Metal Hard Mater. **36**, 300–312 (2013)
36. Hu, X., et al.: Kinetics, isotherm and chemical speciation analysis of Hg(II) adsorption over oxygen-containing MXene adsorbent. Chemosphere (Elsevier) **278**, 130206 (2021)
37. Huang, R., et al.: Self-assembled Ti_3C_2/MWCNTs nanocomposites modified glassy carbon electrode for electrochemical simultaneous detection of hydroquinone and catechol. Ecotoxicol. Environ. Saf. **184**, 109619 (2019)
38. Huang, R., et al.: Electrostatically assembling 2D hierarchical Nb_2CT_x and zifs-derivatives into Zn-Co-NC nanocage for the electrochemical detection of 4-nitrophenol. Sens. Actuators B Chem. **338**, 129828 (2021)
39. Hui, X., et al.: A nanocomposite-decorated laser-induced graphene-based multi-functional hybrid sensor for simultaneous detection of water contaminants. Anal. Chim. Acta **1209**, 339872 (2022)
40. Hussein, D., et al.: Overview of structural, electronic, elastic, thermal, optical, and nuclear properties of Zr_2AC (A = Al, Si, P, S, Ge, As, Se In, Sn, Tl, and Pb) MAX phases: a brief review. Heliyon e18303 (2023)
41. Ihsanullah, I., et al.: Emerging MXenes: revolutionizing oily wastewater treatment—A comprehensive and critical review. Sep. Purif. Technol. (Elsevier) **329**, 125181 (2024)
42. Ishfaq, M., et al.: Synthesis and characterization of a novel MnO_2@MXene-based 2D/3D hierarchical Z-scheme for sustainable environmental applications. Ceram. Int. (Elsevier) **50**(6), 9801–9810 (2024)
43. Jun, B.M., et al.: Effective removal of Pb(ii) from synthetic wastewater using Ti_3C_2T: X MXene. Environ. Sci.: Water Res. Technol. R. Soc. Chem. **6**(1), 173–180 (2020)
44. Kamakshi, P., et al.: Enhancing photocatalytic applications with SnO_2/Ti_3C_2 MXene nanocomposite: synthesis and characterization. Chem. Phys. Lett. (Elsevier) **833**, 140945 (2023)
45. Kamarulzaman, M.K., et al.: Improvement in stability and thermophysical properties of CNC-MXene nanolubricant for tribology application. J. Mol. Liquids **381** (2023)
46. Karataş, Y., et al.: Synthesis and characterization of Pd doped MXene for hydrogen production from the hydrolysis of methylamine borane: effect of cryogenic treatment. J. Energy Inst. 1–9 (2023)

47. Karimi-Maleh, H., et al.: Recent advances in carbon nanomaterials-based electrochemical sensors for food azo dyes detection. Food Chem. Toxicol. **164**, 11296 (2022)
48. Kashif, S., et al.: Development of MOF-MXene composite for the removal of dyes and antibiotic. Diam. Relat. Mater. **136**, 110023 (2023)
49. Kasinathan, K., et al.: Synergistically self-assembled 2D nanosheets of MXene@MOF derived CoW-LDH into 3D frameworks functionalized with chitosan for improved skin wound healing. Chem. Eng. J. **482** (2024)
50. Kealey, G.P.: Carbon monoxide toxicity. J. Burn Care Res. **30**, 146–147 (2009)
51. Khan, M.S., et al.: Heavy metals: biological importance and detoxification strategies. J. Biorem. Biodegrad. **7**(2), 334 (2016)
52. Khazaei, M., et al.: Two-dimensional molybdenum carbides: potential thermoelectric materials of the MXene family. Phys. Chem. Chem. Phys. **16**(17), 7841–7849 (2014)
53. Khazaei, M., et al.: Electronic properties and applications of MXenes: a theoretical review. J. Mater. Chem. **5**(10), 2488–2503 (2017)
54. Kim, S.J., et al.: Metallic $Ti_3C_2T_x$ MXene gas sensors with ultrahigh signal-to-noise ratio. Am. Chem. Soc. Nano **12**(2), 986–993 (2018)
55. Kim, H., Alshareef, H.N.: MXetronics: MXene-enabled electronic and photonic devices. Am. Chem. Soc. Mater. Lett. **2**(1), 55–70 (2020)
56. Kim, S.H., Rhee, M.S.: Environment-friendly mild heat and relative humidity treatment protects sprout seeds (radish, mung bean, mustard, and alfalfa) against various foodborne pathogens. Food Control **93**, 17–22 (2018)
57. Koh, H.J., et al.: Enhanced selectivity of MXene gas sensors through metal ion intercalation: in situ X-ray diffraction study. Am. Chem. Soc. Sens. **4**(5), 1365–1372 (2019)
58. Kumar, J., et al.: Methods of synthesis, characteristics, and environmental applications of MXene: a comprehensive review. Chemosphere **286**, 131607 (2022)
59. Kumar, Y.R., et al.: Chapter 7—MXenes and their composites: emerging materials for gas sensing and biosensing. In: Sadasivuni, K.K., Deshmukh, K., Pasha, S.K.K., Kovářík, T. (eds.) MXenes and Their Composites, pp. 241–279. Elsevier (2022)
60. Lee, E., et al.: Two-dimensional vanadium carbide MXene for gas sensors with ultrahigh sensitivity toward nonpolar gases. Am. Chem. Soc. Sens. **4**(6), 1603–1611 (2019)
61. Lee, J.T., et al.: Covalent surface modification of $Ti_3C_2T_x$ MXene with chemically active polymeric ligands producing highly conductive and ordered microstructure films. Am. Chem. Soc. Nano **15**, 19600–19612 (2021)
62. Lee, S., et al.: An electrochemical sensor for capsaicin based on two-dimensional titanium carbide (MXene)-doped titania-Nafion composite film. Microchem. J. **185**, 108216 (2023)
63. Lee, E., Kim, D.J.: Recent exploration of two-dimensional MXenes for gas sensing: from a theoretical to an experimental view. J. Electrochem. Soc. **167**(3), 037515 (2020)
64. Li, Z., et al.: Synthesis and thermal stability of two-dimensional carbide MXene Ti_3C_2. Mater. Sci. Eng. B **19**, 33–40 (2015)
65. Li, T., et al.: Two-dimensional materials fluorine-free synthesis of high-purity $Ti_3C_2T_x$ (T= -OH, -O) via alkali treatment. Angew. Chem. **130**, 6223–6227 (2018)
66. Li, C., et al.: Non-covalent and covalent immobilization of papain onto Ti_3C_2 MXene nanosheets. Enzyme Microb. Technol. **148**, 109817 (2021)
67. Li, Q., et al.: MXenes-based adsorbents for environmental remediation. Sep. Purif. Technol. **342**, 126982 (2024)
68. Li, J., et al.: Synthesis of MXene-Based functional coatings on rigid polyurethane foam surfaces: a comparative study of layer-by-layer self-assembly and hydrothermal methods. Polym. Degrad. Stability **220** (2024)
69. Liao, M., et al.: MXenes as emerging adsorbents for removal of environmental pollutants. Sci. Total Environ. (2024)
70. Lin, Z., et al.: Microstructures and theoretical bulk modulus of layered ternary tantalum aluminum carbides. J. Am. Ceram. Soc. **89**(12), 3765–3769 (2006)
71. Liu, J., et al.: Hydrophobic, flexible, and lightweight MXene foams for high-performance electromagnetic-interference shielding. Adv. Mater. **29**(38), 1–6 (2017)

72. Liu, X., et al.: Two-dimensional nanostructured materials for gas sensing. Adv. Func. Mater. **27**(37), 1702168 (2017)
73. Liu, R., et al.: Large-scale two-dimensional titanium carbide MXene as SERS-active substrate for reliable and sensitive detection of organic pollutants. Spectrochim. Acta Part A Mol. Biomol. Spectrosc. **236**, 118336 (2020)
74. Liu, M., et al.: Au-decorated Ln_2O_3 nanospheres/exfoliated $Ti_3C_2T_x$ MXene nanosheets for highly sensitive formaldehyde gas sensing at room temperature. Appl. Surf. Sci. **605**, 154839 (2022)
75. Lqbal, A., et al.: Improving oxidation stability of 2D MXenes: synthesis, storage media, and conditions. Nano Converg. **8**(9), 1–22 (2021)
76. Ma, X., et al.: A comprehensive review of MXene-based nanofluids: preparation, stability, physical properties, and applications. J. Mol. Liquids **365** (2022)
77. Mashhadizadeh, M.H., et al.: Synthesis and characterization of N-MPG/CuS flower-like/ MXene to modify a screen-printed carbon electrode for electrochemical determination of nalbuphine. J. Electroanal. Chem. (Elsevier) **957**, 118130 (2024)
78. Mashtalir, O., et al.: Kinetics of aluminum extraction from Ti_3AlC_2 in hydrofluoric acid. Mater. Chem. Phys. **139**(1), 147–152 (2013)
79. McDonald, B.C., et al.: Volatile chemical products emerging as largest petrochemical source of urban organic emissions. Science **359**, 760–764 (2018)
80. Megbenu, H.K., et al.: Photocatalytic and adsorption performance of MXene@Ag/cryogel composites for sulfamethoxazole and mercury removal from water matrices. Environ. Technol. Innov. **32**, 103350 (2023)
81. Mohanapriya, D., et al.: Design and engineering of 2D MXenes for point-of-care electrochemical detection of bioactive analytes and environmental pollutants. Coord. Chem. Rev. **507** (2024)
82. Mozafari, M., Soroush, M.: Surface functionalization of MXenes. Mater. Adv. **2**, 7277–7307 (2021)
83. Mustafa, G., et al.: Water pollutants and nanosensors. Aquananotechnology (Elsevier) 105–133 (2021)
84. Mustafa, M.R.U., et al.: Nanoscale zero-valent iron (NZVI) intercalated MXene ($Ti_3C_2T_x$) for removal of polycyclic aromatic hydrocarbons in wastewater. J. Environ. Chem. Eng. **12**(2) (2024)
85. Najibikhah, P., Rahbar-Kelishami, A.: Preparation of cationic surfactant modified two-dimensional (2D) multi-layered $Ti_3C_2T_x$ MXene for methyl orange removal from aqueous solution: kinetic, equilibrium, and adsorption mechanisms. Chemosphere **350** (2024)
86. Nazemi, H., et al.: Advanced micro- and nano-gas sensor technology: a review. Sensors **19**(6), 1285 (2019)
87. Neolaka, Y.A., et al.: Adsorption of methyl red from aqueous solution using Bali cow bones (Bos javanicus domesticus) hydrochar powder. Results Eng. **17**, 100824 (2023)
88. Nezami, S., et al.: Application of titanium carbide/nitride (MXene)-based NPs in adsorption of radionuclides and heavy metal ions for wastewater remediation: a review. Case Stud. Chem. Environ. Eng. **7** (2023)
89. Nimis, P., et al.: Monitoring Lichens. Springer Netherland, Dordrecht (2002)
90. Numan, A., et al.: Microwave-assisted rapid MAX phase etching and delamination: a paradigm shift in MXene synthesis. Mater. Chem. Phys. **288**, 126429 (2022)
91. Oyehan, T., et al.: MXenes: synthesis, properties, and applications for sustainable energy and environment. Appl. Mater. Today **35**, 101993 (2023)
92. Panda, S., et al.: MXene based emerging materials for supercapacitor applications: recent advances, challenges, and future perspectives. Coord. Chem. Rev. **462**, 214518 (2022)
93. Parra-Muñoz, N., et al.: Covalent functionalization of MXenes for tribological purposes—A critical review. Adv. Coll. Interface Sci. **309**, 102792 (2022)
94. Pazniak, H., et al.: Partially oxidized $Ti_3C_2T_x$ MXenes for fast and selective detection of organic Vapors at part-per-million concentrations. Am. Chem. Soc. Appl. Nano Mater. **3**(4), 3195–3204 (2020)

95. Pei, Y., et al.: $Ti_3C_2T_X$ MXene for sensing applications: recent progress, design principles, and future perspectives. Am. Chem. Soc. Nano **15**(3), 3996–4017 (2023)
96. Phoon, B.L., et al.: Conventional and emerging technologies for removal of antibiotics from wastewater. J. Hazard. Mater. **400**, 122961 (2020)
97. Qin, Y., et al.: Enhanced NH_3 sensing performance at ppb level derived from Ti_3C_2T-supported ZnTi-LDHs nanocomposite with similar metal-semiconductor heterostructure. Sens. Actuators B Chem. **352**, 131077 (2022)
98. Ranjith, K.S., et al.: Hybridized 1D–2D $MnMoO_4$–MXene nanocomposites as high-performing electrochemical sensing platform for the sensitive detection of dihydroxybenzene isomers in wastewater samples. J. Hazard. Mater. **421**, 126775 (2022)
99. Rasheed, P.A., et al.: Ultra-sensitive electrocatalytic detection of bromate in drinking water based on Nafion/$Ti_3C_2T_x$ (MXene) modified glassy carbon electrode. Sens. Actuators B Chem. **265**, 652–659 (2018)
100. Rasheed, P.A., et al.: Platinum nanoparticles/$Ti_3C_2T_x$ (MXene) composite for the effectual electrochemical sensing of Bisphenol A in aqueous media. J. Electroanal. Chem. **880**, 114934 (2021)
101. Rasheed, T.: MXenes as emerging two-dimensional analytical modalities for potential recognition of hazardous environmental contaminants. Mater. Today Chem. **24**, 100859 (2022)
102. Rayhan, M.A., et al.: First-principles study of Vickers hardness and thermodynamic properties of Ti_3SnC_2 polymorphs (2015). arXiv:1505.03666
103. Reddy, M.S.B., Aich, S.: Recent progress in surface and heterointerface engineering of 2D MXenes for gas sensing applications. Coord. Chem. Rev. **500**, 215542 (2024)
104. Riley, D.P., Kis, E.H.: The design of crystalline precursors for the synthesis of $M_{n-1}AX_n$ phases and their application to Ti_3AlC_2. J. Am. Ceram. Soc. **90**(7), 2231–2235 (2007)
105. Ronchi, R., et al.: Synthesis, structure, properties and applications of MXenes: current status and perspectives. Ceram. Int. **45**(15), 18167–18188 (2019)
106. Saeed, M., et al.: Synthesis, characterization and applications of silylation based grafted bentonites for the removal of Sudan dyes: isothermal, kinetic and thermodynamic studies. Microporous Mesoporous Mater. **291**, 109697 (2020)
107. Salim, O., et al.: Introduction to MXenes: synthesis and characteristics. Mater. Today Chem. **14**, 100191 (2019)
108. Sangamithirai, D., Ramanathan, S.: Electrochemical sensing platform for the detection of nitroaromatics using g-C_3N_4/V_2O_5 nanocomposites modified glassy carbon electrode. Electrochim. Acta **434**, 141308 (2022)
109. Sheikh, Z.A., et al.: Transition metal chalcogenides, MXene, and their hybrids: an emerging electrochemical capacitor electrodes. J. Energy Storage **71**, 107997 (2023)
110. Shekhirev, M., et al.: Characterization of MXenes at every step, from their precursors to single flakes and assembled films. Progr. Mater. Sci. (Elsevier) **120**, 100757 (2021)
111. Sun, Y., Li, Y.: Potential environmental applications of MXenes: a critical review. Chemosphere **271**, 129578 (2021)
112. Szuplewska, A., et al.: The 10th anniversary of MXenes: challenges and prospects for their surface modification toward future biotechnological applications. Adv. Drug Deliv. Rev. **182**, 114099 (2022)
113. Tan, Q., et al.: Recent progress in additive manufacturing of bulk MAX phase components: a review. J. Mater. Sci. Technol. **131**, 30–47 (2022)
114. Tang, J., et al.: Tuning the electrochemical performance of titanium carbide MXene by controllable in situ anodic oxidation. Angew. Chem. **131**(49), 18013–18019 (2019)
115. Tawalbeh, M., et al.: MXenes and MXene-based materials for removal of pharmaceutical compounds from wastewater: critical review. Environ. Res. (Elsevier) **228**, 115919 (2023)
116. Thomas, T., Bowen, C.R.: Effect of particle size on the formation of Ti_2AlC using combustion synthesis. Ceram. Int. **42**(3), 4150–4157 (2016)
117. Thurakkal, S., Zhang, X.: Noncovalent functionalization of $Ti_3C_2T_X$ using cationic porphyrins with enhanced stability against oxidation. Mater. Chem. Front. **6**, 561–569 (2021)

118. Tripathi, S., et al.: Elevating selective ethanol detection based on unlocking the potential of accordion structured MXene. Mater. Chem. Phys. **313**, 128780 (2024)
119. Tu, X., et al.: MXene/carbon nanohorn/β-cyclodextrin-Metal-organic frameworks as high-performance electrochemical sensing platform for sensitive detection of carbendazim pesticide. J. Hazard. Mater. **396**, 122776 (2020)
120. Tumrani, S.H., et al.: Synergistic integration of few-layer thick MXenes and small Pd nanocubes for enhanced electrochemical nitrofurantoin detection: implications in pharmaceutical pollutant monitoring. J. Environ. Chem. Eng. **11**(6), 111152 (2023)
121. Uddin, M.M., et al.: Structural, elastic, electronic and optical properties of metastable MAX phase Ti_5SiC_4 compound. Indian J. Pure Appl. Phys. **54**(6), 386–390 (2016)
122. Verger, L., et al.: Overview of the synthesis of MXenes and other ultrathin 2D transition metal carbides and nitrides. Curr. Opin. Solid State Mater. Sci. **23**, 149–163 (2019)
123. Vilian, A.E., et al.: Polypyrrole-MXene supported gold nanoparticles for the trace-level detection of nitrofurantoin. Chem. Eng. J. **454**, 139980 (2023)
124. Vilian, A.E., et al.: Palladium nanoparticles anchored MoS_2-MXene composite modified electrode for rapid sensing of toxic bisphenol A in aqueous media. Chem. Eng. J. 144017 (2023)
125. Wang, G., et al.: Fast-response humidity sensor based on laser printing for respiration monitoring. R. Soc. Chem. Adv. **10**(15), 8910–8916 (2020)
126. Wang, X., et al.: A novel electrochemical sensor based on TiO_2-$Ti_3C_2T_X$/CTAB/chitosan composite for the detection of nitrite. Electrochim. Acta **359**, 136938 (2020)
127. Wang, X., et al.: $Ti_3C_2T_x$/PEDOT: PSS hybrid materials for room-temperature methanol sensor. Chin. Chem. Lett. **31**(4), 1018–1021 (2020)
128. Wang, D., et al.: Direct synthesis and chemical vapor deposition of 2D carbide and nitride MXenes. Science **379**, 1242–1247 (2023)
129. Wang, L., et al.: Ultra-high performance humidity sensor enabled by a self-assembled CuO/ $Ti_3C_2T_X$ MXene. R. Soc. Chem. Adv. **13**(9), 6264–6273 (2023)
130. Wang, R., et al.: Ultra-strong adsorption of organic dyes and antibiotic onto the alk-MXene/ ZIF adsorbents with a specific intercalation structure. Chem. Eng. J. **485** (2024)
131. Wasnik, K., et al.: MXenes: Advances in the synthesis and application in supercapacitors and batteries. J. Mater. Res. **37**, 3865–3889 (2022)
132. Wen, L., et al.: A novel electrochemical sensor for simultaneous detection of Cd^{2+} and Pb^{2+} by MXene aerogel-CuO/carbon cloth flexible electrode based on oxygen vacancy and bismuth film. Sci. Total Environ. **851**, 158325 (2022)
133. Wong, A.J., et al.: Fluoride-free synthesis and long-term stabilization of MXenes. J. Mater. Res. **37**, 3988–3997 (2022)
134. Wu, L., et al.: 2D transition metal carbide MXene as a robust biosensing platform for enzyme immobilization and ultrasensitive detection of phenol. Biosens. Bioelectron. **107**, 69–75 (2018)
135. Wu, M., et al.: Ti_3C_2 MXene-based sensors with high selectivity for NH_3 detection at room temperature. Am. Chem. Soc. Sens. **4**(10), 2763–2770 (2019)
136. Wu, J., et al.: High performance humidity sensing property of $Ti_3C_2T_x$ MXene-derived $Ti_3C_2T_x/K_2Ti_4O_9$ composites. Sens. Actuators B Chem. **326**, 128969 (2021)
137. Wu, Z., et al.: Pod-inspired MXene/porous carbon microspheres with ultrahigh adsorption capacity towards crystal violet. Chem. Eng. J. (Elsevier) **426**, 130776 (2021)
138. Wu, D., et al.: Insight into the adsorption and co-adsorption of PPCPs on $Ti_3C_2T_x$ MXene: behaviors, mechanisms and application potentials. Surf. Interfaces **45**, 103853 (2024)
139. Xie, H., et al.: Electrostatic self-assembly of $Ti_3C_2T_x$ MXene and gold nanorods as an efficient surface-enhanced raman scattering platform for reliable and high-sensitivity determination of organic pollutants. Am. Chem. Soc. Sens. **4**(9), 2303–2310 (2019)
140. Xie, Y., et al.: Facile synthesis of MXene/electrochemically reduced graphene oxide composites and their application for electrochemical sensing of carbendazim. J. Electrochem. Soc. **166**(16), 1673 (2019)

141. Xu, Q., et al.: H_2S sensing under various humidity conditions with Ag nanoparticle functionalized $Ti_3C_2T_x$ MXene field-effect transistors. J. Hazard. Mater. **424**, 127492 (2022)
142. Xu, Z., et al.: $Ti_3C_2T_x$ MXene modified polyethylene fibers for enhancing interface properties of strain-hardening cementitious composites. Cement Concrete Compos. **145** (2024)
143. Yang, Z., et al.: Flexible resistive NO_2 gas sensor of three-dimensional crumpled MXene $Ti_3C_2T_x$/ZnO spheres for room temperature application. Sens. Actuators B Chem. **326**, 128828 (2021)
144. Yang, Z., et al.: Nb_2C MXene self-assembled Au nanoparticles simultaneously based on electromagnetic enhancement and charge transfer for surface enhanced Raman scattering. Spectrochimica Acta—Part A: Mol. Biomol. Spectrosc. **299**, 1–9 (2023)
145. Yousefi, H., et al.: Intelligent food packaging: a review of smart sensing technologies for monitoring food quality. Am. Chem. Soc. Sens. **4**(4), 808–821 (2019)
146. Zamhuri, A., et al.: MXene in the lens of biomedical engineering: synthesis, applications and future outlook. Biomed. Eng. Online **20**, 1–24 (2021)
147. Zhan, X., et al.: MXene and MXene-based composites: synthesis, properties and environment-related applications. Nanoscale Horizons **5**(2), 235–258 (2020)
148. Zhang, F., et al.: Plasma-enhanced pulsed-laser deposition of single-crystalline Mo_2C ultrathin superconducting films. Phys. Rev. Mater. **1**(3), 034002 (2017)
149. Zhang, G., et al.: Synthesis of amino-functionalized $Ti_3C_2T_x$ MXene by alkalization-grafting modification for efficient lead adsorption. Chem. Commun. **56**(76), 11283–11286 (2020)
150. Zhang, Y., et al.: Highly sensitive and selective NO_2 sensor of alkalized V_2CT MXene driven by interlayer swelling. Sens. Actuators B Chem. **344**, 130150 (2021)
151. Zhang, Z., et al.: On the formation mechanisms and properties of MAX phases: a review. J. Eur. Ceram. Soc. **41**(7), 3851–3878 (2021)
152. Zhao, F., et al.: Self-reduction bimetallic nanoparticles on ultrathin MXene nanosheets as functional platform for pesticide sensing. J. Hazard. Mater. **384**, 121358 (2020)
153. Zhou, L., et al.: Acetylcholinesterase/chitosan-transition metal carbides nanocomposites-based biosensor for the organophosphate pesticides detection. Biochem. Eng. J. **128**, 243–249 (2017)
154. Zhu, X., et al.: Alkaline intercalation of Ti_3C_2 MXene for simultaneous electrochemical detection of Cd(II), Pb(II), Cu(II) and Hg(II). Electrochim. Acta **248**, 46–57 (2017)
155. Zhu, X., et al.: Portable wireless intelligent sensing of ultra-trace phytoregulator α-naphthalene acetic acid using self-assembled phosphorene/Ti_3C_2-MXene nanohybrid with high ambient stability on laser induced porous graphene as nanozyme flexible electrode. Biosens. Bioelectron. **179**, 113062 (2021)
156. Zou, S., et al.: Enhanced gas sensing properties at low working temperature of iron molybdate/MXene composite. J. Alloy. Compd. **817**, 152785 (2020)

MXene and Its Modification for Supercapacitor Application

N. Anwesha, Srikanta Moharana, and Ram Naresh Mahaling

Abstract MXenes, a class of two-dimensional transition metal carbides and nitrides, have emerged as promising materials for supercapacitor applications due to their high electrical conductivity and hydrophilic surface chemistry. This typical chapter examines recent advances in MXene modification strategies, including surface functionalization, intercalation, and composite formation, to enhance their electrochemical performance. These modifications significantly improve specific capacitance, cycling stability, and rate capability. Particular emphasis is placed on the correlation between structural modifications and electrochemical properties, providing insights for the future development of high-performance MXene-based supercapacitors.

Keywords MXene · Supercapacitor · Surface modification · Electrochemical energy storage · 2D materials · Transition metal carbides · Composite materials · Surface functionalization

1 Introduction

Energy storage systems have become increasingly crucial in addressing global energy challenges and supporting sustainable development [1]. Among various energy storage technologies, supercapacitors have garnered significant attention due to their high-power density, rapid charge–discharge capabilities, and exceptional cycling stability. The quest for advanced electrode materials has led to the emergence of MXenes, a relatively new family of two-dimensional transition metal carbides and nitrides, as promising candidates for supercapacitor applications [2].

N. Anwesha · S. Moharana (✉)
Department of Chemistry, School of Applied Sciences, Centurion University of Technology and Management, Bhubaneswar, Odisha, India
e-mail: srikantanit@gmail.com

R. N. Mahaling (✉)
Laboratory of Polymeric and Materials Chemistry, School of Chemistry, Sambalpur University, Jyoti Vihar, Burla, Odisha, India
e-mail: ramnareshmahaling@gmail.com

© The Author(s), under exclusive license to Springer Nature Singapore Pte Ltd. 2025

S. Moharana and R. N. Mahaling (eds.), *Surface Engineering of MXenes*, Engineering Materials, https://doi.org/10.1007/978-981-96-3640-2_9

MXenes, with the general formula Mn+1XnTx (where M represents transition metals, X denotes carbon or nitrogen, and T stands for surface terminations), were first discovered in 2011. Their unique structure combines metallic conductivity with hydrophilic surfaces, making them particularly suitable for electrochemical energy storage. For example, Ti_3C_2Tx-MXene is produced by etching Ti_3AlC_2 in a LiF/HCl solution, where Tx represents surface termination groups whose composition depends on the etching conditions. These surface terminations influence the hydrophilic/hydrophobic balance, affecting electrode–electrolyte interactions. Particularly, Ti_3C_2Tx-MXene exhibits exceptional metallic conductivity (up to ~6700 S cm^{-1}), facilitating efficient electron transport [3]. The layered structure of MXenes facilitates ion intercalation, while their high electrical conductivity enables rapid electron transport, both essential properties for supercapacitor performance. Since graphene's discovery in 2004, numerous two-dimensional (2D) materials have emerged, including hexagonal boron nitride (h-BN), transition metal dichalcogenides (TMDs), transition metal oxides, and single-element compounds such as silicene, borophene, and phosphorene. Among these, graphene and TMDs have dominated research applications, leading the field in terms of practical implementation and scientific investigation. The properties and potential of these materials have opened new avenues in materials science and nanotechnology [4]. Graphene is a remarkable material consisting of a single layer of carbon atoms arranged in a honeycomb lattice through sp^2 bonding. This two-dimensional structure serves as the fundamental building block for various carbon allotropes, including graphite, carbon nanotubes, and fullerenes. While graphite has been known as a natural mineral for five centuries, research into intercalation compounds—where chemical species are inserted between graphite layers—has expanded our understanding of its properties and potential applications. The theoretical framework for graphene was established in 1940, laying the groundwork for understanding both graphite and graphene's unique properties. This theoretical foundation has proven crucial for developing practical applications. Today, graphene demonstrates remarkable versatility in various applications, particularly in membrane-based technologies for gas separation and water purification. Its incorporation into hydrogels and aerogels has also shown promise for energy storage applications and environmental remediation, such as dye adsorption from wastewater [5]. However, transition metal dichalcogenides (TMDs) represent a distinct class of 2D materials with the chemical formula MX_2, where M represents transition metals like Mo, Nb, or W, and X represents chalcogens (S, Se, Te). Unlike graphene, which is composed solely of carbon and exhibits semimetallic behavior with a near-zero bandgap, TMDs offer remarkable compositional diversity and feature substantial bandgaps ranging from 1 to 2 eV. This fundamental difference in electronic structure makes TMDs particularly valuable for applications requiring semiconducting properties [6]. For instance, in 2011, Naguib and his co-workers [7] discovered a groundbreaking method to synthesize two-dimensional transition metal carbides, later termed MXenes, through selective etching of "A" elements from MAX phases. These MXenes, named to reflect their structural similarity to graphene, are derived from MAX phases—ternary layered compounds with the formula Mn+1AXn. In this formula, M represents transition metals (such as Mo,

Ti, Zr, or Cr), A denotes elements from groups 13–16 (including Al, Ga, Ge, and Si), and X represents carbon, nitrogen, or their combinations. The value of n typically ranges from 1 to 3, determining the number of layers in the structure. This discovery opened new possibilities in the field of 2D materials, offering unique properties distinct from other known 2D compounds [8]. The MAX phases exhibit a distinctive layered hexagonal structure where transition metal (M) layers are densely packed with X atoms occupying octahedral sites between them. The A-element atoms serve as connectors between these M-X layers [9]. The bonding characteristics within these structures are complex: M-A bonds demonstrate metallic character, while M-X bonds show a unique combination of ionic, metallic, and covalent bonding types. This intricate bonding arrangement contributes to the structural stability and unique properties of MAX phases [10].

MXenes demonstrate remarkable versatility due to their tunable chemical and structural properties, metallic conductivity, and abundant surface functional groups. These materials have found applications across diverse fields, including energy storage/conversion, electromagnetic shielding, sensing, biomedical applications (imaging and therapy), environmental remediation (water purification and gas separation), and catalysis. Through various synthetic approaches, researchers have developed porous MXene structures with optimized topologies, leading to significant performance improvements across these applications [11]. In supercapacitor applications, the restacking of MXene flakes during electrode preparation poses a significant challenge, as it diminishes the accessible electroactive surface area and impedes electrolyte ion diffusion through the electrode structure. This restacking phenomenon directly impacts the supercapacitor's performance by reducing both energy and power density. To enhance supercapacitor performance, researchers must develop strategies to prevent restacking by modifying MXene layer morphology and texture, thereby maintaining the high surface area and facilitating efficient ion transport through the electrode material [12]. This typical chapter focuses on recent developments in MXene modification strategies and their impact on supercapacitor performance. Understanding these modification approaches and their mechanisms is crucial for designing next-generation energy storage devices that can meet increasingly demanding requirements for both high energy and power densities.

2 Fundamentals of MXenes

2.1 Structure and Composition

The MXenes family immediately benefits from any development in the synthesis of MAX phases or the growth of their structural and compositional variety; in fact, the desire to create new MXenes has driven many recent investigations on MAX phases. For many years, MAX phases containing several transition metals in solid solutions have been recognized. Here are a few instances of its MXenes, which were only recently discovered [13]. An ordered double transition metal MAX phase, Cr_2TiAlC_2, was discovered in 2014. The Ti atoms dominate the center layer (2a Wyckoff sites) in an out-of-plane ordered MAX phase (P63/mmc space group), but the Cr atoms nearly fill the outside layers of M (4f Wyckoff sites) in the M_3C_2 blocks. These MAX phases' layered architectures provide them with a special blend of the advantages of metals and ceramics. It also exhibits outstanding resistance to heat corrosion. A Cr–C sublayer appears under the Al_2O_3 scale due to the continual consumption of Al during the oxidation of Cr_2AlC bulk with general grains, which breaks down into Cr–C binary compounds [14]. Anasori and his co-workers have reported several significant findings regarding out-of-plane ordered MXenes. Beyond discovering three novel structures (Mo_2TiC_2Tx, $Mo_2Ti_2C_3Tx$, and Cr_2TiC_2Tx), they identified their corresponding MAX phase precursors Mo_2TiAlC_2 and $Mo_2Ti_2AlC_3$. Their calculations revealed that Mo_2TiC_2 exhibits the lowest energy stability when configured with a Mo–Ti–Mo stacking arrangement. Energy calculations also showed that the total energy of Mo_2TiC_2 increases approximately linearly with an increase in the concentration of Mo atoms in the middle layer [15].

In 2017, Meshkian and his co-workers synthes ized and characterized out-of-plane ordered Mo_2ScAlC_2 and successfully etched it to produce its corresponding MXene Mo_2ScC_2Tx [16]. They theoretically predicted and experimentally verified the existence of Mo_2ScAlC_2, a quaternary MAX phase with out-of-plane chemical ordering consisting of a Sc atomic layer between two Mo-C layers. Through selective Al etching, they transformed this MAX phase into Mo_2ScC_2, marking the first successful synthesis of a Sc-containing MXene [16]. Tao and colleagues synthesized $(Mo_{2/3}Sc_{1/3})2AlC$, a novel ordered double transition metal carbide with in-plane ordered Mo and Sc atoms in the transition metal carbide sheet. During etching, both Al and Sc atoms were removed, resulting in Mo4/3CTx MXene featuring ordered vacancies at the original Sc sites. Figure 1 illustrates the transformation from MAX to MXene, with the right panel highlighting the divacancy configuration. This 2D material exhibits exceptional properties, including one of the highest specific capacitances and conductivities among MXenes. The incorporation of ordered divacancies represents a significant advancement in property tuning of 2D materials, with the additional benefit of reduced mass density [17].

Persson and his co-workers have demonstrated selective etching behavior in (Mo2/3Y1/3)2AlC (as shown in Fig. 2), where modified etching conditions removed only Al layers while preserving the in-plane ordered double M structure ((Mo2/3Y1/3)2CTx MXene). In contrast, etching of (Mo2/3Sc1/3)2AlC removed both Al and Sc atoms due to Sc's weaker bonding and greater distance from the C plane, similar to Y atom's behavior. This selective etching produced Mo1.33C, a vacancy-ordered MXene exhibiting exceptional volumetric capacitance ($\approx$1150 F cm^{-3}) and high conductivity [18]. Deysher and his co-workers have synthesized $(Mo_{0.8}V_{0.2})5AlC_4$, a novel n = 4 MAX phase, using powder metallurgy at 1650 °C. They successfully etched this material to form $(Mo_{0.8}V_{0.2})5C_4Tx$ MXene, expanding the compositional range of both MAX phases and MXenes [19].

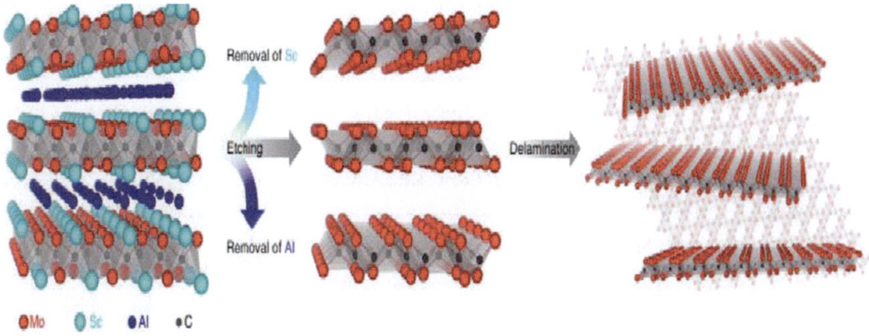

Fig. 1 Structural evolution from (Mo2/3Sc1/3)2AlC MAX phase to Mo1.33C MXene. The schematic shows initial in-plane ordered MAX phase (left), post-etching intermediate (center), and final delaminated MXene with ordered divacancies (right, with top-view inset) [17]. Copyright 2017. Open access. Reproduced with permission from Springer Nature

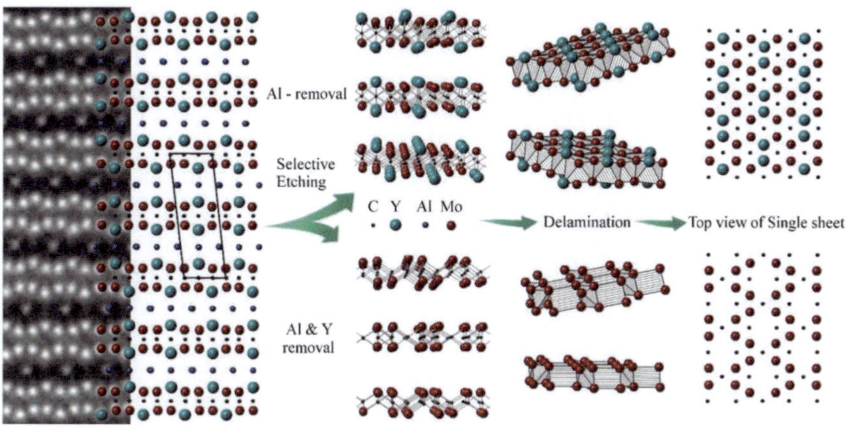

Fig. 2 STEM characterization and structural models of (Mo2/3Y1/3)2AlC. *Left* STEM image along [110] zone axis with overlaid structural model. *Middle* Selective etching outcomes showing Al-only removal (top) versus Al and Y removal (bottom). *Right* Top-view representations of resulting structures [18]. Copyright 2018. Reproduced with permission from Wiley & Sons

2.2 Classification of MXenes

MXenes encompass diverse categories based on their structural and compositional characteristics. Their stoichiometric classifications include M_2X, M_3X_2, M_4X_3, and M_5X_4 formulations, where M represents transition metals and X denotes carbon or nitrogen. Surface terminations like fluorine (–F), oxygen (–O), hydroxyl (–OH), and their combinations significantly influence MXene properties through their effects on electronic structure, surface chemistry, and interlayer interactions. Figure 3 shows the classification scheme of MXenes into seven major categories based on composition and structure, illustrating the diverse family of 2D transition metal carbides and nitrides [20].

- **MXene-conducting polymers**: Large organic molecules known as polymers are made up of repeating units called monomers, which may be found in both natural and artificial forms. It has been demonstrated that conducting polymers offer unique qualities, including flexibility, economic viability, ease of synthesis, and adjustable intrinsic properties for biosensing and other uses. In addition to their conduction characteristics, conducting polymers also display the physical, chemical, and biological characteristics of other nanocomponents, as well as certain unexpected characteristics brought about by the components' synergistic effects [21].
- **MXene-ionic liquid**: The characteristics of ionic liquids, which are organic salts, include improved catalytic performance, redox stability, biocompatibility, a greater range of solubilities, and high ionic conductivity. A top view of the M_3X_2 monolayer, three representative structures of the MAX and MXene phases, the molecular structure model of the single-layer M_3X_2Tz, and the elements in periodic tables utilized to construct MAX phases are all displayed in Fig. 4 [22]. Figure 4a–c displays the typical structure of the MAX and MXene phases as well as the potential component atoms. Different terminal groups can form bonds with interfacial M atoms by eliminating the A atom layers (Fig. 4d). Consequently, the concentration of etchants and etching temperature can significantly affect the as-synthesized MXene's quality and characteristics [20].
- **MXene-perylenediimide**: Perylenediimide (PDI), an electroactive material featuring a planar backbone, enhances MXene's electronic properties through quinone functional group interactions. This combination promotes electronic conjugation and increases charge transport mobility [23].
- **MXene-metal organic**: Metal–organic frameworks (MOFs) have attracted significant research interest due to their inherent porosity, high surface area, and

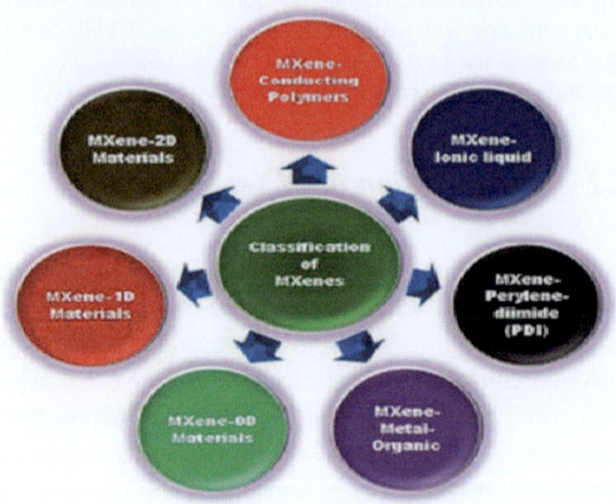

Fig. 3 Classification of MXenes [20]. Copyright 2025. Reproduced with permission from Elsevier

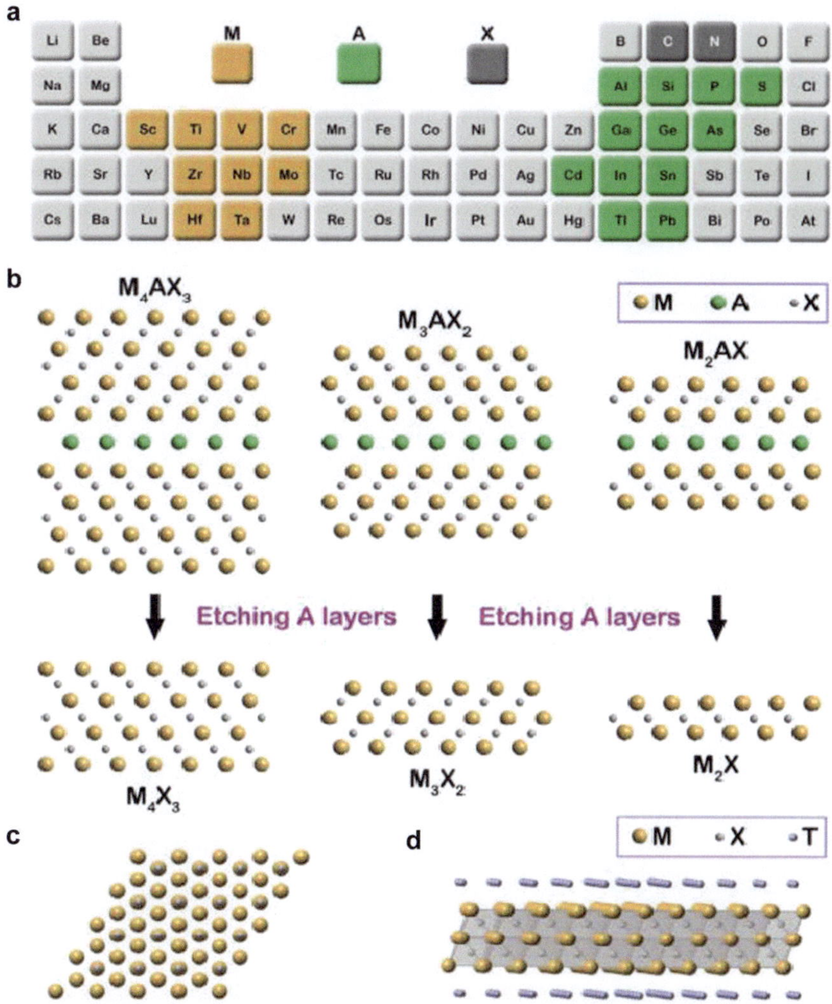

Fig. 4 Compositional and structural elements of MAX phases and MXenes. **a** Elemental composition map showing MAX phase constituent elements. **b** Representative structures of MAX phases and their derived MXenes. **c** Top view of M_3X_2 monolayer structure. **d** Molecular model showing surface-terminated M_3X_2Tz single layer [22]. Copyright 2020. Open access. Reproduced with permission from Wiley

tunable structural and chemical properties, finding applications in gas storage/ separation, sensing, and catalysis. However, converting traditional bulk MOFs into 2D nanosheets and thin films remains a significant challenge [20]. This is because these MOF topologies can open up new possibilities for sensors, electronics, and other device applications. Particularly, the terminal atoms on the surface of MXene are powerful electron acceptors due to their low work function and high electronegativity. This is advantageous for the deprotonation of organic ligands and, consequently, the bridging with the underlying metal atoms. In the meanwhile, the protonated ligands may find sufficient attack sites on the atomically thin 2D MXene's ample accessible surfaces, which might speed up the process while maintaining the underlying 2D structure. However, vanadium-based metal–organic frameworks (V-PMOF) were synthesized using V_2CTx MXene as a metal source, together with H_2TCPP ligand (meso-tetra(4-carboxyl-phenyl) porphyrin). The process involved first creating V_2CTx MXene through Al etching, followed by delamination using TMAOH (tetramethylammonium hydroxide) intercalation (as shown in Fig. 5a). The final V-PMOF, analogous to Al-PMOF, was produced via the solvothermal treatment (Fig. 5b) of the delaminated V_2CTx with H_2TCPP [24].

- **MXene-0D materials**: Generally speaking, 0D materials are composed of three-dimensional particles bigger than 100 nm, such as nanoclusters, nanoparticles, nanodots, and nanospheres. When creating devices, a 0D-2D material combination exhibits superior chemistry, performance, and stability and has been investigated for a range of biological applications [25].
- **MXene-1D materials**: Nanotubes, nanowires, and nanofibers make up the 1D nanostructure. Through the formation of a stable complex, they lessen MXene restacking [26].
- **MXene-2D materials**: The atoms of 2D materials, including graphene, phosphorene, silicon dioxide, molybdenum disulfide, and others, are organized in layers

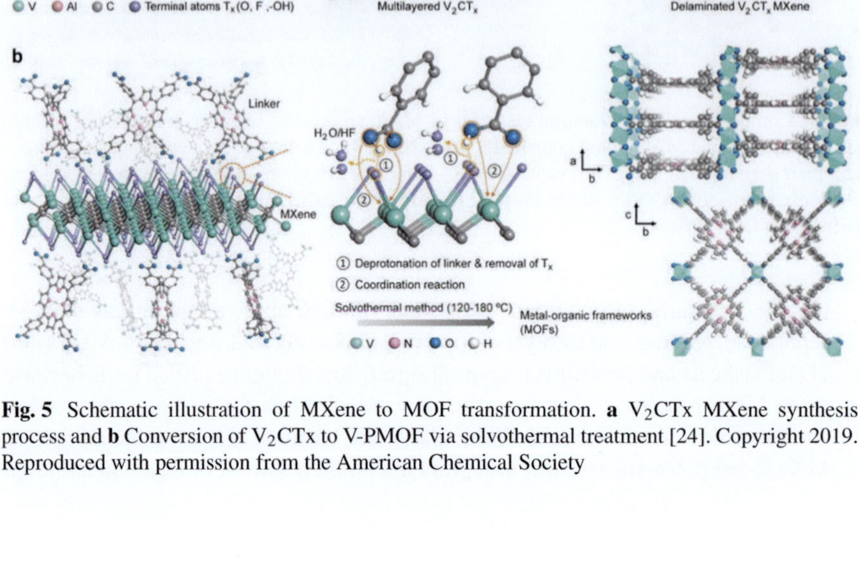

Fig. 5 Schematic illustration of MXene to MOF transformation. **a** V_2CTx MXene synthesis process and **b** Conversion of V_2CTx to V-PMOF via solvothermal treatment [24]. Copyright 2019. Reproduced with permission from the American Chemical Society

that vary in thickness from 1 to 10. MXene is a 2D material that may be mixed with other 2D materials to create compounds with improved qualities [27].

3 Properties

MXenes share several properties with clay materials, particularly in their physical and chemical characteristics. Since their discovery, both theoretical and experimental studies have focused on understanding their optical and electrical properties, which we will examine in detail [28].

3.1 *Optical Properties*

MXene colloidal suspensions can be processed into highly transparent spincast films when sufficiently thin, with Ti_3C_2Tz films exhibiting 97% transmission per nanometer in the visible spectrum [28]. A study of Ti_3C_2Tx films' transport properties revealed exceptional conductivity characteristics (as shown in Fig. 6). Sheet conductance measurements plotted against AFM-determined thickness showed a bulk conductivity of 6,600 S cm^{-1} for spin-cast films, surpassing previously reported solution-processed films made by vacuum filtering or rolling. This conductivity matches that of Ti_3C_2Tx produced by HF etching of physical-vapor-deposited Ti_3AlC_2 films. Notably, films maintained linear conductivity with thickness even below 10 nm, demonstrating continuous nanosheet networks across centimeter-scale areas [29].

The optical properties of MXenes, including transmittance, absorption, and reflection, can be theoretically analyzed through calculations of the imaginary component of the dielectric function tensor across different photon wavelengths [30]. Theoretical studies have examined the optical properties of functionalized Ti_2C and Ti_3C MXenes with –F, –OH, and –O surface terminations. Results show that –F and –OH functionalized Ti_2C and Ti_3C exhibit reduced in-plane absorption coefficients compared to bare and O-terminated structures across infrared to UV wavelengths. These findings predict that –F and –OH terminated Ti_2CT_2 and $Ti_3C_2T_2$ should appear white [31]. The optical response of materials directly correlates with their electrical and structural properties. For MXenes like Ti_3C_2Tx, surface terminations (Tx) significantly influence both structure and electrical characteristics, thereby affecting optical properties. These relationships can be theoretically analyzed through the dielectric constant $\omega(\varepsilon)$, which comprises real ($\omega(\varepsilon)$Re) and imaginary ($\omega(\varepsilon)$Im) components [32]. Berdiyorov used computational methods to analyze how functionalization affects Ti_3C_2Tx's optical properties, finding that surface functionalization doubled the static dielectric function value. Optical behavior varies among Ti_3C_2 MXenes with different surface terminations. Oxygen-terminated samples show higher visible spectrum absorption than pure Ti_3C_2 (as shown in Fig. 7a), while fluorinated surfaces demonstrate reduced

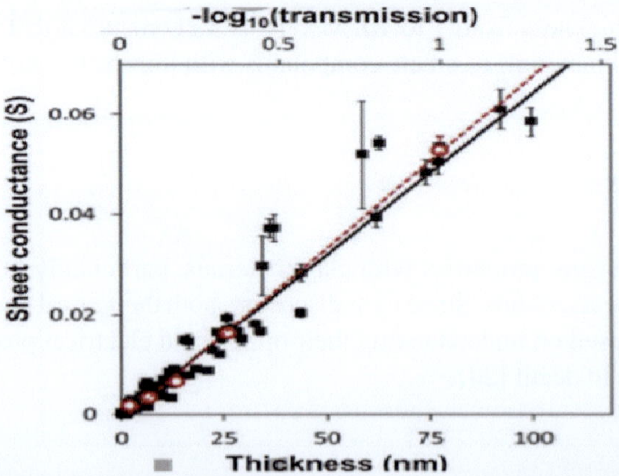

Fig. 6 Transport properties of Ti_3C_2Tx spin-cast films. Sheet conductance plotted against film thickness (AFM measurements, red circles) and optical transmission (black squares). The dashed red line shows a linear fit with conductivity $\sigma = 6600$ S cm^{-1}. The black line represents fit for films on various substrates with $\sigma = 6500 \pm 800$ S cm^{-1} [29]. Copyright 2016. Reproduced with permission from Wiley & Sons

Fig. 7 Calculated absorption spectra of Ti_3C_2Tx MXene with different surface terminations. **a** Visible spectrum range. **b** Ultraviolet spectrum range [33]. Copyright 2016. Open access. Reproduced with permission from AIP Publishing

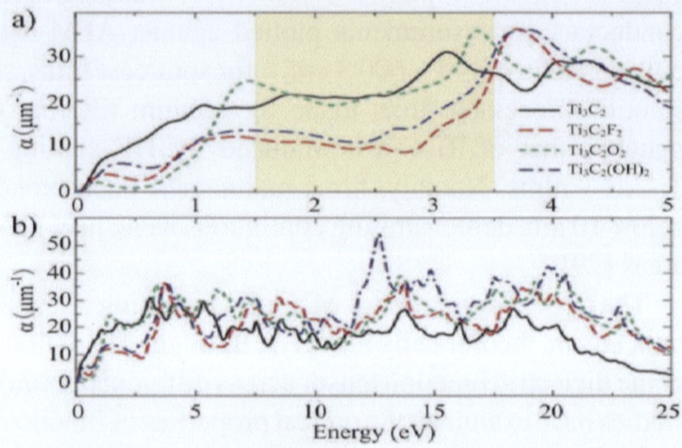

absorption, due to oxygen atoms' contributions to the density of states near the Fermi level. All surface-terminated variants exhibit enhanced UV spectrum reflectance and absorption compared to pure Ti_3C_2. However, hydroxylated and fluorinated terminations display lower visible spectrum absorption and reflectivity, resulting in greater transparency than pristine Ti_3C_2 (Fig. 7b) [33].

3.2 Electrical Properties

Pristine MXenes exhibit metallic conductivity similar to their MAX phase precursors. Surface functionalization can modify their electronic structure, converting some variants into semiconductors [34]. Most functionalized MXenes maintain metallic behavior, but specific compositions develop semiconducting properties when surface-terminated. Using general gradient approximation (GGA) calculations, Sc_2CT_2 shows bandgaps of 1.03 eV (F), 0.45 eV (OH), and 1.8 eV (O). Other semiconducting MXenes include Ti_2CO_2 (0.24 eV), Zr_2CO_2 (0.88 eV), and Hf_2CO_2 (1.0 eV). However, F and OH surface terminations similarly affect the MXene electronic structure since each accepts one electron from the surface. O terminations differ by requiring two surface electrons for stability. Electronic structure studies of functionalized and pristine MXenes reveal the mechanisms behind semiconductor transformation. Calculations show all pure M_2X systems (X = C or N) are metallic, with the Fermi energy residing in the transition metal M's d bands. Surface functionalization with F, OH, or O creates new bands below the Fermi energy through hybridization between F/O p orbitals and M d orbitals. In Ti_2CO_2, Zr_2CO_2, Hf_2CO_2, and Sc_2CT_2, this shifts the Fermi energy into a gap between M d and X p bands, inducing semiconducting behavior. Figure 8 demonstrates this electronic structure modification through band structures and density of states for Ti_2C, Ti_2CF_2, and Ti_2CO_2 [34].

Figure 9 demonstrates the mechanism of large polaron formation in MXene materials. A layered MXene device is shown with electrical contacts, where pump-probe spectroscopy is used to study electron behavior. The key feature is the large polaron formation, depicted by an electron (blue) surrounded by a cloud of positive and negative charges (red/blue dots) within a dotted boundary. This represents the electron-lattice interaction characteristic of MXenes. The experimental setup uses a pump beam to excite electrons and a probe beam to measure the response, with arrows indicating electron transport through the material. This configuration enables studying the electronic and transport properties of MXene-based devices [35].

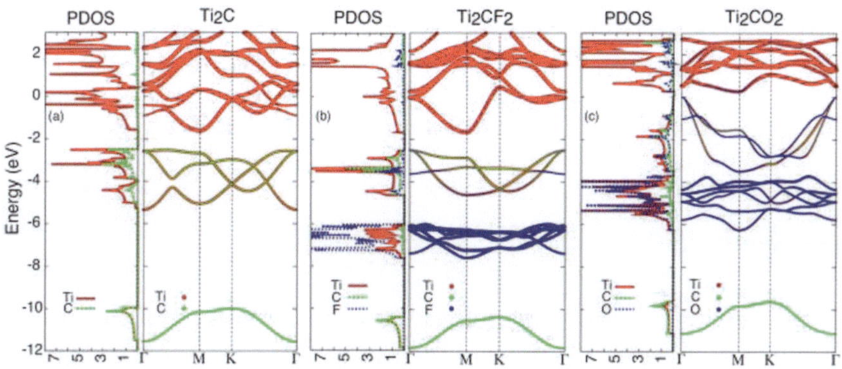

Fig. 8 Electronic structure of Ti_2C-based MXenes. Band structures and projected density of states for **a** pristine Ti_2C, **b** fluorinated Ti_2CF_2, and **c** oxidized Ti_2CO_2 [34]. Copyright 2017. Reproduced with permission from the Royal Society of Chemistry

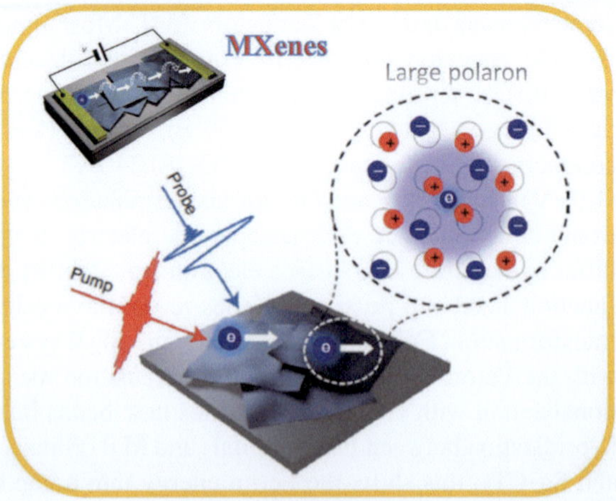

Fig. 9 Formation mechanism of large polarons in MXene showing device structure, experimental setup with pump-probe configuration, and resulting charge distribution pattern in the polaronic state [35]. Copyright 2025. Reproduced with permission from Elsevier

The frequency-dependent conductivity of composites containing MXene and MXene@PDA fillers is presented in Fig. 10a–c. The conductivity shows a positive correlation with increasing frequency, following a pattern similar to pure PVDF, suggesting the composites retain some insulating characteristics. The higher filler content leads to increased conductivity due to MXene's inherent electrical conductivity and the formation of conductive pathways. However, the PDA coating on MXene surfaces slightly reduces the composite conductivity owing to the insulating nature of PDA [36].

MXene electrical properties can be tuned through functional group modification, stoichiometry adjustment, and solid solution formation. Their conductivities exceed carbon nanotubes and reduced graphene oxide, matching multilayer graphene (resistances 22–339 Ω, varying with the "n" index and composition). Ti_3C_2Tx shows conductivities from 850–9880 S cm^{-1} (as shown in Fig. 11), influenced by defect concentration, surface groups, delamination yield, d-spacing, and flake size. Lower HF concentrations and shorter etching times produce MXenes with fewer defects and larger lateral sizes, resulting in higher conductivity [37].

3.3 Synthesis Methods

MXenes are created using a variety of synthesis techniques, such as chemical vapor deposition, physical vapor deposition, plasma-enhanced pulsed laser deposition, and the chemical etching approach. The chemical vapor deposition method is one of the several bottom-up techniques that have been documented in the literature to create MXenes without surface termination. Because of its simplicity in synthesis, convenience of handling in the lab, and inexpensive preparation, the selective chemical etching method (also known as the top-down approach) is the most often used technique among researchers. For example, the morphology of MXene may be adjusted

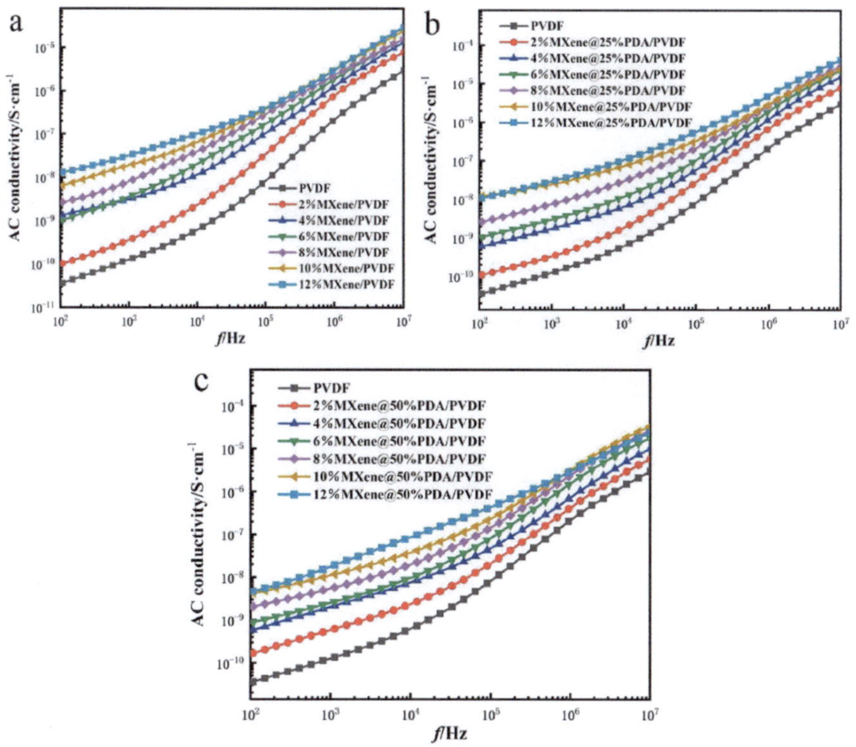

Fig. 10 AC conductivity behavior of PVDF composites with varying filler compositions: **a** MXene/PVDF, **b** MXene@25%PDA/PVDF, and **c** MXene@50%PDA/PVDF [36]. Copyright 2024. Reproduced with permission from Elsevier

by the etching conditions. When intense HF was employed, accordion-like particles were seen; they were absent at softer etching conditions. Additionally, a greater etchant concentration. By employing less hazardous, eco-friendly chemicals, several researchers have attempted to make MXenes more ecologically friendly. This might lead to easier synthesis of MXenes and more chances in the future. Furthermore, many MXenes are still not accessible using the following synthesis techniques. Therefore, it is necessary to investigate novel and distinct synthesis techniques for the production of unidentified MXenes. The following sections cover the two most often used methods [38]. Figure 12 shows the schematic representation of various synthetic routes for MXene preparation [39]. Since 2011, MXene synthesis advancements have led to increased applications (Fig. 13) in energy storage, water purification, and sensing, improving properties and performance [40].

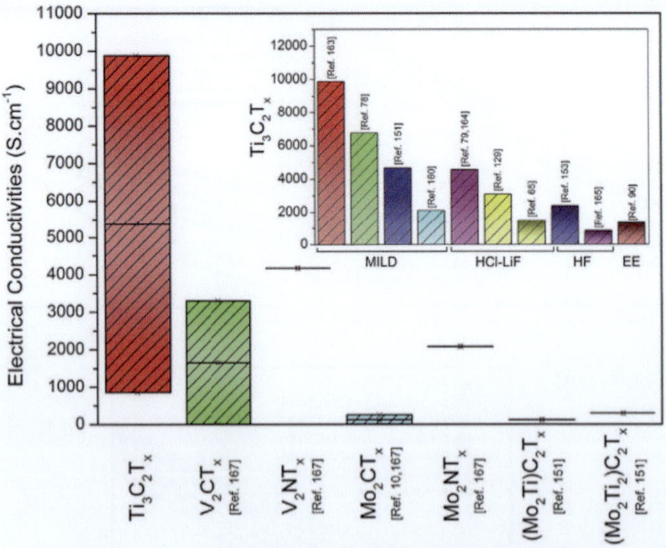

Fig. 11 Comparative electrical conductivities of MXenes variation with different etching protocols [37]. Copyright 2019. Reproduced with permission from Elsevier

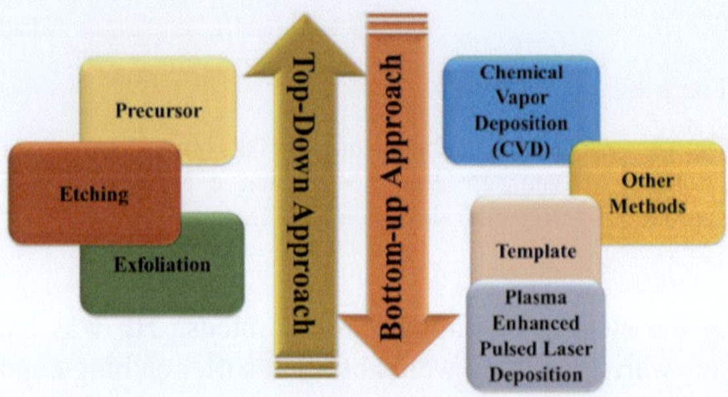

Fig. 12 Schematic representation of various synthetic routes for MXene [39]. Copyright 2023. Open access. Reproduced with permission from MDPI

Top-Down Approach

The synthesis of MXenes through top-down approaches relies on specific precursor materials, primarily MAX phases such as Ti_3AlC_2 and Ti_2AlC, and non-MAX structures including $Zr_3Al_3C_5$, $Hf3(Al, Si)_4C_6$, and $ScAl_3C_3$. MAX phases remain the predominant precursor choice due to their unique bonding characteristics, where M–X bonds exhibit greater strength than M–A bonds. During synthesis, selective etching of the A element occurs through M–A bond cleavage, resulting in multi-layered MXene structures. These layers can then be delaminated to produce single-layer MXenes, which spontaneously form surface terminal groups (−OH, O, F) that stabilize the structure. While mechanical exfoliation is commonly employed for producing 2D materials, the strong M–A bonds in MXene precursors resist mechanical cleavage alone. Consequently, chemical etching serves as the primary synthesis

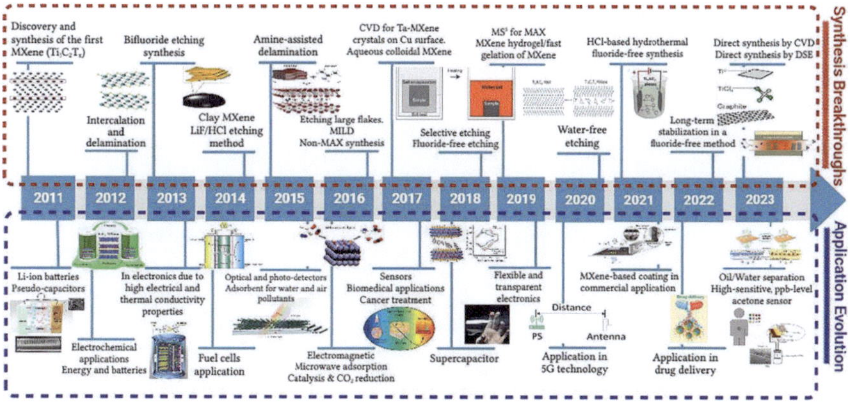

Fig. 13 Chronological development of MXene synthesis methods and applications from 2011 to present [40]. Copyright 2023. Reproduced with permission from Elsevier

method, with mechanical processes reserved for single-layer production. Various etching techniques have been developed to break M–A bonds, including fluoride-based, alkali-based, electrochemical, Lewis's acid molten salt, and photo-Fenton methods (as shown in Fig. 14) [41].

Bottom-Up Approach

The bottom-up synthesis (Fig. 14) serves as an alternative to the conventional top-down approach, particularly for producing nanocrystals that resist traditional exfoliation methods. This synthetic strategy begins with small organic or inorganic molecular precursors or atoms, which then crystallize and self-assemble into ordered two-dimensional layers. Chemical vapor deposition (CVD) exemplifies this bottom-up approach [41, 42]. CVD is a flexible method that uses gaseous precursors interacting with a substrate to create films, heterostructures, and whole devices. The synthetic alternative for MXenes has not yet been made accessible, although CVD can be used to generate transition-metal carbides and nitrides like Mo_2C, Mo_2N, and TiCx. CVD allows for the direct synthesis of MXenes, paving the way toward novel MXene geometries with exposed catalytically active edges and more accessible surfaces. MXene synthesis via CVD was achieved by exposing a Ti surface to a mixture of CH4 and $TiCl_4$ gases diluted in Ar at 950 °C (Fig. 15a). The resulting product, designated as CVD-Ti_2CCl_2, was characterized after 15 min of exposure using XRD (Fig. 15b). Rietveld refinement revealed lattice parameters (a = 3.2225(2) Å and c = 8.7658(8) Å) in close agreement with previously reported values for Ti_2CCl_2 MXene. The phase purity of Ti_2CCl_2 MXene was further confirmed through Raman spectroscopy (Fig. 15c). The detailed structural and compositional analysis using high-resolution EDX and STEM-EELS validated the stoichiometry and crystallinity of CVD-Ti_2CCl_2. STEM imaging showed an interlayer spacing of 0.88 ± 0.05 nm between Ti_2CCl_2 MXene layers. SEM imaging revealed a substrate uniformly covered with a corrugated layer of Ti_2CCl_2 (Fig. 15d, e). This unique

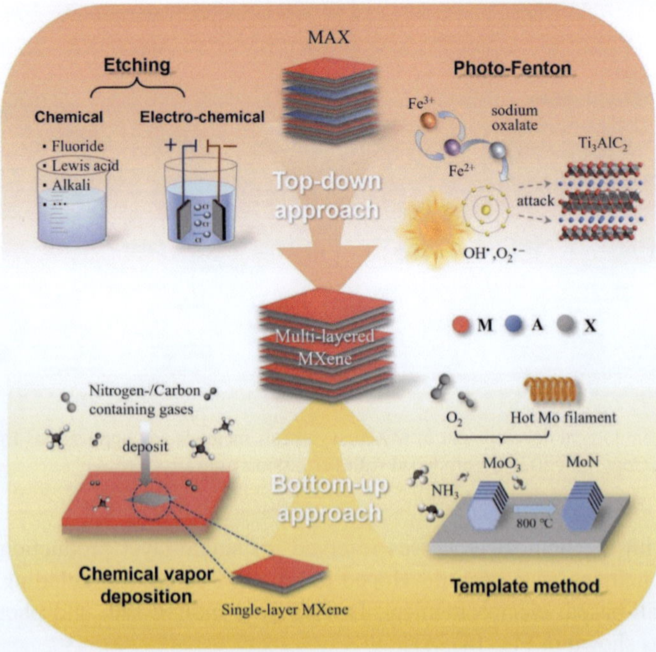

Fig. 14 Synthesis techniques of MXenes [41]. Copyright 2024. Reproduced with permission from Elsevier

morphology, where Ti_2CCl_2 MXene sheets grow perpendicular to the substrate, forming a carpet-like structure, would be difficult to achieve using conventional synthesis methods. The vertical alignment of these sheets presents advantages for applications such as supercapacitors, where efficient ion intercalation is crucial, surpassing the potential of other CVD-grown 2D materials like MoS_2 [43].

4 Surface Modification Strategies

The surface of MXenes, rich in –OH and –O containing groups, provides active sites for covalent functionalization through surface terminations, surface-active initiators, small molecules, and polymers. The hydrophilic nature of MXenes facilitates the incorporation of appropriate functional groups to enhance their performance [44]. Figure 16 illustrates the range of modification strategies developed to enhance MXene properties, encompassing surface functionalization, composite blending, and exfoliation techniques [41].

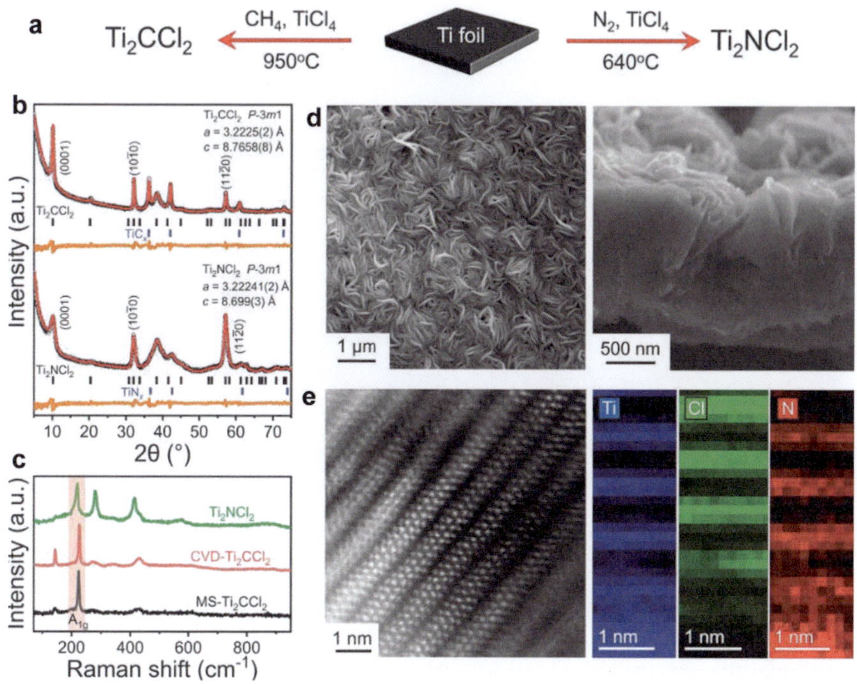

Fig. 15 CVD growth of MXenes. **a** The CVD reactions are shown schematically. **b** Rietveld refinement and XRD patterns for CVD-Ti_2CCl_2 and CVD-Ti_2NCl_2. **c** Comparing the Raman spectra of CVD-Ti_2CCl_2 and CVD-Ti_2NCl_2 MXenes to a conventional MS-Ti_2CCl_2 MXene, which was created by etching the Ti_2AlC MAX phase with molten salt of $CdCl_2$. **d** CVD-Ti_2CCl_2 frontal and cross-sectional SEM pictures and **e** EELS elemental mapping and high-resolution HAADF pictures of CVD-Ti_2NCl_2 [43]. Copyright 2023. Open access. Reproduced with permission from Science

4.1 Chemical Modification Methods

The surface termination of MXene sheets is determined during the MAX phase etching process (as shown in Fig. 17a). While aqueous solutions typically result in mixed surface terminations of Cl, O, and OH groups, recent advances in electrochemical and hydrothermal methods have enabled MAX phase etching without using HF solutions. For example, etching Ti_3AlC_2 MAX phase in molten $ZnCl_2$ and other Lewis acidic molten salts at 500 °C yields $Ti_3C_2Cl_2$ MXene with pure chlorine termination. The molten salt etching approach was successfully extended beyond chlorides by using Lewis acidic $CdBr_2$, enabling the first synthesis of bromine-terminated $Ti_3C_2Br_2$ and Ti_2CBr_2 MXenes (Fig. 17b, c). These halide-terminated MXenes, particularly the bromine-terminated variants, demonstrate exceptional capability for a new class of surface reactions where halide ions can be exchanged for other atoms and functional groups. These exchange reactions enable unprecedented control over MXene surface chemistry, structure, and properties. The outer transition-metal atoms

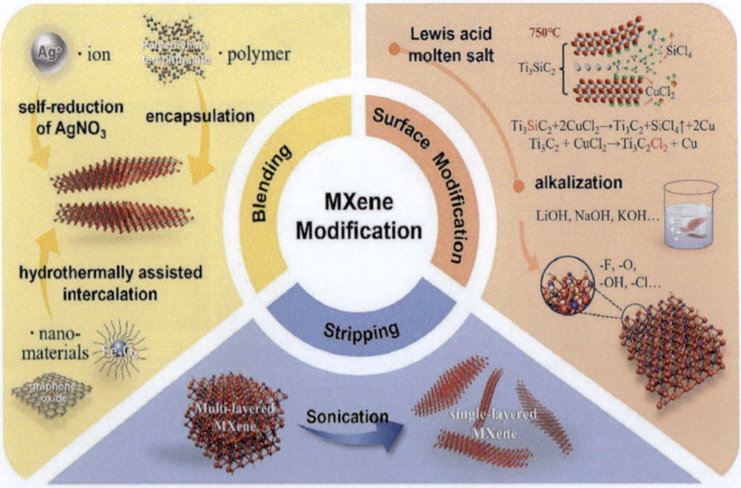

Fig. 16 Various surface modification techniques of MXenes [41]. Copyright 2024. Reproduced with permission from Elsevier

(Ti, Mo, Nb, and V) of MXene sheets form relatively weak M-Cl and M-Br bonds, in contrast to the stronger M–F and M–OH bonds typically found in MXenes with Tx surface groups. While the strong Ti–F and Ti–O bonds make post-synthetic covalent surface modifications challenging, the labile surface bonding in chlorine- and bromine-terminated MXenes makes them versatile building blocks for further chemical transformations [45]. MXene surface exchange reactions typically require temperatures ranging from 300° to 600 °C, making conventional solvents impractical. Molten alkali metal halides serve as ideal solvents due to their wide electrochemical windows, high solubility for various ionic species, and exceptional thermal stability at elevated temperatures. For example, when $Ti_3C_2Br_2$ MXene (Fig. 17b) was dispersed in a CsBr-KBr-LiBr eutectic mixture (melting point: 236 °C), reactions with Li_2Te and Li_2S yielded Ti_3C_2Te (Fig. 17d) and Ti_3C_2S (Fig. 17e) MXenes, respectively. Similarly, reactions of $Ti_3C_2Cl_2$ and $Ti_3C_2Br_2$ with Li_2Se, Li_2O, and $NaNH_2$ produced Ti_3C_2Se, Ti_3C_2O, and $Ti_3C_2(NH)$ MXenes. Moreover, High-angle annular dark-field (HAADF) imaging reveals the atomic structure of $Ti_3C_2Br_2$ MXene, clearly delineating the surface vacancies created by the systematic removal of bromine terminations (Fig. 17f). Further treatment of the multilayered Ti_3C_2Tn MXenes (where T = Cl, S, or NH) with n-butyl lithium (n-BuLi) resulted in Li^+ intercalated sheets bearing negative surface charges [45].

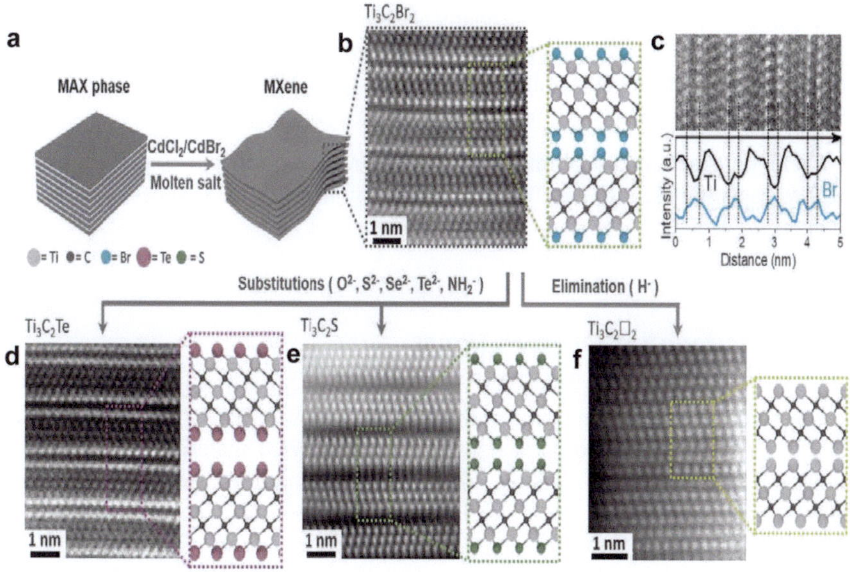

Fig. 17 Chemical modification of MXene surfaces in molten salt systems. **a** Schematic of MAX phase etching in Lewis acidic molten salts, **b** HAADF atomic-resolution imaging of $Ti_3C_2Br_2$ MXene obtained from Ti_3AlC_2 MAX phase etching in molten $CdBr_2$, **c** EDX line scan analysis of $Ti_3C_2Br_2$ MXene, **d** HAADF imaging of Ti_3C_2Te, **e** Ti_3C_2S MXenes from bromine exchange reactions, and **f** HAADF image of $Ti_3C_2Br_2$ MXene showing surface vacancies after Br elimination [45]. Copyright 2020. Open access. Reproduced with permission from Science

4.2 *Physical Modification Approaches*

Physical modification strategies play a crucial role in enhancing MXene's performance in supercapacitor applications by addressing the restacking issue and improving accessibility [46]. These approaches focus on structural manipulation without altering the chemical composition of MXenes.

Key physical modification methods include:

1. **Spacer Introduction**: The incorporation of nanoscale spacers between MXene layers prevents restacking and maintains interlayer spacing. Common spacers include carbon nanotubes (CNTs), graphene, and metal oxide nanoparticles. These spacers create stable channels for electrolyte ion transport while maintaining electrical conductivity [47].
2. **Architectural Engineering**: Creating hierarchical structures through techniques like freeze-drying or template-assisted synthesis produces 3D architectures with interconnected porous networks. These structures significantly increase the accessible surface area and facilitate rapid ion transport throughout the electrode material [48].

3. **Delamination Processing**: Mechanical delamination techniques, such as sonication and shear mixing, help separate MXene layers and reduce their tendency to restack. This process creates single or few-layer MXene sheets with enhanced surface area and improved accessibility to electrolyte ions [49].

These physical modifications have demonstrated significant improvements in specific capacitance, rate capability, and cycling stability of MXene-based supercapacitors.

4.3 Functionalization Techniques

Self-initiated photo-grafting and photopolymerization (SIPGP) represent one of several surface modification strategies for MXene-based materials. Compared to traditional polymerization methods, SIPGP offers a simplified approach that eliminates the need for initiators, anchoring layers, or self-assembled monolayers (SAMs). The technique has demonstrated broad applicability across various substrates, including diamond, graphene, carbon nanotubes, and silicon-based materials [50]. Chen et al. demonstrated a reliable method for grafting poly(2-(dimethylamino) ethylmethacrylate) (PDMAEMA) brushes onto MXene surfaces using SIPGP [51]. As shown in Fig. 18, V_2C MXenes were synthesized by treating V2AlC particles with hydrofluoric acid at room temperature. The hydroxyl groups on the MXene surface served as photo-active sites for the continuous growth and amplification of polymer brushes through SIPGP. The resulting V_2C@PDMAEMA hybrid materials exhibited dual-responsive behavior to both temperature and CO_2 stimuli. By modulating temperature or introducing/removing CO_2, the researchers could control the dispersion state and conductivity of these hybrid MXene materials in aqueous environments, demonstrating their potential as "smart" responsive materials [51].

Wang et al. have demonstrated that aryl diazonium salts can form stable Ti–O–C bonds through covalent attachment to the MXene surface. Notably, the introduction of aryl diazonium salts increases the interlayer spacing of MXene-based materials. As illustrated in Fig. 19, the process begins with Na^+ ion intercalation of MXenes, followed by direct immobilization of phenylsulfonic groups on the MXene surface through aryl diazonium chemistry. These chemical modifications, combining diazonium ion intercalation and electroactive group grafting, enhance the supercapacitive performance of the MXenes. While the application of aryl diazonium chemistry for MXene surface modification remains in its early stages, this approach shows considerable promise for attaching diverse functional groups and molecules to MXene-based materials [52].

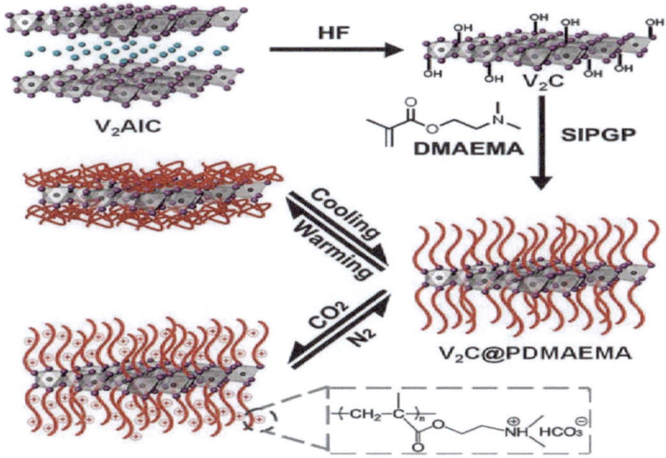

Fig. 18 Schematic representation of smart V_2C@PDMAEMA hybrid system synthesis, showing the stepwise process from V_2AlC MAX phase etching to PDMAEMA polymer brush grafting via SIPGP onto V_2C MXene surfaces [51]. Copyright 2015. Reproduced with permission from the Royal Society of Chemistry

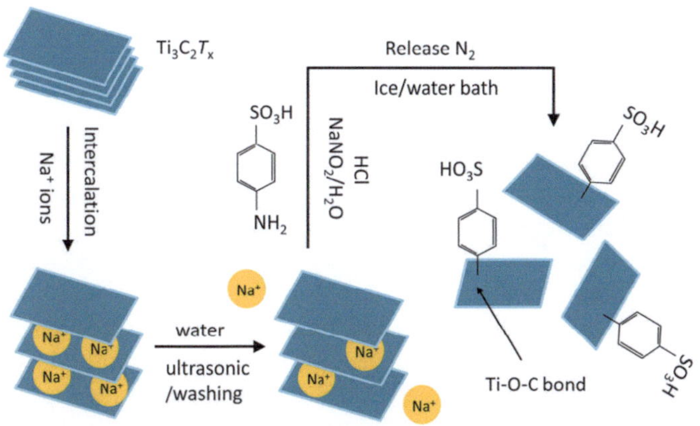

Fig. 19 Surface functionalization of Ti_3C_2 MXene through diazonium chemistry: Na+ ion pre-intercalation followed by aryl group grafting and exfoliation into individual sheets via sonication [52]. Copyright 2018. Reproduced with permission from Elsevier

5 MXene-Based Composites

MXenes can be combined with various materials to create composites that utilize their unique properties, often demonstrating enhanced performance compared to their constituents, with primary categories including:

5.1 Carbon-MXene Hybrids

Carbon-based materials exist in various forms, including diamond, graphite, and amorphous structures, exhibiting exceptional mechanical, electrical, optical, and physicochemical properties. These characteristics make carbon-based components valuable additives for MXene-based composites, enhancing their conductivity, flexibility, and electrochemical performance. Among these carbon materials, carbon nanotubes (CNTs), graphene (G), and mesoporous carbon have attracted particular attention in recent research [53]. Figure 20a depicts the synthesis route for Ti_3C_2Tx HCF production. MXene was selected as the multilayer platform due to its morphology-controlled synthesis capabilities. The HF etching process generates a distinctive accordion-like structure with consistent interlayer spacing, which enhances electromagnetic wave reflection. The template synthesis method utilizes SiO2 nanobeads as scaffolds to transform RF into porous carbon foam through carbonization and etching processes, as illustrated in Fig. 20b. The pore architecture in the carbon foam can be precisely controlled through the ordered arrangement of SiO_2 nanobeads. The resultant porous structure exhibits ultralow density and spherical voids that enhance electromagnetic wave reflection and absorption. The integration of micrometer-scale layered Ti_3C_2Tx MXene with mesoscopic porous carbon preserves MXene's characteristic accordion-like morphology, as shown in Fig. 20c, d. Elemental analysis confirms the complete removal of SiO2 templates through etching, as demonstrated in Fig. 20e.

MXene–Graphene Composites

The unique combination of high conductivity, flexibility, and large surface area has made graphene an attractive component for composite materials, leading to extensive research in this field. These materials not only maintain graphene's intrinsic properties but also preserve its structural integrity, leading to enhanced performance [53]. Xu et al. successfully synthesized a Mo_2C/graphene composite using chemical vapor deposition (CVD) (as shown in Fig. 21). During the reaction process, 2D Mo_2C layers grow in situ on the graphene surface. The resulting Mo_2C/graphene composite demonstrates exceptional performance, combining the superconducting properties of 2D Mo_2C MXene with the superior qualities of high-quality graphene [55].

Figure 22 depicts the schematic preparation procedure for the Mo_2C/N-doped graphene (N-G) composite. A Mo_2C/N-G composite has also been produced by sintering a MoO_3/graphene oxide (MoO_3/GO) composite with a specific nitrogen source (dicyandiamide). First, MoO_3 nanoribbons with a size range of 10–200 nm were created by hydrothermal treatment in order to prepare the Mo_2C nanoribbon/N-G film. After that, a very homogenous combination was created by immediately ultrasonically mixing the produced MoO_3 nanoribbons with a graphene oxide (GO) solution. This mixture was subsequently filtered into a self-standing film [56].

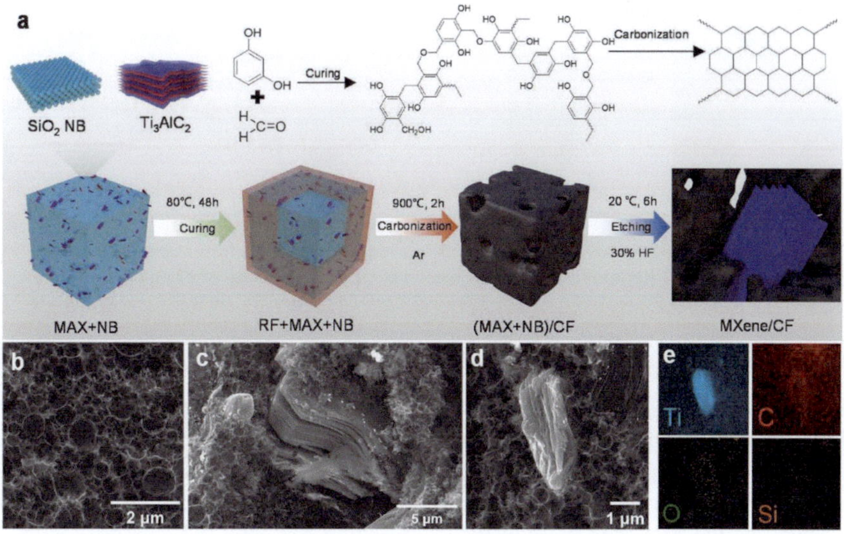

Fig. 20 Synthesis process and structural characterization of hierarchical carbon foam (HCF): **a** Schematic representation of HCF fabrication process, **b** SEM micrograph of carbon foam (CF) template, **c**, **d** SEM images showing HCF morphology at different magnifications, and **e** EDS elemental mapping [54]. Copyright 2024. Reproduced with permission from Elsevier

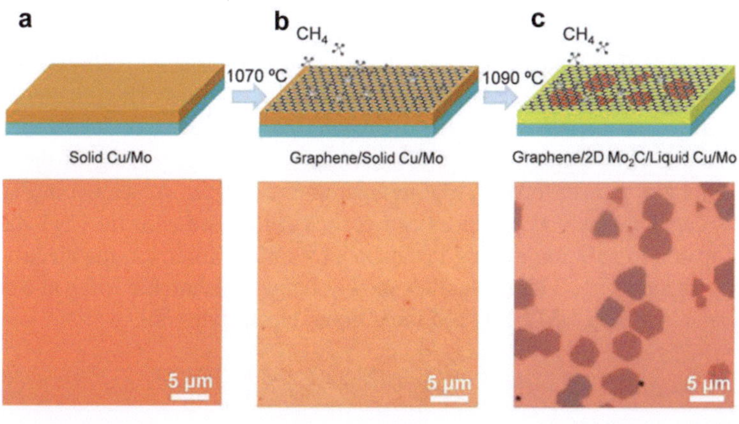

Fig. 21 Stepwise CVD synthesis of graphene/2D α-Mo_2C heterostructures: **a** Initial Cu/Mo bilayer substrate configuration, **b** CVD growth of graphene layer at 1070 °C on solid Cu/Mo substrate, and **c** Formation of 2D α-Mo_2C crystals beneath graphene on liquid Cu/Mo substrate at 1090 °C [55]. Copyright 2017. Reproduced with permission from American Chemical Society

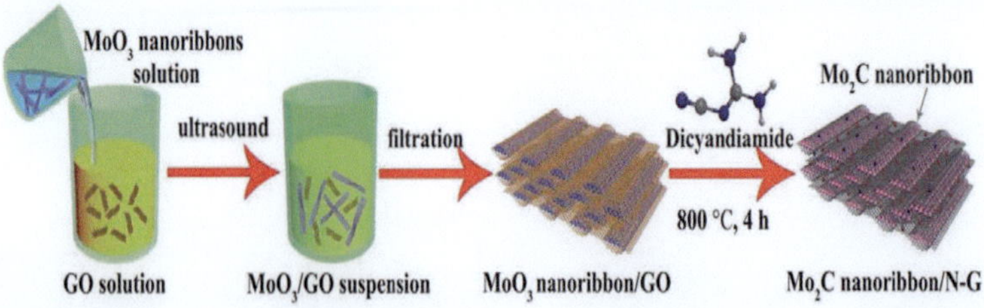

Fig. 22 Diagrammatic representation of the Mo_2C nanoribbon/N-G film manufacturing process [56]. Copyright 2017. Reproduced with permission from Royal Society of Chemsitry

MXene-Carbon Nanotubes

Carbon nanotubes have attracted significant research attention due to their exceptional mechanical properties, chemical stability, and electrical conductivity. CNTs effectively prevent graphene restacking and serve as spacers between MXene layers when compounded with MXene, preventing reaggregation. Solution mixing has emerged as the preferred method for synthesizing MXene/CNT composites [57]. Among these, Ti_3C_2Tx/CNT and Nb_2CTx/CNT composites are the most extensively studied. The surface functionalization of MXenes results in negative surface charges, facilitating composite formation with either positively charged or neutral CNTs through solution mixing [58]. Beyond ex-situ mixing methods, MXene/CNT composites can also be prepared through in situ growth approaches. Li et al. demonstrated the successful synthesis of Ti_3C_2Tx/CNT composites using CVD, where the Ni^{2+} catalyst is first deposited on exfoliated Ti_3C_2Tx surfaces, followed by CNT growth from a C_2H_4 carbon source under catalytic conditions. The process is illustrated in Fig. 23a, b [59]. In another approach, the Co^{2+} catalyst is adsorbed onto polydopamine (PDA)-masked Ti_3C_2Tx surfaces, with urea serving as the carbon source for CNT growth through high-temperature catalysis. As shown in Fig. 23c, the Co^{2+} catalysis method produces composites with more uniform CNT distribution compared to solution mixing techniques [59, 60]. Moreover, the electrochemical performance (Fig. 23d) evaluation shows charge–discharge cycling at a current density of 1 A g^{-1} in 1 M EMITFSI ionic liquid electrolyte [60].

5.2 *Polymer-MXene Composites*

The integration of MXenes with polymeric materials has emerged as a promising strategy for developing functional composites that combine the unique properties of both components. These hybrid materials leverage MXene's electrical conductivity, mechanical strength, and surface chemistry while benefiting from the processability, flexibility, and versatile functionality of polymers. The resulting composites demonstrate enhanced performance in various applications, including electromagnetic interference shielding, energy storage, and sensing devices [61]. For instance,

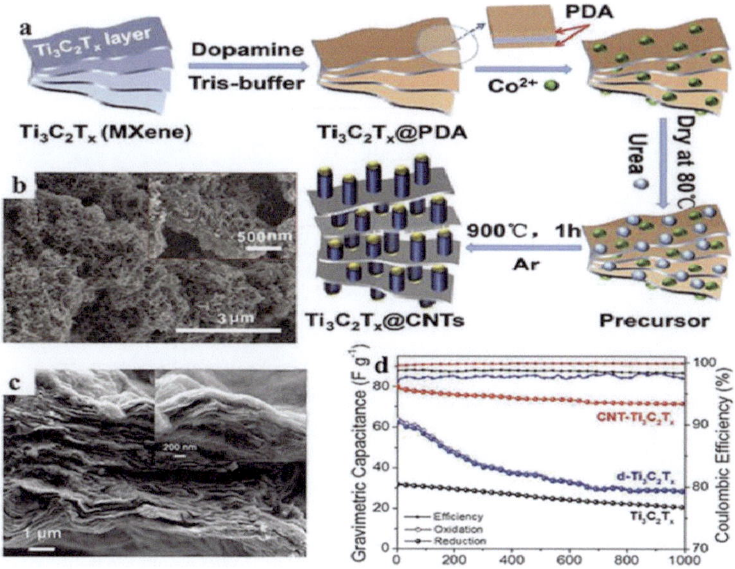

Fig. 23 Ti_3C_2Tx/CNT composite synthesis and characterization: **a** Schematic illustration of 3D networked Ti_3C_2Tx/CNT composite synthesis, **b** SEM micrograph of in-situ grown Ti_3C_2Tx/CNT composite [59]. Copyright 2017. Reproduced with permission from Elsevier. **c** SEM image of Ti_3C_2Tx/CNT composite prepared by ultrasound-assisted solution mixing, and **d** Cycling stability test at 1 A g^{-1} in 1 M EMITFSI electrolyte [60]. Copyright 2016. Reproduced with permission from Elsevier

Xu et al. developed polyvinyl alcohol (PVA)/MXene composite foams and films using a simple freeze-drying technique. The composites achieved a specific EMI shielding effectiveness (SSEt) of 5136 dB cm^2 g^{-1} with just 0.15 vol% MXene content. As shown in Fig. 24a, the synthesis of f-Ti_2CTx/PVA foam and films begins with the selective etching of Al atoms from the Ti_2AlC MAX phase using a hydrochloric acid (HCl) and lithium fluoride (LiF) solution to produce few-layered Ti_2CTx (f-Ti_2CTx). The precursor hydrosol was then formed by mixing poly(vinyl alcohol) (PVA) solution with f-Ti_2CTx solution, as illustrated in Fig. 24b. The final f-Ti_2CTx/PVA foam (22.86 mm × 10.16 mm × 10.00 mm) was produced through freeze-drying in a custom-made hollow square Teflon tube with a copper plate at the bottom, as depicted in Fig. 24c. The resulting f-Ti_2CTx/PVA composite foam exhibits strong enough mechanical properties and can support a 140,000 mg metal block, which is more than 5000 times its weight, due to the strong hydrogen bonding between PVA molecular chains and surface moieties (Tx: a mixture of –OH, –O, and – F) of f-Ti2CTx (Fig. 24f). As seen in Fig. 24d, the f-Ti_2CTx/PVA composite foam may be compressed to create the f-Ti_2CTx/PVA composite film. Figure 24g illustrates the composite film's flexibility, which is also a result of strong hydrogen bonding [62].

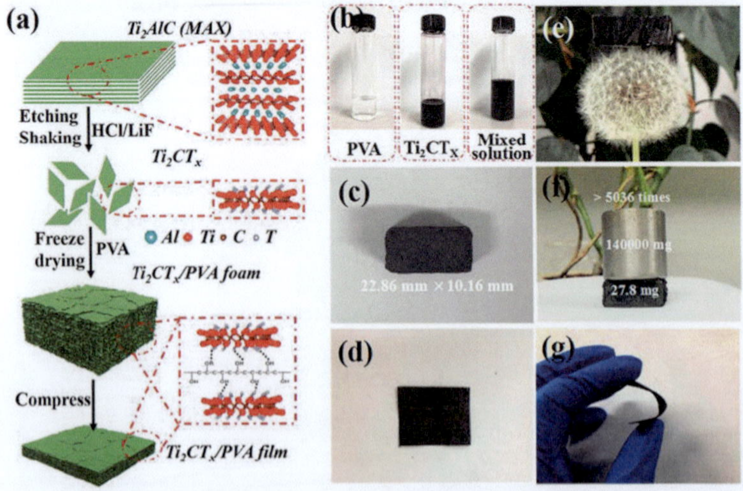

Fig. 24 **a** A diagram showing how f-Ti_2CTx/PVA composite foam and film are prepared, **b** PVA solution bottle, pure black f-Ti_2CTx solution bottle, and f-Ti_2CTx/PVA solution bottle. **c** f-Ti_2CTx/PVA foam photograph, **d** A typical f-Ti_2CTx/PVA film image, **e** An ultralow density piece of f-Ti_2CTx/PVA foam standing on a dandelion, **f** An example of f-Ti_2CTx/PVA foam that can sustain almost 5000 times its weight and **g** Flexible f-Ti_2CTx/PVA film photograph [62]. Copyright 2019. Reproduced with permission from American Chemical Society

MXene-Wrapped Sponge

MXene-wrapped sponges are a novel class of three-dimensional structures that combine the special qualities of MXene with the porosity and flexibility of sponge substrates. Usually, MXene nanosheets are applied to the sponge scaffold by a straightforward immersion or dip-coating procedure to create these composites. Yue and his coworkers have reported a dip-coating technique to create MXene-wrapped sponge with noticeably improved piezoresistive performance. The MXene-sponge synthesis method is shown in detail in Fig. 25a. To absorb the 2D nanosheet, the sanitized melamine sponge was first soaked in the MXene solution. Following complete submersion, the sponge was taken out of the MXene solution and dried for 24 h at 80 °C in a vacuum oven. Ultimately, it created the 3D network structure, with the sponge's 3D skeleton being anchored by a thin layer of MXene. MXene is a 2D substance that adheres to the sponge well due to its wide contact area and robust van der Waals interaction. MXene sponge is a promising contender for a pressure-sensing material because of its 3D porous structure, low density, high elasticity, and excellent mechanical strength. Figure 25b, c provides a schematic representation of the MSP-based piezoresistive sensor production method. First, using the practical inkjet printing template sacrifice technique that we previously described, a pair of interdigitated electrodes was deposited onto a piece of thin polyimide (PI) film. Second, PVA NWs were uniformly deposited using the electrospinning technique on the surface of the PI substrate as a spacer. Thirdly, a segment of MXene-sponge

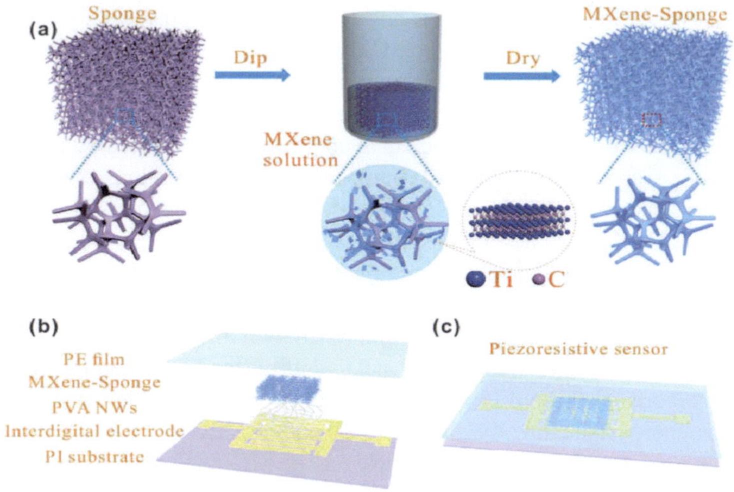

Fig. 25 Fabrication and assembly of MXene-based sensors: **a** Process flow diagram for MXene-sponge synthesis. **b**, **c** Schematic illustration of sensor fabrication using MXene-Sponge/PVA nanowires (NWs) [63]. Copyright 2018. Reproduced with permission from Elsevier

was positioned on the PVA NWs network as it was prepared. In order to ensure the MXene-sponge and interdigitated electrodes made a conformal contact, a thin PE (thermoplastic polymer) sheet was finally glued and packed on it [63].

5.3 *Metal Oxide-MXene Composites*

Metal oxide-MXene composites represent a significant advancement in supercapacitor electrode materials, combining the high conductivity of MXenes with the exceptional pseudocapacitive properties of metal oxides. The integration of metal oxides such as RuO_2, MnO_2, Fe_2O_3, and NiO with MXenes creates synergistic effects that enhance both energy storage capacity and rate capability [64]. These composites address the limitations of specific components—the relatively low capacitance of pristine MXenes and the poor conductivity of metal oxides. The synthesis of metal oxide-MXene composites typically involves controlled growth of metal oxide nanostructures on MXene surfaces through hydrothermal methods, electrodeposition, or in-situ oxidation processes [65]. The unique layered structure of MXenes provides an ideal platform for anchoring metal oxide particles, while their high conductivity facilitates rapid electron transport throughout the composite structure. The interface between MXenes and metal oxides plays a crucial role in determining the composite's performance, influencing charge transfer kinetics and stability during cycling [66]. Manganese oxide (MnO_2)-MXene composites have shown particularly promising results due to their cost-effectiveness and environmental compatibility. The incorporation of MnO_2 nanostructures onto MXene surfaces significantly enhances pseudocapacitive charge storage while maintaining good rate capability [67]. Similarly, ruthenium oxide (RuO_2)-MXene composites exhibit exceptional specific capacitance and cycling stability, though their practical applications are limited by cost considerations [68]. Geng et al. synthesized composite materials combining metal–organic framework-derived Cr_2O_3 with MXene. Figure 26 presents the microstructural analysis of MC-2 using transmission electron microscopy (TEM). The TEM images in Fig. 26a reveal MXene's layered structure alongside agglomerated nanoparticles. High-angle annular dark-field (HAADF) imaging and energy-dispersive X-ray spectroscopy (EDX) mapping (Fig. 26b–f), combined with SEM observations (Fig. 26c, d), confirm MXene's oxidation to TiO_2 and C-dots during hydrothermal processing. The selected area electron diffraction (SAED) pattern in Fig. 26g exhibits distinct diffraction points, indicating the single-crystal nature of the Cr_2O_3/MXene composites. High-resolution TEM (HRTEM) analysis of MC-2 (Fig. 26h) reveals lattice spacings corresponding to both Cr_2O_3 crystal planes (d = 0.2479 nm for (110) and d = 0.3631 nm for (012)) and TiO_2 crystal planes (d = 0.2729 nm for (200) and d = 0.3465 nm for (111)). These observations confirm the formation of multiple heterojunctions: MXene/TiO_2 and Cr_2O_3/TiO_2 p-n heterojunctions, as well as MXene/Cr_2O_3 p-p heterojunctions in the composite structure. Figure 27 presents scanning electron microscopy (SEM) analysis of Cr_2O_3, MC-2, and MC-2* samples. The Cr_2O_3 nanoparticles (Fig. 27a, b) show significant agglomeration and stacking, which reduces gas-sensing active sites and impedes gas molecule diffusion. SEM imaging of MC-2 (Fig. 27c, d) reveals MXene's lamellar structure decorated with numerous nanoparticles. During hydrothermal processing, MXene partially oxidizes to TiO_2 and C-dots, as confirmed by EDX mapping (Fig. 27c–f). While some MXene layers remain intact, most nanoparticles consist of Cr_2O_3 and C-dot-doped TiO_2. The preferential growth of nanoparticles along MXene layer edges helps mitigate particle agglomeration and stacking. MC-2* samples (Fig. 27e, f) demonstrate uniform attachment of Cr_2O_3 nanoparticles across the MXene surface [69].

5.4 *Other Hybrid Systems*

Besides metal oxide composites, MXenes have been successfully integrated with various other materials to create hybrid systems with enhanced supercapacitor performance. Conducting polymers, particularly polyaniline (PANI), polypyrrole (PPy), and poly(3,4-ethylenedioxythiophene) (PEDOT), form effective composites with MXenes [70]. These polymer-MXene hybrids combine the high pseudocapacitance of conducting polymers with MXene's excellent conductivity and mechanical stability, resulting in improved energy storage capabilities and cycling performance [71]. Metal sulfides and phosphides have emerged as promising components in MXene hybrid systems. Materials such as NiS_2, CoS_2, and Ni_2P exhibit high theoretical capacitance and can be grown directly on MXene surfaces. These hybrid

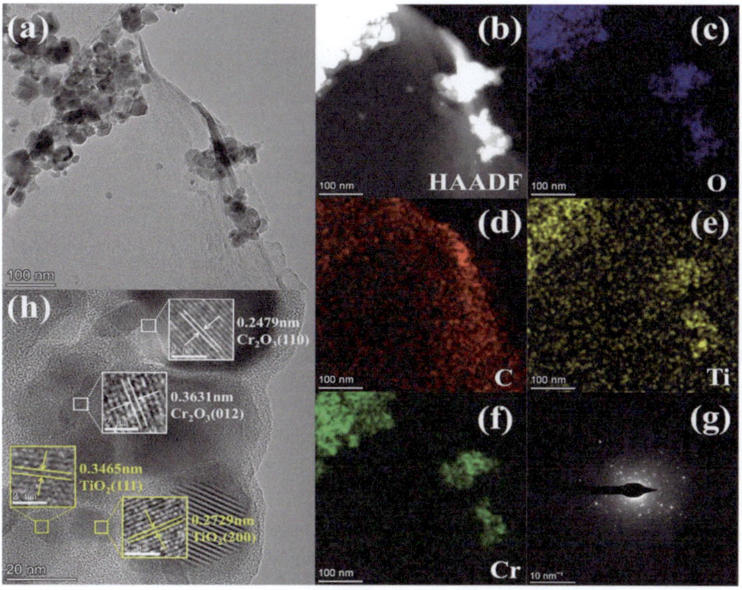

Fig. 26 Microstructural characterization of MC-2 composite: **a** TEM micrograph showing overall morphology, **b–f** High-angle annular dark-field (HAADF) imaging with corresponding elemental mapping by EDX, **g** Selected area electron diffraction (SAED) pattern revealing crystalline structure, and **h** High-resolution TEM image showing lattice fringes and heterojunction interfaces [69]. Copyright 2024. Reproduced with permission from Elsevier

Fig. 27 Comparative SEM analysis at different magnifications: **a**, **b** Pure Cr_2O_3 nanoparticles showing agglomeration, **c**, **d** MC-2 composite revealing MXene layers with attached nanoparticles, and **e**, **f** MC-2* demonstrating uniform distribution of Cr2O3 on MXene surfaces [69]. Copyright 2024. Reproduced with permission from Elsevier

Table 1 Overview of MXene-based hybrid systems

Hybrid system	Composition	Key features	Performance metrics	Applications	References
MXene/ metal oxides	Ti_3C_2Tx/ metal oxide	Enhanced conductivity, increased surface area	Specific capacitance, cycling stability	Supercapacitors	[74]
MXene/ polymers	Ti_3C_2Tx/ Conductive polymers	Improved flexibility, stability	Mechanical strength, conductivity	Flexible electronics	[75]
MXene/ carbon materials	Ti_3C_2Tx/ CNT, graphene	High electrical conductivity, large surface area	Power density, energy density	Energy storage	[76]
MXene/ quantum dots	Ti_3C_2Tx/QDs	Enhanced optoelectronic properties	Photocatalytic activity	Sensors	[77]
MXene/ MOFs	Ti_3C_2Tx/ MOF composites	High porosity, selective absorption	Surface area, selectivity	Gas separation	[78]

structures benefit from enhanced charge transfer kinetics and increased active sites for energy storage [72]. Carbon-based materials, including carbon nanotubes (CNTs), graphene, and activated carbon, form another important category of MXene hybrids. These carbon-MXene composites address the restacking issue while providing enhanced conductivity and surface area. CNTs and graphene sheets act as spacers between MXene layers, maintaining accessible channels for electrolyte ion transport. The synergistic effects in these hybrids lead to improved rate capability and long-term stability [73]. Metal–organic frameworks (MOFs) combined with MXenes create unique hybrid architectures with high surface area and tunable pore structures [72]. The integration of MOFs introduces additional active sites while maintaining the electrical conductivity provided by the MXene network. These hybrids demonstrate enhanced capacitance through the combination of electrical double-layer capacitance and pseudocapacitance mechanisms [73]. Table 1 shows MXene-based hybrid systems and their potential applications.

6 Supercapacitor Applications

Supercapacitors employing MXene-based electrodes operate through both electrical double-layer capacitance (EDLC) and pseudocapacitive mechanisms. The high electrical conductivity of MXenes facilitates rapid electron transport, while their 2D structure and surface chemistry enable efficient ion adsorption and surface redox reactions. The presence of surface functional groups (Tx) on MXene surfaces contributes significantly to pseudocapacitive charge storage through fast, reversible redox reactions. The interlayer spacing between MXene sheets plays a crucial role in facilitating electrolyte ion transport and maximizing the accessible surface area for charge storage. However, supercapacitors are classified into symmetric and asymmetric types based on their electrode design. Symmetric supercapacitors utilize identical materials and structures for both electrodes, while asymmetric configurations employ different materials with distinct properties for each electrode. Figure 28 illustrates the development of MXene-based electrodes in both symmetric and asymmetric configurations. The incorporation of transition metal dichalcogenides and carbides enables the creation of heterostructures, which is crucial for enhancing their performance capabilities [79].

The cyclic voltammetry (CV) curves measured at 5 mV/s are presented in Fig. 29a. The electrical double layer (EDL) exhibits a rectangular shape in neutral solution, with pronounced edge effects visible in the CV curves for all samples at low scan rates. The pristine MXene shows significant capacitance degradation at higher scan rates (200 and 500 mV/s), characterized by distortion of the CV curve shape. Figure 29b demonstrates the relationship between specific capacitance (F/g) and scan rate for

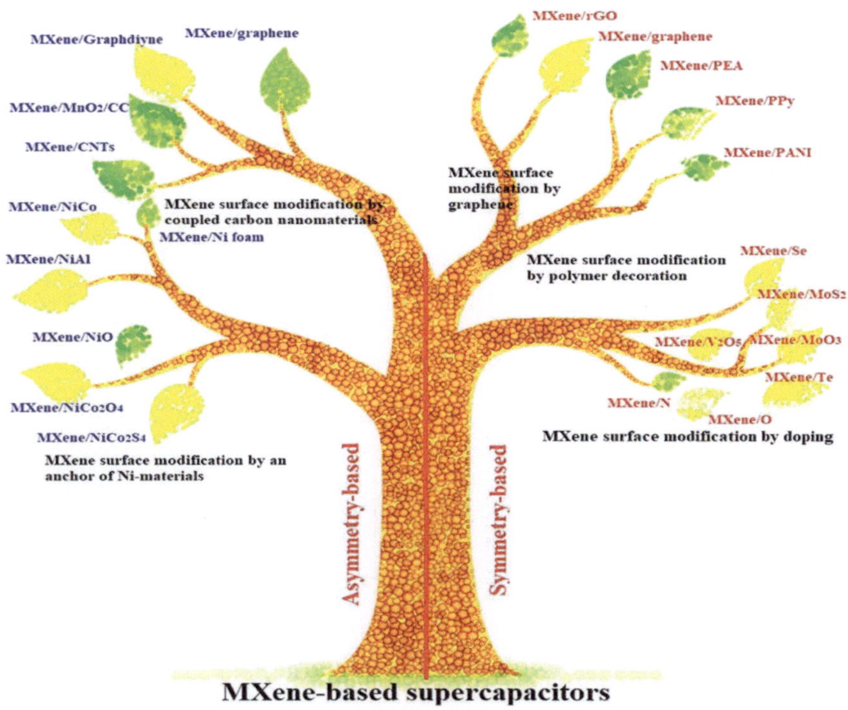

Fig. 28 Schematic representation of symmetric electrodes using identical MXene materials and asymmetric electrodes incorporating different MXene-based heterostructures [79]. Copyright 2024. Open access. Reproduced with permission from Elsevier

2-l-MXene, ranging from low to high scan rates. The electrochemical impedance spectroscopy (EIS) analysis results are illustrated in Fig. 29c. Figure 29d illustrates the rate capability with specific capacitance measured across scan rates from 5 to 2000 mV/s. While pristine MXene exhibits poor capacitance retention of only 33%, the modified 1.5-l-MXene and 2-l-MXene demonstrate superior performance with capacitance retention of 80 and 84.4%, respectively. Figure 29e presents the chronopotentiometry (CP) measurements at 10 A/g, comparing pristine and laser-treated MXene films, confirming similar enhancement patterns observed in acidic solution. The 1.5-l-MXene and 2-l-MXene samples demonstrate superior capacity retention, delivering 31.4 and 34 mAh/g at 10 A/g from initial values of 35.7 and 37.9 mAh/g at 1 A/g, respectively. In contrast, the pristine MXene shows significant capacity loss, dropping from 26.8 mAh/g at 1 A/g to 15.1 mAh/g at 10 A/g [80].

Figure 30a presents cyclic voltammetry (CV) curves for CoMn-OH@Ti_3C_2Tx and AC electrodes within voltage windows of 0 to 0.6 and −1 to 0 V, respectively, indicating a potential operating voltage window of 1.6 V. The CV curves at varying scan rates (10–100 mV s^{-1}) and galvanostatic charge–discharge (GCD) curves in Fig. 30b, c demonstrate non-linear, symmetrical characteristics across current densities of 1–10 A g^{-1}, indicating fast charge/discharge response, reversibility, and effective capacitive behavior of the NiCoMn-OH@Ti_3C_2Tx//AC asymmetric supercapacitor (ASC). Figure 30d shows the specific capacity derived from GCD curves: 223.8, 208, 180, 169.9, 141.4, and 89.9 C g^{-1} at current densities of 1, 2, 3, 5, 8, and 10 A g^{-1}, respectively. The capacity decrease at higher current densities results from incomplete electrode material polarization. The Ragone plot in Fig. 8e displays energy

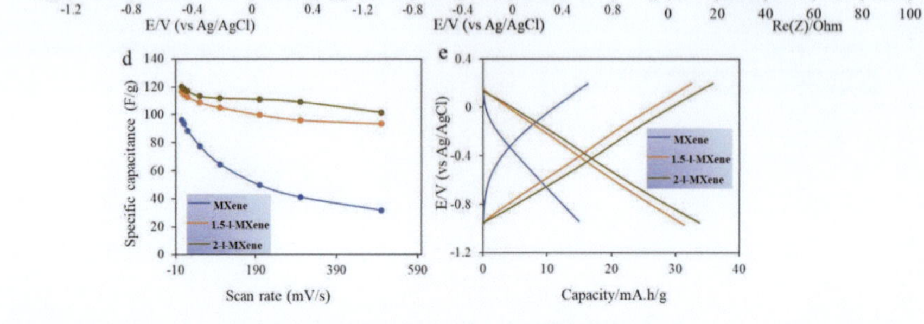

Fig. 29 Comparative electrochemical characterization of pristine and laser-treated MXene films in 7 M LiCl electrolyte: **a** Cyclic voltammetry curves at 5 mV/s scan rate, **b** Rate-dependent specific capacitance of 2-l-MXene, **c** Electrochemical impedance spectroscopy (EIS) analysis, **d** Scan rate dependence of specific capacitance across all samples, and **e** Chronopotentiometry measurements at 10 A/g current density [80]. Copyright 2024. Reproduced with permission from Elsevier

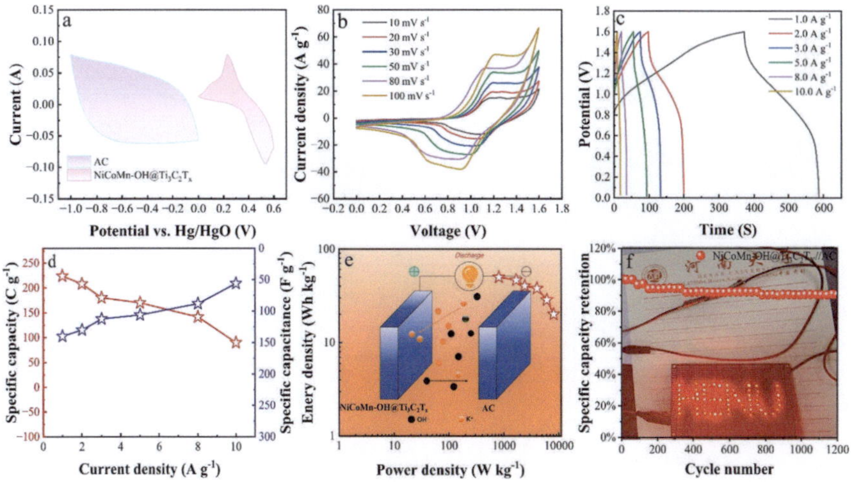

Fig. 30 Electrochemical characterization of NiCoMn-OH@Ti_3C_2Tx/NF//AC asymmetric supercapacitor: **a** Comparative CV curves of individual electrodes at 60 mV s^{-1}, **b** Rate-dependent CV profiles, **c** Galvanostatic charge–discharge curves at various current densities, **d** Specific capacitance versus current density (1–10 A g^{-1}), **e** Energy and power density relationship (Ragone plot), and **f** Long-term cycling stability assessment [81]. Copyright 2025. Reproduced with permission from Elsevier

density versus power density, with the device achieving maximum values of 49.7 and 800 W kg^{-1} at 1 A g^{-1}. Figure 30f demonstrates cycling stability, showing 90.1% capacity retention after 1200 cycles, with an inset showing LED illumination powered by the device for 5 min, confirming its practical applicability [81].

A symmetric solid device (SSD) was fabricated using two GMF3 electrodes (2×2 cm^2 each) separated by PVA-KOH gel electrolyte, forming a GMF3//GMF3 configuration, as depicted in Fig. 31a. Comprehensive electrochemical testing included cyclic voltammetry (CV), galvanostatic charge–discharge (GCD), electrochemical impedance spectroscopy (EIS), and cycling stability measurements. Device voltage optimization through CV at 100 mV/s achieved successful operation up to 1 V (Fig. 31b). CV measurements at scan rates from 10 to 100 mV/s demonstrated consistent curve shapes, indicating robust rate capability (Fig. 11c). GCD testing at current densities between 7 and 10 mA/cm^2 revealed a specific capacitance of 79.33 F/g at 7 mA/cm^2, with corresponding energy and power densities of 11.01 Wh/kg and 583.33 W/kg, respectively (Fig. 31d). Long-term stability testing showed 88.67% capacitance retention after 2000 GCD cycles at 10 mA/cm^2, with slight performance degradation attributed to active material deterioration (Fig. 31e). EIS analysis over 0.1 Hz to 100 kHz yielded low series resistance (Rs $\approx 1\ \Omega$) and charge transfer resistance (Rct $\approx 6\ \Omega$), confirming efficient charge transport properties. These results demonstrate the promising potential of this symmetric solid device configuration for supercapacitor applications [82].

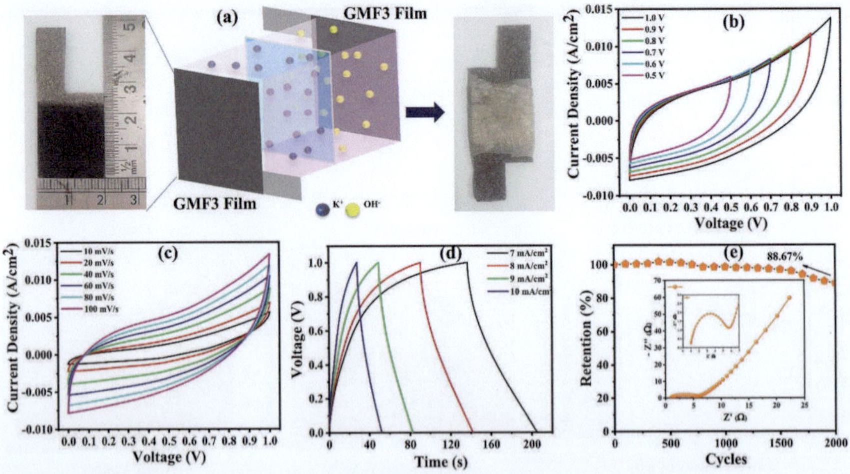

Fig. 31 Performance characteristics of GMF3//GMF3 symmetric solid device (SSD): **a** Device assembly schematic, **b** Operating voltage window optimization at 100 mV/s, **c** Rate-dependent cyclic voltammetry curves (10–100 mV/s), **d** Galvanostatic charge–discharge profiles at varying current densities (7–10 mA/cm^2), **e** Long-term cycling stability over 10,000 cycles with inset showing electrochemical impedance spectroscopy (Nyquist plot) [82]. Copyright 2025. Reproduced with permission from Elsevier

7 Challenges and Future Prospects

MXene-based supercapacitor development faces several significant challenges that hinder widespread commercial adoption. The synthesis of high-quality MXenes remains complex and time-consuming, with challenges in controlling surface terminations and maintaining consistency across batches. Scalability issues persist in production processes, particularly in maintaining uniform properties when scaling up from laboratory to industrial quantities. Additionally, the long-term stability of MXene-based electrodes under various operating conditions requires further improvement, especially regarding oxidation stability and performance degradation over extended cycling. However, the unique properties of MXene-based supercapacitors are driving their exploration in emerging applications. MXene's mechanical flexibility and high conductivity make it a promising material for flexible and wearable electronics, renewable energy systems, smart textiles, biomedical devices, and autonomous systems. Moreover, the future research focuses on efficient synthesis methods, eco-friendly etching processes, optimization of MXene composition, advanced characterization techniques, and innovative electrode architectures and electrolyte systems to enhance performance. The commercial potential of MXene-based supercapacitors depends on cost-effective production methods, market adoption, standardization, quality control measures, environmental considerations, and strategic partnerships between research institutions and industry players to ensure long-term commercial viability.

8 Summary and Conclusions

MXenes have emerged as promising materials for next-generation supercapacitor applications, demonstrating exceptional properties through their unique 2D structure and versatile surface chemistry. Their development since discovery in 2011 has shown remarkable progress in addressing energy storage challenges through various modification strategies and hybrid systems development. Physical and chemical modifications have successfully addressed key challenges such as restacking and limited ion accessibility, while maintaining MXene's intrinsic advantages of high conductivity and surface area. The development of hybrid systems, particularly with metal oxides, conducting polymers, and carbon materials, has demonstrated significant improvements in electrochemical performance metrics, including specific capacitance, rate capability, and cycling stability. The electrochemical performance of MXene-based supercapacitors has shown continuous improvement through optimization of electrode design, electrolyte selection, and fabrication processes. These advances have positioned MXene-based devices as strong candidates for applications ranging from portable electronics to grid-scale energy storage. Despite these achievements, several challenges remain, including scalable synthesis, long-term stability, and cost-effective production. Future research aims to develop efficient synthesis methods, explore new compositions, and comprehend fundamental mechanisms. The commercial viability of MXene-based supercapacitors depends on addressing these challenges while maintaining their superior performance characteristics. The field of MXene-based supercapacitors continues to evolve rapidly, with ongoing research addressing current limitations and exploring new applications. The successful integration of these materials into commercial devices could significantly impact the future of energy storage technology, particularly in applications requiring high power density and rapid charge–discharge capabilities.

Acknowledgements The authors gratefully acknowledge the support provided by Centurion University of Technology and Management, Odisha, India, for carrying out the present research work.

Conflicts of Interest The author(s) declare no conflict of interest.

References

1. Kumar, Y.A., Reddy, G.R., Ramachandran, T., Kulurumotlakatla, D.K., Abd-Rabboh, H.S., Hafez, A.A.A., Rao, S.S., Joo, S.W.: Supercharging the future: MOF-2D MXenes supercapacitors for sustainable energy storage. J. Energy Storage **80**, 110303 (2024)
2. Ali, S., Qaisar, M.A.F., Javed, M.S., Umer, K., Alarfaji, S.S., Mateen, M., Chhattal, M., Ali, S., Parkash, A., Tamang, T.L., Qi, J.: Two-dimensional MXene based innovative electrode materials for supercapacitors: recent advances and prospects. Fuel **377**, 132783 (2024)
3. Zhu, Y., Rajoua, K., Le Vot, S., Fontaine, O., Simon, P., Favier, F.: Modifications of MXene layers for supercapacitors. Nano Energy **73**, 104734 (2020)

4. Salim, O., Mahmoud, K.A., Pant, K.K., Joshi, R.K.: Introduction to MXenes: synthesis and characteristics. Mater. Today Chem. **14**, 100191 (2019)
5. Kuila, T., Bose, S., Mishra, A.K., Khanra, P., Kim, N.H., Lee, J.H.: Chemical functionalization of graphene and its applications. Prog. Mater Sci. **57**(7), 1061–1105 (2012)
6. Wang, Q.H., Kalantar-Zadeh, K., Kis, A., Coleman, J.N., Strano, M.S.: Electronics and optoelectronics of two-dimensional transition metal dichalcogenides. Nat. Nanotechnol. **7**(11), 699–712 (2012)
7. Naguib, M., Kurtoglu, M., Presser, V., Lu, J., Niu, J., Heon, M., Hultman, L., Gogotsi, Y., Barsoum, M.W.: Two-dimensional nanocrystals produced by exfoliation of Ti_3AlC_2. In: MXenes, pp. 15–29. Jenny Stanford Publishing (2023)
8. Magnuson, M., Mattesini, M.: Chemical bonding and electronic-structure in MAX phases as viewed by X-ray spectroscopy and density functional theory. Thin Solid Films **621**, 108–130 (2017)
9. Barsoum, M.W.: The MN+ 1AXN phases: a new class of solids: thermodynamically stable nanolaminates. Prog. Solid State Chem. **28**(1–4), 201–281 (2000)
10. Sun, Z., Music, D., Ahuja, R., Li, S., Schneider, J.M.: Bonding and classification of nanolayered ternary carbides. Phys. Rev. B—Condens. Matter Mater. Phys. **70**(9), 092102 (2004)
11. Bu, F., Zagho, M.M., Ibrahim, Y., Ma, B., Elzatahry, A., Zhao, D.: Porous MXenes: Synthesis, structures, and applications. Nano Today **30**, 100803 (2020)
12. Mendoza-Sánchez, B., Gogotsi, Y.: Synthesis of two-dimensional materials for capacitive energy storage. Adv. Mater. **28**(29), 6104–6135 (2016)
13. Naguib, M., Barsoum, M.W., Gogotsi, Y.: Ten years of progress in the synthesis and development of MXenes. Adv. Mater. **33**(39), 2103393 (2021)
14. Liu, Z., Zheng, L., Sun, L., Qian, Y., Wang, J., Li, M.: (Cr 2/3 Ti 1/3) 3 AlC 2 and (Cr 5/8 Ti 3/ 8) 4 AlC 3: new MAX-phase Compounds in Ti–Cr–Al–C System. J. Am. Ceram. Soc. **97**(1), 67–69 (2014)
15. Anasori, B., Xie, Y., Beidaghi, M., Lu, J., Hosler, B.C., Hultman, L., Kent, P.R., Gogotsi, Y., Barsoum, M.W.: Two-dimensional, ordered, double transition metals carbides (MXenes). ACS Nano **9**(10), 9507–9516 (2015)
16. Meshkian, R., Tao, Q., Dahlqvist, M., Lu, J., Hultman, L., Rosen, J.: Theoretical stability and materials synthesis of a chemically ordered MAX phase, Mo_2ScAlC_2, and its two-dimensional derivate Mo_2ScC_2 MXene. Acta Mater. **125**, 476–480 (2017)
17. Tao, Q., Dahlqvist, M., Lu, J., Kota, S., Meshkian, R., Halim, J., Palisaitis, J., Hultman, L., Barsoum, M.W., Persson, P.O., Rosen, J.: Two-dimensional Mo1. 33C MXene with divacancy ordering prepared from parent 3D laminate with in-plane chemical ordering. Nat. Commun. **8**(1), 14949 (2017)
18. Persson, I., El Ghazaly, A., Tao, Q., Halim, J., Kota, S., Darakchieva, V., Palisaitis, J., Barsoum, M.W., Rosen, J., Persson, P.O.: Tailoring structure, composition, and energy storage properties of MXenes from selective etching of in-plane, chemically ordered MAX phases. Small **14**(17), 1703676 (2018)
19. Deysher, G., Shuck, C.E., Hantanasirisakul, K., Frey, N.C., Foucher, A.C., Maleski, K., Sarycheva, A., Shenoy, V.B., Stach, E.A., Anasori, B., Gogotsi, Y.: Synthesis of Mo_4VAlC_4 MAX phase and two-dimensional Mo_4VC_4 MXene with five atomic layers of transition metals. ACS Nano **14**(1), 204–217 (2019)
20. Banjare, M.K., Behera, K., Banjare, R.K., Pandey, S., Ghosh, K.K.: Challenges and future prospective of MXenes and MXene-based hybrid nanostructures. In: MXenes as Emerging Modalities for Environmental and Sensing Applications, pp. 253–265. Elsevier (2025)
21. Lu, X., Zhang, W., Wang, C., Wen, T.C., Wei, Y.: One-dimensional conducting polymer nanocomposites: synthesis, properties and applications. Prog. Polym. Sci. **36**(5), 671–712 (2011)
22. Luo, J., Matios, E., Wang, H., Tao, X., Li, W.: Interfacial structure design of MXene-based nanomaterials for electrochemical energy storage and conversion. InfoMat **2**(6), 1057–1076 (2020)

23. Amara, U., Mehran, M.T., Sarfaraz, B., Mahmood, K., Hayat, A., Nasir, M., Riaz, S., Nawaz, M.H.: Perylene diimide/MXene-modified graphitic pencil electrode-based electrochemical sensor for dopamine detection. Microchim. Acta **188**(7), 230 (2021)
24. Wu, H., Almalki, M., Xu, X., Lei, Y., Ming, F., Mallick, A., Roddatis, V., Lopatin, S., Shekhah, O., Eddaoudi, M., Alshareef, H.N.: MXene derived metal–organic frameworks. J. Am. Chem. Soc. **141**(51), 20037–20042 (2019)
25. Khan, K., Tareen, A.K., Aslam, M., Wang, R., Zhang, Y., Mahmood, A., Ouyang, Z., Zhang, H., Guo, Z.: Recent developments in emerging two-dimensional materials and their applications. J. Mater. Chem. C **8**(2), 387–440 (2020)
26. Samtham, M., Singh, D., Hareesh, K., Devan, R.S.: Perspectives of conducting polymer nanostructures for high-performance electrochemical capacitors. J. Energy Storage **51**, 104418 (2022)
27. Chen, W., Liu, P., Liu, Y., Liu, Z.: Recent advances in Two-dimensional Ti_3C_2Tx MXene for flame retardant polymer materials. Chem. Eng. J. **446**, 137239 (2022)
28. Verger, L., Natu, V., Carey, M., Barsoum, M.W.: MXenes: an introduction of their synthesis, select properties, and applications. Trends Chem. **1**(7), 656–669 (2019)
29. Dillon, A.D., Ghidiu, M.J., Krick, A.L., Griggs, J., May, S.J., Gogotsi, Y., Barsoum, M.W., Fafarman, A.T.: Highly conductive optical quality solution-processed films of 2D titanium carbide. Adv. Func. Mater. **26**(23), 4162–4168 (2016)
30. Wang, Y., Xu, Y., Hu, M., Ling, H., Zhu, X.: MXenes: focus on optical and electronic properties and corresponding applications. Nanophotonics **9**(7), 1601–1620 (2020)
31. Bai, Y., Zhou, K., Srikanth, N., Pang, J.H., He, X., Wang, R.: Dependence of elastic and optical properties on surface terminated groups in two-dimensional MXene monolayers: a first-principles study. RSC Adv. **6**(42), 35731–35739 (2016)
32. Fu, B., Sun, J., Wang, C., Shang, C., Xu, L., Li, J., Zhang, H.: MXenes: synthesis, optical properties, and applications in ultrafast photonics. Small **17**(11), 2006054 (2021)
33. Berdiyorov, G.R.: Optical properties of functionalized Ti3C2T2 (T= F, O, OH) MXene: first-principles calculations. AIP Adv. **6**(5), 055105 (2016)
34. Khazaei, M., Ranjbar, A., Arai, M., Sasaki, T., Yunoki, S.: Electronic properties and applications of MXenes: a theoretical review. J. Mater. Chem. C **5**(10), 2488–2503 (2017)
35. Song, B., Shi, C., Qu, Y., Liu, H., Yang, H., Wu, X., Liu, X.: The electrical properties and charge transport mechanism of MXenes. Chin. Chem. Lett. 110823 (2025)
36. Li, Y., Wang, J., Chen, H., Sun, H., Xu, H., Yang, D.: Preparation and studies of MXene@ PDA/PVDF composites with improved dielectric properties. Polymer **307**, 127239 (2024)
37. Ronchi, R.M., Arantes, J.T., Santos, S.F.: Synthesis, structure, properties and applications of MXenes: current status and perspectives. Ceram. Int. **45**(15), 18167–18188 (2019)
38. Shukla, V.: The tunable electric and magnetic properties of 2D MXenes and their potential applications. Mater. Adv. **1**(9), 3104–3121 (2020)
39. Awan, H.T.A., Kumar, L., Wong, W.P., Walvekar, R., Khalid, M.: Recent progress and challenges in MXene-based phase change material for solar and thermal energy applications. Energies **16**(4), 1977 (2023)
40. Oyehan, T.A., Salami, B.A., Abdulrasheed, A.A., Hambali, H.U., Gbadamosi, A., Valsami-Jones, E., Saleh, T.A.: MXenes: Synthesis, properties, and applications for sustainable energy and environment. Appl. Mater. Today **35**, 101993 (2023)
41. Liao, M., Zheng, Z., Jiang, H., Ma, M., Wang, L., Wang, Y., Zhuang, S.: MXenes as emerging adsorbents for removal of environmental pollutants. Sci. Total Environ. **912**, 169014 (2024)
42. Fadahunsi, A.A., Li, C., Khan, M.I., Ding, W.: MXenes: state-of-the-art synthesis, composites and bioapplications. J. Mater. Chem. B **10**(23), 4331–4345 (2022)
43. Wang, D., Zhou, C., Filatov, A.S., Cho, W., Lagunas, F., Wang, M., Vaikuntanathan, S., Liu, C., Klie, R.F., Talapin, D.V.: Direct synthesis and chemical vapor deposition of 2D carbide and nitride MXenes. Science **379**(6638), 1242–1247 (2023)
44. Mozafari, M., Soroush, M.: Surface functionalization of MXenes. Mater. Adv. **2**(22), 7277–7307 (2021)

45. Kamysbayev, V., Filatov, A.S., Hu, H., Rui, X., Lagunas, F., Wang, D., Klie, R.F., Talapin, D.V.: Covalent surface modifications and superconductivity of two-dimensional metal carbide MXenes. Science **369**(6506), 979–983 (2020)
46. Jayakumar, S., Santhosh, P.C., Ramakrishna, S., Radhamani, A.V.: 2D (Ti3C2Tx) MXene: a comprehensive review of advancements in synthesis protocols, applications in supercapacitors, sustainability targets and future prospects. J. Energy Storage **97**, 112741 (2024)
47. Taufiq Musa, M., Shaari, N., Kamarudin, S.K.: Carbon nanotube, graphene oxide and montmorillonite as conductive fillers in polymer electrolyte membrane for fuel cell: an overview. Int. J. Energy Res. **45**(2), 1309–1346 (2021)
48. Wu, L., Li, Y., Fu, Z., Su, B.L.: Hierarchically structured porous materials: synthesis strategies and applications in energy storage. Natl. Sci. Rev. **7**(11), 1667–1701 (2020)
49. Pandey, M., Deshmukh, K., Raman, A., Asok, A., Appukuttan, S., Suman, G.R.: Prospects of MXene and graphene for energy storage and conversion. Renew. Sustain. Energy Rev. **189**, 114030 (2024)
50. Huang, H., Jiang, R., Feng, Y., Ouyang, H., Zhou, N., Zhang, X., Wei, Y.: Recent development and prospects of surface modification and biomedical applications of MXenes. Nanoscale **12**(3), 1325–1338 (2020)
51. Chen, J., Chen, K., Tong, D., Huang, Y., Zhang, J., Xue, J., Huang, Q., Chen, T.: CO_2 and temperature dual responsive "Smart" MXene phases. Chem. Commun. **51**(2), 314–317 (2015)
52. Wang, H., Zhang, J., Wu, Y., Huang, H., Jiang, Q.: Chemically functionalized two-dimensional titanium carbide MXene by in situ grafting-intercalating with diazonium ions to enhance supercapacitive performance. J. Phys. Chem. Solids **115**, 172–179 (2018)
53. Yang, J., Bao, W., Jaumaux, P., Zhang, S., Wang, C., Wang, G.: MXene-based composites: synthesis and applications in rechargeable batteries and supercapacitors. Adv. Mater. Interfaces **6**(8), 1802004 (2019)
54. Lu, J., Xu, L., Xie, C., Wei, Q., Jiang, Q., Yao, G., Han, Z.: Porous-multilayered Ti3C2Tx MXene hybrid carbon foams for tunable and efficient electromagnetic wave absorption. Carbon **229**, 119477 (2024)
55. Xu, C., Song, S., Liu, Z., Chen, L., Wang, L., Fan, D., Kang, N., Ma, X., Cheng, H.M., Ren, W.: Strongly coupled high-quality graphene/2D superconducting Mo_2C vertical heterostructures with aligned orientation. ACS Nano **11**(6), 5906–5914 (2017)
56. Gao, J., Cheng, Z., Shao, C., Zhao, Y., Zhang, Z., Qu, L.: A 2D free-standing film-inspired electrocatalyst for highly efficient hydrogen production. J. Mater. Chem. A **5**(24), 12027–12033 (2017)
57. Yu, L.P., Zhou, X.H., Lu, L., Xu, L., Wang, F.J.: MXene/carbon nanotube hybrids: synthesis, structures, properties, and applications. Chemsuschem **14**(23), 5079–5111 (2021)
58. Mohajer, F., Ziarani, G.M., Badiei, A., Iravani, S., Varma, R.S.: MXene-carbon nanotube composites: properties and applications. Nanomaterials **13**(2), 345 (2023)
59. Li, X., Zhu, J., Wang, L., Wu, W., Fang, Y.: In–situ growth of carbon nanotubes on two–dimensional titanium carbide for enhanced electrochemical performance. Electrochim. Acta **258**, 291–301 (2017)
60. Dall'Agnese, Y., Rozier, P., Taberna, P.L., Gogotsi, Y., Simon, P.: Capacitance of two-dimensional titanium carbide (MXene) and MXene/carbon nanotube composites in organic electrolytes. J. Power Sour. **306**, 510–515 (2016)
61. Das, P., Marvi, P.K., Ganguly, S., Tang, X., Wang, B., Srinivasan, S., Rajabzadeh, A.R., Rosenkranz, A.: MXene-based elastomer mimetic stretchable sensors: design, properties, and applications. Nano-Micro Lett. **16**(1), 135 (2024)
62. Xu, H., Yin, X., Li, X., Li, M., Liang, S., Zhang, L., Cheng, L.: Lightweight Ti2CT x MXene/ poly (vinyl alcohol) composite foams for electromagnetic wave shielding with absorption-dominated feature. ACS Appl. Mater. Interfaces. **11**(10), 10198–10207 (2019)
63. Yue, Y., Liu, N., Liu, W., Li, M., Ma, Y., Luo, C., Wang, S., Rao, J., Hu, X., Su, J., Zhang, Z.: 3D hybrid porous Mxene-sponge network and its application in piezoresistive sensor. Nano Energy **50**, 79–87 (2018)

64. Kitchamsetti, N., Cho, J.S.: A roadmap of recent advances in MXene@ MOF hybrids, its derived composites: Synthesis, properties, and their utilization as an electrode for supercapacitors, rechargeable batteries and electrocatalysis. J. Energy Storage **80**, 110293 (2024)
65. Karmakar, M., Swain, M., Punyasloka, S., Mondal, B., Noriyoshi, M., Ghoroi, C.: Enhanced capacitance of MXene synthesized through safer route and its composite with amino graphene oxide. Mater. Chem. Phys. **331**, 130187 (2025)
66. Ravichandran, A.: MXene-based hybrid for electrochemical supercapacitor applications. In: MXenes as Emerging Modalities for Environmental and Sensing Applications, pp. 279–293. Elsevier (2025)
67. Zhang, Y., Wang, Y., Huo, H., Zhang, P., Zhang, X., Che, H., Mu, J., Zhang, Z., Li, P., Wu, C.: Dimension-combined nanoarchitectonics of MXene/MnO2/Ag ternary heterostructure as electrodes for enhanced energy storage performance. J. Alloy. Compd. **1010**, 177459 (2025)
68. Zhang, Y., Zhang, Z., Yu, Z., Addad, A., Wang, Q., Roussel, P., Szunerits, S., Boukherroub, R.: Ruthenium oxide nanoparticles immobilized on Ti3C2 MXene nanosheets for boosting seawater electrolysis. ACS Appl. Mater. Interfaces. **15**(50), 58345–58355 (2023)
69. Geng, W., Song, P., Xie, T., Duan, L.: Enhanced n-butanol sensing performance of metal-organic frameworks-derived Cr2O3/MXene composites. Appl. Surf. Sci. 160571 (2024)
70. Nezafat, Z., Mirzaei, A., Nasrollahzadeh, M., Javanshir, S.: Recent advances in MXene-based composite materials with conducting polymers for supercapacitors. J. Energy Storage **86**, 111215 (2024)
71. Wang, Y., Wang, Y.: Recent progress in MXene layers materials for supercapacitors: high-performance electrodes. SmartMat **4**(1), e1130 (2023)
72. Hanan, A., Lakhan, M.N., Bibi, F., Khan, A., Soomro, I.A., Hussain, A., Aftab, U.: MOFs coupled transition metals, graphene, and MXenes: emerging electrocatalysts for hydrogen evolution reaction. Chem. Eng. J. 148776 (2024)
73. Thomas, S.A., Patra, A., Al-Shehri, B.M., Selvaraj, M., Aravind, A., Rout, C.S.: MXene based hybrid materials for supercapacitors: recent developments and future perspectives. J. Energy Storage **55**, 105765 (2022)
74. Jiang, X., Bai, J., Wijerathne, B., Zhou, Q., Zhang, F., Liao, T., Sun, Z.: 3D Printing MXene-based electrodes for supercapacitors. Chem. Asian J. e202400568 (2024)
75. Manibalan, K., Chen, J.T.: Recent progress on MXene–polymer composites for soft electronics applications in sensing and biosensing: a review. J. Mater. Chem. A (2024)
76. Jaberi, S.S., Asen, P., Esfandiar, A., Tolstoy, V.P.: MXene/carbon hybrid nanostructures and heteroatom-doped derivatives for enhanced electrochemical energy storage. J. Energy Storage **90**, 111751 (2024)
77. Karimi, S., Amjadi, M.: Investigation of the chemiluminescence behavior of Ti3C2 MXene quantum dots as a sensing platform for uric acid. Microchem. J. **207**, 112068 (2024)
78. Li, R., Fu, X., Liu, G., Li, J., Zhou, G., Liu, G., Jin, W.: Room-temperature in situ synthesis of MOF@ MXene membrane for efficient hydrogen purification. J. Memb. Sci. **664**, 121097 (2022)
79. Pour, G.B., Aval, L.F.: Recent advances in supercapacitors based on MXene surface modification: a review of symmetric and asymmetric electrodes material. Results Eng **24**, 103045 (2024)
80. Dutta, A., Porat, H., Goldreich, A., Yadgarov, L., Kafizas, A., Shpigel, N., Borenstein, A.: Laser exfoliated 2D MXene for supercapacitor applications. Chem. Eng. J. **500**, 157342 (2024)
81. Shu, P., Luo, X., Jiang, R., Yue, G., Gao, Y., Huo, J., Dong, C., Tan, F.: Design of NiCoMn-OH@ Ti_3C_2Tx composite electrode with hollow rhombic dodecahedral structure for efficient supercapacitor. J. Energy Storage **107**, 114949 (2025)
82. Chavan, P.P., Babar, U.D., Redekar, R.S., Chougale, A.D., Tarwal, N.L., Kamble, P.D.: Exploring the influence of precursor concentration on hydrothermally synthesized rGO/ $MnFe_2O_4$ composite for a high-performance supercapacitor. J. Energy Storage **109**, 115215 (2025)

Application of MXenes for Perovskite Solar Cells

Armeen Singh, Rohit Yadav, Ramesh Banoth, and Kusum Kumari

Abstract Energy requirements are increasing day by day and efforts are being made to combat the use of fossil fuels for raising energy needs. The use of renewable energy sources has gained a boost in recent years. Photovoltaics technology is being employed to extract energy from the sun. Photovoltaic technology captures sunlight and converts it into electrical energy. It plays a key role in harnessing renewable solar power for various applications. The conventional solar cell technology is associated with problems of high production cost and efficiency. Third generation solar cells are gaining popularity due to their low manufacturing cost, flexibility, and potential to increase efficiency. Perovskite solar cells (PSCs) have been studied a lot and efforts are being made to improve their efficiency from 25.7% as reported in 2021 as they have the potential to overcome the Shockey Queisser limit. They are seen as a promising alternative to traditional solar technologies. MXenes have a unique two-dimensional layered structure, composed of transition metal carbides, nitrides, or carbonitrides. MXenes offer excellent electrical conductivity, high carrier mobility, and tunable work function, making them ideal for use in PSCs as additives in the absorber layer, as well as in ETLs, HTLs, and electrodes. Integrating MXenes into perovskite solar cells enhances conductivity and charge transport, improving overall device efficiency. The following study explores the use of MXenes in perovskite solar cells and their effect on the efficiency of PSCs.

Keywords MXenes · Perovskite solar cells · Energy conversion · 2D materials

A. Singh
Department of Electronics and Communication Engineering, Punjab Engineering College, Chandigarh, Punjab, India

R. Yadav · R. Banoth · K. Kumari (✉)
Energy Materials and Photovoltaics Laboratory, Department of Physics, National Institute of Technology, Warangal, Telangana, India
e-mail: kusum@nitw.ac.in

© The Author(s), under exclusive license to Springer Nature Singapore Pte Ltd. 2025
S. Moharana and R. N. Mahaling (eds.), *Surface Engineering of MXenes*, Engineering Materials, https://doi.org/10.1007/978-981-96-3640-2_10

1 Introduction

1.1 Overview of Perovskite Solar Cells

Definition and Significance:

The Sun offers an abundant and renewable energy source that can significantly reduce our reliance on fossil fuels. Harnessing solar energy is key to achieving a sustainable and clean energy future. Solar technology plays a crucial role in directly converting sunlight into electrical energy through photovoltaic cells. Solar energy stands out as a superior renewable resource to wind, biomass, and hydroelectricity due to its widespread availability, scalability, and minimal environmental impact [1–3]. Unlike wind and hydroelectricity, which depend on specific geographic conditions, solar energy can be harnessed virtually anywhere on Earth. Additionally, solar panels are easy to install on a variety of surfaces and have low operational costs, unlike biomass, which requires a consistent supply of organic material and in transportation to plants need fossil fuels like petrol, and diesel [4, 5]. Among various photovoltaic technologies, Solar cells are categorized into three generations (as shown in Fig. 1), each with distinct characteristics and power conversion efficiency (PCE) levels. The first generation includes wafer-based solar cells, such as monocrystalline and polycrystalline silicon cells. First-generation solar cells achieve a peak efficiency of 26.1%, are more efficient, and have a longer lifespan, though they are expensive to produce [6, 7]. The second generation encompasses thin-film solar cells, including CIGS (Copper Indium Gallium Selenide), CdTe/CdS (Cadmium Telluride/Cadmium Sulfide), and amorphous silicon cells [8, 9]. These cells boast an efficiency of up to 23% and exhibit high light absorption; however, they degrade quickly and are less efficient overall. The third generation focuses on advanced materials and technologies, featuring perovskite solar cells (PSCs), organic solar cells, quantum dot solar cells (QDSCs), and dye-sensitized solar cells (DSSCs) with efficiencies of around 18.1%. Third-generation cells are noted for their low cost and potential to overcome the Shockley-Queisser limit of 31–41% power efficiency, representing a significant advancement in solar technology. Perovskite solar cells (PSCs) have rapidly gained prominence, with a PCE of 25.7% reported in 2021 [10, 11]. This efficiency is nearing the theoretical limit of about 31%, and recent research suggests that halide-based PSCs could potentially exceed the Shockley-Queisser limit of 33.7% for single-junction solar cells [12, 13]. The appeal of PSCs lies not only in their high efficiency but also in their cost-effectiveness and ease of manufacturing, largely due to solution-based processing methods.

Perovskite materials such as $MAPbI_3$, $FAPbI_3$, and $CsPbI_3$ have been the focus of extensive research because of their tunable bandgap, high light absorption capabilities, efficient charge carrier mobility, and long diffusion lengths. The general chemical formula for perovskite compounds is ABX_3, where "A" and "B" are cations and "X" is an anion. Despite these promising features, the stability of PSCs remains

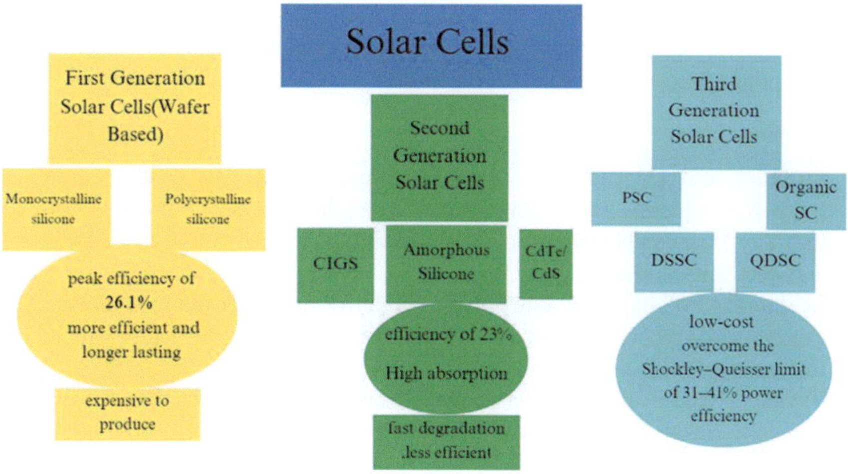

Fig. 1 Types of solar cells

a significant challenge, largely due to the inherent instability of perovskite materials. Efforts to improve stability and performance have included approaches like passivation techniques, grain size optimization, and advanced interface engineering. Perovskite solar cells (PSCs) are a class of photovoltaic devices that use perovskite-structured materials as the light-absorbing layer. Named after the mineral perovskite ($CaTiO_3$), these materials are characterized by their unique crystal structure, which is crucial for their efficient light absorption and charge transport properties. The advent of perovskite solar cells has revolutionized the field of photovoltaics due to their high power conversion efficiencies, low production costs, and ease of fabrication [14].

Key Advantages:

This level of efficiency is comparable to that of traditional silicon-based solar cells, which have been the industry standard for decades. The high efficiency of perovskite solar cells is attributed to their excellent light absorption properties, tunable bandgap, and high charge-carrier mobility, which allow them to convert sunlight into electricity with minimal losses. This rapid improvement in efficiency has positioned perovskite solar cells as a leading contender in the quest for more efficient, next-generation solar technologies.

- **High Efficiency**: Perovskite solar cells have achieved rapid advancements in efficiency, reaching over 25% in laboratory settings, which is comparable to traditional silicon-based solar cells.
- **Flexible and Lightweight**: Unlike rigid silicon cells, perovskite solar cells can be deposited on flexible substrates, making them suitable for a range of applications including flexible and lightweight devices.

- **Low-Cost Manufacturing**: The fabrication processes for perovskite solar cells are relatively simple and cost-effective, allowing for potential large-scale production at lower costs [15, 16].

Challenges:

- **Stability and Degradation**: One of the major challenges is the stability of perovskite materials under environmental conditions. Degradation due to moisture, oxygen, and light exposure can significantly affect the long-term performance and reliability of these cells.
- **Scalability**: While perovskite solar cells perform well at the laboratory scale, scaling up the production process to commercial levels poses difficulties, including maintaining uniform quality and performance.
- **Toxicity**: Lead is often used in perovskites which make them toxic. Many other rear earth metals also result in toxicity [17, 18].

1.2 Introduction to MXenes

Definition and Properties:

In the realm of advanced materials, two-dimensional (2D) MXenes are a class of two-dimensional inorganic compounds, that consist of atomically thin layers of transition metal carbides, nitrides, or carbonitrides, particularly Titanium Carbide $Ti_3C_2Ti_x$, have emerged as promising candidates for enhancing the performance of PSCs. MXenes are a novel class of two-dimensional materials, derived from MAX phases, composed of transition metals, carbon or nitrogen, and surface termination groups like hydroxyl or fluorine. These materials are produced by selectively etching the "A" element from their layered ceramic parent compounds, resulting in a structure that offers high electrical conductivity, mechanical strength, and versatility in surface chemistry. Since their discovery in 2011, MXenes have shown promise in various applications, including energy storage, catalysis, sensing, and water purification, due to their unique combination of properties. The name "MXene" is derived from "M" (the transition metal), "X" (carbon or nitrogen), and the suffix "ene," which is used to denote their two-dimensional nature, similar to graphene [19, 20].

Historical Background and Discovery:

- **Discovery**: MXenes were first synthesized by researchers at Drexel University, and since their discovery, they have attracted significant attention due to their exceptional electrical conductivity, high surface area, and diverse chemical properties.
- **Development**: The exploration of MXenes has rapidly expanded, revealing a broad range of compositions and potential applications in various fields, including energy storage, catalysis, and electronics.

1.3 Relevance of MXenes in Solar Cells

Overview of Their Potential in Enhancing Perovskite Solar Cells

Since their introduction into perovskite solar cells in 2018, MXenes have been recognized for their excellent electrical conductivity, high carrier mobility, and tunable work function. These properties make them suitable for a wide range of applications within PSCs, including their use as additives in the absorber layer, as well as in electron transport layers (ETL), hole transport layers (HTL), and electrodes. Recent studies have shown that incorporating $Ti_3C_2Ti_x$ MXene into PSCs can significantly enhance the efficiency of these devices. By adjusting the work function of both the perovskite absorber and the TiO_2 ETL, MXenes facilitate better charge extraction and improve the overall energy conversion process. This modification can lead to substantial improvements in PCE and reduce hysteresis, which is often observed in PSCs. The unique properties of MXenes, such as their high transparency and mechanical strength, further contribute to their potential in advancing photovoltaic technology [21, 22].

Potential Benefits:

- **Enhanced Conductivity**: MXenes exhibit high electrical conductivity, which can improve charge transport and collection in solar cells, potentially enhancing overall efficiency.
- **High Surface Area**: The large surface area of MXenes can facilitate better interaction with the perovskite layer, potentially improving charge extraction and reducing recombination losses.
- **Flexibility and Mechanical Strength**: MXenes can add mechanical stability and flexibility to the solar cell structure, which is advantageous for the development of flexible and durable solar panels [23].

Application Prospects:

The remarkable electrical conductivity of MXenes facilitates efficient charge transport within the solar cells, potentially leading to higher power conversion efficiencies. Additionally, the hydrophilic nature of MXenes can be leveraged to improve the adhesion and uniformity of perovskite layers, thereby enhancing the overall structural integrity of the device. Moreover, MXenes can act as effective barrier layers, protecting the perovskite material from environmental factors like humidity and oxygen, which are known to degrade perovskite films over time.

In the following sections, we will delve into the intrinsic properties of MXenes, including their synthesis methods, chemical structure, and functionalization techniques. We will also review the fundamentals of perovskite solar cells, covering their operational principles, material composition, and existing challenges. Finally, we will explore the synergies between MXenes and perovskite materials, discussing how the integration of MXenes can lead to advancements in the performance, stability, and manufacturability of perovskite solar cells. By understanding these interactions, we can better appreciate the potential of MXene-enhanced PSCs in paving the way for more efficient, durable, and scalable solar energy solutions.

2 MXenes: Properties and Synthesis

2.1 Structural Properties

Layered Structure and Composition:

MXenes are two-dimensional (2D) materials that are derived from their precursor MAX phases, which are a family of layered ceramics. The general formula for a MAX phase is $M_{n+1}AX_n$, where:

- **M** represents a transition metal (such as titanium, vanadium, or molybdenum).
- **A** is an element like aluminium, silicon, or gallium.
- **X** is carbon (C) or nitrogen (N).

To synthesize MXenes, the "A" layer is selectively removed from the MAX phase using a chemical etching process, typically involving hydrofluoric acid (HF) or a mixture of other acids. This removal leaves behind a layered structure composed of $M_{n+1}X_n$ layers. These layers are held together by weak van der Waals forces, similar to other 2D materials like graphene, which allows them to be easily exfoliated into individual sheets. The thickness of these sheets typically ranges from just a few nanometers, with a high aspect ratio, meaning they are much wider than they are thick. This 2D structure is fundamental to the unique properties of MXenes, such as their high surface area, excellent mechanical strength, and potential for high electrical conductivity.

Surface Terminations:

One of the defining features of MXenes is the presence of surface termination groups, which are functional groups that attach to the surface of the $M_{n+1}X_n$ layers during the synthesis process. Common surface terminations include:

- **–O (oxygen)**
- **–OH (hydroxyl)**
- **–F (fluorine)**

These terminations are introduced during the etching process and can be further modified through post-synthesis treatments. The nature and density of these termination groups have a significant impact on the properties of MXenes. For instance, they influence the material's hydrophilicity (how easily it interacts with water), chemical reactivity, electrical conductivity, and even its mechanical properties. The ability to tune these surface terminations makes MXenes highly versatile for different applications. For example, the hydrophilic nature of certain terminations allows MXenes to disperse easily in water, making them suitable for use in aqueous environments, while specific terminations can enhance the material's catalytic activity or selectivity when used in sensors or energy storage devices.

This tunable surface chemistry is one of the reasons why MXenes are considered highly promising for a wide range of applications, including energy storage, water purification, catalysis, and electromagnetic interference shielding. The ability to control and modify the surface properties of MXenes opens up possibilities for customizing the material to meet the specific needs of different technological applications [24].

Surface Chemistry and Modification:

- **Surface Functionalization**: The surface chemistry of MXenes can be tuned by modifying the termination groups or introducing additional functional groups. This tunability allows for the optimization of MXenes for specific applications, including their integration into perovskite solar cells.
- **Interaction with Other Materials**: The ability to modify the surface of MXenes enhances their compatibility with various materials, enabling effective coupling with perovskite layers and improving overall device performance [25].

2.2 Electrical and Optical Properties

Electrical Conductivity: MXenes are renowned for their exceptional electrical conductivity, a characteristic largely attributed to their unique metallic bonding and layered structure. This high level of conductivity is a result of the free movement of electrons within the layers of MXenes, which resembles the behaviour of metals. This property is particularly advantageous in the realm of electronic devices, including solar cells, where efficient charge transport is crucial for maximizing device performance. The outstanding conductivity of MXenes enables them to serve as effective charge transport layers or conductive additives in perovskite solar cells, thereby reducing resistive losses and enhancing the overall PCE. Additionally, their conductive nature makes them suitable for other applications like supercapacitors, batteries, and electromagnetic shielding, where rapid electron transfer is essential [26–28].

Band Structure: The electronic band structure of MXenes is highly versatile and can be finely tuned by altering their composition and surface terminations. MXenes typically consist of transition metals such as titanium, vanadium, or molybdenum, combined with carbon or nitrogen, forming a layered structure. The electronic properties of MXenes, including their bandgap, can be adjusted by varying the transition metal elements, the carbon-to-nitrogen ratio, or by introducing different surface functional groups such as –OH, –F, or –O. This tunability allows MXenes to be tailored for specific applications by optimizing their electronic properties, such as their work function, band alignment, and carrier mobility. For instance, in the context of perovskite solar cells, adjusting the band structure of MXenes can help align the energy levels with the perovskite layer, facilitate efficient charge transfer, and minimize energy losses. This ability to customize the band structure makes MXenes a highly adaptable material, suitable for a wide range of applications, from energy storage and conversion to sensing and catalysis [29].

Optical Properties: MXenes are distinguished by their strong optical absorption, particularly in the visible and near-infrared regions of the electromagnetic spectrum. This characteristic arises from the intrinsic electronic structure of MXenes, where the interaction between light and the material's surface electrons leads to significant absorption across a broad range of wavelengths. The ability of MXenes to absorb light so effectively makes them highly advantageous in photovoltaic applications, including perovskite solar cells, where maximizing light absorption is crucial for enhancing PCE. By incorporating MXenes into the active layers of perovskite solar cells, it is possible to increase the amount of light that is absorbed and converted into electrical energy, thereby improving the overall efficiency of the device. Moreover, the strong optical properties of MXenes are not only beneficial for solar energy applications but also for photothermal therapy, photodetectors, and other optoelectronic devices where efficient light absorption is required [26, 27].

Reflection and Transmission: The reflective and transmissive properties of MXenes can be finely tuned by modifying their surface chemistry, such as by altering the types and densities of functional groups (e.g., –OH, –O, –F), and by controlling the thickness and stacking order of MXene layers. This ability to adjust reflection and transmission is particularly useful in optimizing the optical performance of MXenes for specific applications. For instance, in perovskite solar cells, MXenes can be engineered to minimize unwanted reflection and maximize light transmission into the active layer, ensuring that more photons are absorbed by the perovskite material rather than being lost to reflection or passing through the device. Additionally, the layered structure of MXenes allows for the creation of thin films with varying degrees of transparency and reflectivity, making them ideal candidates for use as antireflective coatings, light-trapping layers, or transparent conductive electrodes in solar cells. This versatility in manipulating the optical properties of MXenes extends their applicability beyond photovoltaics, making them suitable for use in smart windows, displays, and other technologies where control over light transmission and reflection is critical [30, 31].

2.3 Synthesis Methods

- **Chemical Etching**: The most common method (Fig. 2) for synthesizing MXenes involves selective chemical etching of the A element from the MAX phase. This process typically uses strong acids or other etchants to remove the A element, leaving behind the MXene layers.
- **Exfoliation Techniques**: Once synthesized, MXenes can be further exfoliated into thinner layers or nanosheets using techniques such as mechanical exfoliation

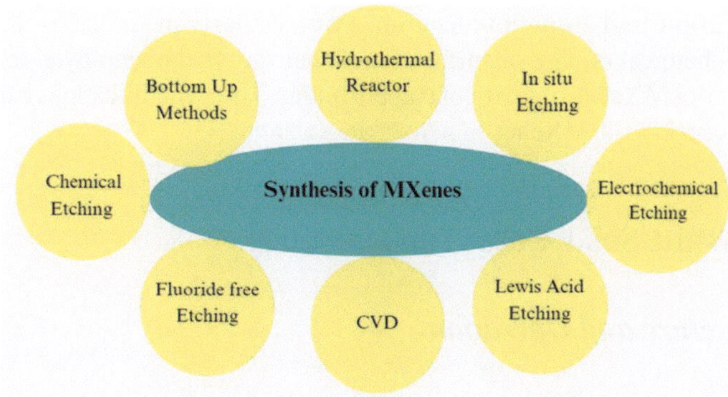

Fig. 2 Common methods for synthesis of MXenes

or chemical exfoliation. These methods are crucial for obtaining MXenes with desirable properties for specific applications [32].

Other Methods of Synthesis:

- **Synthesis from Precursors**: MXenes can also be synthesized directly from precursor materials using methods such as solid-state reactions or chemical vapor deposition (CVD). These techniques offer alternative routes for fabricating MXenes with controlled properties and structures. The chemical etching process begins with the preparation of the MAX phase material, which is typically in bulk form. The material is then exposed to a strong etching solution, commonly consisting of concentrated acids or other potent chemical reagents. For instance, hydrochloric acid (HCl) or a mixture of hydrofluoric acid (HF) and other acids can be used, depending on the specific MAX phase and desired MXene.The etching solution selectively targets and removes the A element (often aluminum) from the MAX phase, while leaving the M and X elements behind. This selective removal is facilitated by the high reactivity of the etching solution with the A element, which effectively dissolves it and separates it from the rest of the material. The result of this process is the formation of a layered MXene structure, which consists of stacked MXene sheets.Following the etching process, the MXene material is typically washed and purified to remove any residual etching chemicals and to stabilize the MXenes. This purification step is crucial to ensure that the MXenes are free from contaminants that could affect their properties.After etching and purification, the MXenes are often further processed into thin films or dispersions for use in various applications. The chemical etching method allows for precise control over the composition and structure of the MXenes, making it possible to tailor their properties for specific uses. This flexibility is one of the reasons why MXenes are such a versatile material, with applications ranging from energy storage and conversion to sensors and catalysis [32, 33].

- **Modification and Functionalization**: Post-synthesis modification techniques, such as chemical doping or surface functionalization, are employed to tailor the properties of MXenes for particular applications. These modifications can enhance the performance of MXenes in perovskite solar cells.

3 Perovskite Solar Cells: Fundamentals

3.1 Structure and Operation

Perovskite solar cells (PSCs) are constructed with a unique layered architecture that contributes to their high efficiency. The typical structure of a perovskite solar cell includes the following layers:

Perovskite Light-Absorbing Layer: The perovskite light-absorbing layer is the heart of the perovskite solar cell, responsible for absorbing sunlight and generating electron–hole pairs (excitons). This layer is typically composed of a perovskite material with the general chemical formula ABX_3. In this formula, **A** is an organic cation such as methylammonium (MA^+), formamidinium (FA^+), or cesium (Cs^+); **B** is a metal cation like lead (Pb^{2+}) or tin (Sn^{2+}); and **X** is a halide anion such as iodine (I^-), bromine (Br^-), or chlorine (Cl^-). The ideal perovskite material must exhibit strong light absorption across the solar spectrum, a suitable bandgap (typically around 1.5 eV for optimal efficiency), high exciton binding energy, and efficient charge generation. Additionally, the material should possess good crystallinity and minimal defects to enhance the photovoltaic performance and stability of the solar cell.

Electron Transport Layer (ETL): The ETL is positioned between the perovskite layer and the cathode. Its primary role is to facilitate the efficient extraction and transfer of electrons from the perovskite layer to the electrode while blocking the movement of holes, preventing recombination losses. For a material to be used as an ETL, it must have a high electron mobility to ensure rapid electron transport, a suitable conduction band alignment with the perovskite layer to enable efficient electron extraction, and good transparency to allow maximum light to reach the perovskite layer. It should also be chemically stable and compatible with the adjacent layers to maintain the overall device stability. Common ETL materials include titanium dioxide (TiO_2), zinc oxide (ZnO), and tin oxide (SnO_2), which are known for their high electron mobility, appropriate band alignment, and stability.

Hole Transport Layer (HTL): The HTL is located between the perovskite layer and the anode. Its function is to transport holes from the perovskite layer to the electrode while blocking electrons, thus reducing recombination and enhancing the overall efficiency of the solar cell. To be effective, an HTL material must possess high hole mobility for efficient charge transport, suitable valence band alignment with the perovskite layer to facilitate hole extraction, and excellent transparency to ensure that incident light reaches the perovskite layer. Additionally, it should exhibit good chemical stability and form a stable interface with both the perovskite and electrode layers. Common HTL materials include spiro-OMeTAD and poly(3,4-ethylenedioxythiophene) polystyrene sulfonate (PEDOT), both of which offer good hole transport properties, appropriate energy level alignment, and stability.

Electrodes: The electrodes, comprising the anode and cathode, are the final layers in the solar cell structure and are responsible for collecting the free charge carriers (electrons and holes) and completing the electrical circuit (as shown in Fig. 3). The **cathode** typically consists of metals such as gold (Au), silver (Ag), or aluminum (Al), which have high electrical conductivity, good work function alignment with the ETL, and excellent stability. The **anode** is usually made from transparent conductive oxides (TCOs) like indium tin oxide (ITO) or fluorine-doped tin oxide (FTO), which combine high transparency with good electrical conductivity, ensuring efficient light transmission and charge collection. For both electrodes, the materials must maintain good electrical contact with the adjacent layers, exhibit minimal resistive losses, and be stable under operational conditions to ensure long-term device performance [34, 35].

Charge Generation and Transport Mechanisms:

Exciton Generation: When sunlight illuminates the perovskite layer in a solar cell, the energy from the light is absorbed by the perovskite material. This absorption process excites the electrons within the material, causing them to move from the valence band to the conduction band, thereby creating electron–hole pairs known as excitons. These excitons are the initial step in converting solar energy into electrical energy. The efficiency of exciton generation is critical for the overall performance of the solar cell, as it determines how much of the absorbed light can be converted into usable electrical energy.

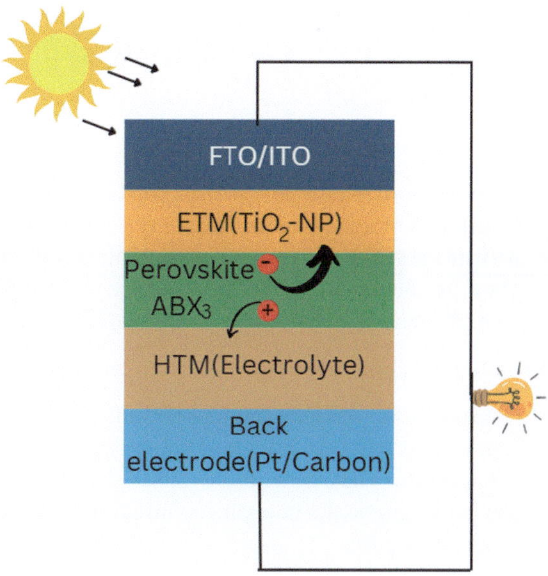

Fig. 3 Basic structure of perovskite solar cell

Charge Separation: After the excitons are generated, the next crucial step is their separation into free charge carriers free electrons and holes. This separation occurs at the interface between the perovskite layer and the adjacent charge transport layers: the ETL and HTL. The energy difference at these interfaces facilitates the dissociation of excitons into individual electrons and holes, preventing them from recombining and instead driving them toward their respective transport layers. Efficient charge separation is essential for minimizing recombination losses and maximizing the flow of charge carriers within the solar cell.

Charge Transport: Once separated, the free electrons and holes need to be transported to the electrodes for collection. Electrons travel through the ETL toward the cathode, while holes move through the HTL toward the anode. The efficiency of charge transport depends on the conductivity and mobility of the materials used in the ETL and HTL, as well as the alignment of their energy levels with the perovskite layer. Properly designed transport layers ensure that the electrons and holes reach the electrodes quickly and with minimal energy loss, which is crucial for maintaining high PCE in the solar cell.

Charge Collection: The final step in the operation of a perovskite solar cell is the collection of the separated charges at the electrodes. The cathode collects the free electrons, while the anode collects the holes, completing the electrical circuit. This process generates an electric current that can be harnessed to power electronic devices or stored in batteries for later use. The efficiency of charge collection depends on the effectiveness of the electrode materials and their interfaces with the transport layers. Optimizing this step ensures that the maximum amount of generated electricity is collected and used, contributing to the overall efficiency and performance of the solar cell [36, 37].

3.2 *Types of Perovskite Materials*

Organic–Inorganic Hybrid Perovskites:

- **Composition**: These perovskites combine organic cations (e.g., methylammonium) with inorganic metal halides (e.g., lead iodide).
- **Advantages**: They offer excellent light absorption and charge transport properties. They are the most commonly used type of perovskite in solar cells due to their high efficiency and ease of fabrication.

All-Inorganic Perovskites:

- **Composition**: All-inorganic perovskites are composed entirely of inorganic components (e.g., caesium lead halides).
- **Advantages**: They generally exhibit better thermal and environmental stability compared to organic–inorganic hybrids, making them more durable for long-term applications [38].

3.3 Current Challenges

Stability and Degradation Issues:

Environmental Degradation: Perovskite materials, while promising in terms of efficiency, are notably sensitive to environmental factors such as moisture, oxygen, and ultraviolet (UV) light. Exposure to moisture can lead to hydrolysis of the perovskite material, causing degradation and loss of its photovoltaic properties. Oxygen, especially when combined with moisture, can result in oxidative degradation of the perovskite layer, further compromising performance. UV light can also cause photodegradation, where the high-energy photons break down the perovskite structure, leading to a decrease in efficiency and the potential for device failure. The cumulative effect of these environmental factors is a reduction in the operational lifespan and reliability of perovskite solar cells, posing significant challenges for their long-term use and commercialization.

Thermal Stability: Thermal stability is another critical concern for perovskite solar cells. High temperatures can induce phase transitions or cause thermal decomposition of the perovskite materials. During phase transitions, the perovskite structure may change, affecting its electronic and optical properties, which in turn impacts the overall efficiency of the solar cells. Prolonged exposure to elevated temperatures can lead to the decomposition of the perovskite material, resulting in the loss of its light-absorbing capabilities and structural integrity. This thermal instability poses a challenge for the practical deployment of perovskite solar cells, particularly in environments with fluctuating or high temperatures. Addressing these thermal stability issues is crucial for enhancing the durability and performance of perovskite solar cells under real-world operating conditions [39, 40].

Scalability and Commercialization:

Manufacturing Scale-Up: Scaling up perovskite solar cell production from laboratory to industrial levels presents several challenges. One major issue is ensuring uniform quality and reproducibility of perovskite materials across large batches. Variability in material properties, such as composition and crystal structure, can affect the performance and reliability of the solar cells. Achieving consistent manufacturing conditions, controlling environmental factors, and maintaining process precision are critical for successful scale-up.

Cost and Integration: The commercial viability of perovskite solar cells is influenced by the cost of raw materials and processing techniques. While perovskite materials themselves are relatively inexpensive, the overall cost can be affected by the need for specialized equipment and processing steps. Additionally, integrating perovskite solar cells with existing technologies and infrastructure poses challenges. This includes developing compatible interfaces with current solar cell technologies and adapting manufacturing processes to ensure compatibility with established production lines. Addressing these cost and integration issues is essential for widespread adoption and commercialization of perovskite solar cells.

3.4 Advantages of PSCs

- **High PCE**: PSCs have rapidly achieved high power conversion efficiencies, surpassing 25% in a relatively short period. This efficiency rivals that of traditional silicon-based solar cells, making PSCs highly competitive in the solar energy market.
- **Low Cost of Materials and Production**: The materials used in perovskite solar cells are relatively inexpensive compared to those in silicon solar cells. Additionally, the manufacturing process for PSCs is less energy-intensive and can be done using solution-based methods like spin coating or inkjet printing, which further reduces costs.
- **Versatility in Material Composition**: Perovskite materials can be easily tuned by altering their chemical composition, allowing for customization of properties such as bandgap and absorption spectrum. This tunability enables the optimization of PSCs for different applications, including tandem solar cells and flexible electronics.
- **Thin-Film Technology**: PSCs can be fabricated as thin films, which are lightweight and flexible. This flexibility allows for the development of innovative applications, such as building-integrated photovoltaics (BIPV), portable solar panels, and wearable solar devices.
- **Broad Absorption Spectrum**: Perovskite materials have a broad absorption spectrum, efficiently capturing light across the visible and near-infrared regions. This broad absorption enhances the light-harvesting capability of PSCs, contributing to their high efficiency [17, 41–43].

4 Integration of MXenes in Perovskite Solar Cells

4.1 Role of MXenes in Enhancing Performance

Charge Transport and Collection:

- **Improved Conductivity**: MXenes exhibit exceptional electrical conductivity due to their metallic bonding and layered structure. This high conductivity can enhance the charge transport within the perovskite layer and between different layers of the solar cell. By integrating MXenes, the efficiency of charge collection can be significantly improved, reducing resistive losses and increasing the overall PCE of the perovskite solar cells.
- **Electron and Hole Transport**: MXenes can be incorporated into the electron transport layer (ETL) or hole transport layer (HTL) to facilitate better charge separation and transport (as shown in Fig. 4). The high conductivity and large surface area of MXenes can lead to improved interface engineering and reduce charge recombination losses, contributing to higher device efficiency.

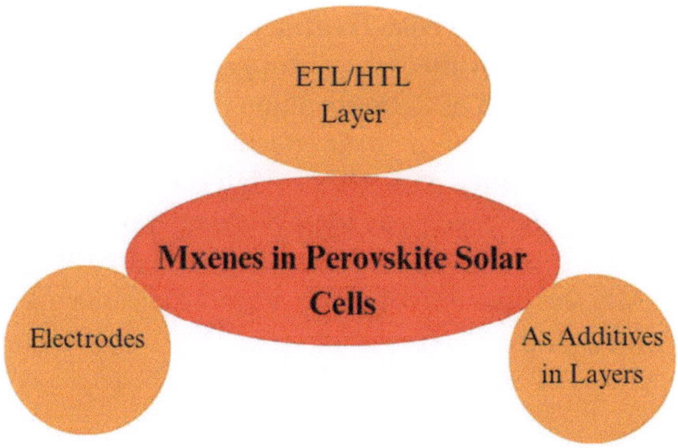

Fig. 4 MXenes in perovskite solar cells

Interface Engineering and Contact Improvement:

- **Optimized Interfaces**: The interface between the perovskite layer and the ETL or HTL is crucial for efficient charge transfer. MXenes can be used to modify these interfaces, improving the contact quality and reducing energy barriers for charge extraction. This can result in better alignment of energy levels and enhanced charge injection and extraction processes.
- **Surface Functionalization**: The surface chemistry of MXenes can be tuned to interact favorably with the perovskite material, leading to better adhesion and reduced interfacial resistance. Functionalized MXenes can also provide additional sites for charge transfer, further enhancing the overall performance of the solar cell.

4.2 Device Fabrication Techniques

Methods of Incorporating MXenes into Perovskite Cells:

- **Layer-by-Layer Deposition**: MXenes can be deposited using layer-by-layer techniques to create thin films or coatings that are integrated into the ETL or HTL. This method allows for precise control over the thickness and uniformity of the MXene layer.
- **Mixed-Material Approaches**: MXenes can be mixed with other materials, such as organic or inorganic semiconductors, to form composite layers. These composites can be used as part of the ETL or HTL, combining the beneficial properties of both MXenes and other materials.
- **Blending with Perovskite**: MXenes can also be blended directly with the perovskite layer to improve charge transport and stability. This approach involves incorporating MXenes into the perovskite precursor solution or depositing them as a separate layer in contact with the perovskite.

Processing Conditions and Optimization:

- **Annealing and Post-Treatment**: Processing conditions such as annealing or post-treatment can affect the performance of MXene-enhanced perovskite solar cells. Optimization of these conditions is essential to achieve the desired properties of MXenes and ensure their effective integration into the solar cell structure.
- **Compatibility with Other Materials**: The choice of fabrication methods and processing conditions must be compatible with the other materials used in the solar cell. Careful optimization is required to ensure that the incorporation of MXenes does not negatively impact the performance or stability of the perovskite layer.

4.3 Performance Metrics

- **PCE**: The incorporation of MXenes into perovskite solar cells has the potential to increase PCE by improving charge transport and reducing losses. Performance metrics such as short-circuit current density (J_{SC}), open-circuit voltage (V_{OC}), and fill factor (FF) are used to evaluate the impact of MXenes on overall device efficiency.
- **Light Absorption and Utilization**: MXenes can enhance light absorption and utilization by modifying optical properties and improving light harvesting. This can lead to higher photocurrent generation and better overall performance of the solar cells.

Stability and Durability Enhancements:

- **Environmental Stability**: MXenes may contribute to improved environmental stability by providing protective layers or modifying interfaces. Enhanced stability can lead to longer operational lifetimes and better performance retention over time.
- **Mechanical Robustness**: The incorporation of MXenes can improve the mechanical robustness of perovskite solar cells, making them more resilient to mechanical stresses and environmental factors.

4.4 MXenes in PSCs

MXenes as Additives:

The performance of a perovskite solar cell (PSC) is primarily influenced by the light-absorbing perovskite film layer, which generates charge carriers (electrons and holes) (Fig. 5 and Table 1). These charge carriers are then separated and collected at the respective electrodes. The quality of the perovskite film is crucial, as it is determined by crystal size and grain boundaries. To achieve optimal charge transfer and high performance, the perovskite film should have large crystal sizes and minimal grain boundaries. Researchers have explored the use of $Ti_3C_2T_X$ MXene in perovskite solar cells, finding that its addition improves performance in several ways. $Ti_3C_2T_X$ helps to retard the nucleation of $MAPbI_3$, resulting in larger crystal sizes, and acts as an "electron bridge," enhancing electron transfer. This has led to an increase in PCE. $Ti_3C_2T_x$ also optimizes the work function of $MAPbI_3$ films, improving energy-level alignment and boosting PCE. In other studies, $Ti_3C_2T_X$ was used to decorate $MAPbBr_3$ nanocrystals, creating a heterostructure that improves electron injection and overall solar cell performance. Recent advancements include the use of ultrathin $Ti_3C_2T_X$ quantum dots to enhance CsFAMA perovskite absorbers and the perovskite/TiO_2 ETL interface, resulting in a high hysteresis-free PCE and improved stability. $Ti_3C_2T_X$ has also been incorporated into electron transport layers (ETLs) and hole transport layers (HTLs), showing benefits like improved electron transport, hole blocking, and increased PCE.

Additionally, $Ti_3C_2T_x$ has been used to modify ZnO ETLs, creating hybrid structures that enhance charge transport and stability, leading to better performance and durability in various organic solar cells [44].

MXenes as ETL:

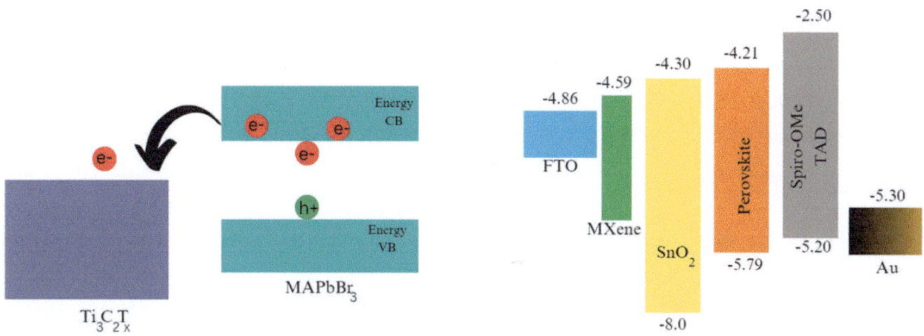

Fig. 5 The alignment of energy levels and electron transfer between $MAPbBr_3$ crystals and the coated $Ti_3C_2T_x$ nanosheets

Table 1 MXenes in perovskite solar cells

MXene used as	Advantage
Electrode	(a) high transparency, remarkable flexibility, conductivity and tunable work functions are well (b) forming hybrid electrodes when paired with other conductive nanomaterials like carbon nanotubes or metallic nanowires
Additives	(a) Enhancing electron transfer, similar to creating an "electron bridge" (b) Improving the crystallinity of perovskite materials (c) Adjusting the work function of carrier transport materials and other properties like conductivity (d) Passivating surfaces and engineering interfaces
ETL/HTL	(a) Work functions and carrier conduction properties that can be easily adjusted

Researchers have explored various approaches to enhance PSCs using MXenes. One study used $Ti_3C_2T_X$ MXene nanosheets as an ETL in planar PSCs processed at low temperatures. Applying a UV-ozone treatment improved the $Ti_3C_2T_X$ layer by increasing surface Ti–O bonds, which accelerated electron transport. This treatment boosted the PSC's PCE from 5 to 17.2% and reduced recombination at the ETL/perovskite interface. Another study involved Nb_2CT_X MXene nanosheets, where the work function was adjusted by replacing –F groups with $–NH_2$ groups through hydrazine treatment. This modification improved the crystallization of the perovskite precursor and led to high-quality films. The PSC with these treated MXene ETLs achieved a maximum PCE of 21.79%, with flexible and large-area devices showing PCEs of 19.2 and 18.3%, respectively. The devices retained 93% of their initial PCE after 1500 h of storage. Despite these advancements, $Ti_3C_2T_X$ MXene-based PSCs still lag behind traditional TiO_2- or SnO_2-based devices. Issues at the MXene/perovskite interface persist. One approach to address these challenges involved using oxygen plasma treatment to enhance the $Ti_3C_2T_X$ surface, which improved electron transport and device stability. This treatment led to a high PCE of 18.9% and good stability, with devices retaining 90% of their initial PCE after 750 h.

Additionally, oxidizing $Ti_3C_2T_X$ can adjust its characteristics for use as an ETL in low-temperature processed PSCs. Oxidation produces Ti–O bonds, which reduce defects and improve energy level matching between the perovskite and ETL layers. A hybrid of oxidized and pristine $Ti_3C_2T_X$ achieved a PCE of 18.3%, demonstrating the potential of MXene-derived materials in enhancing PSC performance.

MXenes as HTL:

PSCs, HTL is crucial for controlling the crystallization of the perovskite film and the efficiency of hole transfer at the perovskite/HTL interface. While organic hole transport materials (HTMs) such as N, N′-bis(3-methylphenyl)-N, N′-bis(phenyl)benzidine, poly(trimethylene terephthalate-co-trimethylene isophthalate)-Terephthalic acid (PTTI-TPA), triphenylamine-based HTM with a pyridine core, and pyridine-based polymer semiconductors have shown success, their complex synthesis, purification, and high costs hinder their mass production. In contrast, the common inorganic HTM, NiOx, requires annealing at temperatures above 200 °C, making it unsuitable for flexible devices. Thus, there is a need for low-cost, low-temperature processed HTMs with excellent electrical properties. Nb_2CT_X MXene is emerging as a promising HTM due to its good conductivity and adjustable work function. Compared to $Ti_3C_2T_X$ MXene, Nb_2CT_X has three atomic layers, providing a larger specific surface area and increased contact area with the perovskite layer, which enhances charge transfer in PSCs. The work function of Nb_2CT_X MXene can be controlled by modifying the surface functional groups. For instance, first-principles calculations show that the work function of pristine Nb_2CT_X with –O, –OH, and –F groups is about 4.7 eV, which is insufficient for effective hole transport. Among these, O-terminated Nb_2CT_X typically shows a higher work function due to differences in surface dipole moments and charge transfer. Recent work involved treating Nb_2CT_X MXene with oxygen plasma, which increased its work function from 4.68 to 5.04 eV by enhancing the O functional groups on its surface. This treatment improved the electron extraction from the perovskite layer to the HTL, resulting in a higher PCE of 20.74% compared to 18.1% with untreated Nb_2CT_X and 15.5% without Nb_2CT_X HTL. The enhanced external quantum efficiency (EQE) further indicates better charge separation and collection. Flexible and large-area PSCs with oxygen plasma-treated Nb_2CT_X HTLs achieved PCEs of 17.3% and 17.9%, respectively, and demonstrated excellent long-term stability and thermal resistance.

Additionally, using a silane coupling agent (SCA) can improve interface bonding and reduce costs by increasing corrosion resistance. In 2022, $Ti_3C_2T_X$ MXene nanosheets were sprayed with varying concentrations of SCA to enhance the HTL. The resulting $Ti_3C_2T_X$ films had improved hole transfer paths, increased hole mobility from 1.30×10^{-6} to 3.08×10^{-7} cm^2 V^{-1} s^{-1}, and reduced transfer resistance from 369.3 to 194.1 Ω at the HTL/perovskite interface, leading to better performance.

MXene as Electrode:

The commercialization of PSCs is hindered by the high costs of traditional hole transport materials (HTMs) like Spiro-OMeTAD, PTAA, and NiO_x, as well as noble metal electrodes such as gold and silver. Coal-based carbon electrodes offer a more affordable alternative to both HTMs and noble metals. However, PSCs with carbon electrodes still exhibit lower PCEs compared to conventional devices. Researchers have explored using 2D MXene materials, such as Ti_3C_2, as back electrodes in HTM and noble-metal-free PSCs. By employing a hot-pressing technique at 85 °C and 0.4 MPa, a seamless interface between the MAPbI3 perovskite layer and Ti_3C_2 was achieved. Ti_3C_2's work function aligns well with the valence band of MAPbI, facilitating efficient hole transfer from the perovskite layer to the Ti_3C_2 electrode and electron transfer from the TiO_2 layer to the FTO electrode. Increasing the Ti_3C_2 electrode thickness from 280 to 330 μm improved charge transport by reducing the square resistance. However, further increases in thickness led to higher resistance and decreased photovoltaic performance. The optimized Ti_3C_2 electrode PSC achieved a PCE of 13.83%, with improved voltage and current density values, and demonstrated better reproducibility and stability compared to conventional carbon electrode-based PSCs. Carbon-based PSCs are cost-effective and stable, but their performance generally lags behind traditional structures. Commercial carbon paste often leads to point-to-point contact with the perovskite layer, resulting in pinholes that impede carrier transport. To enhance performance, researchers incorporated commercial carbon nanotubes (CNTs) and 2D Ti_3C_2 MXene into a mixed carbon electrode for perovskite-based PSCs. This mixed electrode creates a network structure with multi-dimensional charge transfer paths, as MXene nanosheets fill the voids in the carbon powder, significantly boosting conductivity and improving carrier transport.

4.5 MXene as Interfacial Layer

Interface Between Perovskite and ETL:

Enhancing PSCs involves optimizing device design and interfaces to improve efficiency. Strategies like using $Ti_3C_2T_X$. MXenes have increased the work function, boosting PCE from 15.6 to 20% by enhancing light absorption and interface alignment. Incorporating $Ti_3C_2T_X$ quantum dots and Cu1.8S nanocrystals improved film quality and charge extraction, raising PCE from 18.3 to 21.6%. $Ti_3C_2Cl_X$ MXenes reduced lattice strain in CsPbBr3 films, achieving a PCE of 11.08% with a high open-circuit voltage (V_{OC}). Combining MXenes with single-walled carbon nanotubes (SWCNTs) and using SnO2-MXene composites further enhanced conductivity and stability, with PCEs reaching 21.42 and 23.07%, respectively. These advancements highlight MXenes' potential in significantly improving PSC performance and durability.

Interface Between Perovskite and Electrode:

MXenes are emerging as promising materials for photovoltaic applications due to their high transparency, adjustable work functions (ranging from 2.14 to 5.65 eV), and metallic conductivity. Chen et al. were the first to integrate 2D Ti_3C_2. MXene nanosheets as an interlayer in all-inorganic $CsPbBr_3$ perovskite solar cells. The Ti_3C_2 MXene interlayer was applied on the FTO/c-TiO_2/$CsPbBr_3$ substrate using a spin-coating method.

The selected area electron diffraction (SAED) pattern of Ti_3C_2 MXene nanosheets reveals a few well-defined crystalline layers. The device configuration used was FTO/TiO_2/$CsPbBr_3$/Ti_3C_2 MXene/Carbon. This arrangement helps in achieving better alignment of energy levels at the interface, which reduces energy-level mismatches, accelerates hole extraction, and minimizes recombination at the perovskite/carbon electrode interface. Moreover, the surface functional groups of the Ti_3C_2 MXene enhance interactions with the under-coordinated lead atoms, reducing deep trap defects in the $CsPbBr_3$ films. As a result, the device with the Ti_3C_2 MXene interlayer exhibited a notable initial PCE of approximately 9% and demonstrated exceptional long-term stability, retaining its performance for over 1900 h in a humid environment and more than 600 h under elevated temperatures.

Interface Between Perovskite and HTL:

$CsPbI_3$ inorganic perovskites are promising for their thermal stability and suitable band gap for tandem solar cells, but they face challenges such as low PCE and high open-circuit voltage (V_{OC}) loss due to nonradiative recombination. To address these issues, $Ti_3C_2F_X$ monolayer quantum dots (MQDs) were synthesized and used as interface passivators. These MQDs enhance performance by tuning energy levels, reducing defect-induced recombination, and forming a barrier layer. With $Ti_3C_2F_X$ MQDs, $CsPbI_3$ PSCs achieved a PCE of 20.4%, a V_{OC} of 1.22 V, and a high fill factor of 81.5%. The devices also demonstrated excellent stability, retaining 93% of their initial efficiency after 600 h in ambient air. Additionally, FA-based perovskites, which struggle with phase instability, were improved using halogen-terminated nano-MXenes. These nano-MXenes stabilize the perovskite lattice, enhance energy level alignment, and reduce charge transfer barriers. The resulting device reached a record PCE of 24.2%, with exceptional stability, maintaining over 90% of its initial efficiency after 1000 h of continuous illumination and high temperatures [45, 46].

Interface Between ETLs and Electrodes:

The high annealing temperatures, photo-instability of TiO_2, and aggregation issues with SnO_2 pose challenges for scaling up PSCs. The energy-level mismatch between SnO_2's conduction band minimum and FTO's work function exacerbates charge extraction and hysteresis problems. Wang et al. addressed this by using 2D carbide MXenes to enhance the electron mobility and charge transfer of SnO_2 ETLs. The MXene-modified SnO_2 improved perovskite film growth orientation, reduced trap density, and decreased non-radiative recombination, leading to a PSC with a PCE of 20.7%, minimal hysteresis, and 90% of its initial efficiency after 3 months. Yang et al. further advanced this by incorporating in-situ oxidized $Ti_3C_2T_X$ (O-$Ti_3C_2T_X$) MXene, which adjusted its properties from metallic to semiconducting. This modification enhanced the electron mobility of the SnO_2 ETL, improved electron extraction, and increased the PCE from 17.7 to 20.1%, while preserving 80% of its initial efficiency in air.

5 Fabrication Techniques for MXene-Enhanced Perovskite Solar Cells

5.1 *Synthesis and Integration of MXenes*

Synthesis of MXenes:

- **Chemical Etching**: MXenes are commonly synthesized from their MAX phase precursors using selective chemical etching (as shown in Fig. 6). This process involves treating the MAX phase material with a strong acid or other etchants to remove the 'A' element, leaving behind a 2D MXene structure. This method allows for the production of MXenes with controlled composition and surface properties.
- **Exfoliation Techniques**: After the initial synthesis, MXenes can be further exfoliated into nanosheets using methods such as mechanical exfoliation or chemical exfoliation. Exfoliated MXenes are then used to fabricate thin films or composites that can be integrated into perovskite solar cells.

Integration into Perovskite Solar Cells:

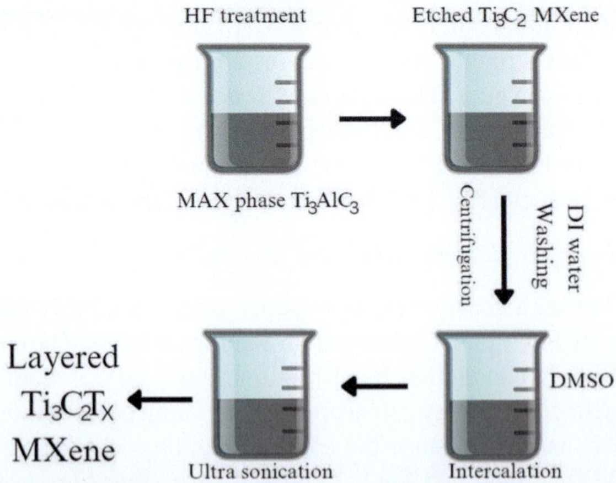

Fig. 6 Preparation of MXenes using chemical etching

- **Layer-by-Layer Deposition**: MXenes can be deposited onto substrates or between layers of the perovskite solar cell using techniques such as spin-coating, dip-coating, or spray-coating. This method allows for precise control over the thickness and uniformity of the MXene layer.
- **Mixed-Material Approaches**: MXenes can be combined with other materials, such as organic or inorganic semiconductors, to form composite layers. These composites can be used as part of the electron transport layer (ETL) or hole transport layer (HTL), leveraging the benefits of both MXenes and other materials.

5.2 *Thin Film and Composite Fabrication*

Thin Film Deposition Techniques:

- **Chemical Vapor Deposition (CVD)**: CVD is a widely used technique for depositing thin films of MXenes onto substrates. This method involves the chemical reaction of gaseous precursors to form a solid MXene layer. CVD offers high-quality films with excellent uniformity and control over thickness.
- **Sputtering**: Sputtering is another technique used to deposit thin films of MXenes. This method involves bombarding a target material with high-energy ions to release atoms, which then deposit onto a substrate. Sputtering can be used to create thin films with good adhesion and uniformity.
- **Atomic Layer Deposition (ALD)**: ALD allows for the deposition of thin, conformal layers of MXenes with precise control over thickness. This technique involves alternating exposures to different precursor gases, leading to layer-by-layer growth.

Composite Fabrication Techniques:

- **Sol–Gel Method**: MXenes can be combined with sol–gel precursors to form composite materials. The sol–gel method involves the transformation of a sol (a liquid containing dispersed particles) into a gel (a network of interconnected particles) through chemical reactions. This method is useful for creating MXene-based composites with tailored properties.
- **Electrochemical Methods**: Electrochemical deposition can be used to incorporate MXenes into composites. This technique involves applying an electric current to drive the deposition of MXenes onto a substrate, forming a composite layer with controlled composition and structure.

5.3 Interface Engineering and Optimization

Improving Adhesion and Electrical Contact:

- **Surface Functionalization**: Functionalizing the surface of MXenes with various chemical groups can improve adhesion to other materials and optimize electrical contact. This surface modification enhances the compatibility of MXenes with perovskite layers and other components of the solar cell.
- **Interface Modifications**: Modifying the interfaces between MXenes and other layers in the solar cell can reduce energy barriers and improve charge transfer. Techniques such as chemical treatment or the introduction of interfacial layers can enhance the performance of MXene-enhanced solar cells.

Process Optimization:

- **Annealing and Post-Treatment**: Annealing or post-treatment processes can be used to optimize the properties of MXene layers and composites. These processes can improve the crystallinity, electrical conductivity, and overall performance of the MXene-based layers.
- **Compatibility with Perovskite Materials**: Ensuring that MXenes are compatible with perovskite materials and other components is crucial for achieving optimal device performance. Compatibility testing and optimization are necessary to address any issues related to chemical interactions or mechanical stresses.

5.4 Scaling up and Industrial Considerations

Challenges in Scaling up:

- **Manufacturing Uniformity**: Scaling up the production of MXenes and their integration into perovskite solar cells requires maintaining uniformity and consistency in the fabrication process. Ensuring that MXene layers are consistently applied and perform reliably at larger scales is a key challenge.

- **Cost-Effectiveness**: The cost of synthesizing and integrating MXenes needs to be evaluated in the context of large-scale production. Cost-effective methods for synthesizing MXenes and incorporating them into perovskite solar cells are essential for commercial viability.

Industrial Production Techniques:

- **Roll-to-Roll Processing**: Roll-to-roll processing techniques can be employed for the large-scale fabrication of MXene-enhanced perovskite solar cells. This approach involves continuously processing flexible substrates to produce solar cells efficiently and cost-effectively.
- **Automation and Quality Control**: Implementing automated processes and quality control measures is crucial for scaling up production. Automation can improve precision and reproducibility, while quality control ensures that the final products meet performance and reliability standards.

6 Electrochemical Performance of MXene-Enhanced Perovskite Solar Cells

6.1 Power Conversion Efficiency (PCE)

- **PCE** is a key performance metric for solar cells, representing the ratio of the electrical power output to the incident solar power. For perovskite solar cells, enhancing PCE is crucial for maximizing energy yield and making the technology commercially viable.

Impact of MXenes:

Improved Charge Transport: The high electrical conductivity of MXenes significantly enhances charge transport within the perovskite layer and across its interfaces. This improved conductivity reduces resistive losses, allowing for more efficient charge flow, which directly contributes to higher PCE. By facilitating smoother and faster charge movement, MXenes help in minimizing energy dissipation, thereby optimizing the overall performance of the solar cell.

Optimized Interfaces: MXenes play a crucial role in refining the interfaces between the perovskite layer and the charge transport layers, namely the electron transport layer (ETL) and hole transport layer (HTL) (as shown in Tables 2, 3, 4 and 5). By improving the quality of these interfaces, MXenes help to lower energy barriers, which enhances the efficiency of charge extraction and collection. This optimization leads to more effective separation and transport of charge carriers, further boosting the efficiency and stability of the perovskite solar cells.

Evaluation Methods:

Table 2 MXenes as HTL/HTL additives

Device	J_{SC} [mAcm^{-2}]	V_{OC} [V]	FF [%]	PCE [%]	Refs.
ITO/Nb_2CT_x/$MAPbI_3$/PCBM/Ag	23.06	1.128	79.75	20.74	[47]
ITO/ $Ti_3C_2T_x$-SCA/$MAPbI_3$/PC61BM/ Ag	21.30	0.87	73.47	13.65	[47]
ITO/Mo_2C-CNT@PEDOT:PSS/ $MAPbI_3$/PCBM/LiF/Al	22.219	0.881	62	12.14	[48]
FTO/TiO_2/$CsPbBr_3$/$Ti_3C_2T_x$/Carbon	8.54	1.44	73.04	9.03	[49]

Table 3 MXenes as additives layer in PSCs

Device	J_{SC} [mAcm^{-1}]	V_{OC} [V]	FF [%]	PCE [%]	Refs.
ITO/SnO_2/$MAPbI_3$:$Ti_3C_2T_x$/ Spiro-OMeTAD/Au	22.26	1.03	76.00	17.41	[50]
FTO/c-TiO_2/m-TiO_2/ CsFAMA-TQD/ Spiro-OMeTAD:Cu1.8S/Au	24.12	1.14	78.70	21.64	[51]
ITO/SnO_2-Ti_3C_2/$MAPbI_3$/ Spiro-OMeTAD/Ag	23.14	1.06	75.00	18.34	[52]
FTO/TiO_2:SnO_2:$Ti_3C_2T_x$/ $(FAPbI3)_{0.97}$ $(MAPbBr3)_{0.03}$/ Spiro-OMeTAD/Au	22.03	1.10	77.78	18.90	[53]
FTO/ c-TiO_2:$Ti_3C_2T_{x/m-}$$TiO_2$:$Ti_3C_2T_x$/ Ti_3C-T_x/$MAPbI_3$:$Ti_3C_2T_x$/Spiro OMeTAD/Au	23.82	1.09	77.60	20.14	[54]
ITO/PEDOT:PSS:$Ti_3C_2T_x$/ PBDB-T:ITIC/PFN-Br/Al	17.08	0.91	70.93	11.02	[55]
ITO/PEDOT:PSS:$Ti_3C_2T_x$/ PM6:Y6/PFN-Br/Al	25.63	0.83	68.40	14.55	[55]
ITO/ZnO:Ti_3C_2T-/PBDB-T:ITIC/ MoO_3/Ag	18.63	0.93	70.39	12.20	[56]
ITO/ZnO:$Ti_3C_2T_x$/PM_6:Y_6/MoO_3/ Ag	26.38	0.83	75.40	16.51	[56]

J-V Curve Analysis: The current–voltage (J-V) characteristics of MXene-enhanced perovskite solar cells are analyzed under standard test conditions (AM 1.5G, 100 mW/cm^2). This analysis involves measuring the solar cell's electrical output in response to varying voltages, which generates the J-V curve. From this curve, key performance parameters such as short-circuit current density (J_{SC}), open-circuit voltage (V_{OC}), and fill factor (FF) are extracted. These parameters are essential for calculating the PCE of the solar cell, providing a comprehensive understanding of the cell's performance and efficiency.

Table 4 MXenes as electrodes

Device	J_{SC} [$mAcm^{-1}$]	V_{OC} [V]	FF [%]	PCE [%]	Refs.
$FTO/TiO_2/MAPbI_3/Ti_3C_2T_x$	22.96	0.95	63.00	13.83	[57]
FTO/c-TiO2/CPbBr3/ Carbon:CNT:$Ti_3C_2T_x$	7.16	1.357	72.97	7.09	[57]
Al/PrC_{60}MA/PTB_7-Th: PC71BM/PEDOT:PSS/ MXene-AgNW-PUA	14.62	0.79	61.00	7.16	[58]
Al/PDINO/PC71BM/PBDB-T: ITIC/PEDOT:PSS/ MXene-AgNW-PUA	13.98	0.86	64.00	7.70	[58]
Al/PDINO/ PBDB-T:ITIC:PC71BM/ PEDOT:PSS/ MXene-AgNW-PUA	14.85	0.88	63.00	8.30	[58]
Glass/$Ti_3C_2T_x$/PEDOT:PSS/ PM6:Y6/PFN-Br/Al	24.97	0.84	64.90	13.62	[59]
$Ti_3C_2T_x$/n+-Si/n-Si/p+-Si/ Al_2O_3/SiN_x/Ag/Al	36.7	0.54	57.99	11.47	[60]

Table 5 MXenes as ETL

Device	J_{SC} [$mAcm^{-1}$]	V_{OC} [V]	FF [%]	PCE [%]	Refs.
$ITO/Ti_3C_2T_x/CH_3NH_3PbI_3$/ Spiro-OMeTAD/Ag	22.63	1.08	70.00	17.17	[60]
FTO/MXene/SnO_2/perovskite/ Spiro-OMeTAD/Au	24.34	1.11	76.40	20.65	[61]
ITO/UH-$Ti_3C_2T_x$/ PBDB-T:ITIC/MoO_3/Al	17.36	0.87	60.00	9.06	[62]
FTO/SnO_2-Ti_3C_2 MXene/ $MAPbI_3$/Spiro-OMeTAD/Ag	23.44	1.07	75.86	18.84	[63]
FTO/TiO_2:SnO_2:$Ti_3C_2T_x$/ $(FAPbI_3)_{0.97}(MAPbBr_3)_{0.03}$/ Spiro-OMeTAD/Au	22.03	1.10	77.78	18.9	[64]
ITO/T-Nb_2CT_x/ $FA_{0.85}Cs_{0.15}PbI_3$/ Spiro-OMeTAD/Ag	24.55	1.117	73.79	20.23	[65]
ITO/HO-$Ti_3C_2T_x$@$Ti_3C_2T_x$/ $MAPbI_3$/Spiro-OMeTAD/Ag	23.11	1.07	74	18.29	[54]
FTO/c-TiO_2/m-TiO_2/$Ti_3C_2T_x$ QD/CsFAMAIBr perovskite/ Spiro-OMeTAD/Au	23.51	1.12	74.62	19.65	[51]
ITO/NiO/$Ti_3C_2T_x$ doped-MAPbI3/MXene-PCBM/BCP/ Ag	22.88	1.09	77	19.20	[53]

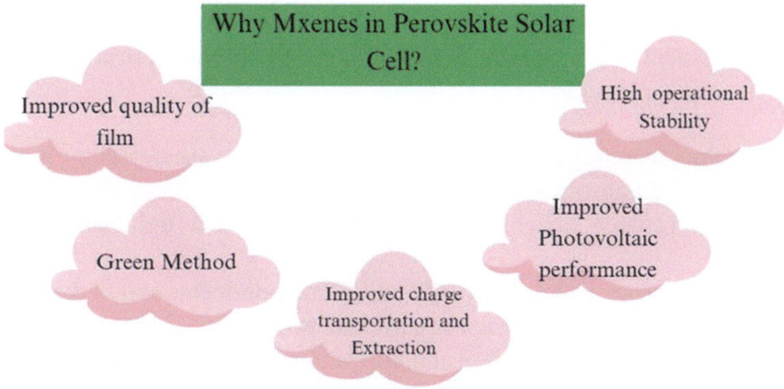

Fig. 7 Importance of MXenes in perovskite solar cell

6.2 *Stability and Durability*

Long-Term Stability:

- **Environmental Stability**: The inclusion of MXenes can improve the environmental stability of perovskite solar cells by providing protective layers or modifying interfaces (as shown in Fig. 7). Enhanced resistance to moisture, oxygen, and UV light can contribute to longer operational lifetimes.
- **Thermal Stability**: MXenes may also enhance the thermal stability of perovskite solar cells, reducing degradation at elevated temperatures and ensuring reliable performance over time.

Testing Methods:

- **Accelerated Aging Tests**: Stability is assessed through accelerated aging tests, which involve exposing the solar cells to harsh conditions (e.g., high humidity, elevated temperatures) to simulate long-term use.
- **Cycling Stability Tests**: Repeated cycling tests, including charge–discharge cycles, evaluate the durability and performance retention of MXene-enhanced perovskite solar cells.

6.3 *Light Absorption and Utilization*

Enhanced Light Absorption:

- **Optical Properties of MXenes**: MXenes have unique optical properties, including high absorption in the visible and near-infrared regions. Incorporating MXenes into perovskite solar cells can enhance light absorption and utilization, leading to increased photocurrent generation.

- **Light Scattering Effects**: MXenes can also influence light scattering within the solar cell, potentially improving light trapping and enhancing overall efficiency.

Measurement Techniques:

- **UV–Vis Spectroscopy**: Ultraviolet–visible (UV–Vis) spectroscopy is used to measure the light absorption spectra of MXene-enhanced perovskite solar cells. The spectra provide insights into how MXenes affect light absorption and utilization.
- **External Quantum Efficiency (EQE)**: EQE measurements assess the efficiency of photon-to-electron conversion at different wavelengths, providing a detailed understanding of light absorption and utilization across the solar cell spectrum.

6.4 Charge Transport and Recombination

Charge Transport Properties:

- **Electrical Conductivity**: The high conductivity of MXenes enhances charge transport within the solar cell, reducing resistive losses and improving overall efficiency.
- **Mobility and Lifetime**: MXenes can influence charge carrier mobility and lifetime, affecting how efficiently electrons and holes are transported through the solar cell.

Recombination Dynamics:

- **Reduced Recombination Losses**: MXenes can help reduce charge recombination losses by improving the interface between the perovskite layer and the transport layers, leading to enhanced performance.
- **Recombination Measurements**: Techniques such as time-resolved photoluminescence (TRPL) and space-charge-limited current (SCLC) measurements are used to study charge recombination dynamics in MXene-enhanced perovskite solar cells.

6.5 Comparative Analysis

Performance Comparison:

- **MXene versus Non-MXene Devices**: Comparative analysis of MXene-enhanced perovskite solar cells versus conventional devices without MXenes can highlight the improvements in efficiency, stability, and performance metrics.
- **Benchmarking Against Other Additives**: The performance of MXene-enhanced solar cells can also be compared to those with other additives or enhancements to evaluate their relative effectiveness (as shown in Table 3).

7 Applications of MXene-Enhanced Perovskite Solar Cells

7.1 Portable Electronics

- **Compact Energy Solutions**: MXene-enhanced perovskite solar cells offer compact and high-efficiency energy solutions for portable electronic devices such as smartphones, tablets, and wearables. Their improved efficiency and lightweight nature make them ideal for integrating into portable devices where space and weight are critical factors.
- **Extended Battery Life**: The integration of these advanced solar cells can extend the battery life of portable electronics by providing supplementary power from ambient light sources. This can reduce the frequency of charging and enhance the usability of devices during extended outdoor activities or in low-light conditions.
- **Wearable Solar Devices**: Examples include wearable solar-powered chargers or integrated solar panels in smartwatches and fitness trackers, which use MXene-enhanced perovskite solar cells to provide efficient energy harvesting in a small form factor.
- **Flexible Solar Fabrics**: MXene-enhanced perovskite solar cells are also being utilized in flexible solar fabrics, which can be integrated into clothing or outdoor gear. These fabrics incorporate lightweight, flexible solar panels that harness solar energy, making them ideal for applications such as powering portable electronic devices or providing supplemental energy in outdoor environments. This application demonstrates the versatility of MXene-enhanced perovskite solar cells in creating innovative and functional energy solutions in wearable technology.
- **Solar-Powered Smart Glasses**: Another example is the use of MXene-enhanced perovskite solar cells in solar-powered smart glasses. These glasses integrate compact, lightweight solar panels into their frames to power built-in electronics, such as augmented reality (AR) displays, communication systems, or sensors. This application leverages the high efficiency and flexibility of MXene-enhanced perovskite cells to provide a seamless energy source for advanced wearable tech while maintaining a sleek, unobtrusive design.

7.2 Electric Vehicles (EVs)

- **Supplemental Power Generation**: MXene-enhanced perovskite solar cells can be integrated into the design of electric vehicles to provide supplemental power for auxiliary systems, such as climate control and infotainment systems. This can help reduce the load on the main battery and improve overall vehicle efficiency.
- **Charging Assistance**: Solar cells can also contribute to charging assistance, offering additional energy to recharge the vehicle's battery when parked in sunlight. This feature can be particularly beneficial for extending the range of

electric vehicles and reducing dependence on conventional charging infrastructure.

- **Solar Panels on Roofs**: Incorporating flexible or semi-transparent MXene-enhanced solar panels on the roofs or surfaces of electric vehicles can harness solar energy while maintaining the vehicle's aerodynamic design and aesthetics.

7.3 *Renewable Energy Systems*

- **Solar Energy Harvesting**: MXene-enhanced perovskite solar cells can be used in renewable energy systems to improve the efficiency of solar energy harvesting. Their high efficiency and stability make them suitable for deployment in large-scale solar farms and residential solar panels.
- **Hybrid Systems**: These solar cells can also be integrated into hybrid renewable energy systems that combine solar power with other energy sources, such as wind or hydroelectric power, to create more reliable and consistent energy solutions.
- **Enhanced Efficiency**: Incorporating MXenes into solar panels can significantly boost their overall efficiency by improving charge transport and light absorption. This enhancement leads to higher energy output and more effective utilization of available sunlight, contributing to more efficient and cost-effective solar energy systems.
- **Flexible and Lightweight Panels**: MXene-enhanced solar cells offer the advantage of being flexible and lightweight. This flexibility allows for innovative applications, such as integrating solar panels into building materials or creating portable and adaptable solar solutions. These properties expand the potential uses of solar technology, making it suitable for a wider range of applications and environments.

7.4 *Industrial Power Backup*

- **Emergency Power Supply**: MXene-enhanced perovskite solar cells can provide reliable power backup solutions for industrial applications. Their high efficiency and durability make them suitable for emergency power supplies in critical infrastructure, such as data centers, communication networks, and manufacturing facilities.
- **Reducing Downtime**: By integrating these solar cells into industrial power systems, companies can reduce downtime and maintain operational continuity during power outages or fluctuations. This can lead to increased reliability and reduced costs associated with power disruptions.
- **Solar Power Integration:** MXene-enhanced perovskite solar panels can be seamlessly integrated into existing power backup systems, providing a sustainable and cost-effective solution for maintaining power during outages. This integration

enhances the reliability of backup systems and reduces dependence on traditional energy sources.

- **Energy Storage Solutions:** Combining MXene-enhanced solar cells with advanced energy storage systems, such as high-capacity batteries or capacitors, creates a robust backup power solution. This setup ensures an uninterrupted power supply, enhances energy security, and optimizes the use of harvested solar energy by storing excess power for future use.

8 Challenges and Future Directions

$Ti_3C_2T_X$ MXene has seen significant developments in solar cell applications since its first introduction in 2018. It primarily functions as an additive, electrode, or charge transport layer in various solar cell types, with perovskite and organic solar cells being the most common. The exploration of MXene's potential in solar cells is still in its early stages, with much room for improvement in efficiency and stability. The properties of MXenes, such as morphology, conductivity, and stability, are highly sensitive to fabrication processes, highlighting the need for controlled, scalable, and cost-effective production methods. Further research is needed to optimize MXene's material properties, enhance device performance, and address challenges like air stability and F-free synthesis. The vast family of MXenes holds promise for future breakthroughs in photovoltaic devices and other optoelectronic applications, driven by a deeper understanding of their properties and innovative approaches to their integration.

8.1 Scalability of Synthesis

- **Material Production**: Scaling up the production of MXenes involves challenges related to the synthesis of large quantities of high-quality materials. Ensuring uniformity and consistency in MXene properties at a larger scale is crucial for maintaining performance in commercial applications.
- **Integration with Perovskites**: Integrating MXenes into perovskite solar cells on a large scale requires scalable methods for deposition and processing. Techniques used in laboratory settings may not always translate directly to industrial-scale production, leading to potential issues with uniformity and efficiency.
- **Developing Scalable Methods**: Research into scalable synthesis techniques for MXenes, such as continuous-flow processes or large-scale chemical vapor deposition (CVD), could help address these challenges. Optimizing these methods for consistency and cost-effectiveness is essential for commercial viability.
- **Optimizing Integration Techniques**: Advancements in deposition and integration techniques, such as roll-to-roll processing or automated manufacturing

processes, could improve the scalability and efficiency of incorporating MXenes into perovskite solar cells.

8.2 Stability and Durability

Environmental Sensitivity: MXenes and perovskite materials are highly sensitive to environmental factors, including moisture, oxygen, and ultraviolet (UV) light. MXenes, particularly when exposed to air, can undergo oxidation, leading to increased resistance and diminished electrical performance. Similarly, perovskite materials are known to degrade rapidly when exposed to moisture and UV light, which can affect their stability and efficiency. To ensure the long-term stability and durability of MXene-enhanced perovskite solar cells, it is crucial to develop effective strategies to protect these materials from environmental stresses. This may involve implementing robust encapsulation techniques, selecting appropriate passivation materials, and designing devices that are resistant to environmental degradation.

Material Degradation: The degradation of both MXenes and perovskite materials over time poses significant challenges for the longevity and performance of solar cells. Exposure to harsh environmental conditions, such as high humidity, elevated temperatures, and intense UV radiation, can accelerate the degradation processes. For MXenes, oxidation and surface contamination can impair electrical conductivity and alter their properties. Perovskites can experience phase transitions, loss of crystallinity, and formation of unwanted by-products, all of which contribute to performance degradation. To mitigate these issues, it is essential to explore advanced stabilization techniques, such as developing new material formulations, optimizing processing conditions, and incorporating protective coatings. Enhancing the resistance of both MXenes and perovskites to degradation will be key to achieving more reliable and durable solar cells.

- **Enhancing Material Stability**: Research into protective coatings and stabilizing agents for MXenes and perovskites can help mitigate degradation and enhance the longevity of the solar cells. Developing materials with intrinsic stability and resistance to environmental factors is also important.
- **Long-Term Testing**: Implementing comprehensive long-term testing protocols to assess the durability and performance of MXene-enhanced perovskite solar cells under real-world conditions can provide valuable insights for improving their stability.

8.3 Cost-Effectiveness

- **Material Costs**: The synthesis and processing of MXenes and perovskite materials can be expensive. Reducing the cost of raw materials and production processes is crucial for making MXene-enhanced perovskite solar cells economically viable.
- **Manufacturing Costs**: The cost of integrating MXenes into perovskite solar cells, including deposition techniques and quality control, can impact the overall cost of the final product. Optimizing manufacturing processes to reduce costs while maintaining quality is essential.
- **Developing Cost-Effective Methods**: Innovations in synthesis and processing techniques that lower production costs without compromising quality are important. Exploring alternative materials and scalable methods could also contribute to cost reduction.
- **Economies of Scale**: Achieving economies of scale through large-scale production and automation can help reduce costs and make MXene-enhanced perovskite solar cells more competitive in the market.

8.4 Emerging Trends

New Materials and Innovations:

Advanced MXenes: Research into novel MXene compositions and structures has the potential to yield enhanced properties and performance for perovskite solar cells. Exploring MXenes with varying chemical compositions or hybrid structures, such as those combining different transition metals or surface terminations, could lead to improved electrical conductivity, stability, and compatibility with perovskite materials. This could result in more efficient charge transport, reduced recombination losses, and overall better solar cell performance.

Innovative Fabrication Techniques: Emerging fabrication methods, such as 3D printing and advanced coating techniques, offer new opportunities for integrating MXenes into perovskite solar cells. 3D printing could enable the creation of complex, custom-designed structures and multi-layered devices, potentially leading to novel device architectures that enhance light absorption and charge extraction. Advanced coating methods, including spray coating and vapor deposition, might allow for more precise and uniform application of MXenes, improving their interaction with perovskite layers and optimizing device performance. These innovations could facilitate the development of scalable, cost-effective production processes and lead to significant advancements in solar cell technology.

Future Directions:

- **Exploring Hybrid Materials**: Combining MXenes with other advanced materials, such as nanomaterials or organic semiconductors, could lead to new hybrid systems with improved performance and functionality.
- **Investigating New Applications**: Beyond traditional solar cells, exploring new applications for MXene-enhanced perovskite materials, such as flexible electronics or integrated energy systems, could open up new opportunities for innovation.

8.5 *Environmental and Safety Considerations*

Eco-Friendly Materials and Processes:

- **Sustainable Practices**: Ensuring that the synthesis and processing of MXenes and perovskites are environmentally friendly is important for reducing the ecological impact of solar cell production. Research into sustainable practices and the use of eco-friendly materials is essential.
- **Safe Disposal**: Proper disposal and recycling of MXenes and perovskite materials are necessary to address potential environmental and health concerns. Developing safe disposal methods and recycling processes is crucial for minimizing the impact of end-of-life solar cells.

Future Directions:

- **Green Chemistry**: Advancing green chemistry approaches in the synthesis and processing of MXenes and perovskites can help reduce environmental impact. This includes using non-toxic chemicals and minimizing waste.
- **Recycling and Reuse**: Developing effective recycling and reuse strategies for MXene-enhanced perovskite solar cells can contribute to a circular economy and reduce the environmental footprint of solar technologies.

Summary:

The challenges of MXene-enhanced perovskite solar cells include scaling up synthesis, ensuring stability, reducing costs, and addressing environmental and safety concerns. Scaling requires efficient, cost-effective production methods, while stability is challenged by MXenes' susceptibility to oxidation and degradation. Cost reduction is vital for commercial viability, and environmental safety must be considered due to potential toxicity. Future efforts should focus on developing scalable and affordable synthesis techniques, improving material stability through advanced passivation, exploring new materials and fabrication methods, and adopting sustainable practices. By overcoming these obstacles, MXene-enhanced perovskite solar cells could significantly advance solar energy technology, leading to broader applications and a more sustainable future.

References

1. Maka, A.O.M.: Solar energy technology and its roles in sustainable development, 476–483 (2022)
2. Azarpour, A., Suhaimi, S.: A review on the drawbacks of renewable energy as a promising, 317–328 (2013). https://doi.org/10.1007/s13369-012-0436-6
3. Patriche, C., Bandoc, G., Pr, R.: Spatial assessment of solar energy potential at global scale. A geographical approach **209** (2019). https://doi.org/10.1016/j.jclepro.2018.10.239
4. Rahman, A., Farrok, O., Haque, M.: Environmental impact of renewable energy source based electrical power plants: solar, wind, hydroelectric, biomass, geothermal, tidal, ocean, and osmotic. Renew. Sustain. Energy Rev. **161**, 112279 (2022). https://doi.org/10.1016/j.rser.2022.112279
5. Armaroli, N., Balzani, V.: The future of energy supply: challenges and opportunities, 52–66 (2007). https://doi.org/10.1002/anie.200602373
6. Pastuszak, J., Photovoltaic cell generations and current research directions for their development (2022)
7. Nazir, S., Ali, A., Aftab, A., Muqeet, H.A., Mirsaeidi, S., Zhang, J.: Techno-economic and environmental perspectives of solar cell technologies: a comprehensive review (2023)
8. Olorunfemi, T.R., Nwulu, N.I., Ismail, S.O.: Composites as candidate materials for photovoltaic cells ABSTRACT. Adv. Mater. Process. Technol. **8**(4), 2378–2397 (2022). https://doi.org/10.1080/2374068X.2022.2046328
9. Sharma, P., Goyal, P.: Materials today: proceedings evolution of PV technology from conventional to nano-materials. Mater. Today Proc. **28**, 1593–1597 (2020). https://doi.org/10.1016/j.matpr.2020.04.846
10. Fazal, M.A., Rubaiee, S.: Progress of PV cell technology: feasibility of building materials, cost, performance, and stability. Sol. Energy **258**, 203–219 (2023). https://doi.org/10.1016/j.solener.2023.04.066
11. Cells, P.S.: Advancements in photovoltaic cell materials: silicon, organic, and perovskite solar cells (2024)
12. Shah, N., et al.: A review of third generation solar cells (2023)
13. Fakharuddin, A., Schmidt-mende, L.: Hybrid organic/inorganic and perovskite solar cells (2018)
14. Veldhuis, S.A., et al.: Perovskite materials for light-emitting diodes and lasers (2016). https://doi.org/10.1002/adma.201600669
15. Yang, G., et al.: Stable and low-photovoltage-loss perovskite solar cells by multifunctional passivation. Nat. Photon. **15** (2021). https://doi.org/10.1038/s41566-021-00829-4
16. Elumalai, N.K., Mahmud, A., Wang, D., Uddin, A.: Perovskite solar cells: progress and advancements (2016). https://doi.org/10.3390/en9110861
17. Saliba, M., Buonassisi, T., Grätzel, M., Abate, A., Tress, W., Hagfeldt, A.: Promises and challenges of perovskite solar cells. **744**, 739–744 (2017)
18. Rao, M.K., Sangeetha, D.N., Selvakumar, M., Sudhakar, Y.N., Mahesha, M.G.: Review on persistent challenges of perovskite solar cells ' stability. Sol. Energy **218**, 469–491 (2021). https://doi.org/10.1016/j.solener.2021.03.005
19. Hasnain, M., Hussain, I.: A review on catalyst convergence: unleashing the potential of MXenes for CO2 electrochemical reduction into high-value liquid product nano materials science a review on catalyst convergence: unleashing the potential of MXenes for CO2 electrochemical reduction into high-value liquid product (2024). https://doi.org/10.1016/j.nanoms.2024.06.006
20. Hussain, T., Chandio, I., Ali, A., Hyder, A., Memon, A.: Nanoscale Recent developments of artificial intelligence in MXene-based devices: from synthesis to applications (2024). https://doi.org/10.1039/d4nr03050h
21. Alhamada, T.F., Hanim, M.A.A., Jung, D.W., Nuraini, A.A., Hasan, W.Z.W.: A brief review of the role of 2D Mxene nanosheets toward solar cells efficiency improvement, 1–18 (2021)

22. Yang, L., Li, P., Ma, J., Zhang, X., Wang, X., Liu, Y.: MXenes for perovskite solar cells: progress and prospects. J. Energy Chem. **81**, 443–461 (2023). https://doi.org/10.1016/j.jechem.2023.02.041
23. Aftab, S., et al.: Two-dimensional MXene incorporating for electron and hole transport in high-performance perovskite solar cells. Mater. Today Energy **36**, 101366 (2023). https://doi.org/10.1016/j.mtener.2023.101366
24. Natu, V., Barsoum, M.W.: MXene surface terminations: a perspective (2023). https://doi.org/10.1021/acs.jpcc.3c04324
25. Bj, J., Rosen, J.: Functionalizing MXenes by tailoring surface terminations in different chemical environments. https://doi.org/10.1021/acs.chemmater.1c01264
26. Chaudhuri, K., et al.: Optical properties of MXenes optical properties of 2D materials, 327–346 (2019)
27. Jiang, X., Kuklin, A.V., Baev, A., Ge, Y., Ågren, H.: Two-dimensional MXenes: from morphological to optical, electric, and magnetic properties and applications. Phys. Rep. **848**, 1–58 (2020). https://doi.org/10.1016/j.physrep.2019.12.006
28. Wang, Y., Xu, Y., Hu, M., Ling, H., Zhu, X.: Review article MXenes: focus on optical and electronic properties and corresponding applications **9**(7), 1601–1620 (2020)
29. Guo, J., Sun, Y., Liu, B., Zhang, Q., Peng, Q.: Two-dimensional scandium-based carbides (MXene): band gap modulation and optical properties. J. Alloys Compd. **712**, 752–759 (2017). https://doi.org/10.1016/j.jallcom.2017.04.149
30. Fu, B., et al.: MXenes: synthesis, optical properties, and applications in ultrafast photonics **2006054**, 1–23 (2021). https://doi.org/10.1002/smll.202006054
31. Han, M., et al.: Tailoring electronic and optical properties of MXenes through forming solid solutions (2020). https://doi.org/10.1021/jacs.0c07395
32. Salim, O., Mahmoud, K.A., Pant, K.K., Joshi, R.K.: Introduction to MXenes: synthesis and characteristics. Mater. Today Chem. **14**, 100191 (2019). https://doi.org/10.1016/j.mtchem.2019.08.010
33. Wei, Y., Zhang, P., Soomro R.A., Zhu, Q., Xu, B.: Advances in the synthesis of 2D MXenes **2103148**, 1–30. https://doi.org/10.1002/adma.202103148
34. Green, M.A., Ho-baillie, A., Snaith, H.J.: The emergence of perovskite solar cells **8** (2014). https://doi.org/10.1038/nphoton.2014.134.
35. Jung, H.S., Park, N.: Perovskite solar cells: from materials to devices, 1–16 (2014). https://doi.org/10.1002/smll.201402767.
36. Kim, J.Y., Lee, J., Jung, H.S., Shin, H., Park, N.: High-efficiency perovskite solar cells (2020). https://doi.org/10.1021/acs.chemrev.0c00107
37. Zuo, C., Bolink, H.J., Han, H., Huang, J., Cahen, D.: Advances in perovskite solar cells, 1–16 (2016). https://doi.org/10.1002/advs.201500324
38. Zhang, H., Park, N.: DeCarbon progress and issues in p-i-n type perovskite solar cells. DeCarbon **3**(2023), 100025 (2024). https://doi.org/10.1016/j.decarb.2023.100025
39. Rong, Y., et al.: Challenges for commercializing perovskite solar cells (2018). https://doi.org/10.1126/science.aat8235
40. Wang, D., Wright, M., Elumalai, N.K., Uddin, A.: Solar energy materials & solar cells stability of perovskite solar cells. Sol. Energy Mater. Sol. Cells **147**, 255–275 (2016). https://doi.org/10.1016/j.solmat.2015.12.025
41. Park, N.: Perovskite solar cells: an emerging photovoltaic technology. Biochem. Pharmacol. **18**(2), 65–72 (2015). https://doi.org/10.1016/j.mattod.2014.07.007
42. Ahmed, M.I., Habib, A., Javaid, S.S.: Perovskite solar cells: potentials , challenges , and opportunities (2015). https://doi.org/10.1155/2015/592308
43. Li, L., Zhang, S., Yang, Z., Esso, E., Berthold, S., Chen, W.: Recent advances of flexible perovskite solar cells. J. Energy Chem. **27**(3), 673–689 (2018). https://doi.org/10.1016/j.jechem.2018.01.003
44. Asgharizadeh, S., Azadi, S.K.: Additive MXene and dominant recombination channel in perovskite solar cells. Sol. Energy **241**, 720–727 (2022). https://doi.org/10.1016/j.solener.2022.07.001

45. Niu, Y., et al.: Nano energy Nb2C MXenes modified SnO2 as high quality electron transfer layer for efficient and stability perovskite solar cells. Nano Energy **89**(PB), 106455 (2021). https://doi.org/10.1016/j.nanoen.2021.106455
46. Palei, S., Murali, G., Kim, C.H., In, I., Lee, S.Y.: A review on interface engineering of MXenes for perovskite solar cells **15**(1) (2023). https://doi.org/10.1007/s40820-023-01083-9
47. Du, J.B., Yang, L., Jin, X., Liu, C.L., Wang, H.H., Wang, X.F.: Spray deposition of vinyl tris(2-methoxyethoxy) silane-doped Ti3C2Tx MXene hole transporting layer for planar perovskite solar cells. J. Alloys Compd. **900**, 163372 (2022). https://doi.org/10.1016/j.jallcom.2021.163372
48. Hussain, S., et al.: Characteristics of Mo2C-CNTs hybrid blended hole transport layer in the perovskite solar cells and X-ray detectors. J. Alloys Compd. **885**, 161039 (2021). https://doi.org/10.1016/j.jallcom.2021.161039
49. Chen, T., et al.: Accelerating hole extraction by inserting 2D Ti3C2-MXene interlayer to all inorganic perovskite solar cells with long-term stability. J. Mater. Chem. A **7**(36), 20597–20603 (2019). https://doi.org/10.1039/c9ta06035a
50. Guo, Z., et al.: High electrical conductivity 2D MXene serves as additive of perovskite for efficient solar cells. Small **14**(47), 1–8 (2018). https://doi.org/10.1002/smll.201802738
51. Chen, X., et al.: Dual interfacial modification engineering with 2D MXene quantum dots and copper sulphide nanocrystals enabled high-performance perovskite solar cells. Adv. Funct. Mater. **30**(30), 1–11 (2020). https://doi.org/10.1002/adfm.202003295
52. Yang, L., et al.: SnO2-Ti3C2 MXene electron transport layers for perovskite solar cells. J. Mater. Chem. A **7**(10), 5635–5642 (2019). https://doi.org/10.1039/c8ta12140k
53. Huang, L., Zhou, X., Xue, R., Xu, P., Wang, S., Xu, C.: Low-temperature growing anatase—multi-dimensional Heterojunctions at MXene conductive network for high-efficient perovskite solar cells. Nano-Micro Lett. **12**(1), 1–19 (2020). https://doi.org/10.1007/s40820-020-0379-5
54. Agresti, A., et al.: Titanium-carbide MXenes for work function and interface engineering in perovskite solar cells. Nat Mater **18** (2019). https://doi.org/10.1038/s41563-019-0478-1
55. Manuscript, A.: Mater. Chem. C (2020). https://doi.org/10.1039/D0TC00075B
56. Hou, C., Yu, H.: ZnO/Ti3C2Tx monolayer electron transport layers with enhanced conductivity for highly efficient inverted polymer solar cells. Chem. Eng. J. **407**, 127192 (2021). https://doi.org/10.1016/j.cej.2020.127192
57. Cao, J., et al.: RSC advances free perovskite solar cells: 2D MXenes electrodes †, 34152–34157 (2019). https://doi.org/10.1039/c9ra06091j
58. Tang, H., Feng, H., Wang, H., Wan, X., Liang, J.: Highly conducting MXene—silver nanowire transparent electrodes for flexible organic solar cells (2019). https://doi.org/10.1021/acsami.9b04113
59. Qin, L., et al.: Supercapacitor based on water-processed MXene, 5467–5475 (2020). https://doi.org/10.1039/d0ta00687d
60. Fu, H., et al.: MXene-contacted silicon solar cells with 11.5% efficiency **1900180**, 1–9 (2019). https://doi.org/10.1002/aenm.201900180
61. Zhang, J., et al.: MXene-modulated electrode/SnO2 interface boosting charge transport in perovskite solar cells (2020). https://doi.org/10.1021/acsami.0c17338
62. Manuscript, A.: Mater. Chem. A (2019). https://doi.org/10.1039/C9TA01195A.
63. Bati, A.S.R., et al.: 1D-2D synergistic MXene-nanotubes hybrids for efficient perovskite solar cells **2**, 1–9 (2021). https://doi.org/10.1002/smll.202101925
64. Wang, J., et al.: Plasma Oxidized Ti3C2T (2021). https://doi.org/10.1021/acsami.1c07146
65. Yang, L., et al.: Performance improvement of MXene-based perovskite solar cells upon property transition from metallic to semiconductive by oxidation of Ti3C2Tx in air † 5016–5025 (2021). https://doi.org/10.1039/d0ta11397b

MXene and Its Composites for Battery Applications

Chandan Kumar Maity, Khusboo Kumari, Salim Hassan Siddiki, and Ganesh Chandra Nayak

Abstract The family of 2D transition metals carbides, carbonitrides, and nitrides—communally known as MXenes—have been explored dramatically since Ti_3C_2 was first discovered in 2011. MXenes' hydrophilic surfaces are a result of surface functionalities like fluorine, oxygen, or hydroxyls. MXenes' versatile surface chemistry allows for customization for a variety of applications, like lubrication, photo-, electro-, and chemical catalysis, energy storage, composite reinforcement, shielding against electromagnetic interferences, reinforcements, biosensors, and water purifications. Thus, this chapter provides a thorough discussion of the usage of MXenes and their composites in rechargeable batteries including the evaluation of their efficacy. On the other hand, MXenes suffer from low electrochemical kinetics and capacity degradation due to self-stacking and agglomeration. Therefore, it is crucial to modify MXene using different nanomaterials. Consequently, MXene-based composites can provide more control over the chemical and physical characteristics of MXene because of their surface terminal groups. The current developments in MXene and its composite nanoarchitectures, which were created for various rechargeable batteries to improve electrochemical properties, problems, and scientific solutions, were covered in this chapter.

Keyword MXene · MXene-based composites · Energy storage · Rechargeable batteries

1 Introduction

One of the main struggles facing humanity in the last few years is the energy crisis. Modern research is focused on developing renewable and sustainable energy sources in order to solve such a predicament. Secondary batteries exhibit high potentiality in the wide field of energy storages in this regard. Li-ion batteries (LIBs) are the most widely used type among them [1, 2]. Nonetheless, current research focuses

C. K. Maity · K. Kumari · S. H. Siddiki · G. C. Nayak (✉)
Department of Chemistry and Chemical Biology, IIT (ISM) Dhanbad, Dhanbad, India
e-mail: gcnayak@iitism.ac.in

© The Author(s), under exclusive license to Springer Nature Singapore Pte Ltd. 2025

S. Moharana and R. N. Mahaling (eds.), *Surface Engineering of MXenes*, Engineering Materials, https://doi.org/10.1007/978-981-96-3640-2_11

on investigating alternative metal-based batteries, such as Na-ion, Li–S, and Mg-ion batteries [3–5]. Other battery types, such as metal-air batteries have also been developed in addition to metal-ion batteries [6, 7]. These two classes of energy storage technologies are regarded as the forerunners of contemporary energy research, even though their charge storing methods are different.

Due to their rich physicochemical properties, 2D materials have amassed a substantial research attention in the last several years. With the discovery of new material types, the family of 2D materials is growing daily. The 2D materials' remarkable properties, such as their high conducting properties, surface areas, heightened mechanical strength, and robust flexibility, are noteworthy. MXenes, a two-dimensional (2D) substance, are composed of carbonitrides, carbides, and transition metal nitrides [8]. Since the initial study on this substance was published in 2011, it has been broadly studied for a wide range of purposes [9]. Moreover, it has been claimed that MXenes work well in LIB, supercapacitor, catalysis, sensors, and further applications [10–15]. Some characteristics of MXenes, including as their rich organic bonding, hydrophilicity, high electrical conductivity, abundance of surface functions, and improved surface areas, are responsible for their many applications [11].

The electrolyte, cathode, and anode are the main parts of rechargeable batteries. An external circuit allows electrons to transport from anode to cathode during the discharge cycle, powering the attached device. Ions flow via the electrolyte simultaneously to speed up the electrochemical processes at the cathode and anode. The electrical current required to power gadgets is produced by this method [16]. Different types of rechargeable batteries involve different specialized systems and materials. Lithium ions, for instance, are shuttled between the anode and cathode of LIBs, which are commonly found in portable gadgets [16, 17]. The mobility of metal hydride or cadmium ions is involved in nickel-metal hydride [18] and nickel–cadmium [19] batteries, respectively. The electrolyte's composition greatly affects the battery's performance properties. It might be a liquid, gel, or solid.

Because of its longer lifespan, reduced environmental impact, and increased energy density, LIBs are widely used today [11]. Because there are less resources available to make LIBs, their costs have gone up recently. Researchers have invented other types of batteries, including lithium-oxygen batteries (LOBs), lithium-sulfur batteries (LSBs), and lithium-metal batteries (LMBs), to evade the ongoing price spike in LIBs [20]. Zinc-ion batteries (ZIBs) [11], potassium-ion batteries (PIBs) [11, 21], sodium-ion batteries (SIBs) [22], magnesium-ion batteries (MgIBs) [23], and so on are some more examples. These recently created energy storage systems (ESS) do have certain drawbacks, though, for which researchers are still looking for fixes. For example, it is necessary to have an appropriate electrode with minimal diffusion barriers and ion migration potential. It's interesting to note that MXenes, a promising electrode, serves as the electrode for the recently created ESS. In addition to the previously listed benefits, MXene is shown to be helpful because, among other things, it can restrict the metal dendrites growth and lessen the shuttling impact of polysulfide [11].

MXenes has been found to have certain disadvantages, such as limited flexibility, restacking tendency, and instability in oxygenated settings, despite its many uses in ESS. Consequently, significant researches have been made to further develop MXenes [11]. Lately, scientists have started to combine other appropriate components to create a new substance called a composite, which has better qualities. Consequently, MXenes have been combined with perovskite [24], nanomaterials [25], metal–organic frameworks [26], conducting polymers [27], and other materials in an effort to find high-quality MXenes. A comprehensive review of MXene, its electrochemistry, the ways in which it can be physically and chemically altered, and the numerous MXene composites for rechargeable batteries are given in this chapter. Additionally, we evaluated the efficacy of MXene composites in comparison to unmodified MXenes and had a strong discussion about their potential in rechargeable batteries. Lastly, a future research focus was placed on the perspective and present difficulties facing the use of MXene composites in rechargeable batteries.

2 MXene and Its Advantages

$M_{n+1}X_nT_x$ (n = 1–3) is the representation for MXenes, where M stands for transition metals (such as Ti, Cr, Mo, V, etc.), X for elements (such as C, N, etc.), and T_x for terminal groups (such as –O, –OH, –F, etc.) (Fig. 1a, b) [28]. Typically, the "A" layer in the MAX phase is specifically removed in order to create MXenes. Prof. Yury Gogotsi of Drexel University in the USA founded the MXene era in 2011 [9]. Based on Ti, the first MXene was found. Nonetheless, other transition metals-based MXenes have also been produced in less than ten years due to the quick advancement of MXene research. The initial synthetic method relied on etching with HF assistance. The HF-free etching method was unveiled in 2015. Later on, reports of further etching techniques included electrochemical, alkali, molten salt, and others. In addition, in 2021 the ionic liquid-assisted etching method was introduced [29]. Furthermore, MXene derivatives with diverse functions, such as 1D MXene nanoribbons, 2D MXene nanosheets, 3D MXene aerogels, and 0D MXene quantum dots, have been discovered for diverse applications [30]. $Ti_3C_2T_x$ was the first member of the MXene family, and other transition metal-based MXenes such as V_2CT_x, Mo_2CT_x, Nb_2CT_x, Mn_2CT_x, etc. were discovered as the family was being further enlarged. Additional bimetallic MXenes have also been investigated, such as Cr_2TiC_2, Mo_2TiC_2, $Mo_2Ti_2C_3$, etc. Nitrides and carbonitrides such as Ti_2NT_x, $VTi_4N_3T_x$, $MnTi_4N_3T_x$, etc. have also been found in addition to the usual carbide MXenes (Fig. 1a, b). Doping and defects can enhance the properties of two-dimensional materials. Heteroatom-doped/substituted MXenes and vacancy-induced MXenes are being investigated to improve the optical, electrical, electrochemical, and magnetic properties of MXenes, much like other 2D materials [31]. There are many chances for scientists to go deeper into this research area because MXene research is still in its early stages.

MXene, graphene, and MoS_2 belong to the same 2D family class. On the other hand, their inherent qualities and structural features differ. The main variations are enumerated below:

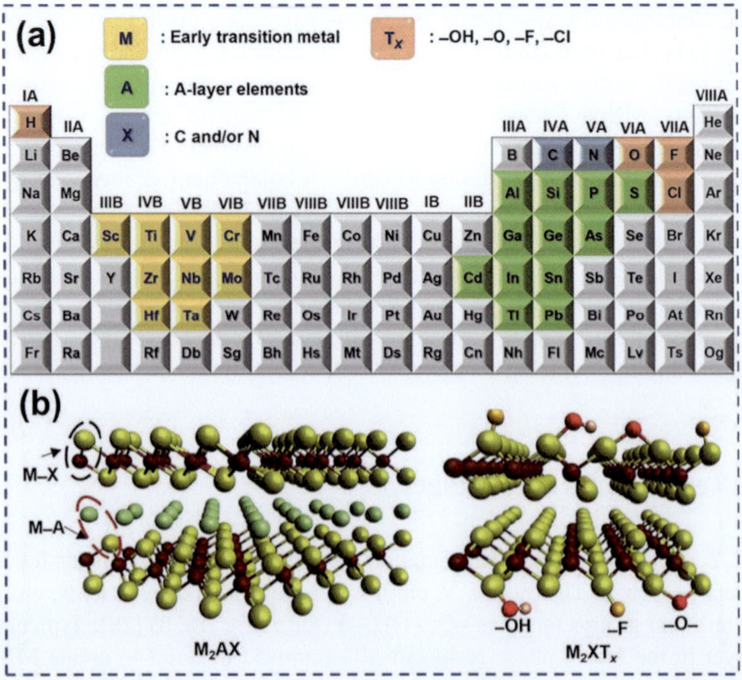

Fig. 1 **a** A periodic table with colors designating the constituent elements of MAX and MXene and **b** Representative structures of MXene (M_2XT_x) and MAX (M_2AX) phases. Redrawn with permission from Ref. [28]

i. Although graphene is hydrophobic by nature, MXene has strong hydrophilicity, suggesting that it can be used in aqueous solutions. However, it is discovered that MoS_2's metallic phase is hydrophilic.
ii. MXenes include a particular class of materials that includes functional groups like –OH, –F, and –O as well as transition metals and N/C. But the only element in graphene is carbon. Though certain oxygen functions are present in the comparable reduced graphene oxide. Conversely, MoS_2 is composed of S–Mo–S alternating layers.
iii. Graphene and MoS_2 possess lower electrical conductivity than MXenes. The conductivity of MXenes materials was clearly ten times higher than that of reduced graphene oxide sheet [32, 33].
iv. MXene's higher mechanical capabilities over other 2D counterparts are demonstrated by its stiffer nature compared to both graphene and MoS_2 [34, 35].

3 Stability and Electrochemical Conductivity of MXene

MXenes are becoming more and more popular for many useful purposes. Their unpredictability, though, is a recurring barrier. Their microstructure is altered and their electrochemical characteristics are diminished as a result of their fast oxidation caused by atomic-scale flaws and storage conditions [11]. Furthermore, MXenes readily react to generate more stable metal oxides when they mix with trace oxygen (even in a safe atmosphere) and oxygen-containing groups that are attached to the surface in both ambient and aquatic conditions. Among the MXenes materials that are most frequently utilized, $Ti_3C_2T_x$-MXene, for instance, oxidizes in the presence of air easily, sometimes to the point where their T–C bonds break and TiO_2 particles and carbon are formed. Through the Reactive force field (ReaxFF) molecular dynamics modeling, Lotfi et al. examined the level of oxidation for Ti_3C_2 MXene in the presence of H_2O_2, wet air (water and oxygen), and dry air (oxygen), and at varied oxidation ranges from 1000 to 3000 K [36]. Figure 2 illustrates that the rate of oxidation for $Ti_3C_2T_x$ MXene was higher in H_2O_2 than in wet air at 1000 K and 1500 Kps, whereas there was no discernible oxidation effect when exposed to dry air [36]. According to a different study, MXenes change structurally when exposed to various oxidative environments. This is because Ti atom migrates from the center of MXene to their surfaces, functionalizing the surfaces with –OH and –O– groups [37]. The oxidations that started on the edges of the MXenes flakes and eventually expanded to the whole surface was documented by Naguib et al. [38]. Stability is said to be improved by surface functionalization or coating, freeze drying, pH adjustment, MAX phase synthesis optimization, appropriate storage conditions, etc. To comprehend the stabilities and structures of terminated-MXenes more clearly, significant numbers of researches have been performed to evaluate the effects and configurations of different functional moieties. The stability of $Ti_3C_2T_x$ in the presence of anhydrous isopropanol and water saturated with argon or air was examined by Huang et al. [39]. The results indicated fast oxidation in the presence of water and oxygen. When there was simply oxygen or water present, the rate of deterioration dropped significantly. MXenes required 41 days in the presence of argon and water; and 2026 days in the presence of oxygen and isopropanol to degrade. On the other hand, MXene's two-dimensional structure and remarkable mechanical properties influence it as a promising option for applications requiring flexible energy storage. MXenes' two-dimensional, flexible lamellar structure makes them a viable option for "micro-electronic" system devices, the demand for which has sharply increased recently. These devices need to be small and portable.

According to reports, MXenes exhibit cost-effective production, hydrophilicity, and greater electric conductivities ranging from metallic to semiconductor than multilayered graphene, carbon nanotubes, and reduced graphene oxide materials [11]. Additionally, a number of papers demonstrate that simulated conductivities are often significantly higher than those found through experimentation [40]. Simple MXenes like $Ti_{n+1}X_n$ have metallic characteristics that decrease with increasing value of "n" because of the extra Ti–X bonds. Changing the chemical on the surface is one way

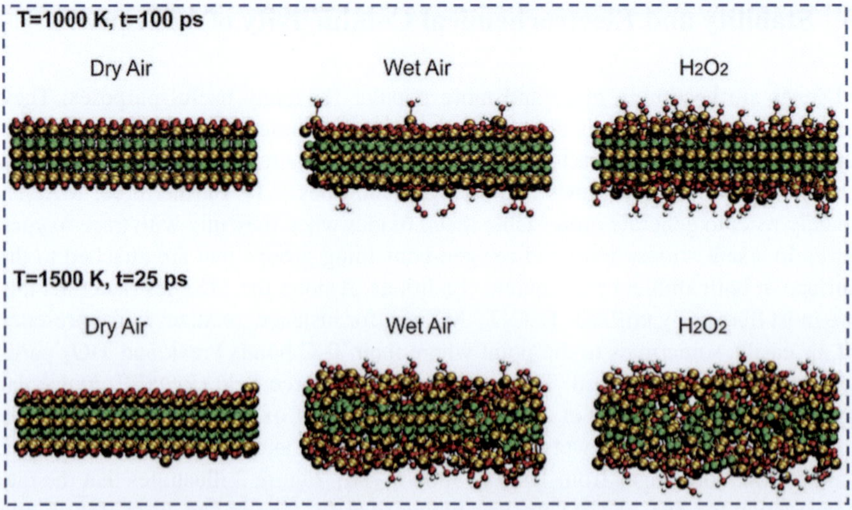

Fig. 2 Ti_3C_2 MXene structures' rates of oxidation at different oxidation simulation temperatures (1000 and 1500 K) in the presence of H_2O_2, dry air, and wet air (O: red, C: green, and Ti: tan). Redrawn with permission from Ref. [36]

to increase their conductivity. For instance, adding –F or –Cl groups in place of the surface –OH groups can increase electrical conductivity. MXenes' conductivity can also be increased by doping them with heteroatoms like S or N, or by generating surface defects like dislocations or vacancies, among other methods. Intercalation also can enhance conductivity and it is appropriate for multilayers MXenes. Hart et al. have demonstrated that intercalation can cause a material's nature to change from metallic to semiconductor-like by raising interlayer distance and, as a result, interlayer resistance [41]. Additionally, they demonstrated that the termination, intercalation, and MXene chemistry all trigger to conductivity for every multilayer MXene sample. Moreover, the electronic conductivity of MXenes with comparatively low etching interval and HF concentration levels was increased due to their reduced flaws and bigger lateral diameters (for example, larger flake sizes produced conductivities that were five times higher than those of small-sized MXenes). Their conductivities can be impacted by environmental humidity, which highlights the necessity for relative humidity sensing materials. Improving the electrical characteristics of a surface with thermal and/or alkaline treatments works effectively. They demonstrate the two orders of magnitude growth in conductivities. Functional groups (especially –F) and intercalated molecules are the source of this rise and can be added or removed [42].

4 Electrochemistry of MXene in Favour of Battery Application

The main application for MXene and its derivatives has been as energy-storing electrode materials. In ionic liquid, aqueous, and organic electrolytes as well as in a hybrid method combining the capacitive and pseudocapacitive electrochemical reactions, steady potential output and quick charge–discharge have been achieved. Because of the quick charge/discharge capabilities of pseudocapacitive MXene, pure MXene electrodes typically lack a stable potential window with minimal voltage decay [43–45]. Several MXene modification techniques, including pre-intercalation, surface chemistry tuning, and, phase composition tuning, cannot stimulate chemically stable M-X bonds or electrochemically inactive surface terminations [46, 47]. As of right now, with the emergence of MXenes with similar halogenated surface terminations, such as $Ti_3C_2Br_2$ and $Ti_3C_2I_2$, starts a different MXene redox process [43, 48]. When $Ti_3C_2Br_2$ and $Ti_3C_2I_2$ cathodes are charged, the surface functional halogens T_x (Br, I) absorb electrons and become elemental [43]. Surface terminations function as electrochemically active halogen sources during the reaction, according to the typical conversion type properties of halogen batteries [49]. These MXenes contribute to surface diffusion type and pseudocapacitive charge storage, which leads to a pronounced redox peak, high energy density, and a constant discharge potential window (1.6 V for $Ti_3C_2Br_2$ and 1.1 V for $Ti_3C_2I_2$ vs. Zn^{2+}/Zn) (Fig. 3a) [43, 49, 50]. A further avenue of investigation concerning the surface chemistry of MXene centers on Lewis acid-base chemistry. The properties of the etchant influence the surface responses. Several Lewis acid molten salts, such as $CdBr_2$ and $ZnCl_2$, have been used to create MXene with regulated surface chemistry. Covalent surface modification has been used to create termination with stoichiometric chalcogens including Te, Se, S, and O in a variety of MXene systems, such as Nb_2CO_x, Ti_3C_2S, Ti_3C_2Se, Ti_3C_2Te, Nb_2CS_2, and Nb_2CSe. Furthermore, enhanced redox kinetics, stability, and reversibility may be advantageous due to the active functional groups connected within the nano space [51, 52].

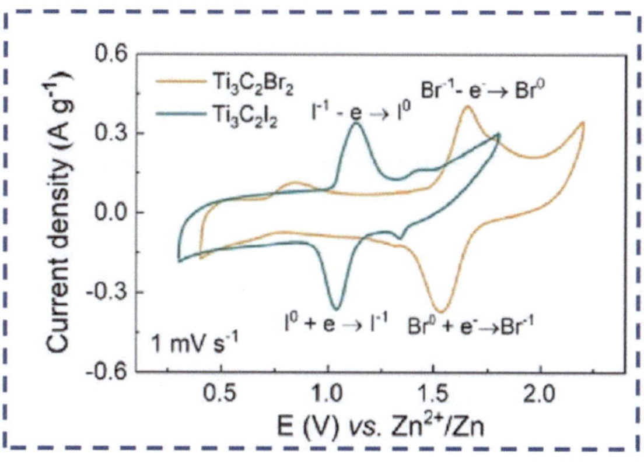

Fig. 3 CV profile of $Ti_3C_2I_2$, and $Ti_3C_2Br_2$ cathodes with Zn anode and Zn-based electrolytes showing redox performance. Redrawn with permission from Ref. [43]

Similar to carbon supercapacitors in aqueous electrolytes like Na_2SO_4 and organic electrolytes like tetraethylammonium tetrafluoroborate ($TEABF_4$) in acetonitrile, MXenes exhibit a double layer charge storage mechanism; however, their double layer capacitive performance is limited by their specific surface area, which ranges from 100 to 100 m^2/g [53]. In a number of electrolytes, MXene furthermore exhibits pseudocapacitive performance, in which ions interleave into the interlayer space as opposed to the inner lattice [27]. The spacing between the MXene sheets is typically feasible for the reversible deintercalation/intercalation of various ions, like non-metallic anion/cations, metallic cations, and their hybrids in both dissolved and undissolved states [54–56]. In contrast to the Faradaic diffusion-controlled reactions of typical battery electrode, the pseudocapacitive contributions of MXene are controlled by ion diffusion channels and surface redox reactions [57, 58]. The combination of this mechanism with MXene's high electrical conductivity leads to a high capacity for storing charge. The rate capacity is contingent upon the type of MXene and electrolyte selected. The intrinsic properties of the ions and their interactions with one another, such as their surface energy, electronegativity, valence state, ionization energy, and radius, also affect MXene's electrochemical efficiency [59–61]. Through hydrogen bonding, electrostatic contact, or string ionic bond formation, the intercalated ions aid in the electron mobilization of MXene [60]. The surface functional groups (T_x) and outer M layers, which control ion shuttling, are mainly accountable for the interactions between ions and the MXene framework [62, 63]. Organic electrolytes have slower ion diffusion and desolvation, but they also have faster charge transfer, a larger capacity than their solvated counterparts, and less volume change [58]. Large interlayer gaps lower the barrier to ion transport and are kinetically advantageous for improving the ion storage capacity [64]. Simple atomic structures of MXene generally show higher capacitance than their complicated equivalents, such as Ti_2CT_x vs. $Ti_3C_2T_x$ [65].

Theoretically, the cation's radius has a more direct impact on capacitive performance than its valence states [66]. Li^+ and Mg^{2+} are more difficult to intercalate than K^+ and Na^+. [56, 66] displays the voltage window vs. theoretical capacity of MXene electrodes intercalated with various ions. Multivalent heavy metal ions like Ca^{2+} and Al^{3+} can be accommodated and provide substantial capacity by adjusting the interlayer gap in MXenes [67]

5 MXene-Based Composites for Battery Application

High energy density batteries that can perform better than the batteries that are already in use are being advanced with respect to the growing need for electric vehicles and wearable microelectronics [68, 69]. Years of research and development have gone into discovering new electrode materials, battery topologies, and electrolytes in an attempt to produce batteries with improved performance and economic viability [70, 71].

Because of their exceptional characteristics, like their excellent electrical conductivity, enlarged surface area, safety, significant interlayer spacing, and remarkable thermal stability, recently, 2D MXenes have fascinated attentions as essential parts for rechargeable batteries electrodes. It suggests that high performing MXene-based composites must be used as energy storing technologies in order to integrate renewable energy sources into practical applications [11]. Due to their excellent qualities—rapid charge/discharge rates, high specific capacitance, remarkable cycling stability, remarkable chemical stability, exceptional electrical conductivity, and high surface area—MXene composites have revealed enormous promises as electrode material for rechargeable battery [11].

MXene composites are used in rechargeable batteries for the following purposes:

i. Battery separator: Coatings made of MXene composites have been utilized to advance the thermal stability and safety of batteries. By reducing the polysulfide shuttle effect, MXene-based coatings can accelerate the efficiencies of Li–S battery [72].
ii. Current collector: MXene composites are being used as current collectors for Li–S battery. MXene-based current collectors can accelerate the electrochemical activity of the batteries by promoting improved electron and ion mobility [72].
iii. Anode material: LIBs with MXene-based anodes had increased battery capacity, shown exceptional rate capability, and extended cycle stability [73].
iv. As cathode in Li–S battery: By enhancing the usage of active materials and minimizing the shuttle effect, the overall performance of the battery can be improved [74].

The function of various MXene composite forms in various battery and fabrication applications is examined in the section that follows.

5.1 MXene-Nanocarbon Based Composites

Different types of nanocarbons like carbon nanotube (CNT), reduced graphene oxide, graphene oxide, carbon dots etc. are the promising materials for the fabrication of electrode materials for electrochemical energy storage devices [75–81]. Porous CNT/Ti_3C_2 MXene composite paper electrode was developed by Xie et al. for sodium-based energy storage device [80]. They formed a heterostructure between the positively charged 1D CNT and negatively charged 2D MXene nanosheets via electrostatic contact. This composite created a unique porosity structure that facilitated ions availability to the electrolyte and electrode transportation by stopping the MXene nanosheets from restacking. MXene-based composite demonstrated remarkable rate capability, a large volumetric capacity, and outstanding cyclic stability when utilized for sodium-ion storage as freestanding electrode. Specifically, the film having porosity displayed 421 mA h/cm^3 volumetric capacity at 20 mA/g [80]. Mashtalir et al. utilized the i-PrA intercalation process to synthesize Nb_2C flakes, and vacuum filtering was used to create flexible MXene/CNT electrodes [81]. The synthesized

MXene/CNT composite exhibited better electrochemical and mechanical properties and provided 400 mA h/g Li-ion storage capacity at 0.5 C, which was higher compared to MXene without CNTs [81]. Lin et al. revealed an innovative way to enhance the Li-ion storing performance for layered $Ti_3C_2T_x$ MXene [82]. In order to build conducting "CNF bridge" between two $Ti_3C_2T_x$ nanosheets space, the plan was carried out utilizing the CVD growth technique. Li-ions diffusion into the MXene nanosheets space was made easier by the of vine-like CNFs that surround the $Ti_3C_2T_x$ nanosheets. The graphic illustration for the synthesis of MXene/CNF nanocomposite and the corresponding SEM image of MXene/CNF are represented in Fig. 4a, b, correspondingly. Finally, even at a high rate of 100 C, the composite anode of CNFs and $Ti_3C_2T_x$ exhibits exceptional long-term stability, as shown by electrochemical data. Additionally, it possesses a nearly twice as large reversible capacity than pure $Ti_3C_2T_x$, with 320 mA h/g capacity at 1 C (Fig. 4c) [82]. The goal of Chen et al.'s work was to create a sandwich-like structure (MGL) made of intercalated layered double hydroxide (LDH), graphene oxide (GO), and MXene [83]. By using the components' conductive interface, this novel configuration seeks to reveal more active sites and avoid MXene stacking problems. By efficiently reducing the morphological collapse of LDH, the MGL structure increases its specific capacity. The materials' electron density rises and charge transmission is enhanced when a thin coating of GO is used. At a current density of 1 A/g, the combination of these parameters results in an extraordinary specific capacity of 241.9 mA h/g and a remarkably high cyclic stability of around 90.9% [83]. But the primary issue with 2D MXene flakes is that they tend to self-restack, which limits the application of functionalized MXene flakes in conducting electrodes. The MXene flakes' two-dimensional structure is the root cause of self-restacking. According to Zahra et al., multiwall carbon nanotubes and V_2CT_x MXene (V_2CT_x/MWCNT) composite can be used as a feasible material for electrodes in energy-storage device [84]. With an increased value of 62.5 A h/g specific capacity and gravimetric capacitance of 1842 F/g at 2 mV/s, the hybrid electrode (V_2CT_x/MWCNT) showed improved electrochemical activity. In addition, the electrode showed well rate performance, durability, and a 94% retention capacity during 10,000-cycle lifespan [84]. Zhang et al. produced the three-dimensional carbon-coated MXene structures by carbonizing $Ti_3C_2T_x$ nanosheets following dopamine self-polymerization across their surface [85].

Consequently, the hybrid exhibited superior capacity and rate capability as anode material for Na- and Li-ion storing. The Mo_2CT_x MXene/graphitic carbon nitride/ (g-C_3N_4/Mo_2CT_x) composite was easily synthesized through an in-situ process by cleverly using the urea to g-C_3N_4 conversion to both produce a protective layer and reach the desired amount of exfoliation for Mo_2CT_x MXene [86]. In addition to providing additional MXene layers to enhance lithium storage, the g-C_3N_4/Mo_2CT_x makes use of the unique structure to shorten electrons and ions dispersion length to expedite charge transportation and to decrease electrode polarization. The theoretically optimized structure of g-C_3N_4/Mo_2CT_x is portrayed in Fig. 4d. Furthermore, it's possible that the in-situ coating of g-C_3N_4 layer will shield the Mo_2CT_x surface terminations from the electrolyte. As a result, the g-C_3N_4/Mo_2CT_x has excellent cyclic stability (229.6 mA h/g after 1800 cycles at 2 A/g, as shown in Fig. 4e), strong

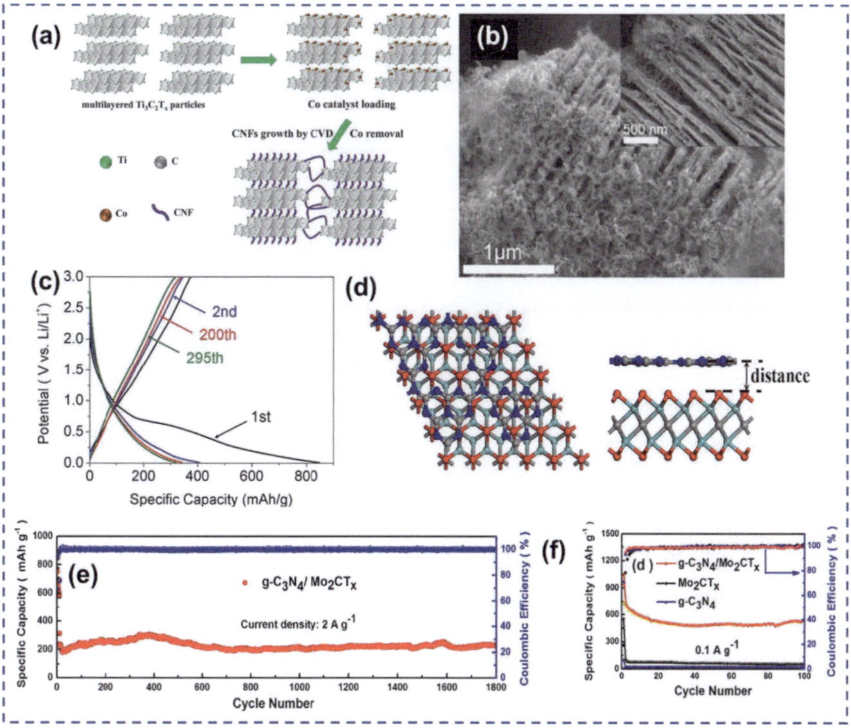

Fig. 4 **a** Diagram illustrating the process of creating MXene/CNF hybrid particles, **b** SEM image of CNF/MXene nanohybrid, **c** Charging/discharging curves of the MXene/CNF nanohybrid at 1 C for Li-storage, **d** Top and side views of the g-C_3N_4/Mo_2CO_2 optimized structure. The atoms C, N, O, and Mo are represented by the gray, blue, red, and green balls, respectively, **e** Cycling stability performance of g-C_3N_4/Mo_2CT_x at 2 A/g, and **f** Cyclic performances of g-C_3N_4/Mo_2CT_x, Mo_2CT_x and g-C_3N_4 at 0.1 A/g. **a–c** Reprinted with permission from Ref. [82] and **d–f** Reused with permission from Ref. [86]

cyclic performance (reversible capacity of 528.5 mA h/g after 100 cycles at 0.1 A/g, as illustrated in Fig. 4f), and a decent Coulombic efficiency of 70.8% [86]. This methodology offers a workable method for creating multilayer structures based on MXene that can be utilized to the energy storing industry and other related areas.

Thus, the combination of nanocarbons with MXenes enhanced the MXenes' conductivity and reaction kinetics while also making it easier for them to retain charges, which is useful for energy storage in next-generation smart devices.

5.2 MXene-Metal Oxides/Chalcogenides Composites

MXenes are able to improve Na^+ storing capacity and stability through the interlayers spacing expansions and structural designs, although more remarkable performance is still expected. Because of their tunable surface properties, researchers combine high-capacity materials with MXenes to generate composites, which are useful for acquiring high power and energy. In this respect, some composites like $Ti_3C_2T_x$/$FePS_3$ [87], Sb_2O_3/$Ti_3C_2T_x$ [88], $Ti_3C_2T_x$/$CoSe_2$ [89], and SnS/$Ti_3C_2T_x$ [90], are painstakingly built for Na-ion storage. The oxides, composite oxides, and tin-based alloys that are also utilized as anode material for lithium-based batteries are seen to be potential alternatives for enhanced energy storage. As the structural frameworks of Li battery, Wang et al. pioneered the utilization of conducting networks composed of Ti_3C_2-MXene and amorphous carbon coated tin oxide [91]. Hydrothermal and calcination processes were used to create (C@SnO_2/Ti_3C_2) composite using carbon-coated kernel-like SnO_2 with 2D layered Ti_3C_2-MXenes coupling. Synthesized C@SnO_2/Ti_3C_2 nanocomposites exhibited greater cyclic stability and a significantly high diffusion coefficient for Li^+, having a lesser charge transfer resistance as compared with SnO_2/Ti_3C_2 and Ti_3C_2-MXenes. Above all, at 100 mA/g current density, C@SnO_2/Ti_3C_2 composite showed a remarkable initial capacity of 1531.5 mA h/g. [91]. Li et al. developed a 3D spring-roll hybrid (C–Fe_3O_4/Ti_3C_2) using carbon-coated Fe_3O_4 wrapped with Ti_3C_2 MXene layers to avoid the inherent limitations and increase the stability of Fe_3O_4 [92]. The 3D C–Fe_3O_4/Ti_3C_2 composite, which was used in a LIB as the anode, receives the high electrochemical activity of Fe_3O_4 and demonstrates exceptional ionic and electrical conductivity. The C–Fe_3O_4/Ti_3C_2 anode showed superior electrochemical efficiencies and a longer cycle life at higher current densities. At 1 A/g, a reversible capacity value of 997 mA h/g was observed [92]. Using a simple solution-phase process, the Sb_2O_3/MXene ($Ti_3C_2T_x$) composite was prepared [88]. For sodium-ion cells, the composite showed enhanced electrochemical performances and structural stability. The homogeneous entrapment of Sb_2O_3 nanoparticles in the MXene 3D networks is responsible for the hierarchical structures of the as-synthesized Sb_2O_3/$Ti_3C_2T_x$ composite. The MXene ($Ti_3C_2T_x$) networks offer a highly efficient way to buffer the expansion of Sb_2O_3 and accelerate the movement of electrons during Na^+ extraction and insertion [88]. Using oleylamine and the interfacial self-assembly method, several sandwich-structured hybrid arrangements were produced by Zou et al. [93]. Hydrogen bonding holds 2D $Ti_3C_2T_x$ nanosheets and 1D Co–NiS nanorods together closely at the cyclohexane and solution interface. Electrochemical experiments show that the hybrid anode has an excellent cycling stability (409 mA h/g after 100 cycles) at 0.1 A/g and a high Na-ion storage capacity of 541 mA h/g [93]. Zhang et al. invented stretchable Na^+ capacitor with AC/MXene and elastomeric $MoSe_2$/MXene electrodes [94]. According to the study, the electrodes exhibited remarkable performances in terms of stretchability, stability (after 1000 cycles 95%), and sodium ion storage (420 mA h/g at 5 A/g) [94].

5.3 MXene-Conducting Polymer Composites

Because of their exceptional ability to carry charges due to significant π-electron delocalization, conducting polymers (CPs) are particularly attractive as potential contenders for energy storing in rechargeable metal-ion batteries (MIBs) [28]. Specifically, through the discriminate binding of anions and cations to π-conjugated systems, the stability of the overall charged states can be improved, which is facilitated by Lewis acid-base interactions [28]. The de-intercalation of tiny cations serves as an example of how specialized charge storage routes can be enabled by strategically designing redox processes through the use of anionic backbones and cation grafting [28]. MXene and CPs have different benefits, but their stand-alone applications in ion storage have very small intrinsic capacities, which restricts development. Optimizing electrochemical performance requires the synergistic interaction of CP and MXene to overcome these constraints. The CP matrix contains metal-based chemicals that reduce MXene restacking while improving structural stiffness, flexibility, and cycling efficiency. MXene's distinct flexibility and interlayer spacing allow for volume variations in the composite, which guarantees stability throughout cycling [28]. Finally, MXene-CP derivatives provide a larger surface area, which helps to enhance the interaction between electrolyte and electrode material, reduce restacking problems, and facilitate ion diffusion. Figure 5a shows the MXenes' evolutionary advancement in battery applications (LIBs, MgIBs, ZIBs, Li–S battery, PIB etc.) with different conjugated conducting polymers like polypyrrole (PPy), polyaniline (PANI), Poly(ethylene oxide) (PEO), polydopamine (PDA), poly(diallyl dimethyl ammonium chloride) (C-PDDA), and Poly(glutamic acid) (PGA) [28].

Because of their exceptional host–guest chemistry, strong redox-active and pseudocapacitive performances, and adjustable structure, MXene-CPs are indispensable parts of metal-ion electrodes [28]. Lui et al. used mechanical interaction techniques to create a sulfonic group-grafted $Ti_3C_2T_x$ and PANI (S-$Ti_3C_2T_x$/PANI) as an illustration of the application of MXene-CP in the cathode fabrication for ZIBs [95]. MXene was grafted with $–SO_3H$ groups during the synthesis process to increase its compatibility with PANI and lessen shrinking and swelling throughout consecutive cycles. It was capable to attain a high capacity of 262 mA h/g, while the cathode based on MXene-CP maintained 100% columbic efficiency and even after 5000 cycles a specific capacity of 103 mA h/g [95]. Redox-active metal centers and metal-ion-stabilizing designs can raise the volume of metal ions, which raises the specific capacity. Since PPy is synthetically diverse among CPs, it is the most commonly used in hybridization. There are a number of ways to combine PPy and MXene, including electrochemical polymerization, mechanical intercalation, and oxidant-free and oxidative in-situ polymerization [95]. The structural-characteristic relationship of MXene-CPs was thoroughly reviewed by Adekoya et al., who also pointed out that MXene-PPy interactions significantly contribute to enhanced thermal and electrical belongings as well as synergetic pseudocapacitive nature [96]. An urchin-like 3D architecture was fabricated by Zhang et al. as an example of how MXene-PPy can be used on the anode side as a host for Li-ion [97]. This was

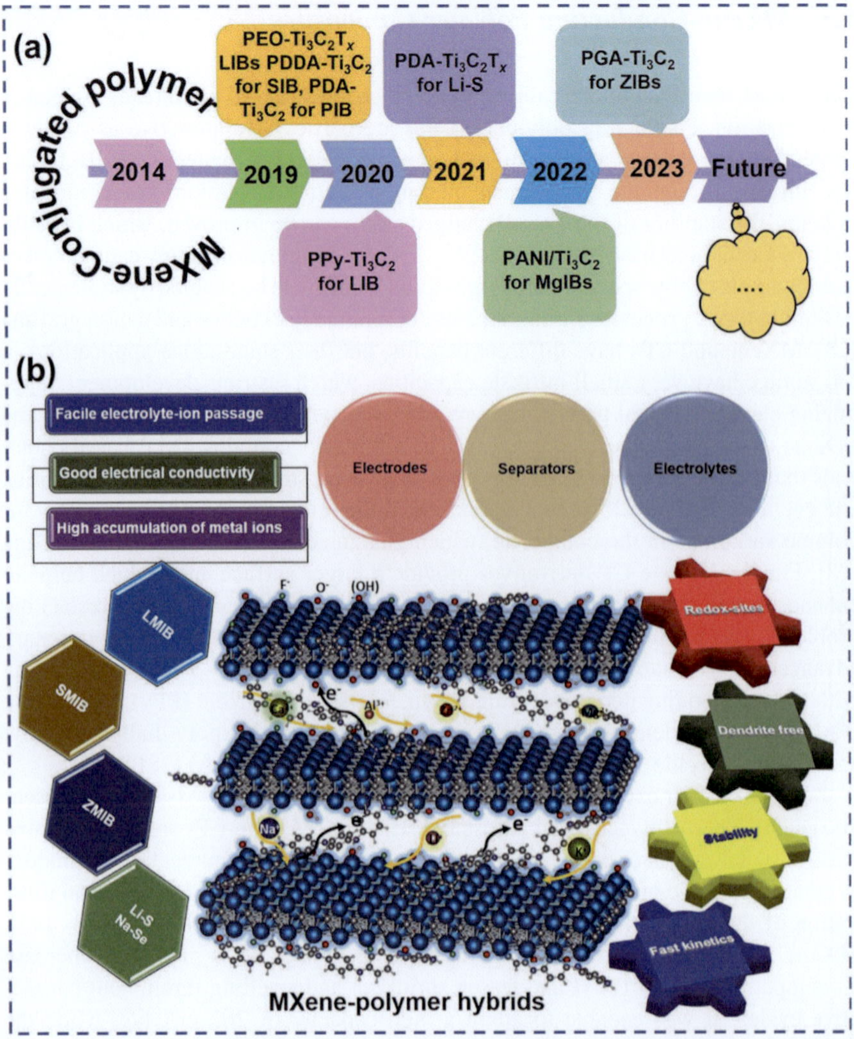

Fig. 5 **a** MXene-CP composites' evolutionary advancement in battery applications and **b** Illustration of the MXene-CP nanoarchitectures' functional advantages. Redrawn with permission from Ref. [28]

accomplished by vulcanization after in-situ polymerization of PPy using initiator on alkalized $Ti_3C_2T_x$ [97]. Using a hydrothermal method, Ding et al. created a P-SnO_2/Ti_3C_2 composite for SIBs' anode by in-situ polymerizing pyrrole monomers on $Ti_3C_2T_x$ in conjunction with $SnCl_2$ [98]. SnO_2 nanocrystals were uniformly fixed on the MXene surface by Sn–O–Ti bonds, and they were positioned between PPy and $Ti_3C_2T_x$ nanosheets. This arrangement made it easier to store Na^+ and stopped the $Ti_3C_2T_x$ nanosheets from restacking. The SnO_2 nano-crystals are contained by the conductive PPy outer layer, which prevents the active particles and electrolyte from coming into direct contact and creates a steady electrochemical connection. With a capacity retention of 57% at 2 A/g and a high reversible specific capacity of 325.6 mA h/g generated at 100 mA/g after 200 cycles, the P-SnO_2/Ti_3C_2 composite showed exceptional rate capability, cycling stability, and high reversible capacity as a SIB anode [98]. As a result of the charge-enriched method, Zhang et al. created an interfacial layer of MXene-PPy on a Zn anode that was connected by N-Zn bonds [99]. Using this unique anode design, the MXene-PPy layers aid in the accumulation of charge concentrations and promote the ion flux and electric field homogenization. At 1 A/g, after 700 cycles, their analysis revealed dendrite-free performance with a high specific capacity of 170 mA h/g and a steady cycle life of up to 3000 cycles at 10 A/g [99]. Cheng et al. investigated the creation of an electrode for LIBs by combining $V_4C_3T_x$ MXene with poly(o-phenylenediamine) (PoPD) [100]. Along with remarkable cycle stability (1000 cycles) and rate performance (10 A/g), the $V_4C_3T_x$/PoPD blend has a noteworthy capacity of 561.2 mA h/g at 0.1 A/g [100]. MXene-CP is also shown to be useful in the future as SIBs' cathode material. As their experiment shows, Gao et al. successfully attained both ionic and electron transportation through in-situ intercalation of polyanthraquinone sulfide (PAQS) on the $Ti_3C_2T_x$ MXene sheets (PAQS@MXene) [101]. Having a remarkable reversible capacity of 231 mA h/g at 100 mA/g, the PAQS@MXene composite as a cathode demonstrated remarkable Na-storage performance in SIBs and superb rate capability. Along with, having a remarkable capacity of 144.7 mA h/g at 2000 mA/g, the PAQS@MXene composite outperformed than bare PAQS cathode under similar condition [101]. Ling and colleagues synthesized a three-dimensional MXene-based material ($Ti_3C_2T_x$/C-PDDA) by intercalating $Ti_3C_2T_x$ nanosheets with a blend of poly(diallyl dimethyl ammonium chloride) (C-PDDA) and functionalized active carbon. As cathode at 0.2 C, $Ti_3C_2T_x$/C-PDDA@S showed superior reversible specific capacity of 1016.8 mA h/g and long-cyclic stability (0.075% capacity degradation per cycle at 1 C), indicating that the 3D MXene-based S hosts reveal the potentiality for high-performing Li–S batteries [102].

Another practical application of MXenes is the addition of MXenes to the conventional polypropylene (PP) separators to improve LIBs' electrochemical performances. For instance, the $LiNi_{0.8}Co_{0.1}Mn_{0.1}O_2$ (NCM811) cathode's wettability and stability were increased when MXene ink was sprayed over a PP membrane to create a PP/MXene hybrid separator for LIBs [103]. Yang et al. produced a vacuum-filtered PP separator for Li–S batteries that was entrenched through mechanical intercalation with carbon nanotube (CNT)-modified $Ti_3C_2T_x$ MXene and deoxyribonucleic acid (DNA) [104]. The hybrid separator's thermotolerance (150 °C) and nonflammability were enhanced when the CNT-DNA structure associated with the MXene by coordination and π-conjugation, therefore inhibiting the polysulfide shuttle effect. Furthermore, at a rate of 1 C and after 200 cycles, the DNA-CNT-MXene separator demonstrated exceptional electrochemical efficacy having a specific capacity of 592 mA h/g [104]. This was due to its enhanced electrolyte wettability. It is shown that functionalizing the separator with MXene-CP is not only a successful method for binding of discharge species and ion percolation, but also the corrective actions for battery safeties and thermal runaway preventions.

The long-term stability of MIBs depends on solid-state electrolytes with strong ionic conductivities and superior mechanical characteristics that prevent dendritic development. The creation of composite polymeric electrolytes with hydrophilic surfaces and MXene is a promising avenue for solid-state battery technology [28]. This is because the hydrophilic surface can improve the interface between the polymer chain and MXene; and MXenes show tunable surface chemistry and superior interlayer spacing that can promote ionic conductivity. Poly(methyl acrylate) (PMA) was grafted onto $Ti_3C_2T_x$ in the presence of poly(vinylidene fluoride-co-hexafluoropropylene) (PVHF) by Zhi et al. to create an improved solid-state electrolyte (SSE) for ZIBs [105]. The presence of F–H linkages between PVHF and PMA caused the MXene matrix to plasticize equally. This solid-state electrolyte demonstrated strong ionic conductivity (2.69×10^{-4} S/cm), enhanced stability, and a good transference number for Zn^{2+} at room temperature (RT), indicating that it is a possible solution for multiple fundamental ZIB issues. In LIBs, an electrolyte made of $Ti_3C_2T_x$/lithium bis(tri fluoromethanesulfonyl)imide (LiTFSI)/Poly(ethylene oxide) (PEO) produced by mechanical blending has been employed [106]. Better LIB performance is obtained via MXene's inhibition of PEO crystallization, acceleration of polymer chain dynamics, and enhancement of ionic conductivity in the electrolyte. With 3.6 weight percent MXene, the developed composite polymer electrolyte (CPE) outperforms previously published 1D and 0D nanofiller CPE. At room temperature, it showed outstanding ionic conductivity (2.2×10^{-5} S/m), 0.18 transference number for Li^+, and up to 4.7 V wide range of potential stability. After over 50 cycles, a $LiFePO_4$/Li battery with this CPE of 1.5 weight percent MXene demonstrated a consistent capacity of 140 mA h/g at 60 °C [106]. With the help of a dual needle electrostatic atomization method, Kou et al. created a laminar MXene-PEO layer that allowed for the simultaneous insertion of PEO/LiTFSI and homogenous polymerization of the MXene layer [107]. The electrolyte anode interface displayed a high concentration of Li^+ due to the enhancement of Li^+ in the MXenes functional layers and its resistance to TFSI ion. The MXene layer promoted uniform Li^+ dispersion with a decreased Li deposition/nucleation overpotential by improving the electric field distribution in comparison to PEO electrolyte only. The PEO-MXene electrolyte showed a high capacity of 142.8 mA h/g after 140 cycles at 60 °C and excellent dendritic-free cyclic stability, despite only possessing a PEO layer [107]. Figure 5b shows the multifunctional properties of the MXene-CP composites for enhancing the performance and stability of the electrolytes, separators, and electrodes for battery technologies [28].

5.4 MOF/MXene Composites

Significantly functional tunability, high crystallinity, large surface area (1000–10,000 m^2/g), and remarkable porosity are among the main structural characteristics of metal–organic frameworks (MOFs). These characteristics are essential for uses including energy conversion and storage and gas sensors [108]. In contrast to their naked MOF structure, treatment of MOF under suitable temperature can produce a variety of derivatives, such as hetero atom coordinated single atom moieties, stable carbon matric hybrids, and meal clusters/oxides. For instance, using Zn-MOF/Co-MOF, Zhou et al. produced Co/CoO@NSC catalyst. This catalyst's superb aqueous zinc air battery performance, ORR, and OER activity were made possible by its intact metal@carbon structures and high porosity [109]. Gu et al. employed a highly conductive carbon made from ZIF-8 as a suitable anode material for Na^+ capacitor [110]. Comparing this carbon generated from MOF to commercially available activated carbon (898 m^2/g), the latter has a lower surface area (1259 m^2/g). This means that more ions can be adsorbed onto the carbon, increasing its capacity. Despite the fact that MOFs have many benefits, their low mechanical resistance, structural porosity, and electrical conductivity lead to poor coulombic efficiency, fragility, and poor mechanical resistance, which prevents them from being directly used as catalytic and electrode materials [108]. MXenes, on the other hand, have not found extensive industrial utilization in energy storage technologies like batteries. These individual MXene and MOF, however, suffer from problems with agglomeration and restacking that reduce their electrochemical energy storage, conversion capacity, and active metal sites [108]. Crucially, it is difficult to create structures having multidimensionality using these. For these reasons, MXene/MOF nanohybrids are gaining attention. (1) Porous high-surface-area nanoarchitectures can be produced by inserting porous MOFs in between 2D MXene nanosheets [108]. (2) The ion/electron transit is accelerated by the nanoarchitectures' constituent parts working together to impede the accumulation and restacking of the MXene nanosheets [108]. (3) By increasing the quantity of operational sites revealed by small branches, the interlayer gaps can store a significant quantity of charges [108]. (4) The nanohybrid materials that were created are flexible, and it is possible to lessen the electrode stress that is produced when charging and discharging [108]. (5) MXene-enclosed porous metal oxides produced from MOFs show improved mechanical and chemical stability [108]. (6) Electrochemical energy storage systems are best suited for nanoarchitectures with high packing density [108]. Because MOFs exist in two forms—their original phases and those generated from transition-metal/metal-oxide derivatives—the dimensions of the MOF and MXene are thought to be crucial when building MOF/MXene nanoarchitectures. By adjusting their functionalities, morphologies, sizes, and porosity, MOFs' flexibility and stiffness can permit remarkable features [108]. Because MXenes have high d-spacing between the layers and a lot of terminal functional groups having negative charges on their surface, they can also react easily with different molecules or structures.

Through interlayer hydrogen bonding between Ti_3C_2 nanosheets and NiCo-MOF, Liu et al. produced 3D porous Ti_3C_2/NiCo-MOF composite, which was then used as a LIB anode [10]. The Ti_3C_2/NiCo-MOF architecture attained a good rate performance (256 mA h/g at 1 A/g current density) and a high reversible capacity (after 300 cycles 402 mA h/g). Sun et al. explored an in-situ solvothermal process to design a MXene/Ni-ferrocene based MOF (NF-MOF/MXene) composite in order to create stable anode for Li^+ storage [111]. The combined structure demonstrated significantly more Li^+ storage capacity than the separate parts. Liu et al. synthesized CoP–N doped carbon (NC)@MXene by treating ZIF-67/$Ti_3C_2T_x$ derivatives to phosphorylation having a sandwich-like structure (Fig. 6a) [112]. To further enhance the Li/Na-ion storage capacity, the resultant CoP-NC supplied active sites and assisted in extending the interlayers gap between the MXene nanosheets. Following 500 and 2000 cycles at 0.5 and 1 A/g, the CoP-NC@MXene system demonstrated 101.6 and 147.8 mA h/g as SIB and LIB anode, respectively (Fig. 6b, c) [112]. A simple, scalable method based on solvothermal techniques and electrostatic interactions has been utilized to generate a 3D hierarchical nMOF-867/$Ti_3C_2T_x$ (referred to as Zr-MOF/$Ti_3C_2T_x$) system (Fig. 6d) [113]. Through the stable double bonds (Zr–S and Li–N), the polysulfides in the heterostructure were chemically trapped by the nMOF-867 particles having high porosity that resulted from the Lewis acid-base reactions between the Zr-MOF and Li polysulfides. The sulfur host for Li–S batteries, nMOF-867/$Ti_3C_2T_x$, demonstrated remarkable reversible capacity (1302 mA h/g at 0.2 C) and outstanding rate capability (581 mA h/g at 4 C current rate) [113]. In-situ method at room-temperature combined with phosphorization allowed Zong et al. to fabricate a MOF-derived CoP/Ti_3C_2 MXene nanohybrid for LIB, wherein MXene layers and ZIF-67 particles self-assembled and converted into CoP@Ti_3C_2 MXene [114]. Because of its distinct dual functions, the MOFs-CoP@Ti_3C_2 MXene system demonstrated enhanced Li-storage capacity. In order to intercalate Cu-HHTP, a 2D MOF, into the layers of the V_2CT_x MXene for Zn-ion batteries (ZIBs), Wang et al. investigated solution-phase MOF assembly [115]. In order to optimize electron conduction and avoid accumulation on nanosheets during successive discharging and charging, MOF sheets were alternately stacked between MXene sheets in the final Cu-HHTP/MXene architecture. With 92.5% capacity retention after 1000 successive cycles, the Cu-HHTP/MXene heterostructure demonstrated long-term cyclic stability and 260.1 mA h/g capacity at 0.1 A/g [115]. Similarly, Ti_3C_2 sheets were used to encapsulate Ni-naphthalene dicarboxylic acid nanosheets, which were then used in LIBs. These devices demonstrated respectable long-term cycling performance, with a capacity of 310 mA h/g after 500 cycles at 1 A/g, and a reversible specific capacity of 579.8 mA h/g at a current density of 100 mA/g over 100 cycles [116]. In order to promote OER and serve as an energy storage material (zinc air battery), Zhao et al. developed a novel catalyst known as 2D $Ti_3C_2T_x$/Co–1,4-benzenedicarboxylate nanoarchitecture [117]. The NiCoS/$Ti_3C_2T_x$ nanoarchitecture was coupled with Pt/C to create a zinc-air battery in the lab. The Zn plate was used as the anode, and this was loaded onto Ni foam to form the cathode. The device demonstrated decent stability, and the charge–discharge voltage gap (0.7 V) was smaller than that of the noble-metal/air cathode [118].

Specifically created MXene/MOF nanohybrids with remarkable features have been highlighted for battery use as a new class of cutting-edge developing materials. For MXene/MOF nanohybrids with their electrochemical energy-conversion and storage properties and varied morphological structures, various synthesis procedures

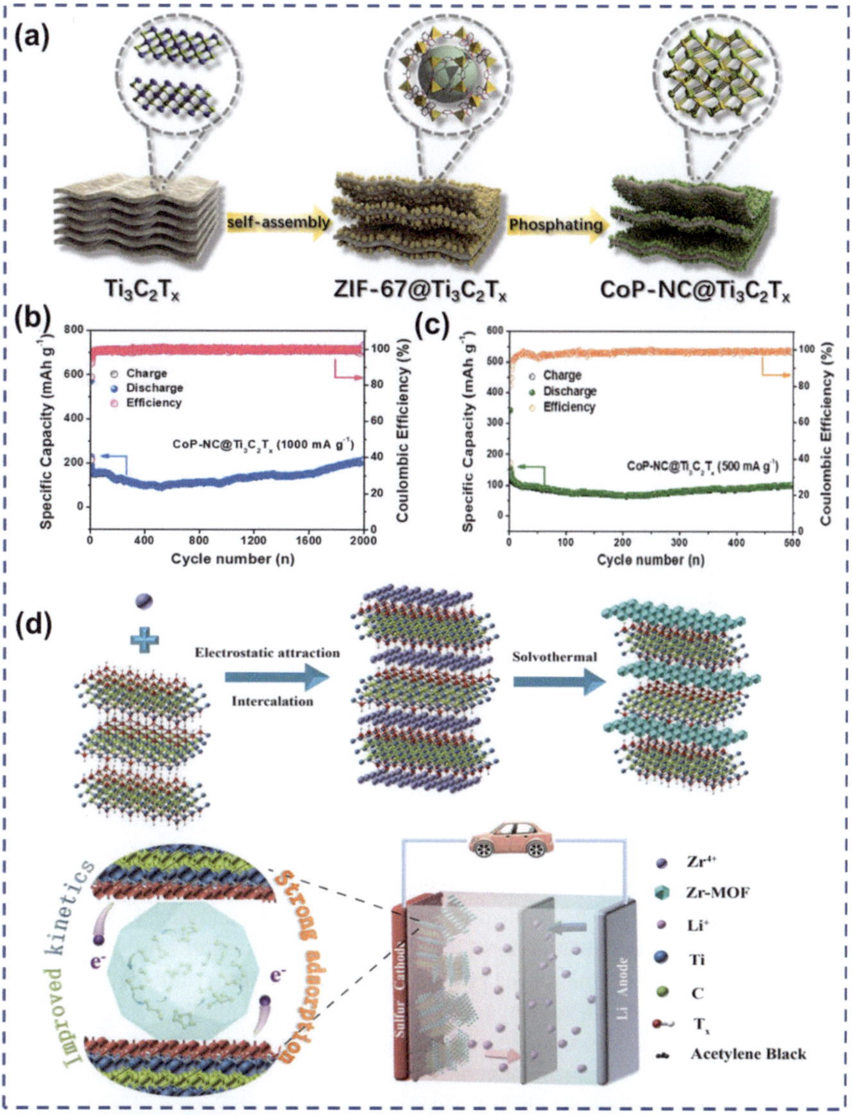

Fig. 6 **a** Diagrammatic representation of the CoP-NC@$Ti_3C_2T_x$ synthesis, **b** Extended cycle performance of CoP-NC@$Ti_3C_2T_x$ at 1000 mA/g current density for LIB, **c** CoP-NC@$Ti_3C_2T_x$ long-cycling performance at 500 mA/g current density for SIB, and **d** Schematic representation of the Zr-MOF/$Ti_3C_2T_x$ composite preparation process and the Zr-MOF/$Ti_3C_2T_x$/S composite operation in lithium-sulfur batteries. **a–c** Reprinted with permission from Ref. [112] and **d** Redrawn with permission from Ref. [113]

have been described in detail. The 3D, 2D, 1D, and 0D MOF/MXene nanoarchitectures with the following intriguing properties are created by adjusting the synthesis conditions:

1. High conductivity
2. Considerable storage capacity
3. Redox-active sites
4. Tunable porosity
5. Economic benefits.

6 3D Printed MXene-Based Electrodes for Batteries

The assembly of energy storage devices using conventional engineering techniques is far more complicated and expensive, with fewer geometrical variations and morphologies for electrode materials, electrolyte types, and cell packaging designs. Thicker batteries filled with more active material have been produced with high areal capacity and energy density; nevertheless, ions need to travel a greater distance in thick 2D electrodes, hence 3D architectures with shorter diffusion routes and less resistance are required [119–121]. As a technologically advanced and remarkable feature for microscale scope and artistic variation, 3D printed batteries have emerged. The most recent advancements in 3D-printed MXenes-based batteries are outlined in this section. We'll talk about printing designs, battery configurations, and performance in order to improve future battery technologies and address issues and inspirations.

MXene is a formidable competitor in the quest for batteries, having bravely made its way into the LIB systems. In order to create a Li anode free of dendrites with minimal over potential (~10 mV), Shen et al. reported using an extrusion-type 3D printing process [122]. The team produced a highly concentrated (~300 mg/ml) MXene ink with a high shear stress (102 Pa) and elastic modulus (104 Pa), which makes it a potential ideal ink for 3D printing different structures (lattices and arrays). Li nucleation is aided by these arrays and lattices, as Li grows parallel to the filaments. Moreover, the uniform dispersion of the Li-ion electric field and flux diminishes the propensity for dendritic development throughout the plating and stripping procedure. The MXene arrays' vast gaps allow for the formation of cobblestone-like Li, which results in a high areal capacity of 10–20 mA h/cm^2. Furthermore, during 1200 h of cycling, no Li dendrites were detected in the MXene arrays. Al was etched in the LiF/HCl mixture to create MXene flakes with a thickness of roughly 3 nm. MXene was added at a concentration of 300 mg/ml to produce a highly concentrated ink, but the result was a diluted, low viscosity dispersion of MXene ink. Such an extremely thick ink had a high viscosity and shear-thinning. It was determined that the shear stress was 102 Pa and the storage modulus was 104. The MXene ink's rheological nature allows for the layer-by-layer printing of continuous filament into pre-designed lattices and arrays. After being freeze-dried, the printed arrays were employed at various current densities for the Li plating and stripping procedure. Ma et al. developed a conductive and porous MXene skeleton for the stable Li anode and $LiFePO_4$ for the cathode using a 3D printed approach [123]. For printing 3D patterned electrodes, highly concentrated MXene ink with a viscosity of 105 Pa-s and a stress yield of 3990 Pa is advantageous. In addition, MXene's lithiophilic F and O groups react with Li ions, causing Li to develop uniformly and preventing dendritic formation. MXene's 3D printed electrodes exhibit a very consistent rate performance after one hour of Li plating, as well as a stable coulombic efficiency (CE) of 98% for 100 cycles. Li stripping and plating were studied by fabricating Li foils and MXene@Li symmetric cells. Multitasking screen printable MXene inks for lithium ion microbatteries and micro supercapacitors were reported by Zheng et al. [124]. MXene and active materials combine to form the semi-solid MXene lithium ion microbatteries, which have a high areal energy density of 154 μW h/cm^2. The MXene current collectors were printed on A4 size lithium iron phosphate (LFP) and aqueous MXene-lithium titanate (LTO) in order to build MXene microbatteries. Step-by-step printing of inks was done on the MXene current collectors, as illustrated in Fig. 7a. The entire printing process was carried out without the use of binders, organic solvents, or any conductive additives, making it environmentally benign. The constructed microbattery was electrochemically tested in an ionogel electrolyte which brought 88 μW h/cm^2 high areal discharge capacitance and capacity retention of 82% up to 1000 cycles.

In parallel with the widely used Li batteries, Li–S batteries have developed into the next generation of energy storage technologies. It has been discovered that the Li–S batteries' fixed operation prevents the disorderly growth of Li dendrites, which lowers anodic and cathodic performance [125]. The positive electrode is designated as S, and the negative electrode is designated as Li in the extremely basic Li–S cell structure [126]. Its mode of operation, redox reactions, and intercalation reactions set it apart from typical Li ions. Since the reaction in Li–S batteries is topotactic at both electrodes, the greatest capacity that has been observed and published to date is 180 mA h/g, and the storage capacity is limited to approximately 300 mA h/g [69]. But even with all of the Li–S cell's benefits, there are too many drawbacks for real-world application. The unchecked development of dendrites at the anode and inadequate cathode loading act are the real uses of these batteries. A drawback of S or S compounds is their insulating nature, which necessitates intimate contact with conductive additives in order for a reversible electrochemical reaction to occur. Consequently, the Li–S batteries' cycling performance and output capacity are unreasonable despite their diverse designs. According to Wei et al. [127], MO_x-MXene heterostructures can serve as hosts for Li–S batteries. S cathodes were created using the 3D printing method with various S loadings. The methods used to synthesize the 3D printed (3DP) electrode with S loading is represented in Fig. 7b. With VO_x-V_2C serving as the hosts and the 3D printed cathodes' open and porous structure, 9.74 mA h/cm^2 of areal capacity at 0.05 C was achieved with significant S loading.

The same research team revealed a 3D printed porous network of nitrogen-doped Ti_3C_2 MXene that prevents dendrite growth in Li–S batteries while also acting as an effective host for S [128]. The N-p$Ti_3C_2T_x$ template, which is produced to act as

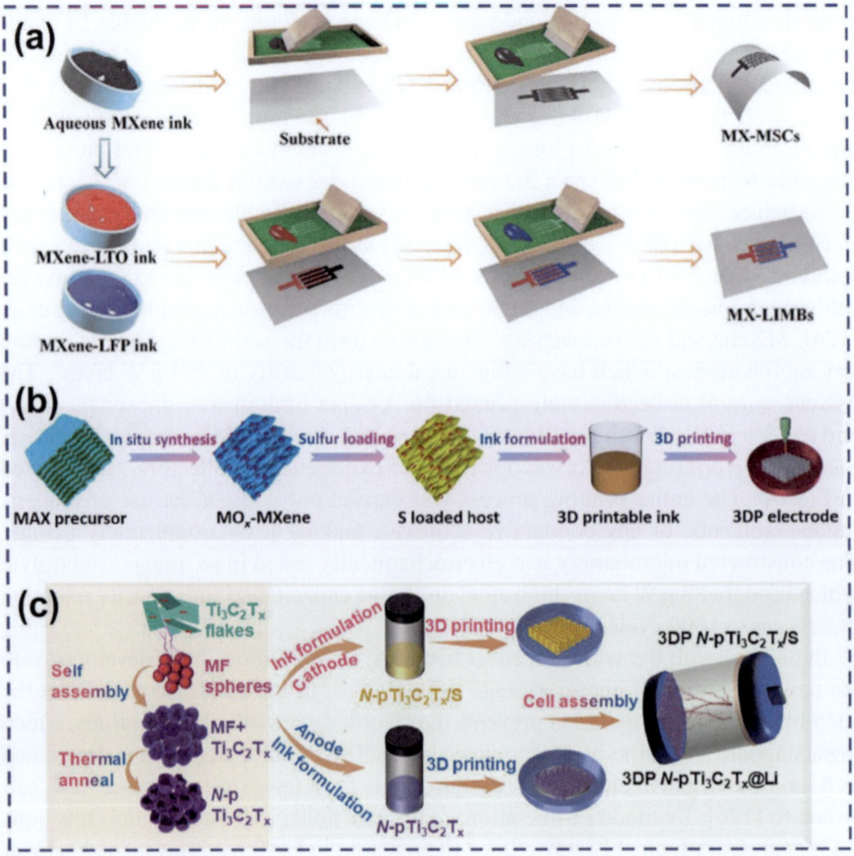

Fig. 7 **a** Diagram showing how MXene MSCs and microbatteries are made (by screen printing LFP and LTO one at a time). Redrawn with permission from Ref. [124]. **b** Diagram for building a 3D printed S electrode. Redrawn with permission from Ref. [127]. **c** Schematic of the Li–S battery dual electrodes preparation by 3D printing of N-p$Ti_3C_2T_x$. Redrawn with permission from Ref. [128]

a porous structure for the S cathode, is included in the 3D printed design (Fig. 7c). This porous structure has several micro, meso, and macro channels that facilitate easy charge transportation and provide ample sites for redox reactions involving sulfur. The bare *N*-doped sites help the soluble polysulfides to convert more easily. The anode site is the 3D printed N-p$Ti_3C_2T_x$ with a unique interlayer architecture. Overall, with a S loading of 7.56 mg and a decay capacity of 0.06% each cycle, the whole cell 3D printed Li–S battery (3DP N-p$Ti_3C_2T_x$/S||3DP N-p$Ti_3C_2T_x$@Li) can run continuously for more than 250 cycles. The investigation of 3D-printed MXenes for high-performance Li–S batteries remains highly motivated and challenging. The electrochemical performances can be improved by fine-tuning and modifying the chemical and physical characteristics of the electrode materials and by using theoretical calculations with assistance.

The SIBs operate on the same concept like LIBs using Na-ion plating and stripping between the anode and cathode to facilitate charging and discharging. Dendrite generation is a problem for SIBs, but researchers are having trouble creating electrodes that can withstand testing conditions at high current densities while still having a high capacity. Since MXene and its related hybrids are rarely used in SIB applications, a thorough understanding of this unique 2D material is required. Numerous research teams worldwide have already conducted some ground-breaking studies demonstrating its effectiveness in Na^+ ion battery applications [129–131].

Wang et al. reported used direct-ink writing printing to create a 3D porous sodiophilic microgrid aerogel of V_2CT_x/rGO-CNT [132]. It has been demonstrated that the V_2CT_x/rGO-CNT microgrid is a superior matrix for Na metal anodes. Reduced current density and an abundance of nucleation sites for Na deposition can be found in areas with high specific surface area. As shown in Fig. 8a, this porous architecture improves ion transport and reduces electrode volume expansion by enhancing electrolyte infiltration and ensuring close adhesion between the electrolyte and active materials. The high binding energy of the sodiophilic MXene (V_2CT_x) nanoflakes with sodium ions stimulates the nucleation and deposition of Na, as supported by data from in-situ TEM, ex-situ SEM, DFT calculations, and simulations. With a coulombic efficiency of 99.54%, the electrode V_2CT_x/rGO-CNT has a good cycle capacity of 3000 h. Moreover, it has a high areal capacity of 50 mA h/cm^2 and can sustain stability for 900 h at a current density of 500 mA/cm^2. The whole battery performance of Na@V_2CT_x/rGO-CNT||NVP@C-rGO is displayed in Fig. 8b–f, together with an LED that glows to verify the real-time application.

7 Conclusions and Future Perspectives

All things considered, MXene composites in rechargeable batteries present a bright future because they represent a previously unheard-of breakthrough in energy storage with higher sustainability and efficiency. This is because of their superior conductivity and surface area, which improve battery performance. When utilized as a part for rechargeable battery, composites of MXene have the potential to offer greater energy density, quicker charging rates, and longer cycle life. Because of its ability to efficiently hold ions, MXene is essential to the advancement of next-generation battery technology.

However, MXene and composites have some drawbacks and further research in this area can be concentrated on solving the following issues.

1. Most of the MXene materials are synthesized through strong-acid-assisted etching processes, which causes major safety issues. In this aspect, green scalable synthetic approaches are highly essential.

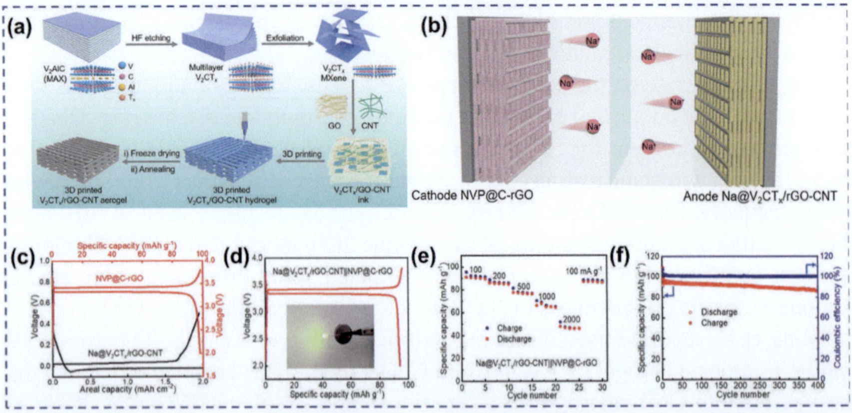

Fig. 8 **a** Graphic for the preparation of 3D printed V_2CT_x/rGO-CNT; Na@V_2CT_x/rGO-CNT||NVP@C-rGO full battery performances: **b** Graphic illustrating the full battery, **c** GCD of the cathode (NVP@C-rGO) and anode (Na@V_2CT_x/rGO-CNT), **d** full battery GCD curve at 100 mA/g (inset: LED lighted by battery), **e** full battery rate performance at current density ranging from 100 to 2000 mA/g, and **f** Coulombic efficiency and cyclic stability of full cell at 100 mA/g. Redrawn with permission from Ref. [132]

2. The stability of MXenes in an oxygen atmosphere is very poor. MXenes are easily oxidized in aqueous dispersion, which also hampers the industrial application of such materials.
3. Ti-based MXenes are the basis of the majority of MXene works. While some MXene materials are theoretically anticipated and others have been found, they have not yet been thoroughly investigated. Certain MXenes have limited energy-related applications because of their low conductivity. Future studies ought to concentrate more on identifying and fine-tuning the characteristics and uses of these MXene compounds.

Acknowledgements The authors would like to acknowledge the IIT (ISM), Dhanbad, India.

References

1. Li, M., Lu, J., Chen, Z., Amine, K.: 30 years of lithium-ion batteries. Adv. Mater. **30**, 1800561 (2018)
2. Etacheri, V., Marom, R., Elazari, R., Salitra, G., Aurbach, D.: Challenges in the development of advanced Li-ion batteries: a review. Energy Environ. Sci. **4**, 3243–3262 (2011)
3. Winter, M., Barnett, B., Xu, K.: Before Li ion batteries. Chem. Rev. **118**, 11433–11456 (2018)
4. Loaiza, L.C., Monconduit, L., Seznec, V.: Si and Ge-based anode materials for Li-, Na-, and K-ion batteries: a perspective from structure to electrochemical mechanism. Small **16**, 1905260 (2020)

5. Bashir, T., Zhou, S., Yang, S., Ismail, S.A., Ali, T., Wang, H., Zhao, J., Gao, L.: Progress in 3D-MXene electrodes for lithium/sodium/potassium/magnesium/zinc/aluminum-ion batteries. Electrochem. Energy Rev. **6**, 5 (2023)
6. Wang, H.F., Wang, X.X., Li, F., Xu, J.J.: Fundamental understanding and construction of solid-state Li− air batteries. Small Sci. **2**, 2200005 (2022)
7. Wu, Z., Tian, Y., Chen, H., Wang, L., Qian, S., Wu, T., Zhang, S., Lu, J.: Evolving aprotic Li–air batteries. Chem. Soc. Rev. **51**, 8045–8101 (2022)
8. Zhang, W., Ji, X.X., Ma, M.G.: Emerging MXene/cellulose composites: design strategies and diverse applications. Chem. Eng. J. **458**, 141402 (2023)
9. Naguib, M., Kurtoglu, M., Presser, V., Lu, J., Niu, J., Heon, M., Hultman, L., Gogotsi, Y., Barsoum, M.W.: Two-dimensional nanocrystals produced by exfoliation of Ti_3AlC_2. Adv. Mater. **23**, 4248–4253 (2011)
10. Liu, Y., He, Y., Vargun, E., Plachy, T., Saha, P., Cheng, Q.: 3D porous Ti_3C_2 MXene/NiCo-MOF composites for enhanced lithium storage. Nanomaterials **10**, 695 (2020)
11. Alli, Y.A., Bamisaye, A., Nancy, P., Zachariah, S.M., Oladoye, P., Bankole, O.M., Akamo, D.O., Chkirida, S., Anuar, H., Thomas, S.: MXene composites: properties, synthesis and its emerging application in rechargeable batteries. J. Energy Storage **77**, 109954 (2024)
12. De, S., Acharya, S., Maity, C.K., Nayak, G.C.: Boron nitride/$Ti_3C_2T_x$ MXene nanosheet/WS_2 nanostructure ternary composites for all-solid-state flexible asymmetric supercapacitors. ACS Appl. Nano Mater. **6**, 11175–11186 (2023)
13. De, S., Maity, C.K., Kim, M.J., Nayak, G.C.: Tin (IV) selenide anchored-biowaste derived porous carbon-$Ti_3C_2T_x$ (MXene) nanohybrid: an ionic electrolyte enhanced high performing flexible supercapacitor electrode. Electrochim. Acta **463**, 142811 (2023)
14. Moniruzzaman, M., Maity, C.K., De, S., Kim, M.J., Kim, J.: SnO2 nanosphere/carbon dot-embedded $Ti_3C_2T_x$ MXene nanocomposites for high-performance binder-free asymmetric supercapacitor electrodes. ACS Appl. Nano Mater. **7**, 6636–6649 (2024)
15. De, S., Acharya, S., Maity, C.K., Sahoo, S., Nayak, G.C.: MXene ($Ti_3C_2T_x$)-/amine-functionalized graphene-supported self-assembled Co9S8 nanoflower for ultrastable hybrid supercapacitor. Ind. Eng. Chem. Res. **61**, 7727–7738 (2022)
16. Xiao, P., Yun, X., Chen, Y., Guo, X., Gao, P., Zhou, G., Zheng, C.: Insights into the solvation chemistry in liquid electrolytes for lithium-based rechargeable batteries. Chem. Soc. Rev. (2023)
17. Liu, Y.K., Zhao, C.Z., Du, J., Zhang, X.Q., Chen, A.B., Zhang, Q.: Research progresses of liquid electrolytes in lithium-ion batteries. Small **19**, 2205315 (2023)
18. Nei, J., Wang, M.: Hydrogen storage alloy development for wide operating temperature nickel-metal hydride battery applications. Int. J. Hydrog. Energy (2023)
19. Łukomska, A., Wiśniewska, A., Dąbrowski, Z., Kolasa, D., Lach, J., Wróbel, K., Domańska, U.: New method for recovery of nickel and cadmium from the" black mass" of spent Ni–Cd batteries by solvent extraction. J. Mol. Liq. **357**, 119087 (2022)
20. Wu, Y., Sun, Y., Zheng, J., Rong, J., Li, H., Niu, L.: Exploring MXene-based materials for next-generation rechargeable batteries. J. Phys.: Energy **3**, 032009 (2021)
21. Shu, W., Han, C., Wang, X.: Prussian blue analogues cathodes for nonaqueous potassium-ion batteries: past, present, and future. Adv. Func. Mater. **34**, 2309636 (2024)
22. Mathiyalagan, K., Shin, D., Lee, Y.C.: Difficulties, strategies, and recent research and development of layered sodium transition metal oxide cathode materials for high-energy sodium-ion batteries. J. Energy Chem. (2023)
23. Man, Y., Jaumaux, P., Xu, Y., Fei, Y., Mo, X., Wang, G., Zhou, X.: Research development on electrolytes for magnesium-ion batteries. Sci. Bull. (2023)
24. Wu, C., Fang, W., Cheng, Q., Wan, J., Wen, R., Wang, Y., Song, Y., Li, M.: MXene-regulated perovskite vertical growth for high-performance solar cells. Angew. Chem. **134**, e202210970 (2022)
25. Perera, A.A.P.R., Madhushani, K.A.U., Punchihewa, B.T., Kumar, A., Gupta, R.K.: MXene-based nanomaterials for multifunctional applications. Materials **16**, 1138 (2023)

26. Fan, W.K., Sherryna, A., Tahir, M.: Advances in titanium carbide (Ti3C2Tx) MXenes and their metal-organic framework (MOF)-based nanotextures for solar energy applications: a review. ACS Omega **7**, 38158–38192 (2022)
27. De, S., Maity, C.K., Sahoo, S., Nayak, G.C.: Polyindole booster for $Ti_3C_2T_x$ MXene based symmetric and asymmetric supercapacitor devices. ACS Appl. Energy Mater. **4**, 3712–3723 (2021)
28. Vallem, S., Venkateswarlu, S., Song, S., Kim, J., Li, Y., Bae, J.: Flexible MXene-conjugated polymer nanoarchitectures: current developments and future frontiers in battery technology. Coord. Chem. Rev. **510**, 215778 (2024)
29. Wei, Y., Zhang, P., Soomro, R.A., Zhu, Q., Xu, B.: Advances in the synthesis of 2D MXenes. Adv. Mater. **33**, 2103148 (2021)
30. Sui, J., Chen, X., Li, Y., Peng, W., Zhang, F., Fan, X.: MXene derivatives: synthesis and applications in energy convention and storage. RSC Adv. **11**, 16065–16082 (2021)
31. Gao, L., Bao, W., Kuklin, A.V., Mei, S., Zhang, H., Ågren, H.: Hetero-MXenes: theory, synthesis, and emerging applications. Adv. Mater. **33**, 2004129 (2021)
32. Kumar, P., Yu, S., Shahzad, F., Hong, S.M., Kim, Y.H., Koo, C.M.: Ultrahigh electrically and thermally conductive self-aligned graphene/polymer composites using large-area reduced graphene oxides. Carbon **101**, 120–128 (2016)
33. Kumar, P., Shahzad, F., Yu, S., Hong, S.M., Kim, Y.H., Koo, C.M.: Large-area reduced graphene oxide thin film with excellent thermal conductivity and electromagnetic interference shielding effectiveness. Carbon **94**, 494–500 (2015)
34. Wan, S., Li, X., Wang, Y., Chen, Y., Xie, X., Yang, R., Tomsia, A.P., Jiang, L., Cheng, Q.: Strong sequentially bridged MXene sheets. Proc. Natl. Acad. Sci. **117**, 27154–27161 (2020)
35. Cao, H.: Beyond graphene and boron nitride: why MXene can be used in composite for corrosion protection on metals?. Compos. Part B: Eng. 111168 (2023)
36. Lotfi, R., Naguib, M., Yilmaz, D.E., Nanda, J., Van Duin, A.C.: A comparative study on the oxidation of two-dimensional Ti_3C_2 MXene structures in different environments. J. Mater. Chem. A **6**, 12733–12743 (2018)
37. Tang, X., Zhou, D., Li, P., Guo, X., Sun, B., Liu, H., Yan, K., Gogotsi, Y., Wang, G.: MXene-based dendrite-free potassium metal batteries. Adv. Mater. **32**, 1906739 (2020)
38. Naguib, M., Mashtalir, O., Lukatskaya, M.R., Dyatkin, B., Zhang, C., Presser, V., Gogotsi, Y., Barsoum, M.W.: One-step synthesis of nanocrystalline transition metal oxides on thin sheets of disordered graphitic carbon by oxidation of MXenes. Chem. Commun. **50**, 7420–7423 (2014)
39. Huang, S., Mochalin, V.N.: Hydrolysis of 2D transition-metal carbides (MXenes) in colloidal solutions. Inorg. Chem. **58**, 1958–1966 (2019)
40. Muckley, E.S., Naguib, M., Wang, H.W., Vlcek, L., Osti, N.C., Sacci, R.L., Sang, X., Unocic, R.R., Xie, Y., Tyagi, M., Mamontov, E.: Multimodality of structural, electrical, and gravimetric responses of intercalated MXenes to water. ACS Nano **11**, 11118–11126 (2017)
41. Hart, J.L., Hantanasirisakul, K., Lang, A.C., Anasori, B., Pinto, D., Pivak, Y., van Omme, J.T., May, S.J., Gogotsi, Y., Taheri, M.L.: Control of MXenes' electronic properties through termination and intercalation. Nat. Commun. **10**, 522 (2019)
42. Khazaei, M., Ranjbar, A., Arai, M., Sasaki, T., Yunoki, S.: Electronic properties and applications of MXenes: a theoretical review. J. Mater. Chem. C **5**, 2488–2503 (2017)
43. Li, M., Li, X., Qin, G., Luo, K., Lu, J., Li, Y., Liang, G., Huang, Z., Zhou, J., Hultman, L., Eklund, P.: Halogenated Ti_3C_2 MXenes with electrochemically active terminals for high-performance zinc ion batteries. ACS Nano **15**, 1077–1085 (2021)
44. Wang, X., Mathis, T.S., Sun, Y., Tsai, W.Y., Shpigel, N., Shao, H., Zhang, D., Hantanasirisakul, K., Malchik, F., Balke, N., Jiang, D.E.: Titanium carbide MXene shows an electrochemical anomaly in water-in-salt electrolytes. ACS Nano **15**, 15274–15284 (2021)
45. Choi, C., Ashby, D.S., Butts, D.M., DeBlock, R.H., Wei, Q., Lau, J., Dunn, B.: Achieving high energy density and high power density with pseudocapacitive materials. Nat. Rev. Mater. **5**, 5–19 (2020)

46. Garg, R., Agarwal, A., Agarwal, M.: A review on MXene for energy storage application: effect of interlayer distance. Mater. Res. Express **7**, 022001 (2020)
47. Pang, J., Mendes, R.G., Bachmatiuk, A., Zhao, L., Ta, H.Q., Gemming, T., Liu, H., Liu, Z., Rummeli, M.H.: Applications of 2D MXenes in energy conversion and storage systems. Chem. Soc. Rev. **48**, 72–133 (2019)
48. Kamysbayev, V., Filatov, A.S., Hu, H., Rui, X., Lagunas, F., Wang, D., Klie, R.F., Talapin, D.V.: Covalent surface modifications and superconductivity of two-dimensional metal carbide MXenes. Science **369**, 979–983 (2020)
49. Yang, C., Chen, J., Ji, X., Pollard, T.P., Lü, X., Sun, C.J., Hou, S., Liu, Q., Liu, C., Qing, T., Wang, Y.: Aqueous Li-ion battery enabled by halogen conversion–intercalation chemistry in graphite. Nature **569**, 245–250 (2019)
50. Liu, Z., Huang, Y., Huang, Y., Yang, Q., Li, X., Huang, Z., Zhi, C.: Voltage issue of aqueous rechargeable metal-ion batteries. Chem. Soc. Rev. **49**, 180–232 (2020)
51. Li, Y., Shao, H., Lin, Z., Lu, J., Liu, L., Duployer, B., Persson, P.O., Eklund, P., Hultman, L., Li, M., Chen, K.: A general Lewis acidic etching route for preparing MXenes with enhanced electrochemical performance in non-aqueous electrolyte. Nat. Mater. **19**, 894–899 (2020)
52. Li, D., Chen, X., Xiang, P., Du, H., Xiao, B.: Chalcogenated-$Ti_3C_2X_2$ MXene ($X = O$, S, Se and Te) as a high-performance anode material for Li-ion batteries. Appl. Surf. Sci. **501**, 144221 (2020)
53. Ma, R., Chen, Z., Zhao, D., Zhang, X., Zhuo, J., Yin, Y., Wang, X., Yang, G., Yi, F.: $Ti_3C_2T_x$ MXene for electrode materials of supercapacitors. J. Mater. Chem. A **9**, 11501–11529 (2021)
54. Carbide, T.D.T.: Cation intercalation and high volumetric capacitance of. Science **1241488**, 341 (2013)
55. Shekhirev, M., Shuck, C.E., Sarycheva, A., Gogotsi, Y.: Characterization of MXenes at every step, from their precursors to single flakes and assembled films. Prog. Mater Sci. **120**, 100757 (2021)
56. Li, X., Huang, Z., Shuck, C.E., Liang, G., Gogotsi, Y., Zhi, C.: MXene chemistry, electrochemistry and energy storage applications. Nat. Rev. Chem. **6**, 389–404 (2022)
57. Gogotsi, Y., Penner, R.M.: Energy storage in nanomaterials–capacitive, pseudocapacitive, or battery-like? ACS Nano **12**, 2081–2083 (2018)
58. Wang, X., Mathis, T.S., Li, K., Lin, Z., Vlcek, L., Torita, T., Osti, N.C., Hatter, C., Urbankowski, P., Sarycheva, A., Tyagi, M.: Influences from solvents on charge storage in titanium carbide MXenes. Nat. Energy **4**, 241–248 (2019)
59. Liang, G., Mo, F., Ji, X., Zhi, C.: Non-metallic charge carriers for aqueous batteries. Nat. Rev. Mater. **6**, 109–123 (2021)
60. Gao, Q., Sun, W., Ilani-Kashkouli, P., Tselev, A., Kent, P.R., Kabengi, N., Naguib, M., Alhabeb, M., Tsai, W.Y., Baddorf, A.P., Huang, J.: Tracking ion intercalation into layered Ti_3C_2 MXene films across length scales. Energy Environ. Sci. **13**, 2549–2558 (2020)
61. Shpigel, N., Levi, M.D., Sigalov, S., Mathis, T.S., Gogotsi, Y., Aurbach, D.: Direct assessment of nanoconfined water in 2D Ti_3C_2 electrode interspaces by a surface acoustic technique. J. Am. Chem. Soc. **140**, 8910–8917 (2018)
62. Griffith, K.J., Hope, M.A., Reeves, P.J., Anayee, M., Gogotsi, Y., Grey, C.P.: Bulk and surface chemistry of the niobium MAX and MXene phases from multinuclear solid-state NMR spectroscopy. J. Am. Chem. Soc. **142**, 18924–18935 (2020)
63. Li, N., Meng, Q., Zhu, X., Li, Z., Ma, J., Huang, C., Song, J., Fan, J.: Lattice constant-dependent anchoring effect of MXenes for lithium–sulfur (Li–S) batteries: a DFT study. Nanoscale **11**, 8485–8493 (2019)
64. Gao, X., Du, X., Mathis, T.S., Zhang, M., Wang, X., Shui, J., Gogotsi, Y., Xu, M.: Maximizing ion accessibility in MXene-knotted carbon nanotube composite electrodes for high-rate electrochemical energy storage. Nat. Commun. **11**, 6160 (2020)
65. Wang, L., Han, M., Shuck, C.E., Wang, X., Gogotsi, Y.: Adjustable electrochemical properties of solid-solution MXenes. Nano Energy **88**, 106308 (2021)
66. Eames, C., Islam, M.S.: Ion intercalation into two-dimensional transition-metal carbides: global screening for new high-capacity battery materials. J. Am. Chem. Soc. **136**, 16270–16276 (2014)

67. Aslam, M.K., Niu, Y., Xu, M.: MXenes for non-lithium-ion (Na, K, Ca, Mg, and Al) batteries and supercapacitors. Adv. Energy Mater. **11**, 2000681 (2021)
68. Schmuch, R., Wagner, R., Hörpel, G., Placke, T., Winter, M.: Performance and cost of materials for lithium-based rechargeable automotive batteries. Nat. Energy **3**, 267–278 (2018)
69. Kang, K., Meng, Y.S., Breger, J., Grey, C.P., Ceder, G.: Electrodes with high power and high capacity for rechargeable lithium batteries. Science **311**, 977–980 (2006)
70. Jin, T., Wang, Y., Hui, Z., Qie, B., Li, A., Paley, D., Xu, B., Wang, X., Chitu, A., Zhai, H., Gong, T.: Nonflammable, low-cost, and fluorine-free solvent for liquid electrolyte of rechargeable lithium metal batteries. ACS Appl. Mater. Interfaces **11**, 17333–17340 (2019)
71. Nelson, J., Misra, S., Yang, Y., Jackson, A., Liu, Y., Wang, H., Dai, H., Andrews, J.C., Cui, Y., Toney, M.F.: In operando X-ray diffraction and transmission X-ray microscopy of lithium sulfur batteries. J. Am. Chem. Soc. **134**, 6337–6343 (2012)
72. Wang, Z., Jiang, H., Zhang, Y., An, Y., Wei, C., Tan, L., Xiong, S., Qian, Y., Feng, J.: Application of 2D MXene in organic electrode materials for rechargeable batteries: recent progress and perspectives. Adv. Func. Mater. **33**, 2210184 (2023)
73. Guo, M., Zhong, S., Xu, T., Huang, Y., Xia, G., Zhang, T., Yu, X.: 3D hollow MXene (Ti_3C_2)/reduced graphene oxide hybrid nanospheres for high-performance Li-ion storage. J. Mater. Chem. A **9**, 23841–23849 (2021)
74. Jiang, G., Zheng, N., Chen, X., Ding, G., Li, Y., Sun, F., Li, Y.: In-situ decoration of MOF-derived carbon on nitrogen-doped ultrathin MXene nanosheets to multifunctionalize separators for stable Li–S batteries. Chem. Eng. J. **373**, 1309–1318 (2019)
75. Maity, C.K., Hatui, G., Sahoo, S., Saren, P., Nayak, G.C.: Boron nitride based ternary nanocomposites with different carbonaceous materials decorated by polyaniline for supercapacitor application. ChemistrySelect **4**, 3672–3680 (2019)
76. Maity, C.K., Hatui, G., Tiwari, S.K., Udayabhanu, G., Pathak, D.D., Nayak, G.C., Verma, K.: One pot solvothermal synthesis of novel solid state N-doped TiO2/n-Gr for efficient energy storage devices. Vacuum **164**, 88–97 (2019)
77. Maity, C.K., De, S., Acharya, S., Siddiki, S.H., Sahoo, S., Verma, K., Thakur, V.K., Nayak, G.C.: Copper oxide stabilized oxy-functionalized boron nitride-carbon nanotube nanohybrid: an ultra-stable electrode for flexible asymmetric supercapacitor device in ionic electrolyte. J. Energy Storage **56**, 105928 (2022)
78. Maity, C.K., De, S., Panigrahi, A., Acharya, S., Verma, K., Kim, M.J., Nayak, G.C.: Aerosol derived carbon dots decorated boron nitride supported Zn-doped MoS_2 for high performing flexible asymmetric supercapacitor. Compos. B Eng. **264**, 110887 (2023)
79. Maity, C.K., Hatui, G., Verma, K., Udayabhanu, G., Pathak, D.D., Nayak, G.C.: Single pot fabrication of N doped reduced GO (N-rGO)/ZnO–CuO nanocomposite as an efficient electrode material for supercapacitor application. Vacuum **157**, 145–154 (2018)
80. Xie, X., Zhao, M.Q., Anasori, B., Maleski, K., Ren, C.E., Li, J., Byles, B.W., Pomerantseva, E., Wang, G., Gogotsi, Y.: Porous heterostructured MXene/carbon nanotube composite paper with high volumetric capacity for sodium-based energy storage devices. Nano Energy **26**, 513–523 (2016)
81. Mashtalir, O., Lukatskaya, M.R., Zhao, M.Q., Barsoum, M.W., Gogotsi, Y.: Amine-assisted delamination of Nb_2C MXene for Li-ion energy storage devices. In: MXenes, pp. 401–414. Jenny Stanford Publishing (2015)
82. Lin, Z., Sun, D., Huang, Q., Yang, J., Barsoum, M.W., Yan, X.: Carbon nanofiber bridged two-dimensional titanium carbide as a superior anode for lithium-ion batteries. J. Mater. Chem. A **3**, 14096–14100 (2015)
83. Chen, M., Chen, J., Tan, X., Yang, W., Zou, H., Chen, S.: Facile self-assembly of sandwich-like MXene/graphene oxide/nickel–manganese layered double hydroxide nanocomposite for high performance supercapacitor. J. Energy Storage **44**, 103456 (2021)
84. Zahra, S.A., Anasori, B., Iqbal, M.Z., Ravaux, F., Al Tarawneh, M., Rizwan, S.: Enhanced electrochemical performance of vanadium carbide MXene composites for supercapacitors. APL Mater. **10** (2022)

85. Zhang, P., Soomro, R.A., Guan, Z., Sun, N., Xu, B.: 3D carbon-coated MXene architectures with high and ultrafast lithium/sodium-ion storage. Energy Storage Mater. **29**, 163–171 (2020)
86. Wan, L., Tang, Y., Chen, L., Wang, K., Zhang, J., Gao, Y., Lee, J.Y., Lu, T., Xu, X., Li, J., Zheng, Y.: In-situ construction of g-C_3N_4/Mo_2CT_x hybrid for superior lithium storage with significantly improved Coulombic efficiency and cycling stability. Chem. Eng. J. **410**, 128349 (2021)
87. Ding, Y., Chen, Y., Xu, N., Lian, X., Li, L., Hu, Y., Peng, S.: Facile synthesis of $FePS_3$ nanosheets@MXene composite as a high-performance anode material for sodium storage. Nano-Micro Lett. **12**, 1–12 (2020)
88. Guo, X., Xie, X., Choi, S., Zhao, Y., Liu, H., Wang, C., Chang, S., Wang, G.: Sb_2O_3/MXene ($Ti_3C_2T_x$) hybrid anode materials with enhanced performance for sodium-ion batteries. J. Mater. Chem. A **5**, 12445–12452 (2017)
89. Xu, E., Li, P., Quan, J., Zhu, H., Wang, L., Chang, Y., Sun, Z., Chen, L., Yu, D., Jiang, Y.: Dimensional gradient structure of $CoSe_2$@CNTs–MXene anode assisted by ether for high-capacity, stable sodium storage. Nano-Micro Lett. **13**, 1–14 (2021)
90. Zhang, Y., Guo, B., Hu, L., Xu, Q., Li, Y., Liu, D., Xu, M.: Synthesis of SnS nanoparticle-modified MXene ($Ti_3C_2T_x$) composites for enhanced sodium storage. J. Alloy. Compd. **732**, 448–453 (2018)
91. Wang, Z., Wang, F., Liu, K., Zhu, J., Waras, A.: Carbon-coated SnO2/Ti3C2 composites with enhanced lithium storage performance. J. Nanomater. **2019**, 9082132 (2019)
92. Li, Q., Zhou, J., Li, F., Sun, Z.: Spring-roll-like Ti 3 C 2 MXene/carbon-coated Fe_3O_4 composite as a long-life Li-ion storage material. Inorgan. Chem. Front. **7**, 3491–3499 (2020)
93. Zou, Z., Wang, Q., Yan, J., Zhu, K., Ye, K., Wang, G., Cao, D.: Versatile interfacial self-assembly of $Ti_3C_2T_x$ MXene based composites with enhanced kinetics for superior lithium and sodium storage. ACS Nano **15**, 12140–12150 (2021)
94. Zhang, X., Chen, S., Cai, J., King, S., Liu, C., Yuan, W., Wang, R., Wang, G.: Pre-strain accommodation enabled multi-dimensionally and hierarchically elastomeric $MoSe_2$/MXene and AC/MXene electrodes for stretchable sodium ion capacitors. J. Alloy. Compd. **935**, 168065 (2023)
95. Liu, Y., Dai, Z., Zhang, W., Jiang, Y., Peng, J., Wu, D., Chen, B., Wei, W., Chen, X., Liu, Z., Wang, Z.: Sulfonic-group-grafted $Ti_3C_2T_x$ MXene: a silver bullet to settle the instability of polyaniline toward high-performance Zn-Ion batteries. ACS Nano **15**, 9065–9075 (2021)
96. Adekoya, G.J., Adekoya, O.C., Sadiku, R.E., Hamam, Y., Ray, S.S.: Applications of MXene-containing polypyrrole nanocomposites in electrochemical energy storage and conversion. ACS Omega **7**, 39498–39519 (2022)
97. Zhang, X., Li, J., Li, J., Han, L., Lu, T., Zhang, X., Zhu, G., Pan, L.: 3D TiO_2@nitrogen-doped carbon/Fe7S8 composite derived from polypyrrole-encapsulated alkalized MXene as anode material for high-performance lithium-ion batteries. Chem. Eng. J. **385**, 123394 (2020)
98. Ding, J., Tang, C., Zhu, G., He, F., Du, A., Wu, M., Zhang, H.: Ultrasmall SnO_2 nanocrystals sandwiched into polypyrrole and $Ti_3C_2T_x$ MXene for highly effective sodium storage. Mater. Chem. Front. **5**, 825–833 (2021)
99. Zhang, Y., Cao, Z., Liu, S., Du, Z., Cui, Y., Gu, J., Shi, Y., Li, B., Yang, S.: Charge-enriched strategy based on MXene-based polypyrrole layers toward dendrite-free zinc metal anodes. Adv. Energy Mater. **12**, 2103979 (2022)
100. Cheng, Y., Li, Y., Yang, L., Yin, S.: Poly (o-phenylenediamine)-decorated $V_4C_3T_x$ MXene/poly (o-phenylenediamine) blends as electrode materials to boost storage capacity for supercapacitors and lithium-ion batteries. ACS Appl. Nano Mater. **6**, 9186–9196 (2023)
101. Gao, Y., Xue, P., Ji, L., Pan, X., Chen, L., Guo, W., Tang, M., Wang, C., Wang, Z.: Interfacial self-assembly of organics/MXene hybrid cathodes toward high-rate-performance sodium ion batteries. ACS Appl. Mater. Interfaces **14**, 8036–8047 (2022)
102. Zhang, L., Bi, J., Zhao, Z., Wang, Y., Mu, D., Wu, B.: Sulfur@ Self-assembly 3D MXene hybrid cathode material for lithium-sulfur batteries. Electrochim. Acta **370**, 137759 (2021)
103. Rao, Q.S., Liao, S.Y., Huang, X.W., Li, Y.Z., Liu, Y.D. and Min, Y.G.: Assembly of MXene/PP separator and its enhancement for Ni-Rich $LiNi_{0.8}Co_{0.1}Mn_{0.1}O_2$ electrochemical performance. Polymers **12**, 2192 (2020)

104. Li, Y., Li, M., Zhu, Y.C., Song, S., Li, S.N., Aarons, J., Tang, L.C., Bae, J.: Polysulfide-inhibiting, thermotolerant and nonflammable separators enabled by DNA co-assembled CNT/MXene networks for stable high-safety Li–S batteries. Compos. B Eng. **251**, 110465 (2023)
105. Chen, Z., Li, X., Wang, D., Yang, Q., Ma, L., Huang, Z., Liang, G., Chen, A., Guo, Y., Dong, B., Huang, X.: Grafted MXene/polymer electrolyte for high performance solid zinc batteries with enhanced shelf life at low/high temperatures. Energy Environ. Sci. **14**, 3492–3501 (2021)
106. Pan, Q., Zheng, Y., Kota, S., Huang, W., Wang, S., Qi, H., Kim, S., Tu, Y., Barsoum, M.W., Li, C.Y.: 2D MXene-containing polymer electrolytes for all-solid-state lithium metal batteries. Nanoscale Adv. **1**, 395–402 (2019)
107. Kou, W., Zhang, Y., Wu, W., Guo, Z., Hua, Q., Wang, J.: Thin polymer electrolyte with MXene functional layer for uniform Li^+ deposition in all-solid-state lithium battery. Green Energy Environ. **9**, 71–80 (2024)
108. Venkateswarlu, S., Vallem, S., Umer, M., Jyothi, N.V.V., Babu, A.G., Govindaraju, S., Son, Y., Kim, M.J., Yoon, M.: Recent progress on MOF/MXene nanoarchitectures: a new era in coordination chemistry for energy storage and conversion. J. Energy Chem. (2023)
109. Zhou, D., Fu, H., Long, J., Shen, K., Gou, X.: Novel fusiform core-shell-MOF derived intact metal@carbon composite: an efficient cathode catalyst for aqueous and solid-state Zn–air batteries. J. Energy Chem. **64**, 385–394 (2022)
110. Gu, H., Kong, L., Cui, H., Zhou, X., Xie, Z., Zhou, Z.: Fabricating high-performance sodium ion capacitors with P2-$Na_{0.67}Co_{0.5}Mn_{0.5}O_2$ and MOF-derived carbon. J. Energy Chem. **28**, 79–84 (2019)
111. Sun, L., Wang, H., Zhai, S., Sun, J., Fang, X., Yang, H., Zhai, D., Liu, C., Deng, W.Q., Wu, H.: Dual-conductive metal-organic framework@MXene heterogeneity stabilizes lithium-ion storage. J. Energy Chem. **76**, 368–376 (2023)
112. Liu, X., Liu, F., Zhao, X., Fan, L.Z.: Constructing MOF-derived CoP-NC@ MXene sandwich-like composite by in-situ intercalation for enhanced lithium and sodium storage. J. Materiomics **8**, 30–37 (2022)
113. Wen, C., Guo, D., Zheng, X., Li, H., Sun, G.: Hierarchical *n*MOF-867/MXene nanocomposite for chemical adsorption of polysulfides in lithium-sulfur batteries. ACS Appl. Energy Mater. **4**, 8231–8241 (2021)
114. Zong, H., Hu, L., Wang, Z., Qi, R., Yu, K., Zhu, Z.: Metal-organic frameworks-derived CoP anchored on MXene toward an efficient bifunctional electrode with enhanced lithium storage. Chem. Eng. J. **416**, 129102 (2021)
115. Wang, Y., Song, J., Wong, W.Y.: Constructing 2D sandwich-like MOF/MXene heterostructures for durable and fast aqueous zinc-ion batteries. Angew. Chem. **135**, e202218343 (2023)
116. Shi, Y., Zhu, G., Guo, X., Jing, Q., Pang, H., Zhang, Y.: Three-dimensional MXene-encapsulated porous Ni-NDC nanosheets as anodes for enhanced lithium-ion batteries. Nano Res. **16**, 2528–2535 (2023)
117. Zhao, L., Dong, B., Li, S., Zhou, L., Lai, L., Wang, Z., Zhao, S., Han, M., Gao, K., Lu, M., Xie, X.: Interdiffusion reaction-assisted hybridization of two-dimensional metal–organic frameworks and $Ti_3C_2T_x$ nanosheets for electrocatalytic oxygen evolution. ACS Nano **11**, 5800–5807 (2017)
118. Zou, H., He, B., Kuang, P., Yu, J., Fan, K.: Metal–organic framework-derived nickel–cobalt sulfide on ultrathin mxene nanosheets for electrocatalytic oxygen evolution. ACS Appl. Mater. Interfaces **10**, 22311–22319 (2018)
119. Wang, L., Wu, Z., Zou, J., Gao, P., Niu, X., Li, H., Chen, L.: Li-free cathode materials for high energy density lithium batteries. Joule **3**, 2086–2102 (2019)
120. Zhou, Q., Ma, J., Dong, S., Li, X., Cui, G.: Intermolecular chemistry in solid polymer electrolytes for high-energy-density lithium batteries. Adv. Mater. **31**, 1902029 (2019)
121. Niu, Y., Feng, W., Lei, Z., Hu, W., Zheng, X., Su, W., Zhang, L.: MXene surface-attached Ni2P on lithium-sulfur battery catalytic effect. J. Electroan. Chem. **946**, 117743 (2023)
122. Shen, K., Li, B., Yang, S.: 3D printing dendrite-free lithium anodes based on the nucleated MXene arrays. Energy Storage Mater. **24**, 670–675 (2020)

123. Ma, J., Zheng, S., Zhou, F., Zhu, Y., Das, P., Huang, R., Zhang, L., Wang, X., Wang, H., Cui, Y., Wu, Z.S.: All 3D printing lithium metal batteries with hierarchically and conductively porous skeleton for ultrahigh areal energy density. Energy Storage Mater. **54**, 304–312 (2023)
124. Zheng, S., Wang, H., Das, P., Zhang, Y., Cao, Y., Ma, J., Liu, S., Wu, Z.S.: Multitasking MXene inks enable high-performance printable microelectrochemical energy storage devices for all-flexible self-powered integrated systems. Adv. Mater. **33**, 2005449 (2021)
125. Ji, X., Lee, K.T., Nazar, L.F.: A highly ordered nanostructured carbon–sulphur cathode for lithium–sulphur batteries. Nat. Mater. **8**, 500–506 (2009)
126. Rauh, R.D., Abraham, K.M., Pearson, G.F., Surprenant, J.K., Brummer, S.B.: A lithium/dissolved sulfur battery with an organic electrolyte. J. Electrochem. Soc. **126**, 523 (1979)
127. Wei, C., Tian, M., Wang, M., Shi, Z., Yu, L., Li, S., Fan, Z., Yang, R., Sun, J.: Universal in situ crafted MO_x-MXene heterostructures as heavy and multifunctional hosts for 3D-printed Li–S batteries. ACS Nano **14**, 16073–16084 (2020)
128. Wei, C., Tian, M., Fan, Z., Yu, L., Song, Y., Yang, X., Shi, Z., Wang, M., Yang, R., Sun, J.: Concurrent realization of dendrite-free anode and high-loading cathode via 3D printed N–Ti_3C_2 MXene framework toward advanced Li–S full batteries. Energy Storage Mater. **41**, 141–151 (2021)
129. Fang, Y., Zhang, Y., Zhu, K., Lian, R., Gao, Y., Yin, J., Ye, K., Cheng, K., Yan, J., Wang, G., Wei, Y.: Lithiophilic three-dimensional porous $Ti_3C_2T_x$-rGO membrane as a stable scaffold for safe alkali metal (Li or Na) anodes. ACS Nano **13**, 14319–14328 (2019)
130. He, X., Jin, S., Miao, L., Cai, Y., Hou, Y., Li, H., Zhang, K., Yan, Z., Chen, J.: A 3D hydroxylated MXene/carbon nanotubes composite as a scaffold for dendrite-free sodium-metal electrodes. Angew. Chem. Int. Ed. **59**, 16705–16711 (2020)
131. Luo, J., Wang, C., Wang, H., Hu, X., Matios, E., Lu, X., Zhang, W., Tao, X., Li, W.: Pillared MXene with ultralarge interlayer spacing as a stable matrix for high performance sodium metal anodes. Adv. Func. Mater. **29**, 1805946 (2019)
132. Wang, Z., Huang, Z., Wang, H., Li, W., Wang, B., Xu, J., Xu, T., Zang, J., Kong, D., Li, X., Yang, H.Y.: 3D-printed sodiophilic V_2CT_x/rGO-CNT MXene microgrid aerogel for stable Na metal anode with high areal capacity. ACS Nano **16**, 9105–9116 (2022)

MXene as Nanofillers for Polymer Composites

Pooja Mohapatra, Shreelata Behera, Saleja Sahoo, Annada Mishra, Lipsa Shubhadarshinee, Bigyan Ranjan Jali, Aruna Kumar Barick, and Priyaranjan Mohapatra

Abstract Incorporation of MXene nanofillers into the polymer matrices represents a burgeoning area of research with promising implications for enhancing overall material properties. The distinctive structural and functional characteristics of MXene and MXene-derived materials facilitate seamless integration with polymer matrices, addressing the elevated performance demands in electrical, thermal, mechanical, and various other domains. Moreover, this combination yields remarkable flexibility, mechanical resilience, and tensile strength, particularly evident in the fabrication of flexible supercapacitors. Polymers stand out as an excellent option for designing MXene composites, given their adaptability, compatibility, and cost-effectiveness. Various types of MXene nanofillers are scrutinized, and their integration techniques into polymer matrices are critically evaluated, including mechanical mixing, solution mixing, and in situ polymerization. The mechanical properties of MXene-based polymer composites are rigorously assessed, focusing on tensile strength, flexural strength, impact resistance, fatigue behavior, and comparative analyses with other nanofillers. Furthermore, the book chapter scrutinizes the thermal and electrical enhancements facilitated by MXene, delving into thermal conductivity, electrical conductivity, and dielectric properties. The role of MXene in various applications, including aerospace, automotive, electronics, and biomedical fields, is critically discussed, shedding light on both successes and challenges in real world scenarios. Environmental considerations are rigorously addressed, including exploring the eco-friendly aspects of MXene, recycling issues, and conducting a life cycle assessment. Through a meticulous analysis of existing literature, this study aims to provide insights into the current limitations, propose potential solutions, and forecast emerging trends in MXene reinforced polymer composites, employing both critical and statistical methodologies.

Keywords MXene · Polymers · Composites · Mechanical properties · Thermal conductivity · Electrical conductivity · Supercapacitors

---

P. Mohapatra · S. Behera · S. Sahoo · A. Mishra · L. Shubhadarshinee · B. R. Jali · A. K. Barick (✉) · P. Mohapatra
Department of Chemistry, Veer Surendra Sai University of Technology, Siddhi Vihar, Burla, Sambalpur, Odisha, India
e-mail: akbarick_chem@vssut.ac.in

© The Author(s), under exclusive license to Springer Nature Singapore Pte Ltd. 2025
S. Moharana and R. N. Mahaling (eds.), *Surface Engineering of MXenes*, Engineering Materials, https://doi.org/10.1007/978-981-96-3640-2_12

1 Introduction

Since their synthesis by Naguib et al. from the selective etching and exfoliation of the MAX phases in 2011 at Drexel University, USA, the MXenes, which are two-dimensional (2D) nanosheets of most transition metal carbides, nitrides, or carbonitrides with graphene-like nanostructure, have drawn a great deal of interest [1]. MXenes' extraordinary material properties, including their metallic electrical conductivity, flexibility, tunable band gap, high charge carrier mobility, stiffness and strength, anticorrosion, chemical resistance, electronic, tribological, magnetic, thermal, optical, electrochemical, barrier, and other unique characteristics, make them a promising candidate for application in several fields, such as energy storage and conversion, optoelectronics, electromagnetic interference (EMI) shielding, catalysis, sensing, biomedicine, environmental remediation, and among others [2]. MXenes encounter various processing challenges that hinder their development, such as restacking or union of nanolayers caused by weak van der Waals force and strong hydrogen bonding, limited flexibility, and oxidation or surface defects [3]. As a result, MXenes are incorporated into different polymer matrices to prevent their layers from stacking on top of each other and to ensure uniform dispersion of the nanolayers within the polymer matrix. This greatly improves various properties such as structural, electrical, tribological, thermal, mechanical, biocompatibility, and other characteristics [4–6]. Nanofillers are crucial in enhancing the properties of composite materials by improving their mechanical strength, thermal stability, electrical conductivity, and barrier properties. Due to their large surface area and nanoscale dimensions, nanofillers can effectively reinforce the composite matrix, distributing stress more evenly and preventing crack propagation. They also improve thermal conductivity and resistance, increase electrical conductivity in insulating matrices, and enhance chemical resistance and durability. Additionally, nanofillers can modify optical properties and reduce weight while maintaining strength, making them ideal for applications in aerospace, automotive, electronics, and packaging, where performance and material efficiency are critical.

MXene-based polymer composites can be fabricated through ultrasonic mixing, solution casting, melt blending, emulsion method, vacuum filtration, freeze-drying, etc. Consequently, MXenes-reinforced thermosetting and thermoplastic polymer composites can be applicable in anti-corrosive coating [7], dielectric, EMI shielding [8, 9], fire retardant [10], anti-wear coating, high strength coating, opto/spintronic, sensing, wearable devices [11], biomedical [12], separating membranes [13], environmental remediation [14], air quality monitoring [15], energy storage devices like supercapacitors and batteries [16–18], etc.

The primary obstacles to produce polymer/MXene composites are identifying appropriate processing settings and achieving precise control over the orientation and arrangement of the MXene nanosheets inside the polymer matrix. This book chapter provides a concise overview of the latest knowledge on the links between processing, structure, properties, and applications of polymer/MXene composites. These composites consist of MXene nanosheets that are evenly distributed throughout the polymer matrix. Consequently, MXenes are reinforced into the various polymer matrices to restrict their restacking and homogeneous exfoliation of the nanolayer within the polymer matrix, which significantly enhances several characteristics like structural, electrical, thermal, mechanical, etc., resulting in the design of energy storage devices like supercapacitors, batteries, etc. A brief description is given in Table 1, which is explained later in this study.

The superior 2D flake structure of MXenes, along with their rich functional groups, grants them a wide range of tunable properties, making them highly advantageous in various fields. Additionally, these unique structural and functional characteristics allow MXenes and MXene-based materials to easily integrate with various polymer matrices like polyaniline (PAni), polypyrrole (PPy), polyindole (PIn), polythiophene (PTh), polyvinylidene fluoride (PVDF), and poly(3,4-ethylenedioxythiophene) (PEDOT), among others, to fulfill high-performance requirements in electrical, thermal, mechanical, and other domains. Various synthesis methods for MXenes have been devised to improve their uses, including those involving hydrofluoric acid, fluorine-containing acid solutions, and salts. Furthermore, various modification techniques are reviewed, focusing on the huge surface terminations of MXenes. The preparation techniques and performance of polymer/MXene composites are investigated in order to analyze the effect of MXenes on the performance of these composites. Additionally, the possible use of polymer/MXene composites with diverse characteristics are studied to assess their potential applications. Finally, the current status of MXene development is discussed, along with the challenges and prospects for the advancement of polymer/MXene composites.

Considering the exceptional performance of MXenes, it is indisputable that MXenes will have a crucial impact on polymer composites and offer significant opportunities for the advancement of high-performance composite materials [16]. Though MXenes exhibit excellent performances, they face different processing challenges, including restacking/union of nanolayers due to weak van der Waals forces of interactions and hydrogen bonding, low flexibility, and oxidation/surface defects, which substantially inhibit their development.

Although MXenes and their composites have shown excellent results in energy storage devices, the tendency of MXene layers to restack hinders electrolyte penetration and reduces the effectiveness of their ion adsorption and desorption sites. To address this issue, developing MXene polymer composites presents a novel strategy that prevents the restacking of MXene nanosheets and integrates functional groups, thus enhancing the electrochemical characteristics of the material. This study examines MXene and its polymer based composites for designing supercapacitors in energy storage devices. An overview of various methods was highlighted for utilizing MXene and polymers, along with a discussion of their structural, electrical, thermal, and mechanical properties. Additionally, the use of MXene and polymer materials in supercapacitors, as well as the future challenges and potential developments, are thoroughly reviewed [29].

Table 1 Representation of synthesis procedures, properties, and applications of polymer/MXene composites

Sl. No	Types of polymers	Methods of preparation	Properties	Applications	References
1	PAni	In situ intercalation polymerization	Coating resistance, charge transfer resistance, impedance modulus, and lower electrical conductivity, and corrosion protection	Anticorrosion performance of epoxy coating	[19]
2	PPy	One-step co-electrodeposition	Electron transfer and ionic diffusion	Electrochemical energy storage	[20]
3	Functionalised-PEDOT	Embedding method	–	Wearable and non-invasive monitoring of sweat glucose	[21]
4	PIn	Polymerisation	Large proton storage capacity, excellent cycle stability, corrosion inhibition efficiency, enhanced redox capability, and structural stability	Superior proton storage and corrosion protection	[22]
5	PTh	Interfacial polymerization	Electro-catalytic activity, and reduction in charge transfer resistance	Dye-sensitized solar cells	[23]
6	PAni	Hydrothermal	High specific capacity, high capacitance retention, and high cycle lifespan	Advanced asymmetric supercapacitor	[24]
7	PEDOT	Self-assembly	Superior capacitive performance and flat surface	Supercapacitors	[25]

(continued)

Table 1 (continued)

Sl. No	Types of polymers	Methods of preparation	Properties	Applications	References
8	PPy	In situ polymerization	High desalination capacity and excellent cycling stability	Enhanced sodium storage capacity for efficient hybrid capacitive deionization	[26]
9	PAni	Steam induced foaming	High sensitivity, rapid response, and recovery time, and outstanding fatigue resistance	Pressure sensors	[27]
10	PPy (Sulfur-doped MXene/PPy)	Hydrothermal, In situ polymerization	High power density and maximum output voltage	Extracellular electron transfer process	[28]

2 Polymer/MXene Composites

2.1 *Synthesis of Polymer/MXene Composites*

Awan et al. described MXene and polymer-based composites as bulk materials in which both components are uniformly distributed, often as fillers or structural elements. The methods for fabricating these composites include in situ polymerization, hot or dry pressing, emulsion mixing, and solution blending. Solution blending is the most commonly employed technique for synthesizing polymer/MXene composites. The authors also discussed various techniques for creating composites using MXene and polymers [30].

Kauser et al. highlighted that because of their large surface area, unique graphene-like morphology, electrical conductivity, and mechanical strength, MXenes are often utilized in combination with different polymers [31]. Furthermore, polymer/MXene composites exhibit potential tribological properties. Ti_3C_2, a representative MXene, is commonly utilized to create polymer-based composites. Jin et al. developed a polyvinyl alcohol (PVA)/$Ti_3C_2T_x$ composite [32]. Figure 1 represents the fabrication of PVA/MXene composites using a multilayered casting technique, where the MXene layers are continuously formed from an aqueous dispersion, and the polymer layers are generated at 45 °C to create a multilayered structure.

Lim et al. synthesized PAni/$Ti_3C_2T_x$ nanoarchitecture through a simple and cost-effective self-alignment process that combines 1D PAni nanofibers with 2D $Ti_3C_2T_x$ nanosheets. This method results in significantly improved specific capacitance, an exceptionally high-rate capability, and unique cycling stability, offering a valuable

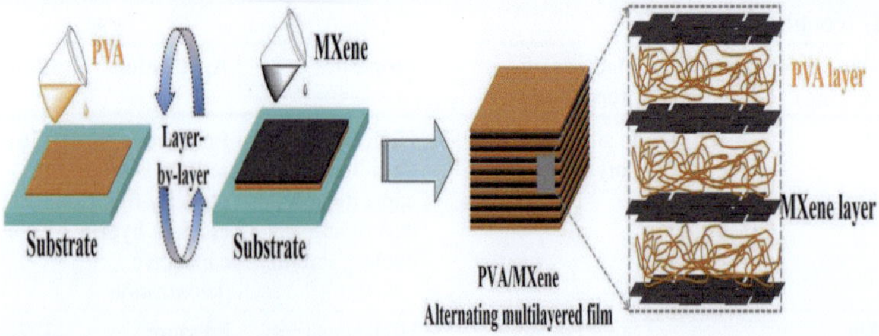

Fig. 1 Fabrication of PVA/MXene film using multilayered casting [32]. Copyright 2019. Reproduced with permission from Elsevier

approach for designing superior energy storage devices with both huge energy and power density [18, 33]. In another study done by Ezika et al. MXene/PAni nanoarchitectures were synthesized using a hydrothermal method, where chain-like PAni structures were anchored onto the surface and wrapped with PAni nanofibers. The chain-like PAni was a structural link and conducting pathway, effectively connecting nearby MXene layers and enhancing charge transfer between the various MXene sheets [18]. The interaction between amine (–NH) groups and $Ti_3C_2T_x$ in the pyrrole ring promotes the formation of an alternating sheet structure through hydrogen bonding that is valuable for the directional growth of PPy. The intercalation of PPy increases the interlayer spacing of the $Ti_3C_2T_x$, as shown in Fig. 2, resulting in a uniformly arranged structure that preserves the high conductivity of $Ti_3C_2T_x$ and facilitates the penetration of electrolytic ions [18, 34].

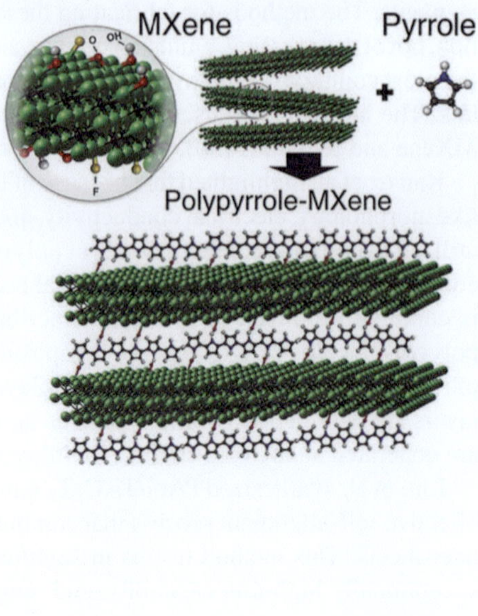

Fig. 2 Schematic representation of fabrication of PPy/MXene nanoarchitectures [34]. Copyright 2015. Reproduced with permission from John Wiley and Sons

3 Polymer-MXene Interactions

The interactions between MXenes and polymers in composites are fundamental to their performance and functionality across various applications. These interactions can be categorized into several types, such physical, chemical, interfacial, mechanical, and electrochemical interactions.

3.1 *Physical Interactions*

In polymer/MXene composites, van der Waals forces act a fundamental factor in maintaining the structural integrity of the composite. These weak but essential forces facilitate the uniform dispersion of MXene nanosheets within the polymer matrix, ensuring consistent mechanical, thermal, and electrical properties throughout the material. Hydrogen bonding stabilizes the composite when polymers with functional groups like hydroxyl or amine groups are involved. These groups can form strong hydrogen bonds with the surface functionalities of MXenes (such as –OH or –F), leading to improved stability and dispersion within the polymer matrix. This combination of van der Waals forces and hydrogen bonding is critical in enhancing the overall performance and uniformity of polymer/MXene composites [31, 35, 36].

3.2 *Chemical Interactions*

In polymer/MXene composites, covalent bonding can significantly enhance the mechanical and thermal properties of the material. This type of bonding occurs when chemical modifications are applied to either the MXenes or the polymer chains, allowing for the formation of strong, durable covalent bonds between the two components. Such covalent linkages result in improved load transfer within the composite, thereby increasing its structural integrity and thermal stability. Electrostatic interactions also play a crucial role, particularly given that MXenes often carry surface charges due to their inherent surface chemistry. These charges enable electrostatic interactions with oppositely charged polymer matrices, which can further enhance the compatibility and stability of the composite. This electrostatic compatibility is essential for maintaining the homogeneous dispersion of MXenes throughout the polymer matrix, thereby contributing to the overall performance of the composite [14].

3.3 Interface Interactions

In polymer/MXene composites, the interaction at the interface between MXenes and the polymer matrix is crucial for the composite's overall performance. Good interfacial adhesion enhances load transfer from the polymer to the MXenes, thereby improving the mechanical strength and structural integrity of the composite. This strong adhesion ensures that the MXenes are well-integrated within the polymer matrix, leading to a more robust material. Additionally, the conformation of the polymer at the MXene surface plays a significant role in determining the composite's behavior. The polymer can adopt various conformations, such as forming a protective layer around the MXenes, which can influence the reactivity and stability of the composite by shielding the MXenes from environmental factors. Together, these factors contribute to the improved functionality and durability of polymer/MXene composites [37, 38].

3.4 Mechanical Interactions

MXenes play a crucial role in reinforcing polymer matrices through mechanical interactions, significantly enhancing the strength and stiffness of the composite. This reinforcement is particularly vital for applications requiring high structural integrity and durability. Additionally, when MXenes interact with flexible polymers, they contribute to the flexibility and stretchability of the composite. This combination makes polymer/MXene composites especially suitable for applications in wearable electronics and flexible devices, where both robustness and adaptability are essential [39].

3.5 Electrochemical Interactions

In energy storage applications like supercapacitors and batteries, the interaction between MXenes and polymers affects the charge transport properties. MXenes contribute to high electrical conductivity, while the polymer matrix can influence the overall electrochemical performance of the composite [40].

4 Characterization of Polymer/MXene Composites

The structural and morphological analysis of polymer/MXene composites are two critical aspects of material characterization that provide comprehensive insights into the composition, distribution, and interactions between the polymer matrix and MXene fillers. Here is a more detailed explanation of each type of analysis [41].

4.1 *Structural Analysis*

Structural analysis of polymer/MXene composites focuses on understanding the internal arrangement, chemical structure, and interactions between polymers and MXenes at atomic, molecular, and crystalline levels. This type of analysis is crucial for determining how MXenes are incorporated into the polymer matrix, the nature of bonding, and the composite's overall structural integrity. The primary objective is to characterize these composites' internal structure, chemical bonding, crystallinity, and phase composition.

Various analytical techniques are employed for this purpose. X-ray diffraction (XRD) is employed to find out the crystalline structure and phase composition of both the polymer and MXenes, providing insights into the interlayer spacing in MXenes and any changes due to the interaction or intercalation within the polymer matrix. Fourier transform infrared (FTIR) spectroscopy identifies functional groups and chemical bonds within the composite, detecting changes in bonding that indicate interactions like hydrogen bonding or covalent bonds between MXenes and polymer chains. Raman spectroscopy provides information on molecular vibrations, which can reveal stress distributions within the composite and changes in the MXene structure due to polymer interactions. X-ray photoelectron spectroscopy (XPS) analyzes the elemental composition and oxidation states of elements within the composite, detecting surface chemistry changes such as oxidation or functionalization of MXenes, which are crucial for understanding interactions with the polymer. Nuclear magnetic resonance (NMR) spectroscopy offers insights into the local chemical environment of nuclei in the polymer matrix and MXenes, helping to understand molecular-level interactions and dynamics. Together, these techniques provide a comprehensive understanding of the structural characteristics of polymer/MXene composites, which is crucial for optimizing their properties for various applications. It facilitates identifying how MXenes interact with the polymer at a molecular level, which can affect properties like thermal stability, chemical resistance, and mechanical strength.

4.2 Morphological Analysis

Morphological analysis is crucial for examining the external structure, shape, size, and distribution of components such as MXenes and polymers within a composite material. This analysis is essential for understanding the dispersion, alignment, and overall morphology of MXenes within the polymer matrix. Scanning electron microscopy (SEM) is carried out to detect the surface morphology and overall distribution of MXene sheets in the polymer matrix, providing high-resolution images that can reveal the homogeneity of the composite, as well as any agglomeration of MXenes and phase separation between MXenes and polymers. Transmission electron microscopy (TEM) offers even higher resolution imaging, allowing for detailed visualization of the internal microstructure, including the thickness, shape, and orientation of MXene sheets within the polymer matrix. TEM also reveals the nanoscale dispersion and interfacial characteristics between MXenes and polymers. Atomic force microscopy (AFM) is used to study the surface topography and roughness of the composite, offering insights into the height and thickness of MXene sheets on the composite surface, thus enhancing the understanding of surface morphology at the nanoscale. Environmental scanning electron microscopy (ESEM) provides the ability to observe the composite morphology under varying environmental conditions, such as changes in humidity or temperature, which aids in assessing the morphological stability and durability of the composite under realistic conditions. Although nanoindentation is primarily a mechanical testing technique, it also provides insights into the morphological characteristics of the material by revealing its response to localized pressure, thereby helping to assess the uniformity of the composite structure, especially in MXene-rich regions.

5 Properties of Polymer/MXene Composites

MXenes exhibit a range of unique properties that make them highly effective as nanofillers in polymer composites. One of the most notable properties of MXenes is their high electrical conductivity, which significantly enhances the conductivity of polymer matrices, making them suitable for applications in electronics, energy storage devices, and EMI shielding. Additionally, MXenes possess excellent thermal conductivity, which, when incorporated into polymers, improves heat dissipation and thermal management in applications such as flexible electronics and high-power devices. Liu et al. explained it by giving an example of PVA/MXene from the thermogravimetric (TG) curve, where the increase in thermal stability of the polymer/MXene composite can be seen in the temperature range of 30–950 °C [42]. The mechanical strength of MXenes is another key property, contributing to the enhanced durability, toughness, and stiffness of polymer composites. The large surface area and layered structure of MXenes, combined with the presence of abundant surface functional groups, allow for strong interfacial interactions with polymer matrices. This not only leads to better load transfer, enhancing mechanical properties but also ensures good dispersion within the polymer matrix [43]. The hydrophilic nature of MXenes, due to their surface chemistry, further aids in their uniform distribution, particularly in polar polymers, thereby improving the overall performance of composites in terms of mechanical strength and barrier properties [44]. MXenes also exhibit excellent barrier properties due to their 2D layered structure, which creates a tortuous path for gas and liquid molecules, which significantly enhances the barrier capabilities of polymer composites that make them ideal for applications such as packaging materials and protective coatings [45]. MXenes contribute to enhanced energy storage capabilities when integrated into polymer composites that makes them highly applicable for flexible energy storage devices, including supercapacitors and wearable electronics [45].

Additionally, the ability of MXenes to provide superior EMI shielding due to their high electrical conductivity, positions them as essential materials for shielding sensitive electronic devices from electromagnetic radiation. The EMI shielding effectiveness (EMI SE) of both MXene and MXene/polyurethane (PU) composites was studied, as illustrated in Fig. 3a. The MXene, known for its superior electrical conductivity, attained the maximum EMI SE of 61 dB in the frequency ranging from 8.2 to 12.4 GHz. However, the EMI SE of the MXene/PU composites gradually reduced with increase in MXene content due to the insulating behaviour of the PU matrix. To better understand the EMI shielding mechanism, the total shielding effectiveness (S_T), absorption effectiveness (S_A), and reflection effectiveness (S_R) were separately analyzed, as depicted in Fig. 3b. It is important to note that S_A was significantly higher than S_R, suggesting that the absorption is the primary shielding mechanism, which can be partially explained by the layered microstructure of the MXene/PU composites. The possible mechanism of EMI shielding of the MXene/PU composite is shown in Fig. 3c. When incident electromagnetic radiation hit the surface of the MXene/PU composite, a portion of the radiations is instantaneously reflected due to the impedance difference between the MXene that contains a high concentration of free electrons and the surrounding air. To assess the EMI SE of composites, the parameter SSE/t (specific shielding effectiveness per unit thickness) is used, as presented in Fig. 3d. The SSE/t magnitudes of the MXene/PU composites are decreased with an increase in the MXene content [46].

Varghese et al. conducted a comparative study on the electrochemical properties of neat MXene, MXene-PPy, and MXene-PAni electrodes across various scan rates and current densities, focusing specifically on a scan rate of 20 mV/s and a current density of 1 A/g. As shown in Fig. 4a, the cyclic voltammetry (CV) profile of neat MXene displays an insignificant response. In contrast, the MXene-PAni electrode shows a significantly bigger loop area and a superior current response at the same scan rate, indicating a substantially greater capacitance compared to MXene-PPy. Figure 4b, d illustrate the variation in specific capacitance (C_{sp}) with increasing scan rates and current densities for all three materials, which indicate that the value of C_{sp} decreases both at higher scan rates and current densities due to the hindered ion accessibility and controlled ion transport that attributed to the slower kinetics. Notably, MXene-PAni demonstrates superior rate capability (90%) compared to the

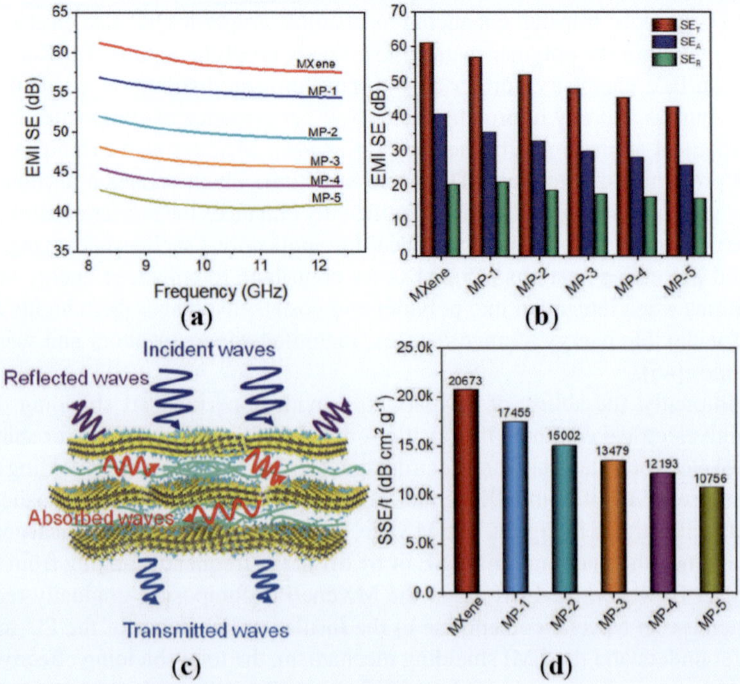

Fig. 3 **a** Variation of EMI SE with applied frequency, **b** S_T, S_A, and S_R values for pure MXene and MXene/PU composite films with different MXene loadings, **c** proposed mechanism of EMI shielding, and **d** SSE/t values for pure MXene and MXene/PU composite films with different MXene loadings [46]. Copyright 2024. Reproduced with permission from Elsevier

neat MXene and MXene-PPy, which show rate capabilities of 77% and 80%, respectively. The improved capacitance of MXene-PAni electrodes is due to large amount of active sites provided via self-assembly of the PAni molecular chains inside the MXene nanolayers. Consequently, MXene-PAni shows a longer charge and discharge time under the same current density, achieving a specific capacitance of 430 F/g, which surpasses the specific capacitances of MXene-PPy and neat MXene, measured at 305 and 105 F/g, respectively, at 1 A/g from galvanostatic charge discharge (GCD) curves, as shown in Fig. 4c [47].

In some cases, MXenes also demonstrate biocompatibility and antimicrobial properties, broadening their potential for use in biomedical applications, such as tissue engineering, biosensors, and antimicrobial coatings. Overall, the combination of electrical, thermal, mechanical, and barrier properties, along with biocompatibility and flame retardancy, makes MXenes highly versatile and valuable nanofillers for advanced polymer composites in a wide range of industries.

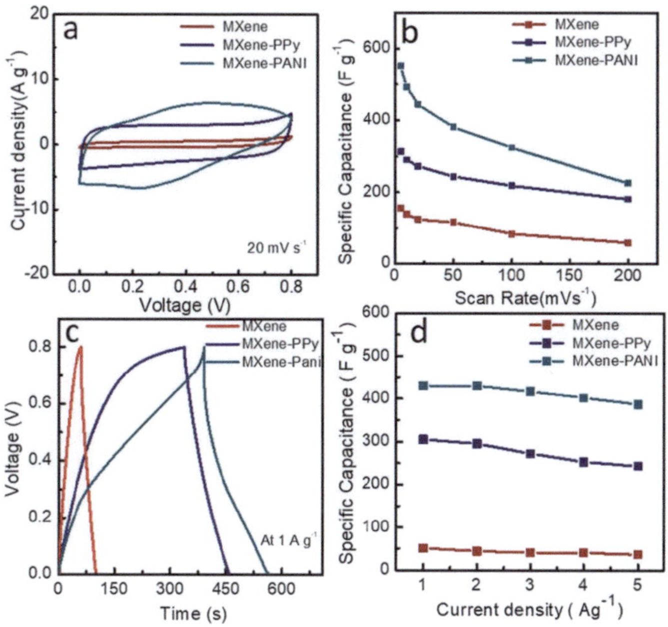

Fig. 4 **a** CV and **b** GCD curves at 20 mV s^{-1} and 1 A g^{-1}, respectively, variation of specific capacitance with various. **c** Scan rates and **d** Current densities for pure MXene, MXene-PPy, and MXene-PANI electrodes [47]. Copyright 2023. Reproduced with permission from Elsevier

6 Applications of Polymer/MXene Composites

Polymer/MXene composites are gaining attention for their wide range of applications due to the unique combination of properties provided by polymers and MXenes. MXenes offer excellent electrical conductivity, thermal stability, and mechanical strength. When combined with polymers, which offer flexibility, ease of processing, and chemical resistance, these composites can be used in several advanced technologies [48]. These composites exhibit outstanding properties, including efficient radiation-to-heat conversion, superior EMI shielding, and enhanced energy storage, outperforming other composites. They find applications across various fields, including chemical, electrical, mechanical, materials, environmental, and biomedical engineering, as well as in medicine [49]. Here are some of the applications of polymer/MXene composites.

6.1 Energy Storage Devices

Polymer/MXene composites are highly suitable for energy storage applications due to the synergistic properties of both materials. In supercapacitors, MXenes provide high surface area, electrical conductivity, and efficient ion transport, making these composites ideal for electrodes, while the polymer matrix adds mechanical flexibility and durability, enhancing overall performance. Similarly, in lithium-ion and sodium-ion batteries, MXenes improve electron transport and charge storage, and the polymer contributes to mechanical integrity and processability. Moreover, the flexibility of these composites allows their use in wearable energy storage systems, such as flexible batteries and supercapacitors, enabling advanced portable technologies [30].

6.2 EMI Shielding

Owing to the excellent electrical conductivity and large surface area of MXenes, polymer/MXene composites are excellent for EMI shielding in electronics. These composites can block electromagnetic radiation effectively and are lighter than traditional metal-based shielding materials. Their lightweight and flexible nature also makes them suitable for shielding, as well as portable and wearable electronics [50]. Figure 5 shows various structured polymer/MXene composites used for EMI shielding [51].

6.3 Sensors

Polymer/MXene composites excel in sensor applications due to their unique properties. In strain sensors, the high electrical conductivity and sensitivity of MXenes enable accurate detection of resistance changes caused by mechanical deformation, making them ideal for soft robotics, wearable health monitoring, and human–machine interfaces. For gas and vapor sensors, MXenes offer a large surface area and tunable surface chemistry, enhancing the detection of gases like ammonia, methane, and CO_2, which is crucial for environmental monitoring [52]. In biosensors, these composites provide a conducive platform for biomolecule immobilization, while the polymer matrix ensures biocompatibility and flexibility, making them effective for detecting biomolecules in biomedical applications.

6.4 Biomedical Applications

Polymer/MXene composites are effective for biomedical fields like drug delivery and tissue engineering. The high surface area and biocompatibility of MXenes enable precise and controlled drug delivery systems, with the polymer matrix providing mechanical flexibility and degradability to efficiently release therapeutic agents. Additionally, the conductive properties of MXenes make these composites ideal for

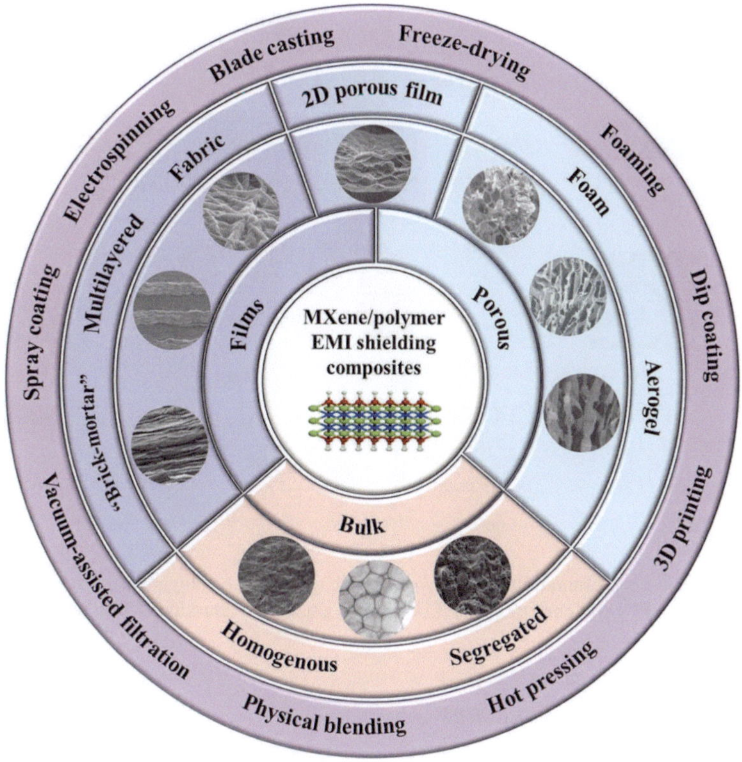

Fig. 5 Schematic representation of polymer/MXene composites for EMI shielding [51]. Copyright 2024. Reproduced with permission from Elsevier

tissue scaffolds, particularly in neural and cardiac tissue engineering, where electrical stimulation is crucial for promoting cell growth and development [53].

Rizi et al. demonstrated that the PPy/MXene biosensor exhibited excellent sensitivity, stability, and selectivity. This biosensor was highly effective in detecting Koch's bacillus (*Mycobacterium tuberculosis*) in human sputum samples, achieving high detection recovery rates between 90.52 and 100.8%. The results indicated that the peak current intensity of methylene blue (MB) decreased with an increase in DNA template content caused by the formation of DNA duplex, suggesting that hybridization originated on the surface of biosensors. Differential pulse voltammetry (DPV) study were performed over a potential range from 0.4 to 0.1 V, with a potential step of 5 mV, a modulation time of 0.02 s, and a time interval of 0.0329 s, in phosphate buffer saline (PBS) ($p^H = 7.0$, 0.05 M) with three repetitions. The design and sensing mechanism of the *M. tuberculosis* genosensor is shown in Fig. 6 [54].

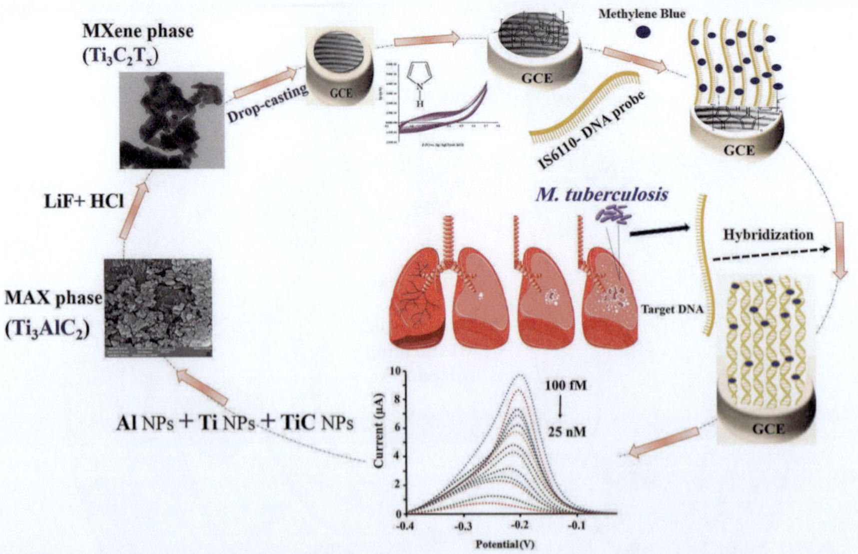

Fig. 6 Schematic representation of electrochemical biosensor for mycobacterium tuberculosis detection using a PPy/MXene composite [54]. Copyright 2022. Reproduced with permission from Elsevier

6.5 Anticorrosion Coatings

Polymer/MXene composite coatings have recently emerged as a promising solution for metal anticorrosion because of their large aspect ratio and excellent impermeability. Nevertheless, practical applications are significantly hindered by challenges such as poor dispersion, oxidation, and sedimentation of MXenes within the resin matrix, which are commonly faced with curing methods that can be overcome by the addition of polymers [55]. The negative surface charge of MXene facilitates its effective dispersion in aqueous and other polar solvents. This broad solubility, coupled with excellent interfacial interactions, adjustable surface functionality, and stability when combined with organic or polymeric materials, along with the layered structure of MXene, makes it a highly suitable material for anticorrosion coatings [7]. MXenes exhibit excellent chemical resistance, and when combined with polymers, can prevent corrosion of metal surfaces due to the formation of protective coatings. These coatings can be applied to metal structures in harsh environments, such as marine and industrial sites. Figure 7 shows the protection of surfaces from corrosion using MXene-based coatings.

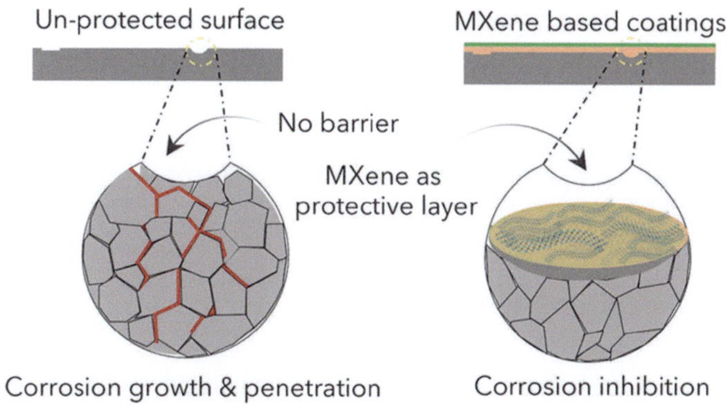

Fig. 7 Schematic representation of protection from corrosion using MXene-based coatings [7]. Copyright 2022. Reproduced with permission from American Chemical Society

7 Comparative Analysis with Other Nanofillers

MXenes, as a class of 2D materials, exhibit unique properties when used as nanofillers in polymer composites. A comparative analysis between MXenes and other commonly used nanofillers such as graphene, carbon nanotubes (CNTs), nanoclays, and metal oxides highlights key differences and similarities in their performance across various properties.

7.1 *Electrical Conductivity*

Electrical conductivity is the measurement of a material's ability to transport electric charge under an applied voltage. It depends on the presence and mobility of charge carriers, such as electrons or ions, and is influenced by the material's structure, temperature, and impurities. High conductive materials allow efficient current flow, while low conductive materials resist it. This section focuses on comparing the mechanical properties of different materials [56, 57].

MXenes: MXenes, particularly $Ti_3C_2T_x$, are highly conductive due to their metallic nature, which makes them superior to many other nanofillers for applications in energy storage, EMI shielding, and sensors. Their surface terminations can also be tuned to optimize conductivity while maintaining good dispersion.

Graphene: Graphene has excellent electrical conductivity and is often used in electronic and conductive composites. However, its production is more complex, and achieving good dispersion in polymer matrices can be challenging.

CNTs: CNTs are known for their exceptional conductivity and are widely used in conductive polymers, though the entanglement of CNTs can lead to aggregation, reducing their effectiveness.

Metal Oxides (e.g., TiO_2): Metal oxides generally have lower electrical conductivity and are used more for optical and thermal properties than electrical enhancement.

7.2 Mechanical Properties

Mechanical properties of a material define its behavior under applied forces, including strength, elasticity, toughness, hardness, and ductility. These properties determine how the material deforms, resists stress, absorbs energy, or breaks, influencing its performance in structural and functional applications. Here, a comparison of the mechanical properties of various materials are reported [16, 58, 59].

MXenes: MXenes significantly enhance the mechanical properties of polymer composites due to their layered structure, strong interfacial adhesion, and ability to distribute stress effectively within the matrix. However, at higher loadings, agglomeration may reduce the reinforcing effect.

Graphene: Graphene nanofillers also provide excellent mechanical reinforcement, improving tensile strength, toughness, and Young's modulus. Its high aspect ratio helps efficiently transfer load across the matrix.

CNTs: CNTs improve mechanical properties due to their tubular structure, high aspect ratio, and strong van der Waals interactions, but achieving a uniform dispersion remains challenging, which can hinder reinforcement effects.

Nanoclays: Nanoclays such as montmorillonite (MMT) offer good mechanical reinforcement and improve barrier properties due to their high surface area, though their mechanical enhancements are generally inferior to MXenes, graphene, and CNTs.

7.3 Thermal Conductivity

The thermal conductivity of a material measures its ability to transfer heat through conduction. It quantifies how efficiently heat flows from a high to low temperature region, influenced by the material's structure and bonding. High thermal conductive materials transfer heat quickly, while low conductive materials act as insulators. In this analysis, a comparison of thermal properties of different materials are highlighted [60, 61].

MXenes: MXenes exhibit moderate thermal conductivity compared to graphene and CNTs. Their tunable surface chemistry allows for good interfacial bonding, which can enhance heat transfer in polymer composites.

Graphene: Graphene has one of the highest thermal conductivities among nanofillers, making it ideal for applications in thermal management, heat dissipation, and conductive coatings.

CNTs: Similar to graphene, CNTs offer excellent thermal conductivity and are used in composites designed for heat dissipation in electronics or aerospace.

Nanoclays and Metal Oxides: Metal oxides like alumina or nanoclays provide modest improvements in thermal conductivity but are generally less effective compared to MXenes, graphene, and CNTs.

7.4 Barrier Properties

Barrier properties refer to a material's ability to resist the permeation of gases, liquids, or other substances, influencing its effectiveness for protective, packaging, or insulation applications. The barrier properties of different materials are mentioned below [62].

MXenes: MXenes have shown potential for improving gas and moisture barrier properties in polymer composites. Their 2D layered structure can create a tortuous path, impeding the diffusion of gases, making them suitable for packaging and protective coatings.

Graphene and CNTs: Both graphene and CNTs can improve barrier properties, but their performance is generally less effective than nanoclays and MXenes, especially for gas permeability.

Nanoclays: Nanoclays, due to their platelet structure, are highly effective in enhancing the barrier properties of polymers. They have been widely used in food packaging and environmental protection applications.

7.5 Ease of Functionalization

MXenes: One of the advantages of MXenes over other nanofillers is their tunable surface chemistry. The surface terminations (–OH, –O, –F) on MXenes enable easier functionalization, leading to better compatibility and dispersion within different polymer matrices.

Graphene: Functionalizing graphene can be challenging due to its stable structure, though modifications such as oxygen-containing groups (graphene oxide) can improve dispersion and compatibility.

CNTs: CNT functionalization is typically necessary to improve their dispersion and compatibility with polymer matrices, but this process can affect their properties, especially conductivity.

Nanoclays and Metal Oxides: These fillers are often surface-modified to enhance compatibility with polymers, but the process is less versatile compared to MXenes.

7.6 Cost and Scalability

MXenes: MXenes are currently more expensive and less scalable than traditional nanofillers like nanoclays and metal oxides, due to the complex and hazardous synthesis processes involved in their production. Their high cost limits large-scale applications.

Graphene: The cost of graphene has been gradually decreasing with advancements in synthesis methods, but it remains expensive compared to nanoclays and metal oxides.

CNTs: CNTs, particularly single-walled variants, are expensive to produce, though multi-walled CNTs are becoming more affordable.

Nanoclays and Metal Oxides: These materials are relatively inexpensive and widely available, making them highly scalable for commercial applications.

7.7 Environmental and Safety Concerns

MXenes: The environmental impact and safety of MXenes, particularly during synthesis and use, are not fully understood. Their potential toxicity and environmental degradation require more investigation.

Graphene and CNTs: Both graphene and CNTs have raised concerns regarding health and environmental safety, especially in the form of nanoparticles that can be inhaled or persist in the environment.

Metal Oxides and Nanoclays: These fillers are generally considered safer and have been widely used without significant environmental concerns.

MXenes offer unique advantages over other nanofillers, particularly in terms of electrical conductivity, tunable surface chemistry, and multifunctionality, causing them suitable for uses in energy storage, sensors, and advanced electronics. However, challenges such as oxidation stability, cost, and scalability currently limit their widespread use. In contrast, fillers like graphene and CNTs excel in conductivity and mechanical reinforcement but face similar dispersion challenges. Nanoclays and metal oxides provide cost-effective solutions for barrier and mechanical property enhancements but lack the multifunctionality and high conductivity of MXenes, CNTs, and graphene.

8 Challenges and Limitations

MXenes, despite their promising potential as nanofillers for polymer composites, face several scientific challenges and limitations. A primary issue is their susceptibility to oxidation, particularly in ambient environments, which results in the degradation of their electrical conductivity and mechanical properties over time. This makes it difficult to maintain their effectiveness in long-term applications [63, 64]. Another challenge is achieving uniform dispersion of MXenes within polymer matrices, as they tend to agglomerate, especially at higher concentrations. Poor dispersion leads to weaker interfacial interactions between MXenes and polymers, negatively impacting the composite's overall mechanical, thermal, and electrical properties [65, 66].

Furthermore, the surface chemistry of MXenes is complex due to their variety of surface terminal functional groups, such as –OH, –O, and –F which can be difficult to control. This variability poses challenges in ensuring consistent interaction with different polymers, complicating the fabrication process [67, 68]. On the synthesis side, current methods, particularly fluoride-based etching, face limitations in scalability, safety, and environmental impact. These processes are often hazardous and expensive, restricting the large-scale production of MXenes. Additionally, the high cost and limited availability of raw MAX phase materials further hinder MXenes' commercial viability as nanofillers [69].

The long-term stability of MXenes within polymer matrices also remains an area of concern. Over time, changes in the polymer/MXene interface or within the MXenes themselves could result in performance degradation, particularly under harsh environmental conditions. Lastly, the environmental and health impacts of MXenes are not fully understood. Their potential toxicity and environmental degradation, especially in consumer products, raise concerns that must be addressed to ensure safe and sustainable use. These challenges must be overcome for MXenes to achieve their full potential as high-performance nanofillers for polymer composites across various industries.

9 Conclusions and Future Scopes

MXenes are novel 2D materials that have demonstrated significant capability as nanofillers for enhancing the performance of polymer composites. Their unique structure, high surface area, electrical conductivity, mechanical strength, and tunable surface chemistry make them suitable for uses in different fields. Integrating MXenes into polymers improves properties such as mechanical strength, electrical conductivity, thermal stability, and barrier properties, thereby broadening the applicability of polymer composites in fields like energy storage, EMI shielding, sensors, and filtration membranes. MXenes also offer advantages over traditional fillers such as graphene or CNTs due to their ease of surface functionalization, enabling better dispersion and compatibility within polymer matrices. However, challenges such as agglomeration at high filler loadings, stability in ambient conditions (particularly oxidation susceptibility), and the cost-effective synthesis of MXenes remain critical barriers to widespread adoption.

Improving the oxidation resistance and stability of MXenes is crucial for expanding their use in long-term applications like structural materials and outdoor electronics. Research will focus on surface modification and encapsulation techniques to mitigate environmental degradation. Additionally, optimizing scalable and cost-effective synthesis methods will be key to meeting industrial demand, focusing on developing green and safer etching processes to minimize environmental impact. Tailored surface functionalization will also be vital, as it enhances MXene compatibility with different polymer systems, enabling the creation of specialized composites for diverse applications.

References

1. Naguib, M., Kurtoglu, M., Presser, V., Lu, J., Niu, J., Heon, M., Hultman, L., Gogotsi, Y., Barsoum, M.W.: Two-dimensional nanocrystals produced by exfoliation of Ti_3AlC_2. Adv. Mater. **23**(37), 4248–4253 (2011)
2. Chen, X., Zhao, Y., Li, L., Wang, Y., Wang, J., Xiong, J., Du, S., Zhang, P., Shi, X., Yu, J.: MXene/polymer nanocomposites: preparation, properties, and applications. Polym. Rev. **61**(1), 80–115 (2021)
3. Abdah, M.A.A.M., Awan, H.T.A., Mehar, M., Mustafa, M.N., Walvekar, R., Alam, M.W., Khalid, M., Umapathi, R., Chaudhary, V.: Advancements in MXene-polymer composites for high-performance supercapacitor applications. J. Energy Storage **63**, 106942 (2023)
4. Hadi, Z., Yeganeh, J.K., Munir, M.T., Zare, Y., Rhee, K.Y.: An innovative model for electrical conductivity of MXene polymer nanocomposites by interphase and tunneling characteristics. Compos. A Appl. Sci. Manuf. **186**, 108422 (2024)
5. Sahoo, P., Shubhadarshinee, L., Mohapatra, P., Mohanty, P., Jali, B.R., Mohapatra, P., Barick, A.K.: Processing, morphology, mechanical, and electrical properties, and applications of thermoplastic polymer/MXenes nanocomposites. In: MXene-Filled Polymer Nanocomposites, pp. 43–69. CRC Press (2022)
6. Malaki, M., Varma, R.S.: Mechanotribological aspects of MXene-reinforced nanocomposites. Adv. Mater. **32**(38), 2003154 (2020)
7. Amin, I., Brekel, H.V.D., Nemani, K., Batyrev, E., de Vooys, A., van der Weijde, H., Anasori, B., Shiju, N.R.: $Ti_3C_2T_x$ MXene polymer composites for anticorrosion: an overview and perspective. ACS Appl. Mater. Interfaces **14**(38), 43749–43758 (2022)
8. Idumah, C.I., Iwuchukwu, F.U., Okoye, I., Ogbu, J.E.: Construction, characterization, properties and electromagnetic interference applications of MXene polymeric nanoarchitectures. Polym.-Plastics Technol. Mater. **63**(1), 1–22 (2024)
9. Idumah, C.I.: Recent advancements in electromagnetic interference shielding of polymer and MXene nanocomposites. Polym.-Plastics Technol. Mater. **62**(1), 19–53 (2023)
10. Idumah, C.I.: Recently emerging advancements in thermal conductivity and flame retardancy of MXene polymeric nanoarchitectures. Polym.-Plastics Technol. Mater. **62**(4), 510–546 (2023)
11. Ma, C., Ma, M.G., Si, C., Ji, X.X., Wan, P.: Flexible MXene-based composites for wearable devices. Adv. Func. Mater. **31**(22), 2009524 (2021)
12. Parajuli, D., Murali, N., KC, D., Karki, B., Samatha, K., Kim, A.A., Park, M., Pant, B.: Advancements in MXene-polymer nanocomposites in energy storage and biomedical applications. Polymers **14**(16), 3433 (2022)

13. Karahan, H.E., Goh, K., Zhang, C., Yang, E., Yıldırım, C., Chuah, C.Y., Ahunbay, M.G., Lee, J., Tantekin-Ersolmaz, ŞB., Chen, Y., Bae, T.H.: MXene materials for designing advanced separation membranes. Adv. Mater. **32**(29), 1906697 (2020)
14. Khosla, A., Sonu, Awan, H.T.A., Singh, K., Gaurav, Walvekar, R., Zhao, Z., Kaushik, A., Khalid, M., Chaudhary, V.: Emergence of MXene and MXene–polymer hybrid membranes as future-environmental remediation strategies. Adv. Sci. **9**(36), 2203527 (2022)
15. Chaudhary, V., Ashraf, N., Khalid, M., Walvekar, R., Yang, Y., Kaushik, A., Mishra, Y.K.: Emergence of MXene–polymer hybrid nanocomposites as high-performance next-generation chemiresistors for efficient air quality monitoring. Adv. Func. Mater. **32**(33), 2112913 (2022)
16. Gong, K., Zhou, K., Qian, X., Shi, C., Yu, B.: MXene as emerging nanofillers for high-performance polymer composites: a review. Compos. B Eng. **217**, 108867 (2021)
17. Aghamohammadi, H., Amousa, N., Eslami-Farsani, R.: Recent advances in developing the MXene/polymer nanocomposites with multiple properties: a review study. Synth. Met. **273**, 116695 (2021)
18. Ezika, A.C., Sadiku, E.R., Idumah, C.I., Ray, S.S., Adekoya, G.J., Odera, R.S.: Recently emerging trends in MXene hybrid conductive polymer energy storage nanoarchitectures. Polym.-Plastics Technol. Mater. **61**(8), 861–887 (2022)
19. Li, C., Xu, J., Xu, Q., Xue, G., Yu, H., Wang, X., Lu, J., Cui, G., Gu, G.: Synthesis of Ti_3C_2 MXene@PANI composites for excellent anticorrosion performance of waterborne epoxy coating. Prog. Org. Coat. **165**, 106673 (2022)
20. Jian, X., He, M., Chen, L., Zhang, M.M., Li, R., Gao, L.J., Fu, F., Liang, Z.H.: Three-dimensional carambola-like MXene/polypyrrole composite produced by one-step co-electrodeposition method for electrochemical energy storage. Electrochim. Acta **318**, 820–827 (2019)
21. Pan, Y., He, M., Wu, J., Qi, H., Cheng, Y.: One-step synthesis of MXene-functionalized PEDOT: PSS conductive polymer hydrogels for wearable and noninvasive monitoring of sweat glucose. Sens. Actuators B Chem. **401**, 135055 (2024)
22. Chen, N., He, J., Xuan, H., Jin, J., Yu, K., Shi, M., Yan, C.: Dual-functional polyindole/MXene composite for superior proton storage and corrosion protection. Compos. B Eng. **270**, 111145 (2024)
23. Nagalingam, S.P., Pandiaraj, S., Alzahrani, K.E., Alodhayb, A.N., Grace, A.N.: Fabrication of a free-standing $Ti_3C_2T_x$-PTh counter electrode via interfacial polymerization for dye-sensitized solar cells. RSC Adv. **14**(33), 24000–24009 (2024)
24. Li, Y., Kamdem, P., Jin, X.J.: Hierarchical architecture of MXene/PANI hybrid electrode for advanced asymmetric supercapacitors. J. Alloys Compd. **850**, 156608 (2021)
25. Ma, Y., Xu, K., Liu, X., Yao, S., Li, X., Si, Y., Li, X.: 3D porous PEDOT/MXene scaffold toward high-performance supercapacitors. Chem. Eng. J. 154348 (2024)
26. Tang, J., Chen, Z., Chen, Y., Xu, X., Zhu, J., Lu, T., Pan, L.: In situ constructed $Ti_3C_2T_x$ MXene/polypyrrole composite with enhanced sodium storage capacity for efficient hybrid capacitive deionization. J. Polym. Sci. **60**(21), 3035–3042 (2022)
27. Yin, T., Cheng, Y., Hou, Y., Sun, L., Ma, Y., Su, J., Zhang, Z., Liu, N., Li, L., Gao, Y.: 3D porous structure in MXene/PANI foam for a high-performance flexible pressure sensor. Small **18**(48), 2204806 (2022)
28. Bi, C., Wen, Q., Chen, Y., Xu, H.: Boosting power output in microbial fuel cell with Sulfur-doped MXene/polypyrrole hydrogel anode for enhanced extracellular electron transfer process. J. Appl. Electrochem. (2024). https://doi.org/10.1007/s10800-024-02137-5
29. Jimmy, J., Kandasubramanian, B.: MXene functionalized polymer composites: synthesis and applications. Eur. Polymer J. **122**, 109367 (2020)
30. Awan, H.T.A., Abdah, M.A.A.M., Mehar, M., Walvekar, R., Chaudhary, V., Khalid, M., Khosla, A.: MXene-polymer hybrid composites for advanced energy storage: Insights into supercapacitors and batteries. J. Energy Storage **95**, 112449 (2024)
31. Kausar, A.: Polymer/MXene nanocomposite—A new age for advanced materials. Polym.-Plastics Technol. Mater. **60**(13), 1377–1392 (2021)

32. Jin, X., Wang, J., Dai, L., Liu, X., Li, L., Yang, Y., Cao, Y., Wang, W., Wu, H., Guo, S.: Flame-retardant poly (vinyl alcohol)/MXene multilayered films with outstanding electromagnetic interference shielding and thermal conductive performances. Chem. Eng. J. **380**, 122475 (2020)
33. Lim, K.R.G., Handoko, A.D., Nemani, S.K., Wyatt, B., Jiang, H.Y., Tang, J., Anasori, B., Seh, Z.W.: Rational design of two-dimensional transition metal carbide/nitride (MXene) hybrids and nanocomposites for catalytic energy storage and conversion. ACS Nano **14**(9), 10834–10864 (2020)
34. Boota, M., Anasori, B., Voigt, C., Zhao, M.Q., Barsoum, M.W., Gogotsi, Y.: Pseudocapacitive electrodes produced by oxidant-free polymerization of pyrrole between the layers of 2D titanium carbide (MXene). Adv. Mater. **28**(7), 1517–1522 (2016)
35. Li, H., Wang, Q., Hong, X.: Self-healing polymer/MXene composites for healthcare applications. Fibers Polym. (2024). https://doi.org/10.1007/s12221-024-00658-6
36. Zhan, X., Si, C., Zhou, J., Sun, Z.: MXene and MXene-based composites: synthesis, properties and environment-related applications. Nanoscale Horizons **5**(2), 235–258 (2020)
37. Parajuli, D.: MXenes-polymer nanocomposites for biomedical applications: fundamentals and future perspectives. Front. Chem. **12**, 1400375 (2024)
38. Lee, C.H., Khalina, A., Lee, S.H.: Importance of interfacial adhesion condition on characterization of plant-fiber-reinforced polymer composites: a review. Polymers **13**(3), 438 (2021)
39. Cui, Y., Zhu, J., Tong, H., Zou, R.: Advanced perspectives on MXene composite nanomaterials: Types synthetic methods, thermal energy utilization and 3D-printed techniques. Iscience **26**(1), 105824 (2023)
40. Bilibana, M.P.: Electrochemical properties of MXenes and applications. Adv. Sens. Energy Mater. **2**(4), 100080 (2023)
41. Assad, H., Fatma, I., Sharma, P.K., Kumar, A.: Structural, morphological and tribological properties of polymer/MXene Composites. In: MXene Reinforced Polymer Composites: fabrication, Characterization and Applications, pp. 221–255. Scrivener Publishing (2024)
42. Liu, R., Li, W.: High-thermal-stability and high-thermal-conductivity $Ti_3C_2T_x$ MXene/poly (vinyl alcohol) (PVA) composites. ACS Omega **3**(3), 2609–2617 (2018)
43. Gao, L., Li, C., Huang, W., Mei, S., Lin, H., Ou, Q., Zhang, Y., Guo, J., Zhang, F., Xu, S., Zhang, H.: MXene/polymer membranes: synthesis, properties, and emerging applications. Chem. Mater. **32**(5), 1703–1747 (2020)
44. Ghanbari, R., Zare, E.N.: Engineered MXene-polymer composites for water remediation: promises, challenges, and future perspective. Coord. Chem. Rev. **518**, 216089 (2024)
45. Champagne, A., Charlier, J.C.: Physical properties of 2D MXenes: from a theoretical perspective. J. Phys.: Mater. **3**(3), 032006 (2020)
46. Huang, C., Huang, R., Cheng, Y., Zhao, L., Hu, N., Wei, Q.: Strong and flexible MXene-based nanocomposite films for atomic oxygen resistance and electromagnetic interference shielding. Compos. Sci. Technol. **253**, 110665 (2024)
47. Varghese, S.M., Mohan, V.V., Suresh, S., Gowd, E.B., Rakhi, R.B.: Synergistically modified $Ti_3C_2T_x$ MXene conducting polymer nanocomposites as efficient electrode materials for supercapacitors. J. Alloy. Compd. **973**, 172923 (2024)
48. He, S., Sun, X., Zhang, H., Yuan, C., Wei, Y., Li, J.: Preparation strategies and applications of MXene-polymer composites: a review. Macromol. Rapid Commun. **42**(19), 2100324 (2021)
49. Riazi, H., Nemani, S.K., Grady, M.C., Anasori, B., Soroush, M.: Ti_3C_2 MXene–polymer nanocomposites and their applications. J. Mater. Chem. A **9**(13), 8051–8098 (2021)
50. Tan, H., Gou, J., Zhang, X., Ding, L., Wang, H.: Sandwich-structured $Ti_3C_2T_x$-MXene/reduced-graphene-oxide composite membranes for high-performance electromagnetic interference and infrared shielding. J. Membr. Sci. **675**, 121560 (2023)
51. Gao, Q., Wang, X., Schubert, D.W., Liu, X.: Review on polymer/MXene composites for electromagnetic interference shielding applications. Adv. Nanocompos. **1**(1), 52–76 (2023)
52. Chaudhary, V.: MXene–Polymer nanocomposites for gas-and vapor-sensing applications. In: MXene-Filled Polymer Nanocomposites, pp. 211–234. CRC Press (2022)

53. George, S.M., Kandasubramanian, B.: Advancements in MXene-polymer composites for various biomedical applications. Ceram. Int. **46**(7), 8522–8535 (2020)
54. Rizi, K.S., Hatamluyi, B., Darroudi, M., Meshkat, Z., Aryan, E., Soleimanpour, S., Rezayi, M.: PCR-free electrochemical genosensor for Mycobacterium tuberculosis complex detection based on two-dimensional Ti_3C_2 MXene-polypyrrole signal amplification. Microchem. J. **179**, 107467 (2022)
55. Zhang, Y., Chen, C., Chen, Z., Zhang, T., Wang, Y., Cao, S., Ma, J.: Superior anticorrosion performance of well-dispersed MXene-polymer composite coatings enabled by covalent modification and ambient electron-beam curing. ACS Appl. Mater. Interfaces **15**(8), 11099–11110 (2023)
56. Irfan, M.S., Ali, M.A., Khan, T., Anwer, S., Liao, K., Umer, R.: MXene and graphene coated multifunctional fiber reinforced aerospace composites with sensing and EMI shielding abilities. Compos. A Appl. Sci. Manuf. **165**, 107351 (2023)
57. Zarepour, A., Ahmadi, S., Rabiee, N., Zarrabi, A., Iravani, S.: Self-healing MXene-and graphene-based composites: properties and applications. Nano-Micro Lett. **15**(1), 100 (2023)
58. Adstedt, K., Buxton, M.L., Henderson, L.C., Hayne, D.J., Nepal, D., Gogotsi, Y., Tsukruk, V.V.: 2D graphene oxide and MXene nanosheets at carbon fiber surfaces. Carbon **203**, 161–171 (2023)
59. Gautam, R., Marriwala, N., Devi, R.: A review: Study of MXene and graphene together. Meas.: Sens. **25**, 100592 (2023)
60. Ruiz-Hitzky, E., Ruiz-Garcia, C.: MXenes versus clays: emerging and traditional 2D layered nanoarchitectonics. Nanoscale **15**(47), 8959–18979 (2023)
61. Zhang, Z., Cao, H., Quan, Y., Ma, R., Pentzer, E.B., Green, M.J., Wang, Q.: Thermal stability and flammability studies of MXene–organic hybrid polystyrene nanocomposites. Polymers **14**(6), 1213 (2022)
62. Cao, F., Zhang, Y., Wang, H., Khan, K., Tareen, A.K., Qian, W., Zhang, H., Ågren, H.: Recent advances in oxidation stable chemistry of 2D MXenes. Adv. Mater. **34**(13), 2107554 (2022)
63. Kumar, S., Kumari, N., Singh, T., Seo, Y.: Shielding 2D MXenes against oxidative degradation: recent advances, factors and preventive measures. J. Mater. Chem. C **12**(23), 8243–8281 (2024)
64. Lee, Y., Kim, S.J., Kim, Y.J., Lim, Y., Chae, Y., Lee, B.J., Kim, Y.T., Han, H., Gogotsi, Y., Ahn, C.W.: Oxidation-resistant titanium carbide MXene films. J. Mater. Chem. A **8**(2), 573–581 (2020)
65. Han, X., Qiu, X., Zong, M., Hao, J.: Assembled MXene macrostructures for multifunctional polymer nanocomposites. Small Struct. **4**(10), 2300090 (2023)
66. He, X., Cui, C., Chen, Y., Zhang, L., Sheng, X., Xie, D.: MXene and polymer collision: sparking the future of high-performance multifunctional coatings. Adv. Funct. Mater. 2409675 (2024). https://doi.org/10.1002/adfm.202409675
67. Rahman, M., Al Mamun, M.S.: Future prospects of MXenes: synthesis, functionalization, properties, and application in field effect transistors. Nanoscale Adv. **6**(2), 367–385 (2024)
68. Matias, M.L., Pereira, C., Almeida, H.V., Jana, S., Panigrahi, S., Menda, U.D., Nunes, D., Fortunato, E., Martins, R., Nandy, S.: 3D printed MXene architectures for a plethora of smart applications. Mater. Today Adv. **23**, 100512 (2024)
69. Abid, Z., Rafiq, K., Aslam, A., Jin, R., Hussain, E.: Scope, evaluation and current perspectives of MXene synthesis strategies for state of the art applications. J. Mater. Chem. A **12**, 7351–7395 (2024)

MXene and Its Composites for Biomedical Applications

Kathellen Pintado Esteves, Ana Paula de Oliveira Lopes Antunes, Isaac dos Santos Nunes, Ananda Fagundes Guarda, Eliezer Quadro Oreste, and Daiane Dias

Abstract MXene is a two-dimensional layered material composed of carbides and nitrides of transition metals, which have been gaining attention for use in developing detection platforms. The impressive progress can be attributed to its exceptional electrical conductivity, which is higher than all synthetic two-dimensional materials. MXenes combine extraordinary chemical and mechanical stability, good hydrophilicity, and dispersion. Also, the surface of these materials presents many covalent and ionic bonds that can help in the performance. Some modifications are available to control oxidation, expand the distance between layers, and functionalize the surface for better sensing uses. MXenes are easily functionalized, presenting a wide range of surface states that are helpful for biomedical sensing. They have already been applied to detection platforms' biosensors, electrochemical and wearable sensors, polymers, MOF, metals, hydroxides, sulfides, and others. The applications of these sensors are diverse, and the literature reported their use in health monitoring to identify volatile organic compounds associated with human diseases and recognize biomolecules, pressure, electroencephalograms, and strain sensors. Therefore, this chapter describes the use of MXenes in (bio)sensors for biomedical applications, providing an overview of recent innovations. Additionally, we highlight the challenges and opportunities regarding the use of these materials.

Keyword MXenes · Composites · Diseases · Biosensors

K. P. Esteves · A. P. de Oliveira Lopes Antunes · I. dos Santos Nunes · A. F. Guarda · E. Q. Oreste · D. Dias (✉)
Laboratório de Eletroespectro Analítica (LEEA), Escola de Química e Alimentos, Universidade Federal do Rio Grande, Rio Grande, RS, Brazil
e-mail: daianezd@gmail.com; daianedias@furg.br

I. dos Santos Nunes
Federal Institute of Education, Science and Technology of São Paulo, Suzano Campus, Suzano, SP, Brazil

© The Author(s), under exclusive license to Springer Nature Singapore Pte Ltd. 2025
S. Moharana and R. N. Mahaling (eds.), *Surface Engineering of MXenes*, Engineering Materials, https://doi.org/10.1007/978-981-96-3640-2_13

1 Introduction

MXenes are two-dimensional materials with exciting characteristics, such as high thermal and electrical conductivity and melting and boiling points [91]. This material is synthesized by selectively etching a layer from another three-dimensional material, called the MAX phase [78]. MAX phases have a chemical composition represented by the general formula $M_{n+1}AX_n$, where n usually ranges from 1 to 4 [43]. Within the term MAX, the letter M represents an early transition metal, A represents a group A element of the periodic table, mainly groups 13–16, and X represents carbon and/or nitrogen [34].

The structure of MAX phases, a three-dimensional material, consists of octahedra (XM_6) with shared edges, stacked and interleaved with flat layers of element A in its pure form [49]. Since the bonds between M–X are stronger than those between M–A, the layers of substance A can be chemically and selectively removed without destroying the M–X bonds, resulting in a two-dimensional material with weakly bonded layers of $M_{n+1}X_n$ [128]. This material is named MXene to emphasize the removal of substance A from the MAX phase and its structure like graphene [43].

During the formation process, the surfaces of the $M_{n+1}X_n$ material are coated with hydrophilic functional groups such as oxygen, hydroxyl, and/or fluorine, resulting in the general formula of $M_{n+1}X_nT_x$, where T_x represents the functional groups located on the material surface [128]. Furthermore, MXenes can undergo surface modifications to enhance material characteristics [90] or be utilized to form composites due to their ability to interact with different materials [128].

The variety of MAX phases and, consequently, of MXenes impart exciting characteristics to this material. These characteristics, combined with the layered structure morphology, facilitate surface modification and the formation of composites with other materials, further expanding the materials' application horizon [128].

MXenes are materials with several properties that make them highly desirable for various applications. These properties include high surface area, active sites, easy functionalization, biocompatibility, electrical conductivity, and good mechanical properties at the nanoscale [17, 1]. These remarkable properties of MXenes make MXenes excellent for biomedical applications. However, the chemical stability of MXenes may be affected by the environment, such as the presence of water and dissolved hydrogen, which can lead to oxide formation [76]. MXenes are often modified and combined with other materials to improve their performance and form composites.

To ensure that MXene has been successfully formed, the X-ray Diffraction (XRD) pattern before and after the process can be used to detect the presence of characteristic peaks from crystals with the A element (MAX). If the characteristic peaks from the MAX phase are missing, MXene has been formed. X-ray Photoelectron Spectroscopy (XPS) can also confirm the presence of MXene by detecting the characteristic signal of the MAX phase that is absent in MXene samples. This technique is also used to identify functional groups and to confirm any modifications or composite formation. Scanning Electron Microscopy (SEM) and Transmission Electron Microscopy

(TEM) are also helpful in evaluating the MXene formation, presenting exfoliated structures, and assessing modifications and composites with MXenes.

In the field of application of MXenes to sensors, biochemical sensors, and wearable biosensors are a class of devices that have gained prominence among researchers and institutions due to the integration between the properties of MXenes and the eventual advancement of these devices regarding constant monitoring and invasive of health. Many pioneering works have been dedicated to this feat, using this material in health monitoring studies in general, with emphasis on prognosis, diagnosis, or even treatment of diseases. Research groups have focused on developing these biosensors based on analytical sensitivity and patient comfort. The MXene composite sensors are usually evaluated based on their biocompatibility, mechanical properties, and electrical conductivity in detecting chemical components or physical aspects like moves. Many innovative ideas have been using bodily fluids such as sweat, tears, saliva, and others for detection through the most varied mechanisms, emphasizing electrochemicals and considering, for example, the significant advantages of amperometric biosensors, which will be discussed in this chapter.

2 MXenes, Its Composites and Biomedical Applications

As previously cited, MXenes exhibit interesting characteristics for various applications, such as good electrical and thermal conductivity, ductility, and high melting and boiling points [91]. Moreover, due to their layered two-dimensional nature, it is possible to form composites with other materials [114]. The newly formed material has different characteristics from the starting materials, significantly expanding its applications [128].

Therefore, we will further discuss the composites formed with MXenes, namely MXenes/polymer, MXene/MOF-based composites, MXene/metal materials-based composites, MXene/hydroxide-based composites, MXene/oxide-based composites, MXene/sulfide-based composites, and MXene/CNT-based composites.

2.1 MXenes/Polymer Composites

The IUPAC can define polymers as high molecular weight molecules whose structure essentially relies on the repetition of a low molecular weight unit [81]. The long polymeric chains provide high versatility to the material, including its use in composites [36]. Polymeric matrices used in composites are characterized by resistance to different environments, such as abrasives and corrosives [84]. They also possess fatigue resistance [61] and fracture [35]. Their use is also reported in the literature as suitable for composite coating, characterizing the material as a hydrophobic surface [5], to increase the refractive index in polystyrene [24], or even in nanoparticle stabilization [92]. Due to these applications, polymeric materials become strong candidates for incorporating and forming composites with MXenes [36].

MXenes/polymer composites exhibit properties that combine the polymer's inherent characteristics, such as flexibility [28], with metal-like features, such as ion channels and high electrical conductivity [14]. Additionally, the insertion of polymers between the two-dimensional layers of MXenes prevents their aggregation and enhances material stability [36].

In summary, mechanisms for forming MXenes/polymer composites may include physical mixing, surface modification, and in situ polymerization [29].

He and colleagues report that the most common method for preparing MXene/polymer composites is directly physically mixing the components [27]. This preparation approach is widely reported in the literature due to the polar functional groups of MXenes, such as hydroxyl, which can form hydrogen bonds with other polar functional groups of polymers, such as carboxylic and amino groups [23]. The hydrogen bonding between MXene and the polymer makes the composite material more stable and uniform [29].

Examples of this type of preparation are reported with a hydrophilic polysaccharide with many oxygenated functional groups, such as xanthan. Due to its strong hydrogen bonding between xanthan's functional groups and MXene, the resulting composite exhibits higher strength than MXene films, thanks to the homogeneous distribution of xanthan between MXene layers. The MXene suspension ($Ti_3C_2T_x$) was mixed with xanthan solution under stirring for 2 h, followed by filtration and drying at room temperature [97]. More details of the process can be seen in Fig. 1.

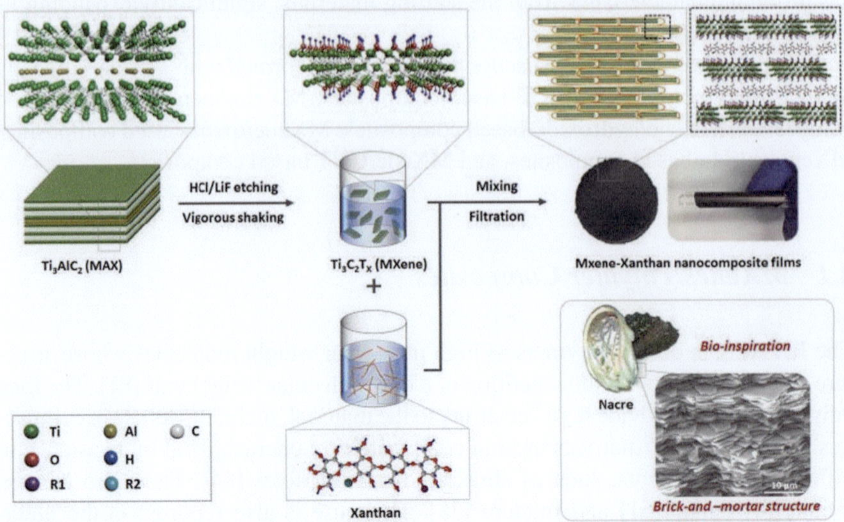

Fig. 1 Representation of the preparation of $Ti_3C_2T_x$ MXene-xanthan composites using the physical mixing method. Reprinted with permission from [97]. Copyright 2024, Elsevier

Although physical mixing easily achieves the synthesis of MXene/polymer composites, some limitations still need to be overcome [27]. These occur due to the absence of chemical bonds between the polymeric layer and the MXene, which makes the mechanical and resistance properties of the compound formed through this process poor [96].

Additionally, MXenes exhibit charged groups on their surface, resulting in the incompatibility of using hydrophobic polymers [111]. One way to select the hydrophobic or hydrophilic character of the MXene surface is through surface modification [29]. This process can occur through methods such as anchoring through Ti–N or Ti–O–C bonds [64], esterification reaction [25], grafting functional alkoxysilanes through Ti–O–Si bonds [26], among others. The representation of the surface modification of an MXene through an esterification reaction is presented in Fig. 2.

Zhang, Ji and Ma have shown that cellulose has garnered attention recently as a material for producing composites with MXene due to its abundance, low cost, biodegradability, and biocompatibility [134]. By combining these two materials, the poor mechanical properties of MXene on the macroscale are compensated for, leading to improved performance of composite materials. Wan and co-workers have developed a highly conductive and transparent hydrogel with an adhesion function using 2D MXene with polydopamine (PDA) [103]. The polymer was designed to be used as a corrosion protector. The authors noted that PDA caused MXene to flocculate and precipitate immediately through electrostatic adsorption and cross-linking. 1D cellulose nanocrystals (CNCs) were employed as a steric stabilizer before

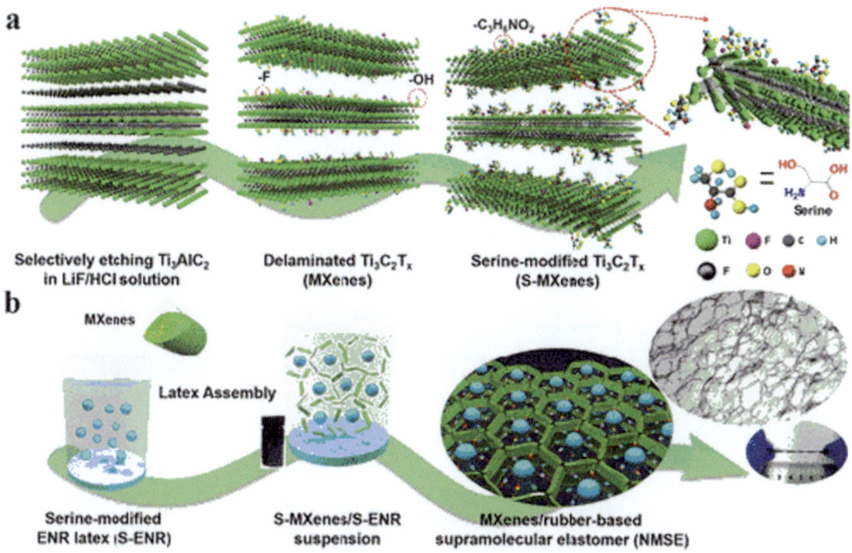

Fig. 2 Representation of the change in the surface of MXene nanosheets through the occurrence of an esterification reaction. Reprinted with permission from [25]. Copyright 2024, ACS Publications

PDA coating to prevent agglomeration. The results showed that the material has outstanding water dispersibility and anti-oxidation stability.

Yu and co-workers produced cellulose/MXene composite by spraying an MXene nanosheet suspension in cellulose nonwovens [126]. The authors have reported that the presence of the MXene layer has enhanced the interface between cellulose and inorganic materials for biomedical applications. On the other hand, Zhang, Ji, and Ma present that composite fibers based in MXene/cellulose enable weaving that offers the possibility to obtain flexible devices [131].

MXene composites have been reported to contain not only cellulose but also other materials, such as chitosan, which is particularly attractive for developing MXene-based composites due to its biocompatible, antibacterial, and surface-modifiable properties [88]. Additionally, other natural polymers like chitin [40], rubber [120], and starch [85] have also been found in composites with MXenes.

Lastly, forming polymers directly on the surface of MXenes (in situ) emerges as an efficient way to form MXene/polymer composites [72]. This procedure begins with the monomer insertion into the MXene surface through intermolecular interactions, such as hydrogen bonding and/or electrostatic interactions [66]. Different polymerization methods can be carried out once this procedure is performed on the MXene structure [27]. This composite formation strategy can contribute to improving the spacing of MXene layers and the distribution of the matrix in the polymeric phase, promoting the stability of the composites [23]. Figure 3 illustrates in situ polymerization for preparing the composite formed by incorporating Polyaniline (PANI) and/or TiO_2 into the layers of $Ti_3C_2T_x$ [65].

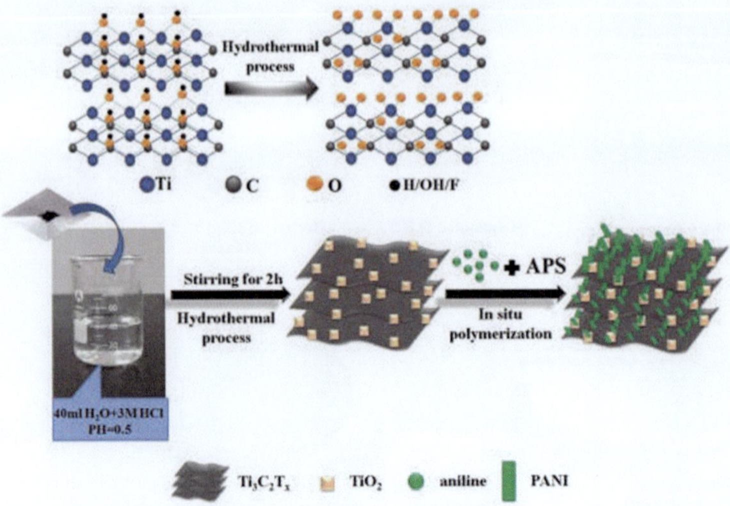

Fig. 3 Representation of the in-situ polymerization of the PANI@TiO_2/$Ti_3C_2T_x$ composite. Reprinted with permission from [65]. Copyright 2024, Elsevier

2.2 MXene/MOF-Based Composites

Metal–organic frameworks (MOFs) are materials with high surface area and porosity formed from metal ions with organic ligands [19]. However, their low electrical conductivity [115] and low chemical stability [29] are reported as limitations for specific applications.

One way to overcome these limitations is by combining MOFs with other materials to form composites, aiming to maintain the exciting characteristics of MOFs while using the attractive attributes of different materials [38]. MXenes have become an excellent option for forming composites with MOFs due to their exceptional electrical conductivity, hydrophilic surface, and transition metals with redox activity [137]. Additionally, MXenes have adjustable spacing between their two-dimensional layers that accommodate different intercalating molecules, making them an ideal material for composing and optimizing MOF performance [71].

Due to the surface being composed of many metals with redox activity, external functional groups in the MXene structure can react with oxygen from the environment and form stable oxides [135]. To increase the stability of these materials for subsequent composite formation, vacuum drying or freeze drying is suggested [4]. After this step, two main synthesis strategies for MXene/MOF-based composites are reported in the literature: in situ and ex-situ [137].

In situ synthesis is one of the most widely used methods to obtain MXene/MOF-based composites due to its simplicity and ease of use [122]. In this process, MXenes are added to solutions of metal ions and fully dissolved organic ligands, which should have the same synthesis conditions as MOFs. The room-temperature diffusion method is one of the most cost-effective and straightforward techniques [137].

The synthesis of hybrid nanosheets Ni-MOF/$Ti_3C_2T_x$ is reported by adding MXene ($Ti_3C_2T_x$) to a solution of nickel (II) with terephthalic acid containing Co^{2+} ions, Ni^{2+}, 2-methylimidazole and hexadecyltrimethylammonium bromide (CTAB) under stirring and room temperature to obtain CoNi-ZIF-67/$Ti_3C_2T_x$ [86].

Yu and co-workers prepared an organic–inorganic composite by treating nonwoven cellulose with MXene [126]. The material was soaked in a Zn^{2+} solution to promote the in-situ growth of zeolitic imidazolate framework-8 (ZIF-8). Treating cellulose with MXene endows the material with outstanding synergistic photothermal/photodynamic therapy (PTT/PDT) activity and adjustable electromagnetic interference (EMI). MOF (ZIF-8) in the bilayer structure protected MXene from oxidation, enhancing its stability. The composite was tested for bactericidal efficacy, and after 5 min of NIR (300 mW cm^2) exposure, it exhibited over 99.99% against *Escherichia Coli* (*E. coli*) and *Staphylococcus aureus (S. aureus)*. In situ preparation of MXene/MOF was also reported for sensing acetone and triethylamine [32].

Another example of a composition synthesized in situ is CoNi-ZIF-67@$Ti_3C_2T_x$, which, in short, is prepared through the in-situ growth of the CoNi-ZIF-67 compound on MXene $Ti_3C_2T_x$ by an occurrence of coprecipitation [112]. The scheme of this occurrence can be seen in Fig. 4.

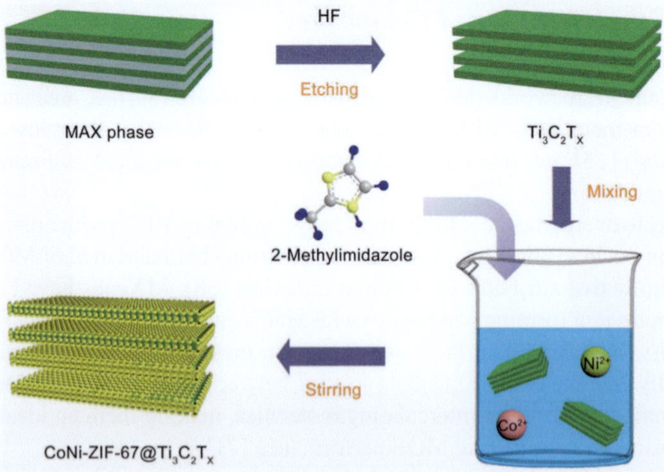

Fig. 4 Representation of the synthesis of the compound CoNi-ZIF-67@$Ti_3C_2T_x$. Reprinted with permission from [112]. Copyright 2024, MDPI

On the other hand, ex-situ synthesis involves synthesizing the composite using methods not influenced by the MOF synthesis environment, such as simple or physical mixing, self-assembly, and electrochemical methods [137].

Direct or physical mixing involves mixing the MOF and MXene to obtain the composite. An example of this simple synthesis is the formation of the MIL-100(Fe)/Ti_3C_2 composite by mixing the MOF (MIL-100) with different amounts of MXene ($Ti_3C_2T_x$) [108]. More details of this synthesis can be seen in Fig. 5.

In this sense, Chen and co-workers prepared a composite of MXene/Ce-MOF by mixing appropriate amounts of the two materials to disperse them in a dimethyl formamide (DMF) solution, resulting in a Ce-MOF/$Ti_3C_2T_x$ MXene solution [16]. This solution was then placed on a glassy carbon electrode (GCE) to fabricate a modified sensor for L-Tryptophan detection that exhibited a wide linear range, low detection limit, excellent selectivity, good reproducibility, and long-term stability.

The self-assembly method involves electrostatic attraction and other intermolecular interactions that spontaneously form and assist in the ordered assembly of the composite [137]. A new composite was formed through hydrogen bonding between

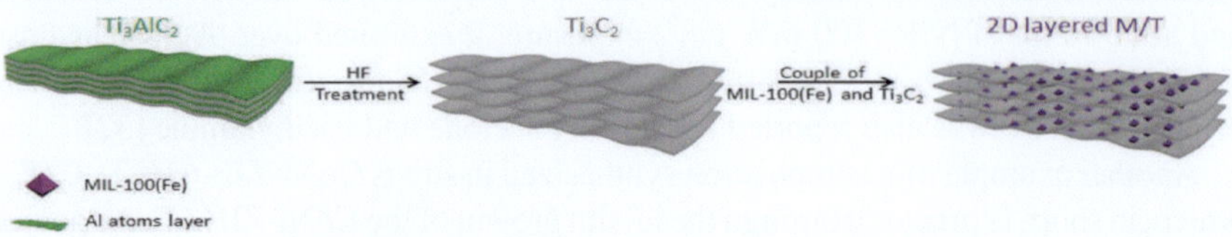

Fig. 5 Representation of the compound MIL-100(Fe)/Ti_3C_2 synthesis. Reprinted with permission from [109]. Copyright 2024, ACS Publications

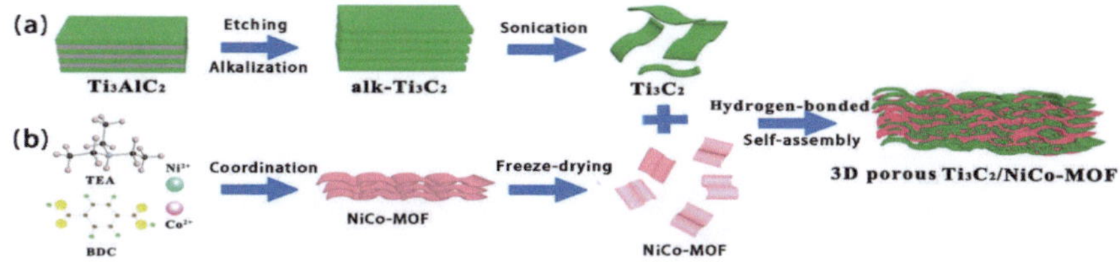

Fig. 6 Representation of the synthesis of the compound Ti_3C_2/NiCo-MOF. Reprinted with permission from [53]. Copyright 2024, MDPI

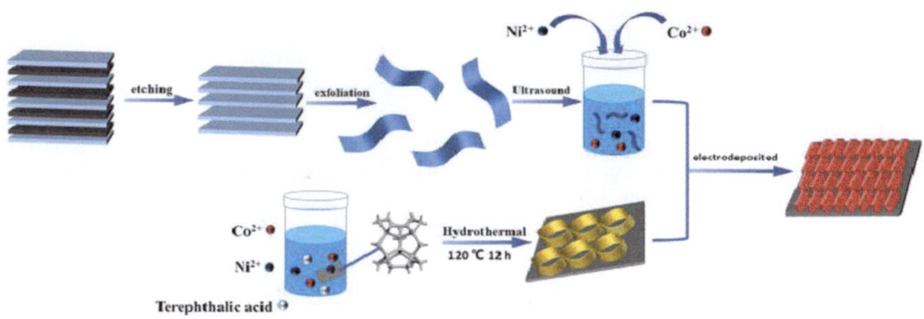

Fig. 7 Representation of the synthesis of the compound MXene-Ni-Co@NiCo-MOF/NF. Reprinted with permission from [105] Copyright 2024, Elsevier

MXene ($Ti_3C_2T_x$) and NoCo-MOF layers, resulting in a three-dimensional and porous structure [54]. Figure 6 shows more details of this synthesis.

Lastly, the electrochemical synthesis method involves the application of electric current or voltage to drive synthesis [137]. The MXene-Ni-Co@NiCo-MOF/NF composite synthesis is reported in the literature from preparing the MOF (NiCo-MOF) on nickel foam. This MOF is used as a precursor for the composite, and the solution containing a mixture of metals (Ni^{2+} and CO^{2+}) and MXene is used as the electrolyte for electrochemical deposition and composite formation [105]. More details of this synthesis can be seen in Fig. 7.

2.3 MXene/Metal Materials Based Composite

Metals to modify the surface of MXenes aims to decrease layer stacking and facilitate electron transfer within the material [118].

Generally, metals-modified MXenes are synthesized by mixing the components under agitation and/or sonication, followed by filtration [42]. The general representation of this type of synthesis can be seen in Fig. 8.

An example of this process is the synthesis of Au/$Ti_3C_2T_x$, where the gold metal is first reduced by a reducing agent under agitation. The MXene ($Ti_3C_2T_x$) is added

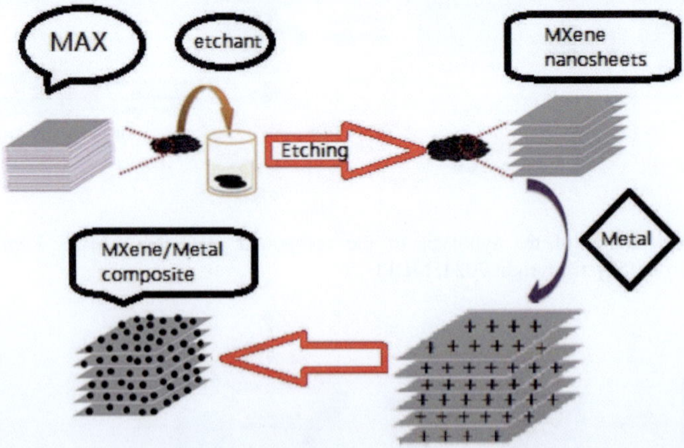

Fig. 8 General representation of the synthesis of metal/MXene composite. Reprinted with permission from [42]. Copyright 2024, MDPI

to the metal solution and subjected to ultrasound for 30 min. The resulting solution is then filtered, and the solid material formed is dried in an oven at 80 °C [87].

In addition, Talreja and co-workers prepared Ti_3AlC_2 MXene and modified this material with copper (Cu) [98]. The MXene was obtained through an etching process with HF. Cu-MXene was synthesized by mixing the prepared MXene with $CuSO_4$ in DI water. The suspension was then sonicated for 3 h at 25 °C to incorporate Cu metal particles within the interlayer spacing of MXene. The authors aimed to develop antibacterial materials. Both MXene and Cu-MXene showed high biocompatibility in erythrocyte cells. Regarding antibacterial activity, Cu-MXene responded more to *E. coli* and *S. aureus*. The authors concluded that incorporating Cu in the MXene matrix enhanced the antibacterial activity, making the material suitable for biomedical applications.

Another example of MXene modification using metals is the use of metallic nanoparticles, such as Ag [117], Au [127], and Pt [63], or even bimetallic nanoparticles [18].

2.4 MXene/Hydroxide-Based Materials

The formation of MXene and hydroxide composites is attractive for developing multifunctional materials with specific properties [37]. Recently, studies have reported the use of layered double hydroxides (LDHs) in composite formation, which are two-dimensional nanostructured materials arranged vertically, having the formula $[M^{2+}{}_{1\text{-}X}M^{3+}{}_{x}(OH)_2]^{x+}(A^{n-})_{x/n} \cdot mH_2O$, where A^{n-} is an anion, M^{2+} is a divalent cation (e.g., M = Mg, Fe, Co, Cu, Ni, or Zn), and M^{3+} is a trivalent cation (e.g., Al, Cr, Ga, Mn, or Fe) [52].

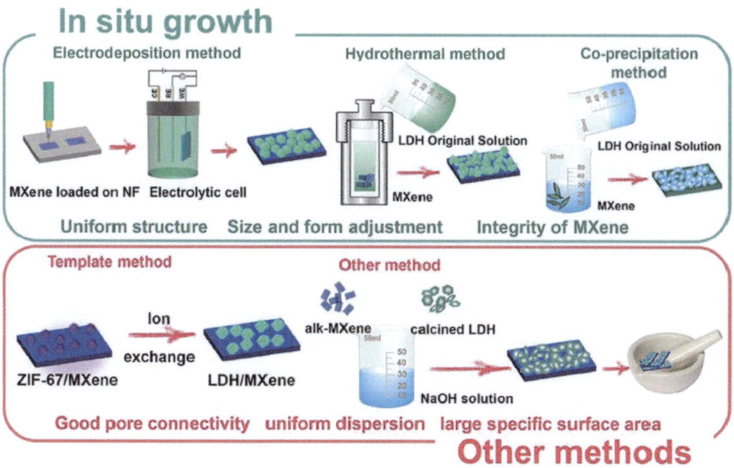

Fig. 9 Obtaining MXene/LDH composite by different methods. Reprinted with permission from [37] Copyright 2024, Elsevier

A large specific surface area and negatively charged surface can be characteristic of MXenes, making them an excellent substrate for the in-situ formation of LDH crystals [37]. Another way to obtain this composite is through a self-assembly approach, where MXene nanosheets (containing negative charge) and LDH (containing positive charge) are attracted by electrostatic forces [37, 80]. This approach can be achieved through a template method, obtaining a composite in an effectively controlled area [37]. Figure 9 illustrates a scheme for obtaining MXene/LDH composites using different methods.

In the electrodeposition method, the LDH flows towards the MXene upon applying an electric field, resulting in the uniform crystallization of LDH on the surface of MXene [37, 79]. Another commonly employed method is the hydrothermal method, which involves the reaction between the materials in a closed container at elevated temperatures (near the boiling point), followed by subsequent LDH crystallization as the temperature of the reaction system decreases [10, 37]. As for the co-precipitation method, the starting materials are mixed (under atmospheric pressure conditions and temperatures below 100 °C) for the subsequent crystallization of LDH, facilitated by functional groups and negative charge on the surface of MXene, which exhibit strong electrostatic force interactions with cations [37]. Through these methods, it is possible to obtain good stability and conductivity composites, owing to the excellent properties and interfacial interactions between LDH and MXene [37].

The materials used to form this type of composite (MXene and LDH) possess distinct physical and chemical properties but can be complementary [3]. For example, the excellent electrochemical activity of LDH, as well as the metallic conductivity, trivial bandgaps, and hydrophilic property of the functionalized surface of MXene, ensure the attainment of a material with these intrinsic benefits, potentially exhibiting improved performance in specific applications compared to the use of unmodified MXene alone [37].

2.5 MXene/Oxide-Based Materials

Composites based on MXene and oxides exhibit excellent performance, mainly when applied as sensors, due to their high electrical conductivity and electrochemical stability [128]. Combining MXene with different metal oxides can obtain this composite type, significantly enhancing their properties [56].

The fabrication of MXene/oxide composites can be achieved through various methods, which directly influence the material's properties. A simple fabrication method is a chemical deposition, as it utilizes relatively low temperature and pressure conditions, allowing for oxide deposition on the surface of MXene [55]. Another possible approach is the use of the hydrothermal method, which is widely employed in composite fabrication and employs high temperature and pressure, resulting in the precipitation of oxide nanostructures on MXene, which typically exhibit good crystallinity, controllable nanoparticle size and high dispersibility [57, 60].

Regarding this, Ulas and co-workers synthesized $BaMnO_3$ nanoparticle-doped MXene using a combination of HF etching and the hydrothermal method [101]. They prepared oxide nanoparticles by mixing $BaCl_2$ and $MnCl_2 \cdot 4H_2O$ aqueous solutions with $Ti_2C_2X_2$ MXene colloidal suspension. The resulting mixture was then subjected to a hydrothermal reaction in an autoclave at 150 °C for 3 h. The particles were washed until reaching a neutral pH and then dried. The process is illustrated in a scheme presented in Fig. 10. Characterization was performed using Scanning Electron Microscopy and X-ray diffraction, confirming the presence of oxide doping on the MXene surface. The composite was assessed for hydrazine electrooxidation, exhibiting lower charge transfer resistance and higher electrocatalytic activity than pristine MXene.

Another approach is solvothermal, a method very similar to hydrothermal. Still, it occurs in non-aqueous solutions, as under these conditions, unwanted reactions induced by H_2O or O_2 are avoided [60]. Approaches involving physical methods, such as vacuum filtration, ultrasonication, freeze-drying, and self-assembly, can also be employed for composite fabrication [55].

Recent studies demonstrate the fabrication and application of composites formed by MXene and graphene oxides (GO) [128]. This material possesses a 2D laminar structure, containing oxygen groups in its composition, such as hydroxyl, carboxyl, and epoxy groups located on GO nanosheets' edges and basal planes [59]. MXene/GO composites exhibit essential characteristics, such as good crystallinity, uniform distribution of metals, spacing between internal layers, and interconnected network structure, which qualify this material for potential applications in adsorption, detection, energy storage, and others [13]. Additionally, electronic, electrical, magnetic, and mechanical properties can be attributed, depending on the starting material used

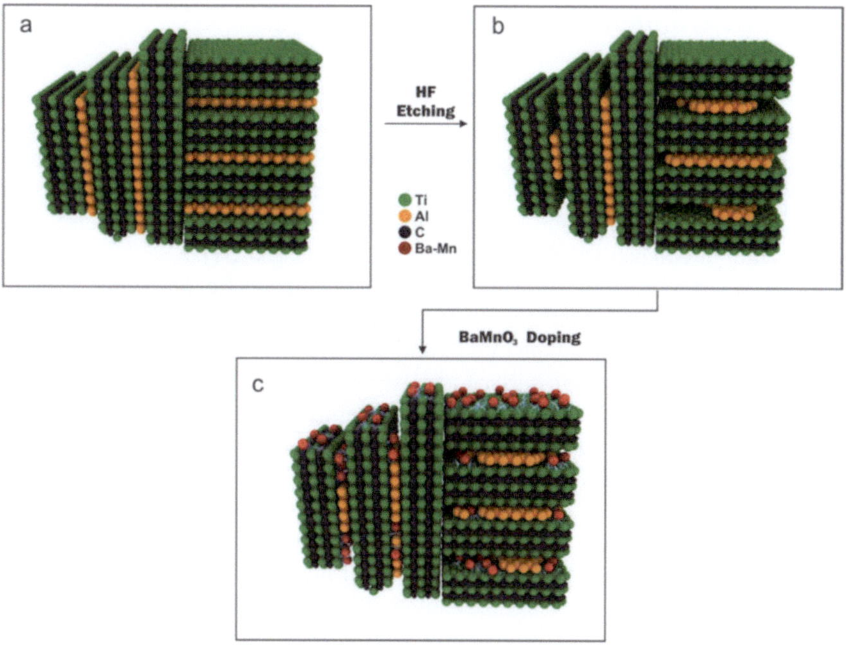

Fig. 10 Etching mechanism of MAX phases with HF solution. Illustration of **a** Ti_3AlC_2 MAX phase before etching, **b** the removal of Al between the layers to obtain MXene, and **c** $BaMnO_3$ nanoparticles doping. Reprinted with permission from [101]. Copyright 2024, Elsevier

for composite synthesis [13]. Figure 11 illustrates a schematic of synthesizing the starting materials (MXene and GO) and the MXene/GO composite.

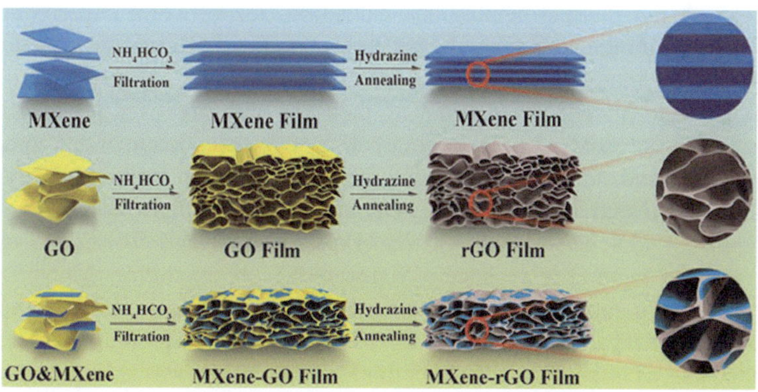

Fig. 11 Scheme for obtaining MXene, GO, and MXene/GO composite. Licensed by Creative Commons adapted from previously published by [68] Copyright 2024, Elsevier

2.6 MXene/Sulfide-Based Materials

Composites based on metal sulfide and MXene are obtained to enhance the materials' performance in different application scenarios, primarily to address inconveniences such as sheet aggregation and partial surface oxidation observed in unmodified MXenes [51]. The use of this composite is promising, especially concerning energy storage and conversion technologies, due to its synergistic effect from the combination of conductive MXenes and metal sulfides [20, 51].

Metal sulfides possess attractive electrical and magnetic properties, making them widely employed in sensor devices. Examples include molybdenum disulfide (MoS_2), vanadium disulfide (VS_2), tungsten disulfide (WS_2), and tin disulfide (SnS_2), characterized by high surface area, strong interlayer interactions, and electron transport [20]. Various methods can be employed for obtaining MXene and metal sulfide composites, such as hydrothermal, solvothermal, sol–gel, calcination, ultrasonication, and others [20].

The hydrothermal method is based on the principle previously described for MXene/oxide composites. This method is widely employed and advantageous for preparing unstable crystalline phases at their melting point [20]. The solvothermal method is also widely used for obtaining this composite type, which utilizes organic solutions and results in materials with different properties and morphologies [20, 130]. Another commonly employed method is sol–gel, based on reactions where primary homogeneous molecules (sol) become infinite and three-dimensional molecules called gel [8]. This method allows for controlling the size and porosity of the material, along with other advantages such as simplicity, relatively lower cost compared to other methods, and excellent performance at low temperatures [20]. However, some drawbacks are highlighted in the sol–gel method, such as the long-required time and potentially toxic organic solvents [20, 136]. Other methods employed in obtaining the composite include calcination and ultrasonication, which use thermal heating and cavitation (respectively) under controlled conditions, obtaining materials with different properties [20].

Studies show that MXene/sulfide-based materials exhibit enhanced characteristics compared to MXene alone, allowing their use with improved performance in various application fields [20]. However, scaling up the production of this composite for commercial and industrial applications is a challenge to overcome, as the methods used have the disadvantage of low yield and relatively high cost [20, 84]. As an example of MXene/sulfide-based materials, Hussain and co-workers utilized a hydrothermal method to produce a Manganese niobium sulfide (MnNbS) for a MnNbS-MXene composite electrode, as can be seen in Fig. 12a [33]. The composite, consisting of equal parts of sulfide and MXene dispersed in water, was ultrasonicated in an autoclave, as seen in Fig. 12b. After this the material was washed until neutral pH and then dried. The nanocomposite was designed to improve electrochemical characteristics to detect hydrogen peroxide (H_2O_2). The composite exhibited superior properties compared to MXene and MnBsS electrodes. The authors suggest that this discovery offers promise for practical applications in developing high-performance electrode materials to detect cancerous cells in the body. These findings illustrate the potential for various applications through modifying and combining MXenes into composites.

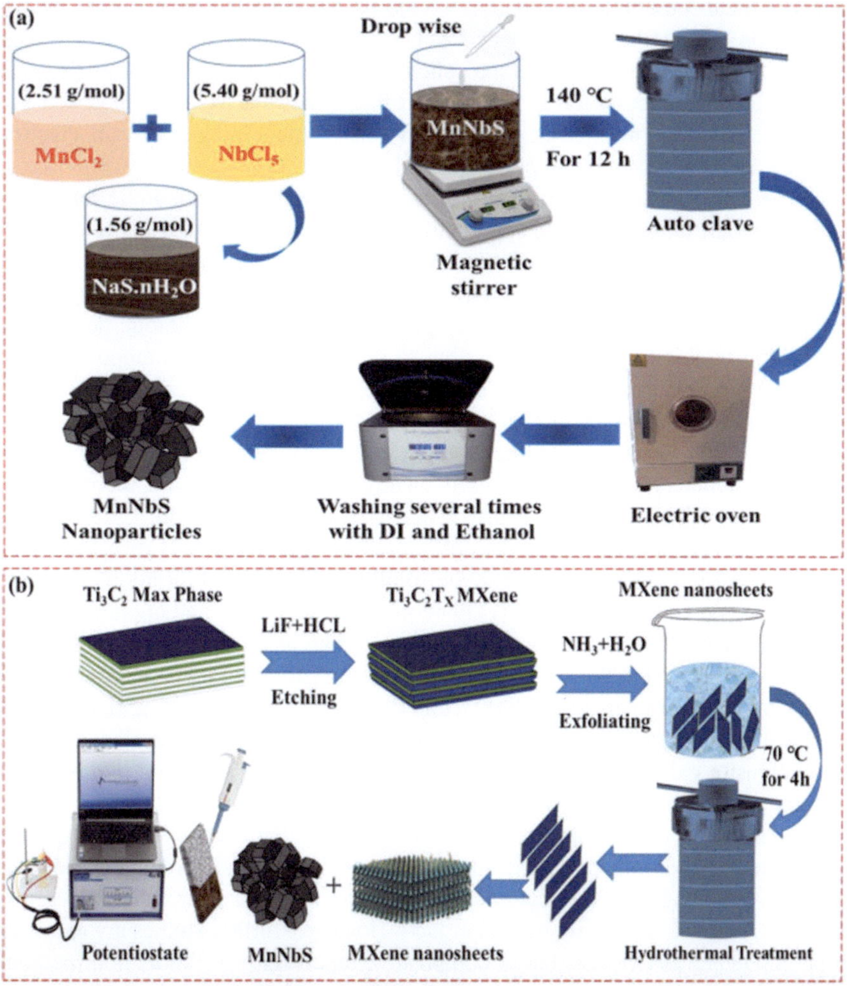

Fig. 12 Schemes to illustrate the synthesis of **a** MnNbS and **b** MnNbS@MXene. Reprinted with permission from [33] Copyright 2024, Elsevier

2.7 MXene/CNT-Based Materials

Composites based on MXene and carbon nanotubes (CNTs) also exhibit enhanced performance compared to MXene alone [75, 128]. CNTs consist of curled graphene sheets and can be single-walled (SWCNTs) or multi-walled (MWCNTs), which is based on the number of graphene layers [70]. Comparing both types of CNTs, SWCNTs have a simpler structure and a chemically inert surface, while MWCNTs are more chemically reactive. CNTs function as bridges between MXene nanosheets in a composite structure, promoting a significant improvement in ion diffusion, electron transport, and electrochemical performance [11].

One way to obtain this type of composite is through the electrophoretic deposition method, which is based on the application of an electric field to a stable colloidal suspension between two electrodes, offering advantages such as the use of simple instrumentation, high deposition rate and excellent uniformity [123]. Figure 13 depicts a scheme for obtaining the composite through electrophoretic deposition. In addition to this method, it is possible to obtain MXene/CNT composite using co-deposition, self-assembly, chemical vapor deposition, hydrothermal process, and microwave-assisted processes [75].

In their study, He and co-workers introduced a simple method to create a wearable motion detector [30]. They used electrospinning to prepare thermoplastic polyurethane (TPU) nanofiber films, which were then coated with CNT/MXene/CNT layers. This coating process involved alternating ultrasonically dispersed CNT diluted solution and a few-layer MXene solution with vacuum filtration, as shown in Fig. 14. The CNT/MXene/CNT/TPU composite was dried and applied between copper electrodes using conductive silver paste. The electrode showed high sensitivity, low detection limit, fast response time, and high durability. The authors demonstrated that the strain sensors produced from PU can effectively detect a wide range of human strains, including oversized, medium, and small strain ranges. Additionally, the composite exhibited inherent antimicrobial properties, attending requirements for contact with the human body.

Due to their enhanced properties and unique characteristics, MXene/CNT composites hold promising applications in biomedicine, catalysis and electrocatalysis, energy storage, sensor manufacturing, electromagnetic interference shielding, pollutant removal, and water treatment [75]. However, further studies are expected to be conducted on MXene-based composites, reporting new fabrication methods (especially on a larger scale) and expanding the knowledge of their properties and applications [11, 75].

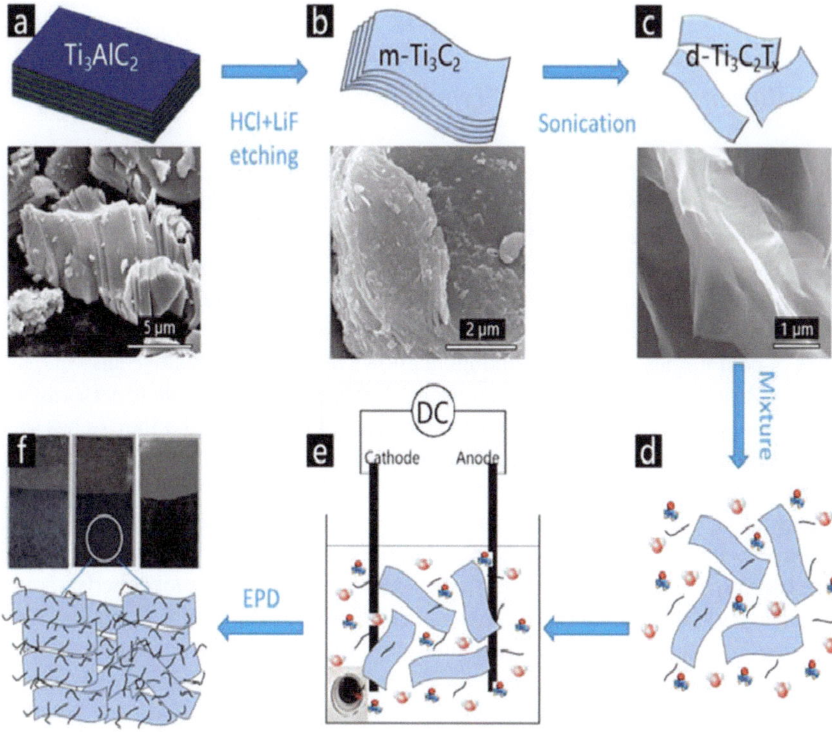

Fig. 13 **a** Morphology of MXene Ti_3AlC_2; **b** morphology of m-Ti_3C_2; **c** morphology of d-Ti_3C_2; **d** solution mixture of deionized water and acetone with d-Ti_3C_2 and CNTs; **e** electrophoretic deposition process; **f** sketch and photographs of d-Ti_3C_2 film, Ti_3C_2/CNTs film, and CNTs. Reprinted with permission from [123]. Copyright 2024, Elsevier

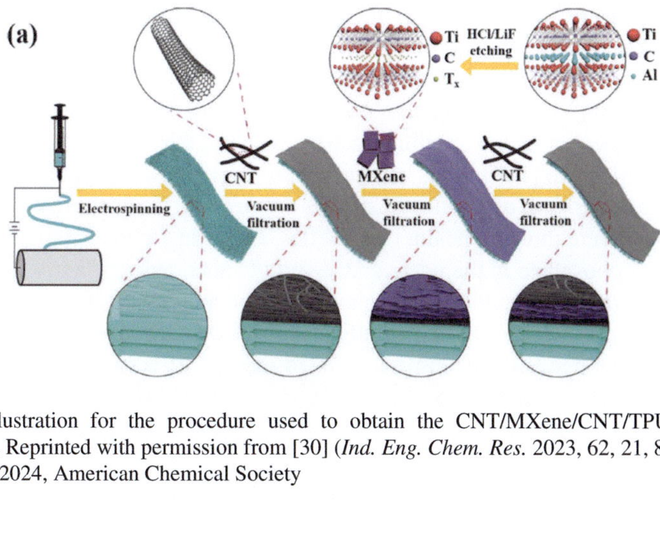

Fig. 14 Illustration for the procedure used to obtain the CNT/MXene/CNT/TPU nanofiber composite. Reprinted with permission from [30] (*Ind. Eng. Chem. Res.* 2023, 62, 21, 8327–8338). Copyright 2024, American Chemical Society

3 MXenes for Biomedical Sensing

3.1 Biosensor: DNA/RNA Hybridizations

Currently, detection technologies that are fast, selective, and sensitive have become increasingly emerging, especially when talking about a pandemic scenario [48]. In this context, diagnosing pathogens based on genomes and their hybridizations, DNA-DNA and DNA-RNA, have been seen as a means with great potential for performative and complete diagnoses [83]. It is known that even today, the real-time quantitative polymerase chain reaction (PCR is the standard method for diagnosing a series of pathologies, however, although it allows reliable and sensitive answers, it is wholly restricted to laboratories and professionals, which makes their performance in rapid diagnosis positions unfeasible [133]. Therefore, alternative approaches such as microfluidic platforms, colorimetric, and electrochemical methods have aroused considerable interest [83].

Considering sensitivity and practicality, electrochemical methods, among those mentioned, are the most used [83]. Furthermore, the advantage of miniaturization makes this type of sensor a promising path [7]. However, these are not directly applied to the detection of a single copy, for example, of DNA, in biological samples [110]. Thus, much research has been carried out to amplify electrochemical signals, making such biosensors more selective and sensitive, capable of detecting analytes at very low concentrations and, therefore, facilitating the diagnosis of diseases in their initial stage [83, 102].

One of the approaches adopted to provide greater sensitivity to these biosensors is using conductive materials, such as MXenes [9]. In addition to its flexibility and high electrical conductivity, a series of ligands are commonly integrated into this system and precisely onto this nanomaterial's surface, producing hybrids with the genotypic structure and contributing to signal improvement [39]. However, since a MXene-genone interaction can be formed without the use of a specific ligand, the construction of biosensors is widely explored due to ease [9, 39, 107].

In this context, Wang and collaborators developed an electrochemical biosensor based on MXene nanosheets modified with tetrahedral DNA nanostructures for the detection of gliotoxin [108]. As is known, gliotoxin is a toxic metabolite biosynthesized during the growth of the fungal species *Aspergillus fumigatus* (*A. fumigatus*), considered a severe threat to health [108]. Thus, according to the authors, due to the abundance of phosphate in the nucleic acid and the abundance of a transition metal in MXene, it was believed that strong chelation-type interactions between the metal group and the phosphate could efficiently adsorb DNA nanostructures. In this way, a broad electroactive surface, and a facilitated method of constructing a biosensor are provided, which does not involve complex modifications. The manufactured biosensor enabled, across a range of 5 to 1×10^4 pmol L^{-1}, promoted a LOD of 5 pmol L^{-1} for gliotoxin.

At the same time, Chen and collaborators proposed the rapid detection of the SARS-CoV-2 nucleocapsid gene through a chemiresistive device developed based on the functionalization of MXene with DNA probe molecules [15]. According to the authors, the respective functionalization occurred through non-covalent adsorption, which made it possible to eliminate costly steps involving other ligands. The recognition mechanism was based on nucleic acid hybridization and chemiresistive transduction, and it is further reported that the spacing between MXene layers would serve as sieving channels, which would function as a kind of host for organic molecules and ions, highlighting an advantage for biomolecular sensing. From the studies, therefore, a sensitive and rapid ability to detect the analyte with a LOD of 105 copies/mL was observed, in addition to a high selectivity when tested against SARS-CoV-1 and Middle East respiratory syndrome.

As seen in Chen and collaborators' work and their chemical resistive devices, although electrochemical sensors are the most used in terms of the facilities already mentioned in this item, other types of sensors have also been used to diagnose diseases. The same can be seen in the work of Tao and collaborators, where the authors, instead of expensive instruments and laborious procedures, chose to develop a colorimetric biosensor based on clustered systems of regularly interspaced short palindromic repeats (CRISPR-Cas12a) using modified MXene modified with DNA-Ag/Pt nanohybrids for detection of Hepatitis B virus DNA in human serum samples [99]. The proposed method and device proved sensitive and selective, and its stability and reliability were further confirmed by the recovery of 99.5 to 101.3% in the respective samples. In Fig. 15, it is possible to verify the integration of the colorimetric device with a smartphone and the change in color of the biosensor in the presence of different concentrations of virus DNA.

Wang and collaborators also relied on other mechanisms to propose a biosensor, in this case in question, capable of simultaneously mapping dual breast biomarkers in living cells [106]. To this end, the authors developed a MXene-based nonaprobe functionalized with chimeric DNA for detection of the transmembrane glycoprotein mucin 1 (MUC1) and cytoplasmic microRNA-21 (miR-21) through electrochemiluminescence. In the presence of MUC1, recognition between the DNA-MXene probe

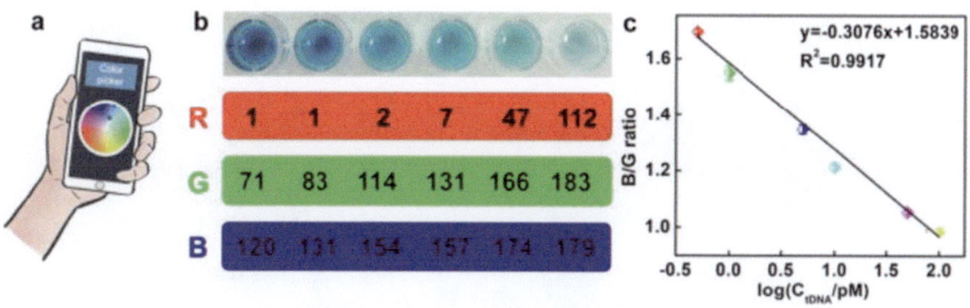

Fig. 15 **a** Integration of the biosensor with a smartphone; **b** photographic image of the color change of the sensor for concentrations between 0.5 and 100 pmol L^{-1} of Hepatitis B virus DNA; **c** analytical curve for respective concentration range. Reprinted with permission from [99]. Copyright 2024, Elsevier

and the biomarker induces the separation of the aptamer called TAMRA, which triggers a red fluorescence. In contrast, in the face of miR-21, hybridization with the probe increases the green fluorescence, enabling detection. Therefore, the developed biosensor enabled in vitro detection at concentrations in the nanomolar range for both biomarkers.

On the other hand, as mentioned, some authors choose to insert ligands in the construction of sensors to improve sensitivity and/or selectivity. Thus, for Liu and collaborators, the detection of Pb^{2+} was carried out using a biosensor based on a glassy carbon electrode (CGE) modified with MXene and GR5 DNAzyme, a type of RNA cleavage particular for Pb^{2+} [57]. For this construction, an aliquot of a previously prepared MXene dispersion in Nafion was deposited on the surface of the GCE, followed by the deposition of an aliquot of a solution of GR5 DNAzyme. Cyclic voltammetry was used to evaluate the parameters and detect Pb^{2+}. The authors discuss that in the presence of Pb^{2+}, in which the GR5 DNAzyme has high specificity, its cleavage occurs activated by a riboadenine site in the DNA strand, which leads to the unwinding of the double helix. Due to this, the adsorptive property of MXene is stimulated about GR5 DNAzyme, favoring the intercalation of ions and inducing an increase in the Pb^{2+} oxidation peak current and, consequently, greater sensitivity. Thinking about the bioaccumulation of Pb^{2+} through the food chain, which could threaten human health, the authors applied the sensor developed to detect the respective analyte in chicken liver tissue, reaching LOD of 0.1 nmol L^{-1}. Figures 16 and 17 show a representative scheme of the preparation and application of the sensor, the voltammetric responses at different concentrations of Pb^{2+}, and the analytical curve.

However, among the applications found in the literature, it is worth highlighting due to the numerous varieties of biosensors developed for diagnosing COVID-19.

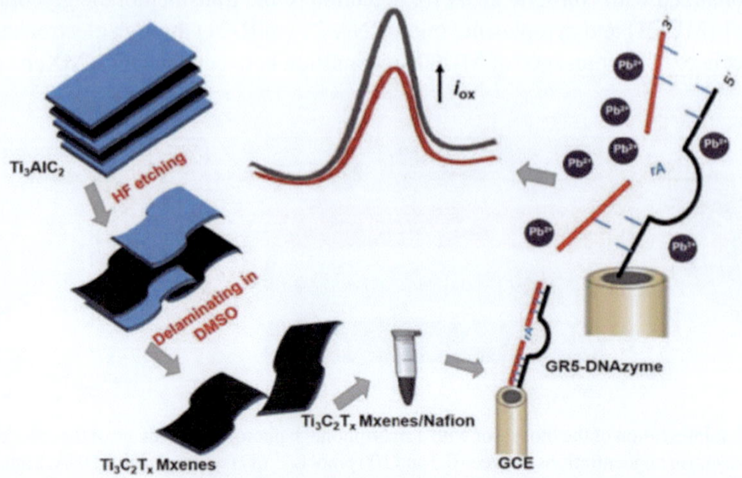

Fig. 16 Representation of the preparation and application of the GR5 DNAzyme/MXene/GCE sensor. Reprinted with permission from [57]. Copyright 2024, Elsevier

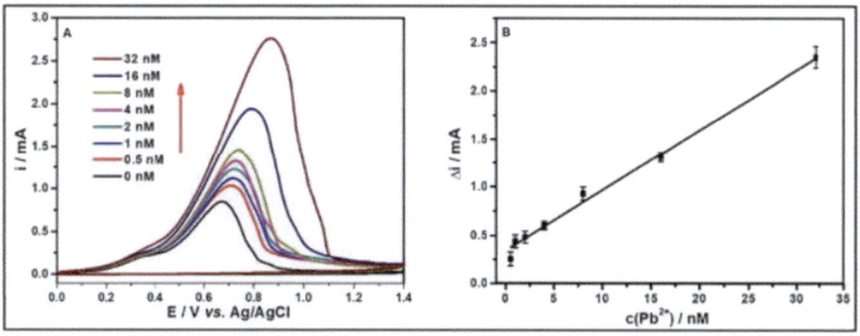

Fig. 17 CV responses of the GR5 DNAzyme/MXene/GCE sensor for different concentrations of Pb^{2+} and calibration curve. Reprinted with permission from [55]. Copyright 2024, Elsevier

In this context, it is known that the COVID-19 pandemic was, despite being a tragic period in our history, a strong incentive for the manufacture of new technologies aimed at clinical diagnosis [100]. This is because, in more than 7 million deaths, the urgency for a quick and efficient diagnosis has become the focus of many researchers [100]. In this sense, Bolourinezhad and collaborators, faced with the highlighted need, developed a fast, precise, and economical electrochemical sensor based on MXene nanosheets and platinum carbon, bovine serum albumin (BSA), and ssDNA for detection in nasopharyngeal, saline and serum samples. According to the authors, this combination was proposed to promote signal amplification after a hybridization reaction of the DNA probe and the target oligonucleotide specific to the COVID-19 virus. Thus, differential pulse voltammetry (DPV) achieved an LOD of 0.4 amol L^{-1}. Furthermore, specificity and sensitivity studies proposed against 192 clinical samples analyzed by PCR showed 100% accuracy and 97.87% specificity.

With a similar aim, Bharti and collaborators proposed a sensor based on a screen-printed electrode (SPE) modified with MXene nanosheets and a DNA probe functionalized with an amino group. While CV and DPV were used to evaluate the electrochemical composting, define the parameters, and optimize the method, electrochemical impedance spectroscopy was used to evaluate the biosensor's analytical response. The biosensor presented LOD of 0.004 pmol L^{-1} and selectivity and stability for 40 days [6].

3.2 *Wearable Biosensors*

Lately, scientists have been exhaustively exploring the potential of MXenes incorporated into biosensors for detecting various biomarkers, monitoring the body, and diagnosing diseases [44]. Among the different types of sensors, MXenes have gained considerable prominence in modifying wearable biosensors to make them more sensitive and selective and meet some mechanical needs [44]. In this context, wearable biosensors are nothing more than versatile and flexible detection devices that have excellent mechanical compliance and sensitivity and can be attached to the human body, either superficially or internally, enabling the emergence of interactive platforms for monitoring physiological activity, health management, disease prevention, diagnosis, and others [31, 121]. Due to rapid growth in recent years, many studies have been carried out on the preparation, design, applications, and improvement of the properties of these sensors to make them more adaptable to the skin, making them light, flexible, mechanically robust, and friendly [67]. Thus, researchers are looking for new materials capable of meeting this need because such devices require flexible electronics that at the same time present good electrical performance, which is not possible with conventional silicon bases [89]. Over the years, it was realized that due to their high electrical conductivity, wearable biosensors based on MXenes allow the detection of biomarkers with high sensitivity due to efficient signal transduction. Furthermore, the large surface area of this material allows ample space for immobilizing these marker biomolecules [41, 44].

However, given some disadvantageous properties of MXenes, such as poor mechanical properties, ease of restacking, low stability in an oxygen atmosphere, and others, a bias that has been widely explored is the functionalization of this material, that is, the insertion of functional groups on its surface, which allows greater specificity and selectivity due to specific binding to target biomarkers that are present in very low concentrations. This aims to minimize false positives, increase monitoring accuracy, and enable these biosensors to act in complex biological environments [14]. However, introducing functional groups may generate non-specific bonds, interfering with the molecular discrimination capacity of the biosensor. Thus, one of the main challenges is promoting efficient bioconjugation between the functionalized surface of MXene and the biomarkers [17, 44]. It is noticeable that the integration of MXenes-based biosensors into wearable diagnostic platforms requires extraordinarily cautious and calculated steps since ensuring biocompatibility without losing the inherent properties of a sensor is crucial for inserting these devices into medical and clinical environments [44]. Therefore, in Table 1, some works represent the integration among wearable platforms, MXenes, and health monitoring.

According to the work presented in Table 1, electrochemical devices are intensely involved in health monitoring. This fact is in line with what is reported in the literature since great efforts in the field of research worldwide are focusing on developing, validating, and commercializing wearable electrochemical biosensors [73]. The great advantage of using this type of sensing mechanism is, in addition to sensitivity, the possibility of processing chemical analyses in a non-invasive way [73]. Additionally, electrochemical devices provide greater integration and robustness, as it becomes possible to eliminate optical components and moving parts, such as flow systems, microfluidic components, pumps, and others, enabling miniaturization simplification. In this context, Zhang and collaborators developed a highly sensitive, flexible, and fully integrated potentiometric electrochemical detection system for monitoring the concentration of K^+ in human sweat [129]. To this end, MWCNT and MXene-based hybrid multidimensional networks were integrated to obtain a high surface activation area and a high charge transfer rate. The choice of carbonaceous

Table 1 MXene-based biosensors for health monitoring in sweat samples

Biosensor	Detection technique	Analyte	LOD	Sensitivity	References
$Ti_3C_2T_x$/PB	Chronoamperometry	Glucose and lactate	3.3×10^{-1} and 6.7×10^{-1} μmol L^{-1}	35.3 and 11.74 μA mmol L^{-1}	[50]
$Ti_3C_2T_x$/MoS_2	CV	Ascorbic acid	4.2 μmol L^{-1}	54.6 μA mmol L^{-1}	[134]
$Ti_3C_2T_x$/AuNPs/ssDNA	OPC-EIS and SWV	Oestradiol	1.4×10^{-7} μmol L^{-1}	NA	[125]
$Ti_3C_2T_x$/MWCNT	Potentiometry	K^+	NA	173 mV dec^{-1}	[132]
$Ti_3C_2T_x$/LBG/PDMS	EIS	Cortisol	8.8×10^{-5} μmol L^{-1}	NA	[94]
$Ti_3C_2T_x$/PEDOT:PSS	EIS	Na^+ and pro-inflammatory cytokine protein	NA	40 and 3.9 mV dec^{-1}	[93]
$Ti_3C_2T_x$/ZnOTPs	Chronoamperometry	Glucose	17 μmol L^{-1}	27.87 and 29.88 μA mmol L^{-1}	[77]
$Ti_3C_2T_x$/L-cys/AuNPs	Amperometry	Cortisol	0.54 ng mL^{-1}	NA	[47]
$Ti_3C_2T_x$/PEDOT:PSS	Amperometry	Glucose	1.9 μmol L^{-1}	21.7 μA mmol L^{-1}	[82]
$Ti_3C_2T_x$/$Pd(OH)_2$	Amperometry	Nicotine	0.027 μmol L^{-1}	0.286 μA μmol L^{-1} cm^2	[69]

NA = Not available; PB = Prussian blue; PVA = Polyvinyl alcohol; AuNPs = Gold nanoparticles; ssDNA = Single-stranded DNA; OPC-EIS = Open circuit potential-electrochemical impedance spectroscopy; SWV = Square wave voltammetry; MWCNT = Multiwalled carbon nanotube; LBG = Laser-burned graphene; PDMS = Polydimethylsiloxan; EIS = Electrochemical impedance spectroscopy; PEDOT:PSS = poly(3,4-ethylenedioxythiophene):poly(styrenensulfonate); ZnOTPs = Zinc oxide tetrapods; PBS = Phosphate-buffered saline

materials to compose the biosensor was also based on the possibility of accelerating the kinetic process of ionic diffusion, which contributes to the electrochemical response. The electrode was then prepared from a flexible PET substrate by a screen-printing process, and valinomycin, a known K^+ carrier, was used to coat the electrode. According to the authors, the manufactured sensor exhibited high sensitivity, presenting a LOD of 63 mV dec^{-1} and an amplified sensitivity of 173 mV dec^{-1}. Figure 18 presents a representation of the detection mechanism and biosensor fabrication.

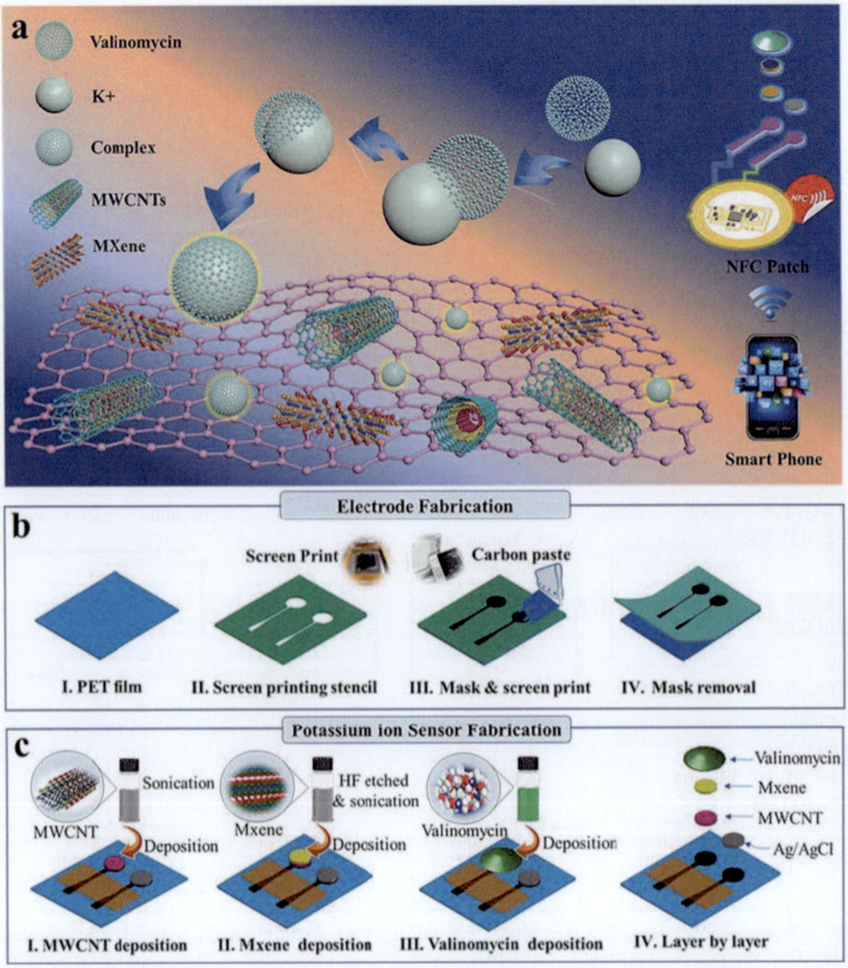

Fig. 18 **a** Representation of the K^+ detection mechanism, **b** printing of the biosensor by screen-printing, and **c** manufacturing of the biosensor. Reprinted with permission from [132]. Copyright 2024, Elsevier Publisher

At the same time, Saleh and collaborators proposed the development of a wearable electrochemical biosensor for multimodal detection of analytes in the skin [93]. To this end, the authors developed an aqueous paint based on MXene, which was integrated into a Na^+ ion-selective membrane and functionalized with antibodies responsive to the pro-inflammatory cytokine protein interferon gamma. The ink was printed on poly(3,4-ethylenedioxythiophene):poly(styrenensulfonate) (PEDOT:PSS) substrates by an inkjet printer in a single step. Electrochemical impedance spectroscopy (EIS) performed detection and the biosensors achieved a sensitivity of 40 mV and 3.9 mV dec^{-1}, respectively.

It is also noted in Table 1 that among the electrochemical techniques applied to these sensors, the amperometry technique gains considerable prominence. Even though the choice of redox enzymes available is limited, which limits the scope of amperometrically detectable metabolites, the wide use of amperometric wearable biosensors is justified due to the rapid and continuous response, which allows prolonged monitoring [73]. Considering that amperometric biosensors are based on current monitoring at a fixed potential, some criteria need to be considered, such as the choice of electrode material, the applied potential and surface modifications [73]. However, when such parameters are well evaluated, the amperometry technique can provide selective detections of different analytes [45, 73]. This was done in the work of Laochai and collaborators when they evaluated the fabrication of a biosensor that integrated gold nanoparticles (AuNPs), MXene, and L-cys [47]. A thread-based biosensor was then fabricated for the non-invasive detection of cortisol in sweat through the immobilization, by coupling agents, of anticortisol on the surface of the modified conductive thread. According to the studies, the authors stated that the choice to use AuNPs and MXene considerably increased the surface area of the electrode, facilitating the immobilization of the anticortisol and, therefore, leading to greater sensitivity of the device. An illustration of the sensor preparation is shown in Fig. 19.

With the ideal conditions optimized, the developed biosensor showed high sensitivity, reproducibility, and stability (>6 weeks), with a LOD of 0.54 ng mL^{-1} and a wide linear range of 5–180 ng mL^{-1}, as can be seen in Fig. 20.

Hydroxides can also be combined with MXene to obtain specific application properties. In a study by Magesh and co-workers, MXene (Ti_3AlC_2) was combined with Palladium Hydroxide ($Pd(OH)_2$) (PHC) to develop a sensor for detecting nicotine in human sweat [69]. The composite was prepared by mixing MXene and PHC-supported carbon in equal proportions and grinding the mixture for 20 min to produce fine particles, as presented in Fig. 21. The powder mixture was then ultrasonicated with diluted Nafion, and the composite was diluted with distilled water. The dispersion was drop-casted over the GCE surface and then dried. The modification of GCE with MXene/PHC composite resulted in a reduced potential for nicotine oxidation and enhanced oxidation peak current compared with GCE and MXene/GCE. Nicotine quantification ranged from 0.25 to 37.5 μM, demonstrating a LOD of 27 nM. The sensor was tested in the presence of various cations and molecules, showing good selectivity as there was no interference during analysis.

Another amperometric biosensor was developed by Lei and collaborators using chronoamperometry as an electrochemical technique to detect glucose and lactate in sweat simultaneously [50]. Although the authors have emphasized that electrochemical detection in sweat by enzyme-based biosensors presents several challenges, mainly due to the easy degradation of enzymes, limited range of sensitivity due to oxygen deficiency in sweat, and the low helpful life of the device, a multifunctional biosensor was developed here from the integration of MXene, Prussian blue (PB) and the enzymes lactate and glucose oxidase. This approach allowed electrochemical sensitivities of 35.3 and 11.74 μA mmol L^{-1} and LODs of 3.3×10^{-1} and $6.7 \times$

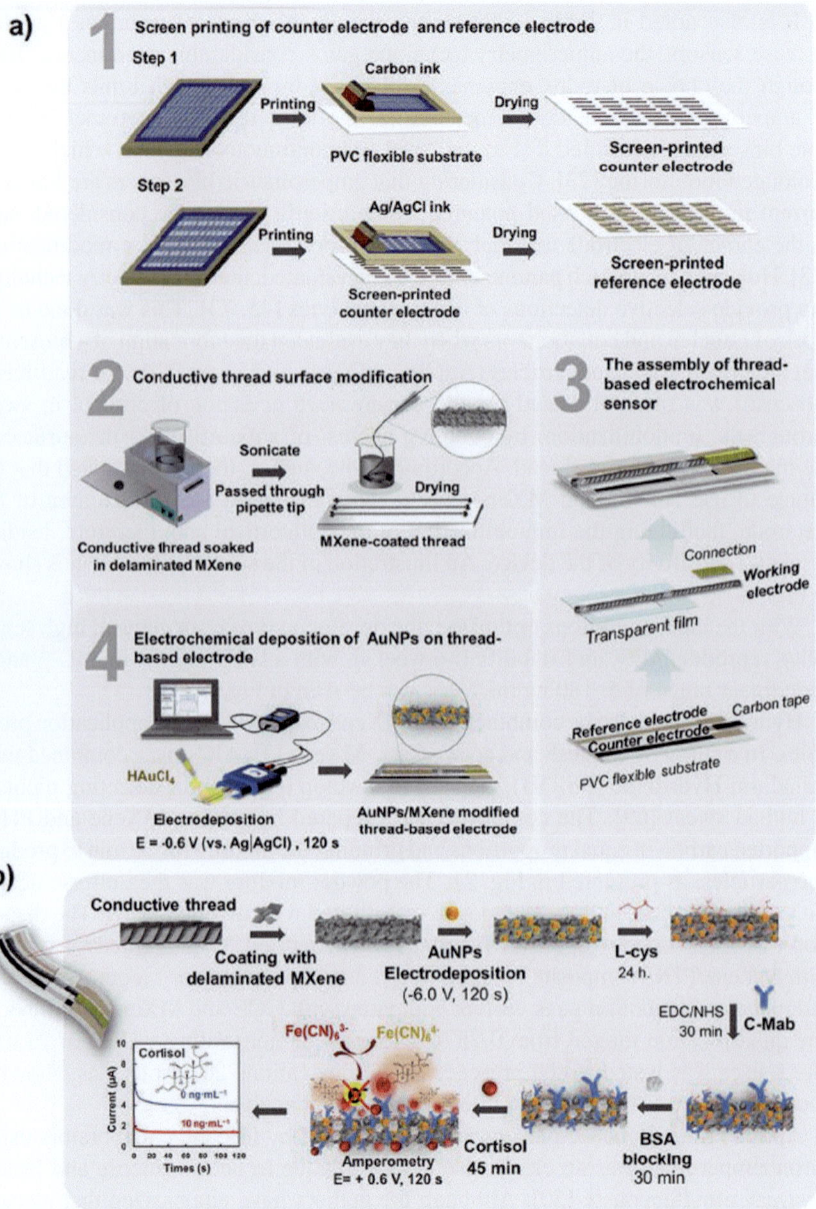

Fig. 19 Representation of the manufacturing steps of the amperometric biosensor for cortisol, where **a** is the electrode printing and **b** the immobilization of the anticortisol and detection mechanism. Reprinted with permission from [47]. Copyright 2024, Elsevier Publisher

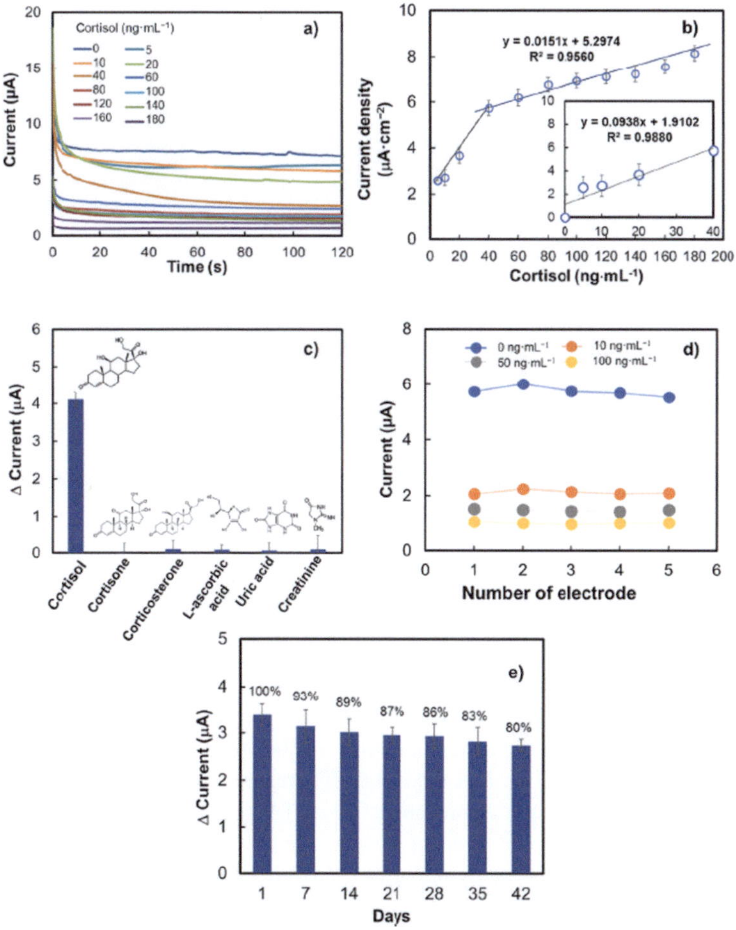

Fig. 20 Amperometric responses for cortisol: 5–180 ng mL^{-1}; constant potential +6 V; time 120 s. **a** Cortisol calibration curve, **b** Study of the effect of interferents (cortisone, corticosterone, L-ascorbic acid, uric acid, and creatinine), **c** Amperometric response for different concentrations of cortisol, **d** Stability study, and **e** Reproducibility study. Reprinted with permission from [47]. Copyright 2024, Elsevier Publisher

10^{-1} μmol L^{-1}, respectively, representing an essential step towards the construction of ultrasensitive enzymatic biosensors applied to the sciences of health.

On the other hand, glucose is a widely investigated analyte when it comes to wearable sensors applied to health monitoring. Much of the motivation behind the development of wearable sensors for this purpose is based on the possibility of manufacturing a sensor that, unlike conventional invasive subcutaneous glucometers, these new sensors enable non-invasive measurements, offering the patient less discomfort, in addition to the promise of providing better glycemic control through constant monitoring [46].

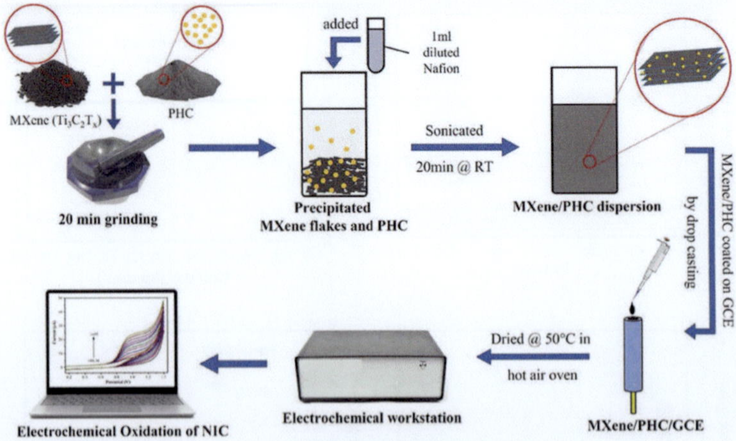

Fig. 21 Scheme of MXene/PHC composite preparation for the sensor development to determine nicotine. Reprinted with permission from [69]. Copyright 2024, MDPI

In this sense, Pan and collaborators sought to develop an electrochemical biosensor based on a hydrogel composed of MXene and PEDOT:PSS for continuous and non-invasive monitoring of glucose in sweat. A schematic of the biosensor preparation and detection can be seen in Fig. 22 [82].

The use of ethylene glycol (EG) in the manufacture of the sensor, as seen in the steps in Fig. 22, promotes porous networks in the material, contributing to film-forming properties, stability, and flexibility. Therefore, the glucose biosensor used the amperometric technique to present LOD of 1.9 μmol L^{-1} and a sensitivity of 21.7 μA mmol L^{-1}.

Myndrul and collaborators, by integrating MXene nanoflakes and ZnO tetrapods, manufactured a chronoamperometric enzymatic biosensor to detect glucose in sweat continuously [77]. The authors state that the high specific area of the tetrapods, combined with the conductivity of MXene, allowed a high sensitivity of 29 μA mmol L^{-1} and LOD of 17 μmol L^{-1}. The data obtained satisfy the proposed application with advanced mechanical stability during sugar intake and physical activity.

In summary, the use of MXene in manufacturing wearable biosensors is a considerable advance regarding health monitoring, considering the need for more sensitive, selective, and flexible biosensors, as discussed. Once the parameters and conditions are optimized, a series of medical studies can be carried out based on the constant and noninvasive monitoring of a series of analytes, this area not being limited to glucose studies.

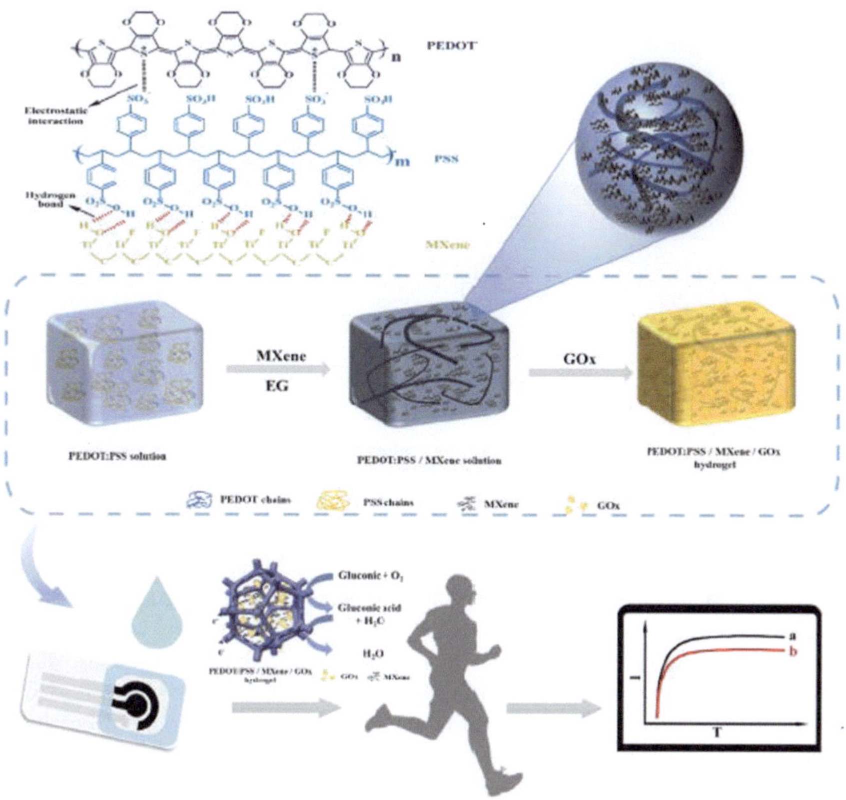

Fig. 22 Representation of the construction of the MXene/PEDOT:PSS biosensor and detection of glucose in sweat. Reprinted with permission from [82]. Copyright 2024, Elsevier Publisher

4 Characterization of MXenes

As seen previously, MXenes as 2D materials are obtained by transforming MAX phases into MXene after the etching process. Simple X-ray diffraction analyses can be used to evaluate MXene formation. Yang and collaborators prepared ultrathin $Ti_3C_2T_x$ MXene sheets from Ti_3AlC_2 MAX precursor [124]. The authors confirmed the formation of the MXene phase by comparing peaks from the MAX phase and ones presented by 2D and ultrathin (UT) MXene, as seen in Fig. 23. MAX phase shows various peaks non-presented in 2D and UT patterns. For example, the peak at ~8.5° represents (0 0 2) of MAX (Ti_3AlC_2) changes to a lower angle (~6.2°) in 2D and UT. This change confirms the removal of Al from the MAX structure and the increase of interlayer spacing by the inception of hydroxyl and fluorine groups after etching. The same approaches to confirm MXene formation were reported by Alhabeb and co-workers [2], Chen and collaborators [16].

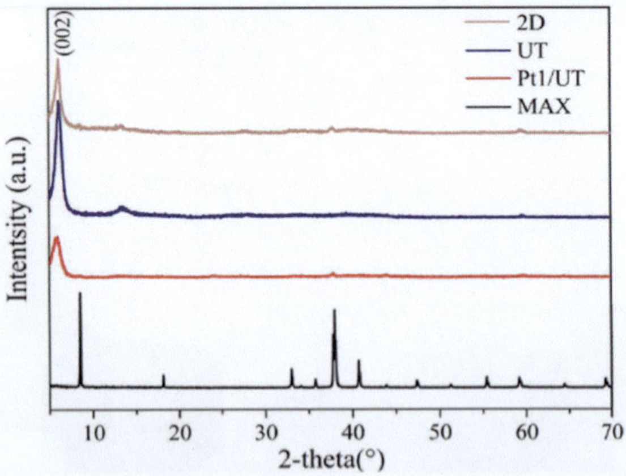

Fig. 23 XRD patterns of MAX phase, 2D, and UT MXenes. License by Creative Commons from [122]. Copyright 2024, Elsevier Publisher

The successful synthesis of MXene sheets can also be verified by Scanning Electron Microscopy (SEM), Transmission Electron Microscopy (TEM), and X-ray Photoelectron Spectroscopy (XPS) [58, 109]. SEM and TEM can reveal the structure of MXenes. In the most expected cases, an SEM micrograph showing an "accordion-like" structure, like Fig. 24a, is considered a successful MXene formation. This characteristic is due to the release of gasses from the MAX phase during the etching process with highly concentrated acid [2]. Nonetheless, not all MXene structures are accordion-like, as seen in Fig. 4b [95], reinforcing the necessity of using multiple coordinated techniques to characterize these materials. SEM was also used to evaluate the differences between 2D and UT MXene, presented in Fig. 24c–f. Authors reported that, by SEM characterization, the thickness of 2D MXene was observed and determined (~500 nm). Still, for the UT MXene for an accurate measure, Atomic Force Microscopy was necessary [124].

TEM is used to confirm MXene formation by the presence of exfoliated particles, allowing the determination of sheet thickness [57]. With high resolution (HRTEM), the technique is used to measure particle size and determine interplanar distance, as presented in Fig. 25. Ulas and co-workers prepared $Ti_3C_2X_2$ MXene supported $BaMnO_3$ and used TEM to characterize the material at low and high magnification before, as can be seen in Fig. 25a, b and after $BaMnO_3$ doping [101]. The formation of $BaMnO_3$/MXene was attributed to the presence of black dots in MXene nanosheets, presented in Fig. 25d, e. Figure 25c presents that the interplanar distance in MXene sheets is 0.175 nm, and for $BaMnO_3$/MXene, Fig. 25c presents an increase to 0.349 and 0.354 nm, which confirms the impregnation of MXene with the nanoparticles.

Xing and co-workers evaluated the formation of MXenes comparing different XPS spectra that demonstrate the removal of Al of MAX, resulting in successful synthesis of 2D Ti_3C_2 MXene with –OH, –F, and –O– group terminations [116]. The peak with a binding energy of 74.9 eV, attributed to Al 2p, decreased significantly after HF-etching and is absent in MXene nanosheets. This finding reinforces MXene formation. XPS is also used to evaluate functional groups on the surface of MXenes [101, 108], determine material composition [113], evaluate oxidation [34],

Fig. 24 SEM images of **a** exfoliated MXene (Reprinted with permission from [22]. Copyright 2024, Elsevier Publisher), **b** Ti_3AlC_2 produced by HF etching (Scale: 5 μm) (Reprinted with permission from [108], Copyright 2024, Elsevier Publisher) and to present the morphology of **c**, **d** 2D MXene and **e**, **f** UT MXene. Licensed by Creative Commons adapted from previously published from [122], Elsevier Publisher

and analyze the formation of composites [62]. Wang and co-workers present that the presence of polar groups determined by XPS spectra in the surface and edges of MXenes plays an exciting role in the excellent dispersibility in water and intense interactions with some polymers [108]. These characteristics may be interesting for sensing applications.

Mechanical properties need to be considered in some cases of MXene applications. MXenes have excellent mechanical properties and are predicted to be better than graphene [21]. However, these remarkable properties are limited to the nanoscale scale because, on the macroscopic scale, there are weak van der Waals interactions between layers, resulting in poor mechanical behavior [17]. Considering that MXenes can be used in composite with various polymers, enhancing mechanical properties, Gao and co-workers reported that MXene is an ideal nanofiller for polymer composites [21].

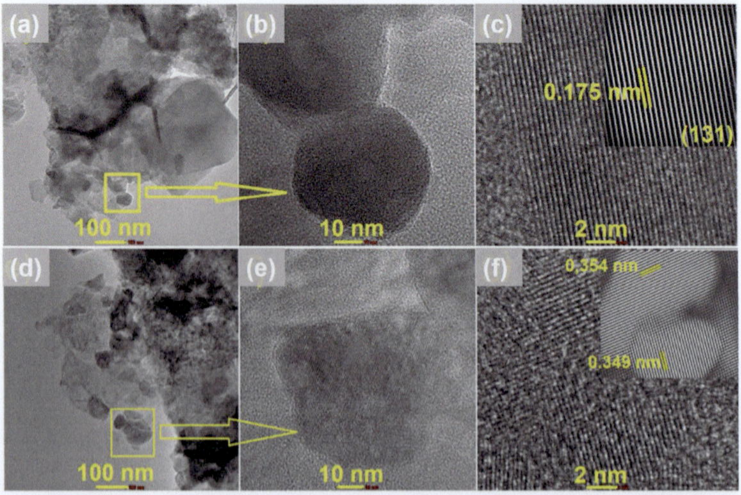

Fig. 25 TEM images of a MXene with **a** low and **b** high magnification, **c** interplanar *d*-space of MXene, TEM images of $BaMnO_3$/MXene at **d** low and **e** high magnification, **f** interplanar *d*-space in $BaMnO_3$/MXene sample. Reprinted and adapted with permission from [101]. Copyright 2024, Elsevier Publisher

The presence of MXene in cellulose composite can also strengthen and toughen nanocomposites [12]. Wan and collaborators also evaluated the mechanical properties of MXene and films with modifications. The results were reported in a tensile stress–strain curve. The authors present that MXene exhibited a tensile strength of 87 $\pm$ 3 MPa, Young's modulus of 6.1 $\pm$ 0.6 GPa, and a toughness of 1.3 $\pm$ 0.1 MJ/m^3. The best result was achieved with modified film with sequentially bridged MXene that presented a tensile strength of 583 $\pm$ 16 MPa, Young's modulus of 27.8 $\pm$ 2.8 GPa, and a toughness of 15.9 $\pm$ 1.0 MJ/m^3 [104]. The superficial modifications induced by hydrogen and covalent bonding agents justified the improvement in mechanical properties.

The mechanical reinforcement made by introducing MXene in a composite matrix is an essential consideration in sensing applications. Cheng and co-workers prepared a self-healing pressure sensor with high-strength MXene sheets with aramid fibers (ANF) [17]. Figure 26 presents that the treatment for hydrogen bonding with AFN resulted in better mechanical behavior. The best result for mechanical strength was achieved with 50% of AFN (114.1 MPa), which is 1240% higher than pristine MXene, as seen in Fig. 26b. This film also presented better properties like toughness and Young's modulus. According to the authors, MXenes have enhanced mechanical properties that make them suitable for flexible pressure sensors. These sensors can resist external damage and exhibit excellent electronic properties.

After producing sensors, Cheng and co-workers assess the application in some applications [17]. By the interests here in biomedical applications and related, Fig. 27 presents the uses in finger, wrist, knee, and ankle bending, as well as for elbow swing

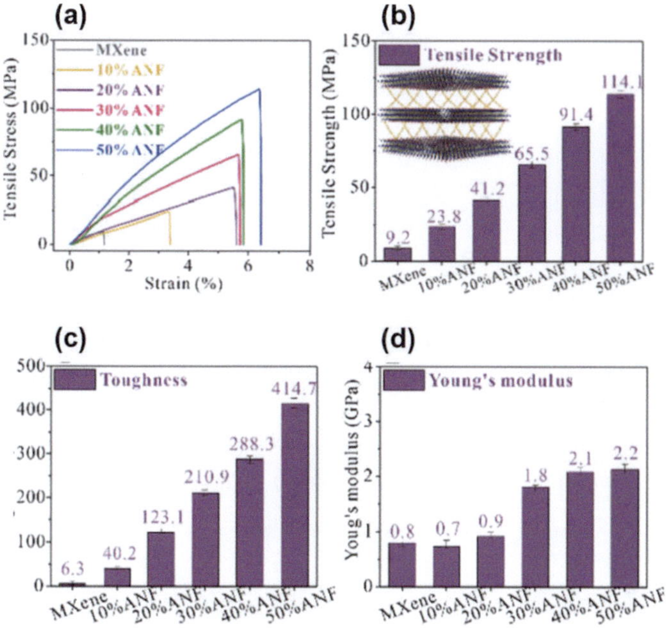

Fig. 26 Results of enhanced mechanical properties of MXene sheets and mechanism of the self-healing pressure sensor for **a** Stress–strain, **b** tensile strength, **c** toughness, and **d** Young's modulus of MXene sheets with different mass ratios of incorporated aramid fibers. Reprinted and adapted with permission from [17]. Copyright 2024, Elsevier

and wrist pulse reported by the authors. For these sensing uses, electrical properties have paramount importance. A computer collector was built to examine the pressure-sensing performances of the high-strength MXene films. The measurements were conducted using a single-axis propulsion device, an ergometer, and an electrical signal-detecting device.

In these sensors, the signal is captured by the relation between the mechanical and electrical conductivity of MXenes. In static mode, there is an internal and inherent resistance to current and conductivity. With external pressure, there is an increased signal (ΔI) relative to the initial one (I_0) by diminishing spacing in material, resulting in $\Delta I/I_0$ shapes presented in Fig. 27. MXenes have excellent electrical conductivity [74], but this property needs to be evaluated when these materials are in composites. As said before, Wan and co-workers produced different films with MXenes. MXene films presented 9468 $\pm$ 115 S/cm conductivity [104]. The insertion of 10 wt. % of sodium carboxy-methyl cellulose (CMC) reduces the conductivity to 6049 $\pm$ 64 S/cm by the insulating characteristic CMC. Some films presented conductivities comparable with pure MXene, resulting in 9491 $\pm$ 95 S/cm justified by the improved alignment, decreased interplanar spacing, and elimination of intercalated water. The values reported by the authors present conductivities that are lesser

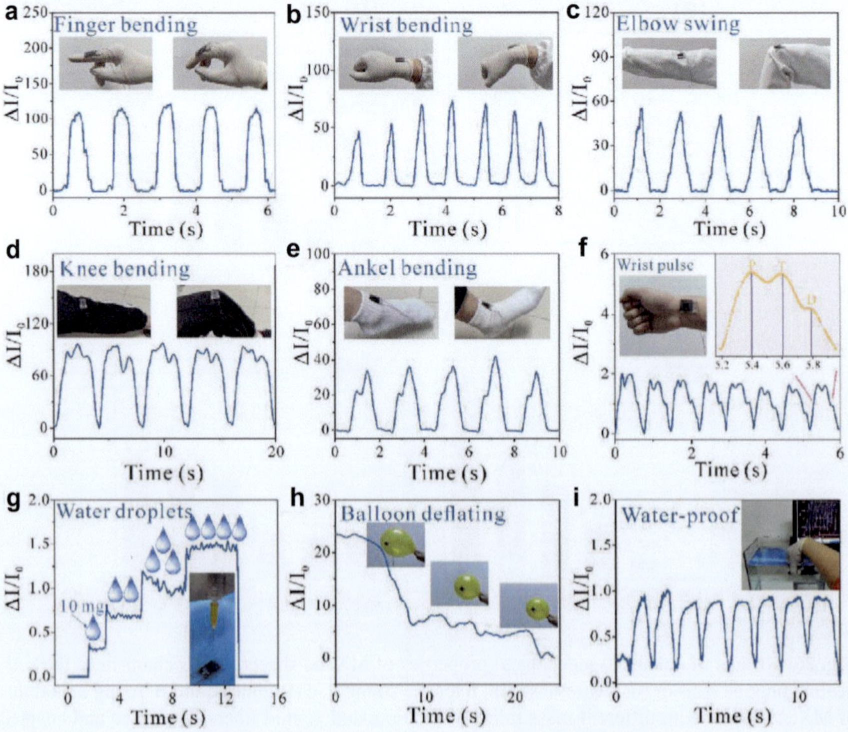

Fig. 27 Signal responses of high-strength MXene used as a pressure sensor to monitor human activities such as **a** finger bending, **b** wrist bending, **c** elbow swing, **d** knee bending, **e** ankle bending, **f** wrist pulse. Reprinted and adapted with permission from [17]. Copyright 2024, Elsevier

than pure MXene until near it. These findings justify the need to evaluate electrical conductivity responses when MXenes are in composites for sensing.

5 Challenges and Perspectives

Different methods can be employed for MXene synthesis, resulting in materials with distinct compositions and electronic, electrochemical, magnetic, and mechanical properties [128]. An advantage of this material is its excellent flexibility, layered structure, and 2D morphology, enabling the formation of various types of composites with other materials such as polymer, MOF, metal, hydroxide, oxide, sulfide, and CNT, thus enhancing the properties of MXene and its performance in various areas, including biomedical applications [119, 128].

One of the main challenges in synthesizing MXene composites with other materials is obtaining them in large quantities to enable their use on an industrial scale [84]. The development of different methods of obtaining these composites, which are simple, efficient, economical, and environmentally friendly on a large scale, is necessary to expand the application segments of these materials. Furthermore, studies involving stability and toxicity in the short and long term of these composites still need to be improved, especially for their use in human biomedical applications [20].

As exemplified, there are several reports on the application of MXene in health monitoring devices, specifically, MXene composites. This chapter introduced some crucial examples highlighting the main techniques and analytes studied. However, in the context of wearable biosensors, even though MXenes are strong candidates for integrating these biosensors, some challenges still need to be overcome, such as improving the material's deficiencies, such as stability and high cost. In addition, optimizing and integrating real-time and easily accessible monitoring platforms associated with these biosensors will promote significant advancement in using MXene for this purpose.

Additionally, the great importance of developing biosensors based on DNA-DNA and DNA-RNA hybridizations is evident, given the gain in flexibility compared to other molecules since a genome is present in all living organisms. Because of this, devices with great potential, as discussed in this chapter, can promote immediate and early diagnoses, with the advancement and insertion of technology always welcome in this scenario.

References

1. Algaradah, M.M.: MXene-based adsorbent materials for pollutants removal from water: current challenges and future prospects. Inor. Chem. Comm. 112113 (2024). https://doi.org/10.1016/j.inoche.2024.112113
2. Alhabeb, M., Maleski, K., Anasori, B., Lelyukh, P., Clark, L., Sin, S., Gogotsi, Y.: Guidelines for synthesis and processing two-dimensional titanium carbide (Ti3C2Tx MXene). Chem. Mater. **29**(18), 7633–7644 (2017). https://doi.org/10.1021/acs.chemmater.7b02847
3. Ali Khan, A., Tahir, M., Khan, N.: Recent developments in titanium carbide (Ti3C2)-based layered double hydroxide (LDH) nanocomposites for energy storage and conversion applications: a minireview and perspectives. Energy Fuels **36**(17), 9821–9843 (2022). https://doi.org/10.1021/acs.energyfuels.2c01752
4. Bao, W., Tang, X., Guo, X., Choi, S., Wang, C., Gogotsi, Y., Wang, G.: Porous cryo-dried MXene for efficient capacitive deionization. Joule **2**(4), 778–787 (2018). https://doi.org/10.1016/j.joule.2018.02.018
5. Basiron, N., Sreekantan, S., Saharudin, K.A., Ahmad, Z.A., Kumaravel, V.: Improved adhesion of nonfluorinated ZnO nanotriangle superhydrophobic layer on glass surface by spray-coating method. J. Nanomater. **2018**,(2018). https://doi.org/10.1155/2018/7824827
6. Bharti, A., Singh, S., Munthala, D., Roy, S., Pojprapai, S., Suksaweang, S., et al.: Development of a nucleic acid-based screen printed electrochemical biosensor using Ti3C2Tx-MXene for the detection of SARS-CoV-2. Microchem. J. **195**, 109521 (2023). https://doi.org/10.1016/j.microc.2023.109521
7. Blair, E.O., Corrigan, D.K.: A review of microfabricated electrochemical biosensors for DNA detection. Biosens. Bioelectron. **134**, 57–67 (2019). https://doi.org/10.1016/j.bios.2019.03.055

8. Bokov, D., Turki Jalil, A., Chupradit, S., Suksatan, W., Javed Ansari, M., Shewael, I.H., Valiev, G.H., Kianfar, E.: Nanomaterial by sol-gel method: synthesis and application. Adv. Mat. Sci. Eng. **2021**(1), 5102014 (2021). https://doi.org/10.1155/2021/5102014
9. Bolourinezhad, M., Rezayi, M., Meshkat, Z., Soleimanpour, S., Mojarrad, M., Aghaee-Bakhtiari, S.H., Taghdisi, S.M.: Design of a rapid electrochemical biosensor based on MXene/Pt/C nanocomposite and DNA/RNA hybridization for the detection of COVID-19. Talanta **265**, 124804 (2023). https://doi.org/10.1016/j.trac.2019.115643
10. Byrappa, K., Adschiri, T.: Hydrothermal technology for nanotechnology. Prog. Cryst. Growth Charact. Mater. **53**, 117–166 (2007). https://doi.org/10.1016/j.pcrysgrow.2007.04.001
11. Cai, Y., Chen, X., Xu, Y., Zhang, Y., Liu, H., Zhang, H., Tang, J.: Ti3C2Tx MXene/carbon composites for advanced supercapacitors: Synthesis, progress, and perspectives. Carbon Energy e501 (2024). https://doi.org/10.1002/cey2.501
12. Cao, W.T., Chen, F.F., Zhu, Y.J., Zhang, Y.G., Jiang, Y.Y., Ma, M.G., Chen, F.: Binary strengthening and toughening of MXene/cellulose nanofiber composite paper with nacre-inspired structure and superior electromagnetic interference shielding properties. ACS Nano **12**(5), 4583–4593 (2018). https://doi.org/10.1021/acsnano.8b00997
13. Chaudhuri, H., Yun, Y.S.: A critical review on the properties and energy storage applications of graphene oxide/layered double hydroxides and graphene oxide/MXenes. J. Power Sour. **564**, 232870 (2023). https://doi.org/10.1016/j.jpowsour.2023.232870
14. Chen, W.Y., Lin, H., Barui, A.K., Gomez, A.M.U., Wendt, M.K., Stanciu, L.A.: DNA-Functionalized Ti3C2Tx MXenes for selective and rapid detection of SARS-CoV-2 nucleocapsid gene. ACS Appl. Nano Mater. **5**(2), 1902–1910 (2021a). https://doi.org/10.1021/acsanm.1c03520
15. Chen, X., Zhao, Y., Li, L., Wang, Y., Wang, J., Xiong, J., Du, S., Zhang, P., Shi, X., Yu, J.: MXene/polymer nanocomposites: preparation, properties, and applications. Polym. Rev. **61**(1), 80–115 (2021b). https://doi.org/10.1080/15583724.2020.1729179
16. Chen, Y., Li, S., Zhang, L., Jing, T., Wang, J., Zhao, L., Fengbo, L., Li, C., Sun, J.: Facile and fast synthesis of three-dimensional Ce-MOF/Ti3C2TX MXene composite for high performance electrochemical sensing of L-Tryptophan. J. Solid State Chem. **308**, 122919 (2022). https://doi.org/10.1016/j.jssc.2022.122919
17. Cheng, Y., Xie, Y., Cao, H., Li, L., Liu, Z., Yan, S., Ma, Y., Liu, W., Yue, Y., Wang, J., Gao, Y., Li, L.: High-strength MXene sheets through interlayer hydrogen bonding for self-healing flexible pressure sensor. Chem. Eng. J. **453**, 139823 (2023). https://doi.org/10.1016/j.cej.2022.139823
18. Ding, R., Jiang, W., Ma, Y., Yang, Q., Han, X., Hou, X.: A highly sensitive MXene/AuPt/AChE-based electrochemical platform for the detection of chlorpyrifos. Microchem. J. **187**, 108425 (2023). https://doi.org/10.1016/j.microc.2023.108425
19. Fan, W.K., Sherryna, A., Tahir, M.: Advances in Titanium Carbide (Ti3C2Tx) MXenes and their metal-organic framework (MOF)-based nanotextures for solar energy applications: a review. ACS Omega **7**(43), 38158–38192 (2022). https://doi.org/10.1021/acsomega.2c05030
20. Fard, N.E., Raeisi, I., Yousefipour, Z., Mosavin, R.: Review of metal sulfide-based MXene nanocomposites for environmental applications, gas sensing, energy storage, and photothermal therapy. ACS Appl. Nano Mater. **7**(6), 5698–5728 (2024). https://doi.org/10.1021/acsanm.3c05455
21. Gao, Q., Feng, M., Li, E., Liu, C., Shen, C., Liu, X.: Mechanical, thermal, and rheological properties of Ti3C2Tx MXene/thermoplastic polyurethane nanocomposites. Macromol. Mater. Eng. **305**(10), 2000343 (2020). https://doi.org/10.1002/mame.202000343
22. Ghani, A.A., Kim, B., Nawaz, M., Devarayapalli, K.C., Lim, Y., Kim, G., Lee, D.S.: Adsorption and electrochemical regeneration of 2D magnetic MXene nanosheets loaded with tetracycline. Chem. Eng. J. **467**, 143473 (2023). https://doi.org/10.1016/j.cej.2023.143473
23. Gong, K., Zhou, K., Qian, X., Shi, C., Yu, B.: MXene as emerging nanofillers for high-performance polymer composites: a review. Compos. B Eng. **217**, 108867 (2021). https://doi.org/10.1016/j.compositesb.2021.108867

24. Guo, Y.B., Yang, L., Wang, D.G.: Preparation and hydrophobic behaviours of polystyrene composite coating. Surf. Eng. **32**(2), 95–101 (2016). https://doi.org/10.1016/j.nanoso.2018.11.001
25. Guo, Q., Zhang, X., Zhao, F., Song, Q., Su, G., Tan, Y., Tao, Q., Tao, Z., Yu, Y., Zhou, Z., Lu, C.: Protein-inspired self-healable Ti_3C_2 MXenes/rubber-based supramolecular elastomer for intelligent sensing. ACS Nano **14**(3), 2788–2797 (2020). https://doi.org/10.1021/acsnano.9b09802
26. Hao, L., Zhang, H., Wu, X., Zhang, J., Wang, J., Li, Y.: Novel thin-film nanocomposite membranes filled with multi-functional Ti3C2Tx nanosheets for task-specific solvent transport. Compos. A Appl. Sci. Manuf. **100**, 139–149 (2017). https://doi.org/10.1016/j.compositesa.2017.05.003
27. He, T., Kong, X.J., Li, J.R.: Chemically stable metal–organic frameworks: rational construction and application expansion. Acc. Chem. Res. **54**(15), 3083–3094 (2021b). https://doi.org/10.1021/acs.accounts.1c00280
28. He, W., Sohn, M., Ma, R., Kang, D.J.: Flexible single-electrode triboelectric nanogenerators with MXene/PDMS composite film for biomechanical motion sensors. Nano Energy **78**, 105383 (2020). https://doi.org/10.1016/j.nanoen.2020.105383
29. He, S., Sun, X., Zhang, H., Yuan, C., Wei, Y., Li, J.: Preparation strategies and applications of MXene-polymer composites: a review. Macromol. Rapid Commun. **42**(19), 2100324 (2021a). https://doi.org/10.1002/marc.202100324
30. He, J., Zou, X., Wang, W., Chen, M., Jiang, S., Cui, C., Tang, H., Yang, L., Guo, R.: Highly stretchable, highly sensitive, and antibacterial electrospun nanofiber strain sensors with low detection limit and stable CNT/MXene/CNT sandwich conductive layers for human motion detection. Ind. Eng. Chem. Res. **62**(21), 8327–8338 (2023)
31. Heikenfeld, J., Jajack, A., Rogers, J., Gutruf, P., Tian, L., Pan, T., Li, R., Khine, M., Kim, J., Wang, J., Wang, J.: Wearable sensors: modalities, challenges, and prospects. Lab Chip **18**(2), 217–248 (2018). https://doi.org/10.1039/C7LC00914C
32. Huang, D., Li, H., Liu, W., Tan, X., Zhao, M., Liu, G.: Engineering metal oxide heterostructures derived from MOFs/MXene hybrids as efficient acetone sensor. J. Alloy. Compd. **985**, 174114 (2024). https://doi.org/10.1016/j.jallcom.2024.174114
33. Hussain, Z., Imran, M., Afzal, A.M., Iqbal, M.W., Mumtaz, S., Munnaf, S.A., Wabaidur, S.M., Fatima, W., Safdar, S., Mumtaz, M.A., Waris, M.H., Ahmad, Z.: High performance MnNbS@MXene hybrid electrode for battery-supercapacitor hybrid device and biomedical applications. Mat. Sci. Semicond. Proc. **173**, 108091 (2024). https://doi.org/10.1016/j.mssp.2023.108091
34. Iqbal, A., Hong, J., Ko, T.Y., Koo, C.M.: Improving oxidation stability of 2D MXenes: synthesis, storage media, and conditions. Nano Converg. **8**(1), 9 (2021). https://doi.org/10.1186/s40580-021-00259-6
35. Jia, J., Sun, X., Lin, X., Shen, X., Mai, Y.W., Kim, J.K.: Exceptional electrical conductivity and fracture resistance of 3D interconnected graphene foam/epoxy composites. ACS Nano **8**(6), 5774–5783 (2014). https://doi.org/10.1021/nn500590g
36. Jimmy, J., Kandasubramanian, B.: MXene functionalized polymer composites: synthesis and applications. Eur. Polymer J. **122**, 109367 (2020). https://doi.org/10.1016/j.eurpolymj.2019.109367
37. Ju, X., Yang, Z., Duan, X., Zhu, H., Cui, B., Wang, J., Du, Y., Liu, Y., Wang, D.: The new emerging MXene/LDHs nanocomposites: a critical review of synthesis strategies, morphological regulation, and application prospects. Appl. Mater. Today **34**, 101908 (2023). https://doi.org/10.1016/j.apmt.2023.101908
38. Kalaj, M., Bentz, K.C., Ayala, S., Jr., Palomba, J.M., Barcus, K.S., Katayama, Y., Cohen, S.M.: MOF-polymer hybrid materials: from simple composites to tailored architectures. Chem. Rev. **120**(16), 8267–8302 (2020). https://doi.org/10.1021/acs.chemrev.9b00575
39. Kalambate, P.K., Gadhari, N.S., Li, X., Rao, Z., Navale, S.T., Shen, Y., et al.: Recent advances in MXene–based electrochemical sensors and biosensors. TrAC, Trends Anal. Chem. **120**, 115643 (2019). https://doi.org/10.1016/j.trac.2019.115643

40. Kasprzak, D., Mayorga-Martinez, C.C., Alduhaish, O., Pumera, M.: Wearable and flexible all-solid-state supercapacitor based on MXene and Chitin. Energy Technol. **11**(3), 2201103 (2023). https://doi.org/10.1002/ente.202201103
41. Khan, R., Andreescu, S.: MXenes-based bioanalytical sensors: design, characterization, and applications. Sensors **20**(18), 5434 (2020). https://doi.org/10.3390/s20185434
42. Khan, M.U., Du, L., Fu, S., Wan, D., Bao, Y., Feng, Q., Grasso, S., Hu, C.: Preparations and applications of MXene–metal composites: a review. Coatings **12**(4), 516 (2022). https://doi.org/10.3390/coatings12040516
43. Khazaei, M., Ranjbar, A., Arai, M., Sasaki, T., Yunoki, S.: Electronic properties and applications of MXenes: a theoretical review. J. Mater. Chem. C **5**(10), 2488–2503 (2017). https://doi.org/10.1039/C7TC00140A
44. Khorsandi, D., Yang, J.W., Ülker, Z., Bayraktaroğlu, K., Zarepour, A., Iravani, S., Khosravi, A.: MXene-based nano (bio) sensors for the detection of biomarkers: a move towards intelligent sensors. Microchem. J. **109874** (2023). https://doi.org/10.1016/j.microc.2023.109874
45. Kim, J., Campbell, A.S., Wang, J.: Wearable non-invasive epidermal glucose sensors: a review. Talanta **177**, 163–170b (2018a). https://doi.org/10.1016/j.talanta.2017.08.077
46. Kim, J., Sempionatto, J.R., Imani, S., Hartel, M.C., Barfidokht, A., Tang, G., Campbell, A.S., Mercier, P.P., Wang, J.: Simultaneous monitoring of sweat and interstitial fluid using a single wearable biosensor platform. Adv. Sci. **5**(10), 1800880a (2018b). https://doi.org/10.1002/advs.201800880
47. Laochai, T., Yukird, J., Promphet, N., Qin, J., Chailapakul, O., Rodthongkum, N.: Non-invasive electrochemical immunosensor for sweat cortisol based on L-cys/AuNPs/MXene modified thread electrode. Biosens. Bioelectron. **203**, 114039 (2022). https://doi.org/10.1016/j.bios.2022.114039
48. Lee, Y., Biswas, S., Koo, M., Kim, H.: Recent development in biocompatible biosensors. J. Sens. Sci. Technol. **32**(6), 403–411 (2023). https://doi.org/10.46670/JSST.2023.32.6.403
49. Lee, I.C., Li, Y.C.E., Thomas, J.L., Lee, M.H., Lin, H.Y.: Recent advances using MXenes in biomedical applications. Mater. Horiz. **11**(4), 876–902 (2024). https://doi.org/10.1039/D3MH01588B
50. Lei, Y., Zhao, W., Zhang, Y., Jiang, Q., He, J.H., Baeumner, A.J., Wolfbeis, O.S., Wang, Z.L., Salama, K.N., Alshareef, H.N.: A MXene-based wearable biosensor system for high-performance in vitro perspiration analysis. Small **15**(19), 1901190 (2019). https://doi.org/10.1002/smll.201901190
51. Li, H., Chen, X., Zalnezhad, E., Hui, K.N., Hui, K.S., Ko, M.J.: 3D hierarchical transition-metal sulfides deposited on MXene as binder-free electrode for high-performance supercapacitors. J. Ind. Eng. Chem. **82**, 309–316 (2020). https://doi.org/10.1016/j.jiec.2019.10.028
52. Li, F., Duan, X.: Applications of layered double hydroxides. In: Duan, X., Evans, D.G. (eds.) Layered Double Hydroxides. Structure and Bonding, p. 119. Springer, Berlin, Heidelberg (2005). https://doi.org/10.1007/430_007
53. Liu, Y., He, Y., Vargun, E., Plachy, T., Saha, P., Cheng, Q.: 3D porous Ti_3C_2 MXene/NiCo-MOF composites for enhanced lithium storage. Nanomaterials **10**(4), 695 (2020). https://doi.org/10.3390/nano10040695
54. Liu, T., Liu, X., Graham, N., Yu, W., Sun, K.: Two-dimensional MXene incorporated graphene oxide composite membrane with enhanced water purification performance. J. Membr. Sci. **593**, 117431 (2020). https://doi.org/10.1016/j.memsci.2019.117431
55. Liu, Y., Qiu, R., Zhang, Z., Chen, D., Gao, Y., Liu, Z., Li, H., Wang, C.: Label-free electrochemical biosensor based on GR5 DNAzyme/Ti3C2Tx MXenes for Pb2+ detection. J. Electroanal. Chem. **905**, 115979 (2022). https://doi.org/10.1016/j.jelechem.2021.115979
56. Liu, S., Tian, X., Zhang, X., Xu, C., Wang, L., Xia, Y.: A green MXene-based organohydrogel with tunable mechanics and freezing tolerance for wearable strain sensors. Chin. Chem. Lett. **33**(4), 2205–2211 (2022). https://doi.org/10.1016/j.cclet.2021.09.063

57. Liu, X., Xu, F., Li, Z., Liu, Z., Yang, W., Zhang, Y., Fan, H., Yang, H.Y.: Design strategy for MXene and metal chalcogenides/oxides hybrids for supercapacitors, secondary batteries and electro/photocatalysis. Coord. Chem. Rev. **464**, 214544 (2022). https://doi.org/10.1016/j.ccr.2022.214544
58. Liu, J., Zhang, H.B., Sun, R., Liu, Y., Liu, Z., Zhou, A., Yu, Z.Z.: Hydrophobic, flexible, and lightweight MXene foams for high-performance electromagnetic-interference shielding. Adv. Mater. **29**(38), 1702367 (2017). https://doi.org/10.1002/adma.201702367
59. Liu, Y., Zhou, W., Teo, W.L., Wang, K., Zhang, L., Zeng, Y., Zhao, Y.: Covalent-organic-framework-based composite materials. Chem **6**(12), 3172–3202 (2020). https://doi.org/10.1016/j.chempr.2020.08.021
60. Liu, Y.T., Zhu, X.D., Pan, L.: Hybrid architectures based on 2D MXenes and low-dimensional inorganic nanostructures: methods, synergies, and energy-related applications. Small **14**(51), 1803632 (2018). https://doi.org/10.1002/smll.201803632
61. Liu, L., Zhu, M., Shi, Y., Xu, X., Ma, Z., Yu, B., Fu, S., Huang, G., Wang, H., Song, P.: Functionalizing MXene towards highly stretchable, ultratough, fatigue-and fire-resistant polymer nanocomposites. Chem. Eng. J. **424**, 130338 (2021). https://doi.org/10.1016/j.cej.2021.130338
62. Liu, W., Li, H., Huang, D., Tan, X., Zhao, M., Cheng, Q., Yi, M., Ren, Y., Liu, G.: Engineering α-MoO3/TiO2 heterostructures derived from MOFs/MXene hybrids for high-performance triethylamine sensor. Chem. Eng. J. **149340** (2024). https://doi.org/10.1016/j.cej.2024.149340
63. Lorencova, L., Bertok, T., Filip, J., Jerigova, M., Velic, D., Kasak, P., Mahmoud, K.A., Tkac, J.: Highly stable Ti_3C_2Tx (MXene)/Pt nanoparticles-modified glassy carbon electrode for H_2O_2 and small molecules sensing applications. Sens. Actuators B Chem. **263**, 360–368 (2018). https://doi.org/10.1016/j.snb.2018.02.124
64. Lu, C., Yang, L., Yan, B., Sun, L., Zhang, P., Zhang, W., Sun, Z.: Nitrogen-doped Ti3C2 MXene: mechanism investigation and electrochemical analysis. Adv. Func. Mater. **30**(47), 2000852 (2020). https://doi.org/10.1002/adfm.202000852
65. Lu, X., Zhu, J., Wu, W., Zhang, B.: Hierarchical architecture of PANI@ TiO2/Ti3C2Tx ternary composite electrode for enhanced electrochemical performance. Electrochim. Acta **228**, 282–289 (2017). https://doi.org/10.1016/j.electacta.2017.01.025
66. Luo, W., Ma, Y., Li, T., Thabet, H.K., Hou, C., Ibrahim, M.M., El-Bahy, S.M., Xu, B.B., Guo, Z.: Overview of MXene/conducting polymer composites for supercapacitors. J. Energy Storage **52**, 105008 (2022). https://doi.org/10.1016/j.est.2022.105008
67. Ma, C., Ma, M.G., Si, C., Ji, X.X., Wan, P.: Flexible MXene-based composites for wearable devices. Adv. Func. Mater. **31**(22), 2009524 (2021). https://doi.org/10.1002/adfm.202009524
68. Ma, Z., Zhou, X., Deng, W., Lei, D., Liu, Z.: 3D porous MXene (Ti3C2)/reduced graphene oxide hybrid films for advanced lithium storage. ACS Appl. Mater. Interfaces **10**(4), 3634–3643 (2018). https://doi.org/10.1021/acsami.7b17386
69. Magesh, V., Sundramoorthy, A.K., Ganapathy, D., Atchudan, R., Arya, S., Alshgari, R.A., Aljuwayid, A.M.: Palladium hydroxide (Pearlman's Catalyst) doped MXene (Ti3C2Tx) composite modified electrode for selective detection of nicotine in human sweat. Biosensors **13**(1), 54 (2022). https://doi.org/10.3390/bios13010054
70. Manikandan, N., VP, S.K., Rathis, G., Shabariganesh, T.K.: Carbon nanotubes and their properties-the review. Mater. Today: Proc. **47**, 4682–4685 (2021). https://doi.org/10.1016/j.matpr.2021.05.543
71. Mathew, A.E., Jose, S., Babu, A.M., Varghese, A.: State of the art MOF-composites and MXene-composites: synthesis, fabrication and diverse applications. Materials Today Chemistry **36**, 101927 (2024). https://doi.org/10.1016/j.mtchem.2024.101927
72. McDaniel, R.M., Carey, M.S., Wilson, O.R., Barsoum, M.W., Magenau, A.J.: Well-dispersed nanocomposites using covalently modified, multilayer, 2D titanium carbide (MXene) and in-situ "Click" polymerization. Chem. Mater. **33**(5), 1648–1656 (2021). https://doi.org/10.1021/acs.chemmater.0c03972
73. Min, J., Sempionatto, J.R., Teymourian, H., Wang, J., Gao, W.: Wearable electrochemical biosensors in North America. Biosens. Bioelectron. **172**, 112750 (2021). https://doi.org/10.1016/j.bios.2020.112750

74. Miranda, A., Halim, J., Barsoum, M.W., Lorke, A.: Electronic properties of freestanding Ti3C2Tx MXene monolayers. Appl. Phys. Lett. **108**(3) (2016). https://doi.org/10.1063/1.4939971
75. Mohajer, F., Ziarani, G.M., Badiei, A., Iravani, S., Varma, R.S.: MXene-carbon nanotube composites: properties and applications. Nanomaterials **13**(2), 345 (2023). https://doi.org/10.3390/nano13020345
76. Mohanapriya, D., Satija, J., Senthilkumar, S., Ponnusamy, V.K., Thenmozhi, K.: Design and engineering of 2D MXenes for point-of-care electrochemical detection of bioactive analytes and environmental pollutants. Coord. Chem. Rev. **507**, 215746 (2024). https://doi.org/10.1016/j.ccr.2024.215746
77. Myndrul, V., Coy, E., Babayevska, N., Zahorodna, V., Balitskyi, V., Baginskiy, I., Gogotsi, O., Bechelany, M., Giardi, M.T., Iatsunskyi, I.: MXene nanoflakes decorating ZnO tetrapods for enhanced performance of skin-attachable stretchable enzymatic electrochemical glucose sensor. Biosens. Bioelectron. **207**, 114141 (2022). https://doi.org/10.1016/j.bios.2022.114141
78. Naguib, M., Barsoum, M.W., Gogotsi, Y.: Ten years of progress in the synthesis and development of MXenes. Adv. Mater. **33**(39), 2103393 (2021). https://doi.org/10.1002/adma.202103393
79. Nguyen, T., Boudard, M., Carmezim, M.J., Montemor, M.F.: Layered $Ni(OH)_2$–$Co(OH)_2$ films prepared by electrodeposition as charge storage electrodes for hybrid supercapacitors. Sci. Rep. **7**, 39980 (2017). https://doi.org/10.1038/srep39980
80. Niu, H., Yang, X., Wang, Q., Jing, X., Cheng, K., Zhu, K., et al.: Electrostatic self-assembly of MXene and edge-rich CoAl layered double hydroxide on molecular-scale with superhigh volumetric performances. J. Energy Chem. **46**, 105–113 (2020). https://doi.org/10.1016/j.jechem.2019.10.023
81. PAC (1996) 68, 2287. 'Glossary of basic terms in polymer science (IUPAC Recommendations 1996)' on page 2289. https://doi.org/10.1351/pac199668122287
82. Pan, Y., He, M., Wu, J., Qi, H., Cheng, Y.: One-step synthesis of MXene-functionalized PEDOT: PSS conductive polymer hydrogels for wearable and noninvasive monitoring of sweat glucose. Sens. Actuators B Chem. **401**, 135055 (2024). https://doi.org/10.1016/j.snb.2023.135055
83. Parihar, A., Yadav, S., Sadique, M.A., Ranjan, P., Kumar, N., Singhal, A., et al.: Internet-of-medical-things integrated point-of-care biosensing devices for infectious diseases: toward better preparedness for futuristic pandemics. Bioeng. Transl. Med. **8**(3), e10481 (2023). https://doi.org/10.1002/btm2.10481
84. Prakash, N.J., Kandasubramanian, B.: Nanocomposites of MXene for industrial applications. J. Alloy Compd. **862**, 158547 (2021). https://doi.org/10.1016/j.jallcom.2020.158547
85. Qi, F., Wang, L., Zhang, Y., Ma, Z., Qiu, H., Gu, J.: Robust Ti3C2Tx MXene/starch derived carbon foam composites for superior EMI shielding and thermal insulation. Mater. Today Phys. **21**, 100512 (2021). https://doi.org/10.1016/j.mtphys.2021.100512
86. Qu, Y., Shi, C., Cao, H., Wang, Y.: Synthesis of Ni-MOF/$Ti_3C_2T_x$ hybrid nanosheets via ultrasonific method for supercapacitor electrodes. Mater. Lett. **280**, 128526 (2020). https://doi.org/10.1016/j.matlet.2020.128526
87. Rakhi, R.B., Nayak, P., Xia, C., Alshareef, H.N.: Novel amperometric glucose biosensor based on MXene nanocomposite. Sci. Rep. **6**(1), 36422 (2016). https://doi.org/10.1038/srep36422
88. Rana, A.K., Gupta, V.K., Hart, P., Scarpa, F., Thakur, V.K.: Sustainable MXene-chitosan/chitin composites for Interdisciplinary applications in water purification, bio-medical, bio-sensing and electronic fields. Mater. Today Sustain. **100671**,(2024). https://doi.org/10.1016/j.mtsust.2024.100671
89. Ray, T., Choi, J., Reeder, J., Lee, S.P., Aranyosi, A.J., Ghaffari, R., et al.: Soft, skin-interfaced wearable systems for sports science and analytics. Curr. Opin. Biomed. Eng. **9**, 47–56 (2019). https://doi.org/10.1016/j.cobme.2019.01.003
90. Riazi, H., Anayee, M., Hantanasirisakul, K., Shamsabadi, A.A., Anasori, B., Gogotsi, Y., Soroush, M.: Surface modification of a MXene by an aminosilane coupling agent. Adv. Mater. Interfaces **7**(6), 1902008 (2020). https://doi.org/10.1002/admi.201902008

91. Ronchi, R.M., Arantes, J.T., Santos, S.F.: Synthesis, structure, properties and applications of MXenes: Current status and perspectives. Ceram. Int. **45**(15), 18167–18188 (2019). https://doi.org/10.1016/j.ceramint.2019.06.114
92. Safari, J., Masouleh, S.F., Zarnegar, Z., Najafabadi, A.E.: Water-dispersible Fe_3O_4 nanoparticles stabilized with a biodegradable amphiphilic copolymer. C. R. Chim. **17**(2), 151–155 (2014). https://doi.org/10.1016/j.crci.2013.08.009
93. Saleh, A., Wustoni, S., Bihar, E., El-Demellawi, J.K., Zhang, Y., Hama, A., Druet, V., Yudhanto, A., Lubineau, G., Alshareef, H.N., Inal, S.: Inkjet-printed Ti3C2Tx MXene electrodes for multimodal cutaneous biosensing. J. Phys.: Mater. **3**(4), 044004 (2020). https://doi.org/10.1088/2515-7639/abb361
94. San Nah, J., Barman, S.C., Zahed, M.A., Sharifuzzaman, M., Yoon, H., Park, C., Yoon, S., Zhang, S., Park, J.Y.: A wearable microfluidics-integrated impedimetric immunosensor based on Ti3C2Tx MXene incorporated laser-burned graphene for noninvasive sweat cortisol detection. Sens. Actuators B Chem. **329**, 129206 (2021). https://doi.org/10.1016/j.snb.2020.129206
95. Shekhirev, M., Shuck, C.E., Sarycheva, A., Gogotsi, Y.: Characterization of MXenes at every step, from their precursors to single flakes and assembled films. Prog. Mater Sci. **120**, 100757 (2021). https://doi.org/10.1016/j.pmatsci.2020.100757
96. Song, P., Liu, B., Qiu, H., Shi, X., Cao, D.: MXenes for polymer matrix electromagnetic interference shielding composites: a review. Compos. Commun. **24**, 100653 (2021). https://doi.org/10.1016/j.coco.2021.100653
97. Sun, Y., Ding, R., Hong, S.Y., Lee, J., Seo, Y.K., Nam, J.D., Suhr, J.: MXene-xanthan nanocomposite films with layered microstructure for electromagnetic interference shielding and Joule heating. Chem. Eng. J. **410**, 128348 (2021). https://doi.org/10.1016/j.cej.2020.128348
98. Talreja, N., Ashfaq, M., Chauhan, D., Mangalaraja, R.V.: Cu-MXene: a potential biocide for the next-generation biomedical application. Mater. Chem. Phys. **294**, 127029 (2023). https://doi.org/10.1016/j.matchemphys.2022.127029
99. Tao, Y., Yi, K., Wang, H., Kim, H.W., Li, K., Zhu, X., Li, M.: CRISPR-Cas12a-regulated DNA adsorption and metallization on MXenes as enhanced enzyme mimics for sensitive colorimetric detection of hepatitis B virus DNA. J. Colloid Interface Sci. **613**, 406–414 (2022). https://doi.org/10.1016/j.jcis.2022.01.038
100. Truong, P. L., Yin, Y., Lee, D., Ko, S.H.: Advancement in COVID-19 detection using nanomaterial-based biosensors. In: Exploration, vol. 3, no. 1, p. 20210232 (2023). https://doi.org/10.1002/EXP.20210232
101. Ulas, B., Çetin, T., Kaya, Ş, Akinay, Y., Kivrak, H.: Novel Ti3C2X2 MXene supported BaMnO3 nanoparticles as hydrazine electrooxidation catalysts. Int. J. Hydrog. Energy **58**, 726–736 (2024). https://doi.org/10.1016/j.ijhydene.2024.01.280
102. Vikrant, K., Bhardwaj, N., Bhardwaj, S.K., Kim, K.-H., Deep, A.: Nanomaterials as efficient platforms for sensing DNA. Biomaterials **214**, 119215 (2019). https://doi.org/10.1016/j.biomaterials.2019.05.026
103. Wan, B., Liu, N., Zhang, Z., Fang, X., Ding, Y., Xiang, H., He, Y., Liu, M., Lin, X., Tang, J., Li, Y., Tang, B., Zhou, G.: Water-dispersible and stable polydopamine coated cellulose nanocrystal-MXene composites for high transparent, adhesive and conductive hydrogels. Carbohyd. Polym. **314**, 120929 (2023). https://doi.org/10.1016/j.carbpol.2023.120929
104. Wan, S., Li, X., Chen, Y., Liu, N., Du, Y., Dou, S., Jiang, L., Cheng, Q.: High-strength scalable MXene films through bridging-induced densification. Science **374**(6563), 96–99 (2021). https://doi.org/10.1126/science.abg2026
105. Wang, J., Gong, J., Zhang, H., Lv, L., Liu, Y., Dai, Y.: Construction of hexagonal nickel-cobalt oxide nanosheets on metal-organic frameworks based on MXene interlayer ion effect for hybrid supercapacitors. J. Alloy. Compd. **870**, 159466 (2021). https://doi.org/10.1016/j.jallcom.2021.159466
106. Wang, S., Wei, S., Wang, S., Zhu, X., Lei, C., Huang, Y., Nie, Z., Yao, S.: Chimeric DNA-functionalized titanium carbide MXenes for simultaneous mapping of dual cancer biomarkers in living cells. Anal. Chem. **91**(2), 1651–1658 (2018). https://doi.org/10.1021/acs.analchem.8b05343

107. Wang, Q.W., Zhang, H.B., Liu, J., Zhao, S., Xie, X., Liu, L., Yang, R., Koratkar, N., Yu, Z.Z.: Multifunctional and water-resistant MXene-decorated polyester textiles with outstanding electromagnetic interference shielding and joule heating performances. Adv. Func. Mater. **29**(7), 1806819 (2019c). https://doi.org/10.1002/adfm.201806819
108. Wang, H., Li, H., Huang, Y., Xiong, M., Wang, F., & Li, C.: A label-free electrochemical biosensor for highly sensitive detection of gliotoxin based on DNA nanostructure/MXene nanocomplexes. Biosens. Bioelectron. **142**, 111531 (2019a). https://doi.org/10.1016/j.bios.2019.111531
109. Wang, H., Zhao, R., Qin, J., Hu, H., Fan, X., Cao, X., Wang, D.: MIL-100 (Fe)/Ti3C2 MXene as a Schottky catalyst with enhanced photocatalytic oxidation for nitrogen fixation activities. ACS Appl. Mater. Interfaces **11**(47), 44249–44262 (2019b). https://doi.org/10.1021/acsami.9b14793
110. Wang, J.: Nanomaterial-based amplified transduction of biomolecular interactions. Small **1**(11), 1036–1043 (2005). https://doi.org/10.1002/smll.200500214
111. Wang, W., Wang, C., Yuen, A.C.Y., Li, A., Lin, B., Yuan, Y., Ma, C., Han, Y., Yeoh, G.H.: 3D MXene frameworks for flame retardant hydrophobic polymer nanocomposites. Compos. A Appl. Sci. Manuf. **173**, 107673 (2023). https://doi.org/10.1016/j.compositesa.2023.107673
112. Wen, Y., Wei, Z., Ma, C., Xing, X., Li, Z., Luo, D.: MXene boosted CoNi-ZIF-67 as highly efficient electrocatalysts for oxygen evolution. Nanomaterials **9**(5), 775 (2019). https://doi.org/10.3390/nano9050775
113. Wen, Q., Yang, M.: Platinum nanoparticles modified MXene for highly sensitive detection of pyruvate. Mater. Lett. **322**(March), 1–4 (2022). https://doi.org/10.1016/j.matlet.2022.132496
114. Wu, S., Wang, H., Li, L., Guo, M., Qi, Z., Zhang, Q., Zhou, Y.: Intercalated MXene-based layered composites: preparation and application. Chin. Chem. Lett. **31**(4), 961–968 (2020). https://doi.org/10.1016/j.cclet.2020.02.046
115. Xie, L.S., Skorupskii, G., Dinca, M.: Electrically conductive metal–organic frameworks. Chem. Rev. **120**(16), 8536–8580 (2020). https://doi.org/10.1021/acs.chemrev.9b00766
116. Xing, C., Chen, S., Liang, X., Liu, Q., Qu, M., Zou, Q., Li, J., Tan, H., Liu, L., Fan, D., Zhang, H.: Two-dimensional MXene (Ti3C2)-integrated cellulose hydrogels: toward smart three-dimensional network nanoplatforms exhibiting light-induced swelling and bimodal photothermal/chemotherapy anticancer activity. ACS Appl. Mater. Interfaces **10**(33), 27631–27643 (2018). https://doi.org/10.1021/acsami.8b08314
117. Xiong, S., He, T., Zhou, T., Wang, D., Wang, Y., Dai, C., Liu, W.: In situ synthesis of MXene/Ag nanocomposites based flexible SERS substrates on PDMS for detection on fruit surfaces. Colloids Surf. A **654**, 130077 (2022). https://doi.org/10.1016/j.colsurfa.2022.130077
118. Xu, J., Li, Y., Yan, F.: Constructed MXene matrix composites as sensing material and its applications: a review. Anal. Chim. Acta **342027**,(2023). https://doi.org/10.1016/j.aca.2023.342027
119. Yang, J., Bao, W., Jaumaux, P., Zhang, S., Wang, C., Wang, G.: MXene-based composites: synthesis and applications in rechargeable batteries and supercapacitors. Adv. Mater. Interfaces **6**(8), 1802004 (2019b). https://doi.org/10.1002/admi.201802004
120. Yang, W., Liu, J.J., Wang, L.L., Wang, W., Yuen, A.C.Y., Peng, S., Yu, B., Lu, H., Yeoh, G.H., Wang, C.H.: Multifunctional MXene/natural rubber composite films with exceptional flexibility and durability. Compos. B Eng. **188**, 107875 (2020). https://doi.org/10.1016/j.compositesb.2020.107875
121. Yang, J.C., Mun, J., Kwon, S.Y., Park, S., Bao, Z., Park, S.: Electronic skin: recent progress and future prospects for skin-attachable devices for health monitoring, robotics, and prosthetics. Adv. Mater. **31**(48), 1904765 (2019a). https://doi.org/10.1002/adma.201904765
122. Yang, H., Zhang, G.X., Zhou, H.J., Sun, Y.Y., Pang, H.: Metal–organic frameworks meet MXene: new opportunities for electrochemical application. Energy Mater. Adv. **4**, 0033 (2023a). https://doi.org/10.34133/energymatadv.0033
123. Yang, L., Zheng, W., Zhang, P., Chen, J., Tian, W.B., Zhang, Y.M., Sun, Z.M.: MXene/CNTs films prepared by electrophoretic deposition for supercapacitor electrodes. J. Electroanal. Chem. **830**, 1–6 (2018). https://doi.org/10.1016/j.jelechem.2018.10.024

124. Yang, Z., Chen, Y., Chen, G., Wang, J., Li, H.: Ultrathin Ti3C2Tx MXene sheets with high electrochemically active area anchored Pt boosting hydrogen evolution. Heliyon **9**(9) (2023b). https://doi.org/10.1016/j.heliyon.2023.e19197.
125. Ye, C., Wang, M., Min, J., Tay, R.Y., Lukas, H., Sempionatto, J.R., Li, J., Xu, C., Gao, W.: A wearable aptamer nanobiosensor for non-invasive female hormone monitoring. Nat. Nanotechnol. **1–8**,(2023). https://doi.org/10.1038/s41565-023-01513-0
126. Yu, Z., Deng, C., Ding, C., Zhang, X., Liu, Y., Liu, C., Lou, Z., Seidi, F., Han, J., Yong, Q., Xiao, H.: Organic-inorganic hybrid ZIF-8/MXene/cellulose-based textiles with improved antibacterial and electromagnetic interference shielding performance. Int. J. Biol. Macromol. **266**, 131080 (2024). https://doi.org/10.1016/j.ijbiomac.2024.131080
127. Yu, Z., Jiang, L., Liu, R., Zhao, W., Yang, Z., Zhang, J., Jin, S.: Versatile self-assembled MXene-Au nanocomposites for SERS detection of bacteria, antibacterial and photothermal sterilization. Chem. Eng. J. **426**, 131914 (2021). https://doi.org/10.1016/j.cej.2021.131914
128. Zhan, X., Si, C., Zhou, J., Sun, Z.: MXene and MXene-based composites: synthesis, properties and environment-related applications. Nanoscale Horizons **5**(2), 235–258 (2020). https://doi.org/10.1039/C9NH00571D
129. Zhang, Z., Duan, X., Jia, D., Zhou, Y., van der Zwaag, S.: On the formation mechanisms and properties of MAX phases: a review. J. Eur. Ceram. Soc. **41**(7), 3851–3878 (2021b). https://doi.org/10.1016/j.jeurceramsoc.2021.02.002
130. Zhang, W., Qi, J., Cao, T., Lei, Z., Ma, Y., Liu, H., Zhu, L., Feng, X., Wei, W., Zhang, H.: A facile method synthesizing marshmallow ZnS grown on Ti3C2 MXene for high-performance asymmetric supercapacitors. J. Energy Storage **50**, 104652 (2022). https://doi.org/10.1016/j.est.2022.104652
131. Zhang, Y., Wang, Z., Liu, X., Liu, Y., Cheng, Y., Cui, D., Chen, F., Cao, W.: A MXene/MoS2 heterostructure based biosensor for accurate sweat ascorbic acid detection. FlatChem **39**, 100503 (2023b). https://doi.org/10.1016/j.flatc.2023.100503
132. Zhang, S., Zahed, M.A., Sharifuzzaman, M., Yoon, S., Hui, X., Barman, S.C., Sharma, S., Yoon, H.S., Park, C., Park, J.Y.: A wearable battery-free wireless and skin-interfaced microfluidics integrated electrochemical sensing patch for on-site biomarkers monitoring in human perspiration. Biosens. Bioelectron. **175**, 112844 (2021a). https://doi.org/10.1016/j.bios.2020.112844
133. Zhang, C., Zheng, X., Zhao, C., Li, Y., Chen, S., Liu, G., Wang, C., Lv, Q., Liu, P., Zheng, Y., Kong, D., Jiang, H., Jiang, Y.: Detection of pathogenic microorganisms from bloodstream infection specimens using TaqMan array card technology. Sci. Rep. **8**,(2018). https://doi.org/10.1038/s41598-018-31200-3
134. Zhang, W., Xing, X.J., Ming, G.M.: Emerging MXene/cellulose composites: design strategies and diverse applications. Chem. Eng. J. **458**(December 2022) (2023a). https://doi.org/10.1016/j.cej.2023.141402
135. Zheng, Z., Guo, C., Wang, E., He, Z., Liang, T., Yang, T., Hou, X.: The oxidation and thermal stability of two-dimensional transition metal carbides and/or carbonitrides (MXenes) and the improvement based on their surface state. Inorg. Chem. Front. **8**(9), 2164–2182 (2021). https://doi.org/10.1039/D1QI00041A
136. Zhou, J., Pei, Z., Sui, Z., Liang, Y., Xu, X., Li, Y., Li, Y., Qiu, J., Chen, Q.: Hierarchical porous and three-dimensional MXene/SiO2 hybrid aerogel through a sol-gel approach for lithium-sulfur batteries. Molecules **27**(20), 7073 (2022). https://doi.org/10.3390/molecules27207073
137. Zhuang, X., Zhang, S., Tang, Y., Yu, F., Li, Z., Pang, H.: Recent progress of MOF/MXene-based composites: synthesis, functionality and application. Coord. Chem. Rev. **490**, 215208 (2023). https://doi.org/10.1016/j.ccr.2023.215208

MXene and Its Composites for Industrial Applications

Priyanka P. Mishra, Ajaya K. Behera, Nigamananda Das, Satyanarayan Sahoo, and Ganesh Chandra Nayak

Abstract MXene, after its discovery in 2011 by Yury Gogotsi and Michel Barsoum, are now center of attraction for research due to their potential applications in various industrial sectors. These are the inorganic materials of molecular formula $M_{n+1}X_n$, which consist of atomically thin layers of transition metal (M)-carbides or nitrides or carbonitrides (X). MXene without surface modification are magnetic and metallic, making them suitable for sensor and energy storage applications. These photothermal materials are promising because they absorb a lot of visible light, have corrosion inhibitors, and are antimicrobial. MXene can be used in a variety of industrial applications, including energy storage devices such as Li-ion batteries, energy storage components, supercapacitors, transparent conducting electrodes, gas sensors, neural electrodes, photonic diodes, electrochromic devices, and triboelectric nanogenerators, as well as composites, photocatalysts for dye degradation, and water purification. To capture the present trend of MXene and MXene-based composites, this chapter will critically analyze their prospective industrial applications.

Keywords MXene · Industrial application · Electronic device · Sensor · Ionic battery

P. P. Mishra · A. K. Behera · N. Das
Department of Chemistry, Utkal University, Bhubaneswar, India

S. Sahoo
Department of Chemistry, Berhampur University, Brahmapur, Odisha, India

G. Ch. Nayak (✉)
Department of Chemistry and Chemical Biology, Indian Institute of Technology (ISM) Dhanbad, Dhanbad, Jharkhand, India
e-mail: gcnayak@iitism.ac.in

© The Author(s), under exclusive license to Springer Nature Singapore Pte Ltd. 2025

S. Moharana and R. N. Mahaling (eds.), *Surface Engineering of MXenes*, Engineering Materials, https://doi.org/10.1007/978-981-96-3640-2_14

1 Introduction

In recent times, two-dimensional materials (2D) such as graphitic carbon nitride (g-C_3N_4), graphene oxide (GO), molybdenum disulfide (MoS_2), and transition metal dichalcogenides (TMDs) have engrossed the attention of researchers owing to their intrinsic properties, easy surface functionalization, and large surface area [26]. Due to these materials' superior electrical and mechanical attributes, researchers are further exploring the analysis, synthesis, surface modification, and application of other 2D materials. In this context, Naguib et al. [30] first discovered the MXene materials from the Ti_3AlC_2 MAX phase, as a new diverse 2D material in the field of electrochemical, photochemical, catalysis, sensing application [30]. MXene was first synthesized by the HF etching ternary metal carbides and nitrides denoted by a universal formula for $M_{n+1}AX_n$. In addition, over thirty distinct varieties of MXene were produced using M and X elements in various proportions, displaying a range of capabilities [28].

MXene is a cutting-edge, two-sided layered material comprising transition metal nitrides, carbides, and carbonitrides. It can be categorized into different types, such as $Ti_3C_2T_x$, V_2CT_x, Mo_2CT_x, and $Ti_5C_2T_X$ [33]. The rapid growth of research on MXene materials can be ascribed to their incomparable electrical conductivity, which is the highest among all two-dimensional materials. Furthermore, the surface of MXene materials features many covalent and ionic bonds. Ionic bonds are caused by strong charge attraction, resulting in a closely packed crystal structure. In contrast, covalent bonds are formed by sharing outer electrons to achieve electronic stability. Similarly to graphene, MXene matrixes also exhibit Van der Waals force, which leads to self-stacking and eventually affects their electrochemical performance [44, 45]. However, MXene materials are highly susceptible to oxidation, which can convert them into TiO_2 [53]. Therefore, incorporating other materials into the MXene matrices can enhance the application area by utilizing synergistic effects. The synergistic effect of MXene-based composite material will improve their performances in numerous arenas such as sensing, energy storage, water remediation, EMI shielding, and catalysis.

2 Types of MXene

MXene is classified into different types as reported in Fig. 1, and discussed in the following subsection.

2.1 MXene-Polymer

Since the 1960s, several additives have been developed to improve the polymer matrix's mechanical, electrical, and catalytic performance [40]. Nevertheless, a new

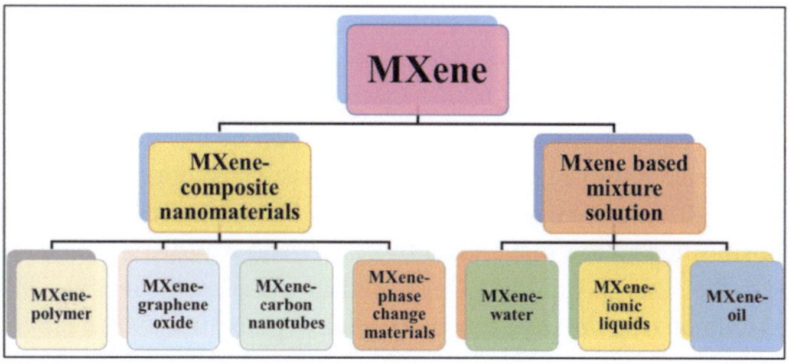

Fig. 1 Classification of MXene

era of polymer composites started with the integration of nanoparticles into the polymer sheets. Clays and other 2D materials, in particular, have significant benefits over their 3D materials due to their large surface area-to-volume ratios along with significant particle dimension ratios. The poor thermal, mechanical, rheological, and gas resistance attributes might be alleviated by dispersing those nanofillers into a polymer matrix [32]. After its discovery, MXene has garnered loads of attention as a nanofiller because of its diverse properties and compositions. The introduction of $Ti_3C_2T_Z$ MXene polymer in 2014, coupled with PVA and PDDA, led to an increase in interest in MXene polymer composites [20]. To improve the mechanical, electrical, and catalytic capabilities of the MXene matrix, other polymer ingredients including polypropylene, PVA, PPy, and PANI were added.

Specifically, Jin et al., synthesized PVA/MXene to introduce multifaceted structures in the MXene layer to increase the thermal conductance. The multidimensional film was provided with excellent EMI SE and TC simultaneously by the MXene layer, which also offered a compact matrix for the transmission of electrons and heat. The composite had an improved electrical resistivity of 716 S/m. In contrast, the multifaceted material had an excessive in-plane temperature factor (TC) of 4.5 W/mK, which was over 23 times higher than the default PVA value. The multifaceted layer of MXene showed excellent anti-dripping efficacy [14]. EMI shielding materials are more appealing than their traditional reflection-dominated equivalents as they reduce the twofold effluence of the reflected electromagnetic (EM) wave. EMI shielding materials such as MXene/PVA composite foams show a shielding efficacy of 5136 dB cm^2 g^{-1}, indicating superb absorption-dominated shielding performance [51]. Electromagnetic interference (EMI) shielding materials based on polymeric materials have excellent properties such as easy processing, superior flexibility, inferior temperature resistance, and low density in comparison to metals. Herein, Lei et al. developed a highly robust, ultrathin, and temperature-steady MXene@ANF film that enhances the mechanical strength and EMI shielding effectiveness up to 8814.5 dB cm^2 g^{-1} and flame retardancy exhibits in the temp. between −1000 °C and 3000 °C. It has also the capability to eliminate electromagnetic waves even in some extreme conditions [18]. To improve MXene's charge storage capacity, Gund et al. incorporated poly (3,4-ethylenedioxythiophene) polystyrene sulfonate (PSS), which offers unparalleled adaptability and endurance up to 30,000 cycles, dependent on the ideal composition and thickness. It is anticipated that these MXene hybrid-based electrochemical capacitors will reduce the effectiveness gap between high frequency and high-capacitance responsiveness in the direction of form-factor-free, modular, and microscopic devices [11]. Introducing robust Ti–O bonds into the MXene's surface without altering the crystal assembly can boost thermal stability. The fabricated composite showed thermal conductivity up to 55.8 W/(m$K), that is greater in comparison to the other metals and 2D materials. Moreover, the increase in the lattice parameter is indicated by the elevation of the peak of phase from 9.6 to 8.6° for MXene and 7.8° for MXene@ PVA [23]. Later on, the introduction of polymeric material through surface modification could also enhance the electrochemical activity. For this context, an MXene@ PANI material has been fabricated via oxidative polymerization to enhance the electrochemical performance through surface modification. The material showed a volume capacitance of 1682.3 F/cm^3 which lasted up to 10,000 cycles [42]. Similarly another polymeric material, polypyrrole has been utilized for the fabrication of MXene-PPy (polypyrrole) composite through the electrochemical polymerization method and determined that the specific capacitance and supercapacitor of the MXene-PPy could reach 458 F/g and 21.61 Wh/kg, respectively, the superior electrochemical performance leads to a steady microstructure and excellent energy storage composite [59, 60].

2.2 *MXene-Graphene Oxide*

The introduction of interfacial interaction to the graphene oxide can improve its conductivity by eliminating the probable accumulation of GO caused by the Vander Waals interaction. The synergistic interaction of MXene and GO results in advancements in many applications which involve stimulating the stability of MXene and protecting it from oxidation. This perilous amendment not only increases the inclusive activity of MXene materials but also confirms their enduring resilience. The MXene/GO composite has been widely utilized in the sensing platform. One significant phenolic environmental pollutant is p-nitrophenol or p-NP. For environmental surveillance, the establishment of new fast assay methods to detect p-NP is essential. The design of an electrochemical sensor based on a $Ti_3C_2T_X$ MXene/Graphene composite has been implemented to detect p-NP in aqueous samples. The presence of intrinsic properties like greater active sites, quick electron migration capability, and superior electro-catalytic activity makes the composite efficient toward the electrochemical responses on the p-NP reduction reaction. The composite showed a linear response in the *p*-NP concentration varying from 1 to 175 $\mu mol \cdot L^{-1}$ having a detection limit of 0.16 $\mu mol \cdot L^{-1}$ having brilliant recyclability [49]. The MXene/GO system has shown great promise as an electrochemical sensing platform because of its enhanced conductivity. For example, MXene/GO composites were built using a top-down methodology to detect bisphenol A electrochemically. The incorporation of MXene molecules is confirmed by the formation of lower oxidation potential at the electrode. This crucial adjustment made the sensor extremely well-suited for the exact detection of bisphenol A [35]. A sustainable approach was used for manufacturing a stacked rGO/MXene-Pd/rGO nanocomposite, which will improve its electrochemical sensing capabilities. Here, the electrocatalyst MXene/N-doped reduced graphene oxide composite was synthesized by Chen et al. for the electrochemical sensing of adrenaline (AD) determination. The doped N in rGO acted as a solid mediator between the GO and MXene through the hydrogen bonding and also acted as a catalyst to oxidize AD. The obtained MXene/N-rGO electrode also demonstrated good stability, repeatability, and reproducibility [7].

2.3 MXene-Metal Nanoparticles

Metal nanoparticles have gained a lot of scrutiny owing to their inimitable electrical, catalytic, optical, and magnetic characteristics. The systematic incorporation of metal nanoparticles into the matrix of MXene resulted in the acceleration of electron migration by minimizing the lamellar arrangements. Yu et al. synthesized MXene-Au nanoparticles using the MXene surface, resulting in a composite that is specifically designed for photothermal sterilization and bacterial detection [56]. For nitrite detection, Feng et al. developed a composite of Au NPs, CS, and MXene. Incorporating CS into the MXene/Au NPs composite improves its film-forming properties while simultaneously promoting biocompatibility [9].

2.4 MXene-Carbon Nanotubes

The mechanical characteristics and alkaline resilience of MXene nanosheets can be improved by incorporating CNTs. For example, after being immersed in alkali solutions for 120 days, three-dimensional (3D) nanofillers of MXene (Ti3C2Tx)/CNTs demonstrated excellent surface roughness and bending strength, along with high stability and better interface adhesion between adjacent fiber. Furthermore, the hybridization of MXene-based membranes with carbon nanotubes (CNTs) has the potential to improve interfacial adhesion. This is because the CNTs' robust π–π interface and van der Waals forces encourage MXene nanosheets to close to improve interfacial bonding forces and anti-swelling properties. To provide appropriate electrical conductivity and facilitate the encapsulation of Cu_2O nanoparticles, Xia et al. proposed CNTs/Cu_2O nanoparticles/MXene ($Ti_3C_2T_x$)-modified electrodes. Alternatively, CNTs were used to increase the susceptibility of these electrochemical sensors, which had adequate stability and a broad linear reactivity range of up to 70 μM and a LOD of 6 nM [50]. To generate solar steam for purifying textile wastewater, Wang et al. developed a unique fabric composed of MXene, CNTs, and Cotton by layer-by-layer fabrication. The effective optical sorption, conversion of light into thermal, and water transport capabilities of the composite fabric were

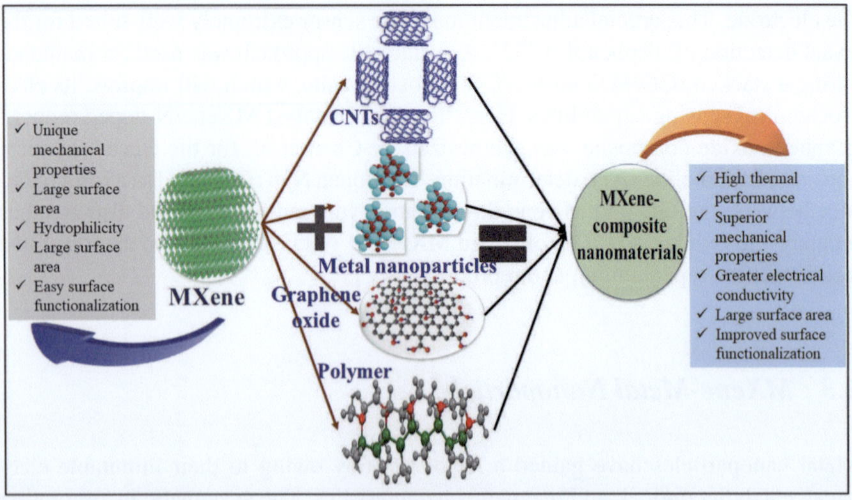

Fig. 2 Schematic diagram of advantages of MXene-composite nanomaterials

facilitated by the high interfacial contacts with excellent recyclability, and textile wastewater evaporation rates exceeded 1.35 kg m^{-2} and >1.16 kg m^{-2} h^{-1} [48, 49]. A schematic presentation of the advantage of MXene-based composites compared to the neat MXene is shown in Fig. 2.

2.5 *MXene-Oil*

To improve thermal efficiency, solar sorption capabilities, and viscosity in a solar PV/T system, olein palm oil (OPO) composite is used as a thermal fluid. Furthermore, owing of their greater surface area and higher thermal conductivity, the 2D components in MXene function as nano-filling agents, facilitating effective heat transfer from fluids. According to Samylingam et al., there is an increase in both liquid layering development and particle mobility leading to a rise in thermal conductivity. A temperature rise can also increase the dynamic viscosity through the hybridization of MXene with olein palm oil. This suggests that MXene@ OPO may find application as a thermal working fluid, particularly at elevated temperature ranges of 60 to 1000 °C [39].

2.6 *MXene-Water*

The thermal and optical conductivity of MXene-water nanofluids can be assessed for photothermal transformation in direct absorption solar collectors. While the graphene nanofluids demonstrated greater thermal conductivity, the MXene nanofluids' optical characteristics are superior because of the MXene nanosheets' LSPR effect. For instance, the photothermal conversion efficacy of MXene nanofluids reached an optimum of 63.35% when the concentration of nanofluid was 20 ppm, which was 4.34% greater than that of graphene nanofluids [48, 49]. The effects of sodium dodecylbenzene sulfonate (SDBS) and cetyltrimethylammonium bromide (CTAB), two distinct types of surfactants, on the absorbance and transmittance spectra rate, were examined by Abdelrazik et al. It is discovered that the transmittance spectra of the MXene nanofluid are significantly influenced by the surfactant type. Furthermore, studies show that MXene disperses more readily in the presence of SDBS and that employing the SDBS surfactant improves the stability of MXene@ water [1].

2.7 MXene-Ionic Liquid

The incorporation of MXene nanoparticles in ionic nanofluids leads to improved thermal conductivity and stability, making them suitable for advanced heat transfer applications.

Nanofluids have garnered considerable attention as greater heat transfer fluids in solar systems over the past several decades. In this study, ionic liquid water nanofluids were articulated with a novel 2D MXene (Ti_3C_2) nanosheet at three varying concentrations (0.05, 0.10, and 0.20 wt %) and their thermophysical and optical properties were measured. The results revealed a significant enhancement of 47% in thermal conductance for the 0.20 wt % MXene loading. Additionally, viscosity was found to increase with the addition of Ti_3C_2 only slightly at varying concentrations. On the other hand, viscosity decreased significantly as the temperature rose from 20 to 60 °C. The IL with water/MXene nanofluid displayed the maximum electrical, thermal, and inclusive heat transfer efficacy than palm oil/MXene, water/alumina, and water. The optimum thermal and electrical efficacy was noted as 81.16% and 13.94% respectively, using the composite. Moreover, the heat transfer constants increased by 12.6% and 2% when compared to water/alumina and palm oil/MXene, respectively [8].

3 Properties of MXene

MXene has inherent qualities that make it an appealing catalyst in the modern world. It can be employed as composites in a variety of industrial applications, regardless of its hydrophilicity or metallic conductivity. Aside from this, their exceptional qualities and distinct structure make them advantageous for a variety of uses, including terahertz shielding, optoelectronics, antibacterial films, flexible and transparent electrodes, thermal heaters, sensors, light-emitting diodes (LEDs), ion battery storage, supercapacitors, water treatment, and additive for lubrication.

3.1 Anti-microbial Efficacy

Multiple 2D materials have been explored to reduce the derogatory impact of microbial agents on the health of human beings. MXene has demonstrated greater antibacterial potency than GO nanosheets. Bacterial cell membrane contact is enhanced by the anionic nature and hydrophilic properties of the MXene surface. The functional groups of MXene moieties form an H-bond with the lipopolysaccharide molecules which causes the inactivation of bacterial cells by inhibiting the uptake of nutrients and consequently obstructs the growth of bacteria. Subsequently, the MXene molecules create a conducting bridge around the lipid molecules which provides the active electrons into the bacterial cell and causes the bacterial cell death. Thus, the greater conducting ability of MXenes makes them superior antimicrobial agents than the GO nanosheets. Similarly, Rasool et al. synthesized a $Ti_3C_2T_x$-PVDF (polyvinylidene fluoride) composite with antimicrobial efficacy. The addition of PVDF material onto the surface of $Ti_3C_2T_x$ increases the aspect ratio, enhances hydrophilicity, and alleviates the existence of large pores in the membrane. There was a lessening in viability and growth inhibition of around 67% B. subtilis, and 73% E. coli [37].

3.2 Mechanical Properties

The mechanical attributes of MXene moieties are significantly influenced by their surface terminus. For instance, compared to OH and F-terminated MXene, the O-terminated MXene exhibits extremely high rigidity. The plausible reason may be the presence of dissimilar lattice coefficients with dissimilar terminations, the O-terminated MXene has smaller lattice constants in comparison to OH and F-modified MXene [58]. Guo et al. explained this by taking the example of Ti_2C, that the modification would decrease Young's modulus of Ti_2C, but the high strain characteristics of Ti_2C will remain higher than bare Ti_2C and even graphene [12]. The breakdown of Ti layers is reduced as the surface terminals act as a buffer layer for Ti_2C under tensile deformation causing the rise in the value of critical strain. The variation in the no. of atomic layers ($n = 1, 2, 3$) in MXene also affects its mechanical attributes. Yorulmaz et al. investigated the elastic constant and Young's modulus of bare MXenes by utilizing classical molecular dynamics. The results showed Ti_2C had the highest Young's modulus between entirely $Ti_{n+1}C_n$. Furthermore, the elastic coefficient of Ti_2C was nearly twofold as high as that of MoS_2 [4].

3.3 Conducting Properties

The researchers investigated the thermal properties of MXene composites based on their conducting properties. They added 5% of single and multi-layer MXenes to ethylene glycol (EG) based nanofluids and examined the thermal physical properties of the mixture. The synthesized composite showed an improvement in thermal conductivity by approximately 64.9% and 53.1%, respectively. MXene also exhibited excellent self-lubricating qualities at volume concentrations below 1%, and the single-layer MXene@EG nanofluid maintained good stability. These enhancements in tensile strength, mechanical properties, self-lubricating properties, and high thermal conductivity make MXene composites highly desirable materials.

3.4 Electronic Properties

MXene possess a diverse range of electronic properties, including metallicity, semiconductivity, and topological insulative, owing to their flexible thickness controllability high compositional variety and numerous surface functionalization potentials. Mostly all the bare and surface-modified MXenes are metallic in nature. The estimated work functions of metallic MXene can fluctuate widely, from 1.8 to 8 eV, depending on their dipole moment [15]. Platinum (Pt) has the highest WFs among all the elements but the WFs of O-terminated MXenes are even higher than Pt. Meanwhile, the WFs of F-terminated MXene vary in between OH- and O-terminated MXene [24].

4 Synthesis of MXene and MXene-Based Composite

The typical procedure to generate MXene involves wet chemically etching MAX precursors using high-temperature sintering and ball milling techniques. The MAX phase has a hexagonal structure representing "M" as the first transition metal, "A" denotes the elements from group IIIA and IVA and X can either be carbon, nitrogen, or both. The general structure of MXene represents covalent/ metallic/ionic bonds between the M-layer and the X-atoms and the A-atom layers interpenetrated between the MX cross-linked layer. However, modifying the surface of MXene by producing an MXene-based composite will improve its chemical composition, thermal physical characteristics, and energy conversion proficiency. Various methods are employed, among which bottom-up and top-down synthesis methods, freeze-drying, direct vacuum impregnation, toluene dilution methods, and two-step techniques, are often utilized in the synthesis procedure. A schematic presentation of synthesis of MXene sheets and MAX phase are shown in Figs. 3 and 4 respectively.

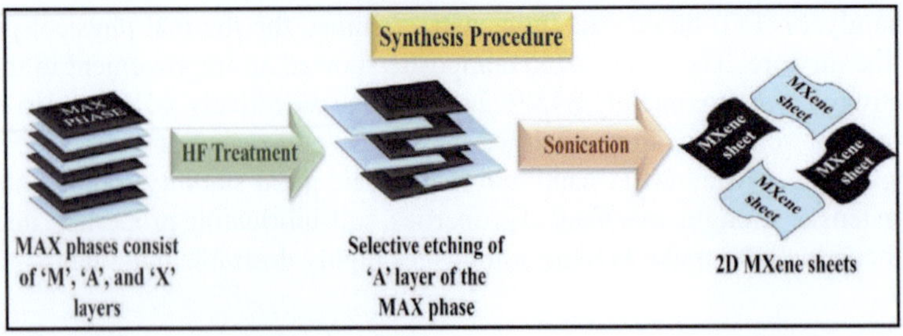

Fig. 3 Synthesis procedure of MXene sheets

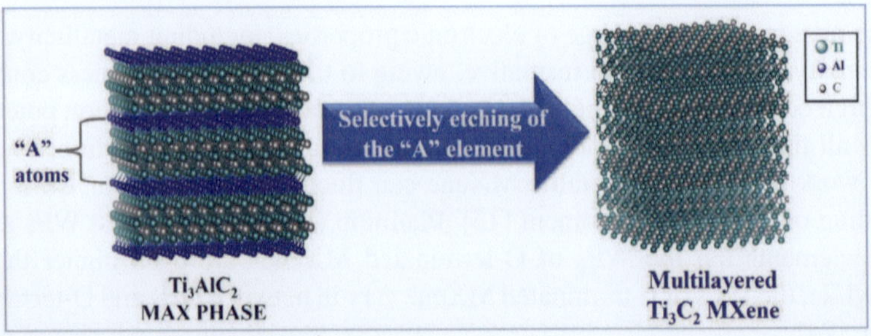

Fig. 4 Schematic presentation of MAX phase

4.1 Bottom-Up Methods

It is noteworthy that the synthesis of MXenes can be attained most efficiently by the use of bottom-up approaches such as chemical vapor deposition (CVD). By combining the sort of substrate, raw ingredients, and reaction conditions, a high-quality batch of MXene sheets can be produced using this process. excellent quality, vast area, huge lateral dimensions, and excellent stability are only a few of the benefits of MXene sheets made via CVD. In this context, Yang et al. employed the CVD technique to synthesize 3D hybrid nanocomposite from the platinum nanoparticles and observed that the composite material exhibits high catalytic efficacy and long-term stability (C. [52]). However, the high reaction time and low yield were the major drawbacks of this method.

4.2 Top-Down Methods

The etching and exfoliation procedure was primarily used in this strategy to synthesize the composite. Several techniques are employed to synthesize MXenes, including, fluoride-salt etching, hydrofluoric (HF) etching, molten salt etching, and fluoride-free etching [57]. Hydrofluoric acid is used in the HF etching process to etch the "A" layer, creating 2D sheets in the process. Subsequently, DMSO is used to delaminate the produced Sheets, transforming MXene's nanocrystal structure into a sheet-like structure. Then the 2D MXene nanosheets interact with the F, OH, O, etc., present in the solution which enhances the bond strengths. The HF-etching method is widely used among researchers and the development of this method is still going on. Additionally, the fluoride-salt etching method which uses diluted hydrochloric acid (HCl) and fluoride-based salts like LiF, CaF, NaF, KF, etc., is a crucial technique for the synthesis of MXene [2]. The intercalation of metal ions into the MXene sheets causes the delamination rapidly. However, scientists are developing an etching method free of fluoride because of the detrimental effects of fluorine, including toxicity problems and low electrical characteristics of MXene [19]. Equations (1) and (2) represent the synthesis process of MXene.

$$M_{n+1}AX_n + 3HF = M_{n+1}X_n + AlF_3 + \frac{3}{2}H_2 \tag{1}$$

$$M_{n+1}X_n + 2HF = M_{n+1}X_nF_2 + H_2 \tag{2}$$

5 Other Synthesis Methods

In addition to the previously discussed two conventional synthesis techniques, other methods have been employed for synthesizing composite-based-MXene nanomaterials (Fig. 5). These include the freeze-dying method, direct vacuum impregnation method, multi-layered casting method, toluene dilution method, and two-step technique. Specifically, an MXene-PDMS-based nanocomposite synthesized via the vacuum-assisted impregnation method. The synthesized composite showed 14 times higher conductivity than the bare PDMS [46–48]. An MXene@PVA material has been developed by freeze-drying with the incorporation of PEG. The resultant composite has 4.2 times enhanced thermal conductivity than the neat PEG sample. The enhancement in the thermal conductivity can be attributed to the 3D structure of the MXene sheet [29]. Furthermore, another MXene@PVA nanocomposite was fabricated via the multilayer casting method to enhance the thermal conductivity. The obtained multidimensional film exhibited thermal conductivity around 4.57 W/mK, which was approximately 23-fold greater in comparison to the pristine PVA moiety [14]. In addition, a SO@MXene hybrid composite was synthesized by using the two-step approach to increase the performance of thermal and electrical efficacy. The electrical output of the resultant composite has increased up to 15.44% compared to the water-alumina nanofluids [38].

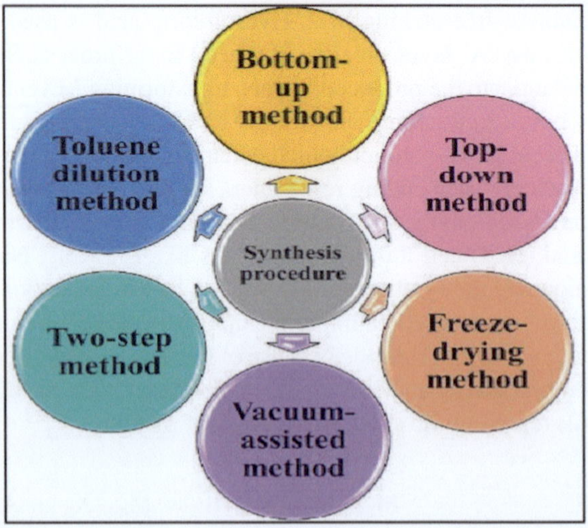

Fig. 5 Synthesis approaches of MXene-based composites

6 Properties

6.1 Conductivity

The high metallic conductivity of MXenes is one of its most intriguing characteristics; spin-cast films of $Ti_3C_2T_z$ have been reported to have conductivities of about 10,000 S/cm [6]. Thus, adding MXene to polymers can make it possible to produce conductive NCs, which can be used in sensing platforms such as flexible strain sensors, selective chemical and gas sensors, energy conversion and storage applications, and shields against electromagnetic and radio frequency interference. The volume fraction of MXene needed to create a percolated network of conducting channels is known as the percolation threshold. In the case of PAM/Ti_3C_2, Naguib et al. initially reported a percolated NC and discovered a temperature dependency of the resistance in addition to a percolation threshold of 1.7%. Shielding against electromagnetic interference (EMI) is one of the main uses for MXene polymer NCs [31].

6.2 Mechanical Properties

Ti_3C_2 MXene is one of the most solution-processable 2D materials with a measured Young's modulus of 333 $\pm$ 30 GPa. On the other hand, Ti_3C_2 was stated to have Young's modulus of 502 GPa; however, lower experimental values are caused due to the imperfection in the surface functionality [21]. The nanocomposites show more elasticity as the filler and matrix rigidity differ more which leads to the transfer of shear deformation and stress to the filler particles. Moreover, strain-induced aligning of the anisotropic particles enables these platelet-like particles to function more like fibers in the conventional fiber-reinforced composite. Tensile strengths and elastic moduli are improved when all of the platelets are oriented perpendicular to the direction of crack propagation [16]. However, this technique is not effective in the case of glassy matrices as they have no strain-induced alignment due to restricted strain. Various studies suggested that the mechanical reinforcement of MXene nanosheets is significantly dependent on the orientation, size, and aspect ratio. The best reinforcement of MXene in polymer NCs is produced when (i) MXene nanosheets are oriented in the direction of the applied load, the MXene, (ii) modulation, and (iii) aspect ratios are high. If the intended NC's moduli and tensile strengths are more important than elongation at break, a high MXene volume fraction should optimize those characteristics. The fact that all of these results are consistent with the basic theories of clay/polymer nano reinforcement emphasizes how much knowledge can be gleaned from this area of study.

6.3 Magnetic Properties

Most 2D materials are typically nonmagnetic, and finding magnetic properties among them can be quite challenging. However, some MXene materials such as Ti_2N, Ti_2C, Cr_2N, Cr_2C, and Mn_2C do possess certain magnetic characteristics. Amid these, Ti_2N, Ti_2C, and Cr_2C are ferromagnetic, while Cr_2N and Mn_2C are antiferromagnetic [10, 41]. Specifically, there has been increasing interest in magnetic MXene with half-metallic materials recently. Half-metallic moieties possess completely spin-polarized electrons at the Fermi level having one metallic and another semiconducting spin channel. Cr_2C is the inaugural member of the MXene clan to exhibit a half-metallic characteristic, demonstrating a half-metallic gap of 2.9 eV. This suggests that the 100% spin-filter efficacy can be sustained across a broad range of bias values. Apart from this, half-metallic attributes have also been predicted in Ti_2C and Ti_2N [41]. As previously discussed, the properties of MXene are greatly influenced by surface modification. The magnetic characteristics of MXene are generally attributed to the presence of unpaired electrons on the surface transition metal moieties, but surface modification can reduce magnetism. However, the magnetism of Cr and Mn-derived MXene cannot be influenced by surface modification. These characteristics of MXene-metal complexes make them promising candidates for various spintronic applications.

7 Application of MXene

Application of MXene and MXene-based composites in different industrial sectors is given in Fig. 6. These include energy storage, biomedical, optical, catalysis, etc. The plausible causes and mechanism of applications of the MXene-based materials are explained below.

7.1 Energy Storage

The growth of energy storage applications is a result of the growing need for renewable energy and the rapid consumption of traditional fossil fuels. Because of their exceptional electrochemical performance, which includes an elevated power density, excellent rate efficiency, remarkable safeguarding, and long cycle life supercapacitors have garnered a lot of attention in this respect. Activated carbon, graphene, and carbon nanotubes are materials with a bulky specific surface area, excellent conductance, and strong chemical resistivity that are frequently used as electrode precursors. The materials' electrochemical activity may be improved by hybridization with the MXene polymer matrix. More precisely, MXene and graphene together can efficiently inhibit self-stacking between layers of graphene and MXene sheets, increasing the specific surface area that is available, expediting the diffusion of electrolyte ions, and ultimately improving the material's activity [25]. Furthermore, by combining the pseudo-capacitance characteristics of MXene with the electric double-layer characteristics of graphene, the composite's electrochemical energy storage will be further enhanced. The high pseudo-capacitance of conjugated polymers, such as polypyrrole and polyaniline, has rendered them preferred options for flexible supercapacitor electrodes [59, 60]. However, the independent conjugated polymer's weak cycle stability and structural instability deteriorate their efficient usage. The high aspect ratio of MXene polymer nanocomposite could surmount this problem.

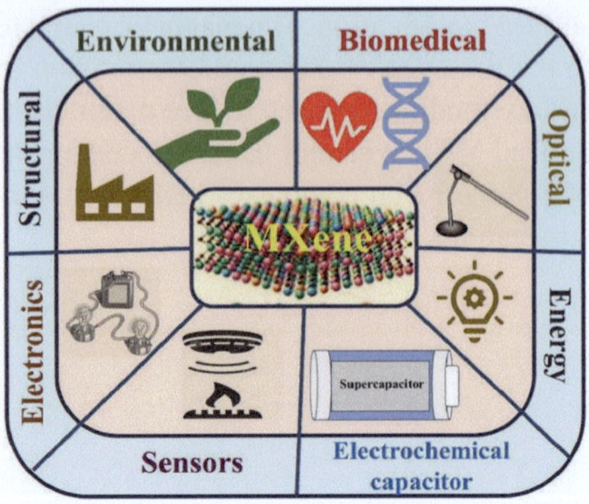

Fig. 6 Applications of MXene-based composite in various fields

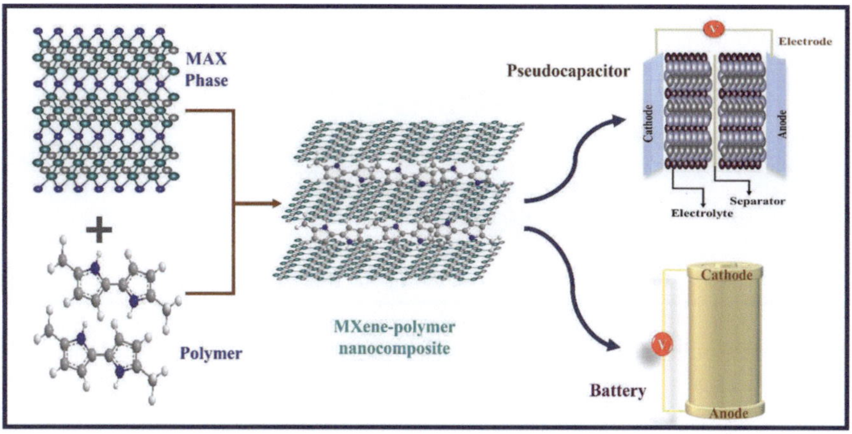

Fig. 7 Energy storage devices using MXene-polymer-based materials

MXene is a potential anode material for batteries due to its layered structure, chemical stability, electrical conductivity qualities, and supercapacitor capabilities. MXenes are competitive with other 2D materials because of their large range of chemical and structural variations. However, because of MXene's high molecular weight, the specific gravimetric capacity of bare MXene can only display a relatively low value. Thus, developing composites based on MXene for electrodes is useful for enhancing electrochemical performance (Fig. 7).

7.2 *Sensors*

2D materials like MXene are commonly used in sensing devices owing to their high conductivity, greater surface area, and biocompatibility. Additionally, MXene has performed exceptionally well in gas and biosensors [55]. In biosensors, selecting the right material for immobilizing active proteins to maintain their bioactivity is crucial. MXene helps immobilize proteins and facilitates direct electron migration between the enzyme and electrode, making it possible to create mediator-free biosensors. Liu et al. utilized $Ti_3C_2T_x$ to immobilize hemoglobin (Hb) in a biosensor for the detection of nitrites [22]. The selected $Ti_3C_2T_X$ showed exceptional biosensing activity with an expanded detection range of 0.5–11,800 mM, with the lowest detection limit at 0.12 mM. More recently, it was discovered that a TiO_2-altered $Ti_3C_2T_x$ nanocomposite can also serve as an immobilization matrix for Hb in a biosensor detecting hydrogen peroxide [44]. TiO_2 provides outstanding chemical stability and biocompatibility, which can protect enzymes. However, experimental results showed that the synthesized composite outperformed $Ti_3C_2T_x$.

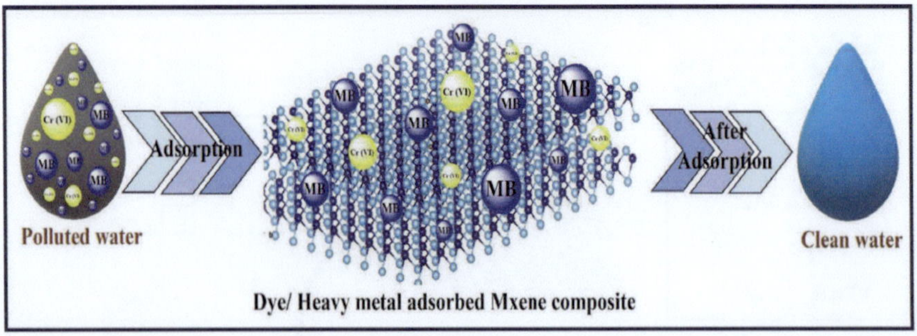

Fig. 8 Schematic illustrations of adsorption of heavy metal ions and dyes by MXene-based adsorbent

7.3 Wastewater Treatment

The exponential growth of industrialization and the escalating human population have resulted in the introduction of a wide array of organic and inorganic pollutants, heavy metals, and aromatic compounds into aquatic habitats, ultimately leading to water pollution. Various methods are used to address this issue, with catalytic degradation and adsorption gaining significant attention in sewage treatment. MXene, a new type of adsorbent with advantages such as ease of preparation, good reducibility, large specific surface area and, are anticipated to play a vital role in sewage treatment. As Ti has a high uptake intensity for heavy metal ions, thus, Ti-comprising materials are capable applicants for removing heavy metal ions from water [17]. Peng et al. have developed an OH-terminated $Ti_3C_2T_x$ with superior adsorption ability for lead (Pb (II)) ions. Alkalization introduces many –OH groups on the surface of $Ti_3C_2T_x$, which benefits the sorption of Pb (II) ions. Notably, 4500 kg of water holding Pb (II) can be treated with just one kilogram of $Ti_3C_2T_x$, and the residual Pb (II) content in the water can validate the intake water standard suggested by the WHO (10 mg L^{-1}) [34]. Moreover, the surface modification of the MXene matrix is also effective in the adsorption of multiple dyes [27]. A schematic illustration of the adsorption of heavy metal ions and dyes by MXene-based adsorbent is shown in Fig. 8. The presence of an adequate number of surface-active sites within the adsorbent ensures the ability to capture and retain dyes and heavy metals, resulting in purified water.

7.4 EMI Shielding

The 2D materials' high conductivity properties make them a key factor for developing EMI shielding materials [13]. Specifically, the 2D MXene's conductivity of up to 20,000 S/cm makes it an ideal choice for this application. Additionally, cost-effective and lightweight conducting polymers can also be utilized as EMI shielding materials. In one study, Cao et al. synthesized a $Ti_3C_2T_X$-cellulose nanofiber to

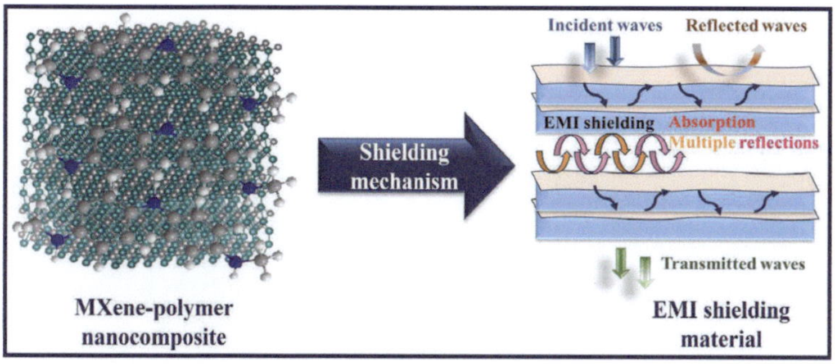

Fig. 9 EMI shielding mechanism of MXene-polymer-based composite

evaluate its mechanical properties. The resulting material exhibited excellent tensile strength and strain of 135.4 MPa and 16.7%, respectively. Its folding endurance was also impressive, at around 14,260 times. Furthermore, the composite demonstrated excellent EMI shielding efficacy and electrical conductivity [5]. Similarly, Yin et al. developed a wearable, lightweight, and durable fabric using MXene-PANI-CF nanocomposites. The fabric showed outstanding performance in EMI shielding [54]. In another study, Wang et al. fabricated MXene/Fe_3O_4@Polyanaline (PANI) using a vacuum-assisted method. The shielding efficacy of this material reached up to 58.8 dB at a thickness of 12.1 μm. Interestingly, the incorporation of Fe_3O_4@PANI into the composite improved both the magnetic loss and impedance-matching characteristics of the material [48, 49]. The incorporation of polymer into the MXene matrix subsequently enhances the mechanical, and tensile strengths which makes them an efficient EMI shielding material which is presented in Fig. 9.

7.5 *Photocatalysis*

With the help of a suitable photocatalyst, water splitting has become a straightforward process that can be achieved directly using solar radiation, without the need for any additional energy sources. This method of water splitting is considerably better than the electrochemical splitting method. In this regard, MXene has emerged as the most promising photocatalyst for the production of hydrogen via water splitting due to its inherent characteristics, such as a narrow band gap, suitable position of CB and VB bands, high conductivity, simple surface functionalization, and lower recombination rates. According to Ran et al., Ti_3C_2Tx is an extremely effective co-catalyst for H_2 production [36]. The incorporation of $Ti_3C_2T_x$ nanoparticles into the surface of the CdS photocatalyst through the hydrothermal method is efficient in generating H_2 production activity of 14,342 mmol h^{-1} g^{-1}, which is more efficient than that of CdS. However, the pristine $Ti_3C_2T_x$ nanoparticles could not show efficacy toward the photocatalytic H_2 production, suggesting that it functions as a co-catalyst rather than

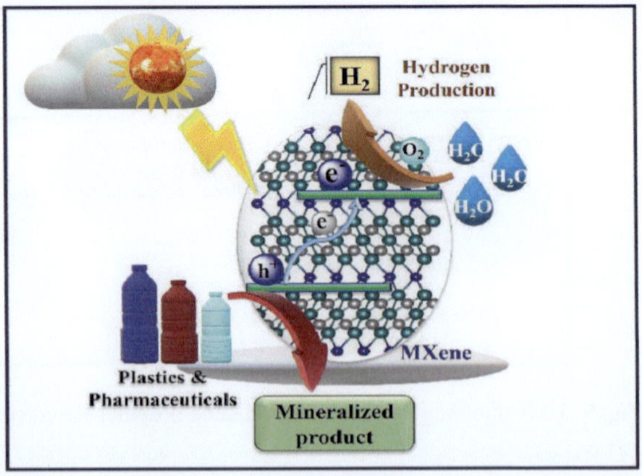

Fig. 10 Plausible mechanism for the H_2 production using MXene-based photocatalyst

a photocatalyst. The possible mechanism for hydrogen production can be explained as follows. First, the incorporation of $Ti_3C_2T_x$ on the CdS surface increases the separation of photoinduced charge carriers and suppresses the recombination of e^-/h^+ pairs. With the exposure of light, the photo-induced electrons migrate from the conduction band (CB) of CdS to $Ti_3C_2T_x$ as the Fermi level of $Ti_3C_2T_x$ is lower than that of the CB of CdS. The exceptional conductivity of $Ti_3C_2T_x$ causes the migration of electrons to its surface-active sites. Second, the hydrothermal treatment leads to the exchange of F-termination into O-termination, which increases the number of O-termination as well as the number of effective active sites. By following a similar mechanism trend various other MXene are also deposited as co-catalysts in the matrix of photocatalyst for the H_2 production [3]. A plausible schematic mechanism of H_2 production is illustrated in Fig. 10. The schematic diagram illustrates the process by which an electron from the conduction band (CB) of MXene material generates hydrogen, while the hole in the valence band (VB) degrades plastic materials, decreasing the recombination of induced charge carriers.

8 Conclusions and Future Perspective

In recent years, the unique properties of MXene have attracted the attention of numerous researchers, with the production of 20 distinct types of MXene materials. This chapter provides a comprehensive summary of the synthesis process for MXene and MXene-based materials, as well as their respective properties, characteristics, and applications. The synthesis process is significantly influenced by factors such as etching, exfoliation, and surface terminations (such as O, F, and OH). The 2D morphology, flexibility, and layered structure of MXenes make them highly suitable for use in composites with other materials, allowing for the optimization of their properties and performance for various applications. The incorporation of CNTs and metal oxide into the MXene matrix can enhance their specific capacity and conductive pathways for the electrons, which enhance transportation of the electrolyte ion, thus improving the overall electrochemical performance of MXenes. Apart from electrochemical applications, MXenes, and their composites can serve as effective catalysts or co-catalysts in a range of areas, including catalysis, sewage treatment, sensors, photocatalytic/electro water splitting, and photocatalytic CO_2 reduction. These qualities make MXene and MXene-based composites viable alternatives to conventional metal-based catalysts. However, compared to other 2D materials such as graphene, MXene is still in its early stages and requires further exploration. It is essential to develop new synthesis approaches for producing eminence MXene with great lateral dimensions, controlled surface terminations, and lesser defects. Furthermore, MXene and MXene-based materials are prone to oxidative degradation due to the abundance of surface functional groups, which results in reduced stability. This limitation restricts their practical application in industrial fields. Developing effective strategies is crucial to prevent oxidative degradation and to improve their application in industrial settings. The greatest challenge that researchers in the field of experimental science face is the ability to create uniform terminations using only one type of functional group. To surmount this obstacle, the synthesis of bare MXene without surface terminations must be carried out using either chemical vapor deposition (CVD) or physical methods. Despite its exceptional performance in various application areas, the underlying physical mechanisms of MXene still need to be thoroughly examined. Additionally, the precise mechanism behind MXene's high sensitivity to ammonia remains a mystery, which hinders the practical implementation of gas sensors based on MXene. Therefore, it is recommended that future research on MXene incorporates both experimental and theoretical investigations.

References

1. Abdelrazik, A.S., Tan, K.H., Aslfattahi, N., Arifutzzaman, A., Saidur, R., Al-Sulaiman, F.A.: Optical, stability and energy performance of water-based MXene nanofluids in hybrid PV/thermal solar systems. Sol. Energy **204**, 32–47 (2020). https://doi.org/10.1016/j.solener.2020.04.063
2. Alhabeb, M., Maleski, K., Anasori, B., Lelyukh, P., Clark, L., Sin, S., Gogotsi, Y.: Guidelines for synthesis and processing of two-dimensional titanium carbide ($Ti_3C_2T_x$ MXene). Chem. Mater. **29**(18), 7633–7644 (2017). https://doi.org/10.1021/acs.chemmater.7b02847
3. An, X., Wang, W., Wang, J., Duan, H., Shi, J., Yu, X.: The synergetic effects of Ti_3C_2 MXene and Pt as co-catalysts for highly efficient photocatalytic hydrogen evolution over g-C_3N_4. Phys. Chem. Chem. Phys. **20**(16), 11405–11411 (2018). https://doi.org/10.1039/C8CP01123K
4. Borysiuk, V.N., Mochalin, V.N., Gogotsi, Y.: Molecular dynamic study of the mechanical properties of two-dimensional titanium carbides $Ti_{n+1}C_n$(MXenes). Nanotechnology **26**(26), 265705 (2015). https://doi.org/10.1088/0957-4484/26/26/265705
5. Cao, W.-T., Chen, F.-F., Zhu, Y.-J., Zhang, Y.-G., Jiang, Y.-Y., Ma, M.-G., Chen, F.: Binary strengthening and toughening of MXene/cellulose nanofiber composite paper with nacre-inspired structure and superior electromagnetic interference shielding properties. ACS Nano **12**(5), 4583–4593 (2018). https://doi.org/10.1021/acsnano.8b00997

6. Carey, M., Barsoum, M.W.: MXene polymer nanocomposites: a review. Mater. Today Adv. **9**, 100120 (2021). https://doi.org/10.1016/j.mtadv.2020.100120
7. Chen, S., Shi, M., Xu, Q., Xu, J., Duan, X., Gao, Y., Lu, L., Gao, F., Wang, X., Yu, Y.: $Ti_3C_2T_x$ MXene/nitrogen-doped reduced graphene oxide composite: A high-performance electrochemical sensing platform for adrenaline detection. Nanotechnology **32**(26), 265501 (2021). https://doi.org/10.1088/1361-6528/abef94
8. Das, L., Habib, K., Saidur, R., Aslfattahi, N., Yahya, S.M., Rubbi, F.: Improved thermophysical properties and energy efficiency of aqueous ionic liquid/MXene nanofluid in a hybrid PV/T solar system. Nanomaterials **10**(7), 1372 (2020). https://doi.org/10.3390/nano10071372
9. Feng, X., Han, G., Cai, J., Wang, X.: Au@Carbon quantum dots-MXene nanocomposite as an electrochemical sensor for sensitive detection of nitrite. J. Colloid Interface Sci. **607**, 1313–1322 (2022). https://doi.org/10.1016/j.jcis.2021.09.036
10. Gao, G., Ding, G., Li, J., Yao, K., Wu, M., Qian, M.: Monolayer MXenes: promising half-metals and spin gapless semiconductors. Nanoscale **8**(16), 8986–8994 (2016). https://doi.org/10.1039/C6NR01333C
11. Gund, G.S., Park, J.H., Harpalsinh, R., Kota, M., Shin, J.H., Kim, T., Gogotsi, Y., Park, H.S.: MXene/polymer hybrid materials for flexible AC-filtering electrochemical capacitors. Joule **3**(1), 164–176 (2019). https://doi.org/10.1016/j.joule.2018.10.017
12. Guo, Z., Zhou, J., Si, C., Sun, Z.: Flexible two-dimensional $Ti_{n+1}C_n$ (n = 1, 2 and 3) and their functionalized MXenes predicted by density functional theories. Phys. Chem. Chem. Phys. **17**(23), 15348–15354 (2015). https://doi.org/10.1039/C5CP00775E
13. Han, M., Shuck, C.E., Rakhmanov, R., Parchment, D., Anasori, B., Koo, C.M., Friedman, G., Gogotsi, Y.: Beyond $Ti_3C_2T_x$: MXenes for electromagnetic interference shielding. ACS Nano **14**(4), 5008–5016 (2020). https://doi.org/10.1021/acsnano.0c01312
14. Jin, X., Wang, J., Dai, L., Liu, X., Li, L., Yang, Y., Cao, Y., Wang, W., Wu, H., Guo, S.: Flame-retardant poly(vinyl alcohol)/MXene multilayered films with outstanding electromagnetic interference shielding and thermal conductive performances. Chem. Eng. J. **380**, 122475 (2020). https://doi.org/10.1016/j.cej.2019.122475
15. Khazaei, M., Arai, M., Sasaki, T., Ranjbar, A., Liang, Y., Yunoki, S.: OH-terminated two-dimensional transition metal carbides and nitrides as ultralow work function materials. Phys. Rev. B **92**(7), 075411 (2015). https://doi.org/10.1103/PhysRevB.92.075411
16. Kim, Y., Kim, Y.K., Kim, S., Harbottle, D., Lee, J.W.: Nanostructured potassium copper hexacyanoferrate-cellulose hydrogel for selective and rapid cesium adsorption. Chem. Eng. J. **313**, 1042–1050 (2017). https://doi.org/10.1016/j.cej.2016.10.136
17. Lee, Y.-C., Yang, J.-W.: Self-assembled flower-like TiO_2 on exfoliated graphite oxide for heavy metal removal. J. Ind. Eng. Chem. **18**(3), 1178–1185 (2012). https://doi.org/10.1016/j.jiec.2012.01.005
18. Lei, C., Zhang, Y., Liu, D., Wu, K., Fu, Q.: Metal-level robust, folding endurance, and highly temperature-stable MXene-based film with engineered aramid nanofiber for extreme-condition electromagnetic interference shielding applications. ACS Appl. Mater. Interfaces **12**(23), 26485–26495 (2020). https://doi.org/10.1021/acsami.0c07387
19. Li, T., Yao, L., Liu, Q., Gu, J., Luo, R., Li, J., Yan, X., Wang, W., Liu, P., Chen, B., Zhang, W., Abbas, W., Naz, R., Zhang, D.: Fluorine-free synthesis of high-purity $Ti_3C_2T_x$ (T=OH, O) via alkali treatment. Angew. Chem. Int. Ed. **57**(21), 6115–6119 (2018). https://doi.org/10.1002/anie.201800887
20. Ling, Z., Ren, C.E., Zhao, M.-Q., Yang, J., Giammarco, J.M., Qiu, J., Barsoum, M.W., Gogotsi, Y.: Flexible and conductive MXene films and nanocomposites with high capacitance. Proc. Natl. Acad. Sci. **111**(47), 16676–16681 (2014). https://doi.org/10.1073/pnas.1414215111
21. Lipatov, A., Lu, H., Alhabeb, M., Anasori, B., Gruverman, A., Gogotsi, Y., Sinitskii, A.: Elastic properties of 2D $Ti_3C_2T_x$ MXene monolayers and bilayers. Sci. Adv. **4**(6), eaat0491 (2018). https://doi.org/10.1126/sciadv.aat0491
22. Liu, R., Li, W.: High-thermal-stability and high-thermal-conductivity $Ti_3C_2T_x$ MXene/poly(vinyl alcohol) (PVA) composites. ACS Omega **3**(3), 2609–2617 (2018). https://doi.org/10.1021/acsomega.7b02001

23. Liu, H., Duan, C., Yang, C., Shen, W., Wang, F., Zhu, Z.: A novel nitrite biosensor based on the direct electrochemistry of hemoglobin immobilized on MXene-Ti_3C_2. Sens. Actuators B Chem. **218**, 60–66 (2015). https://doi.org/10.1016/j.snb.2015.04.090
24. Liu, Y., Xiao, H., Goddard, W.A.: Schottky-barrier-free contacts with two-dimensional semiconductors by surface-engineered MXenes. J. Am. Chem. Soc. **138**(49), 15853–15856 (2016). https://doi.org/10.1021/jacs.6b10834
25. Liu, Y., Yu, J., Guo, D., Li, Z., Su, Y.: $Ti_3C_2T_x$ MXene/graphene nanocomposites: Synthesis and application in electrochemical energy storage. J. Alloy. Compd. **815**, 152403 (2020). https://doi.org/10.1016/j.jallcom.2019.152403
26. Ma, L., Wang, B., Zhao, S., Hao, C., Guo, H., Lei, Q.: The fabrication and electrical properties of polyimide/boron nitride nanosheets composite films. J. Mater. Sci. Mater. Electron. **30**(22), 20302–20310 (2019). https://doi.org/10.1007/s10854-019-02417-3
27. Mashtalir, O., Cook, K.M., Mochalin, V.N., Crowe, M., Barsoum, M.W., Gogotsi, Y.: Dye adsorption and decomposition on two-dimensional titanium carbide in aqueous media. J. Mater. Chem. A **2**(35), 14334–14338 (2014). https://doi.org/10.1039/C4TA02638A
28. Meng, W., Liu, X., Song, H., Xie, Y., Shi, X., Dargusch, M., Chen, Z.-G., Tang, Z., Lu, S.: Advances and challenges in 2D MXenes: from structures to energy storage and conversions. Nano Today **40**, 101273 (2021). https://doi.org/10.1016/j.nantod.2021.101273
29. Mo, Z., Mo, P., Yi, M., Hu, Z., Tan, G., Selim, M.S., Chen, Y., Chen, X., Hao, Z., Wei, X.: $Ti_3C_2T_x$@Polyvinyl alcohol foam-supported phase change materials with simultaneous enhanced thermal conductivity and solar-thermal conversion performance. Sol. Energy Mater. Sol. Cells **219**, 110813 (2021). https://doi.org/10.1016/j.solmat.2020.110813
30. Naguib, M., Kurtoglu, M., Presser, V., Lu, J., Niu, J., Heon, M., Hultman, L., Gogotsi, Y., Barsoum, M.W.: Two-dimensional nanocrystals produced by exfoliation of Ti_3AlC_2. Adv. Mater. **23**(37), 4248–4253 (2011). https://doi.org/10.1002/adma.201102306
31. Naguib, M., Saito, T., Lai, S., Rager, M.S., Aytug, T., Parans Paranthaman, M., Zhao, M.-Q., Gogotsi, Y.: $Ti_3C_2T_x$ (MXene)–polyacrylamide nanocomposite films. RSC Adv. **6**(76), 72069–72073 (2016). https://doi.org/10.1039/C6RA10384G
32. Okada, A., Usuki, A.: Twenty years of polymer-clay nanocomposites. Macromol. Mater. Eng. **291**(12), 1449–1476 (2006). https://doi.org/10.1002/mame.200600260
33. Pang, J., Mendes, R.G., Bachmatiuk, A., Zhao, L., Ta, H.Q., Gemming, T., Liu, H., Liu, Z., Rummeli, M.H.: Applications of 2D MXenes in energy conversion and storage systems. Chem. Soc. Rev. **48**(1), 72–133 (2019). https://doi.org/10.1039/C8CS00324F
34. Peng, Q., Guo, J., Zhang, Q., Xiang, J., Liu, B., Zhou, A., Liu, R., Tian, Y.: Unique lead adsorption behavior of activated hydroxyl group in two-dimensional titanium carbide. J. Am. Chem. Soc. **136**(11), 4113–4116 (2014). https://doi.org/10.1021/ja500506k
35. Rajendran, J., Kannan, T.S., Dhanasekaran, L.S., Murugan, P., Atchudan, R., ALOthman, Z.A., Ouladsmane, M., Sundramoorthy, A.K.: Preparation of 2D graphene/MXene nanocomposite for the electrochemical determination of hazardous bisphenol A in plastic products. Chemosphere **287**, 132106 (2022). https://doi.org/10.1016/j.chemosphere.2021.132106
36. Ran, J., Gao, G., Li, F.-T., Ma, T.-Y., Du, A., Qiao, S.-Z.: Ti_3C_2 MXene co-catalyst on metal sulfide photo-absorbers for enhanced visible-light photocatalytic hydrogen production. Nat. Commun. **8**(1), 13907 (2017). https://doi.org/10.1038/ncomms13907
37. Rasool, K., Mahmoud, K.A., Johnson, D.J., Helal, M., Berdiyorov, G.R., Gogotsi, Y.: Efficient antibacterial membrane based on two-dimensional $Ti_3C_2T_x$ (MXene) nanosheets. Sci. Rep. **7**(1), 1598 (2017). https://doi.org/10.1038/s41598-017-01714-3
38. Rubbi, F., Habib, K., Saidur, R., Aslfattahi, N., Yahya, S.M., Das, L.: Performance optimization of a hybrid PV/T solar system using Soybean oil/MXene nanofluids as a new class of heat transfer fluids. Sol. Energy **208**, 124–138 (2020). https://doi.org/10.1016/j.solener.2020.07.060
39. Samylingam, L., Aslfattahi, N., Saidur, R., Yahya, S.M., Afzal, A., Arifutzzaman, A., Tan, K.H., Kadirgama, K.: Thermal and energy performance improvement of hybrid PV/T system by using olein palm oil with MXene as a new class of heat transfer fluid. Sol. Energy Mater. Sol. Cells **218**, 110754 (2020). https://doi.org/10.1016/j.solmat.2020.110754

40. Scala, E.P.: A brief history of composites in the U.S. the dream and the success. JOM **48**(2), 45–48 (1996). https://doi.org/10.1007/BF03221382
41. Si, C., Zhou, J., Sun, Z.: Half-metallic ferromagnetism and surface functionalization-induced metal-insulator transition in graphene-like two-dimensional Cr_2C crystals. ACS Appl. Mater. Interfaces **7**(31), 17510–17515 (2015). https://doi.org/10.1021/acsami.5b05401
42. VahidMohammadi, A., Moncada, J., Chen, H., Kayali, E., Orangi, J., Carrero, C.A., Beidaghi, M.: Thick and freestanding MXene/PANI pseudocapacitive electrodes with ultrahigh specific capacitance. J. Mater. Chem. A **6**(44), 22123–22133 (2018). https://doi.org/10.1039/C8TA05807E
43. Wang, F., Yang, C., Duan, M., Tang, Y., Zhu, J.: TiO_2 nanoparticle modified organ-like Ti_3C_2 MXene nanocomposite encapsulating hemoglobin for a mediator-free biosensor with excellent performances. Biosens. Bioelectron. **74**, 1022–1028 (2015). https://doi.org/10.1016/j.bios.2015.08.004
44. Wang, Q., Pan, X., Lin, C., Gao, H., Cao, S., Ni, Y., Ma, X.: Modified $Ti_3C_2T_X$ (MXene) nanosheet-catalyzed self-assembled, anti-aggregated, ultra-stretchable, conductive hydrogels for wearable bioelectronics. Chem. Eng. J. **401**, 126129 (2020). https://doi.org/10.1016/j.cej.2020.126129
45. Wang, D., Lin, Y., Hu, D., Jiang, P., Huang, X.: Multifunctional 3D-MXene/PDMS nanocomposites for electrical, thermal and triboelectric applications. Compos. A Appl. Sci. Manuf. **130**, 105754 (2020). https://doi.org/10.1016/j.compositesa.2019.105754
46. Wang, Y., Qi, Q., Fan, J., Wang, W., Yu, D.: Simple and robust MXene/carbon nanotubes/cotton fabrics for textile wastewater purification via solar-driven interfacial water evaporation. Sep. Purif. Technol. **254**, 117615 (2021). https://doi.org/10.1016/j.seppur.2020.117615
47. Wang, Z., Cheng, Z., Xie, L., Hou, X., Fang, C.: Flexible and lightweight $Ti_3C_2T_x$ MXene/Fe_3O_4@PANI composite films for high-performance electromagnetic interference shielding. Ceram. Int. **47**(4), 5747–5757 (2021). https://doi.org/10.1016/j.ceramint.2020.10.161
48. Wang, H., Li, X., Luo, B., Wei, K., Zeng, G.: The MXene/water nanofluids with high stability and photo-thermal conversion for direct absorption solar collectors: a comparative study. Energy **227**, 120483 (2021). https://doi.org/10.1016/j.energy.2021.120483
49. Wang, X., Li, M., Yang, S., Bai, X., Shan, J.: Self-assembled $Ti_3C_2T_X$ MXene/graphene composite for the electrochemical reduction and detection of p-nitrophenol. Microchem. J. **179**, 107473 (2022). https://doi.org/10.1016/j.microc.2022.107473
50. Xia, Y., Hu, X., Liu, Y., Zhao, F., Zeng, B.: Molecularly imprinted ratiometric electrochemical sensor based on carbon nanotubes/cuprous oxide nanoparticles/titanium carbide MXene composite for diethylstilbestrol detection. Microchim. Acta **189**(4), 137 (2022). https://doi.org/10.1007/s00604-022-05249-x
51. Xu, H., Yin, X., Li, X., Li, M., Liang, S., Zhang, L., Cheng, L.: Lightweight Ti_2CT_x MXene/Poly(vinyl alcohol) composite foams for electromagnetic wave shielding with absorption-dominated feature. ACS Appl. Mater. Interfaces **11**(10), 10198–10207 (2019). https://doi.org/10.1021/acsami.8b21671
52. Yang, C., Jiang, Q., Li, W., He, H., Yang, L., Lu, Z., Huang, H.: Ultrafine Pt nanoparticle-decorated 3D hybrid architectures built from reduced graphene oxide and MXene nanosheets for methanol oxidation. Chem. Mater. **31**(22), 9277–9287 (2019). https://doi.org/10.1021/acs.chemmater.9b02115
53. Yang, H., Cheng, L., Yang, Q., Farooq, A., Ying, L., Wang, P., Wang, Z.: Multifunctional and durable thermal management coating from sericin-MXene biohybrid on silk fabric micro-etched by deep eutectic solvent. Appl. Surf. Sci. **623**, 156962 (2023). https://doi.org/10.1016/j.apsusc.2023.156962
54. Yin, G., Wang, Y., Wang, W., Yu, D.: Multilayer structured PANI/MXene/CF fabric for electromagnetic interference shielding constructed by layer-by-layer strategy. Colloids Surf. A **601**, 125047 (2020). https://doi.org/10.1016/j.colsurfa.2020.125047
55. Yousef Elahi, M., Khodadadi, A.A., Mortazavi, Y.: A glucose biosensor based on glucose oxidase immobilized on ZnO/Cu_2O graphene oxide nanocomposite electrode. J. Electrochem. Soc. **161**(5), B81–B87 (2014). https://doi.org/10.1149/2.020405jes

56. Yu, Z., Jiang, L., Liu, R., Zhao, W., Yang, Z., Zhang, J., Jin, S.: Versatile self-assembled MXene-Au nanocomposites for SERS detection of bacteria, antibacterial and photothermal sterilization. Chem. Eng. J. **426**, 131914 (2021). https://doi.org/10.1016/j.cej.2021.131914
57. Zahra, Q.U.A., Ullah, S., Shahzad, F., Qiu, B., Fang, X., Ammar, A., Luo, Z., Abbas Zaidi, S.: MXene-based aptasensors: advances, challenges, and prospects. Prog. Mater. Sci. **129**, 100967 (2022). https://doi.org/10.1016/j.pmatsci.2022.100967
58. Zha, X.-H., Luo, K., Li, Q., Huang, Q., He, J., Wen, X., Du, S.: Role of the surface effect on the structural, electronic and mechanical properties of the carbide MXenes. EPL (Europhys. Lett.) **111**(2), 26007 (2015). https://doi.org/10.1209/0295-5075/111/26007
59. Zhang, A., Liu, R., Tian, J., Huang, W., Liu, J.: MXene-based nanocomposites for energy conversion and storage applications. Chem. Eur. J. **26**(29), 6342–6359 (2020). https://doi.org/10.1002/chem.202000191
60. Zhang, C., Xu, S., Cai, D., Cao, J., Wang, L., Han, W.: Planar supercapacitor with high areal capacitance based on Ti_3C_2/polypyrrole composite film. Electrochim. Acta **330**, 135277 (2020). https://doi.org/10.1016/j.electacta.2019.135277

Surface-Functionalized MXenes for Adsorptive Wastewater Treatment Applications

Adarsh Kumar Arya, Akshay Kapoor, and Ashish Kapoor

Abstract The growing concern over water pollution due to industrialization and urbanization has intensified the need for effective wastewater treatment methods. With its high efficiency, simplicity, and cost-effectiveness, adsorption has emerged as a preferred technique for removing a wide range of contaminants from wastewater. This chapter explores the potential of surface-functionalized MXenes, a novel class of two-dimensional materials, as advanced adsorbents for wastewater treatment applications. MXenes, characterized by their large surface area, high electrical conductivity, and rich surface chemistry, can be tailored through various surface functionalization techniques to enhance their adsorptive properties. The functionalization approaches, including chemical modification with small molecules, polymers, and other agents, allow for tuning MXenes' surface properties, thereby improving their selectivity, capacity, and stability in contaminant removal. The chapter reviews adsorption studies for various contaminants, including heavy metals, dyes, and organic compounds. It discusses the adsorption mechanisms in depth, such as ion exchange, electrostatic interactions, and surface complexation, highlighting how different functional groups on MXenes contribute to efficiently removing specific pollutants. The interplay between the MXene surface chemistry and the nature of the contaminants is thoroughly analyzed to illustrate the versatility and effectiveness of these materials in diverse wastewater treatment scenarios. The challenges associated with using surface-functionalized MXenes in real-world applications are discussed. Further, future directions indicate the need for ongoing research to optimize MXene properties, improve functionalization techniques, and develop scalable production methods.

Keywords MXenes · Surface functionalization · Adsorption · Wastewater treatment

A. K. Arya · A. Kapoor · A. Kapoor (✉)
Department of Chemical Engineering, Harcourt Butler Technical University, Kanpur, Uttar Pradesh, India
e-mail: ashishk@hbtu.ac.in

© The Author(s), under exclusive license to Springer Nature Singapore Pte Ltd. 2025
S. Moharana and R. N. Mahaling (eds.), *Surface Engineering of MXenes*, Engineering Materials, https://doi.org/10.1007/978-981-96-3640-2_15

1 Introduction

The escalating pace of industrialization and urbanization across the globe has led to a significant deterioration in water quality, with severe repercussions for both human health and the environment [1]. Industrial effluents often contain many contaminants, including heavy metals like lead and mercury, organic pollutants such as dyes, pesticides and pharmaceuticals, and other hazardous chemicals [2]. These pollutants, if not adequately treated, can cause detrimental effects on aquatic ecosystems and pose severe health risks to humans through waterborne diseases and bioaccumulation in the food chain [3]. Adequate wastewater treatment is thus imperative to ensure the safety and availability of clean water for diverse uses, including drinking, agriculture, and industrial applications [4]. Moreover, stringent environmental regulations necessitate the removal of these contaminants to safeguard natural water bodies and maintain ecological balance. Wastewater treatment encompasses a range of techniques, each tailored to address specific pollutants. Physical methods, such as sedimentation, filtration, and flotation, are primarily used to remove suspended solids and particulate matter from wastewater. These methods rely on physical forces to separate contaminants from water. Chemical methods, including coagulation, flocculation, and chemical precipitation, involve adding chemicals to induce the aggregation and removal of dissolved inorganic substances and certain organic pollutants [5]. Biological methods utilize microorganisms to degrade organic matter and other pollutants; processes like activated sludge systems and biofilms are commonly employed to break down biodegradable contaminants [6]. Advanced oxidation processes (AOPs) leverage highly reactive oxygen species to degrade complex organic molecules resistant to conventional treatments. Membrane technologies, such as membrane contactors, use semipermeable membranes to separate contaminants based on size and molecular properties [7]. Each of these methods has advantages and limitations, and the treatment choice often depends on the specific characteristics of the treated wastewater and availability of suitable materials for treatment [8].

Among the various wastewater treatment methods, adsorption stands out due to its simplicity, efficiency, and versatility [9]. Adsorption involves the adhesion of pollutants onto the surface of a solid material, known as the adsorbent. This method effectively removes a broad spectrum of contaminants from wastewater, including heavy metals, dyes, and organic compounds. The operational simplicity of adsorption, which does not require complex equipment or stringent conditions, makes it an attractive option for both small-scale and large-scale applications. Additionally, adsorption is often more cost-effective than other treatment methods, involving lower operational and maintenance costs. The selectivity of adsorbents can be tailored through surface modifications, allowing for targeted removal of specific pollutants, thereby enhancing the overall efficiency of the process. These advantages make adsorption a preferred choice in various wastewater treatment scenarios.

Traditional adsorbents, such as activated carbon, clay minerals, and zeolites, have been used in wastewater treatment for decades [10]. However, these materials often face adsorption capacity, selectivity, and regeneration potential limitations. For instance, while activated carbon is effective, its adsorption sites can become saturated quickly, necessitating frequent regeneration or replacement. This has spurred significant research interest in developing novel adsorbents to overcome these limitations. Nanomaterials, in particular, have emerged as promising candidates due to their unique properties. They offer high surface area-to-volume ratios, providing more active sites for pollutant adsorption. Furthermore, the surface chemistry of nanomaterials can be easily tailored through functionalization, enhancing their selectivity and interaction with specific contaminants. The ability to regenerate these materials also adds to their appeal, making them sustainable options for long-term use in wastewater treatment.

Recent advancements in material science have led to the discovery of promising nanomaterials such as MXenes. These materials have shown exceptional promise in diverse applications, including wastewater treatment. This chapter aims to delve deeper into surface-functionalized MXenes, exploring their potential and versatility as advanced adsorbents in adsorptive wastewater treatment.

2 Fundamentals of MXenes

MXene is a material that falls within the transition metal carbides and nitrides category. An uneven distribution of functional groups characterizes it. MXene is a compound with the chemical formula $M_{n+1}X_nT_x$. It is created from the MAX phase ($M_{n+1}AX_n$) by etching, as illustrated in Fig. 1. In the formula, M represents a transition metal, X represents either carbon or nitrogen, T represents functional groups like hydroxyl, oxygen, or fluorine, A represents an element from a specific group, and n is an integer. Oxygen (–O), fluorine (F), and hydroxyl (–OH) functional groups were attached to MXene surfaces. They were identified using energy-dispersive X-ray spectroscopy (EDX) to detect the elemental compositions. The structure of MXene, Ti_3C_2, consists of the intralayer skeleton, interlayer, and surface termination groups. Neutron diffraction has proven that Ti forms an ionic link with C in the interlayer skeleton, influenced by Van der Waals forces and hydrogen bonding with –O or F atoms on the surface [9]. MXene materials have garnered considerable interest due to their diverse compositions, conductive characteristics, and distinctive shape.

MXene is a promising material with several desirable properties for diverse applications [11]. For instance, MXene has excellent UV light reflectivity across the 300 to 500 nm wavelength range [12]. A variety of methods have been explored for synthesizing MXenes, including both top-down and bottom-up methods. Figure 2 shows the various MXene synthesis techniques and etching agents [10].

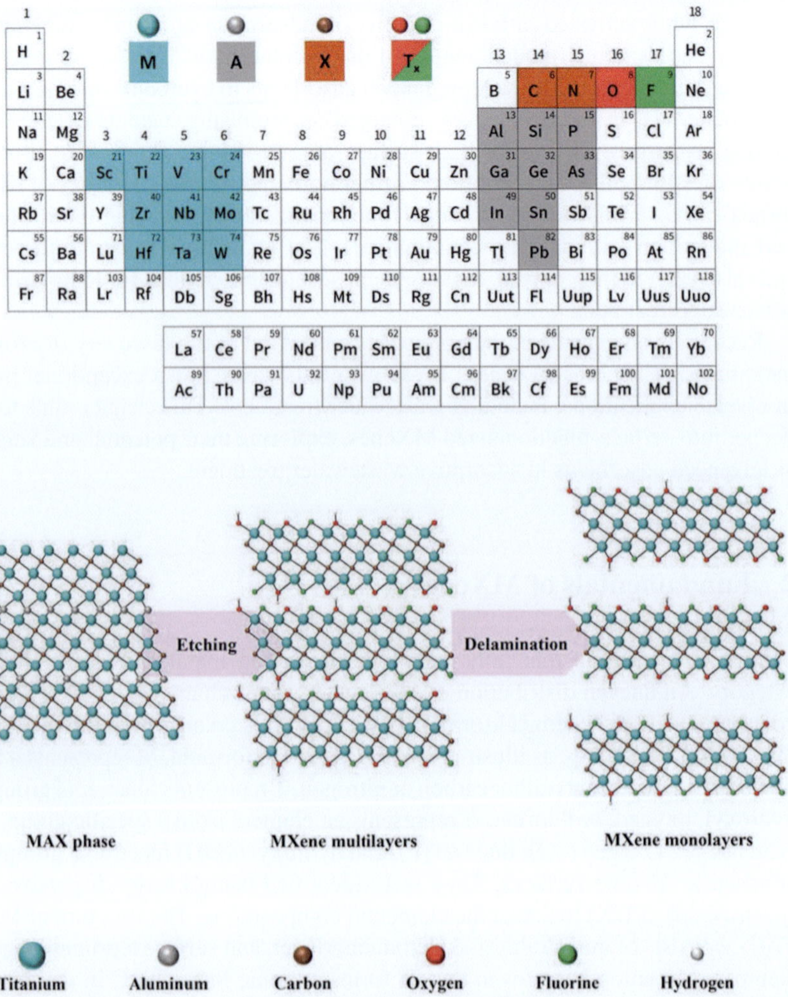

Fig. 1 Overview of MAX phases and their transformation into MXenes. The periodic table indicates elements commonly found in MAX phases, and the schematic demonstrates the synthesis of MXenes through etching and delamination. (Reproduced under CC-BY license from [8])

3 MXenes and Their Surface Functionalization

MXenes are an innovative class of two-dimensional materials derived from the selective etching of layered transition metal carbides, nitrides, or carbonitrides [11]. First discovered in the early 2010s, these materials have the general formula $M_{n+1}X_nT_x$, where M represents an early transition metal (such as titanium, vanadium, or niobium), X denotes carbon and/or nitrogen, and T stands for various surface terminations, including hydroxyl (–OH), oxygen (–O), and fluoride (–F) [12]. The unique

Fig. 2 MXene synthesis techniques and etching agents (Reproduced under CC-BY license from [10])

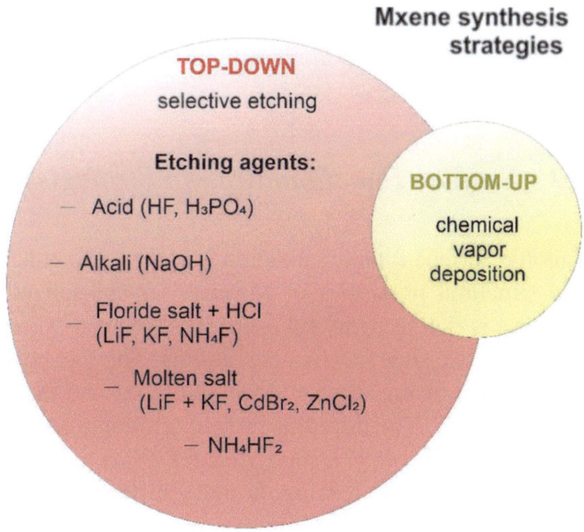

structure of MXenes, consisting of a layered arrangement of metal atoms interspersed with carbon or nitrogen, endows them with a range of exceptional properties that are highly advantageous for several valuable applications, including adsorption. Figure 3 shows the structure diagram of $Ti_3C_2T_x$.

One of the critical features of MXenes is their large surface area, which provides abundant active sites for the adsorption of contaminants [14]. This, coupled with their inherent hydrophilicity and high electrical conductivity, makes MXenes particularly effective in interacting with and removing various pollutants from wastewater. The presence of tunable surface functional groups further enhances their adsorption capabilities, allowing for selective binding to different contaminants.

Surface functionalization of MXenes is a crucial aspect that significantly enhances their performance in wastewater treatment [15]. By modifying the surface chemistry of MXenes, it is possible to tailor their properties to target specific pollutants more effectively. Functionalization can introduce various chemical groups that enhance contaminant interactions through hydrogen bonding, ion exchange, and electrostatic

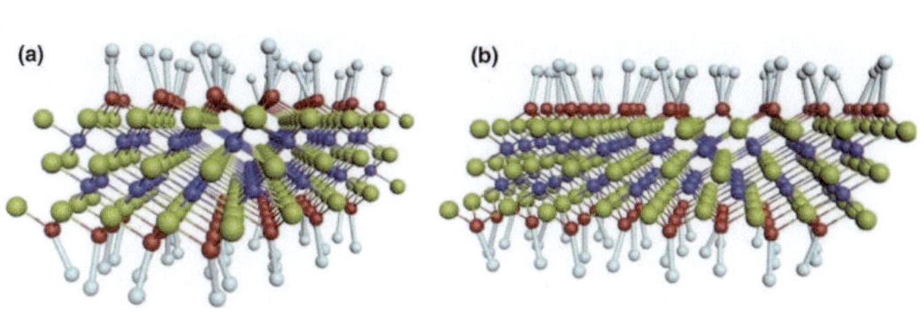

Fig. 3 Structural depiction of $Ti_3C_2T_x$. **a** Side and **b** front views of Ti2C2Tx are presented. Color coding: hydrogen (white), oxygen (red), carbon (blue), titanium (yellow). (Reproduced under CC-BY license from [13]).

attraction [16]. Such modifications can also improve the stability and reusability of MXenes, making them more suitable for practical applications.

3.1 Surface Functionalization Strategies

Tailored surface functionalization is essential to enhance and stabilize the chemical and structural properties of MXenes for specialized applications. The surface of MXenes is rich in hydroxyl (–OH) and oxygen-containing (–O) groups, which can serve as active sites for covalent binding. This can be achieved by controlling surface terminations, using surface-active initiators, and incorporating small molecules and polymers [16]. The different surface terminations and their chemical pathways are illustrated in Fig. 4 [17]. Bare MXenes ($M_{n+1}X_n$) can be produced through exposure to hydrogen gas and heating. Initially, different chemical elements can adsorb from their respective pure diatomic gas phase [$R_2(g)$] to form $M_{n+1}X_n(RH_y)_2$ MXenes. Alternatively, elements may adsorb from their hydrogenated forms [$RH_z(g)$] to form $M_{n+1}X_nR_2$ through one or multiple dehydrogenation steps. $M_{n+1}X_n(RH_y)_2$ can be oxidized into $M_{n+1}X_nR_2$, while $M_{n+1}X_nR_2$ can undergo reduction to $M_{n+1}X_n(RH_y)_2$ under oxygen and hydrogen atmospheres, respectively. Additionally, exposure to oxygen can result in chemical substitution, forming oxygen-terminated $M_{n+1}X_nO_2$ MXenes. The main approaches to functionalization are explained in this section.

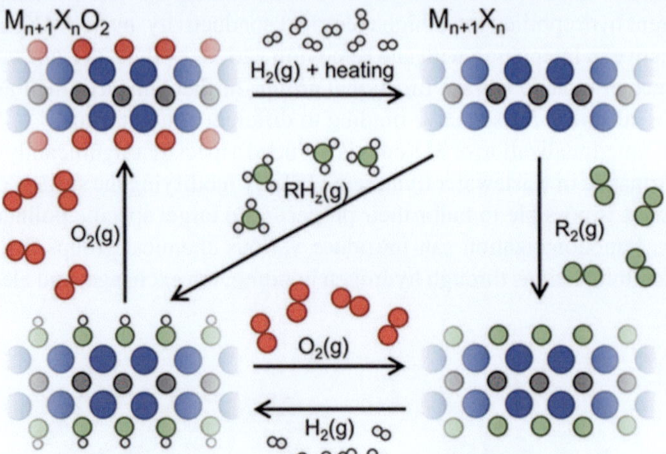

Fig. 4 Various surface terminations of MXenes and their chemical pathways (Reproduced under CC-BY license from [17])

3.2 Control of Surface Terminations

Modifying the surface of MXenes by controlling the T groups allows for the adjustment of their physical properties, including the types and concentrations of electrochemically active sites and their electronic structures [18]. Surface terminations like –O and –F can be readily obtained using conventional etching methods, such as fluoride-containing salts or hydrofluoric acid. Moreover, the concentration and type of surface terminations can be fine-tuned by varying the etching duration and the concentration of etching agents and applying various post-treatments like hydrazine or annealing in different process conditions [19].

3.3 Functionalization Using Small Molecules

The performance attributes of MXenes, including their mechanical stability, electrical properties, and solution stability, can be significantly improved through surface modification with various small molecules [20]. These molecules are not only easy to process but also cost-effective, making them ideal for enhancing the properties of MXenes. The significant surface area and hydrophilic nature of MXenes make them a promising choice for molecular adsorption in aqueous solutions. Recent research has demonstrated that functionalizing the surface of MXene with amino acids and amino acid-like compounds, including glutamic acid, lysine, and l-dopa, can improve urea adsorptive performance [21]. A simplified synthesis technique, minimally intensive layer delamination, was used. This approach leveraged the specific interactions between the functional groups and urea molecules to enhance the adsorption performance.

3.4 Surface-Initiated Polymerization

The surface of MXenes is rich in active functional groups, enabling the integration of polymers through ex-situ blending or in-situ polymerization. Typically, MXene-polymer composites are created using the ex-situ blending method, which offers advantages such as well-defined polymer structures and customizable compositions. The interactions between polymers and MXenes are significantly enhanced through hydrogen bonding and electrostatic interactions. On the other hand, the in situ polymerization approach can result in a covalently bonded network between the MXenes and the polymer. A recent study first synthesized a few-layered MXene, followed by polypyrrole polymerization within the MXene solution to form MXene/PPy composite particles [22]. The study systematically examined the microstructure, adsorption behaviour towards methylene blue and stability during storage. Findings revealed that incorporating PPy nanoparticles facilitated the exfoliation of MXene layers, substantially improving the adsorption capacity of the MXene/PPy composites.

4 Utilizing MXenes for the Adsorption-Based Removal of Pollutants

Adsorbents based on MXene-based materials have recently attracted much attention to reducing wastewater pollution [23, 24]. Large surface areas, hydrophilicity, "electron density", available "adsorption sites", fast ion exchange, and active metallic OH sites are some of the interesting "physicochemical properties of these adsorbents". Photo- and MXenes' strong redox reactivity facilitates electro-catalytic pollution removal [25, 26]. The routine use of MBNs in environmental cleaning is due to their abundance, affordability, safety, and ease of production [26]. MXenes and MXene-based materials have demonstrated versatility in various environmental remediation and monitoring applications [27], as shown in Fig. 5.

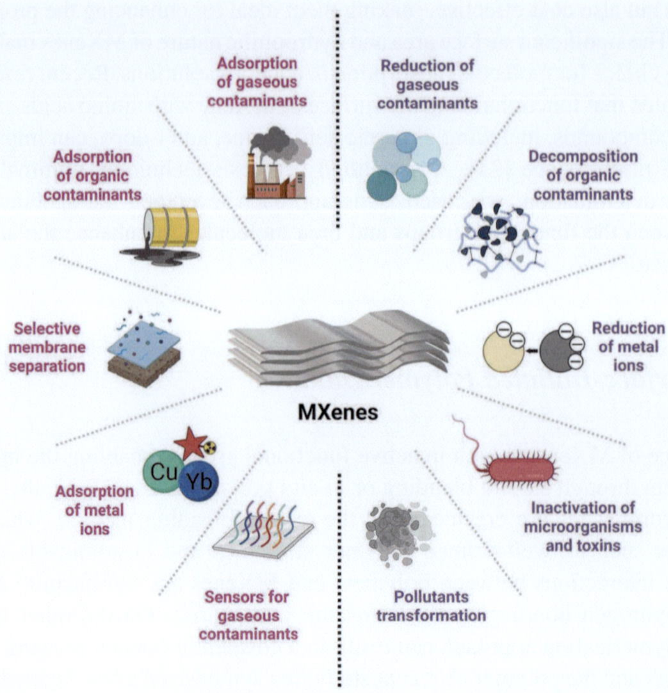

Fig. 5 Multifarious environmental application of MXenes. (Reproduced under CC-BY license from [27])

4.1 Adsorption of Inorganic Contaminants

This section delves into the procedure for removing inorganic contaminants, such as Cu, Hg, Pb, and Cr, from wastewater using MXene-based adsorbents. Mining, dye manufacturing, fertilizer production, battery manufacturing, and metal plating are some of the industrial operations that contribute to the widespread release of lead into the environment. MBNs have diverse uses as very effective adsorbents with a significant ability to absorb lead [26]. Chemical exfoliation of MXene and intercalation of alkali produce 2D alk-MXene (Ti_3C_2 $(OH/ONa)_xF_{2-x}$) with a capacity for adsorption (AC) of 140.10 mg/g, the multifaceted alk-MXene adsorbent shown favourable results in lead degradation. In less than two minutes, adsorption equilibrium was reached. Hydrofluoric acid was used to transform Ti_3AlC_2 into MXene. The interlayer gaps were efficiently filled using a synthetic combination of MXene and bentonite powder, making adsorbing lead easier. The MXene/alginate composite reached equilibrium in about fifteen minutes, and its enhanced adsorption capacity reached 382.70 mg/g [26].

Industries that deal with leather tanners' refractories, the electroplating process, and metallurgy are significant sources of chromium, another inorganic pollutant. Both hexavalent and trivalent chromium (Cr (VI) and Cr (III)) are found in nature. For aquatic creatures, the chromium compounds are an essential nutrient. It becomes dangerous at higher concentrations. A hazardous chemical, hexavalent chromium, is called Cr (VI) [28].

Hence, it is essential to eliminate chromium ions from the surroundings before emission [28]. This section provides a summary of the outcomes achieved by the use of MBNs as absorbents. Creating two-dimensional MXene nanosheets included etching bulk $Ti_3C_2T_x$ using an HF solution and then delaminating it using ultrasonic waves. The $Ti_3C_2T_x$ nanosheets exhibited exceptional removal of Cr (VI) ions from the contaminated aqueous solution. It was shown that Cr (VI) underwent a reduction to Cr (III) that was also subsequently adsorbed [29].

Copper ions are produced by businesses that make paper, medicines, batteries, mining, and fertilizer. People and ecosystems are negatively impacted when these businesses discharge sewage into the natural world [30]. The oxygenated functional groups on MXene nanosheets $Ti_3C_2T_x$ made them an effective copper ion remover. Compared to multilayered MXene, delaminated MXene demonstrated superior performance, according to the demonstrated study [31]. The adsorption capacity was determined to be 78.450 mg/g. In the process 79% or more of the pollutants were captured within 60 s, and the adsorption equilibrium was achieved in less than 3 min. The adsorption effectiveness decreased after three rounds of retention [32]. The MXene Ti_3C_2Tx underwent delamination, and a reaction with the amino acids present in the solution also delaminated the TiO_2 rutile phase. A histidine-functionalized rutile "TiO_2 DL-Ti_3C_2T" hybrid was generated due to the response. The material obtained exhibited a maximum adsorption capacity of about 94.6 mg per gram of activated carbon (AC). Within a few minutes, a significant proportion of 75% of the Cu ions were excluded or left out.

Furthermore, it was discovered that removal efficacy was improved by functionalizing MXene with amino acids [33]. $Ti_3C_2T_x$ powder was subjected to ultrasonication, and a DOPA solution was added while swirling magnetically to produce Ti_3C_2TX-PDOPA composites. With an effectiveness of up to 93.20%, the AC composite attained a specific adsorption rate of 46.60 mg per gram. Pseudo first-order kinetics provides a good description of the adsorption data. The adsorbent performance of Ti_3C_2TX was enhanced due to the functional groups [34]

Among the top ten most lethal compounds is the hazardous element mercury. Burning coal and trash are the main ways mercury gets into the environment. So getting rid of mercury to a tolerable level is essential, if not paramount [28]. Several investigations have employed MBNs to remove mercury from water. The $Ti_3C_2T_x$ MXene nanosheets were modified by adding nano-layered molybdenum disulfide using a combination of hydrothermal treatment and ultra-sonication. With an adsorption rate of 7.160 mmol/g, the composite fulfilled the predictions of the Langmuir adsorption isotherm.

Furthermore, even at parts per billion (ppb) concentrations, metal ions did not impede the elimination of mercury ions [35]. By manipulating the concentrations of Ti_3C_2Tx MXene and sodium alginate, MXene core–shell aerogels (MX-SA) were created. Of all the compositions tested, the "MX-SA4:20" showed the best ability to adsorb mercury ions. The study found that mercury levels had risen to 1.8 ppb after only fifteen minutes. It effectively removes contaminants when the environment is acidic. $Ti_3C_2O_x$, a multilayer oxygen-functionalized Ti_3C_2, may adsorb up to 4806.0 mg/g. At a concentration drop from 10,400 to 33 g/L, oxygen-functionalized multilayered Ti_3C_2 (Ti_3C_2Ox) demonstrated an outstanding degradation efficiency of around 95.6% [36].

4.2 Organic Contaminant Adsorption

MXene-based materials may remove both organic and inorganic contaminants. There is a lot of wastewater from the textile sector that has colours. Multiple investigations have shown that MXenes and materials derived from MXene efficiently absorb and decrease the concentration of the cationic dye MB. Several research have extensively examined methylene blue so far. This text is a discussion of many discoveries. Authors in [37] successfully created porous MXene/single-walled carbon nanotubes using a simple and scalable method. The generated material was used as a dependent electrode to break down MB using AC. Figure 6 illustrates the functional properties and interfacial interactions of MXenes, highlighting their potential for effectively treating aqueous environments contaminated with organic pollutants.

Additionally, MB was degraded by electro-sorption using a high AC of 1068.80 mg/g at an applied voltage of 1.2 V. The highest recorded value for AC was 28,403.70 mg/g. In addition, the pH may be adjusted to remove MO by selective adsorption. A composite of PA and MXene was created hydrothermally, preparing a mixture of MXene and PA at varying percentages. The etching procedure used to develop

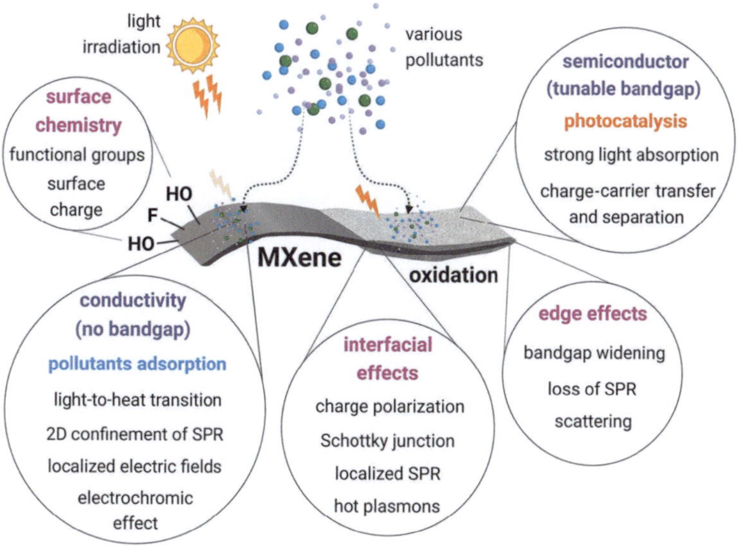

Fig. 6 Schematic representation of the functional attributes and interfacial phenomena of pristine and oxidized MXenes for the removal of organic pollutants via adsorption and photocatalytic processes. (Reproduced under CC-BY license from [27])

MXene also influences the dye adsorption process. To investigate the adsorption of MO and MB dyes, [38] contrasted Ti_3C_2 was produced by hydrothermal synthesis with Ti_3C_2 produced by HF etching. The problem is that no MXene formulation succeeded in adsorbing MO. Hydrothermally produced MXene showed much higher MB adsorption compared to HF-etched MXene. Their high specific surface area, a result of efficient etching and exfoliation, is likely to blame.

The effect of pH on the removal efficiency of MXenes for different aqueous organic dyes was studied by [39]. Figure 7 shows the different organic dyes found in polluted aqueous systems.

Ti_3C_2-MXenes were synthesized by etching Ti_3AlC_2 MAX phase with a concentrated HCl solution containing LiF [39]. XPS analysis confirmed the complete removal of the Al layer (Fig. 8a). XRD patterns showed an increased d-spacing in the etched product compared to the precursor (Fig. 8b), attributed to Li^+ intercalation and the formation of surface hydroxyl groups. SEM imaging revealed a structural transformation from 3 to 2D, indicating successful delamination of Ti_3C_2-MXenes (Fig. 8c and d).

In another study, d-$Ti_3C_2T_x$ nanosheets (MXene) were incorporated into sodium alginate (SA) matrices via electrospinning and subsequent calcium ion crosslinking (Fig. 9), resulting in a series of SA/MXene nanofiber membranes (NMs) [40]. These NMs effectively removed methylene blue from aqueous solutions, demonstrating their potential for water decontamination. SA/MXene nanofibers with 0.74 wt%

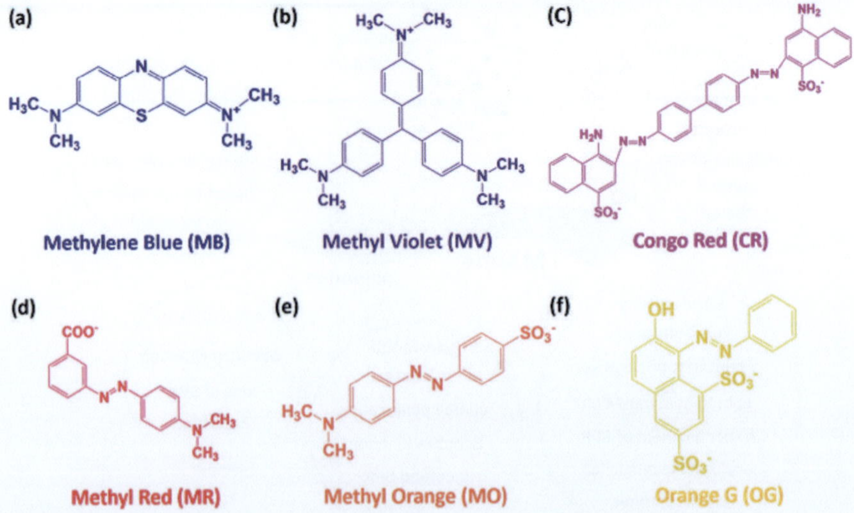

Fig. 7 Commonly occurring dye contaminants in wastewater (Reproduced under CC-BY license from [39])

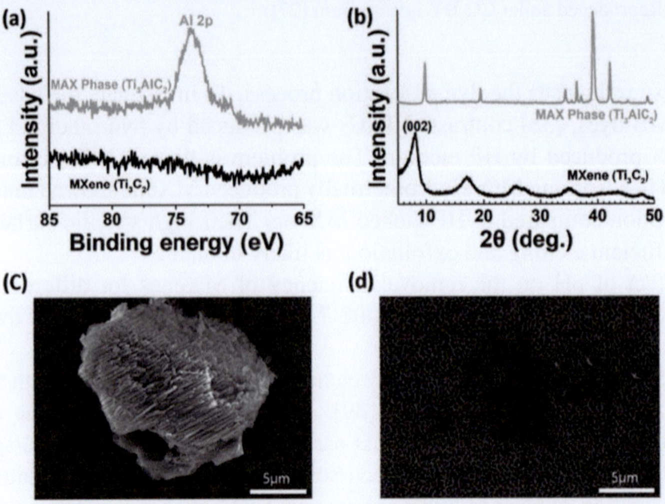

Fig. 8 **a** High-resolution Al 2p XPS spectra of Ti_3AlC_2 MAX phase and Ti_3C_2 MXene, **b** XRD patterns of Ti_3AlC_2 MAX phase and Ti_3C_2 MXene. SEM micrographs of **c** intact Ti_3AlC_2 MAX phase and **d** delaminated Ti_3C_2 MXene. (Reproduced under CC-BY license from [39])

MXene exhibited superior MB adsorption (440 mg/g) compared to other materials tested, such as SA/MXene beads having identical MXene content and pristine MXene. This performance was maintained after 10 cycles, demonstrating the

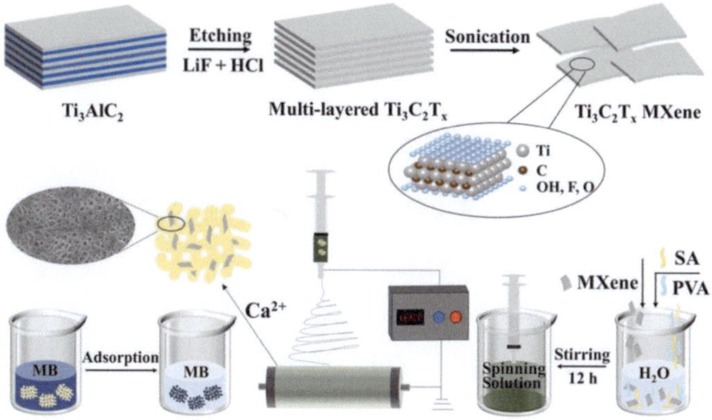

Fig. 9 Schematic representation of the sodium alginate (SA)/MXene nanofiber membrane (NM) fabrication process. Reproduced under (CC-BY license from [40])

reusability of the material and the potential of electrospinning for efficient adsorbent development.

Characterization techniques provide deep insigts into the properties of MXene-based materials. Figure 10 shows the morphologies of MXene-based composite samples as revealed by high-resolution SEM images [41]. The etching process revealed a smooth 2D layer of pure MXene with a lateral dimension of 1–2 μm. There were no imperfections on the surface of the 1D silver nanowire (AgNW), and it was smooth. Last but not least, the MXene sheet showed a uniform distribution of AgNW. The homogeneous deposition of AgNW and successful coordination through hydrogen bonding were both facilitated by the abundance of functional groups in MXene. EDS was used to find out what elements were present and how they were distributed on the surface of the MXene/AgNW. The etching and washing were successful because Al was not detected in the MXene. The oxygen concentration was caused by both the surface chemistry of AgNW and the self-oxidation of MXene during production. The presence of Ag indicates the effective immobilization of AgNW on MXene. Elemental mapping and distribution investigations confirmed Ag, C, F, O, and Ti. A near-identical and complete overlap existed between the SEM pictures and the Ag, O, and Ti mappings. The uniform distribution of AgNW on the MXene sheets was verified by EDS analysis, which showed that the hydrogen bonding was successful. The chemical etching procedure caused the surface to terminate.

Moreover, transmission electron microscopy (TEM) investigation was done to gain further insights into the properties of the synthesized MXen-based material (Fig. 11). The TEM images of AgNWs verified their 1D structure, which was smooth and wire-like. The effective etching of pure MXene was confirmed by TEM, which revealed a sheet-like form. MXene sheets and AgNWs were identifiable in MXene/AgNW. On top of the layered substance, there were several AgNWs. The presence of

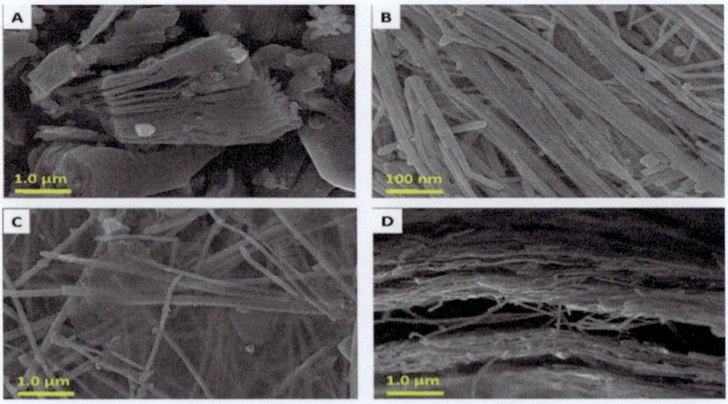

Fig. 10 High-resolution scanning electron micrographs illustrating the morphology of **a** MXene, **b** silver nanowires (AgNWs), and **c**, **d** MXene-AgNW composite. (Reproduced under CC-BY license from [41])

AgNW was confirmed by the elemental structure of the MXene/AgNW composite, which revealed a strong signal for Ag in a magnified dark image.

In another study, low cost MXene synthesis for regerative adsorption of benzene, toluene, ethylbenzene and xylene (BTEX) was demonstrated [42]. Cost-effective titanium oxide replaced high-risk pure titanium as a starting material. Molten salt shielding enabled lower-temperature synthesis of 3D MAX phases, which were subsequently transformed into 2D-MXenes and thoroughly analyzed. SEM analysis was used to get visual confirmation for the successful synthesis of MAX and

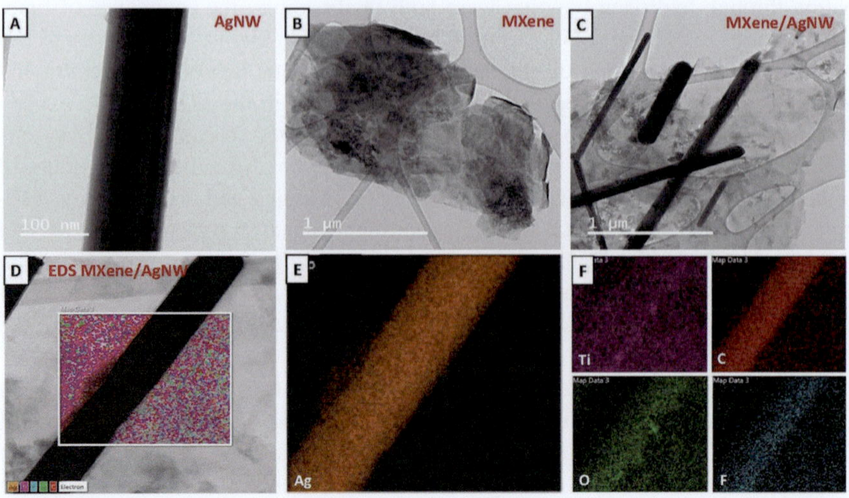

Fig. 11 TEM micrographs of **a** silver nanowire (AgNW), **b** MXene, and **c**, **d** the MXene/AgNW composite. Elemental mapping for the composite is shown in (**e**) and (**f**). (Reproduced under CC-BY license from [41])

MXenes. In general, the morphology of MAX phases shows a solid 3D structure with visible lines from which layers could be formed after being etched. On the other hand, the layered sheet structures (accordion-like) typify the morphology of MXenes. Figure 12 illustrates SEM micrographs of graphite-based and AC-based Ti_3AlC_2 and their respective MXenes. The 3D solid morphologies with compact layers expected in Ti_3AlC_2 confirmed successful MAX formation. Interestingly, the morphologies were very similar, irrespective of the carbon type used. The images showed that the synthesized MXene were multilayer (m-$Ti_3C_2T_x$). In AC-based MXene (MXene-AC), the layers were not so well-separated visibly even though the aqueous MXenes were ultrasonicated at 70% amplitude for 20 min, and tetramethylammonium hydroxide was used as the intercalant.

The mechanistic insights of adsorption onto MXene-based materials are essential to develop effective water purification methodologies. MXenes primarily interact with dye molecules and functional groups on their surface to adsorb dye molecules.

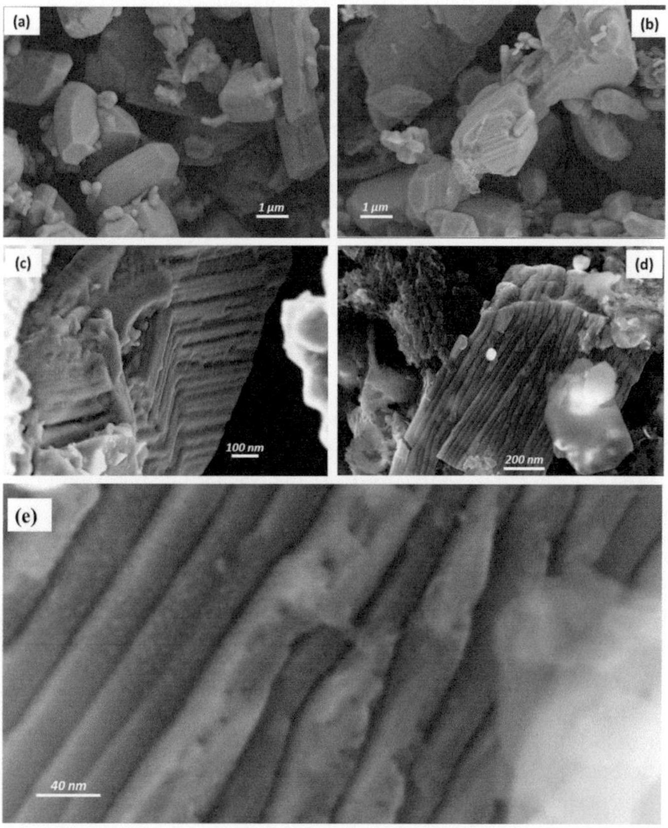

Fig. 12 SEM micrographs of **a** MAX-G, **b** MAX-AC, **c** MXene-G, **d** MXene-AC, and **e** higher magnification SEM image revealing incomplete layer separation. AC denotes activated carbon and G denotes graphite. (Reproduced under CC-BY license from [42])

While surface positive charges on MXene@Fe_3O_4 impede complex formation, MB's electrostatic adsorption helps the process. $Ti_3C_2T_x$ nanosheets may decrease MB and then adsorb it onto the MXene@Fe_3O_4 surface due to interactions among the hydroxyl (OH) group and the titanium (Ti) site. The nitrogen (N) of the MB participates in dipole–dipole hydrogen bonding interactions with the titanium-oxygen-hydrogen (Ti–OH) group, leading to the creation of an OH•••N bond. According to studies by Li et al. [43], three steps are involved in the adsorption of MB on MXene-: surface adsorption, internal particle diffusion, and equilibrium adsorption. The cationic dye MB is removed from MXene ($Ti_3C_2T_x$) by adsorption and photocatalytic degradation [44]. Through a sequence of operations, $Ti_3C_2T_x$ can eradicate MB.

In the first phase, the MB molecules adhere to the surfaces of $Ti_3C_2T_x$, leading to an increase in the arrangement irregularity, which might be attributed to the layered structure or chemical changes in MXene. Oxidation of MXene ($Ti_3C_2T_x$) leads to the synthesis of titania in the final stage. Alternatively, the dye may be broken down by photo-catalysis when exposed to UV light, perhaps due to the production of $Ti(OH)_4$ and TiO_2 on the surfaces of MXene. Due to its excellent efficiency, easy operation, and lack of secondary contaminants, the Fenton technique has attracted huge interest in decontamination processes. Oxidants such as persulfate (PS) and hydrogen peroxide (H_2O_2) are often used in systems that resemble Fenton's. In particular, MXene-based catalysts are very effective for the breakdown of organic pollutants utilizing H_2O_2/PS because they are often conducting and include several active sites.

5 Associated Decontamination Processes Using MXene-Based Materials

The capacity of oxidizing H_2O_2 to remove organic contaminants has attracted much interest. Free radicals are produced by the Fe^{2+} ions, which function as a catalyst [45]. The 2D MXene has electronegativity on the surface and has a lamellar structure. A two-dimensional MXene might have iron particles loaded into its outermost layer and lamellar gaps [46]. Also, MXene can reliably disperse iron nanoparticles so they can't accumulate [36].

Incorporating iron particles into 2D MXene increases the particular surface area and creates a more porous structure than pure 2D MXene. One possible method for activating H_2O_2 is to use the pores of the complex as little reactors. Hydrogen peroxide (H_2O_2) activated by magnetic nanosheets made of zero-valent iron at the (nZVI) and titanium carbide (Ti_3C_2) was able to remove ranitidine with a 91.1% removal efficiency in less than 30 min [47]. Iron species coated with MXene are very effective catalysts for degrading organic pollutants. Most particles are derived from iron, including Fe_3O_4, $MnFe_2O_4$, and nZVI. One important factor determining the activity level of 2D MXene is its weight ratio of the iron species to MXene.

The chemical makeup of 2D MXene was varied by adding or subtracting ferric acetylacetonate [45].

Coating a 2D MXene surface with iron particles might enhance the availability of reaction sites. MXene enhances the electron transport between magnetic particles. In addition, MXene/nZVI nanosheets possess several hydrophilic functional groups, including OH, -COOH, and -CHO, that enhance the activation of H_2O_2 [47].

The degradation of organic pollutants using persulfates (PS) has attracted considerable attention [48]. There are two main types of PS: peroxy-mono sulfate (PMS, HSO_5^-) and perox-disulfate (PDS, $S_2O_8^{2-}$) [49]. Furthermore, according to reference [50], the oxidation potentials of PMS are 1.82, and PDS is 2.01. Both accessible radical and non-radical routes, such as electron transfer, may activate peroxy-onosulfate (PMS) and persulfate (PDS), which can eliminate pollutants [51].

There are several advantages to activating peroxy-monosulfate using molecules based on MXene [52]. Because of its large surface area and layered structure, MXene can immobilize active sites [52]. Moreover, MXene's strong reducibility prevents active site aggregation during catalysis, improving efficiency. As a further advantage, MXene's exceptional specific capacity [53] makes it easy for electrons to move from the interface to the PMS. Peroxy-monosulfate (PMS) is activated when metal atoms are added to it because it increases the electron transfer rate to the surface of MXene[54]. An essential aspect of the manufacturing process was the activation of PMS with the use of nZVIPs/MXene. The interaction between PMS and Fe particles and 2D MoSe caused the lengthening of the lO–O bond. The catalyst was found to have two active centres, as shown by this experiment. Furthermore, PMS reacts well with two-dimensional MXene or Fe particles because of their low energy barrier [55]. As a result of metal ion leaching being less likely, secondary contamination was also less likely to occur.

One potential use of a two-dimensional MXene is as a substrate for metal oxide connections. Making compounds between MXene and metal oxides is easy.

Also, MXene/metal oxide composites exhibit superior catalytic capacities for PS compared to MXene or metal oxides alone. For instance, when comparing the activation of PS with CoFe2O4 alone versus that with MXene added, the latter showed superior results. The result was an increase in the valence cycle processes of Co(II)/Co(III), Fe(II)/Fe(III), and Ti(II)/Ti(IV), which enhanced the formation of a Co/Fe/Ti complex. As a result, electron mobility was amplified, accelerating the pollutants' breakdown[56]. Furthermore, surface functional groups like -O, -OH, and -F interact with metal ions to create stable complexes with metal/Ti, which improve redox reactions [57].

Capacity for activation is proportional to MXene's dimension. The more significant number of exposed active sites in monolayer MXene gives it a more significant PS activation potential than bulk MXene. A tiny "HOMO–LUMO" gap permits electron transport and produces free radicals in monolayer 2D MXene, in contrast to bulk 2D MXene with many layers. According to reported literature [57], the rate of exponential degradation of aqueous 2,4-dichloro phenoxyacetic acid on a single layer of MXene was more than 300 times faster than on bulk MXene. Developing and optimizing a monolayer 2D MXene is essential to improve PMS performance.

Both radicals that are free and non-radical oxidation may result from PDS activation [58]. The remarkable electrical conductivity of materials based on MXene suggests that they may be able to activate PDS and transport electrons. Because of these properties, MXene might be a good choice for water filtration when combined with organic pollutants.

6 Prospects

MXenes and MXene-based materials have captured significant research attention due to their adjustable electrical, optical, mechanical, and electrochemical characteristics, as well as their promising applications in various fields. Increasing efforts are being made to make the process of MXene synthesis more environmentally friendly. Figure 13 depicts the prospects for green synthesis of MXenes using sustainable approaches and their multifaceted applications [59].

Nanomaterials derived from MXene have an active surface area, layered structure, and charged surface that aid in the removal of toxic compounds from solutions in water. Many inorganic and organic compounds, including heavy metals like Pb(II), Cu(II), Cd(II), Cr(IV), and Hg(II), are resistant to breakdown in living organisms and can build up over time. Nanoparticles derived from MXene are ideal for environmental remediation application due to their large surface area and tailored functional groups, allowing them to remove contaminants from effluent effectively.

Wastewater and specific volatile organic compounds induce respiratory infections and other health issues in humans. Nanomaterials based on MXene can remove harmful organic compounds and contaminants from water due to their high adsorption energy and selectivity. The radioactive elements uranium, neptunium, and plutonium

Fig. 13 Sustainable synthesis of MXenes and their promising applications. (Reproduced under CC-BY license from [59])

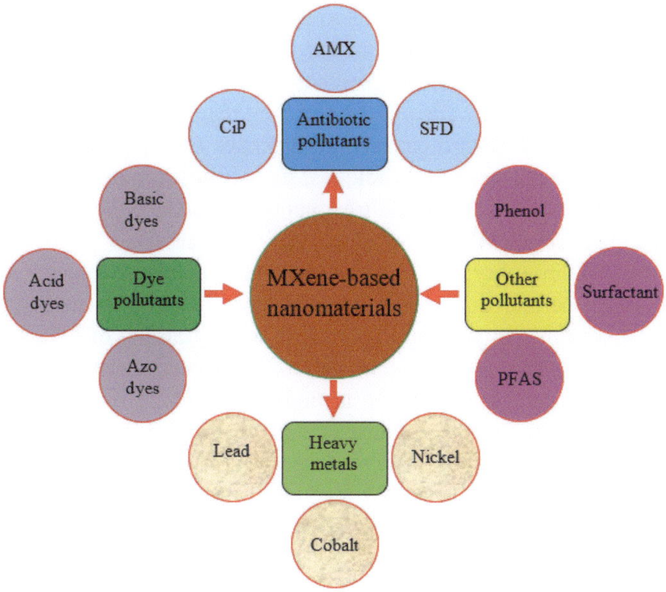

Fig. 14 Scope of MXene-based nanomaterials for wastewater treatment applications. (Reproduced under CC-BY license from [9])

ions pose chemical and radiological risks. Their nuclear waste is a major environmental problem. The extraordinary radiation resistance and good chemical compatibility of nanomaterials based on MXene allow them to be employed to eradicate these radioactive elements. Nanomaterials based on MXene also eliminate pollutants, including phenol, highly charged cations, and aqueous phosphates, because of their active surface morphology. Figure 14 provides an overview of the diverse applications of MXene-based nanomaterials in wastewater treatment [9].

7 Challenges and Recommendations

MXenes, a class of two-dimensional nanomaterials, have shown promise in adsorptive wastewater treatment due to their unique properties, such as high surface area, tunable surface chemistry, and excellent adsorption capabilities. However, there are still challenges to overcome to fully harness the potential of MXenes in this application. Some of the key challenges and recommendations for MXenes in adsorptive wastewater treatment are mentioned below.

i. Selectivity and capacity: One challenge is achieving high selectivity and capacity for specific pollutants. MXenes have shown good adsorption performance for various contaminants, including heavy metals and organic compounds.

However, further research is needed to optimize MXene properties and surface modifications to enhance selectivity and capacity for specific pollutants

ii. Regeneration and reusability: Another challenge is the regeneration and reusability of MXenes as adsorbents. Efficient regeneration methods are crucial to maintain the adsorption capacity of MXenes over multiple cycles. Developing effective regeneration techniques, such as desorption using environmentally friendly solvents or thermal treatment, can enhance the sustainability and cost-effectiveness of MXene-based adsorption processes

iii. Stability and durability: MXenes can undergo structural changes and degradation under certain conditions, affecting their adsorption performance and stability over time. Improving the strength and durability of MXenes through surface modifications, composite materials, or protective coatings can enhance their long-term performance in wastewater treatment

iv. Scale-up and cost: MXene synthesis and integration into practical adsorption systems is challenging. The cost-effectiveness of MXene-based adsorption processes needs to be considered, including the availability and cost of raw materials, synthesis methods, and the overall lifecycle cost of the technology. Further research and development are needed to optimize the synthesis and manufacturing processes to reduce costs and enable large-scale applications

v. Integration with other treatment technologies: MXenes can be integrated with other treatment technologies, such as membrane filtration or advanced oxidation processes, to enhance overall treatment efficiency. Exploring synergistic effects and developing integrated systems can improve the removal efficiency of MXenes and address the limitations of individual treatment methods. For instance, electrified membranes (EMs) offer a promising alternative to traditional water treatment methods [60]. Figure 15 illustrates the contaminant removal mechanisms in an EM-based separation process. Integrating adsorption with these innovative technologies could lead to more efficacious wastewater purification processes.

In conclusion, MXenes hold great potential for adsorptive wastewater treatment. Overcoming challenges related to selectivity, regeneration, stability, scale-up, and integration with other treatment technologies will be crucial for realizing the full benefits of MXenes in this field. Continued research and development efforts are needed to optimize MXene properties, synthesis methods, and system design for efficient and sustainable wastewater treatment.

8 Conclusions

Creating MXenes and MXenes based on nanomaterials was a preferable choice for the future of water remediation. The photodegradation potential of MXene-based nanocomposites to break down harmful environmental contaminants in water was much higher than that of other materials. Environmental photo-catalysis has been

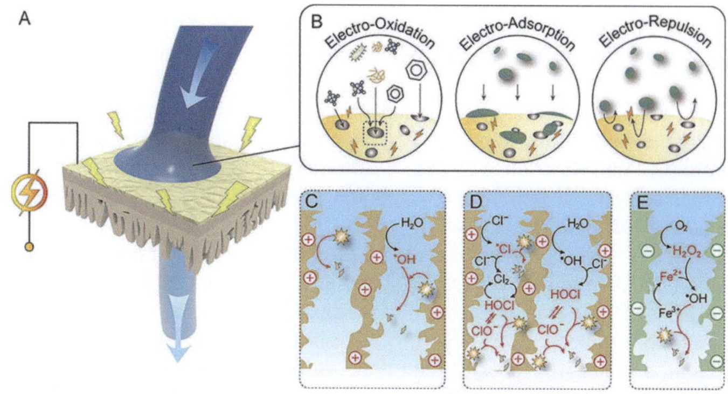

Fig. 15 Synergism of adsorption with electrified membrane (EM) filtration could enhance water purification. **a** Schematic representation of the EM filtration process for contaminant removal. **b** Upon voltage application, the EM interacts with pollutants through a combination of electrooxidation, electrostatic attraction, and repulsion, with electrooxidation being the primary decontamination mechanism. The indirect electrooxidation pathways encompass **c** hydroxyl radical-mediated oxidation, **d** active chlorine-based oxidation, and **e** electro-Fenton reactions. (Reproduced under CC-BY license from [60])

a thriving field for MXene nanostructures because of their exceptional thermal and mechanical durability, high water solubility, ideal thermal and electrical conductivity, and outstanding bandgap. Typically, the advantages of MXene-based composites are twice as significant as those of MXene alone. Due to their structure, which has multiple properties such as a large specific surface area, improved light harvesting capabilities, and shorter diffusion routs for photoexcited ions, MXenes and their composites have been shown in multiple studies to be highly effective at removing heavy metals, dyes, and radionuclides. The promising results of research investigations suggest that these novel materials have a lot of room for advancement despite the many obstacles they face. The exceptional properties of MXene position it as a promising material for transformative advancements in water and environmental remediation. Integrating MXene-based adsorption with complementary technologies can provide comprehensive solutions to address complex pollution challenges.

References

1. Rahman, M., Ishaque, F., Hossain, M.A., Mahdy, I.H., Roy, P.P.: Impact of industrialization and urbanization on water quality of Surma river of Sylhet city. Desalin. Water Treat. **235**, 333–345 (2021). https://doi.org/10.5004/dwt.2021.27661
2. Deniz, F., Dogan, F.: Effective cleaning of a hazardous synthetic triarylmethane-type dye from aquatic environment with a multifunctional waste biomass–based biosorbent. Biomass Convers. Biorefinery **13**(12), 10885–10892 (2022). https://doi.org/10.1007/s13399-021-01995-9
3. Intisar, A., et al.: Occurrence, toxic effects, and mitigation of pesticides as emerging environmental pollutants using robust nanomaterials—a review. Chemosphere **293**, 133538 (2022). https://doi.org/10.1016/j.chemosphere.2022.133538

4. Yadav, S., et al.: Updated review on emerging technologies for PFAS contaminated water treatment. Chem. Eng. Res. Des. **182**, 667–700 (2022). https://doi.org/10.1016/j.cherd.2022.04.009
5. Khor, C.M., et al.: Performance, energy and cost of produced water treatment by chemical and electrochemical coagulation. Water **12**(12), 3426 (2020). https://doi.org/10.3390/w12123426
6. Wu, C., Zhi, D., Yao, B., Zhou, Y., Yang, Y., Zhou, Y.: Immobilization of microbes on biochar for water and soil remediation: a review. Environ. Res. **212**, 113226 (2022). https://doi.org/10.1016/j.envres.2022.113226
7. Poonguzhali, E., Kapoor, A., Prabhakar, S.: Membrane assisted process intensification and optimization for removal and recovery of phenol from industrial effluents. Sep. Purif. Technol. **319**, 124026 (2023). https://doi.org/10.1016/j.seppur.2023.124026
8. Hajian, S., Maddipatla, D., Narakathu, B.B., Atashbar, M.Z.: MXene-based flexible sensors: a review. Front. Sensors **3**(October), 1–24 (2022). https://doi.org/10.3389/fsens.2022.1006749
9. Saafie, N., Zulfiqar, M., Samsudin, M.F.R., Sufian, S.: Current scenario of MXene-based nanomaterials for wastewater remediation: a review. Chemistry (Easton) **4**(4), 1576–1608 (2022). https://doi.org/10.3390/chemistry4040104
10. Pogorielov, M., Smyrnova, K., Kyrylenko, S., Gogotsi, O., Zahorodna, V., Pogrebnjak, A.: Mxenes—a new class of two-dimensional materials: structure, properties and potential applications. Nanomaterials **11**(12) (2021). https://doi.org/10.3390/nano11123412
11. VahidMohammadi, A., Rosen, J., Gogotsi, Y.: The world of two-dimensional carbides and nitrides (MXenes). Science (80) 372(6547) (2021). https://doi.org/10.1126/science.abf1581
12. Siwal, S.S., et al.: Novel synthesis methods and applications of MXene-based nanomaterials (MBNs) for hazardous pollutants degradation: future perspectives. Chemosphere **293**, 133542 (2022). https://doi.org/10.1016/j.chemosphere.2022.133542
13. Li, X., Ran, F., Yang, F., Long, J., Shao, L.: Advances in MXene films: synthesis, assembly, and applications. Trans. Tianjin Univ. **27**(3), 217–247 (2021). https://doi.org/10.1007/s12209-021-00282-y
14. Jeon, M., et al.: A review on MXene-based nanomaterials as adsorbents in aqueous solution. Chemosphere **261**, 127781 (2020). https://doi.org/10.1016/j.chemosphere.2020.127781
15. Liao, M., et al.: MXenes as emerging adsorbents for removal of environmental pollutants. Sci. Total. Environ. **912**, 169014 (2024). https://doi.org/10.1016/j.scitotenv.2023.169014
16. Damptey, L., et al.: Surface functionalized MXenes for wastewater treatment-a comprehensive review. Glob. Challenges (Hoboken, NJ) **6**(6), 2100120 (2022). https://doi.org/10.1002/gch2.202100120
17. Bjork, J., Rosen, J.: Functionalizing MXenes by tailoring surface terminations in different chemical environments. Chem. Mater. (2021). https://doi.org/10.1021/acs.chemmater.1c01264
18. Zhang, T., Chang, L., Xiao, X.: Surface and interface regulation of MXenes: methods and properties. Small Methods **7**(8) (2023). https://doi.org/10.1002/smtd.202201530
19. Kwon, O., Choi, Y., Kang, J., Kim, J.H., Choi, E., Woo, Y.C., Kim, D.W.: A comprehensive review of MXene-based water-treatment membranes and technologies: recent progress and perspectives. Desalination **522**, 11544 (2022)
20. Janica, I., et al.: Covalently functionalized MXenes for highly sensitive humidity sensors. Small Methods **7**(8) (2023). https://doi.org/10.1002/smtd.202201651
21. Yen, Z., Bradley, D.G., Wang, Y., Hanna, J.V., Lam, Y.M.: Organic functionalization of $Ti_3C_2T_x$ MXene for urea adsorption in aqueous solutions. ACS Appl. Nano Mater. **7**(7), 7430–7441 (2024). https://doi.org/10.1021/acsanm.4c00119
22. Shi, X., et al.: Largely enhanced adsorption performance and stability of MXene through in-situ depositing polypyrrole nanoparticles. Sep. Purif. Technol. **287**, 120596 (2022). https://doi.org/10.1016/j.seppur.2022.120596
23. Jawad, A.H., Hameed, B.H., Abdulhameed, A.S.: Synthesis of biohybrid magnetic chitosan-polyvinyl alcohol/MgO nanocomposite blend for remazol brilliant blue R dye adsorption: solo and collective parametric optimization. Polym. Bull. **80**(5), 4927–4947 (2022). https://doi.org/10.1007/s00289-022-04294-z

24. Zain, Z.M., Abdulhameed, A.S., Jawad, A.H., ALOthman, Z.A., Yaseen, Z. M.: A pH-sensitive surface of chitosan/sepiolite clay/algae biocomposite for the removal of malachite green and remazol brilliant blue R dyes: optimization and adsorption mechanism study. J. Polym. Environ. **31**(2), 501–518 (2022). https://doi.org/10.1007/s10924-022-02614-y
25. Schultz, T., et al.: Surface termination dependent work function and electronic properties of $Ti_3C_2T_x$ MXene. Chem. Mater. **31**(17), 6590–6597 (2019). https://doi.org/10.1021/acs.chemmater.9b00414
26. Tunesi, M.M., Soomro, R.A., Han, X., Zhu, Q., Wei, Y., Xu, B.: Application of MXenes in environmental remediation technologies. Nano Converg. **8**(1), 5 (2021). https://doi.org/10.1186/s40580-021-00255-w
27. Bury, D., et al.: Cleaning the environment with MXenes. MRS Bull. **48**(3), 271–282 (2023). https://doi.org/10.1557/s43577-023-00507-6
28. Ibrahim, Y., Kassab, A., Eid, K., Abdullah, A.M., Ozoemena, K.I., Elzatahry, A.: Unveiling fabrication and environmental remediation of MXene-based nanoarchitectures in toxic metals removal from wastewater: strategy and mechanism. Nanomaterials (Basel, Switzerland) **10**(5), 885 (2020). https://doi.org/10.3390/nano10050885
29. Ying, Y., et al.: Two-dimensional titanium carbide for efficiently reductive removal of highly toxic chromium(VI) from water. ACS Appl. Mater. Interfaces **7**(3), 1795–1803 (2015). https://doi.org/10.1021/am5074722
30. Krstić, V., Urošević, T., Pešovski, B.: Corrigendum to 'A review on adsorbents for treatment of water and wastewaters containing copper ions' [Chem. Eng. Sci. **192,** 273–287 (2018)]. Chem. Eng. Sci. **261**, 117987 (2022). https://doi.org/10.1016/j.ces.2022.117987
31. Shahzad, A., et al.: Mercuric ion capturing by recoverable titanium carbide magnetic nanocomposite. J. Hazard. Mater. **344**, 811–818 (2018). https://doi.org/10.1016/j.jhazmat.2017.11.026
32. Shahzad, A., et al.: Two-dimensional $Ti_3C_2T_x$ MXene nanosheets for efficient copper removal from water. ACS Sustain. Chem. Eng. **5**(12), 11481–11488 (2017). https://doi.org/10.1021/acssuschemeng.7b02695
33. Elumalai, S., Yoshimura, M., Ogawa, M.: Simultaneous delamination and rutile formation on the surface of $Ti_3C_2T_x$ MXene for copper adsorption. Chem. Asian J. **15**(7), 1044–1051 (2020). https://doi.org/10.1002/asia.202000090
34. Gan, D., et al.: Bioinspired functionalization of MXenes ($Ti_3C_2T_x$) with amino acids for efficient removal of heavy metal ions. Appl. Surf. Sci. **504**, 144603 (2020). https://doi.org/10.1016/j.apsusc.2019.144603
35. Shahzad, A., Jang, J., Lim, S.-R., Lee, D.S.: Unique selectivity and rapid uptake of molybdenum-disulfide-functionalized MXene nanocomposite for mercury adsorption. Environ. Res. **182**, 109005 (2020). https://doi.org/10.1016/j.envres.2019.109005
36. Shi, W., Fu, Y., Hao, C., Guo, F., Tang, Y.: Heterogeneous photo-Fenton process over magnetically recoverable $MnFe_2O_4$/MXene hierarchical heterostructure for boosted degradation of tetracycline. Mater. Today Commun. **33**, 104449 (2022). https://doi.org/10.1016/j.mtcomm.2022.104449
37. Yao, C., et al.: A facile synthesis of porous MXene-based freestanding film and its spectacular electrosorption performance for organic dyes. Sep. Purif. Technol. **263**, 118365 (2021). https://doi.org/10.1016/j.seppur.2021.118365
38. Peng, C., et al.: A hydrothermal etching route to synthesis of 2D MXene (Ti_3C_2, Nb_2C): enhanced exfoliation and improved adsorption performance. Ceram. Int. **44**(15), 18886–18893 (2018). https://doi.org/10.1016/j.ceramint.2018.07.124
39. Lim, S., et al.: Role of electrostatic interactions in the adsorption of dye molecules by Ti_3C_2-MXenes. RSC Adv. **11**(11), 6201–6211 (2021). https://doi.org/10.1039/d0ra10876f
40. Li, M., Zhang, P., Wang, Q., Yu, N., Zhang, X., Su, S.: Electrospinning novel sodium alginate/MXene nanofiber membranes for effective adsorption of methylene blue. Polymers (Basel) **15**(9), 2110 (2023). https://doi.org/10.3390/polym15092110
41. Mushtaq, S., et al.: MXene/AgNW composite material for selective and efficient removal of radioactive cesium and iodine from water. Sci. Rep. **13**(1), 19696 (2023). https://doi.org/10.1038/s41598-023-47075-y

42. Oyehan, T.A., Ganiyu, S.A., Pfrang, C., Valsami-Jones, E.: Low cost Mxene synthesis for regenerative adsorption of benzene, toluene, ethylbenzene and Xylene (Btex). Elsevier BV (2024). https://doi.org/10.2139/ssrn.4688780
43. Li, K., et al.: Self-assembled MXene-based nanocomposites via layer-by-layer strategy for elevated adsorption capacities. Colloids Surf A Physicochem Eng Asp **553**, 105–113 (2018). https://doi.org/10.1016/j.colsurfa.2018.05.044
44. Mashtalir, O., Cook, K.M., Mochalin, V.N., Crowe, M., Barsoum, M.W., Gogotsi, Y.: Dye adsorption and decomposition on two-dimensional titanium carbide in aqueous media. J. Mater. Chem. A **2**(35), 14334–14338 (2014). https://doi.org/10.1039/c4ta02638a
45. Cui, Y., et al.: A novel one-step strategy for preparation of Fe_3O_4-loaded Ti_3C_2 MXenes with high efficiency for removal organic dyes. Ceram. Int. **46**(8), 11593–11601 (2020). https://doi.org/10.1016/j.ceramint.2020.01.188
46. He, L., et al.: Nanoscale zero-valent iron intercalated 2D titanium carbides for removal of Cr(VI) in aqueous solution and the mechanistic aspect. J. Hazard. Mater. **388**, 121761 (2020). https://doi.org/10.1016/j.jhazmat.2019.121761
47. Ma, Y., Lv, X., Xiong, D., Zhao, X., Zhang, Z.: Catalytic degradation of ranitidine using novel magnetic Ti_3C_2-based MXene nanosheets modified with nanoscale zero-valent iron particles. Appl. Catal. B Environ. **284**, 119720 (2021). https://doi.org/10.1016/j.apcatb.2020.119720
48. Li, N., et al.: Influences and mechanisms of phosphate ions onto persulfate activation and organic degradation in water treatment: a review. Water Res. **222**, 118896 (2022). https://doi.org/10.1016/j.watres.2022.118896
49. Devi, P., Das, U., Dalai, A.K.: In-situ chemical oxidation: Principle and applications of peroxide and persulfate treatments in wastewater systems. Sci. Total. Environ. **571**, 643–657 (2016). https://doi.org/10.1016/j.scitotenv.2016.07.032
50. Peng, W., et al.: Non-radical reactions in persulfate-based homogeneous degradation processes: a review. Chem. Eng. J. **421**, 127818 (2021). https://doi.org/10.1016/j.cej.2020.127818
51. Yan, X., Yue, D., Guo, C., Wang, S., Qian, X., Zhao, Y.: Effective removal of chlorinated organic pollutants by bimetallic iron-nickel sulfide activation of peroxydisulfate. Chinese Chem. Lett. **31**(6), 1535–1539 (2020). https://doi.org/10.1016/j.cclet.2019.11.003
52. Hu, L., et al.: MXene-induced electronic optimization of metal-organic framework-derived CoFe LDH nanosheet arrays for efficient oxygen evolution. Appl. Catal. B Environ. **298**, 120599 (2021). https://doi.org/10.1016/j.apcatb.2021.120599
53. Song, X., et al.: Oligolayered $Ti_3C_2T_x$ MXene towards high performance lithium/sodium storage. Nano Res. **13**(6), 1659–1667 (2020). https://doi.org/10.1007/s12274-020-2789-6
54. Jin, L., You, S., Yao, Y., Chen, H., Wang, Y., Liu, Y.: An electroactive single-atom copper anchored MXene nanohybrid filter for ultrafast water decontamination. J. Mater. Chem. A **9**(46), 25964–25973 (2021). https://doi.org/10.1039/d1ta07396f
55. Ma, Y., et al.: Rapid and long-lasting acceleration of zero-valent iron nanoparticles@Ti_3C_2-based MXene/peroxymonosulfate oxidation with bi-active centers toward ranitidine removal. J. Mater. Chem. A **9**(35), 19817–19833 (2021). https://doi.org/10.1039/d1ta02046c
56. Cai, T., et al.: Ag_3PO_4/Ti_3C_2 MXene interface materials as a Schottky catalyst with enhanced photocatalytic activities and anti-photocorrosion performance. Appl. Catal. B Environ. **239**, 545–554 (2018). https://doi.org/10.1016/j.apcatb.2018.08.053
57. Fayyaz, A., Saravanakumar, K., Talukdar, K., Kim, Y., Yoon, Y., Park, C.M.: Catalytic oxidation of naproxen in cobalt spinel ferrite decorated $Ti_3C_2T_x$ MXene activated persulfate system: mechanisms and pathways. Chem. Eng. J. **407**, 127842 (2021). https://doi.org/10.1016/j.cej.2020.127842
58. Yun, E.-T., Yoo, H.-Y., Bae, H., Kim, H.-I., Lee, J.: Exploring the role of persulfate in the activation process: radical precursor versus electron acceptor. Environ. Sci. Technol. **51**(17), 10090–10099 (2017). https://doi.org/10.1021/acs.est.7b02519

59. Amrillah, T., Abdullah, C.A.C., Hermawan, A., Sari, F.N.I., Alvani, V.N.: Towards greener and more sustainable synthesis of MXenes: a review. Nanomaterials **12**(23) (2022). https://doi.org/10.3390/nano12234280
60. Sun, M., et al.: Electrified membranes for water treatment applications. ACS Es&T Eng. **1**(4), 725–752 (2021). https://doi.org/10.1021/acsestengg.1c00015

Surface Engineering and Processing via MXene for Corrosion Protection Coatings

Bibekananda Nayak, K. Sumiran Dash, Sharmistha Anwar, and Shahid Anwar

Abstract Metals and alloys are crucial part of our day-to-day life, appearing in practically most element of human endeavor. But one of the major complications faced is their subjection toward corrosion. Technologies for corrosion prevention are blooming as this is a major complication faced in numerous industries, including automobile, defence, aerospace, and medical. Consequently, corrosion-protective technologies are crucial for minimizing resource waste. On focusing corrosion protection of the material coating a new novel class of two-dimensional layered material like MXene is currently being considered for potential components of corrosion protection strategy against corrosion. In this book chapter, we are discussing on MXene toward an innovative category of two-dimensional (2D) material, characterized by their layered structure and exceptional performance, rendering them highly promising for diverse applications. Most importantly, the distinctive physical barrier properties of these two-dimensional nanosheets position them as effective barriers against oxygen and hydrogen. MXene often exhibits a good electrical conductivity which can help in creating a conductive protection layer on the substrate, facilitating electrochemical process that inhibits corrosion.

Keywords MXene · Composite coating · Corrosion protection · Surface engineering · Electrochemical properties

---

B. Nayak (✉) · K. S. Dash · S. Anwar (✉)
Materials Chemistry and Interfacial Engineering Department, CSIR-Institute of Minerals and Materials Technology, Bhubaneswar, India
e-mail: shahidanwr@gmail.com

B. Nayak · K. S. Dash · S. Anwar · S. Anwar
Academy of Scientific and Innovative Research (AcSIR) Ghaziabad, Ghaziabad, India

S. Anwar
Pyrometallurgy and Materials Engineering Department, CSIR-Institute of Minerals and Materials Technology, Bhubaneswar, India

© The Author(s), under exclusive license to Springer Nature Singapore Pte Ltd. 2025
S. Moharana and R. N. Mahaling (eds.), *Surface Engineering of MXenes*, Engineering Materials, https://doi.org/10.1007/978-981-96-3640-2_16

1 Introduction

Corrosion is a natural phenomenon that results in materials' deterioration, especially metals deteriorate because of the chemical reactions with their surroundings. It happens when metals interact with elements like oxygen, water, acids, or salts, resulting in the formation of corrosion by-products like rust or tarnish [1]. Corrosion can appear in different ways, from uniform corrosion affecting entire surfaces gradually to localized corrosion, which targets specific areas and can be more harmful. Various factors like temperature, humidity, pH levels, and pollutants can influence how quickly and extensively corrosion occurs. The impacts of corrosion are widespread and can be costly and dangerous [2]. It affects industries across the board, from infrastructure and transportation to manufacturing and electronics, leading to expensive repairs, environmental harm, safety risks, and reduced equipment efficiency. Hence, it's crucial to develop effective strategies to prevent and manage corrosion to safeguard materials and infrastructure integrity in the long run.

For example, when iron (Fe) undergoes or exposed to oxygen and moisture, a chemical reaction occurs and form iron oxide (rust), as demonstrated by the equation:

$$4Fe + 3O_2 \rightarrow 2Fe_2O_3$$

This reaction shows iron with oxygen and water to produce iron oxide, commonly known as rust, which is characterized by its red-brown coloration and often seen on iron surfaces. Similarly, in environments containing acidic substances, metals can experience accelerated corrosion. For example, when iron reacts with hydrochloric acid (HCl), the following chemical reaction occurs:

$$Fe + 2HCl \rightarrow FeCl_2 + H_2$$

This reaction highlights the reaction of iron with hydrochloric acid to generate iron chloride ($FeCl_2$) and hydrogen gas (H_2). These chemical reaction processes involve in corrosion mechanism of metals, where metal atoms lose electrons to form positively charged ions, which then combine with other substances in the environment to produce corrosion by-products [3, 4].

It describes that the two different types of corrosion emphasizes their widespread impact, whether in uniform corrosion, where entire surfaces gradually corrode; galvanic corrosion, which occurs when two different metals are in contact, accelerating degradation; and pitting corrosion, which creates small, localized areas of damage. Corrosion poses significant economic, environmental, and safety challenges. Economically, it requires hefty spending on repairs, maintenance, and premature replacements, creating financial burdens for industries. Environmentally, corrosion leads to pollution through by-products and leaks, endangering ecosystems and human health. Safety risks emerge from weakened structures, increasing the likelihood of accidents and injuries, sometimes even fatalities. Furthermore, corrosion reduces energy efficiency by disrupting the smooth functioning of equipment and systems, leading to higher energy consumption and costs. Addressing corrosion requires comprehensive strategies encompassing coatings, inhibitors, and material selection to mitigate its detrimental effects, ensuring the longevity and reliability of infrastructure, machinery, and other assets of industries [5].

The corrosion of metals presents a longstanding global challenge, resulting in remarkable economic losses and numerous risks. In recent technological applications like gas pipelines and transportation (including cars and trains), corrosion remains a serious issue. Over recent years, numerous methods are being devised to improve corrosion-resistant properties. These include cathodic protection, corrosion inhibitors such as ionic liquids, drugs, and surfactants, and organic coatings. Organic coatings, in particular, are promising due to their simplicity, effectiveness, and cost efficiency. However, they also have drawbacks; for instance, epoxy coatings can develop micropores due to solvent evaporation and cross-linker interaction, allowing electrolytes to penetrate the metal rapidly [6]. Therefore, integrating anticorrosion agents into coating formulations is crucial for achieving highly efficient corrosion protection.

In recent years, significant research focus on the development of "self- protection coatings." These coatings utilize either intrinsic or extrinsic self-protection mechanisms to achieve their intended functionality. Intrinsic self-protection coatings work by disrupting and then reorganizing molecular chains within the material itself. This molecular-level response enables the coating to autonomously repair damage, influencing the inherent properties of the material. However, intrinsic protection coatings often require specialized resin, which can be costly. On the other hand, extrinsic smart coatings incorporate micro- or nano-carrier-loaded materials with active agents, such as corrosion inhibitors. These active agents can be released and adsorbed onto the metal surface either chemically or physically when the corrosion process begins [7]. The released corrosion inhibitors then create a protective layer on the metal surface and effectively protect the damage. Utilizing micro- or nano-carriers presents a promising solution to address challenges associated with directly adding corrosion inhibitors. By embedding corrosion inhibitors into the pores or between the layers of these carriers, they can be released precisely where needed when corrosion occurs. Figure 1 illustrates how coatings containing fillers or carriers loaded with corrosion inhibitors function to reduce the corrosion rate of stainless-steel substrates. In Fig. 1a, without protective coatings, stainless-steel substrate surfaces quickly come into contact with corrosive elements, leading to corrosion within hours. Figure 1b shows the coated stainless-steel surfaces act as barriers, shielding the metal from corrosive ions, but defects and cracks in the coating can still allow penetration by corrosive agents. Figure 1c illustrates addition of fillers helps by extending the diffusion path of corrosive ions, filling voids, and enhancing the coating's mechanical properties. However, all three coating systems lack active inhibition properties, leaving them vulnerable to failure when coating cracks develop. Whereas Fig. 1d, which includes carriers loaded with corrosion inhibitors, not only shares the benefits of the previous coatings but also provides ongoing protection by releasing active corrosion inhibitors on the surface of the bare substrate, even in the presence of coating cracks [8].

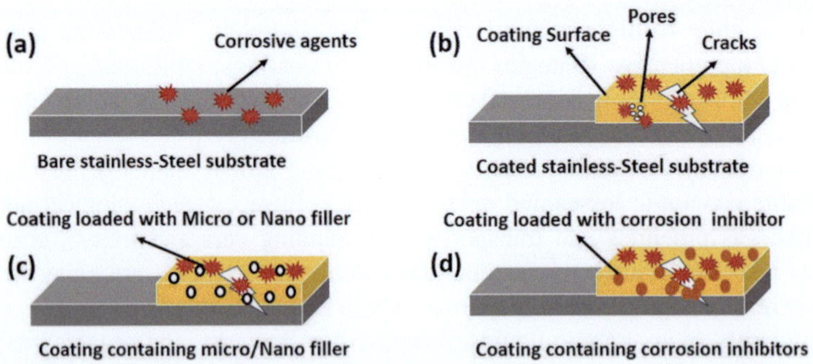

Fig. 1 Schematic representation of the corrosion mechanism on stainless-steel substrate without or with filler carrier coadding and also corrosion inhibitor

2 Introduction to MXene and Its Properties

Inorganic 2D MXenes, similar in structure to graphene, consist of transition metal carbides, nitrides, or carbonitrides. Their general formula is $M_{n+1}X_nT_x$ (where n = 1–3), with M representing an early transition metal (e.g., Sc, Ti, Zr, Hf, V, Nb, Ta, Cr, and Mo), X representing carbon or nitrogen, and T_x denoting surface functional groups (–OH, = O, –F, and occasionally –Cl) that are abundant on the surface of the M-layer during MXene production through chemical etching of the MAX phase. MXenes possess exceptional properties, including a large surface area, excellent biocompatibility, antibacterial properties, high hydrophilicity and electrical conductivity, flexibility in surface functionalization, and targeted therapeutic capabilities. These characteristics have led to the widespread use of MXenes in biomedical applications [9, 10].

One of the most remarkable properties of MXenes is their metallic behavior, characterized by a stable electron density near the Fermi level, similar to their precursor MAX phases. The electronic band structure of MXenes can be adjusted by altering their composition or surface chemistry, achieved through various methods such as using different etchants or controlling intercalated ions, interlayer spacing, defects, and grain boundaries. Additionally, the surface chemistry of MXenes can be modified through techniques like aromatic coupling-diazotization to introduce functional groups. Annealing MXenes under vacuum or inert gas flow can enhance carrier mobility by removing surface terminations and trapped water molecules, thereby significantly increasing electrical conductivity. Intercalating ions such as Li^+, Na^+, or K^+ can also enhance electrical conductivity by boosting carrier density [11]. However, intercalating bulky cations may induce a transition from metallic to semiconductor behavior by expanding interlayer distances and impeding electron hopping between flakes. The electrical conductivity of MXenes is influenced by their microstructure modification. For example, ultrathin, tightly packed MXene films composed of highly aligned flakes exhibit remarkably high electrical conductivity, reaching values as high as 15,100 S/cm [11].

3 Corrosion Mechanism and Challenges

Corrosion occurs through the exchange of electrons between anode and cathode in a conductive solution. At the anode, metal undergoes oxidation, releasing free electrons, while at the cathode, metal accepts these electrons, undergoing reduction [12]. The metal with the higher reduction potential becomes the anode, while the one with lower potential becomes the cathode, resulting in corrosion of the anodic metal. Certain conditions must be met for corrosion to take place.

For corrosion to occur, there must be distinct regions, known as cathodic and anodic areas, between which current can flow through a conducting medium.

At the anode, metal undergoes oxidation, leading to the depletion of the metal surface as it combines with oxygen. Consequently, corrosion always happens at the anode:

$$A \rightarrow A^{n+} + ne^{-}(\text{Oxidation})$$

At the cathode, reduction occurs, during which dissolved oxygen accepts electrons from the anode. This process results in the formation of ions such as OH^- or O^{2-}. The chemical reaction occurs

$$\frac{1}{2}O_2 + H_2O + 2e^- \rightarrow 2OH^-(\text{reduction})$$

$$\frac{1}{2}O_2 + 2e^- \rightarrow O^{2-}(\text{reduction})$$

As ions released from different electrodes combine, a corrosive product forms somewhere between these electrodes and both cathodic and anodic reactions occur.

- **Reaction Occurs at Anode Site**

Corrosion primarily occurs at the anode, where oxidation or the loss of electrons from the metal occurs. Typically, the metal with the higher reduction potential is chosen as the anode. As a result of oxidation, the anodic metal (A) transforms into its respective ions, losing electrons in the process:

$$A \rightarrow A^{n+} + ne^{-}(\text{Oxidation})$$

- **Reaction Occurs at Cathode Site**

At the cathode, reduction occurs, where the metal accepts electrons from the anode. This can happen in two ways, either by the evolution of hydrogen or by the absorption of oxygen.

- **Evolution of Hydrogen**

Hydrogen liberation is a corrosion process that typically occurs in acidic or alkaline environments. In this process, H+ ions from the electrolyte solution accept electrons from the anode, leading to the formation of hydrogen H_2 gas.

$$2H^+ + 2e^- \rightarrow H_2(t)$$

- **Absorption of Oxygen**

This corrosion process involves the absorption of oxygen, which typically occurs in neutral mediums. Often, NaCl is used as a conducting medium. In this scenario, oxygen is absorbed, leading to the formation of OH– ions.

$$\frac{1}{2}O_2 + H_2O + 2e^- \rightarrow 2OH^-$$

Types of corrosion are schematically represented in Fig. 2, and discussed in the following section.

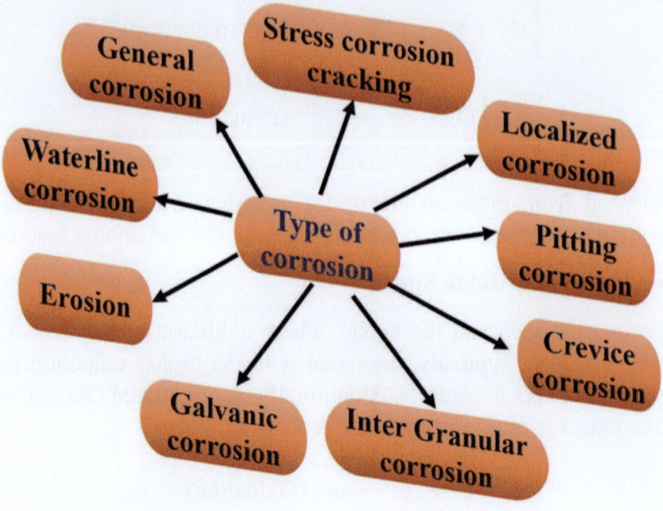

Fig. 2 Schematic representation of type of corrosion

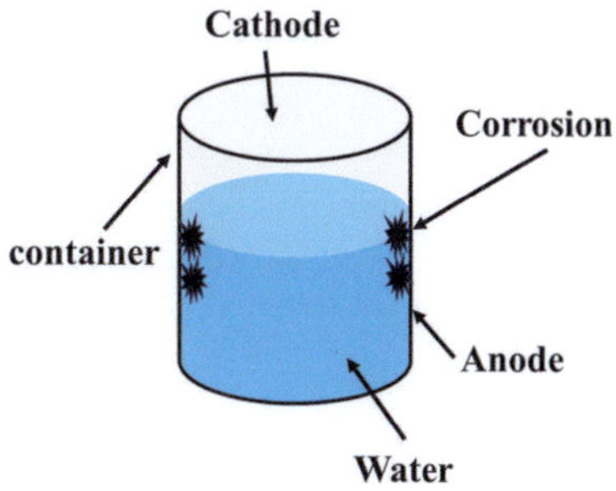

Fig. 3 Schematic representation of waterline corrosion

3.1 Waterline Corrosion

This type of corrosion is commonly observed in metallic tanks. When these tanks are partially filled with water, a unique corrosion process occurs [12]. The area under the water surface, i.e., poorly oxygenated, acts as the anode, while the area above the waterline, where there is a significant amount of oxygen, acts as the cathode. This setup is illustrated in Fig. 3. As a result, the area just below the waterline is susceptible to corrosion, while the area above the waterline is protected from corrosion.

3.2 General Corrosion

This type of corrosion, also referred to as uniform or dry corrosion, gradually deteriorates the entire surface of the metal uniformly. It progresses at a consistent rate across the sterile area. Oxygen plays a significant role in initiating this corrosion process. Commonly affected materials include cast iron and steel, which develop a rust-like appearance when exposed to moist atmospheric conditions [13].

3.3 Stress Corrosion Cracking

Stress corrosion cracking is a concerning phenomenon for a wide range of metal users. It can occur unexpectedly during service conditions, often leading to catastrophic consequences without any prior warning. This type of corrosion occurs when a material, initially in an inert environment, is subjected to stress, resulting in the formation of cracks in a corrosive environment. Stress corrosion cracking can be accelerated by residual internal stresses within the metal or by externally applied stress [12]. This type of corrosion is particularly prevalent at high temperatures and is more commonly observed in alloys than in pure metals. For stress corrosion

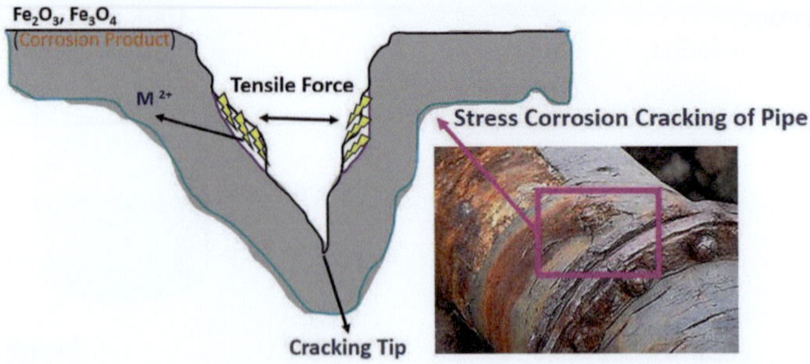

Fig. 4 Schematic representation of stress corrosion cracking. Reproduced from Ref. [13] with permission from Elsevier, Copyright [2008]

cracking to occur, three conditions must be met: the presence of a susceptible material, exposure to a corrosive environment, and the presence of tensile stresses. The schematic representation of stress corrosion cracking on pipe is depicted in Fig. 4.

3.4 Localized Corrosion

Localized corrosion differs significantly from general corrosion. While general corrosion affects relatively larger areas of metal surfaces, localized corrosion targets smaller areas. This type of corrosion occurs when certain spots on a metal surface are more susceptible to corrosion than others in a corrosive environment [12, 13]. These spots undergo corrosion at a faster rate, resulting in partial erosion of the metal surface. The area with limited oxygen supply becomes the anode, where corrosion initiates, while the area with ample oxygen becomes the cathode. The digital overview of localized corrosion is depicted in Fig. 5.

3.5 Pitting Corrosion

Pitting corrosion is a significant form of localized corrosion that begins in a relatively small area on a material's surface. Over time, these areas can grow larger and deeper, creating pits or holes in the surface. In pitting corrosion, pits are formed on the material's surface due to localized chemical reactions. Areas covered by impurities or water tend to have lower oxygen concentrations, making them act as the anode, while other areas act as the cathode [12]. It is believed that electrochemical mechanisms play a key role in the dissolution of the metal during pitting corrosion. Figure 6 illustrates the digital picture of pitting corrosion on a metal chain.

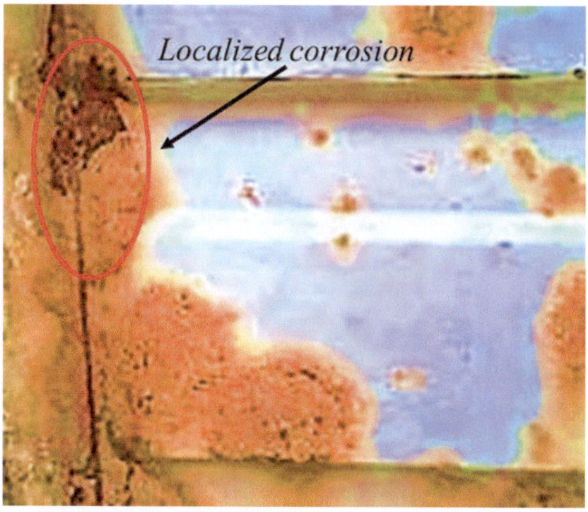

Fig. 5 Digital representation of localized corrosion

Fig. 6 Digital photo of pitting corrosion on metal chain

3.6 Crevice Corrosion

Crevice corrosion is a type of localized corrosion happens in narrow gaps where water gets trapped and oxygen cannot easily reach. This kind of corrosion forms pits in the metal, making it weaker and potentially causing damage. It occurs in places like underwater or in chemical environments. Factors like the type of metal, the shape of the gap, temperature, and the chemicals present affect how bad it gets. It is a big problem in industries like shipbuilding and chemical plants, where there are lots of tight spaces for corrosion to occur. Figure 7 shows the digital photo view of crevice corrosion.

Fig. 7 Digital view of crevice corrosion

3.7 *Intergranular Corrosion*

Intergranular corrosion is a particular type of corrosion that happens within the grain boundaries of a metal, not on its surface. This occurs because these boundaries are more vulnerable to corrosion compared to the actual grains of the metal. Differences in composition or the structure of these grain boundaries can make them more prone to react with corrosive substances [12, 14]. This type of corrosion can weaken the metal and even cause it to fail structurally, particularly in alloys or metals with specific compositions.

Intergranular corrosion often happens in specific types of stainless steels, especially those with high carbon levels or ones that weren't heat treated properly. For instance, sensitized stainless steel can undergo intergranular corrosion because chromium carbide forms along the grain boundaries, causing a drop in corrosion resistance. Additionally, certain aluminum alloys, like those found in aerospace applications, can also experience intergranular corrosion in certain situations. Al-Cu alloy has been extensively studied. It is commonly accepted that the microstructure of the Al–Cu alloy resembles the one depicted in Fig. 8.

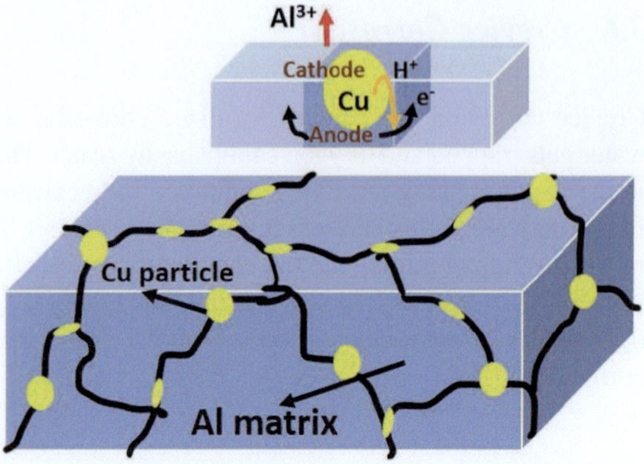

Fig. 8 Schematic representation of the Al–Cu alloy. Reproduced from Ref. [13] with permission from Elsevier, Copyright [2008]

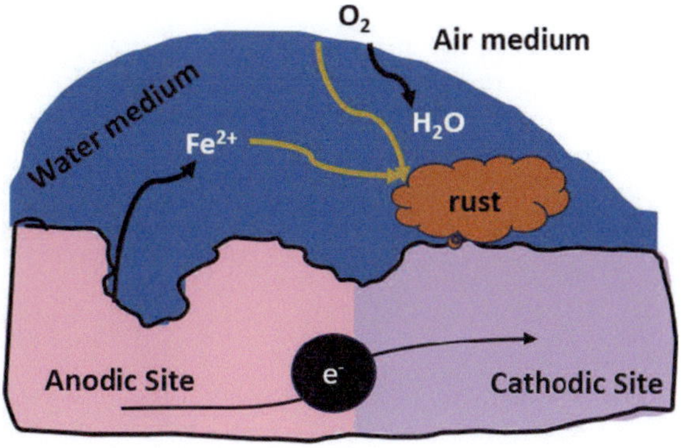

Fig. 9 Schematic representation of galvanic corrosion. Reproduced from Ref. [13] with permission from Elsevier, Copyright [2008]

3.8 Galvanic Corrosion

Galvanic corrosion, also called dissimilar metal corrosion, occurs when two different metals are close to each other and exposed to water or salt water. In this process, one metal corrodes faster (anode) while the other corrodes slower or doesn't corrode at all (cathode). This happens because the metals have different electrical properties, causing electrons to move between them. Galvanic corrosion can cause damage to the metals involved, weaken structures, and even lead to structural failure if not dealt with [12].

The rate of galvanic corrosion depends on the potential difference between the two metals involved. When this difference increases, the corrosion rate also goes up. We can calculate this potential difference using the formula:

$$\text{Potential difference} = \frac{\mathrm{dG}}{\mathrm{nF}}$$

where *dG* represents the Gibbs free energy of the reaction, n is the number of moles of electrons transferred, and F is the Faraday constant, which has a value of 23.06 kcal/volt. This potential difference determines how quickly galvanic corrosion occurs between the metals, as shown in Fig. 9.

3.9 Erosion

Erosion is when the surface of something slowly wears away or gets taken off by outside forces like wind, water, or particles that rub against it. It can happen on its own over time, or people can make it happen faster. Erosion affects lots of different things, like rocks, dirt, metal, and even concrete [13, 15]. How quickly erosion happens depends on how strong and long lasting the outside forces are, as well as

Fig. 10 Digital view of erosion on gas pipe

what kind of material is being worn away. The digital view of erosion on the gas pipe is depicted in Fig. 10.

4 Challenges in Developing Effective Corrosion Protection Coatings

Nobel metals like (Au, Ag, Pt, Pd, Ir) are typically used as corrosion protection system but their high cost makes them impractical for everyday use. Therefore, to protect machinery from corrosion, the alternative metals or alloy or other mechanism must be employed. These approaches are illustrated in Fig. 11.

4.1 Pre-treatment of Coating Substrate

Before applying any protective measures, it is important to thoroughly clean the metal surface. One widely used method is degreasing, which involves using a volatile organic solvent like trichloroethylene to dissolve oily and greasy films on the surface.

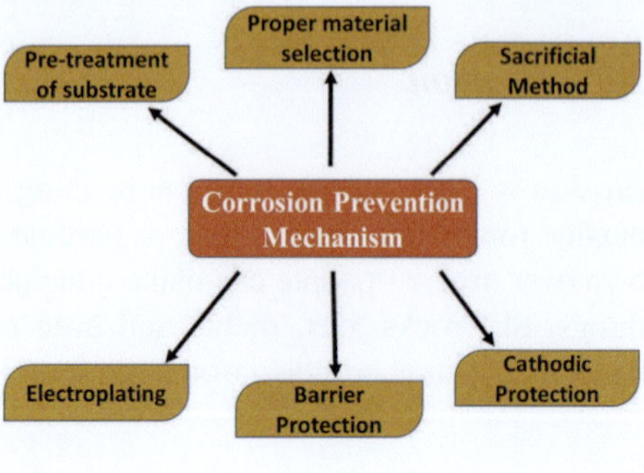

Fig. 11 Schematic representation of corrosion protection mechanism

Acid pickling is another option for removing scale instead of mechanical cleaning [16]. Proper preparation of the metal surface is essential before applying any coatings. Grease and oils also can be washed out with an alkali solution. Another cleaning method involves using a flame from a kerosene burner to remove oil, old paint, and grease by causing a sudden expansion of the metals.

4.2 Proper Material Selection

To minimize corrosion, it is essential to select materials and design with keeping certain principles in mind. Firstly, ensuring proper design involves avoiding contact between dissimilar metals in corrosive environments and, if contact is necessary, ensuring the anode metal has a larger surface area than the cathode metal. Placing dissimilar metals closer together in the electrochemical series and using insulating materials between them can also prevent direct metal-to-metal electrical contact [17]. Additionally, avoiding sharp corners and refraining from painting or coating metals are important measures to prevent rapid localized corrosion. Material selection plays a crucial role, with the ideal material possessing high tensile strength, corrosion resistance, and cost-effectiveness. This selection process typically involves prior experience-based selection and laboratory testing to evaluate suitability under specific conditions.

4.3 Sacrificial Method for Protective Coating

The sacrificial method for protective coating works by applying a layer of a more reactive metal, like zinc or aluminum, onto the surface of another material. This sacrificial metal acts as a sort of shield, sacrificing itself to protect the underlying material from corrosion. When exposed to a corrosive environment, the sacrificial metal corrodes first, preventing the substrate material from corroding. This method is commonly used in industries like marine, automotive, and construction to safeguard steel structures, pipelines, and other metal parts from corrosion damage [12, 16].

4.4 Cathodic Protection

Cathodic protection is a method used to prevent corrosion by turning the metal surface into the cathode of an electrochemical cell. Essentially, it is used to give the metal a protective shield against corrosion. Cathodic protection comes in two main forms: Galvanic Cathodic Protection (GCP) and Impressed Current Cathodic Protection (ICCP). In GCP, a more reactive metal, like zinc or magnesium, is connected to the metal needing protection. This sacrificial metal corrodes instead of the protected metal, effectively shielding it from corrosion. GCP is commonly used for smaller structures such as pipelines, storage tanks, and boat hulls. On the other hand, ICCP involves using an external power source, like a rectifier, to apply a direct current to the metal to be protected [18]. This current counteracts the metal's natural tendency to corrode, providing ongoing protection. ICCP is typically employed for larger structures like ships, offshore platforms, and steel-reinforced concrete structures. Both methods serve to extend the lifespan of metal structures and reduce maintenance costs associated with corrosion.

4.5 Barrier Protection

To prevent corrosion, one effective method is to coat metal surfaces, forming a protective shield that separates them from their environment. A straightforward strategy for corrosion prevention is barrier protection. By introducing a barrier whether through paint, oil, or non-corroding metals, which safeguard metal surfaces from the detrimental effects of air, water, and carbon dioxide. Additionally, certain chemicals can be used to coat the metal surface for protection [19]. Graphene is a commonly used material for barrier protection due to its inert nature, unique structure, and electrical properties. Its tightly bonded carbon atoms form a hexagonal structure, making it resistant to chemical reactions that would otherwise cause corrosion.

4.6 Electroplating

In 1805, Luigi Brugnatelli, an Italian chemist, pioneered the process of electroplating by successfully coating a fine layer of gold onto the silver after several trial-or-error attempts. Electroplating is a technique used to deposit a thin layer of metal onto the surface of a substrate, typically made of a different metal or material. This process occurs by passing an electric current through an electrolyte solution containing metal cations. This method serves various purposes, including decorative finishes and protection against corrosion. For example, chromium plating is commonly used for coating to vehicle wheel rims and gas burners for corrosion resistance, while nickel plating is utilized for decorative purposes and machinery parts [12, 14, 19]. The electroplating process begins with preparing a clean workpiece by removing dirt, grease, and contaminants through various solutions like alkaline cleaners and degreasers. Next, the workpiece is connected to the negative terminal (cathode) of a DC power supply, while the metal to be plated is connected to the positive terminal (anode). An ammeter and voltmeter are used to measure current and voltage, respectively. After thorough polishing, Faraday's law helps determine the coating thickness, ensuring uniformity and continuity for high-quality results.

5 MXene: Properties and Applications in Coating Technology

Coating technology is widely employed for surface protection, but it often faces challenges due to intrinsic defects like cracks and micropores within coating surface, which can compromise long-term effectiveness. Thus, there is a pressing need for the development of stable and durable surface protection technologies. Despite the increasing barrier performance, robust mechanical properties, and structural stability exhibited by 2D materials, composite coatings derived from these materials are emerging as highly promising solutions. These coatings have the capacity to address the requirements of various service conditions and show great potential for safeguarding surfaces.

Yury Gogotsi introduced MXenes—a fascinating class of two-dimensional transition metal carbides or nitrides. Their chemical structure, illustrated in Fig. 12, adheres to the formula $M_{n+1}X_nT_x$. Here, M stands for the transition metal, X represents carbon or nitrogen, and T_x (with x varying) denotes terminations on the transition metal layers. The various terminations present on MXenes (as shown in Fig. 12) enable versatile applications. With functional terminations such as –OH, O, and –F, along with a range of compositions and structures [20].

Furthermore, MXenes boast a pristine structure devoid of defects and showcase remarkable conductivity when compared to graphene oxide. The surface groups, influenced by synthesis techniques and chemical composition, offer exciting possibilities—they can be replaced, modified, or eliminated through additional chemical treatments, providing precise control over MXene surface properties [21]. The intrinsic layered structure of MXenes imparts tuneable properties. In this material, the conductive transition metal cores act as 'electron autobahns,' enabling rapid electron transport, while the interlayers serve as molecular turnstiles, permitting ions to pass through at the sub-nanometer scale between the 2D sheets. By employing techniques such as pre-intercalation, pillaring, and designing hybrids or heterostructures with other 2D materials, the large and adjustable interlayer space can accommodate ions of varying sizes.

Multilayered MXene flakes with diverse compositions—such as Ti_2CT_x, Ti_3CNT_x, Nb_2CT_x, and V_2CT_x—are typically synthesized from their respective MAX phase (Ti_2AlC, Ti_3AlCN, Nb_2AlC, and V_2AlC) through wet chemical etching method using hydrofluoric acid (HF) or a combination of LiF and HCl, followed by appropriate exfoliation techniques. The term 'T_x' denotes surface terminations (e.g., –OH, –F, and =O) introduced during the etching process, which selectively removes the 'A' layer (like Al) from the MAX phase structure. Additionally, innovative methods like CVD, such as salt-solution-based acoustic synthesis system and chemical scissor-mediated structural editing, have been proposed for MXene fabrication. Each of these novel approaches and discoveries has expanded the versatility, diversity, and practical uses of MXenes. As a result, the physical structure and surface properties of MXenes can be finely tuned through various preparation methods and post-chemical modifications, opening up new approach for their

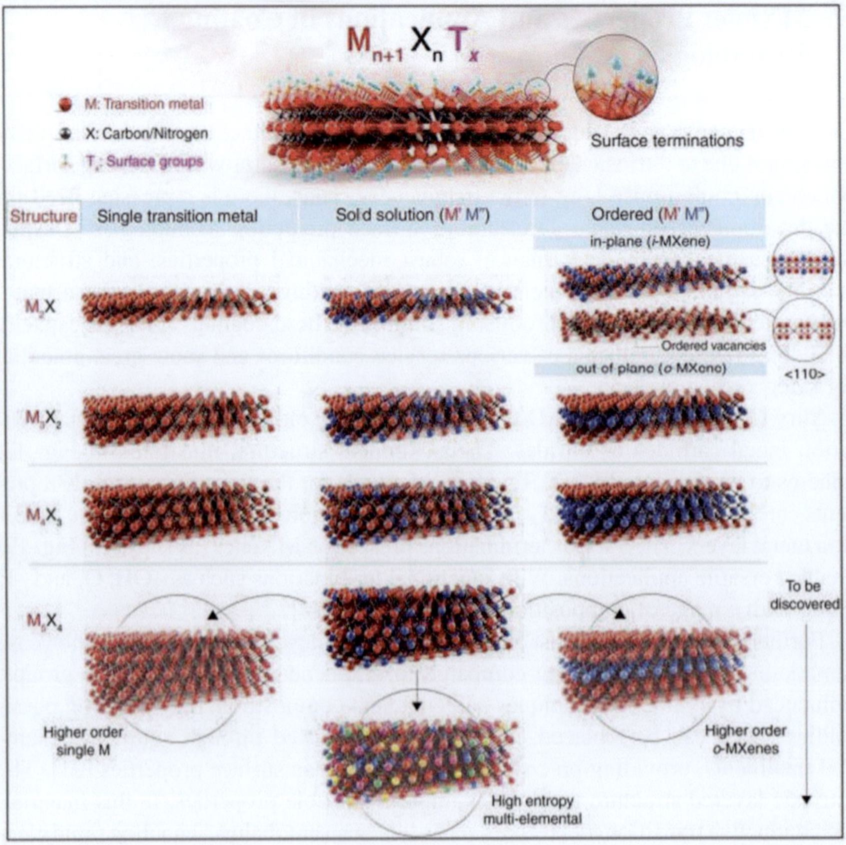

Fig. 12 Schematic representation of MXene structure. Reproduced from Ref. [20] with permission from Elsevier, Copyright [2024]

multifunctional applications [10]. The properties of the MXene are schematically illustrated in Fig. 13.

5.1 Electrical Properties

The standout feature of MXenes is their metallic behavior, which resembles their precursor MAX phases. This behavior is characterized by a consistent electron density close to the Fermi level. The electronic band structure of MXenes can be fine-tuned by adjusting factors like surface chemistry (e.g., using different etchants like LiF/HCL or HF treatment), composition, intercalated ions, defects, interlayer spacing, and grain boundaries. Furthermore, the surface chemistry of MXenes can be modified using techniques such as aromatic coupling-diazotization to attach sulfonic

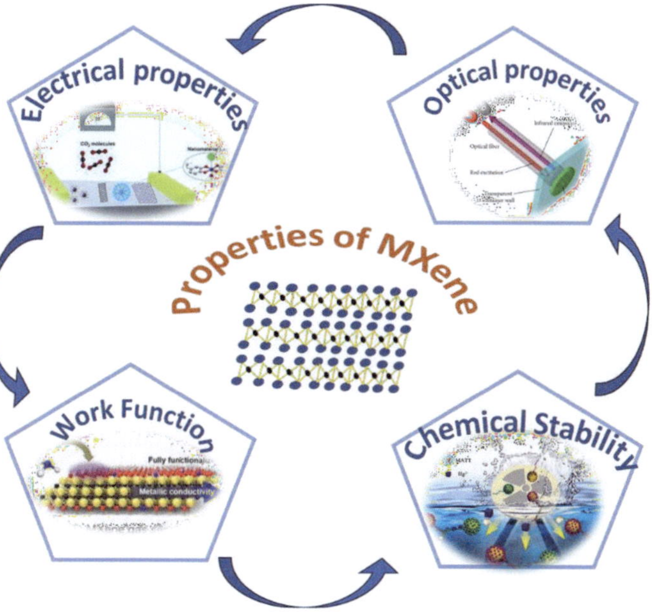

Fig. 13 Schematical representation of properties of the MXene

groups. Annealing MXenes under vacuum or annealing under the presence of inert gas, such as argon or nitrogen, can eliminate surface terminations and also water molecules trapped between adjacent MXene layers increase the electron mobility up to 10–15 times which improve the electrical conductivity [22].

In experiments, $Ti_3C_2T_x$ films have demonstrated superior electrical conductivity compared to other 2D materials like exfoliated graphene and metal sulfides/hydroxides. It is important to mention that the choice of etching method influences the electrical conductivity of $Ti_3C_2T_x$ stronger etchants resulting in higher defect densities in the MXene samples. For example, $Ti_3C_2T_x$ powders etched with HF show moderate electrical conductivities, typically less than 1000 S/cm [10].

5.2 *Optical Properties*

While separating multilayered $Ti_3C_2T_x$ into a few layers suspended in water, an observable Tyndall effect occurs. UV–Vis spectroscopy of ultrathin films composed of these delaminated flakes reveals a transmittance of up to 97%, indicating continuous coverage by single-layer $Ti_3C_2T_x$ nanosheets when supported on a substrate. This implies that a monolayer of $Ti_3C_2T_x$ results in approximately a 3% loss in transmittance, a value similar to graphene (which experiences a 2.3% loss per layer). Taking into account that a monolayer of $Ti_3C_2T_x$ has a thickness four times greater than that of graphene (1.5 nm versus 0.34 nm), it implies that $Ti_3C_2T_x$ may possess superior optical transmittance and more favorable optoelectronic properties compared to graphene. In the UV–Vis spectrum, a broad absorption peak around 800 nm is typically observed in $Ti_3C_2T_x$ MXene, attributed to its inherent out-of-plane interbond transition [23].

5.3 Work Function

Surface groups on $Ti_3C_2T_x$ play a key role in adjusting their work function by influencing the shift in the Fermi level and electron redistribution. When $Ti_3C_2T_x$ are terminated with pure –O groups, they tend to have a higher work function compared to those terminated with pure –F groups. Additionally, bare $Ti_3C_2T_x$ and those terminated with pure –OH groups follow in sequence. By incorporating a higher percentage of –OH functionalities through surface modifications, $Ti_3C_2T_x$ with lower work functions can be achieved. The work function of $Ti_3C_2T_x$ MXene can be lifted between 3.9 and 4.8 eV by altering the annealing temperature, which correlates with the surface termination groups [24].

5.4 Chemical Stability

$Ti_3C_2T_x$ are generally chemically stable, especially when not exposed to oxidizing environments. However, they can undergo oxidation when in contact with water and dissolved oxygen, leading to their deterioration over time. To prevent this, $Ti_3C_2T_x$ dispersions are often stored in environments devoid of oxygen, such as argon-filled containers, and at low temperatures to extend their shelf life. Additionally, freezing $Ti_3C_2T_x$ dispersions at temperatures below −20 °C has been found to preserve $Ti_3C_2T_x$ flakes for up to 2 years. Nevertheless, the stability of $Ti_3C_2T_x$ can vary depending on factors like how they are synthesized, the surface treatments applied, and the surrounding environmental conditions. Therefore, proper handling and storage are essential to maintain the chemical stability and functionality of $Ti_3C_2T_x$.

The chemical stability of $Ti_3C_2T_x$ against oxidation, particularly in electrolyte solution, is a significant apprehension. While oxidation or partial oxidation can have applications in energy storage, disposable electronics, and sensing, it is typically undesirable when electrical conductivity is a priority [25]. Pinholes and defects resulting from synthesis can speed up the oxidation process of $Ti_3C_2T_x$, leading to a reduced shelf life. Water and dissolved oxygen are the primary factors driving the oxidation of MXenes. To extend their shelf life, it is recommended to store MXene aqueous dispersions in vials filled with argon and at low temperatures below 5 °C. Freezing these dispersions at temperatures below −20 °C can preserve MXene flakes for up to 2 years. Upon oxidation, methane (CH_4) gas may be released from $Ti_3C_2T_x$

dispersion, while cloudy-white titanium oxides settle at the bottom. In the case of Ti_3CNT_x, the oxidation process can lead to the detection of ammonium gas in the solution.

6 Overview of MXene Coating in Various Applications

Figure 14 represents the development process of anti-corrosion coating of MXene ($Ti_3C_2T_x$). In the year 2019, Fan's research group pioneered the utilization of $Ti_3C_2T_x$ to enhance the corrosion resistance of epoxy coatings [26]. Following this, in 2020, surface modification techniques were introduced to enhance the stability of $Ti_3C_2T_x$. Song et al. then developed bulk $Ti_3C_2T_x$/Cu composites, significantly improving corrosion resistance within metal matrix composites. Additionally, a composite coating with self-healing properties was engineered. In the realm of corrosion protection, a novel technique emerged: crafting an oriented structure in $Ti_3C_2T_x$ through self-assembly triggered by air flow. Moreover, a heterostructure $Ti_3C_2T_x$ coating was proposed to achieve high stability. Excitingly, $Ti_3C_2T_x$-based coatings were also adapted for anti-corrosion applications in unique environments beyond standard NaCl solutions. Since 2019, there has been notable progress in the development of $Ti_3C_2T_x$ composite anti-corrosion coatings [27]. When it comes to MXene, oxidation and degradation are formidable foes, posing serious obstacles to its practical use. The antagonists such as water, oxygen, and temperature are the ones responsible for MXene's woes. These mechanisms have been elucidated through equations proposed in prior research.

The stability of MXene is crucial for its effectiveness in corrosion protection. This section delves into the detailed discussion on how MXene, with surface modifications, serves as fillers in composite coatings to combat corrosion. By addressing MXene oxidation through surface modification, it becomes viable for protecting metals from corrosion. Research on MXene-based anti-corrosion coatings primarily focuses on structuring design and understanding corrosion protection mechanisms. Research on MXene-based anti-corrosion coatings centers on structural design and understanding corrosion protection mechanisms depicted in Fig. 15. Recent investigations have explored methods like controlling orientation of MXene composite coatings and also heterostructure regulation to achieve desired structures. Techniques such as magnetic fields, electrostatic interaction, self-assembly, and electrodeposition have been utilized for this purpose. Additionally, understanding the relationship between structure and performance is vital for designing effective MXene-based anti-corrosion coatings. Factors such as the shear ability of MXene sheets, barrier and labyrinth effects, and minimizing coating defects are essential considerations for applications in anti-wear and anti-corrosion [28].

Despite MXene's potential as a filler in composite coatings for metal surface protection, it differs from graphene and boron nitride. MXene's unique properties are crucial for surface protection, but its susceptibility to oxidation and degradation poses challenges. Recent research has explored various MXene-based coatings

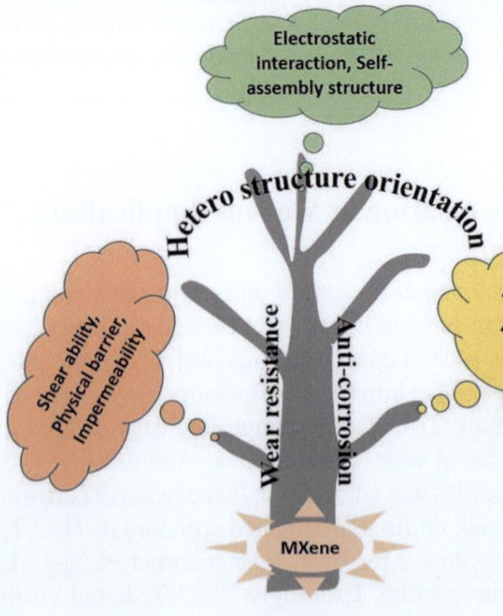

Fig. 14 Schematic illustration of MXene in various applications

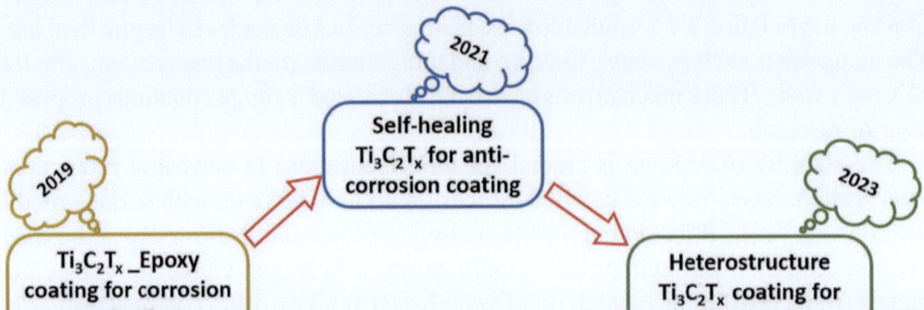

Fig. 15 Schematic illustration of mechanism proposed in prior research

to enhance corrosion and wear resistance, yet a comprehensive review comparing MXene with other 2D materials in anti-corrosive coatings is lacking. This review aims to clarify MXene's advantages over other fillers, summarize recent progress, and provide insights into its protective abilities. By addressing these questions, this review offers guidance for designing high-performance MXene anti-corrosion coatings [29].

7 Surface Engineering Techniques Used in Coating Application of MXene

Each of these techniques offers unique advantages and is suitable for different coating applications, ranging from decorative finishes to functional surface modifications aimed at improving mechanical, chemical, or electrical properties. Over the past few decades, significant research has been conducted on coating with MXene dispersions, along with the development of printing strategies for MXene-based coatings. When thin, continuous conductive films or electrodes are required, direct coating techniques offer promising and versatile options. Techniques such as spray coating, dip coating, blade coating, and spin coating have been explored (Fig. 16). Using these techniques, the MXene based coatings can be easily applied over different substrates or surfaces which are particularly suitable for the fabrication of high-performance, flexible, and foldable devices, making them ideal for advanced applications in next-generation electronics and functional surfaces.

7.1 Spray Coating

Spray coating usually used for thin and uniform layer of material coating onto a surface using a specialized spray gun or nozzle. Typically, the material, which is often in liquid form, is finely atomized into tiny droplets and then directed toward the surface to be coated. This method offers precise control over the dispersion of droplets, enabling accurate and even application. Figure 17 depicted the schematic representation of spray coating. This method offers precise control over the application, ensuring even coverage. It is commonly used in various industries for tasks like painting, coating, and finishing surfaces due to its ability to coat complex shapes and

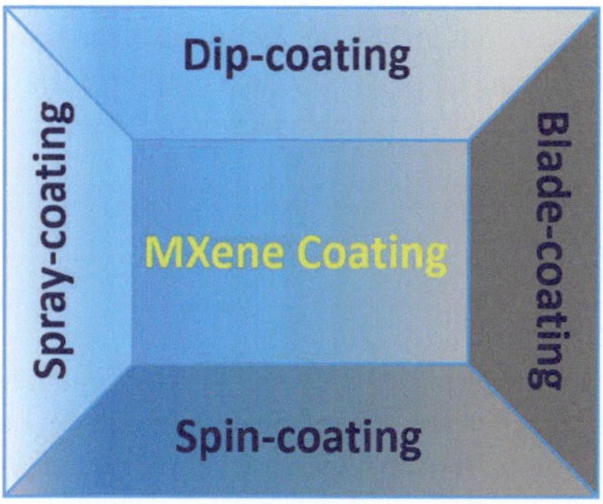

Fig. 16 Schematic representation of MXene coating in different surface engineering coating technology

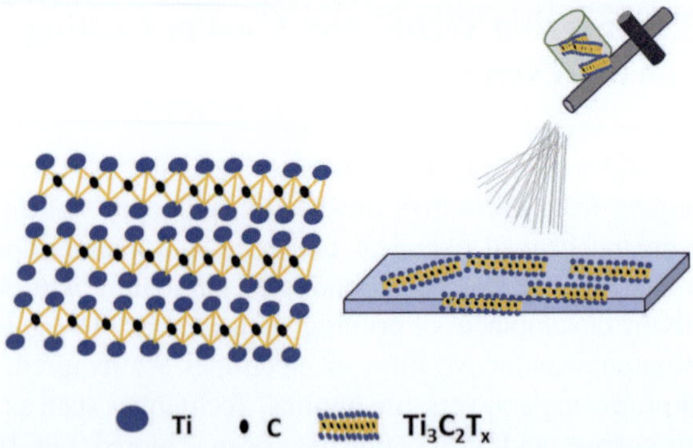

Fig. 17 Schematic representation of spray coating

surfaces effectively [30]. For MXene aqueous dispersion, the concentration range should be 0.1 to 10 mg/ml. In spray coating, MXene dispersions are pressurized and ejected as aerosol droplets by a carrier gas. These droplets, controlled in size and velocity by gas pressure, land on hot substrates. As the solvent evaporates, the droplets merge, leaving behind aligned MXene flakes. Moving the spray gun in different directions helps create a network of MXene flakes, forming continuous thin films or coatings. This method enables rapid fabrication of large-area films varying from nano- to micro-range [31]. The resulting MXene film's uniformity depends on factors like solvent type, gas pressure, temperature, and distance between the substrate and nozzle. Surface optimization techniques like plasma treatment or etching help ensure strong adhesion of MXene films to substrates, even under deformation like bending.

Sprayed MXene thin films have significant applications in wireless communication, where thicker films offer higher reflection coefficients and lower attenuation as shown in Fig. 18a. MXenes, known for their high conductivity and water dispersibility, make them ideal for scalable manufacturing of radio-frequency antennas. Additionally, spray coating enables the easy fabrication of micro-supercapacitors through various techniques like template masking and laser scribing. MXene films serve as both the current collector and active material, enhancing power handling and capacitance in micro-supercapacitors, as shown in Fig. 18b.

7.2 *Dip Coating*

Dip coating involves immersing a substrate into a liquid containing a coating material and then withdrawing it, allowing a thin layer to adhere via capillary action. Film thickness is controlled by withdrawal speed, solution viscosity, and coating cycles. This technique finds widespread use in surface modification, protective coatings, and thin film deposition electronics and optics [30]. MXene colloidal dispersions typically exhibit negative zeta potentials from the rage from −35 to −85 mV, facilitating

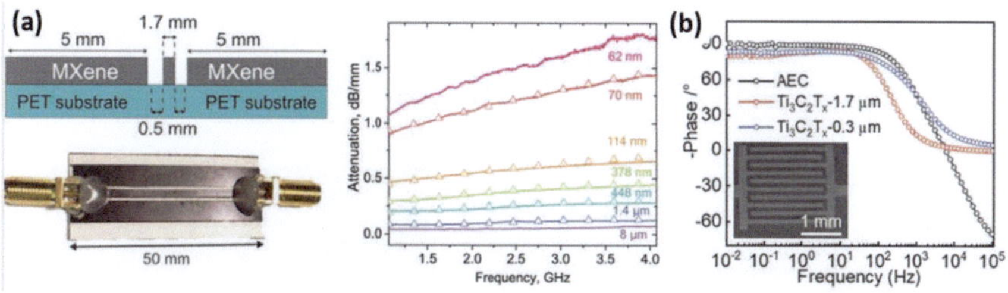

Fig. 18 **a** Wireless communication with schematic and digital photo of the MXene attenuation versus thickness and **b** Bode phase angle plots of $Ti_3C_2T_x$ film-based micro-supercapacitor prepared by spray coating. Reproduced from Ref. [31] with permission from Elsevier, Copyright [2021]

repulsive interactions among the nanosheets and ensuring stable, uniform dispersions suitable for dip coating. The dip coating process involves several sequential steps: dipping the substrate into the dispersion, immersing it for a specific duration, withdrawing at a constant speed, and allowing solvent evaporation, as shown in Fig. 19. This results in the deposition of a thin MXene's thin layer on the substrate, with excess dispersion returning to the bulk solution. Various factors, including MXene concentration, flake quality, pulling speed, immersion duration, and substrate surface, influence the quality and homogeneity of the resulting MXene films. For instance, dipping pre-treated glass in MXene baths with concentrations ranging from 1 to 10 mg/mL and pulling at a constant speed of 2 mm/s, continuous, smooth MXene films with adjustable thicknesses, depending on the concentration and number of dips, are produced. The thickness range is depicted in Fig. 20a and b, from 20 to 150 nm [30, 32].

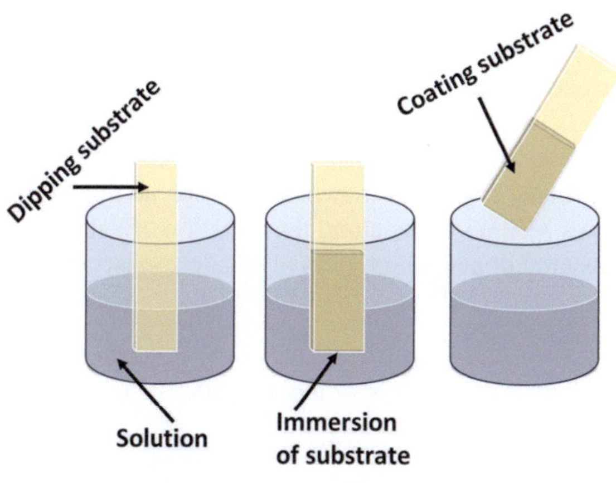

Fig. 19 Schematic representation of dip coating

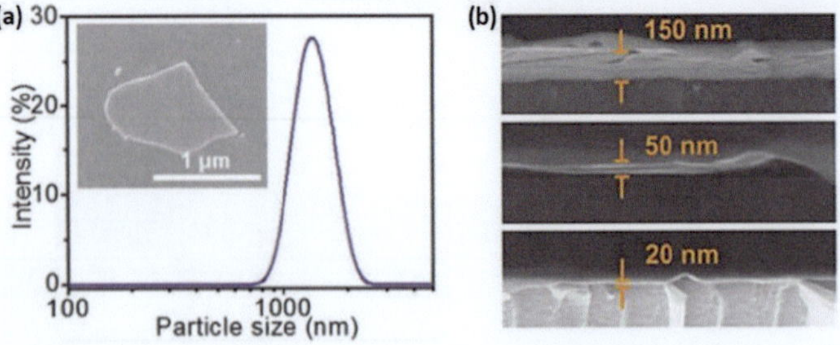

Fig. 20 **a** MXene flake size distribution dispersion plot and **b** Cross-sectional SEM images of dip-coated MXene films on glass substrates. Reproduced from Ref. [31] with permission from Elsevier, Copyright [2021]

7.3 *Blade Coating*

Blade coating is a method where a blade is passed over a substrate to evenly spread a solution across its surface. The gap between the blade and substrate, along with the solution's viscoelastic properties and coating speed, controls the thickness of the wet film. This technique finds applications in various research areas like thin film electronics, battery technology, ceramics, and paints. It is particularly useful for solutions with high viscosity that need thicker films. Although less precise compared to slot die coating, blade coating serves as a prototyping method for large-scale production [30]. Figure 21 shows the schematic overview of blade coating.

Dispersion of MXene with higher concentrations (20–200 mg/mL) or viscosities (up to 2000 mPa.s) is suitable for blade coating, similar to spin coating. Blade coating creates large-area freestanding MXene films with well-oriented flakes, resulting in improved electrical conductivity and mechanical properties. The thickness of the MXene films produced by blade coating depends on the height of the blade and MXene concentration, typically reaching millimeter thicknesses. Concentrated MXene dispersions exhibit shear-thinning behavior, facilitating flowability and processability. However, diluted dispersions with low concentrations (<10 mg/mL) are not suitable for blade coating as they incline to spread thinly over the substrate. Blade coating aligns MXene flakes along the direction of velocity, aided by higher elastic and viscous moduli and a high ratio of elastic-to-viscous moduli, promoting

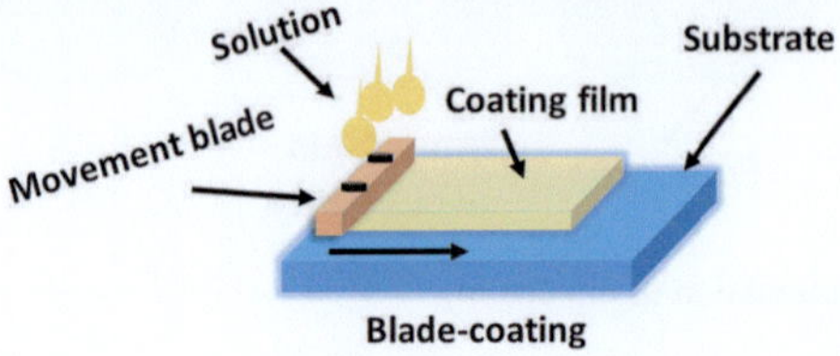

Fig. 21 Schematic representation of blade coating

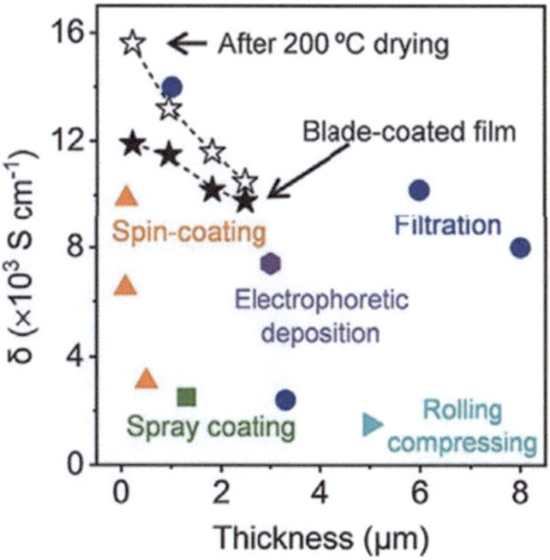

Fig. 22 Comparisons of the electrical conductivity comparison of MXene film through different processing techniques. Reproduced from Ref. [31] with permission from Elsevier, Copyright [2021]

gel state recovery and flake alignment upon solvent evaporation, resulting in large-area conductive MXene films. The electrical conductivity comparison for MXene films through different methods is depicted in Fig. 22.

7.4 Spin Coating

Spin coating is a straightforward technique that demands minimal training. It is capable of producing uniform films, with thicknesses ranging from nanometers to microns, on small and flat substrates. The drying process is expedited by the airflow generated during rotation, resulting in fast drying times. Consequently, post-deposition heat treatment may not always be required to achieve complete drying. In terms of cost-effectiveness, spin coating excels, particularly for producing small batches or individual films. Figure 23 schematically depicted the spin coating technique.

When MXene dispersions reach a medium viscosity (10–200 mPa.s) or concentration (1–20 mg/mL), spin coating becomes the preferred processing method. Spin coating is a frequently employed technique for producing thin, uniform films from dispersions. Initially, the dispersion coats the substrate, spreading as the substrate rotates. As spinning continues, most of the dispersion is removed, leaving behind a thin liquid layer that evaporates rapidly, resulting in homogeneous thin films [30, 31]. The film thickness is determined by factors like MXene concentration, spin speed,

Fig. 23 Schematic illustration of spin coating technique

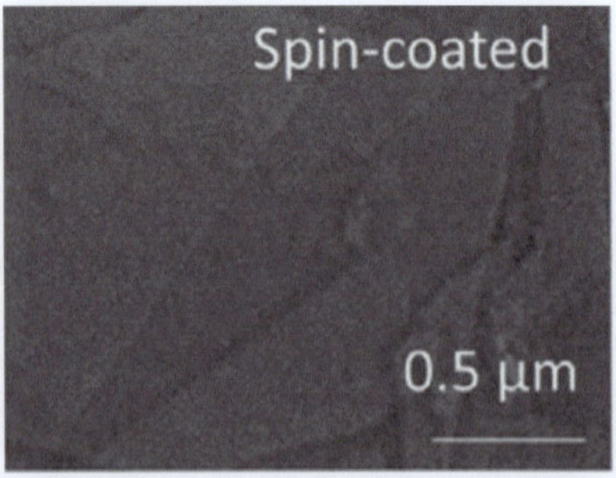

Fig. 24 SEM image of spin-coated MXene film. Reproduced from Ref. [31] with permission from Elsevier, Copyright [2021]

and duration. Higher concentrations yield thicker films compared to more diluted dispersions at the same spin speed. Spin coating can generate films with thicknesses ranging from nanometers to micrometers and various transmittances, making it suitable for lab-scale prototype devices with flat substrates. The SEM image of the spin-coated MXene film is depicted in Fig. 24.

8 Characterization Techniques for Corrosion Protection of MXene-Based Coating

The fabrication of MXene coatings on the X70 steel substrate involves two main steps, as illustrated in Fig. 25. In the first step (Fig. 25a), a colloidal suspension containing MXene nanosheets is prepared by selectively etching the Al interlayer from the MAX precursor, followed by exfoliation into ultrathin flakes.

Figure 26a shows the typical layered structure achieved through micromechanical cleavage of the bulk MAX before etching. Subsequently, in the second step (Fig. 24b), mono- or few-layer freestanding 2D MXene flakes are exfoliated from the multilayered $Ti_3C_2T_x$ with the assistance of sonication. This step demonstrates the effectiveness of Li+ intercalation between the $Ti_3C_2T_x$ layers for delamination [33]. The X-ray diffraction (XRD) patterns depicted in Fig. 26a confirm the formation of MXene. Following etching, the characteristic (104) peak of Ti_3AlC_2 at 39° weakens, while the (002) diffraction peak corresponding to 2D $Ti_3C_2T_x$ shifts to a smaller angle. These changes indicate complete removal of Al atoms from the MAX

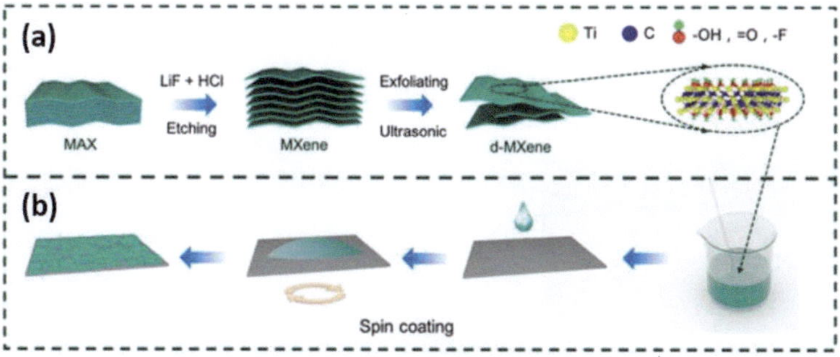

Fig. 25 Schematic illustration of **a** MXene fabrication and **b** MXene spin coating. Reproduced from Ref. [33] with permission from Elsevier, Copyright [2023]

precursor and a surge in interlayer spacing [34]. The Fourier transform infrared (FTIR) spectra in Fig. 26b exhibit peaks at 554 cm^{-1}, 1100 cm^{-1}, 1632 cm^{-1}, and 3445 cm^{-1}, attributed to the stretching vibrations of Ti–O, C–F, C=O, and –OH bonds, respectively, consistent with previous findings.

Potentiodynamic polarization is employed to assess the corrosion behavior of MXene coatings, as depicted in Fig. 27, with Table 1, summarizing the electrochemical properties, notably the corrosion potential (Ecorr) and corrosion current density (Icorr). The initial uncoated X70 steel exhibits an Ecorr of −0.681 V and an Icorr of 5.6 $\mu A{\cdot}cm^{-2}$. Upon coating with MXene (4 mg/mL), a significant positive shift of approximately 170 mV is observed, indicating reduced susceptibility to corrosion after 30 min of immersion. Moreover, the Icorr decreases to 1.5 $\mu A{\cdot}cm^{-2}$, implying protective benefits for the underlying substrate. However, both Ecorr and Icorr decrease with decreasing MXene concentration. This decrease in the anodic current density, indicative of metal dissolution and corrosion rate, suggests that higher defect density in MXene coatings diminishes corrosion resistance.

In 2019, Fan et al. demonstrated the effectiveness of $Ti_3C_2T_x$ nanosheets in enhancing the anti-corrosion properties of epoxy coatings exposed to 3.5 wt% NaCl

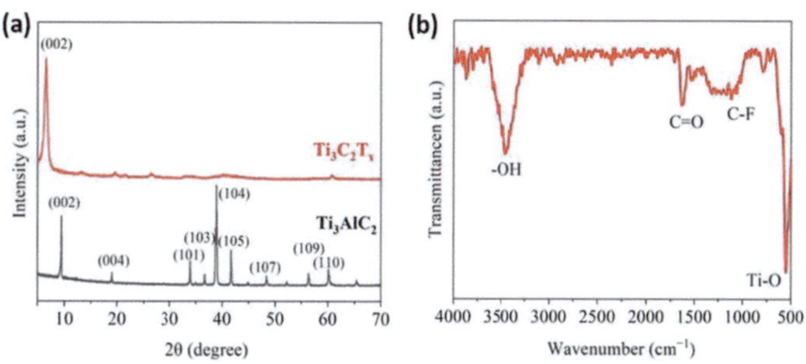

Fig. 26 **a** X-ray diffraction of MAX and MXene phase and **b** FTIR spectroscopy of MXene film Reproduced from Ref. [32] with permission from Elsevier, Copyright [2021]

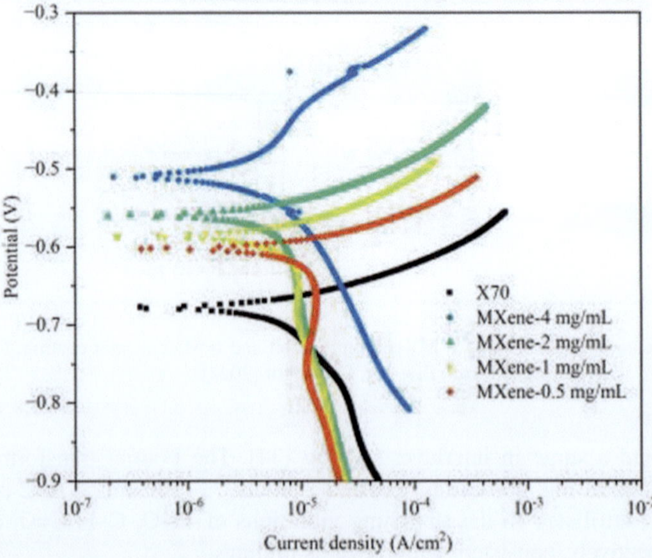

Fig. 27 Potentiodynamic polarization plot of MXene coating. Reproduced from Ref. [33] with permission from Elsevier, Copyright [2023]

Table 1 Electrochemical properties of the MXene coating

Sample name	Ecorr (V versus SCE)	Icorr (μA.cm^{-2})
X70	−0.681	5.6
MXene-4 mg/ml	−0.511	1.5
MXene-2 mg/ml	−0.561	3.3
MXene-1 mg/ml	−0.588	4.2
MXene-0.5 mg/ml	−0.605	5.6

solution. Their study revealed that incorporating MXene into the coating formulation eliminated micropores in the epoxy matrix after curing [31]. With just 1 wt% $Ti_3C_2T_x$, a significant reduction in corrosion current density was achieved, from 1.00 × 10^{-6} A cm^{-2} in pure epoxy coatings to 3.39 × 10^{-8} A cm^{-2}. Moreover, MXene finds application in metal matrix composites, as demonstrated by Song et al., who used $Ti_3C_2T_x$ to reinforce copper matrices, resulting in enhanced corrosion resistance compared to pure copper. Additionally, MXene has proven effective in improving the corrosion-resistant capabilities of metallic coatings. Zhang et al. and Xiao et al. successfully prepared Ni–P–$Ti_3C_2T_x$–Ce and Ni–W–$Ti_3C_2T_x$ composite coatings with enhanced corrosion resistance using electrodeposition methods. The incorporation of MXene facilitated the formation of dense passivation films and contributed to the reinforcement of the coatings. For instance, Ni–P–$Ti_3C_2T_x$–Ce composite coatings exhibited a significant reduction in corrosion current density, from 5.447 × 10^{-4} A cm^{-2} in Ni–P coatings to 5.869 × 10^{-7} A cm^{-2}. Overall, MXene's impermeability and physical barrier properties make it a versatile nanofiller for enhancing corrosion resistance in various coating systems, including organic, metallic, and inorganic–organic hybrid coatings. Metal substrates are susceptible to corrosion when exposed to water or salt solutions. To prevent this, organic coatings are commonly applied to create a protective barrier. In a study by Yan et al., they investigated the anti-corrosive properties of a mixture of few-layer Ti_3C_2 and epoxy resin applied to a steel sheet. It was discovered that adding 1 wt.% of Ti_3C_2 provided the best anti-corrosion performance by effectively covering the pores in the epoxy coating, which act as pathways for corrosive materials. This incorporation of Ti_3C_2 flakes enhanced the coating's ability to resist corrosion. Furthermore, the presence of Ti_3C_2 reduced water absorption in the epoxy coating over time. For instance, while pure epoxy coatings absorbed 0.96 wt.% of water after 96 hours in a 3.5% NaCl solution, those containing 1 wt.% Ti_3C_2 absorbed only 0.23 wt.%. Spray salt tests also confirmed the improved anti-corrosion performance of epoxy coatings with Ti_3C_2, showing significant suppression of metal corrosion over a 15-day period [33].

9 Summary, Performance Evaluation, and Author's View Points for Future Outlook

This section provides a summary overview of MXene-based coatings and offers a comprehensive discussion of their features and mechanisms for corrosion protection. The chapter begins by exploring the development of MXene-based coatings, followed by an examination of MXene's intrinsic surface protection capabilities, including its anti-corrosion properties. Subsequently, it summarizes the properties and aids of MXene in corrosion protection. The main research aspects of MXene-based anti-corrosion coatings are then outlined, covering aspects such as reinforcement effects, surface functionalization, structural design, orientation control, self-healing properties, and corrosion behavior under special conditions. Additionally, this book chapter highlights the diverse roles of MXene nanosheets in anti-corrosion mechanisms.

Despite progress in developing MXene-based coatings and understanding their role in corrosion protection, challenges persist. Prospects for future research in MXene-based anti-corrosion coatings are proposed, including the need to reduce costs, address intrinsic oxidation and degradation issues, develop green and low-cost surface functionalization methods, and achieve large-scale preparation while ensuring uniform dispersion and structure integrity. Maintaining high repeatability, flexibility, and key structural properties such as uniform thickness, good dispersion, and strong adhesion is also identified as crucial areas for improvement in future research endeavors. Figure 28 proposed the perspective of MXene-based anti-corrosion protection outlook.

The practical application of MXene-based coatings introduces several challenges due to the diverse and harsh environments they may encounter, such as seawater, deep sea conditions, acidic solutions, and corrosive gases. Designing coatings that can

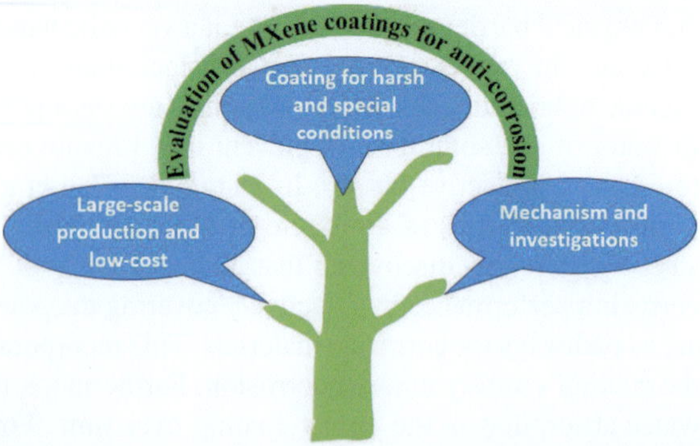

Fig. 28 Overview of opportunities and challenges for MXene-based protective coatings

withstand high pressure, temperature, and concentrations of corrosive species while maintaining stability and impermeability requires careful consideration of factors like thickness, adhesion, compatibility, dispersion, and mechanical resistance. However, current studies often focus on simulated environments, overlooking the complex interactions and coupling effects present in real-world scenarios. Developing environmentally friendly, self-healing, and multifunctional MXene coatings suited for practical applications is thus a pressing need. Understanding the corrosion shielding mechanism of MXene coatings is hindered by the lack of in situ experimental observations and theoretical calculations. Electrochemical results and corrosion morphology provide some insights, direct observation of the corrosion process is challenging. By concluding this our viewpoints on MXene-based corrosion protection coatings have emerged as a promising solution due to their unique properties, including high electrical conductivity, large surface area, and excellent mechanical strength. These coatings have demonstrated significant potential in enhancing the corrosion resistance of various substrates. However, several challenges must be addressed to fully realize their practical applications. One major challenge is the stability of MXene materials under environmental conditions, such as humidity and temperature, which can affect their performance over time. Another concern is the potential toxicity of MXene materials, as their environmental and health impacts are not yet fully understood, necessitating comprehensive studies to ensure their safe use. Additionally, producing MXene materials on a large scale while maintaining consistent quality and performance remains a significant challenge. To overcome these challenges, several research areas and technological advancements should be pursued. Developing surface functionalization techniques can improve the environmental stability of MXene coatings. Conducting extensive research on the biocompatibility and environmental impact of MXene materials is essential to ensure their safe application. Innovating scalable and cost-effective production methods is crucial to facilitate the widespread use of MXene-based coatings. Exploring the development of hybrid coatings that combine MXene with other materials can enhance their protective properties and stability. Performing long-term real-world testing of MXene coatings will help validate their performance under various environmental conditions. Advancements in nanotechnology and material science are expected to play a crucial role in addressing the challenges associated with MXene-based coatings. Innovations in synthesis methods, such as chemical vapor deposition and electrochemical techniques, could lead to more efficient and scalable production processes. Additionally, integrating MXene materials with smart coating technologies could pave the way for developing self-healing and adaptive corrosion protection systems. Additionally, theoretical simulations face difficulties due to the complexity of MXene coatings and the need to consider multiple factors simultaneously. Constructing investigation systems that mimic real-world conditions presents another obstacle in exploring the corrosion protection mechanism comprehensively.

References

1. Revie, R.W.: Corrosion and Corrosion Control: An Introduction to Corrosion Science and Engineering. John Wiley & Sons (2008)
2. Haddadi, S.A., Mehmandar, E., Jabari, H., SA, A.R., Mohammadkhani, R., Yan, N., Arjmand, M.: Zinc-doped silica/polyaniline core/shell nanoparticles towards corrosion protection epoxy nanocomposite coatings. Compos. Part B: Eng. **212**, 108713 (2021)
3. Marinescu, M.: Recent advances in the use of benzimidazoles as corrosion inhibitors. BMC Chem. **13**, 1–21 (2019)
4. Mashuga, M.E., Olasunkanmi, L.O., Ebenso, E.E.: Experimental and theoretical investigation of the inhibitory effect of new pyridazine derivatives for the corrosion of mild steel in 1 M HCl. J. Mol. Struct. **1136**, 127–139 (2017)
5. Ghali, E., Sastri, V.S., Elboujdaini, M.: Corrosion Prevention and Protection: Practical Solutions. John Wiley & Sons (2007)
6. El Ibrahimi, B., Jmiai, A., Bazzi, L., El Issami, S.: Amino acids and their derivatives as corrosion inhibitors for metals and alloys. Arab. J. Chem. **13**(1), 740–771 (2020)
7. Chen, Z., Scharnagl, N., Zheludkevich, M.L., Ying, H., Yang, W.: Micro/nanocontainer-based intelligent coatings: synthesis, performance and applications–a review. Chem. Eng. J. **451**, 138582 (2023)
8. Tian, Z., Li, S., Chen, Y., Li, L., An, Z., Zhang, Y., An, B., et al.: Self-healing coating with a controllable release of corrosion inhibitors by using multifunctional zinc oxide quantum dots as valves. ACS Appl. Mater. Interfaces **14**(41), 47188–47197 (2022)
9. VahidMohammadi, A., Rosen, J., Gogotsi, Y.: The world of two-dimensional carbides and nitrides (MXenes). Science **372**(6547), eabf1581 (2021)
10. Alhabeb, M., Maleski, K., Anasori, B., Lelyukh, P., Clark, L., Sin, S., Gogotsi, Y.: Guidelines for synthesis and processing of two-dimensional titanium carbide ($Ti_3C_2T_x$ MXene). Chem. Mater. **29**(18), 7633–7644 (2017)
11. Naguib, M., Mashtalir, O., Carle, J., Presser, V., Lu, J., Hultman, L., Barsoum, M.W., et al.: Two-dimensional transition metal carbides. ACS Nano **6**(2), 1322–1331 (2012)
12. Harsimran, S., Santosh, K., Rakesh, K.: Overview of corrosion and its control: a critical review. Proc. Eng. Sci **3**(1), 13–24 (2021)
13. Makhlouf, A.S.H., Herrera, V., Muñoz, E.: Corrosion and protection of the metallic structures in the petroleum industry due to corrosion and the techniques for protection. In: Handbook of Materials Failure Analysis, pp. 107–122. Butterworth-Heinemann (2018)
14. Okyere, M.S.: Corrosion Protection for the Oil and Gas Industry: Pipelines, Subsea Equipment, and Structures. CRC Press (2019)
15. Kumar, S., Kumar, M., Handa, A.: High temperature oxidation and erosion-corrosion behaviour of wire arc sprayed Ni-Cr coating on boiler steel. Mater. Res. Express **6**(12), 125533 (2019)
16. Prutton, C.F., Frey, D.R., Turnbull, D., Dlouhy, G.: Corrosion of metals by organic acids in hydrocarbon solvents. Ind. Eng. Chem. **37**(1), 90–100 (1945)

17. Qian, Y., Li, Y., Jungwirth, S., Seely, N., Fang, Y., Shi, X.: The application of anti-corrosion coating for preserving the value of equipment asset in chloride-laden environments: a review. Int. J. Electrochem. Sci. **10**(12), 10756–10780 (2015)
18. Sahoo, P., Das, S.K., Paulo Davim, J.: 3.3 Surface finish coatings. In: Hashmi, M.S.J. (eds.) Comprehensive Materials Finishing (2017)
19. Sharma, V., Kumar, S., Kumar, M., Deepak, D.: High temperature oxidation performance of Ni-Cr-Ti and Ni-5Al coatings. Mater. Today: Proc. **26**, 3397–3406 (2020)
20. Cao, H.: Beyond graphene and boron nitride: why MXene can be used in composite for corrosion protection on metals? Compos. Part B: Eng. 111168 (2023)
21. Naguib, M., Mochalin, V.N., Barsoum, M.W., Gogotsi, Y.: 25th anniversary article: MXenes: a new family of two-dimensional materials. Adv. Mater. **26**(7), 992–1005 (2014)
22. Huang, K., Li, Z., Lin, J., Han, G., Huang, P.: Two-dimensional transition metal carbides and nitrides (MXenes) for biomedical applications. Chem. Soc. Rev. **47**(14), 5109–5124 (2018)
23. El-Demellawi, J.K., Lopatin, S., Yin, J., Mohammed, O.F., Alshareef, H.N.: Tunable multipolar surface plasmons in 2D $Ti_3C_2T_x$ MXene flakes. ACS Nano **12**(8), 8485–8493 (2018)
24. Caffrey, N.M.: Effect of mixed surface terminations on the structural and electrochemical properties of two-dimensional $Ti_3C_2T_2$ and V_2CT_2 MXenes multilayers. Nanoscale **10**(28), 13520–13530 (2018)
25. Wu, Y., Wu, L., Yao, W., Jiang, B., Wu, J., Chen, Y., Pan, F., et al.: Improved corrosion resistance of AZ31 Mg alloy coated with MXenes/MgAl-LDHs composite layer modified with yttrium. Electrochim. Acta **374**, 137913 (2021)
26. Yan, H., Li, W., Li, H., Fan, X., Zhu, M.: Ti_3C_2 MXene nanosheets toward high-performance corrosion inhibitor for epoxy coating. Prog. Org. Coat. **135**, 156–167 (2019)
27. Yan, H., Cai, M., Li, W., Fan, X., Zhu, M.: Amino-functionalized $Ti_3C_2T_x$ with anti-corrosive/wear function for waterborne epoxy coating. J. Mater. Sci. Technol. **54**, 144–159 (2020)
28. Liu, N., Li, Q., Wan, H., Chang, L., Wang, H., Fang, J., Xiao, X., et al.: High-temperature stability in air of $Ti_3C_2T_x$ MXene-based composite with extracted bentonite. Nat. Commun. **13**(1), 5551 (2022)
29. Wu, Y., Wu, L., Zheludkevich, M.L., Chen, Y., Serdechnova, M., Yao, W., Pan, F., et al.: MgAl-$V_2O_7^{4-}$LDHs/(PEI/MXene) 10 composite film for magnesium alloy corrosion protection. J. Mater. Sci. Technol. **91**, 28–39 (2021)
30. Abdolhosseinzadeh, S., Jiang, X., Zhang, H., Qiu, J., Zhang, C.J.: Perspectives on solution processing of two-dimensional MXenes. Mater. Today **48**, 214–240 (2021)
31. Park, T.H., Yu, S., Koo, M., Kim, H., Kim, E.H., Park, J.E., Park, C., et al.: Shape-adaptable 2D titanium carbide (MXene) heater. ACS Nano **13**(6), 6835–6844 (2019)
32. Shi, K., Meng, X., Xiao, S., Chen, G., Wu, H., Zhou, C., Chu, P.K., et al.: MXene coatings: novel hydrogen permeation barriers for pipe steels. Nanomaterials **11**(10), 2737 (2021)
33. Sharma, S., Kumar, A., Ebenso, E.E.: MXenes and MXene-based nanomaterials for corrosion protection. Mater. Lett. **335**, 133789 (2023)
34. Nayak, B., Anwar, S., Roy, A., Anwar, S.: Investigation of 2D-$Ti_3C_2T_X$-loaded PVDF composites for a high-performance flexible piezoelectric energy harvester and wearable sensor. ACS Appl. Electron. Mater. **5**(12), 7045–7060 (2023)

Progress of MXene-Based Materials: Commercialization, Shortcomings, and Multifunctional Applications

D. Parajuli

Abstract MXenes represent a class of two-dimensional ceramic materials, primarily comprising carbides, nitrides, and carbonitrides, with 'MAX' phases as their precursors through the removal of 'A'. Renowned for their conductive, hydrophilic, biocompatible, and adjustable properties, MXenes encounter challenges such as environmental instability, low biodegradability, and complexities in drug release. These issues are effectively addressed through the creation of nanocomposites. The incorporation of MXene terminations onto the polymer post-composite formation enhances functionality. Consequently, there is an augmentation in efficient photo-thermal conversion for cancer treatment, increased antibacterial activity, improved biosensors, and enhanced capabilities in bone regeneration. After the formation of the composite, the hydrophilic surfaces are converted to conducting surfaces near to that of metals. MXenes exhibit compatibility with various materials like ceramics, metals, and polymers in nanocomposite form, yielding superior properties suitable for advanced applications. Composites made of polymer are used extensively for biomedical and energy storage purposes. Similarly, the composite with ferrites is used for dying and shielding purposes which we have studied recently. This chapter also explores the prospects of MXene. Due to their specific properties and controllability from their parent MAX phase, different exotic and flexible applications are leaving a clear picture of the bright and promising future of MXene.

Keywords MXene · Nanocomposites · Biomedical · Energy storage · Sensors · Photocatalysts

D. Parajuli (✉)
Research Center for Applied Science and Technology, Tribhuvan University, Kirtipur 44613, Kathmandu, Nepal
e-mail: deepenparaj@gmail.com

© The Author(s), under exclusive license to Springer Nature Singapore Pte Ltd. 2025
S. Moharana and R. N. Mahaling (eds.), *Surface Engineering of MXenes*, Engineering Materials, https://doi.org/10.1007/978-981-96-3640-2_17

1 Introduction

MXenes are the two-dimensional inorganic compounds characterized by thin carbides, nitrides, or carbonitrides of transitional elements [1]. Renowned for their capacity to accommodate diverse hydrophilic terminations, the discovery of MXenes traces back to 2012, revealing remarkable attributes such as high electrical conductivity, robust mechanical strength, and enduring chemical stability [2]. When synthesized through HF etching, MXenes exhibit an accordion-like morphology in their original state. The nomenclature $M_{n+1}X_nT_x$ denotes MXenes, with T representing a functional group (e.g., O, F, OH, Cl) terminating the MXene surfaces [3]. MXenes with a single transition metal have three main structures: M_2C, M_3C_2, and M_4C_3. These structures arise from selectively removing the A element in the MAX phase. In $M_{n+1}X_nT_x$, A is an element from group 13 or 14, X is C and/or N in the octahedral site [3], and n ranges from 1 to 4 [4]. MXene can also be extracted from layered Mo_2Ga_2C. MAX phases possess a hexagonal layered structure with P63/mmc symmetry. Consequently, MXenes are interleaved between A and M [5, 6]. The transitional metal M_{n+1} can comprise a single type element of or two. Those with two transitional elements (M' and M") are called Double transition metal MXenes. They can be solid solutions or ordered double transition metal (ODTM) MXenes. The ODTM MXene in forms $M'_2\,M''C_2$ or $M'_2\,M''_2C_3$ are Mo_2TiC_2, $Mo_2Ti_2C_3$, Cr_2TiC_2, and Mo_4VC_4, with specific configurations influencing electrochemical properties [7].

Solid-solution MXenes follow general formulas like $(M'_{2-y}M''_y)C$, $(M'_{3-y}M''_y)C_2$, $(M'_{4-y}M''_y)C_3$, or $(M'_{5-y}M''_y)C_4$, involving a random distribution of transition metals, resulting in solid solutions and providing continuous flexibility to customize properties [8]. The etching of Al and Sc atoms from 3D $Mo_{2/3}Sc_{1/3})_2AlC$ gives rise to 2D $Mo_{1.33}C$ sheets [9]. The pie chart given by Anasori and Gogosti about publications of MXenes in different fields is shown in Fig. 1.

2 Synthesis and Characterization

The preparation of MXene is through a scalable top-down approach following selective etching of the MAX phase that exhibits no loss or alteration of properties even with an increased batch size [11, 12]. This synthesis method utilizes the fluoride ion (F−) containing etching solutions like hydrofluoric acid (HF), ammonium bifluoride (NH_4HF_2) [11], and hydrochloric acid (HCl) + lithium fluoride (LiF) [13]. The resulting MXene is with terminations of O, OH, and/or F atoms [14, 15].

The synthesis of MXene from their parent compound sketched by W. Hong et al. 2020 are shown in Fig. 2. Figure (a) represents the A-group layers (illustrated by red spheres) undergoing selective etching, resulting in the formation of three types of mono-M MAX phases, namely M_2AX, M_3AX_2, and M_4AX_3, (b) represents after the selective etching process, MXenes are generated, accompanied by the development of surface terminations (illustrated by yellow spheres) denoted as T and (c) represents

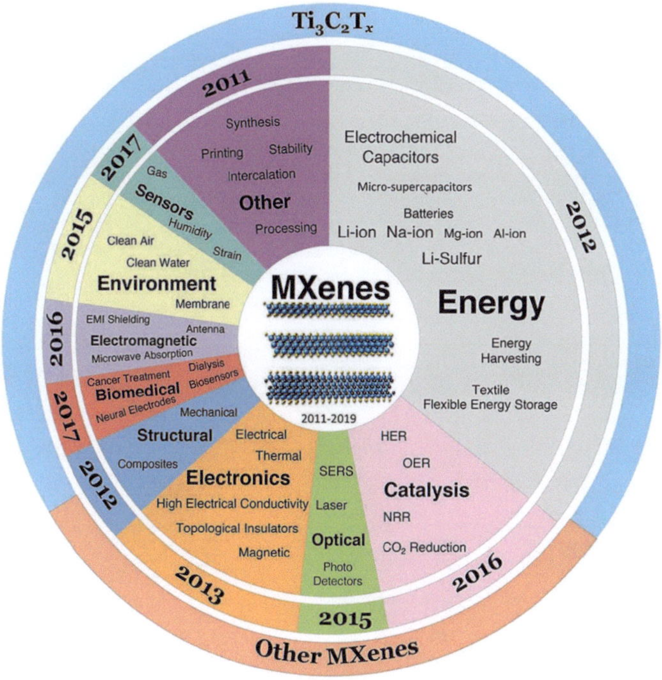

Fig. 1 Applications and properties of MXenes (adopted from [10])

the MAX and MXene phases encompass potential elements for M, A, X, and T. It is noteworthy that MXenes with orders as high as M_5X_4 have been successfully synthesized, although they are not included in the present periodic table. While Mn is present in M sites within MAX phases, MXenes containing manganese (Mn) have yet to be synthesized, as suggested by the horizontally striped background that represents Mn, indicating its absence from successfully developed MXenes.

An alternative synthesis approach involves generating MXene in Lewis acid molten salts, such as $ZnCl_2$, leading to the formation of a Cl-terminated MXene that remains structurally stable up to 750 °C [17, 18]. This method has proven effective in etching various MAX phase members, utilizing molten salts like $CdCl_2$, $FeCl_2$, $CoCl_2$, $CuCl_2$, AgCl, and $NiCl_2$ for precursors with A elements Si, Zn, and Ga [19]. The synthesis of the first reported nitride MXene, Ti_4N_3, differing from carbide MXenes was obtained from Ti_4AlN_3 using a molten eutectic fluoride salt mixture which has fluorides of lithium, sodium, and potassium [20]. MXenes are also obtained through chemical vapor deposition (CVD) processes directly [21].

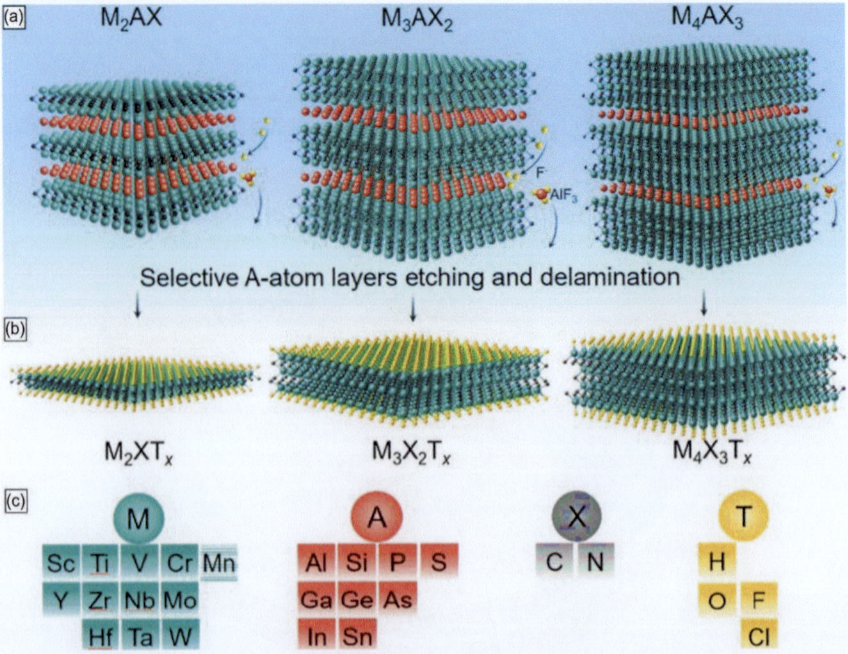

Fig. 2 MXenes precursors and their synthesis **a** MAX phase **b** Terminated MX phase **c** M, A, X, T elements [16]

3 Surface Modification, Intercalation and Delamination

MXenes, 2D transition-metal carbides, exhibit surface transformations by incorporating diverse terminations O, NH, S, Cl, Se, Br, Te, or remaining bare [22]. This transformation involves introducing and removing surface groups [23]. Aryl diazonium salts can be used to prepare MXene from organic molecules [24]. Due to the layered structure of MXenes and the relatively weak interlayer bonding, guest molecules like dimethyl sulfoxide (DMSO), hydrazine (N_2H_4), and urea can be intercalated into MXenes [3, 25]. The wet chemical etching of MXene with hydrofluoric acid and their delamination are shown in Fig. 3.

After HF etching, Ti_3C_2 MXene displays an accordion-like morphology with a weak force which is overcome for delamination. Despite these relatively weak forces, ultrasound treatment yields minimal single-layer flakes. The incorporation of dimethyl sulfoxide (DMSO) into the multi-layered (ML) MXene powders under continuous stirring weakens the bonds and subsequent ultrasound treatment leads to a few layered (FL) ones. Their filtration gives MXene "paper," following a process akin to producing the graphene oxide paper [27].

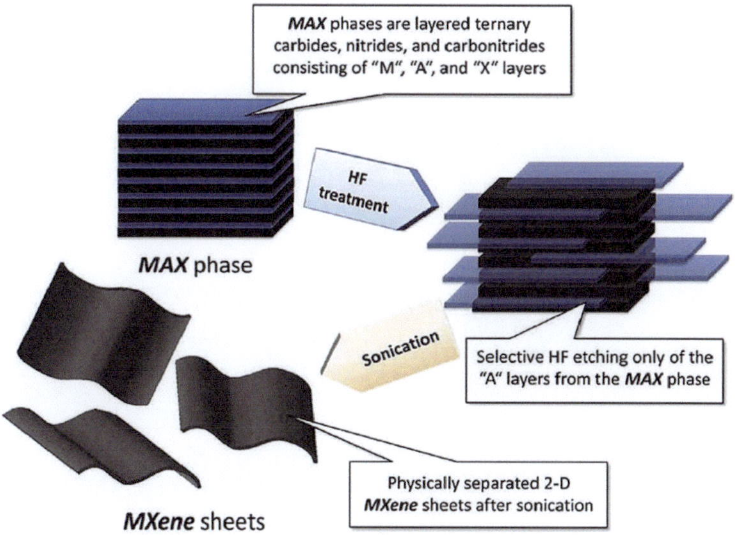

Fig. 3 Wet-chemical etching in hydrofluoric acid. (Reproduced with permission [26]. Copyright 2012, American Chemical Society.)

4 Making MXene Clay

For $Ti_3C_2T_x$ and Ti_2CT_x, the utilization of concentrated hydrofluoric acid in the etching process results in an open, accordion-like morphology with closely spaced layers, a characteristic shared by various MXene compositions. To achieve dispersion in suspension, the material requires pre-intercalation with substances like dimethylsulfoxide. However, when hydrochloric acid and LiF are employed for etching, the morphology becomes more compact, accompanied by a larger inter-layer spacing, likely influenced by the presence of intercalated water [28].

The schematic diagram of the processing of MXene clay is shown in Fig. 4, where the synthesis of clay and electrode preparation is in (a), the TEM image of the flakes is shown in (b), the single- and double-layer flakes are in (c) and (d) respectively and (e) the SEM image. The inset in b is the SAED pattern and in c and d are the respective layers sketches and e shows the flexibility of the film. The 'clay-like' behavior is reminiscent of certain clay minerals such as smectite clays and kaolinite. When hydrated, $Ti_3C_2T_x$ undergoes an expansion in interlayer distance, akin to clay materials [27] which is moldable in any shape and solidified on drying. They are highly conducting and are hydrophilic enough to be separated into single-layer 2D sheets in water. These are used in energy storage devices by rolling the MXene clay.

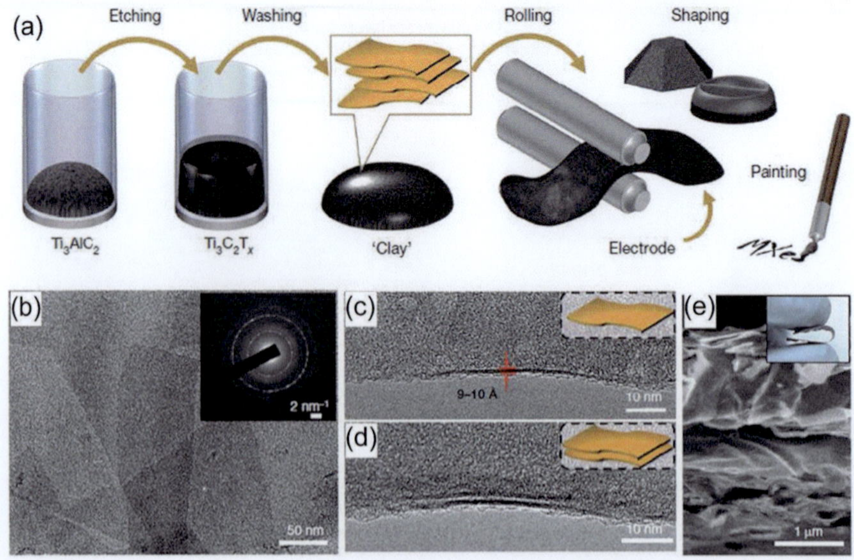

Fig. 4 Schematic diagrams of the processing of MXene Clay **a** electrode preparation **b** microscopy image **c**, **d** MXene layers, and **e** flexible film. Adopted with the permission of ref. [28] © 2014, Nature

5 MXene Processing

MXenes exhibit remarkable adaptability in solution processing, accommodating various solvents [29]. This versatility enables a broad range of deposition techniques [30–32]. Ink-jet printing research has explored both additive and additive-free $Ti_3C_2T_x$ inks [33, 34]. The lateral size of the flake emerges as a critical factor influencing properties, with different synthetic routes yielding varying flake sizes [30, 35]. Notably, using HF as an etchant requires sonication during intercalation and delamination to exfoliate in nanolayers that can be applicable in catalytical, biomedical and electrochemical means. However, the larger flakes are used for electronic and optical purposes. The Minimally Intensive Layer Delamination (MILD) method uses LiF to convert the MAX phase to delamination flakes in a neutral environment [30]. Several post-synthesis techniques have been utilized to tailor the flake size of MXenes, including methods like sonication, differential centrifugation, and density gradient centrifugation [36, 37]. Sonication can significantly reduce flake size, from 4.4 μm (as-synthesized) to around 1.0 μm after 15 min of bath sonication (100 W, 40 kHz), and further down to 350 nm after 3 h of treatment. For finer refinement, probe sonication (8 s ON, 2 s OFF pulse, 250 W) can bring flake size down to approximately 130 nm laterally [36]. Differential centrifugation, also known as cascading centrifugation, allows the sorting of flakes by size by progressively increasing centrifugation speeds, resulting in "large" (800 nm), "medium" (300 nm), and "small" (110 nm) flakes [37]. Additionally, density gradient centrifugation—using a sucrose and water gradient—facilitates the separation of flakes based on lateral dimensions, offering more uniform distributions that can be fine-tuned within a range of 100 to 10 μm, without requiring sonication [36].

6 Properties

MXene monolayers are theoretically expected to demonstrate metallic behavior due to the presence of a high electron density at the Fermi level [2, 38–41]. In MAX phases, the electronic states near the Fermi level are dominated by the M element's 3d orbitals, forming two distinct sub-bands. The first sub-band, primarily comprising hybridized M 3d and A element p orbitals, is close to the Fermi level and associated with M-A bonds. The second sub-band, located between −10 eV and −3 eV, stems from M 3d-C 2p and M 3d-A s orbital hybridization, contributing to M-C bonds. Upon removal of the A element layers, these M 3d states redistribute, forming Ti–Ti metallic bonds, which increases the density of states at the Fermi level by 2.5 to 4.5 times compared to the original MAX phases [2].

Despite these theoretical predictions, experimental data has not fully corroborated the anticipated increase in N(EF), and MXenes tend to exhibit higher resistivities than MAX phases. The energy positions of the O 2p and F 2p bands depend on the bonding environment and surface terminations, such as oxygen or fluorine. For example, in Ti_2CT_x and $Ti_3C_2T_x$, the O 2p band appears around 6 eV from the Fermi level, while the F_2p band is near 9 eV. These positions shift based on the bonding configuration and the adsorption sites of the terminating species. Additionally, heat treatments induce notable changes in Ti–O/F coordination.

Magnetically, theoretical models suggest that surface termination-free MXenes should exhibit magnetic properties [42]. Compounds such as Cr_2C, Cr_2N, and Ta_3C_2 are predicted to be ferromagnetic, while Ti_3C_2 and Ti_3N_2 are expected to exhibit anti-ferromagnetism [43]. However, experimental confirmation of these magnetic behaviors is yet to be realized [2].

7 Optical Properties

The dark membranes consisting of MXenes, such as Ti_3C_2 and Ti_2C, signaling effective absorption visible light are effective promising photo-thermal materials [44, 45]. Notably, the optical characteristics of these MXenes in the infrared (IR) region deviate considerably from their behavior in the visible spectrum [46]. These particles with size above 1.4 μm show robust metallic response to IR light and, consequently, high reflectivity to IR wavelengths due to their negative permittivity. According to Kirchhoff's law of radiation, their low IR absorption leads to a low IR emissivity around 0.1 for certain metals [46]. Materials that appear visually black but demonstrate IR white behavior hold significant promise in various applications.

8 Corrosion Resistivity

Certainly, an expanding body of literature acknowledges MXenes, notably $Ti_3C_2T_x$, as potent corrosion inhibitors. The enhanced corrosion resistance observed in $Ti_3C_2T_x$ MXene is ascribed to the combined benefits of its excellent dispersibility, barrier effect, and the release of corrosion inhibitors [47] due to several factors: (1) Good Dispersibility: MXenes, including $Ti_3C_2T_x$, often exhibit good dispersibility in various solvents. This property allows them to form protective layers on metal surfaces, preventing or inhibiting the corrosive attack. (2) Barrier effect: MXenes can act as a barrier on the metal surface, hindering the penetration of corrosive agents. The ability to form a protective layer can significantly reduce the corrosion rate. (3) Release of corrosion inhibitors: MXenes may release certain ions or functional groups that can act as corrosion inhibitors. These released species can interact with the metal surface, forming a protective layer and mitigating corrosion. It's important to note that the effectiveness of MXenes as corrosion inhibitors can depend on various factors, including the specific MXene composition, surface terminations, and the nature of the corrosive environment. Research in this area indicates the potential of MXenes for corrosion protection in different applications.

9 Biological Properties

In contrast to graphene oxide studied vigorously for its antibacterial properties, Ti_2C MXene does not exhibit antibacterial effects [48]. However from experimental results, Ti_3C_2 MXenes are efficient antibacterial in nature [49] destroying 98% of bacterial cells after exposure to a 200 μg/mL for 4 h [49]. This antibacterial effect is accompanied by observable damage to the cell membrane, resulting in the release of cytoplasmic materials and eventual cell death, providing insights into the antibacterial mechanism [49]. In vitro investigation show their promising use in the biomedical field [50]. The anticancer activity test of Ti_3C_2 MXene on both normal (MRC-5 and HaCaT) and cancerous (A549 and A375) cells shows their higher toxicity against cancerous cells due to the release of oxidative stress and reactive oxygen species (ROS) [50]. Further investigations highlight the potential of Ti_3C_2 MXene as a novel ceramic photothermal agent for cancer therapy [44]. In the context of neuronal biocompatibility, neurons cultured on Ti_3C_2 demonstrate comparable viability to those in control cultures. They exhibit the ability towards functional network and axonal formation, affirming the material's positive neuronal biocompatibility [51].

10 Emerging Applications

The surface terminations of MXenes have exhibited considerable promise across various applications. Some notable areas of application include energy storage applications [52, 53]. MXenes have also shown potential in composites [54, 55], photocatalysis [56], water purification [57], gas sensors [58, 59], transparent conducting electrodes [31], neural electrodes [51], metamaterials [60], surface-enhanced Raman spectroscopy (SERS) substrates [61], photonic diodes [62], electrochromic devices [32], and triboelectric nanogenerators (TENGs) [63]. Some of the emerging applications of MXenes basically in energy storage, water purification, and flexible electronics are shown in the following flowchart Fig. 5.

10.1 Water Purification

MXenes, particularly $Ti_3C_2T_x$ and Ti_2CT_x, have shown promise in water purification applications. The first MXene-based water purification membrane is shown in Fig. 6.

10.1.1 Membrane Application

MXenes, when fabricated into membranes, have demonstrated excellent water purification properties. For example, the water flux due to Ti_3C_2 MXene membranes of 1

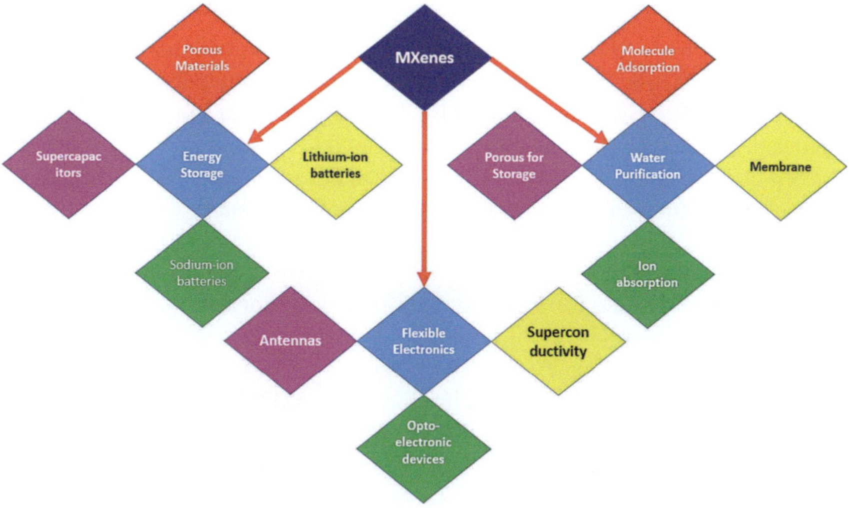

Fig. 5 Some emerging applications of MXenes and their composites

Fig. 6 The first MXene-based-water-purification-membrane through sieving ion by vacuum-assisted filtration. **a** Preparation of $Ti_3C_2T_x$ membrane from Ti_3AlC_2 **b** a photograph, and **c** SEM image of the MXene layer membrane structure (a cross-sectional view). "Credit: Ren et al. The Journal of Physical Chemistry Letters (© 2015 American Chemical Society, open access article)" [57]

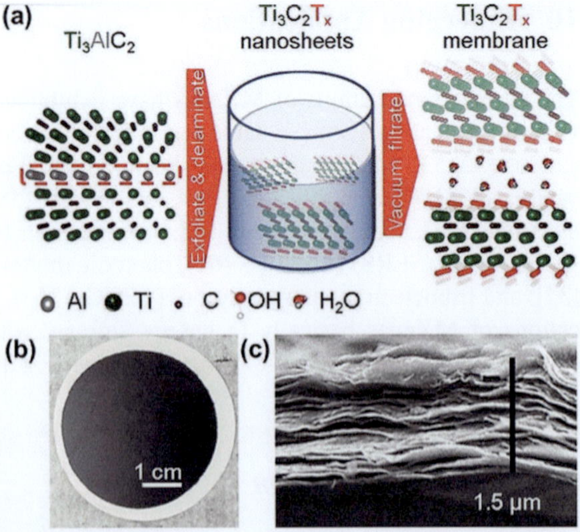

μ has exhibited exceptional water flux, reaching nearly 38 L/(Bar·h·m^2) with selective sieving of salts with size smaller than the interlayer spacing [57]. The sieving is based on the radius of hydration and ion charges demonstrating their potential in the desalination processes.

10.1.2 Ion Absorption in Flow Electrode Capacitive Deionization (FE-CDI)

Ti_3C_2 MXenes have recently demonstrated notable applications in flow-electrode capacitive deionization (FE-CDI) cells that are efficient in removing ammonia from wastewater. In this context, MXene FE-CDI has shown an impressive 100-fold improvement in ion absorption capacity and a tenfold increase in energy efficiency more the electrode based on activated carbon than that of MXene [64]. This suggests potential applications in the removal of specific ions from water.

10.1.3 Porous MXenes for Enhanced Lithium-Ion Storage

Porous MXenes, including Ti_3C_2, Nb_2C, and V_2C, produced through chemical etching, have been utilized in flexible films for improved lithium-ion storage capabilities. These films, especially p- Ti_3C_2/CNT (carbon nanotube) composites, exhibited high lithium-ion storage capacity, cycling stability, and rate performance. While this application is more directly related to energy storage, it underscores the versatility of MXenes in various applications, including those related to water purification.

10.1.4 Adsorption of Molecules

MXenes possess a layered structure, and their surfaces can undergo chemical modifications. This allows for the introduction of various functional groups, enabling MXenes to adsorb and remove contaminants from water. This adsorption capability has been explored for potential use in water purification processes. While research in this area is ongoing, the unique properties of MXenes, such as their tunable surface chemistry, high surface area, and flexibility in processing, make them promising materials for water purification applications.

10.2 Energy Storage

MXenes demonstrate excellent performance in energy storage, and their storing capacity can be further enhanced by incorporating suitable components to form nanocomposites. Critical factors influencing storing capacity include conductivity, cyclability, specific capacitance, capacitive retention, etc. Noteworthy MXenes for Lithium Ion Batteries include V_2CT_x [65], Nb_2CT_x [65], Ti_2CT_x [66], and $Ti_3C_2T_x$ [27]. $Ti_3C_2T_x$ MXene boasts high electrical conductivity (2.4×10^5 Sm^{-1}), comparable to that of multi-layered graphene, exhibiting excellent response to RF fields, charging, and discharging phenomena. Conductive polymers within organic materials offer distinct advantages over inorganic solids, featuring tunable electronic, optical, stable, and conducting properties [67–69]. These materials find applications in various fields, such as smart windows [70], thermoelectrics [71], biosensors [72], corrosion protection [73], light-emitting diodes [74], and energy storage devices [75–77]. The in-situ electrochemical process proves effective for synthesizing such materials [78–82]. However, one-step electroplating (EP) faces challenges in synthesizing organic–inorganic nanocomposites. Overcoming this limitation, i-MAX, a MAX phase with in-plane vacancy ordering, such as $Mo_{1.33}C$, is highly conductive and suitable for supercapacitor applications [9, 83]. Combining $Mo_{1.33}C$ with conducting polymers produces efficient plates for micro-supercapacitors, replacing electroplating with a novel in situ electrochemical polymerization process for the preparation of 2D MXene-doped conductive polymer films [84, 85]. In 2021, Mahmood et al. introduced Vanadium Pentoxide as an efficient electrode material, addressing its lower electrical conductivity by forming nanohybrids with MXene. This resulted in a specific capacity of 768 F/g (at 1 A/g), a specific capacity retention of 93.3% after a 6000 Galvanostatic Charge–Discharge (GDC) test, and increased current density from 1 to 5 A/g, showcasing the advantageous properties of MXenes [86].

The various aspects of energy storage MXene materials and their applications including interdigited electrode are shown in Fig. 7. In **(A)**, a schematic of a layered transition metal oxide structure is shown, highlighting key features such as a redox-active surface, rapid cation intercalation, and a conductive metal carbide core that ensures efficient electron transfer. The material also has tunable surface terminations and adjustable interlayer spacing, enhancing its redox chemistry and ion storage

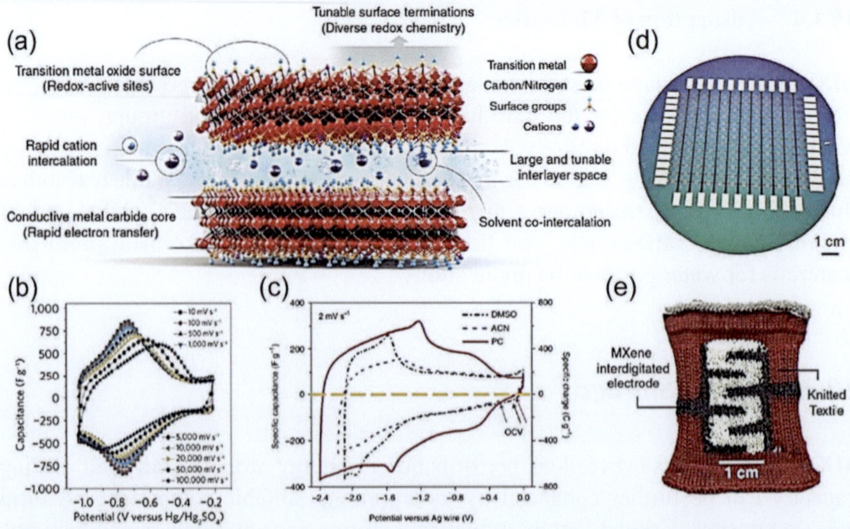

Fig. 7 Advanced layered materials for energy storage. **a** Schematic of a transition metal oxide with tunable surfaces, rapid cation intercalation, and a conductive metal carbide core. **b** Capacitance vs. potential at varying scan rates. **c** Cyclic voltammetry curves showing electrolyte effects on performance. **d** Circular energy storage device with interdigitated electrodes (1 cm diameter). **e** MXene-based knitted textile electrode for flexible energy storage. *Adapted from* [87] –©**Royal Society of Chemistry (RSC)**

capability. In **(B)**, a graph shows the specific capacitance as a function of potential at different scan rates, ranging from 10 to 100,000 mV s^{-1}, illustrating the material's ability to retain high capacitance even at fast charge/discharge rates. **(C)** presents cyclic voltammetry curves for different electrolytes, indicating that propylene carbonate (PC) delivers the best performance in terms of specific capacitance and charge storage compared to DMSO and ACN. **(D)** depicts a circular energy storage device with interdigitated electrodes, demonstrating the material's integration into practical applications for electrochemical testing. Finally, **(E)** showcases a flexible, knitted textile embedded with MXene interdigitated electrodes, suggesting the potential for wearable energy storage devices. Each part of the figure emphasizes these advanced materials' efficiency, flexibility, and versatility in energy storage systems.

10.2.1 Porous Materials

Through a simple room temperature chemical etching process [88], especially porous Ti_3C_2 showcases a more open and larger specific surface area. This material can be further processed into flexible films, either independently or in conjunction with carbon nanotubes (CNTs) [88] forming p-Ti_3C_2/CNT composite showing enhanced storage capacity of lithium-ion, achieving an impressive capacity of up to 1250 $mAhg^{-1}$ at 0.1 C [88].

10.2.2 Lithium-Ion Batteries (LIBs)

MXenes, including V_2CT_x, Nb_2CT_x, Ti_2CT_x, and $Ti_3C_2T_x$, have been comprehensively studied in the context of lithium-ion batteries (LIBs) [27, 65, 66]. In multi-layer form, V_2CT_x MXene has been found to deliver the highest reversible charge storage capacity, reaching 280 $mAhg^{-1}$ at a 1C rate and 125 $mAhg^{-1}$ at 10C. Similarly, multi-layer Nb_2CT_x demonstrates a reversible capacity of 170 $mAhg^{-1}$ at 1C and 110 $mAhg^{-1}$ at 10C, maintaining stability. Although $Ti_3C_2T_x$ in its multi-layer form exhibits a lower capacity, once delaminated via sonication, its performance improves significantly. Delaminated $Ti_3C_2T_x$ paper reaches 410 $mAhg^{-1}$ at a 1C rate and 110 $mAhg^{-1}$ at 36C, attributed to an increased electrochemical surface area. In general, M_2X MXenes tend to have a higher capacity than M_3X_2 or M_4X_3 types under the same current due to their fewer atomic layers. Each MXene also operates within a specific voltage range, which could be useful in battery applications. Interestingly, the experimentally obtained capacity for $Ti_3C_2T_x$ paper surpasses theoretical predictions, suggesting that the underlying charge storage mechanisms need further exploration [89].

10.2.3 Sodium-Ion Batteries (SIBs)

MXenes perform excellently in sodium-ion batteries, where the rapid diffusion of Na + ions on MXene surfaces facilitates fast charging and discharging [90, 91]. It has been observed that two layers of Na + ions can be intercalated between MXene layers [92, 93]. For example, a multilayered negative electrode of Ti_2CT_x demonstrated 175 $mAhg^{-1}$ capacity with excellent rate performance [94]. The surface termination or the transition metal alteration can help in adjusting the insertion potential of Na-ion in MXenes [25, 90]. Furthermore, V_2CT_x is used as the cathode material in batteries [95]. In addition, the higher capacity and longer cyclability is shown by porous MXene-based paper electrodes, making them suitable for applications where compactness is a crucial factor [96].

10.2.4 Supercapacitors

MXenes are being investigated to improve supercapacitor energy density by enhancing their charge storage capabilities. One approach involves increasing the available surface area for redox reactions by expanding the interlayer spacing, though this might reduce electrode density. The synthesis process is critical as it influences surface chemistry, intercalation rates, and overall energy storage capacity. For example, $Ti_3C_2T_x$ MXenes produced using molten salt methods with chlorine surface groups exhibited capacities of 142 $mAhg^{-1}$ at a 13C rate and 75 $mAhg^{-1}$ at a 128C rate, driven by complete Li+ desolvation. In contrast, MXenes synthesized using HF etching showed a lower capacity of 107.2 $mAhg^{-1}$ at a 1C rate [97]. Researchers have also investigated composite $Ti_3C_2T_x$-based electrodes, such as those combined with polymers (PPy, polyaniline) [98, 99], TiO_2, and Fe_2O_3 [100]. Notably, $Ti_3C_2T_x$ hydrogel electrodes achieved volumetric capacitances as high as 1500 F/cm^3 [101].

Supercapacitors built from $Ti_3C_2T_x$ MXene paper in aqueous solutions exhibit impressive cyclability, storing up to 300–400 F/cm^3, three times higher than activated carbon and graphene-based capacitors [102]. Additionally, Ti_3C_2 MXene clay achieved a volumetric capacitance of 900 F/cm^3, maintaining this level in over more than 10,000 cycles without any capacitance loss [103]. For MXene electrodes in lithium-ion electrolytes, solvent choice critically affects ion transport. Using propylene carbonate (PC) facilitated lithium-ion desolvation, enhancing charge storage without significantly increasing electrode volume. This method achieved over double the charge storage density compared to solvents like acetonitrile or dimethyl sulfoxide [104].

Similarly, Sunil Kumar et al. [105] designed a supercapacitor that utilizes nano-layered MXene as the active material with Ni-foil acting as the current collector. Additionally, the surface of the Ni-foil is enhanced with graphene grown via chemical vapor deposition, forming a $Ti_3C_2T_x$/graphene/Ni electrode structure. This surface passivation boosts charge transfer, increasing the capacitance. The graphene-based supercapacitor displays more than 1.5 times the capacitance of those without graphene, achieving a specific capacitance of ~542 F/g at a 5 mV/s scan rate. Cyclic voltammetry shows a quasi-rectangular shape and stable performance over 5000 cycles with excellent retention. [105] The Nyquist plot analysis confirms reduced series resistance, further optimizing the performance of the supercapacitor. Figure 8 outlines the synthesis process of $Ti_3C_2T_x$ MXene and its application in fabricating a supercapacitor. In part (a), the MXene is synthesized through LiF + HCl treatment of the Ti_3AlC_2 MAX phase, followed by washing, centrifugation at different speeds, and subsequent transfer to collect MXene from the supernatant. Part (b) compares earlier synthesis methods, including sonication and filtration, to produce free-standing MXene paper. Part (c) demonstrates the fabrication process of a supercapacitor where MXene slurry is coated onto CVD-grown graphene on Ni-foil, with the final assembly involving a separator soaked in H_2SO_4 between the electrodes [105].

11 Composites

The MXene itself have many limitations which can be overcome by mixing it with several other elements or compounds and enhance their properties. The resulting mixture are termed as composites. There are many composites of MXenes which are widely studied and applied in modern technology. The few-layered Ti_3C_2 nanosheets (FL-Ti_3C_2) exhibit exceptional compatibility with polymers like polyvinyl alcohol

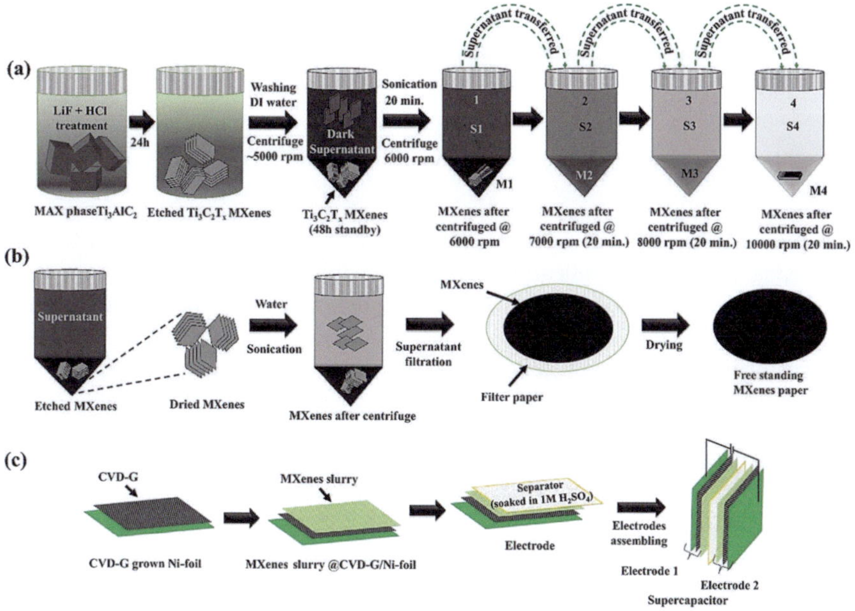

Fig. 8 **a** Synthesis of $Ti_3C_2T_x$ MXene from the supernatant via subsequent transfer, as outlined in this study; **b** a comparison of MXene synthesis methods from earlier research; and **c** the process of supercapacitor fabrication. Adopted from [105] ©Nature

(PVA), forming alternating MXene-PVA composite layers [54, 55, 106, 107]. The electrical conductivity of these composites can be modulated by adjusting the MXene content, which ranges from 40 to 90%, resulting in conductivities between 4×10^{-4} and 220 S/cm. These composites also display enhanced mechanical properties, including tensile strengths up to 400% higher than pure MXene films, and capacitance reaching 500 F/cm^3 [108]. Flexible MXene/graphene supercapacitor electrodes have also been fabricated using electrostatic self-assembly, exhibiting volumetric capacitance of 1040 F/cm^3, with high cycle stability and 61% capacitance retention [109]. Additionally, composite films of MXene and carbon nanomaterials, produced via filtration methods, demonstrate improved supercapacitor performance at high scan rates [110]. MXenes incorporated with carbon nanomaterials facilitate better electrolyte ion diffusion, vital for flexible energy storage devices. Epoxy/MXene composites also exhibit improved mechanical properties, with significant enhancements in tensile strength and modulus, while MXene/C-dot nanocomposites demonstrate synergistic optical and thermal properties [111]. MXene-based sensors have been explored extensively for applications in gas [112] and biological sensing [113], as well as Surface-Enhanced Raman Spectroscopy (SERS) [61, 114], detecting molecules like salicylic acid [115], organic dyes, and biomolecules [116]. Gas sensors based on MXenes have shown high sensitivity and selectivity for detecting ammonia, alcohol, and sulfur dioxide, with potential applications in environmental monitoring and industrial safety [112]. The details are mentioned in the following Sect. 14.

12 Flexible Electronics

The unique electronic and mechanical properties of MXenes make them suitable for flexible electronics applications, such as flexible electrodes and sensors Their 2D layered structure allows for bending and stretching without losing performance, essential for wearable devices. Additionally, MXenes' tunable surface chemistry enhances their sensitivity in detecting gases or biomolecules, while their high surface area ensures fast response times. These properties make MXenes highly suited for advanced applications in energy storage, health diagnostics, and environmental monitoring explained in a different section in this chapter. Some notable flexible electronics are found in the following devices.

12.1 Antennas

Scientists at Drexel University in the United States have engineered spray-on antennas, achieving performance parity with existing antennas in devices such as phones and routers. This breakthrough involves the application of MXenes by coating commonplace objects, thus broadening the scope of applications for the Internet of Things (IoT) [117, 118].

12.2 Optoelectronic Devices

MXene-based SERS (Surface-Enhanced Raman Spectroscopy) platforms have been crafted using a spray-coating technique, facilitating the detection of various dyes, with enhancement factors around 10^6. Titanium carbide MXenes have demonstrated this SERS capability within aqueous colloidal solutions, suggesting promising applications in fields such as biomedicine and environmental monitoring, particularly due to their selective enhancement of positively charged molecules [61]. Moreover, titanium carbide MXenes have also been employed to create transparent conducting electrodes, which exhibit a high light transmittance of approximately 97% per nanometer thickness. The electrode performance is influenced by the MXene composition, as well as the specific methods of synthesis and processing [119].

12.3 Superconductivity

Nb_2C MXenes display superconductivity that is dependent on the surface groups present [22]. The MXene $Ti_3C_2T_x$, one of the most extensively studied MXenes, has been investigated for its superconducting properties. Reports have suggested that, when intercalated or modified, $Ti_3C_2T_x$ may exhibit superconducting behavior. The critical temperature (Tc) for superconductivity in $Ti_3C_2T_x$ can be influenced by factors like intercalation agents, surface terminations, and other structural modifications. Intercalation of various elements or compounds between MXene layers has been explored to induce superconductivity. For example, introducing lithium or other intercalants has been investigated to modify the electronic structure and potentially induce superconducting states. The surface terminations of MXenes can impact their electronic properties. Surface terminations involving functional groups, such as –OH or –F, have been considered in studies aiming to understand and control the superconducting behavior of MXenes. While $Ti_3C_2T_x$ has been a focus, there are efforts to explore other MXenes for potential superconductivity. The field is expanding to include various MXene compositions, each with its unique electronic structure and properties. The potential application of MXenes in superconducting devices and systems could have significant implications for technology, energy transmission, and other fields, making it an exciting area of research within the broader field of two-dimensional materials.

13 Topological Properties

MXenes exhibit interesting topological properties, particularly in the context of condensed matter physics. The topological aspects of MXenes are often related to their electronic band structures, and they have been explored for potential applications in electronics and materials science [120]. Some MAX-phase, Tuned Terminated MXene, and Pure MXene are listed in Table 1. Some key considerations regarding the topological properties of MXenes are illustrated in Figs. 9 and 10.

13.1 Topologically Trivial MXenes

MXenes derive their energy gaps primarily from their surface terminations. For instance, Sc_2CT_2 (T = F, OH, O) exhibits gaps of 1.03, 0.45, and 1.8 eV, respectively, while Ti_2CO_2, Zr_2CO_2, and Hf_2CO_2 have gaps of 0.24 eV, 0.88 eV, and 1.0 eV [40]. Out of these, only $Sc_2C(OH)_2$ displays a direct band gap, with the others having indirect gaps. MXenes containing elements such as Ti, Zr, and Hf exhibit a consistent transition from metallic to semimetallic behavior depending on the surface termination [40]. In semimetallic states, the conduction band minimum and valence band

Table 1 List of some MAX-phase, tuned terminated MXene and pure MXene

S. N	MAX	Terminated MXene	Type	Transition	Pure MXene	Citation
1	–	$M'_2M''C_2O_2$ (M' = Mo, W, Ta; M'' = Ti, Zr, Hf)	Non-trivial	SOC. SM-SC	M_2TiC_2	[121]
2	–	$M'_2M''_2C_3O_2$ (M' = Mo, W, Ta; M'' = Ti, Zr, Hf)	Non-trivial	SOC, M-SM	$M_2Ti_2C_3$	[121]
3	–	M_2CO_2 (M = W, Mo, Cr)	Non-trivial	SOC, M-SC	M_2C	[122]
4	$M_3N_2AlF_2$	$M_3N_2F_2$ (M = Ti, Hf, Zr)	Non-trivial	M,SM,SC-I	M_3N_2	[123, 124]
5	–	$V_2TiC_2X_2$ (X = F, Cl, Br, I, O, H, and OH)	Non-trivial	M–I	V_2TiC_2	[125]
6	–	$M'_2M''C_2F_2$ (M' = V, Nb, Ta; M'' = Ti, Zr, Hf)	Non-trivial	M–I	$M'_2M''C_2$	[125]
7	–	MCO_2 (M = Ti, Zr, Hf)	Non-trivial	M–SC	MC	[126, 127]
8	–	M_2CO_2 (M = Ti, Zr, Hf)	Trivial	M–SC	Ti_2C	[128]
9	–	Ti_2NO_2	Trivial	FMHM–FMSC	Ti_2N	[129]
10	–	Sc_2CX_2(X = O, F, (OH))	Trivial	M–SC	Sc–2C	[130–132]
11	–	$(M'2/3\ M''\ 1/3)_2CO_2$ (M' = Mo, W; M'' = Sc, Y)	Piezoelectric	M-SC	(M'2/3 M'' 1/3)2C	[133]
	M-Metallic, SC-Semiconductor I-Insulator, HM-half metallic, FM-ferromagnetic					

Fig. 9 Edge band structure of $Mo_2TiC_2O_2$ (adopted from [134])

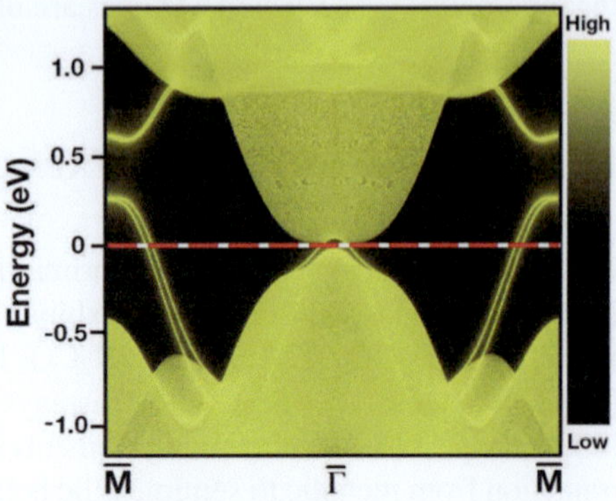

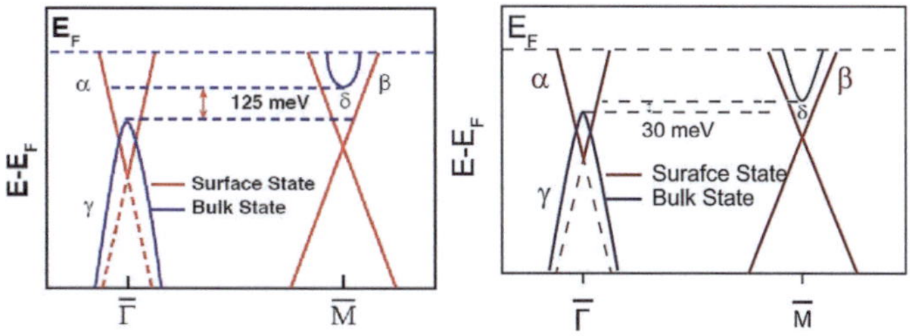

Fig. 10 A schematic of all the bands of Mo_2TiC_2 (left) and $Mo_2Ti_2C_3$ (right) observed along $\overline{\Gamma}-\overline{M}$ direction (adapted from [135])

maximum meet at specific points, leading to parabolic bands. Surface functionalization plays a significant role, in influencing stability and band structure. For example, adding terminations like F, OH, and O alters the electron configuration, with p orbitals hybridizing with the d orbitals of transition metals [40]. This leads to the creation of a semiconducting gap, which can be further adjusted by strain or the application of an electric field [40, 136, 137]. Additionally, specific MXenes under strain, such as the two-dimensional Ti_2NO_2 monolayer, show a transition from half-metal to semiconductor, showcasing potential use in spintronics applications [129]. The list of some MAX phase-tuned terminated MXene, and Pure MXenes extracted from literature [121–133] are listed in Table 1.

13.2 Topologically Non-Trivial MXenes

In the realm of topologically nontrivial semimetals and semiconductors within MXenes, the role of spin–orbit coupling (SOC) emerges as crucial, overshadowing hybridization. Computational studies by Hongming Weng et al. in 2015 using Density Functional Theory (DFT) explored the oxidation of MXenes, revealing that M_2CO_2 ($M = W$, Mo, Cr) serve as 2D Topological Insulators (TIs), with W-based MXene exhibiting the most significant gap of 0.194 eV [122]. These materials are noted for their ease of preparation, stability as antioxidants, and the anticipated enhancement of topological superconductivity and Majorana mode development crucial for spintronic devices. Chen Si and colleagues (2016) projected that a new category of MXenes, $Mo_2MC_2O_2$ ($M = Ti$, Zr, or Hf), containing two transition metals in a structured configuration, would function as strong quantum spin Hall (QSH) insulators. The emergence of the topological gap, particularly at the Γ point, was linked to the influence of spin–orbit coupling (SOC), which arises primarily from the properties of the transition metal (M) atoms within the MXene. This coupling enables the system to maintain topological insulating behavior. Despite smaller gaps compared to Kane-Mele (KM) or Bernevig-Hughes-Zhang (BHZ) models, these MXenes exhibited the QSH effect at room temperature and demonstrated antioxidant properties due to their oxygen-covered surfaces [138].

In 2016, Liu Jiang Zhou and colleagues identified the Quantum Spin Hall Effect (QSHE) and spin textures in transition-metal carbide monolayers (MC) composed of elements such as Zr and Hf. They observed a 54 meV gap in ZrC attributed to band inversion from the d orbitals driven by spin–orbit coupling, while the HfC monolayer showed a 170 meV gap under in-plane strain [126, 127].

Similarly, Khazaei et al. (2017) computationally analyzed oxides of $M'_2\ M''C_2$ (M' = Mo, W; M'' = Ti, Zr, Hf) MXenes, showing that spin–orbit coupling between $dx^2 - y^2$ and dxy orbitals was critical in opening topological gaps. Their work also suggested that W-based MXenes are promising for room-temperature applications due to their larger gaps [121].

Yunye Liang's 2017 work introduced $Ti_3N_2F_2$ as a 2D topological insulator and found $Zr_3N_2F_2$ to be a semimetal with non-trivial topology, with both materials showing transformation into 2D TIs under lattice stretching [30]. In 2018, E. Balci and collaborators described a new kind of band inversion in Si and Ge doped $Sc_2C(OH)_2$ MXenes, emphasizing the importance of SOC for gap formation [132]. In a 2019 study, Mei-Ling Xu examined $V_2TiC_2X_2$ (X = F, Cl, Br, I, O, H, OH), demonstrating their potential as topological insulators with band gaps between 40 and 420 meV, suitable for QSHE applications [125].

Additionally, Parajuli and Samatha (2019) conducted DFT calculations on $M'_2\ M''C_2$ MXenes using the WIEN2K package, re-optimizing their structures and reporting larger gaps compared to Khazaei's results. They synthesized oxides of Mo_2TiC_2 and $Mo_2Ti_2C_3$, analyzing them under ARPES, finding chirality and band gaps of 125 meV and 30 meV, respectively [121]. The Mo_2TiAlC_2 MAX phase was synthesized using a solid-state reaction method, with alloy preparation aided by ball milling. Chemical wet etching with hydrofluoric acid was then employed to remove aluminum [139], yielding $Mo_2TiC_2T_x$. The Tx groups (O, OH, F) were removed via sonication to obtain pure Mo_2TiC_2 MXene. Oxygen adsorption in a PLD chamber resulted in the formation of $Mo_2TiC_2O_2$. The oxidation of Mo_2TiC_2 and the synthesis of $Mo_2TiC_2O_2$ were described in our previous work, where Mo_2TiC_2 was placed in a PLD chamber under specific conditions (750 °C and 0.1 mbar) to allow oxygen to occupy the octahedral sites, producing $Mo_2TiC_2O_2$. This oxygen adsorption was studied using the RHEED technique, as previously reported [140]. The process flow for producing Mo_2TiC_2 flakes, oxidizing them to $Mo_2TiC_2O_2$, and subsequent ARPES characterization is illustrated in Fig. 11.

14 Sensors

The discussion on sensors has become imperative, given their crucial roles in biomedical and energy storage applications. Essential components for the efficient performance of sensors include repeatability, reproducibility, sensitivity, and selectivity over an extended period [141]. MXene stands out as an exceptionally sensitive

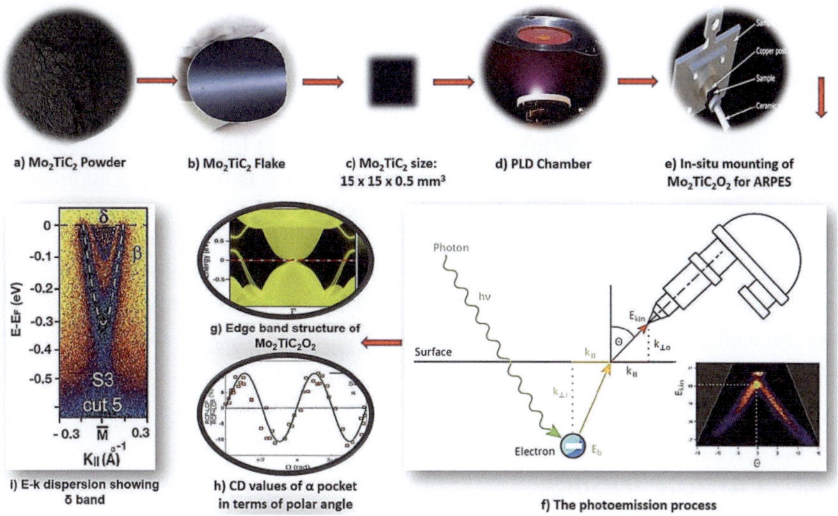

Fig. 11 Flowchart of the sample preparation and characterization for ARPES: **a** Mo_2TiC_2 Powder, **b** Mo_2TiC_2 flake **c** Mo_2TiC_2 of size 15 × 15 × 0.5 mm^3 **d** PLD chamber **e** In-situ mounting of $Mo_2TiC_2O_2$ for ARPES **f** the photoemission process **g** Edge band structure of $Mo_2TiC_2O_2$ **h** CD values of α-pocket in terms of polar angle and **i** E-k dispersion showing δ-band [134]

gas sensor, owing to its terminations that respond even to minute gas concentrations [142]. Notably, MXene sensors can detect ammonia and acetone-like chemicals, serving as indicators for conditions such as ulcers and diabetes. The unique chemical layered structure and porous membrane of MXenes contribute to their outstanding performance. MXene sensors encompass various types, including electrochemical, gas, stress–strain, and photoluminescence sensors. Furthermore, the flexibility of these sensors can be harnessed to create versatile options such as flexible humidity, VOC, mechanical, and parametric sensors [143]. When compared to conductive polymer nanocomposites used for similar purposes, sensors based on MXene-polymer nanocomposites demonstrate significantly higher efficiency than the nanocomposite of conductive polymer [141]. There are two such types of sensors used in biomedical fields:

14.1 Biosensors

Biosensors, which are used to detect specific elements within the human body, typically operate by leveraging both electrical and optical signals. These devices consist of three main components: a sensing element, a transducer, and a signal processor. The sensing component generally contains a biomolecule, such as an enzyme, that reacts with analytes, generating a biochemical signal. This signal is converted into an electrical form by the transducer and further processed for interpretation [144]. For instance, a biosensor designed for glucose detection, developed by Rakhi et al. [145] includes components like GOx (glucose oxidase), Ti_3C_2 MXene, and Nafion, which aid in the conversion of glucose to gluconolactone and hydrogen peroxide, with peroxide oxidation serving as the detection signal [146].

MXene-based polymer nanocomposites show promise in biomedical applications because of their stability, selectivity, biocompatibility, and responsiveness to external stimuli. These composites, particularly in conjunction with polymers like PANI and PVP, have applications ranging from cancer research to photothermal therapy and even kidney stone treatments [147]. The versatility of these nanocomposites in medical applications makes them a key subject for future research into expanded use cases.

14.2 Physical Sensors

Sensors that respond to changes in pressure and tension by altering electrical signals, distinct from the biomolecule-based biosensors, are classified as physical sensors. These sensors find application in human activities through flexible wearables [148–151] as shown in Fig. 12. Enhanced sensitivity and selectivity in various sensors are achieved through improved conductivity, lower bandgap, and sites with higher activity [152–154]. MXene is actively involved in the development of electrochemical, photoluminescence, mechanical, and other types of sensors [150–152]. When combined with polymers, MXene facilitates the creation of flexible sensors for applications such as humidity [151, 155–158], and mechanical sensing [159]. Material impedance and water absorption directly impact sensitivity, especially in humidity sensors utilized across diverse fields like environment monitoring, medical care, food, and metallurgy. In 2017, Sajid's team [158] and Kim's team [157] successfully developed a highly sensitive humidity sensor using a 2D Mo_2C, Cr_3C_2, and polymer (PAM, PVA) composite. Another breakthrough came in 2019 when An et al. [155] identified $Ti_3C_2T_x$/PDAC humidity sensors with repeatable sensing properties, where PDAC (poly diallyl dimethyl ammonium chloride) enhanced electrical properties. The sensitivity of the composite increased to 46% by altering the deposition time, albeit not surpassing other commercially available sensors. Detecting volatile organic compounds (VOCs), toxic pollutants from industries poses another challenge. Yuan et al. [160] coated a PVA/PEI polymer with $Ti_3C_2T_x$ to create a VOCs gas sensor with heightened sensitivity. PVA/PEI aided in the adsorption of polar gas molecules, increasing the specific surface area. Electrical stability was maintained even after 5000 cycles, with Zhang et al. [161] exploring a hydrogel strain sensor using MXene/PVA crystal mud composite. Guo et al. [151] developed a wearable pressure sensor by blending degradable PLA and $Ti_3C_2T_x$, demonstrating exceptional stability over 10,000 compression cycles. Hu et al. [162] utilized Chitosan (CS)/Ti_3C_2 composite for wearable pressure sensors, leveraging CS to connect interlaminar sheets of MXene. Their response time was swift (109.6 ms), with recovery

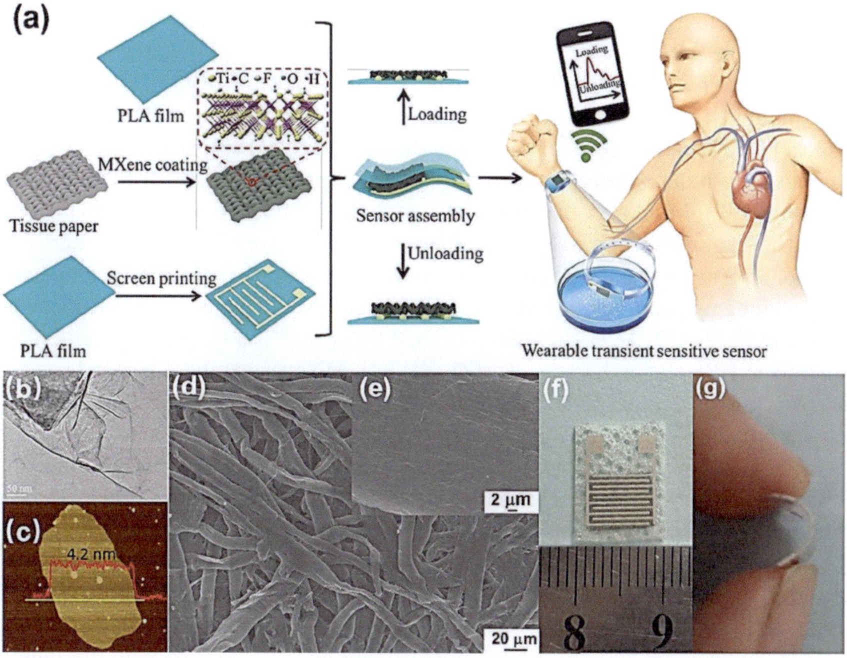

Fig. 12 MXene nanosheets based composites for wearable pressure human sensors (Reprinted with permission from Ref. [151] Copyright 2017 American Chemical Society)

time at 110.6 ms, and exhibited remarkable stability over 150,000 cycles. Li et al. [163] incorporated polyurethane (PU) into Ti_3C_2@CS to create a pressure sensor. These examples showcase the diverse applications of MXene/polymer nanocomposites in various sensor types, but further intensive studies are essential for optimizing and enhancing their applications.

15 Biomedical Applications

MXenes exhibit significant potential in biomedical applications, encompassing drug delivery, biosensing, and imaging, although this field is still in its early stages. The biomedical use of Ti_3C_2-based MXene is pivotal in the biomaterial market for treating, diagnosing, or enhancing tools [164] made from polymer, ceramic, metal, or composite materials [165]. In recent times, nano-medicine-based applications have gained prominence. Ceramic-based MXenes are particularly desirable for biocompatible accessories [166], given their metallic conductivity, tunability, absorbent properties, and non-toxicity to living organisms. However, their stability in the external environment and challenges in controlling their flow for drug delivery, potentially impacting healthy cells or tissues, pose limitations.

These challenges have been addressed by creating composites with various biodegradable polymers such as polyvinylpyrrolidone (PVP) [167, 168], polyethylene glycol (PEG) [169], cellulose [170], polyvinyl alcohol (PVA) [171], soybean phospholipid (SP) [172–176], or inorganic nanoparticles like polyoxometalates (POMs) [177], mesoporous silica nanoparticles (MSNs) [178], etc. These composites not only reduce costs but also minimize the pain associated with traditional stitching [179]. Moreover, these composites serve as effective photothermal agents [36]. The terminations on MXenes influence their electronic behavior, primarily as semiconductors or metallic materials [180–182], imparting antibacterial properties and enabling applications in wastewater treatment [82, 83], cancer diagnostics and treatment [172–175, 183], and photothermal therapy (PTT) through hyperthermia, synergistic PTT/chemotherapy, etc. [71–74, 84]. MXenes exhibit advanced properties such as antibacterial activity, drug delivery, photothermal therapy, tissue culture, bio-sensing, bio-imaging, among others.

In addition to polymer nanocomposites [184], another noteworthy composite is WO_3/MXene, synthesized using a hydrothermal approach, with characterized structural, morphological, spectral, elemental, and superficial properties. The photocatalytic degradation rates for MXene, WO_3/MXene, and WO_3 were 54%, 89%, and 99%, respectively. The addition of WO_3 significantly increased the photocatalytic degradation capacity of MXene. Moreover, these nanocomposites exhibited notable antibacterial activity against positive and negative strains, with the inhibition zone increasing from 7 to 9 mm with concentration. This example highlights the enhancement of MXenes' properties through the formation of nanocomposites [185].

16 Environmental Remediation

MXenes have been explored for their potential in environmental applications, such as water purification and pollutant removal. MXenes, a family of two-dimensional transition metal carbides, nitrides, or carbonitrides, have demonstrated promising properties for environmental remediation. Their unique combination of conductivity, hydrophilicity, and surface chemistry makes them attractive materials for applications ranging from water purification to pollutant removal. Here are some ways MXenes have been explored for environmental remediation:

16.1 Water Purification

MXenes have shown the ability to adsorb a variety of contaminants, including heavy metals, dyes, and organic pollutants, from water. The surface chemistry and high specific surface area of MXenes contribute to their effective adsorption capabilities. MXenes, such as $Ti_3C_2T_x$, have been studied for their affinity to heavy metal ions. The functional groups on MXene surfaces facilitate the chelation of metal ions, leading to their removal from water sources. MXenes have been explored for the removal of dyes from wastewater through adsorption. The high adsorption capacity and rapid kinetics make them promising candidates for treating textile industry effluents. The detail of water purification is described in Sect. 17.10.1.

16.2 Catalysis of Water Treatment

MXenes, particularly those functionalized with hydroxyl or oxygen-containing groups, can serve as catalysts for advanced oxidation processes. These processes involve the generation of reactive oxygen species, such as hydroxyl radicals, to degrade organic pollutants in water.

16.3 Electrochemical Treatment

MXenes have been employed in the development of electrochemical sensors for the detection of environmental pollutants. Their excellent conductivity and electrochemical properties make them suitable for sensing applications. MXenes, especially $Ti_3C_2T_x$, have been investigated for electrochemical reduction processes that can convert toxic pollutants into less harmful or inert species. This approach is useful for degrading pollutants through electrochemical means.

16.4 Antibacterial Applications

MXenes have exhibited antibacterial properties, which can be beneficial for water treatment applications. The interaction of MXenes with bacteria is an area of research, and their antibacterial effects can contribute to the removal of harmful microorganisms from water.

16.5 Adsorption and Removal of Emerging Contaminants

MXenes have shown potential in adsorbing emerging contaminants such as pharmaceuticals and personal care products from water. Their versatility in surface modification allows for tailoring their properties to target specific contaminants. Finally, it's important to note that while MXenes show promise in environmental remediation, the practical implementation of these materials on a larger scale involves addressing challenges such as scalability, stability, and cost-effectiveness. Ongoing research is focused on optimizing MXene properties and exploring their full potential in addressing various environmental challenges.

17 Prospective and Discussion

Prospects for MXene research remain promising, and ongoing studies are likely to uncover new applications and optimize existing ones. Researchers are also working on addressing challenges such as scalability, cost-effectiveness, and long-term stability to facilitate the practical implementation of MXene-based technologies. The focus of current studies lies in a novel era of materials characterized by their assembly from nanoscale constituents, marking a departure from previous material generations spanning the Stone Age to the Silicon Age. Two-dimensional (2D) materials serve as nanometer and subnanometer-thin "bricks" for such assembly. This unique approach allows for the utilization of organic molecules or polymers as mortar, while van der Waals (vdW) or electrostatic forces can facilitate robust bonding between the 2D layers. The realization of this vision hinges on the availability of diverse building blocks with a broad range of physical and chemical properties.

Figure 13 shows the elements contributing to the formatting of MXenes with their positions in the periodic tables. The elements with striped backgrounds are promising to MAX structures for substituting the root elements. The isolation of single layers of graphene and the exploration of unique physics in this atomically thin sp^2 carbon layer ignited interest in other 2D materials. Initially, materials with known van der Waals (vdW) bonds, such as MoS_2 and boron nitride (BN), were the focus of investigation. However, subsequent advancements led to the development of 2D layers from elements like Si, Ge, or B, as well as compounds lacking weakly bonded layered precursors. MXenes belong to this latter category, encompassing carbides, nitrides, oxycarbides, and carbonitrides of early transition metals such as Ti, Nb, Mo, etc. MXenes stand out within the extensive class of 2D materials for several reasons. They exhibit a diverse range of structures with 2–5 layers of early transitional metal atoms (M elements) connected by 1–4 layers of nonmetal atoms ($X = C, N, O$). Additionally, multiple M metals can coexist in one structure, forming ordered MXenes or random solid solutions, including high-entropy MXenes with 3–5 metals. The hexagonal structure of MXene corresponds to a (111) slice of the cubic NaCl structure of bulk cubic carbide/nitride. MXene surfaces can be terminated with oxygen, OH, amines, halogens, chalcogens, and, more recently, antimony and phosphorus.

Figure 14 shows there is a clear upward trend in Google searches for "MXene," reflecting a growing interest in this field as of August 20, 2023. The surge in interest is attributed to the continual addition of new materials to the MXene family and the discovery of exciting properties associated with MXenes. Notably, Asian countries confidently lead in this statistical trend, indicating a strong and active interest in MXene research and developments within the region. The differentiation between i-MXene, representing in-plane ordering, and o-MXene, representing out-of-plane

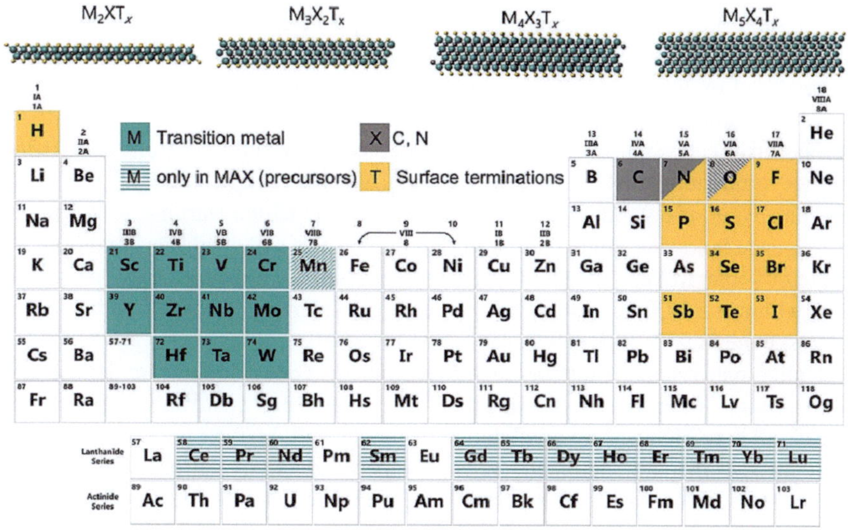

Fig. 13 Periodic table showing elements used to build MXenes

ordering, suggests a nuanced exploration of MXene properties and applications. Researchers and enthusiasts appear to be actively investigating both in-plane and out-of-plane ordering for diverse applications, contributing to the overall rise in interest in MXenes.

The MXene system, consisting of a dozen transition metals, carbon, nitrogen, and various monatomic surface terminations, presents the potential for over a thousand stoichiometric compositions. With solid solutions on M and X sites and mixed surface terminations, an infinite array of 2D materials can be synthesized within

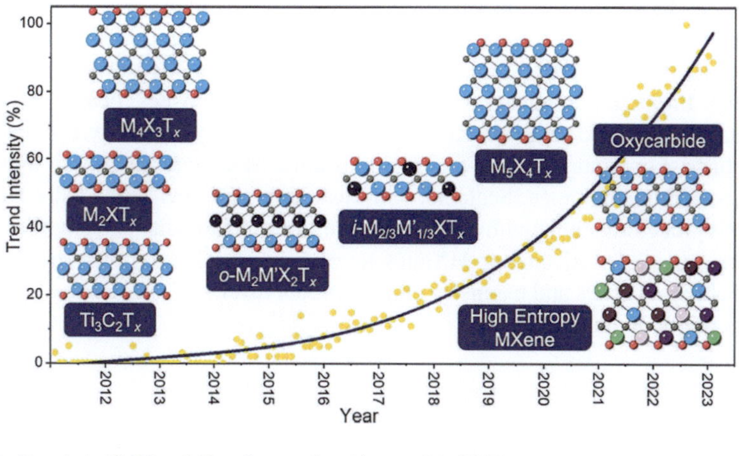

Fig. 14 Trends in "MXene" Google searches (August 20, 2023)

this system, positioning MXenes at the forefront of atomistic design and chemical assembly for new 2D materials. This capability was historically exclusive to the realm of organic molecules, including covalent organic frameworks (COF) and metal—organic frameworks (MOFs). Over 50 MXenes have been reported, with the number likely approaching a hundred when considering solid solutions and distinct surface terminations.

The advent of computational materials science, machine learning, and artificial intelligence holds the promise of predicting properties and synthesis conditions for numerous other MXenes that have yet to be synthesized. The abundance of new materials within the MXene system has garnered considerable attention, with growing interest evident with each new composition discovery.

Unlike many semiconductors and dielectrics, MXenes exhibit rare 2D metallic conductivity. Similarly, unlike graphene, which operates as a zero band gap semiconductor, all bare MXenes are predicted to be metals with high carrier concentrations. MXenes also display chemically tunable superconductivity, surpassing copper in electrical breakdown current while matching graphene. Their O/OH-terminated films surpass reduced graphene oxide (rGO) in conductivity, making MXenes attractive for conductive coatings and 2D/3D printing. O/OH-terminated MXenes' hydrophilic nature facilitates their use in electrochemistry, contributing to applications in energy storage and conversion. They exhibit outstanding mechanical properties, surpassing other solution-processed 2D materials. MXenes interact with the electromagnetic spectrum across UV to microwaves, making them suitable for diverse applications, including photothermal therapy.

The synthesis of new 2D materials within the MXene system is expected to continue, with the recently added $M_5C_4T_x$ family likely expanding to nitrides. The compositional and structural variety of MXenes and related materials extends beyond current descriptions, including possibilities like oxides, nitrides, oxynitrides, and borides. Computational evaluation and experimental testing of these materials' properties will be crucial.

MXenes demonstrate competitiveness in various applications, outperforming gold and silver in printed electronics, surpassing indium–tin–oxide (ITO) in flexible and foldable transparent devices, and excelling in electromagnetic interference (EMI) shielding. Promising applications include structural composites, healthcare, water desalination, chemical catalysis, and sensors. Collaborative efforts between academic and industrial communities are vital for translating MXene research into industrial products and technologies. Scalable, low-cost, safe, and environmentally friendly synthesis processes, along with comprehensive studies of toxicity and environmental fate, are crucial for advancing MXenes toward industrial use, given their unique and tunable properties and extensive compositional diversity. The future of MXenes appears bright and promising.

18 Commercialization and Shortcomings

The commercialization of MXenes, particularly for applications in energy storage, environmental remediation, and flexible electronics, presents both promising opportunities and notable challenges.

18.1 Opportunities for Commercialization

MXenes have demonstrated immense potential due to their unique properties. They are known for their high electrical conductivity, hydrophilicity, and rich surface terminations, making them ideal candidates for supercapacitors, batteries, sensors, and catalysis. The ability of MXenes, such as Ti3_33C2_22Tx_xx, to store and deliver high energy densities in supercapacitors, due to their 2D structure and large surface areas, has drawn significant attention from industries focused on next-generation energy storage devices [186, 187].

Moreover, their tunable surface chemistry allows for enhanced ion intercalation, which is beneficial in battery applications. MXenes' excellent conductivity also positions them as strong contenders in the realm of flexible electronics and wearables [187].

18.2 Shortcomings and Challenges

Despite the potential, several barriers to commercialization exist:

18.2.1 Synthesis and Scalability

One of the primary challenges is the synthesis of MXenes on an industrial scale while maintaining consistent quality. Currently, MXenes are primarily produced using wet chemical etching, often involving hazardous chemicals like hydrofluoric acid (HF), which poses significant safety and environmental risks. Scaling up such processes without compromising the properties of MXenes, such as their surface terminations and structural integrity, is challenging. The etching process must be carefully controlled to avoid defects that can impair conductivity and performance. Additionally, alternative etching processes, such as electrochemical or fluoride-free methods, are being explored but are not yet widely adopted or optimized for large-scale production.

18.2.2 Stability and Durability

Another significant hurdle is the long-term stability of MXenes, especially in applications where they are exposed to air and moisture. MXenes are highly susceptible to oxidation, which leads to the degradation of their electrical and mechanical properties. This sensitivity limits their applicability in practical environments, particularly for devices requiring long-term stability, such as supercapacitors and batteries. Several strategies, including surface passivation, protective coatings, and encapsulation, are being researched to improve the longevity of MXenes under ambient conditions.

18.2.3 Environmental and Safety Concerns

The hazardous nature of the chemicals used in MXene production, especially HF, raises concerns regarding the environmental impact and the safety of workers in an industrial setting. Developing greener, more environmentally friendly synthesis methods is essential to making MXene production safer and more sustainable. Alternative etching agents, such as molten salts or less toxic acids, are being explored to mitigate these issues.

18.2.4 Cost of Production

The cost of producing MXenes is still relatively high due to the complexity of the synthesis and purification processes. Achieving the desired size, quality, and consistency of MXene sheets requires precise control over the etching and exfoliation steps, which adds to the overall cost. As the demand for MXenes increases in industries such as electronics and energy storage, there is a pressing need to develop cost-effective, scalable production techniques.

18.3 Efforts Toward Overcoming Shortcomings

To overcome these commercialization challenges, ongoing research is focused on improving the synthesis process, increasing the yield, and reducing the toxicity of the chemicals used. Researchers are also investigating ways to enhance the stability of MXenes in real-world applications by improving their resistance to oxidation and developing protective coatings.

Moreover, new approaches to reduce production costs, such as automated manufacturing systems and the use of less expensive starting materials, are being tested to make MXenes commercially viable.

18.4 *Future Directions*

With advancements in synthesis techniques and improvements in stability, MXenes are likely to find broader commercial applications. If these challenges can be adequately addressed, MXenes could revolutionize sectors such as renewable energy, flexible electronics, and environmental technology.

19 Conclusion

This chapter has stated the present stage of research on MXenes, a class of 2D ceramic materials, primarily comprising carbides, nitrides, and carbonitrides, derived from 'MAX' phases through the removal of 'A'. They show the conductive, hydrophilic, biocompatible, and adjustable properties as their recognition with some limitations of environmental instability, low biodegradability, and complexities in drug release. In addition, their terminations plays important roles in their tuning properties by the formation of nanocomposites which are highly applicable in photo-thermal conversion efficiency for cancer therapy, increased antibacterial activity, improved biosensors, selectivity, and enhanced capabilities in bone regeneration. The hydrophilic surfaces transition to conductivity within the metallic range post-composite formation. The composites with polymers are used in energy storage and biomedical applications, encompassing antimicrobial agents, anticancer activity, drug delivery, bio-imaging, biosensors, and micro-supercapacitors. Similarly, composites involving ferrites demonstrate efficiency in photocatalytic degradation of hazardous dyes such as Congo Red Dye and addressing invisible pollutants, coupled with applications in advanced electromagnetic absorption. The future of MXene was well stated in this chapter.

References

1. Parajuli, D.: MXenes-polymer nanocomposites for biomedical applications: fundamentals and future perspectives. Front. Chem. **12**, 1400375 (2024). https://doi.org/10.3389/FCHEM.2024.1400375
2. Naguib, M., et al.: Two-dimensional nanocrystals produced by exfoliation of Ti_3AlC_2. Adv. Mater. **23**(37), 4248–4253 (2011). https://doi.org/10.1002/adma.201102306
3. Naguib, M., Mochalin, V.N., Barsoum, M.W., Gogotsi, Y.: 25th anniversary article: MXenes: a new family of two-dimensional materials. Adv. Mater. **26**(7), 992–1005 (2014). https://doi.org/10.1002/adma.201304138
4. Deysher, G., et al.: Synthesis of Mo_4VAlC_4 MAX phase and two-dimensional Mo_4VC_4 MXene with five atomic layers of transition metals. ACS Nano **14**(1), 204–217 (2020). https://doi.org/10.1021/acsnano.9b07708
5. Barsoum, M.W.: MN+1AXN phases: a new class of solids; thermodynamically stable nanolaminates. Prog. Solid State Chem. **28**(1–4), 201–281 (2000). https://doi.org/10.1016/S0079-6786(00)00006-6

6. Sun, Z., Music, D., Ahuja, R., Li, S., Schneider, J.M.: Bonding and classification of nanolayered ternary carbides. Phys. Rev. B Condens. Matter Mater. Phys. **70**(9), 092102 (2004). https://doi.org/10.1103/PhysRevB.70.092102
7. Anasori, B., et al.: Two-dimensional, ordered, double transition metals carbides (MXenes). ACS Nano **9**(10), 9507–9516 (2015). https://doi.org/10.1021/acsnano.5b03591
8. Wang, T., Lang, J., Zhao, Y., Su, Y., Zhao, Y., Wang, X.: Simultaneous doping and heterojunction of silver on $Na_2Ta_2O_6$ nanoparticles for visible light driven photocatalysis: the relationship between tunable optical absorption, defect chemistry and photocatalytic activity. CrystEngComm **17**(35), 6651–6660 (2015). https://doi.org/10.1039/c5ce01125f
9. Tao, Q., et al.: Two-dimensional $Mo_{1.33}C$ MXene with divacancy ordering prepared from parent 3D laminate with in-plane chemical ordering. Nat. Commun. **8**(1), 1–7 (2017). https://doi.org/10.1038/ncomms14949
10. Gogotsi, Y., Anasori, B.: The Rise of MXenes. ACS Nano **13**(8), 84918494 (2019). American Chemical Society. https://doi.org/10.1021/acsnano.9b06394
11. Shuck, C.E., et al.: Scalable Synthesis of $Ti_3C_2T_x$ MXene. Adv. Eng. Mater. **22**(3), 1901241 (2020). https://doi.org/10.1002/ADEM.201901241
12. Carbon-Ukraine—etching reactor for MXene synthesis. https://carbon.org.ua/etching-reactor-for-mxene-synthesis. Accessed 14 Jan 2024
13. Halim, J., et al.: Transparent conductive two-dimensional titanium carbide epitaxial thin films. Chem. Mater. **7**(26), 2374–2381 (2014). https://doi.org/10.1021/cm500641a
14. Halim, J., Cook, K.M., Naguib, M., Gogotsi, Y., Rosén, J., Barsoum, M.: X-ray photoelectron spectroscopy of select multi-layered transition metal carbides (MXenes). Appl. Surf. Sci. **362**, 406–417 (2016). https://doi.org/10.1016/j.apsusc.2015.11.089
15. Harris, K.J., Bugnet, M., Naguib, M., Barsoum, M.W., Goward, G.R.: Direct measurement of surface termination groups and their connectivity in the 2D MXene V_2CT_x using NMR spectroscopy. J. Phys. Chem. C **119**(24), 13713–13720 (2015). https://doi.org/10.1021/acs.jpcc.5b03038
16. Hong, W., Wyatt, B.C., Nemani, S.K., Anasori, B.: Double transition-metal MXenes: atomistic design of two-dimensional carbides and nitrides. MRS Bull. **45**(10), 850–861 (2020). https://doi.org/10.1557/MRS.2020.251/METRICS
17. Li, M., et al.: Element replacement approach by reaction with lewis acidic molten salts to synthesize nanolaminated MAX phases and MXenes. J. Am. Chem. Soc. **141**(11), 4730–4737 (2019). https://doi.org/10.1021/JACS.9B00574/SUPPL_FILE/JA9B00574_SI_001.DOCX
18. Lu, J., et al.: $Ti_{n+1}C_n$ MXene with fully saturated and thermally stable Cl terminations. Nanoscale Adv. **1**(9), 3680–3685 (2019). https://doi.org/10.1039/c9na00324j
19. Li, Y., et al., A general Lewis acidic etching route for preparing MXenes with enhanced electrochemical performance in non-aqueous electrolyte. Nat. Mater. **19**(8), 894–899 (2020). https://doi.org/10.1038/s41563-020-0657-0
20. Urbankowski, P., et al.: Synthesis of two-dimensional titanium nitride Ti_4N_3 (MXene). Nanoscale **8**(22), 11385–11391 (2016). https://doi.org/10.1039/C6NR02253G
21. Wang, D., et al.: Direct synthesis and chemical vapor deposition of 2D carbide and nitride MXenes. Science **379**(6638), 1242–1247 (2023). https://doi.org/10.1126/SCIENCE.ADD9204/SUPPL_FILE/SCIENCE.ADD9204_SM.PDF
22. Kamysbayev, V., et al.: Covalent surface modifications and superconductivity of two-dimensional metal carbide MXenes. Science **369**(6506), 979–983 (2020). https://doi.org/10.1126/SCIENCE.ABA8311/SUPPL_FILE/ABA8311_KAMYSBAYEV_SM.PDF
23. Jeewandara, T.: A new strategy to synthesize 2-D inorganic materials used in capacitors, batteries, and composites. phy.org (2020). https://phys.org/news/2020-07-strategy-d-inorganic-materials-capacitors.html. Accessed 15 Jan 2024
24. Wang, H., et al.: Surface modified MXene Ti_3C_2 multilayers by aryl diazonium salts leading to large-scale delamination. Appl. Surf. Sci. **384**, 287–293 (2016). https://doi.org/10.1016/J.APSUSC.2016.05.060
25. Eames, C., Islam, M.S.: Ion intercalation into two-dimensional transition-metal carbides: global screening for new high-capacity battery materials. J. Am. Chem. Soc. **136**(46), 16270–16276 (2014). https://doi.org/10.1021/JA508154E/SUPPL_FILE/JA508154E_SI_001.PDF

26. Naguib, M., et al.: Two-dimensional transition metal carbides. ACS Nano **6**(2), 1322–1331 (2012). https://doi.org/10.1021/nn204153h
27. Mashtalir, O., et al.: Intercalation and delamination of layered carbides and carbonitrides. Nat. Commun. **4**, 1–7 (2013). https://doi.org/10.1038/ncomms2664
28. Ghidiu, M., Lukatskaya, M.R., Zhao, M.Q., Gogotsi, Y., Barsoum, M.W.: Conductive two-dimensional titanium carbide 'clay' with high volumetric capacitance (2014). https://doi.org/10.1038/nature13970
29. Maleski, K., Mochalin, V.N., Gogotsi, Y.: Dispersions of two-dimensional titanium carbide MXene in organic solvents. Chem. Mater. **29**(4), 1632–1640 (2017). https://doi.org/10.1021/ACS.CHEMMATER.6B04830/SUPPL_FILE/CM6B04830_SI_001.PDF
30. Alhabeb, M., et al.: Guidelines for synthesis and processing of two-dimensional titanium carbide ($Ti_3C_2T_x$ MXene). Chem. Mater. **29**(18), 7633–7644 (2017). https://doi.org/10.1021/acs.chemmater.7b02847
31. Dillon, A.D., et al.: Highly conductive optical quality solution-processed films of 2D titanium carbide. Adv. Func. Mater. **26**(23), 4162–4168 (2016). https://doi.org/10.1002/adfm.201600357
32. Salles, P., Pinto, D., Hantanasirisakul, K., Maleski, K., Shuck, C.E., Gogotsi, Y.: Electrochromic effect in titanium carbide MXene thin films produced by dip-coating. Adv. Func. Mater. **29**(17), 1809223 (2019). https://doi.org/10.1002/ADFM.201809223
33. Zhang, C.J., et al., Additive-free MXene inks and direct printing of micro-supercapacitors. Nat. Commun. **10**(1), 1–9 (2019). https://doi.org/10.1038/s41467-019-09398-1
34. Vural, M., et al.: Inkjet printing of self-assembled 2D titanium carbide and protein electrodes for stimuli-responsive electromagnetic shielding. Adv. Func. Mater. **28**(32), 1801972 (2018). https://doi.org/10.1002/ADFM.201801972
35. Lipatov, A., Alhabeb, M., Lukatskaya, M.R., Boson, A., Gogotsi, Y., Sinitskii, A.: Effect of synthesis on quality, electronic properties and environmental stability of individual monolayer Ti_3C_2 MXene flakes. Adv. Electr. Mater. **2**(12), 1600255 (2016). https://doi.org/10.1002/AELM.201600255
36. Maleski, K., Ren, C.E., Zhao, M.Q., Anasori, B., Gogotsi, Y.: Size-dependent physical and electrochemical properties of two-dimensional MXene flakes. ACS Appl. Mater. Interfaces **10**(29), 24491–24498 (2018). https://doi.org/10.1021/ACSAMI.8B04662/ASSET/IMAGES/MEDIUM/AM-2018-04662N_0007.GIF
37. Zhang, C.J., et al.: Oxidation stability of colloidal two-dimensional titanium carbides (MXenes). Chem. Mater. **29**(11), 4848–4856 (2017). https://doi.org/10.1021/acs.chemmater.7b00745
38. Enyashin, A.N., Ivanovskii, A.L.: Structural and electronic properties and stability of MXenes Ti_2C and Ti_3C_2 functionalized by methoxy groups. J. Phys. Chem. C **117**(26), 13637–13643 (2013). https://doi.org/10.1021/JP401820B/ASSET/IMAGES/MEDIUM/JP-2013-01820B_0005.GIF
39. Tang, Q., Zhou, Z., Shen, P.: Are MXenes promising anode materials for Li ion batteries? Computational studies on electronic properties and Li storage capability of Ti_3C_2 and $Ti_3C_2X_2$ (X = F, OH) monolayer. J. Am. Chem. Soc. **134**(40), 16909–16916 (2012). https://doi.org/10.1021/JA308463R/SUPPL_FILE/JA308463R_SI_001.PDF
40. Khazaei, M., et al.: Novel electronic and magnetic properties of two-dimensional transition metal carbides and nitrides. Adv. Func. Mater. **23**(17), 2185–2192 (2013). https://doi.org/10.1002/adfm.201202502
41. Xie, Y., Kent, P.R.C.: Hybrid density functional study of structural and electronic properties of functionalized $Ti_{n+1}Xn$ (X=C, N) monolayers. Phys. Rev. B Condens. Matter Mater. Phys. **87**(23), 235441 (2013). https://doi.org/10.1103/PhysRevB.87.235441
42. Magnuson, M., Halim, J., Näslund, L.Å.: Chemical bonding in carbide MXene nanosheets. J. Electron Spectrosc. Relat. Phenom. **224**, 27–32 (2018). https://doi.org/10.1016/J.ELSPEC.2017.09.006
43. Magnuson, M., Näslund, L.Å.: Local chemical bonding and structural properties in Ti_3AlC_2MAX phase and Ti_3C_2Tx MXene probed by Ti_1sX-ray absorption spectroscopy.

Phys. Rev. Res. **2**(3), 033516 (2020). https://doi.org/10.1103/PHYSREVRESEARCH.2.033516/FIGURES/6/MEDIUM
44. Lin, H., Wang, X., Yu, L., Chen, Y., Shi, J.: Two-dimensional ultrathin MXene ceramic nanosheets for photothermal conversion. Nano Lett. **17**(1), 384–391 (2017). Accessed 26 Feb 2021. https://doi.org/10.1021/acs.nanolett.6b04339
45. Li, R., Zhang, L., Shi, L., Wang, P.: MXene Ti_3C_2: an effective 2D light-to-heat conversion material. ACS Nano **11**(4), 3752–3759 (2017). Accessed 27 Mar 2021. www.acsnano.org
46. Li, Y., et al.: 2D $Ti_3C_2T_x$ MXenes: visible black but infrared white materials. Adv. Mater. **33**(41), 2103054 (2021). https://doi.org/10.1002/ADMA.202103054
47. Saharudin, M.S., Che Nasir, N.A., Hasbi, S.: Tensile and corrosion resistance studies of MXenes/nanocomposites: a review. Adv. Struct. Mater. **167**, 189–198 (2022). https://doi.org/10.1007/978-3-030-89988-2_14/COVER
48. Jastrzębska, A.M., et al.: The atomic structure of Ti_2C and Ti_3C_2 MXenes is responsible for their antibacterial activity toward E. coli bacteria. J. Mater. Eng. Perform. **28**(3), 1272–1277 (2019). https://doi.org/10.1007/s11665-018-3223-z
49. Rasool, K., Helal, M., Ali, A., Ren, C.E., Gogotsi, Y., Mahmoud, K.A.: Antibacterial activity of $Ti_3C_2T_x$ MXene. ACS Nano **10**(3), 3674–3684 (2016). Accessed 16 Mar 2021. https://doi.org/10.1021/acsnano.6b00181
50. Jastrzębska, A.M., et al.: In vitro studies on cytotoxicity of delaminated Ti_3C_2 MXene. J. Hazard. Mater. **339**, 1–8 (2017). https://doi.org/10.1016/J.JHAZMAT.2017.06.004
51. Driscoll, N., et al.: Two-dimensional Ti_3C_2 MXene for high-resolution neural interfaces. ACS Nano **12**(10), 10419–10429 (2018). https://doi.org/10.1021/ACSNANO.8B06014/SUPPL_FILE/NN8B06014_SI_001.PDF
52. Li, X., Huang, Z., Shuck, C.E., Liang, G., Gogotsi, Y., Zhi, C.: MXene chemistry, electrochemistry and energy storage applications. Nat. Rev. Chem. **6**(6), 389–404 (2022). https://doi.org/10.1038/s41570-022-00384-8
53. Ostadhossein, A., Guo, J., Simeski, F., Ihme, M.: Functionalization of 2D materials for enhancing OER/ORR catalytic activity in Li–oxygen batteries. Commun. Chem. **2**(1), 1–11 (2019). https://doi.org/10.1038/s42004-019-0196-2
54. Parajuli, D., Uppugalla, S., Murali, N., Ramakrishna, A., Suryanarayana, B., Samatha, K.: Synthesis and characterization MXene-ferrite nanocomposites and its application for dying and shielding. Inorg. Chem. Commun. **148**, 110319 (2023). https://doi.org/10.1016/J.INOCHE.2022.110319
55. Parajuli, D., et al.: Advancements in MXene-polymer nanocomposites in energy storage and biomedical applications. Polymers **14**(16), 3433 (2022). https://doi.org/10.3390/POLYM14163433
56. Mashtalir, O., Cook, K.M., Mochalin, V.N., Crowe, M., Barsoum, M.W., Gogotsi, Y.: Dye adsorption and decomposition on two-dimensional titanium carbide in aqueous media. J. Mater. Chem. A **2**(35), 14334–14338 (2014). https://doi.org/10.1039/C4TA02638A
57. Ren, C.E., Hatzell, K.B., Alhabeb, M., Ling, Z., Mahmoud, K.A., Gogotsi, Y.: Charge- and size-selective ion sieving through $Ti_3C_2T_x$ MXene membranes. J. Phys. Chem. Lett. **6**(20), 4026–4031 (2015). https://doi.org/10.1021/ACS.JPCLETT.5B01895/ASSET/IMAGES/LARGE/JZ-2015-018957_0004.JPEG
58. Chen, J., et al.: CO_2 and temperature dual responsive 'smart' MXene phases. Chem. Commun. **51**(2), 314–317 (2014). https://doi.org/10.1039/C4CC07220K
59. Khakbaz, P., et al.: Titanium carbide MXene as NH_3 sensor: realistic first-principles study. J. Phys. Chem. C **123**(49), 29794–29803 (2019). https://doi.org/10.1021/ACS.JPCC.9B09823/ASSET/IMAGES/MEDIUM/JP9B09823_0015.GIF
60. Chaudhuri, K., Alhabeb, M., Wang, Z., Shalaev, V.M., Gogotsi, Y., Boltasseva, A.: Highly broadband absorber using plasmonic titanium carbide (MXene). ACS Photonics **5**(3), 1115–1122 (2018). https://doi.org/10.1021/ACSPHOTONICS.7B01439/SUPPL_FILE/PH7B01439_SI_001.PDF
61. Sarycheva, A., et al.: Two-dimensional titanium carbide (MXene) as surface-enhanced Raman scattering substrate. J. Phys. Chem. C **121**(36), 19983–19988 (2017). https://doi.org/10.1021/ACS.JPCC.7B08180/SUPPL_FILE/JP7B08180_SI_001.PDF

62. Dong, Y., et al.: Saturable absorption in 2D Ti_3C_2 MXene thin films for passive photonic diodes. Adv. Mater. **30**(10), 1705714 (2018). https://doi.org/10.1002/ADMA.201705714
63. Dong, Y., et al.: Metallic MXenes: a new family of materials for flexible triboelectric nanogenerators. Nano Energy **44**, 103–110 (2018). https://doi.org/10.1016/J.NANOEN.2017.11.044
64. Mansoor, N.E., Diaz, L.A., Shuck, C.E., Gogotsi, Y., Lister, T.E., Estrada, D.: Removal and recovery of ammonia from simulated wastewater using $Ti_3C_2T_x$ MXene in flow electrode capacitive deionization. npj Clean Water **5**(1), 1–11 (2022). https://doi.org/10.1038/s41545-022-00164-3
65. Naguib, M., et al.: New two-dimensional niobium and vanadium carbides as promising materials for li-ion batteries. J. Am. Chem. Soc. **135**(43), 15966–15969 (2013). https://doi.org/10.1021/ja405735d
66. Naguib, M., et al.: MXene: a promising transition metal carbide anode for lithium-ion batteries. Electrochem. Commun. **16**(1), 61–64 (2012). https://doi.org/10.1016/j.elecom.2012.01.002
67. Forrest, S.R.: The path to ubiquitous and low-cost organic electronic appliances on plastic (2004). Accessed 16 Mar 2021. www.nature.com/nature
68. Shi, Y., Peng, L., Ding, Y., Zhao, Y., Yu, G.: Nanostructured conductive polymers for advanced energy storage. Chem. Soc. Rev. **44**(19), 6684–6696 (2015). https://doi.org/10.1039/c5cs00362h
69. Reese, C., Roberts, M., Ling, M.M., Bao, Z.: Organic thin film transistors. Mater. Today **7**(9), 20–27 (2004). https://doi.org/10.1016/S1369-7021(04)00398-0
70. Mortimer, R.J.: Electrochromic materials. Chem. Soc. Rev. **26**(3), 147 (1997). https://doi.org/10.1039/cs9972600147
71. Bubnova, O., et al.: Optimization of the thermoelectric figure of merit in the conducting polymer poly(3,4-ethylenedioxythiophene). Nat. Mater. **10**(6), 429–433 (2011). https://doi.org/10.1038/nmat3012
72. Nezakati, T., Seifalian, A., Tan, A., Seifalian, A.M.: Conductive polymers: opportunities and challenges in biomedical applications. Chem. Rev. **118**(14), 6766–6843 (2018). American Chemical Society. https://doi.org/10.1021/acs.chemrev.6b00275
73. Deshpande, P.P., Jadhav, N.G., Gelling, V.J., Sazou, D.: Conducting polymers for corrosion protection: a review. J. Coat. Technol. Res. **11**(4), 473–494 (2014). https://doi.org/10.1007/s11998-014-9586-7
74. Burroughes, J.H., et al.: Light-emitting diodes based on conjugated polymers. Nature **347**(6293), 539–541 (1990). https://doi.org/10.1038/347539a0
75. Simon, P., Gogotsi, Y.: Materials for electrochemical capacitors. Nat. Mater. **7**(11), 845–854 (2008). Nature Publishing Group. https://doi.org/10.1038/nmat2297
76. Meng, C., Maeng, J., John, S.W.M., Irazoqui, P.P.: Ultrasmall integrated 3D micro-supercapacitors solve energy storage for miniature devices. Adv. Energy Mater. **4**(7), 1301269 (2014). https://doi.org/10.1002/aenm.201301269
77. Kurra, N., Hota, M.K., Alshareef, H.N.: Conducting polymer micro-supercapacitors for flexible energy storage and Ac line-filtering. Nano Energy **13**, 500–508 (2015). https://doi.org/10.1016/j.nanoen.2015.03.018
78. Gu, C., et al.: π-conjugated microporous polymer films: designed synthesis, conducting properties, and photoenergy conversions. Angew. Chem. Int. Ed. **54**(46), 13594–13598 (2015). https://doi.org/10.1002/anie.201506570
79. Gu, C., et al.: Color-stable white electroluminescence based on a cross-linked network film prepared by electrochemical copolymerization. Adv. Mater. **22**(24), 2702–2705 (2010). https://doi.org/10.1002/adma.201000347
80. Qin, L., et al.: In situ electrochemical synthesis and deposition of discotic hexa-peri-hexabenzocoronene molecules on electrodes: self-assembled structure, redox properties, and application for supercapacitor. Small **11**(25), 3028–3034 (2015). https://doi.org/10.1002/smll.201403734
81. Gu, C., et al.: Electrochemical route to fabricate film-like conjugated microporous polymers and application for organic electronics. Adv. Mater. **25**(25), 3443–3448 (2013). https://doi.org/10.1002/adma.201300839

82. Gu, C., Fei, T., Yao, L., Lv, Y., Lu, D., Ma, Y.: Multilayer polymer stacking by in situ electrochemical polymerization for color-stable white electroluminescence. Adv. Mater. **23**(4), 527–530 (2011). https://doi.org/10.1002/adma.201003027
83. Dahlqvist, M., Lu, J., Meshkian, R., Tao, Q., Hultman, L., Rosen, J.: Prediction and synthesis of a family of atomic laminate phases with Kagomé-like and in-plane chemical ordering **3**(7), e1700642 (2017). American Association for the Advancement of Science
84. Ghidiu, M., et al.: Synthesis and characterization of two-dimensional Nb_4C_3 (MXene). Chem. Commun. **50**(67), 9517–9520 (2014). https://doi.org/10.1039/c4cc03366c
85. Tran, M.H., et al.: Adding a new member to the MXene family: synthesis, structure, and electrocatalytic activity for the hydrogen evolution reaction of V_4C_3Tx. ACS Appl. Energy Mater. **1**(8), 3908–3914 (2018). https://doi.org/10.1021/ACSAEM.8B00652/ASSET/IMAGES/MEDIUM/AE-2018-00652H_0008.GIF
86. Mahmood, M., et al.: Nanostructured V_2O_5 and its nanohybrid with MXene as an efficient electrode material for electrochemical capacitor applications. Ceram. Int. **48**(2), 2345–2354 (2022). https://doi.org/10.1016/J.CERAMINT.2021.10.014
87. Khan, K., et al.: Recent advances in MXenes: a future of nanotechnologies. J. Mater. Chem. A **11**(37), 19764–19811 (2023). https://doi.org/10.1039/D3TA03069E
88. Ren, C.E., et al.: Porous two-dimensional transition metal carbide (MXene) flakes for high-performance li-ion storage. ChemElectroChem **3**(5), 689–693 (2016). https://doi.org/10.1002/CELC.201600059
89. Xie, Y., et al.: Role of surface structure on li-ion energy storage capacity of two-dimensional transition-metal carbides. J. Am. Chem. Soc. **136**(17), 6385–6394 (2014). https://doi.org/10.1021/ja501520b
90. Yang, E., Ji, H., Kim, J., Kim, H., Jung, Y.: Exploring the possibilities of two-dimensional transition metal carbides as anode materials for sodium batteries. Phys. Chem. Chem. Phys. **17**(7), 5000–5005 (2015). https://doi.org/10.1039/C4CP05140H
91. Er, D., Li, J., Naguib, M., Gogotsi, Y., Shenoy, V.B.: Ti_3C_2 MXene as a high capacity electrode material for metal (Li, Na, K, Ca) ion batteries. ACS Appl. Mater. Interfaces **6**(14), 11173–11179 (2014). https://doi.org/10.1021/AM501144Q/ASSET/IMAGES/MEDIUM/AM-2014-01144Q_0007.GIF
92. Xie, Y., et al.: Prediction and characterization of MXene nanosheet anodes for non-lithium-ion batteries. ACS Nano **8**(9), 9606–9615 (2014). https://doi.org/10.1021/NN503921J/SUPPL_FILE/NN503921J_SI_001.PDF
93. Wang, X., Shen, X., Gao, Y., Wang, Z., Yu, R., Chen, L.: Atomic-scale recognition of surface structure and intercalation mechanism of Ti_3C_2X. J. Am. Chem. Soc. **137**(7), 2715–2721 (2015). https://doi.org/10.1021/JA512820K/SUPPL_FILE/JA512820K_SI_001.PDF
94. Wang, X., et al.: Pseudocapacitance of MXene nanosheets for high-power sodium-ion hybrid capacitors. Nat. Commun. **6**(1), 1–6 (2015). https://doi.org/10.1038/ncomms7544
95. Dall'Agnese, Y., Taberna, P.L., Gogotsi, Y., Simon, P.: Two-dimensional vanadium carbide (MXene) as positive electrode for sodium-ion capacitors. J. Phys. Chem. Lett. **6**(12), 2305–2309 (2015). https://doi.org/10.1021/ACS.JPCLETT.5B00868/ASSET/IMAGES/MEDIUM/JZ-2015-00868R_0006.GIF
96. Xie, X., et al.: Porous heterostructured MXene/carbon nanotube composite paper with high volumetric capacity for sodium-based energy storage devices. Nano Energy **26**, 513–523 (2016). https://doi.org/10.1016/J.NANOEN.2016.06.005
97. Sun, D., Wang, M., Li, Z., Fan, G., Fan, L.Z., Zhou, A.: Two-dimensional Ti_3C_2 as anode material for Li-ion batteries. Electrochem. Commun. **47**, 80–83 (2014). https://doi.org/10.1016/J.ELECOM.2014.07.026
98. Boota, M., Anasori, B., Voigt, C., Zhao, M.Q., Barsoum, M.W., Gogotsi, Y.: Pseudocapacitive electrodes produced by oxidant-free polymerization of pyrrole between the layers of 2D titanium carbide (MXene). Adv. Mater. **28**(7), 1517–1522 (2016). https://doi.org/10.1002/ADMA.201504705
99. Vahidmohammadi, A., et al.: Thick and freestanding MXene/PANI pseudocapacitive electrodes with ultrahigh specific capacitance. J. Mater. Chem. A **6**(44), 22123–22133 (2018). https://doi.org/10.1039/C8TA05807E

100. Zhu, J., Tang, Y., Yang, C., Wang, F., Cao, M.: Composites of TiO_2 nanoparticles deposited on Ti_3C_2 MXene nanosheets with enhanced electrochemical performance. J. Electrochem. Soc. **163**(5), A785–A791 (2016). https://doi.org/10.1149/2.0981605JES/XML
101. Lukatskaya, M.R., et al.: Ultra-high-rate pseudocapacitive energy storage in two-dimensional transition metal carbides. Nat. Energy **2**(8), 1–6 (2017). https://doi.org/10.1038/nenergy.2017.105
102. Lukatskaya, M.R., et al.: Cation intercalation and high volumetric capacitance of two-dimensional titanium carbide. Science **341**(6153), 1502–1505 (2013). https://doi.org/10.1126/SCIENCE.1241488/SUPPL_FILE/LUKATSKAYA.SM.PDF
103. Ghidiu, M., Lukatskaya, M.R., Zhao, M.Q., Gogotsi, Y., Barsoum, M.W.: Conductive two-dimensional titanium carbide 'clay' with high volumetric capacitance. Nature **516**(7529), 78–81 (2015). https://doi.org/10.1038/nature13970
104. Wang, X., et al.: Influences from solvents on charge storage in titanium carbide MXenes. Nat. Energy **4**(3), 241–248 (2019). https://doi.org/10.1038/s41560-019-0339-9
105. Kumar, S., et al.: Supercapacitors based on $Ti_3C_2T_x$ MXene extracted from supernatant and current collectors passivated by CVD-graphene. Sci. Rep. **11**(1), 1–9 (2021). https://doi.org/10.1038/s41598-020-80799-9
106. Parajuli, D.: Ferrite Nanomaterials-Fundamentals & Perspectives, 1st edn. SEND Nepal, Kathmandu, Nepal (2022)
107. Parajuli, D.: MXene Nanocomposites (With Polymers & Ferrites), 1st edn. Send Nepal, Kathmandu, Nepal (2023)
108. Ling, Z., et al.: Flexible and conductive MXene films and nanocomposites with high capacitance. Proc. Natl. Acad. Sci. U.S.A. **111**(47), 16676–16681 (2014). https://doi.org/10.1073/pnas.1414215111
109. Yan, J., et al.: Flexible MXene/graphene films for ultrafast supercapacitors with outstanding volumetric capacitance. Adv. Func. Mater. **27**(30), 1701264 (2017). https://doi.org/10.1002/ADFM.201701264
110. Zhao, M.Q., et al.: Flexible MXene/carbon nanotube composite paper with high volumetric capacitance. Adv. Mater. **27**(2), 339–345 (2015). https://doi.org/10.1002/ADMA.201404140
111. Sreekumar, S., et al.: Thermo-optical characterization of novel MXene/carbon-dot hybrid nanofluid for heat transfer applications. J. Clean. Prod. **434**, 140395 (2024). https://doi.org/10.1016/J.JCLEPRO.2023.140395
112. Kim, S.J., et al.: Metallic $Ti_3C_2T_x$ MXene gas sensors with ultrahigh signal-to-noise ratio. ACS Nano **12**(2), 986–993 (2018). https://doi.org/10.1021/ACSNANO.7B07460/ASSET/IMAGES/NN-2017-07460Z_M001.GIF
113. Ramanavicius, S., Ramanavicius, A.: Progress and insights in the application of MXenes as new 2D nano-materials suitable for biosensors and biofuel cell design. Int. J. Mol. Sci. **21**(23), 9224 (2020). https://doi.org/10.3390/IJMS21239224
114. Adomavičiūtė-Grabusovė, S., Ramanavičius, S., Popov, A., Šablinskas, V., Gogotsi, O., Ramanavičius, A.: Selective enhancement of SERS spectral bands of salicylic acid adsorbate on 2D $Ti_3C_2T_x$-based MXene film. Chemosensors **9**(8), 223 (2021). https://doi.org/10.3390/CHEMOSENSORS9080223
115. Du, M., Yao, Z., Zhou, J., Liu, P., Yao, T., Yao, R.: Design of efficient microwave absorbers based on multi-layered polyaniline nanofibers and polyaniline nanofibers/$Li_{0.35}Zn_{0.3}Fe_{2.35}O_4$ nanocomposite. Synth. Met. **223**, 49–57 (2017). https://doi.org/10.1016/j.synthmet.2016.11.039
116. Peng, Y., et al.: Charge-transfer resonance and electromagnetic enhancement synergistically enabling MXenes with excellent SERS sensitivity for SARS-CoV-2 S protein detection. Nano-Micro Lett. **13**(1), 1–17 (2021). https://doi.org/10.1007/S40820-020-00565-4/FIGURES/6
117. Sarycheva, A., Polemi, A., Liu, Y., Dandekar, K., Anasori, B., Gogotsi, Y.: 2D titanium carbide (MXene) for wireless communication. Sci. Adv. **4**(9) (2018). https://doi.org/10.1126/SCIADV.AAU0920/SUPPL_FILE/AAU0920_SM.PDF
118. Han, M., et al.: Solution-processed $Ti_3C_2T_x$ MXene antennas for radio-frequency communication. Adv. Mater. **33**(1), 2003225 (2021). https://doi.org/10.1002/ADMA.202003225

119. Hantanasirisakul, K., et al.: Effects of synthesis and processing on optoelectronic properties of titanium carbonitride MXene. Chem. Mater. **31**(8), 2941–2951 (2019). https://doi.org/10.1021/ACS.CHEMMATER.9B00401/SUPPL_FILE/CM9B00401_SI_001.PDF
120. Parajuli, D., Samatha, K.: MXene as topological insulator. JETIR **6**(4), 689–706 (2019). https://doi.org/10.1729/Journal.23872
121. Khazaei, M., Ranjbar, A., Arai, M., Yunoki, S.: Topological insulators in the ordered double transition metals M2′ M″ C2 MXenes (M′= Mo, W; M″=Ti, Zr, Hf). Phys. Rev. B **94**(12) (2016). https://doi.org/10.1103/PhysRevB.94.125152
122. Weng, H., et al.: Large-gap two-dimensional topological insulator in oxygen functionalized MXene. Phys. Rev. B Condens. Matter Mater. Phys. **92**(7), 075436 (2015). https://doi.org/10.1103/PhysRevB.92.075436
123. Ivanovskii, A.L., Medvedeva, N.I.: Electronic structure of hexagonal Ti_3AlC_2 and Ti_3AlN_2. Mendeleev Commun. **9**(1), 36–38 (1999). https://doi.org/10.1070/mc1999v009n01abeh001039
124. Liang, Y., et al.: Theoretical prediction of two-dimensional functionalized MXene nitrides as topological insulators. Phys. Rev. B **96**(19), 1–26 (2017). https://doi.org/10.1103/PhysRevB.96.195414
125. Xu, M.L., Huang, Z.Q., Macam, G., Hsu, C.H., Chuang, F.C.: Prediction of large-gap topological insulator in functionalized ordered double transition metal MXene **102**(075306) (2019)
126. Zhang, Z., Liu, X., Yakobson, B.I., Guo, W.: Two-dimensional tetragonal tic monolayer sheet and nanoribbons. J. Am. Chem. Soc. **134**(47), 19326–19329 (2012). https://doi.org/10.1021/ja308576g
127. Zhou, L., et al.: Prediction of the quantum spin Hall effect in monolayers of transition-metal carbides MC (M=Ti, Zr, Hf). 2D Materials **3**(3), 035022 (2016). https://doi.org/10.1088/2053-1583/3/3/035022
128. Yu, X.F., et al.: The band gap modulation of monolayer Ti_2CO_2 by strain. RSC Adv. **5**(39), 30438–30444 (2015). https://doi.org/10.1039/c5ra01586c
129. Chen, W., Li, H.F., Shi, X., Pan, H.: Tension-tailored electronic and magnetic switching of 2D Ti_2NO_2. J. Phys. Chem. C **121**(46), 25729–25735 (2017). https://doi.org/10.1021/acs.jpcc.7b08496
130. Lee, Y., Hwang, Y., Cho, S.B., Chung, Y.C.: Achieving a direct band gap in oxygen functionalized-monolayer scandium carbide by applying an electric field. Phys. Chem. Chem. Phys. **16**(47), 26273–26278 (2014). https://doi.org/10.1039/c4cp03811h
131. Balci, E., Akkuş, Ü.Ö., Berber, S.: Band gap modification in doped MXene: Sc_2CF_2. J. Mater. Chem. C **5**(24), 5956–5961 (2017). https://doi.org/10.1039/c7tc01765k
132. Balci, E., Akkuş, N.Z., Berber, S.: Doped $Sc_2C(OH)_2$ MXene: new type s-pd band inversion topological insulator. J. Phys. Condens. Matter **30**(15) (2018). https://doi.org/10.1088/1361-648X/aab41e
133. Khazaei, M., Wang, V., Sevik, C., Ranjbar, A., Arai, M., Yunoki, S.: Electronic structures of iMAX phases and their two-dimensional derivatives: a family of piezoelectric materials. Phys. Rev. Mater. **2**(7), 1–11 (2018). https://doi.org/10.1103/PhysRevMaterials.2.074002
134. Parajuli, D., Samatha, K.: Topological signatures of $Mo_2TiC_2O_2$. AIP Adv. **14**(1), 035011 (2024). https://doi.org/10.1063/5.0196244
135. Parajuli, D., Samatha, K.: Topological properties of MXenes. In: Mxenes and their composites, pp. 171–199. Elsevier. https://doi.org/10.1016/B978-0-12-823361-0.00015-0
136. Ashton, M., Mathew, K., Hennig, R.G., Sinnott, S.B.: Predicted surface composition and thermodynamic stability of MXenes in solution. J. Phys. Chem. C **120**(6), 3550–3556 (2016). https://doi.org/10.1021/acs.jpcc.5b11887
137. Hu, T., Wang, J., Zhang, H., Li, Z., Hu, M., Wang, X.: Vibrational properties of Ti_3C_2 and $Ti_3C_2T_2$ (T = O, F, OH) monosheets by first-principles calculations: a comparative study. Phys. Chem. Chem. Phys. **17**(15), 9997–10003 (2015). https://doi.org/10.1039/c4cp05666c
138. Si, C., Jin, K.H., Zhou, J., Sun, Z., Liu, F.: Large-gap quantum spin hall state in MXenes: d-band topological order in a triangular lattice. Nano Lett. **16**(10), 6584–6591 (2016). https://doi.org/10.1021/acs.nanolett.6b03118

139. Parajuli, D., Kaphle, G.C., Murali, N., Samatha, K.: Structural identification of cubic aluminum and non-cubic titanium using x-ray diffractometer. J. Lumbini Eng. Coll. **4**(1), 62–71 (2022). https://doi.org/10.3126/LECJ.V4I1.49369
140. Parajuli, D., Devendra, K.C., Reda, T.G., Sravani, G.M., Murali, N., Samatha, K.: RHEED analysis of the oxidized $M'_2M''_x$Xyene sheets by ablated plasma thrust method in pulsed laser deposition chamber. AIP Adv. **11**(11), 115019 (2021). https://doi.org/10.1063/5.0068659
141. Jang, D., Farooq, S.Z., Yoon, H.N., Khalid, H.R.: Design of a highly flexible and sensitive multi-functional polymeric sensor incorporating CNTs and carbonyl iron powder. Compos. Sci. Technol. **207**, 108725 (2021). https://doi.org/10.1016/J.COMPSCITECH.2021.108725
142. MXene—what is it, how it is made, what it is used for. https://www.nanowerk.com/mxene.php. Accessed 16 June 2022
143. Chen, X., et al.: MXene/polymer nanocomposites: preparation, properties, and applications. Polym. Rev., 1–36 (2020). https://doi.org/10.1080/15583724.2020.1729179
144. Prajapati, D.G., Kandasubramanian, B.: Progress in the development of intrinsically conducting polymer composites as biosensors. Macromol. Chem. Phys. **220**(10), 1800561 (2019). https://doi.org/10.1002/macp.201800561
145. Rakhi, R.B., Nayuk, P., Xia, C., Alshareef, H.N.: Novel amperometric glucose biosensor based on MXene nanocomposite. Sci. Rep. **6**(1), 1–10 (2016). https://doi.org/10.1038/srep36422
146. Harper, A., Anderson, M.R.: Electrochemical glucose sensors—developments using electrostatic assembly and carbon nanotubes for biosensor construction. Sensors **10**(9), 8248–8274 (2010). https://doi.org/10.3390/s100908248
147. George, S.M., Kandasubramanian, B.: Advancements in MXene-polymer composites for various biomedical applications. Ceram. Int. **46**(7), 8522–8535 (2020). Elsevier Ltd https://doi.org/10.1016/j.ceramint.2019.12.257
148. Ma, Y., et al.: 3D synergistical MXene/reduced graphene oxide aerogel for a piezoresistive sensor. ACS Nano **12**(4), 3209–3216 (2018). https://doi.org/10.1021/ACSNANO.7B06909
149. Ma, Y., et al.: A highly flexible and sensitive piezoresistive sensor based on MXene with greatly changed interlayer distances. Nat. Commun. **8**(1), 1–8 (2017). https://doi.org/10.1038/s41467-017-01136-9
150. Cai, Y., et al.: Stretchable $Ti_3C_2T_x$ MXene/carbon nanotube composite based strain sensor with ultrahigh sensitivity and tunable sensing range. ACS Nano **12**(1), 56–62 (2018). https://doi.org/10.1021/ACSNANO.7B06251
151. Guo, Y., Zhong, M., Fang, Z., Wan, P., Yu, G.: A wearable transient pressure sensor made with MXene nanosheets for sensitive broad-range human-machine interfacing. Nano Lett. **19**(2), 1143–1150 (2019). https://doi.org/10.1021/ACS.NANOLETT.8B04514
152. Zheng, S., et al.: Ionic liquid pre-intercalated MXene films for ionogel-based flexible micro-supercapacitors with high volumetric energy density. J. Mater. Chem. A **7**(16), 9478–9485 (2019). https://doi.org/10.1039/C9TA02190F
153. Ciou, J.H., Li, S., Lee, P.S.: Ti_3C_2 MXene paper for the effective adsorption and controllable release of aroma molecules. Small **15**(38), 1903281 (2019). https://doi.org/10.1002/SMLL.201903281
154. Zhang, C.J., et al.: Transparent, flexible, and conductive 2D titanium carbide (MXene) films with high volumetric capacitance **29**(36), 1702678 (2017). https://doi.org/10.1002/ADMA.201702678
155. An, H., et al.: Water sorption in MXene/polyelectrolyte multilayers for ultrafast humidity sensing **2**, 43 (2019). https://doi.org/10.1021/acsanm.8b02265
156. Hao, L., Zhang, H., Wu, X., Zhang, J., Wang, J., Li, Y.: Novel thin-film nanocomposite membranes filled with multi-functional $Ti_3C_2T_x$ nanosheets for task-specific solvent transport. Compos. A Appl. Sci. Manuf. **100**, 139–149 (2017). https://doi.org/10.1016/J.COMPOSITESA.2017.05.003
157. Kim, H.B., Sajid, M., Kim, K.T., Hoan Na, K., Choi, K.H.: Linear humidity sensor fabrication using bi-layered active region of transition metal carbide and polymer thin films. Sens. Actuators B **252**, 725–734 (2017). https://doi.org/10.1016/j.snb.2017.06.052

158. Sajid, M., Kim, H.B., Siddiqui, G.U., Na, K.H., Choi, K.H.: Linear bi-layer humidity sensor with tunable response using combinations of molybdenum carbide with polymers. Sens. Actuators, A **262**, 68–77 (2017). https://doi.org/10.1016/J.SNA.2017.05.029
159. An, H., et al.: Surface-agnostic highly stretchable and bendable conductive MXene multilayers. Sci. Adv. **4**(3) (2018). https://doi.org/10.1126/SCIADV.AAQ0118/SUPPL_FILE/AAQ0118_SM.PDF
160. Yuan, W., Yang, K., Peng, H., Li, F., Yin, F.: A flexible VOCs sensor based on a 3D Mxene framework with a high sensing performance. J. Mater. Chem. A **6**(37), 18116–18124 (2018). https://doi.org/10.1039/C8TA06928J
161. Zhang, Y.Z., et al.: MXenes stretch hydrogel sensor performance to new limits. Sci. Adv. **4**(6) (2018). https://doi.org/10.1126/SCIADV.AAT0098/SUPPL_FILE/AAT0098_SM.PDF
162. Hu, Y., et al.: Biomass polymer-assisted fabrication of aerogels from MXenes with ultrahigh compression elasticity and pressure sensitivity. J. Mater. Chem. A **7**(17), 10273–10281 (2019). https://doi.org/10.1039/C9TA01448A
163. Li, X.P., et al.: Highly sensitive, reliable and flexible piezoresistive pressure sensors featuring polyurethane sponge coated with MXene sheets (2019)
164. Markets and markets: Biomaterials market by type of materials (metallic, ceramic, polymers, natural), application (cardiovascular, orthopedic, dental, plastic surgery, wound healing, neurological disorders, tissue engineering, ophthalmology). Global Forecast to 2025 (2019). https://www.marketsandmarkets.com/Market-Reports/biomaterials-393.html.
165. Korde, J.M., Kandasubramanian, B.: Naturally biomimicked smart shape memory hydrogels for biomedical functions. Chem. Eng. J. **379**, 122430 (2020). https://doi.org/10.1016/j.cej.2019.122430
166. Balakrishnan, P., Thomas, S.: Inert ceramics. In: Fundamental Biomaterials: Ceramics, pp. 117–127. Elsevier Inc. (2018)
167. Lin, H., Gao, S., Dai, C., Chen, Y., Shi, J.: A two-dimensional biodegradable niobium carbide (MXene) for photothermal tumor eradication in NIR-I and NIR-II biowindows. J. Am. Chem. Soc. **139**(45), 16235–16247 (2017). Accessed 26 Feb 2021. https://doi.org/10.1021/jacs.7b07818
168. Ren, X., et al.: Highly catalytic niobium carbide (MXene) promotes hematopoietic recovery after radiation by free radical scavenging **13**(6), 6438–6454 (2019). American Chemical Society
169. Szuplewska, A., et al.: 2D Ti_2C (MXene) as a novel highly efficient and selective agent for photothermal therapy. Mater. Sci. Eng. C **98**, 874–886 (2019). https://doi.org/10.1016/j.msec.2019.01.021
170. Xing, C., et al.: Two-dimensional MXene (Ti_3C_2)-integrated cellulose hydrogels: toward smart three-dimensional network nanoplatforms exhibiting light-induced swelling and bimodal photothermal/chemotherapy anticancer activity. ACS Appl. Mater. Interfaces **10**(33), 27631–27643 (2018). https://doi.org/10.1021/acsami.8b08314
171. Feng, W., et al.: Ultrathin molybdenum carbide MXene with fast biodegradability for highly efficient theory-oriented photonic tumor hyperthermia. Adv. Func. Mater. **29**(22), 1901942 (2019). https://doi.org/10.1002/adfm.201901942
172. Liu, Z., et al.: 2D superparamagnetic tantalum carbide composite MXenes for efficient breast-cancer theranostics. Theranostics **8**(6), 1648–1664 (2018)
173. Dai, C., et al.: Two-dimensional tantalum carbide (MXenes) composite nanosheets for multiple imaging-guided photothermal tumor ablation. ACS Nano **11**(12), 12696–12712 (2017). https://doi.org/10.1021/acsnano.7b07241
174. Lin, H., Wang, Y., Gao, S., Chen, Y., Shi, J.: Theranostic 2D tantalum carbide (MXene). Adv. Mater. **30**(4), 1703284 (2018). https://doi.org/10.1002/adma.201703284
175. Han, X., Huang, J, Lin, H., Wang, Z., Li, P., Chen, Y.: 2D ultrathin MXene-based drug-delivery nanoplatform for synergistic photothermal ablation and chemotherapy of cancer **7**(9), 1701394 (2018). Accessed 26 Feb 2021. https://doi.org/10.1002/adhm.201701394
176. Liu, G., et al.: Surface modified Ti_3C_2 MXene nanosheets for tumor targeting photothermal/photodynamic/chemo synergistic therapy. ACS Appl. Mater. Interfaces **9**(46), 40077–40086 (2017). https://doi.org/10.1021/acsami.7b13421

177. Zong, L., Wu, H., Lin, H., Chen, Y.: A polyoxometalate-functionalized two-dimensional titanium carbide composite MXene for effective cancer theranostics **11**(8), 4149–4168 (2018). Accessed 26 Feb 2021. https://doi.org/10.1007/s12274-018-2002-3
178. Li, Z., Zhang, H., Han, J., Chen, Y., Lin, H., Yang, T.: Surface nanopore engineering of 2D MXenes for targeted and synergistic multitherapies of hepatocellular carcinoma. Adv. Mater. **30**(25), 1706981 (2018). https://doi.org/10.1002/adma.201706981
179. Balakrishnan, P., Geethamma, V.G., Sreekala, M.S., Thomas, S.: Polymeric biomaterials: state-of-the-art and new challenges. In: Fundamental Biomaterials: Polymers, pp. 1–20. Elsevier Inc. (2018)
180. Lipatov, A., Alhabeb, M.H., Lukatskaya, M.R., Boson, A.J., Gogotsi, Y.: Effect of synthesis on quality, electronic properties and environmental stability of individual monolayer Ti $_3C_2$ MXene flakes (2016). https://doi.org/10.1002/aelm.201600255
181. Anasori, B., et al.: Control of electronic properties of 2D carbides (MXenes) by manipulating their transition metal layers. Nanoscale Horizons **1**(3), 227–234 (2016). https://doi.org/10.1039/c5nh00125k
182. Feng, L., et al.: Structures and mechanical and electronic properties of the Ti_2CO_2 MXene incorporated with neighboring elements (Sc, V, B and N). J. Electron. Mater. **46**(4), 2460–2466 (2017). https://doi.org/10.1007/s11664-017-5311-5
183. Dai, C., Lin, H., Xu, G., Liu, Z., Wu, R., Chen, Y.: Biocompatible 2D titanium carbide (MXenes) composite nanosheets for pH-responsive MRI-guided tumor hyperthermia. Chem Mater **29**(20), 8637–8652 (2017). Accessed 16 Mar 2021. https://doi.org/10.1021/acs.chemmater.7b02441
184. Parajuli, D., Kaphle, G.C., Samatha, K.: First-principles study of electronic and magnetic properties of anatase and its role in anatase—MXene nanocomposite. J. Nepal Phys. Soc. (2019)
185. Warsi, A.Z., Aziz, F., Zulfiqar, S., Haider, S., Shakir, I., Agboola, P.O.: Synthesis, characterization, photocatalysis, and antibacterial study of WO3, MXene and WO3/MXene nanocomposite. Nanomaterials **12**(4), 713 (2022). https://doi.org/10.3390/NANO12040713
186. He, L., et al.: Advances and challenges in MXene-based electrocatalysts: unlocking the potential for sustainable energy conversion. Mater. Horiz. (2024). https://doi.org/10.1039/D4MH00845F
187. Panda, D., Venkatesh, N., Sakthivel, P.: MXene-based materials for remediation of environmental pollutants. In: MXenes and their Composites: Synthesis, Properties and Potential Applications, pp. 553–594 (2022). https://doi.org/10.1016/B978-0-12-823361-0.00014-9.

MXenes and MXene Based Composite Materials for Tissue Engineering and Cancer Nanotheranostics

Roja Sahu and Srikanta Moharana

Abstract The remarkable versatility of MXenes—a type of two-dimensional (2D) transition metal carbides, nitrides, and carbonitrides—has led to considerable interest in their use in cancer nanotheranostics and tissue engineering. They are perfect candidates for creating cutting-edge biomaterials because of their remarkable physicochemical characteristics, which include high surface area, conductivity, hydrophilicity, and variable surface functionality. MXene nanomaterials have become a rising star in biomedical fields including biological imaging, tumor diagnosis, anti-tumor therapy, biosensor, and tissue engineering in the past few years. The extraordinary capabilities of MXene composite in boosting angiogenesis and tissue remodeling and improving cell adhesion, proliferation, and differentiation are well-established. Their biocompatibility and mechanical durability make them ideal scaffolds for regenerating complicated tissues, such as cardiac and bone tissue. Moreover, the surface functionalization of MXenes allows for the incorporation of bioactive molecules, which further boosts their therapeutic potential. The use of MXenes as multifunctional theranostic platforms has grown popular due to their unique optical and electrical characteristics, enabling cutting-edge imaging modalities such as photoacoustic imaging and magnetic resonance imaging (MRI), as well as in photothermal therapy (PTT), photodynamic therapy (PDT), and drug carrier systems. This review provides a comprehensive update on the recent progress in MXenes and their composites, highlighting their transformative potential in tissue engineering and cancer nanotheranostics while addressing future research directions.

Keywords MXenes · 2D material · Cancer theranostic · Tissue regeneration · Photothermal therapy

R. Sahu (✉)
School of Pharmacy and Life Sciences, Centurion University of Technology and Management, Balangir 767001, Odisha, India
e-mail: rojasahu12@gmail.com

S. Moharana
Department of Chemistry, School of Applied Sciences, Centurion University of Technology and Management, Bhubaneswar 752050, Odisha, India

© The Author(s), under exclusive license to Springer Nature Singapore Pte Ltd. 2025
S. Moharana and R. N. Mahaling (eds.), *Surface Engineering of MXenes*, Engineering Materials, https://doi.org/10.1007/978-981-96-3640-2_18

1 Introduction

The rapid advancement of nanotechnology makes biomedical utilization of nanomaterials a promising area of research [1]. The advent of 2D nanomaterials has given researchers in different sectors new ways to address the increasing need for scientific study. 2D nanomaterials are intriguing due to their distinctive physicochemical and morphological characteristics such as unique electrical properties and broad surface area and are widely used in the field of energy, physics, environmental research, and biomedicine [2]. The biological applications of 2D nanomaterials have long been a focal point of research, encompassing 2D nanomaterials-based therapies, bioimaging, tailored drug delivery systems, and biosensors due to their advantages over conventional pharmaceuticals such as enhanced biocompatibility, facile metabolism, reduced side effects, resistance to photobleaching, and improved targeting capabilities, making their clinical application more promising [3, 4].

Various materials can manifest in their 2D forms, such as graphene [5], germanene [6], silicene [7], and boraphenes [8]. Amongst them, MXene, an inorganic 2D material made up of one or two layers of transition metal carbides, carbonitrides and nitrides bonded together by an element, is a relatively recent addition that has been studied thoroughly [9]. The word "MXene" derives from the Max phase, where "M", "A" and "X" symbolize transition metal atoms; class IIIA or IVA elements (silicon and aluminum); and carbides or nitrides respectively. The general formula for MXenes is represented as $M_{n+1}X_nT_X$, where n ranges from 1 to 3 [8]. MXenes can be synthesized by exfoliating class IIIA or IVA element from "MAX" precursor (as shown in Fig. 1) [10] through various techniques: (i) hydrothermal method (e.g., Ti_3AlC_2) [11]; (ii) chemical vapor deposition (e.g., MoS_2) [12]; (iii) urea glass technique (e.g., Mo_2N) [13]; (iv) molten salt etching (e.g., $Ti_3C_2T_x$) [14]; (v) electrochemical synthesis (e.g., Ti_2CT_X) [15]; (vi) bottom-up chemical vapor deposition (Mo_2C) [16] and (vii) top-down wet etching tactics ($Ti_3C_2T_x$) [10, 17, 18], etc.

MXenes and MXene-based structures, characterized by electronic assemblies, exceptional specific surface area, and readily available nanoscale building blocks, are applicable in various domains including energy storage, electrochemistry, biomedicine, pharmaceuticals, cancer treatment, drug delivery, theranostic nanomedicine, tissue engineering, regenerative medicine, and water purification [19]. Due to the presence of certain functionalities (e.g., −O, −F, and −OH), MXenes exhibit hydrophilic properties, indicating significant potential for surface modification; their extensive surface area enables MXenes to incorporate substantial entities for synergistic treatment, including chemotherapeutic agents, photosensitizers, and immune adjuvants [20]. Researchers have focused a lot of attention on these nanostructures for cancer therapy and diagnostics because of their remarkable photothermal effects and their suitability for multimodal tumor treatment.

$Ti_3C_2T_x$, was the first MXene synthesized from the reaction of Ti_3AlC_2 and hydrofluoric acid (HF) at room temperature by Gogotsi et al. [21]. Afterward, many researchers have studied nearly 250 MXenes containing different metals, C/N pairings, and surface terminations through silico modeling; however, only

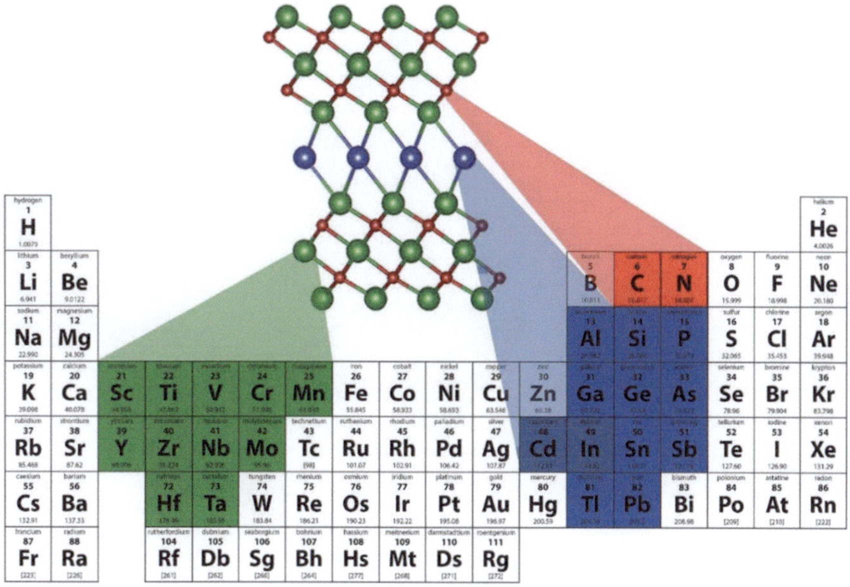

Fig. 1 MAX phases, where M, A, and X signify the transition metal, elements from IIIA or IVA and N or C. Reproduced with permission [18]. Copyright 2019, Elsevier Ltd.

about 30 species of MXenes have been effectively synthesized in the laboratory [22]. Following the initial research, 2D $Ti_3C_2T_x$ was found to possess antibacterial properties [23]. Unlike traditional photosensitizers (PSs), MXene-based materials offer substantial absorption in both NIR-I (650–900 nm) and NIR-II (1000–1700 nm) biowindows with an excellent capability of converting light into heat, making them ideal candidates for photothermal therapy (PTT) and photodynamic therapy (PDT). MXenes has been employed in a range of electrochemical, optical, fluorescent, and biocompatible field-effect transistor biosensors. Further, surface modifications or their integration with other nano-substances can enhance the characteristics of MXenes [24]. Moreover, research has shown that certain MXenes exhibit biocompatibility and non-toxicity to living beings, thus enabling their application in implants [25, 26]. In summary, through diverse surface modifications and functionalizations, MXene-based materials exhibit significant potential for numerous biomedical applications (Fig. 2).

This typical chapter systematically summarises recent advancements in MXene-based materials for cancer theranostics and tissue engineering. We presented a concise overview of its applications in biosensors, cancer diagnosis, implantation, photodynamic treatment, PTT, drug administration, and its role in tissue engineering for specific tissue types in both in vitro and in vivo research. Concludingly, it discusses the present state and potential future directions of clinical application of materials based on MXene.

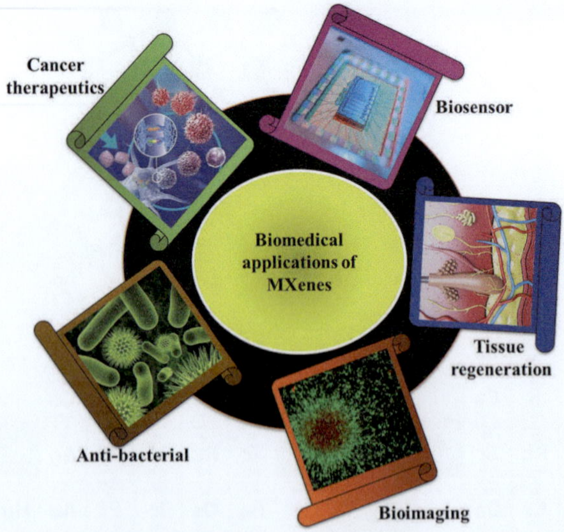

Fig. 2 Biomedical application of MXenes

2 MXenes: Properties Relevant to Biomedical Applications

MXenes demonstrate remarkable potential in various biomedical applications. Their unique characteristics, including superior conductivity, extensive surface area, hydrophilicity, two-dimensional (2D) architecture, and adjustable particle size, have established them as pivotal contributors to the advancement of nanohybrids and nanocomposites. These attributes facilitate techniques to tackle obstacles such as restricted oxidative stability, inadequate target selectivity, and concerns regarding biodegradability and biocompatibility under physiological settings. The mechanical characteristics of MXenes are considerably affected by their surface terminations. Materials with O terminations are quite stiff, whereas those with F or OH terminations are very pliable. It is worth mentioning that the surface-functionalized MXenes showed increased flexibility. By incorporating different types of polymeric materials into MXene-based composites, it is possible to enhance mechanical properties such as toughness, flexibility, tensile strength and compressive strengths [27]. It was demonstrated that MXenes exhibited substantial plasmonic resonances, an extended optical transparency window, remarkable nonlinear optical performance, and considerable optical transparency. Additionally, they have the potential for efficient photothermal conversion, optical responses that may be tuned, and significant surface-sensitive optical properties. By altering the functional groups (for example, functional nanoparticles) that are present on the surfaces of MXenes, it is possible to modify the optical characteristics of these materials [28]. Numerous 2D materials lack magnetic properties, which limits their use in spintronics. Nevertheless, certain

MXenes, including Ti_2N, Cr_2C, and Ti_2C, which exhibit ferromagnetic characteristics, while others, such as Cr_2N and Mn_2C, display antiferromagnetic behaviour. The magnetic properties of MXenes can be easily modified through surface functionalisation [29]. MXenes shown attractive electrical properties, encompassing semiconductivity, topological insulativity, and metallicity; their electrical conductivity is significantly affected by the manufacturing and delamination techniques [30].

The surface of MXene nanosheets serves as a substrate for the development and deposition of diverse chemical and inorganic substances in molecular, chain, and nanoparticle forms. Surface functionalisation and hybridisation can be performed using in situ and ex situ approaches. Alongside their improved physiological characteristics, the surface manipulation techniques can enhance additional attributes of the 2D nanosheets, including conductivity, photosensitivity, and photothermal capacity. To accomplish these objectives, bovine serum albumin [31], poly (ethylene glycol) [32], poly (vinyl alcohol) [33], soybean phospholipid [34] along with nanoparticles of inorganic materials such as manganese oxide) [35], iron oxide [32], gold [36] has been utilized for the functionalization of MXene nanosheets surface. It is also possible to hybridise MXene nanosheets with other 2D nanostructures, such as reduced graphene oxide nanosheets, in order to synergize its thermal characteristics and conductivity [37]. Furthermore, MXenes can be utilised in bioglass frameworks to impart photosensitive qualities to the final scaffold [38]. Diverse approaches have been utilised to fabricate MXene-based nanocomposites into distinct shapes and structures, including membranes, nanofibers, porous materials, polymeric microparticles, and 3D-printed items.

3 Applications of MXene and Its Composites in Tissue Engineering

Tissue engineering represents a cutting-edge biomedical strategy designed to repair and regenerate damaged tissues, offering solutions for patients with organ or tissue loss or failure and enhancing life expectancy. This approach utilizes a combination of scaffolds or biomaterials, cells, and biological molecules to restore or enhance the functionality of tissues and organs [39]. MXenes combine electrical conductivity, mechanical strength, biocompatibility, and functional versatility, making them a highly promising material for advancing tissue engineering technologies (Table 1). Their unique properties allow for applications in scaffold development, drug delivery, wound healing, and even disease modeling, significantly enhancing the effectiveness and adaptability of engineered tissues [40].

Table 1 Application of MXene in tissue regeneration

Target tissue	Type of formulation	Advantages	References
Bone	Bioglass@NbSiR	In vivo studies showed that the BG@NbSiR scaffold might promote bone regeneration, increase immune responses, prevent metastasis when used in conjunction with checkpoint blockade immunotherapy, and remove primary tumours. Furthermore, the scaffold provided a wide variety of tumor-associated antigens in tumour vaccines made from tumour debris following photothermal treatment (PTT)	[41]
	Ultralong hydroxyapatite nanowires/titanium carbide nanocomposite	Cell adhesion, proliferation, and osteogenic differentiation were all improved by its enhanced hydrophilicity and better mechanical characteristics. It greatly enhanced the amount and quality of bone tissue growth in a rat model of calvarial bone deficiency	[42]
	Nb_2C MXene-integrated 3D-printed bone-mimetic scaffold	Niobium carbide MXene facilitated the formation and migration of blood vessels at the defect site, enabling enhanced delivery of oxygen, nutrients, and energy to support the bone repair process	[38]
	Silica@Nb_2C MXene-integrated 3D-printing bioactive glass scaffolds	The composite was engineered with features such as controlled nitric oxide release, efficient photothermal conversion, and the ability to stimulate bone regeneration, offering a versatile biomaterial platform for the multifaceted treatment of bone tumors	[43]
	MXene/hydroxyapatite nanoparticle composite nanofibers	It was revealed that the composite nanofiber based on MXene and hydroxyapatite exhibited a synergistic connection between its photothermal properties and osteogenic features. The bone mesenchymal stem cells were efficiently encouraged in their proliferation and osteogenic differentiation by these nanofibers, which displayed outstanding biocompatibility	[44]
Nerve	$Ti_3C_2T_x$ MXene-Matrigel hydrogel	The 3D $Ti_3C_2T_x$ MXene system may enhance the development of mature synapses and intercellular signalling in spiral ganglionic neurones	[45]

(continued)

Table 1 (continued)

Target tissue	Type of formulation	Advantages	References
	Ti_3C_2 MXene	Primary cortical neurons could adhere, grow, and form neuronal networks on Ti_3C_2 and polystyrene. This material provided a neural interface that has a high-resolution for neuroelectronic devices	[46]
	$Ti_3C_2T_x$ MXene-Matrigel hydrogel	enhanced the efficiency of synapse formation, aided in the development of synaptic connections between hair cells and spiral ganglion neurones, and encouraged the growth, maturation, and production of organoid hair cells	[47]
	GelMA-MXene hydrogel	Successfully restored complete spinal cord transection and established neural networks between the repaired and regenerated nerves	[48]
Skin	Muscle-inspired MXene/PVA hydrogel	The MXene-based hydrogel effectively inhibited bacterial infections without inducing drug resistance. It exhibited outstanding mechanical properties, with a stress capacity of up to 0.5 MPa and a strain limit of 800%. Additionally, the hydrogel could be utilized for localized hyperthermia at infection sites under near-infrared (NIR) laser irradiation. Leveraging the exceptional photothermal performance of MXene, its primary antibacterial mechanism relied on heat generation. The hydrogel demonstrated broad-spectrum antibacterial activity against both gram-positive and gram-negative bacteria	[49]
	Chitosan–hyaluronate hydrogel@ $Ti_3C_2T_x$ MXene nanocomposites	The material exhibited effective antibacterial activity against *Escherichia coli*, *Staphylococcus aureus*, and *Bacillus* species	[50]
	$Ti_3C_2T_x$ MXene nanosheets	It was designed to support the healing of wounds infected with methicillin-resistant *Staphylococcus aureus* (MRSA). It demonstrated potent anti-inflammatory properties, enhanced cell proliferation, and promoted angiogenesis. Additionally, it facilitated the formation of granulation tissue, collagen deposition, vascular endothelial differentiation, and the development of new blood vessels	[51]

(continued)

Table 1 (continued)

Target tissue	Type of formulation	Advantages	References
	MXene–amoxicillin–PVA nanofibrous membrane	The membrane served as both a physical barrier for co-loading amoxicillin and MXene and as an effective antibacterial agent, promoting faster wound healing. This approach provides a foundation for developing innovative wound dressings and antibacterial solutions	[52]
	Nb_2C MXene titanium plate (Nb_2C@TP)-based implant	By neutralising an excessive amount of reactive oxygen species in contaminated microenvironments, the Nb2C@TP medical implant was able to prevent proinflammatory reactions. Moreover, it enhanced tissue regeneration and aided the process of angiogenesis	[26]
Heart	$Ti_3C_2T_x$-PEG composite	3D-printed $Ti_3C_2T_x$-PEG hydrogel aligned the iCMs with an increase in *MYH7*, *TNNT2*, and *SERCA2* expressions	[53]
	Conductive biohybrid platforms of collagen type I and 2D MXene ($Ti_3C_2T_x$)	This study highlights the promising potential of a biohybrid platform composed of collagen type I and MXene ($Ti_3C_2T_x$), which demonstrates excellent electrical properties, enhanced mechanical strength, improved hydrophilicity, and swelling behavior. The platform is highly biocompatible and prevents bacterial adhesion (*S. aureus*), essential qualities for next-generation medical devices. Most notably, it promotes the attachment, elongation, and maturation of iPSC-derived cardiomyocytes when subjected to external electrical stimulation	[54]

3.1 Bone Tissue

The bone matrix comprises organic substances, inorganic minerals, cells, and bioactive components. The inadequate self-healing capacity of bone renders the restoration of defects a significant challenge in clinical practice. Bone transplants have several limitations, including challenges in obtaining sufficient connective tissue and the potential for immunological reactions [55]. A variety of scaffolds and bone substitute biomaterials have been devised to address abnormalities and facilitate the restoration of bone tissues. A fundamental necessity in bone tissue engineering is the fabrication of scaffolds that replicate the essential properties of normal bone, hence facilitating native bone functionality and enhancing the regeneration of injured tissue [56].

Pan et al. documented the novel integration of 2D Ti_3C_2 MXene with three-dimensional (3D) bioactive glass scaffolds (TBGS), thereby facilitating the concurrent management of bone tumours by photonic hyperthermia while enhancing bone tissue regeneration [57] (Fig. 3a–d). To achieve total tumour eradication on in vivo bone-tumor xenografts, the developed TBGS employed the unique characteristic of integrated 2D Ti_3C_2 MXene, which has a high photothermal conversion, in order to eradicate bone tumour by using near-infrared induce photothermal hyperthermia. It is worth noting that the incorporation of 2D Ti_3C_2 MXene into the composite BG scaffolds efficiently speeds up the in vivo formation of new bone tissue. The combination capability of tumour eradication and bone tissue regeneration renders these TBGS exceptionally promising for the management of bone tumours. Yin et al. [38] developed a 2D Nb_2C MXene nanosheet containing BG scaffold and Nb-based species (NBGS) for osteosarcoma treatment (Fig. 4A, B). The integrated 2D Nb_2C-MXene nanosheets exhibit a distinct photonic action in the second near-infrared (NIR-II) biowindow, providing substantial tissue penetration depth, hence enhancing their efficacy in eradicating bone cancer cells. Further, NBGS degradation promotes angiogenesis, thereby accelerating bone repair. Furthermore, the mineralisation of new bone tissue may be aided by the phosphate and calcium produced during scaffold disintegration. Lin et al. [58] synthesised a new 2D niobium carbide (Nb_2C) MXene that efficiently ablated xenograft tumours in mice via exceptional photothermal effect in both the NIR-I (36.4%) and NIR-II (45.65%) windows, coupled with significant light and thermal stability.

In order to encourage the death of tumour cells, combat harmful microorganisms, and boost bone production, Yin et al. [25] generated an SP@MX/GelMA scaffold from MXene nanosheets, GelMA hydrogel, and sulfonated polyetheretherketone (SP). When exposed to 808 nm near-infrared radiation, the synergistic photothermal activity of MXene and PDA results in the efficient removal of osteosarcoma cells on the multifunctional coating by the process of thermal erosion. Tobramycin loading into SP@MX/GelMA also inhibited Gram-negative and Gram-positive bacteria (as shown in Fig. 5).

To direct the process of bone tissue regeneration, another researcher created a Ti_3C_2 MXene film that could differentiate into osteogenic cells in vitro and the in vivo research done on the calvarial defect model on rat demonstrated excellent biocompatibility and significant bone regeneration activity [59]. Moreover, UHAPNWs/MXene nanocomposite membrane prepared by inserting ultralong hydroxyapatite nanowires into the MXene film, can effectively enhance bone tissue formation too, in order to deliver more effective osteoconductivity and osteoinductivity for bone regeneration in dental applications [42] (as shown in Fig. 6a–c). The findings further demonstrated that the mechanical and physical characteristics of the constructed nanocomposite membrane were substantially affected by the concentration of hydroxyapatite nanowires. Additionally, the nanocomposite with 10 wt% hydroxyapatite demonstrated surprisingly good flexibility even after being folded and wrapped around a plastic rod numerous times without causing any damage [42]. Mi et al. [60] developed nanocomposite scaffolds (Ti_3C_2–MXene modified HA/SA) made of Ti_3C_2-MXene modified with hydroxyapatite and sodium alginate using the

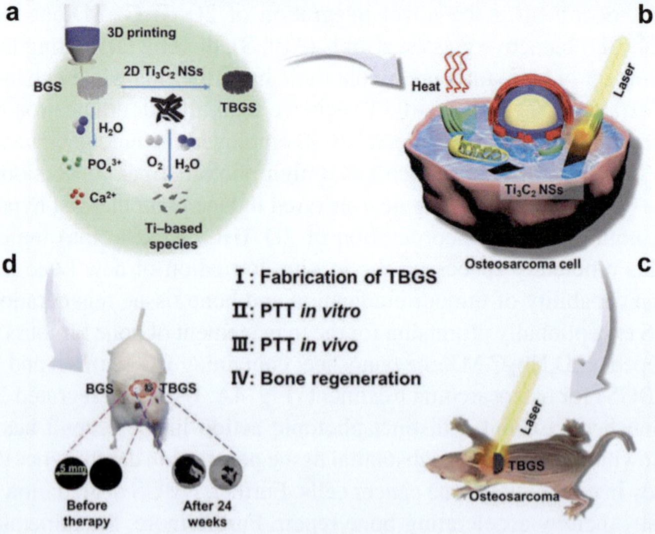

Fig. 3 Schematic depiction of the process for creating TBGS, its application in bone cancer treatment, and subsequent bone tissue regeneration. **a** Steps involved in the preparation of TBGS fabrication, including the 3D printing of pure BGS, amalgamation with Ti_3C_2 MXene, and controlled degradation of Ti_3C_2 MXene on BGS. **b** and **c** Application of TBGS for the eradication of osteosarcoma cells through PTT in both in vitro (**b**) and in vivo (**c**) settings. **d** Restoration of bone tissue and therapeutic outcomes following the implantation of BGS and TBGS. Reproduced with permission [57]. Copyright 2019, Wiley Online Library (Open Access)

extrusion-based 3D printing method. In vitro osteogenecity and biocompatibility study done on bone mesenchymal stem cells (BMSCs) of rat showcased that the rough surface of the scaffolds allows better cell adherence, migration, differentiation, and proliferation, eventually encouraging new bone tissue production. Furthermore, the porosity of the engineered scaffolds facilitated the movement of nutrients, vascular ingrowth, and mesenchymal cell migration, thus creating a conducive osteogenic milieu. This affirms the calvarial bone regeneration and bone healing ability of Ti_3C_2–MXene-modified HA/SA scaffolds.

3.2 *Nerve Tissue*

Despite great advancements in comprehending the biological mechanics of the nervous system, the regeneration and repair of nerve tissues continue to pose significant challenges. This is mainly due to the intricate structure and physiology of the nervous system, which sets it apart from other tissues, with injuries frequently leading

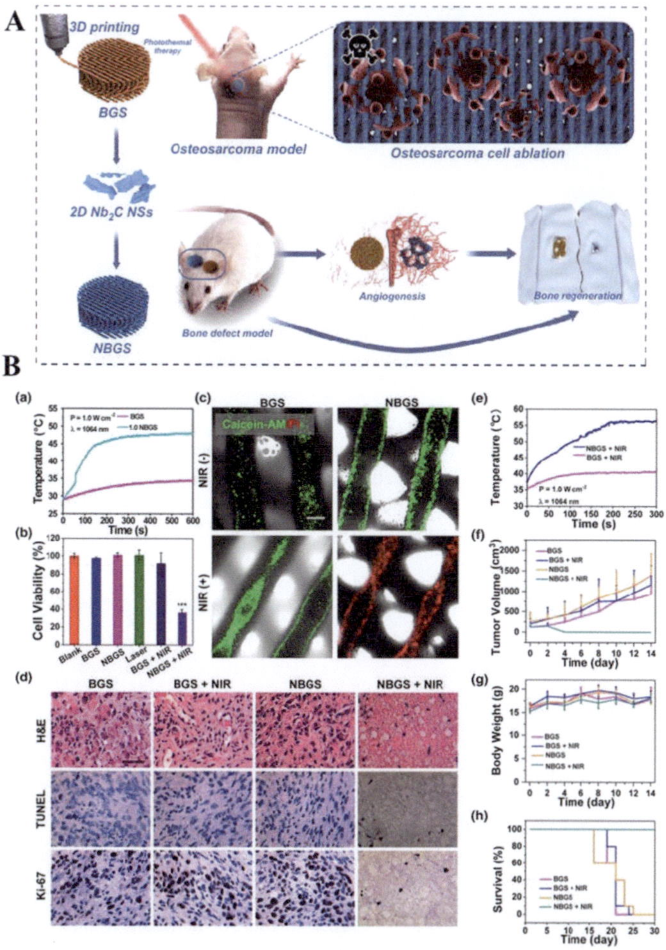

Fig. 4 **A** Schematic representation of PTT for osteosarcoma combined with bone regeneration using NBGS, emphasizing vascularization to facilitate tissue reconstruction. **B** Analysis of the photothermal properties and antitumor effects of BGS and NBGS both in vitro and in vivo. **a** Temperature profiles of 1.0 NBGS and BGS in PBS solution. **b** Viability of Saos-2 cells following various treatments. **c** Confocal laser scanning microscopy (CLSM) images of treated groups (BGS, BGS + NIR, NBGS, NBGS + NIR) showing live and dead cells on the scaffold via Calcein AM/PI staining, along with bright-field visuals. **d** Optical microscopy images of tumor sections stained with H&E, TUNEL, and Ki-67. **e** Tumor temperature trends in mice implanted with 1.0 NBGS and BGS during NIR-II laser irradiation. **f** Tumor volume progression in osteosarcoma-bearing mice subjected to different treatments (n = 5). **g** Body weight variation in treated mice (n = 5). (h) Survival rates across treatment groups (n = 5). Scale bars: 500 μm (**c**) and 100 μm (**d**). Reproduced with permission [38]. Copyright 2021, Springer nature (Open access)

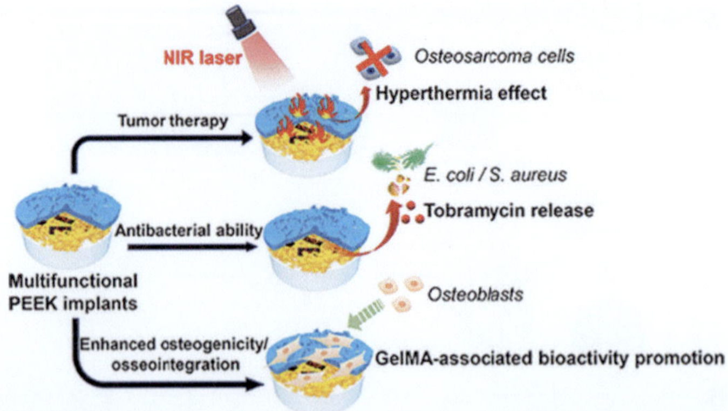

Fig. 5 MXene-based hydrogels enhance the osteogenic capabilities of polyetheretherketone (PEEK) while providing a dual therapeutic approach for osteosarcoma treatment and bacterial infection prevention. This innovative system integrates PTT for tumor elimination and antibacterial action, offering a comprehensive solution for bone regeneration and disease management. Reproduced with permission [25]. Copyright 2020, American Chemical Society

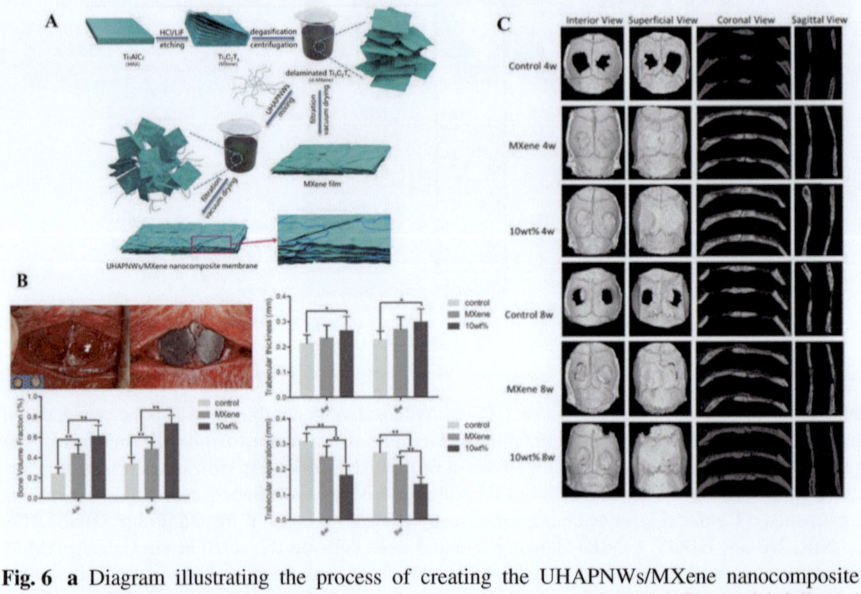

Fig. 6 **a** Diagram illustrating the process of creating the UHAPNWs/MXene nanocomposite membrane. **b** Illustration of the surgical procedure for creating a rat calvarial defect model, followed by coverage of the defect with the respective membranes and the radiographic assessment of bone morphometric parameters. **c** 3D reconstruction images displaying internal and surface views of the bone defect, alongside micro-CT images in coronal and sagittal perspectives in the control, MXene, and UHAPNWs/MXene groups at 4- and 8-weeks post-surgery. Reproduced with permission [42]. Copyright 2021, Elsevier Ltd.

to permanent incapacity. This results in heightened interest in formulating innovative methodologies, including neural tissue engineering and nanotechnology-based methods, for neural tissue regeneration [61].

Electrical conductivity is a key characteristic of MXenes for nervous tissue engineering. Neural tissues work on electrical signaling, thus compounds that replicate or promote these signals are useful for nerve restoration. Recent investigations have revealed that electrical stimulation of MXene-based materials increases brain cell differentiation and synaptic activity [62]. Within the MXene family, $Ti_3C_2T_x$ MXene is applicable for the optical regulation of neural electrical activity, exhibiting great spatiotemporal resolution. Furthermore, $Ti_3C_2T_x$ MXene's photothermal efficacy was studied at the single-flake scale by Wang et al. (2021). They discovered that both films and flakes could stimulate dorsal root ganglion neurons through photothermal means, even at low incident radiation. This method shows promise for creating efficient models of neurological treatment when combined with medicinal and tissue engineering techniques [63]. Li et al. [64] found that electrical stimulation combined with $Ti_3C_2T_x$ MXene could ensure adequate growth, proliferation, and differentiation of neural stem cells (NSCs), signifying that it can enhance NSC maturation (Fig. 7a–f). $Ti_3C_2T_x$ MXenes preferentially enhances voltage-gated Ca^{2+} current, lengthening the neurites. In addition, $Ti_3C_2T_x$ MXene also boosted synaptic release and neuronal differentiation after intense depolarisations, improving synaptic transmission. Using Ti_3C_2 MXene, Driscoll et al. [46] demonstrated that primary cortical neurones could adhere, proliferate, and establish neural networks on top of polystyrene. For neuro-electronic devices, this material offered a high-resolution neural interface. Further, Xiao et al. examined the effects of uncoated $Ti_3C_2T_x$ MXenes on hippocampus cell growth which suggested its role in preservation of the basic physiological activities in neural circuit development. Their efforts paved the way for the development of diagnostic, therapeutic, and research tools for neural repair [65].

MXenes demonstrate superior biocompatibility when suitably treated, facilitating effective interaction with neural cells. The presence of surface functional groups like –O, –OH, and –F on MXenes facilitates cell attachment and proliferation [66]. Recent studies have investigated surface alterations of MXenes to improve their compatibility with neural tissues. Research done by Guo et al. [67] observed that surface modification of $Ti_3C_2T_x$ MXene with laminin improved the adhesion and differentiation of neural stem cells into neurons. In addition, as compared to cells grown on tissue culture polystyrene, neurones produced from NSCs grown on $Ti_3C_2T_x$ MXenes displayed longer neurites, more branches, and tips, and were more actively differentiated, mature, and active (as shown in Fig. 8a–g). This approach highlights the potential of MXenes as bioactive platforms for guiding neural tissue regeneration.

Electrospun nanofibrous scaffolds using MXenes have garnered interest for their capacity to replicate the fibrous architecture of the neural extracellular matrix (ECM) [68]. Diedkova et al. had formed polycaprolactone –$Ti_3C_2T_x$ MXene nanofibrous scaffolds with superior mechanical strength and conductivity, which enabled neuronal alignment and enhanced axonal development (Fig. 9). Moreover, the MXene-infused nanofibers exhibited antibacterial characteristics, diminishing the likelihood of infections at the wound site, hence, showcasing their potential for neural

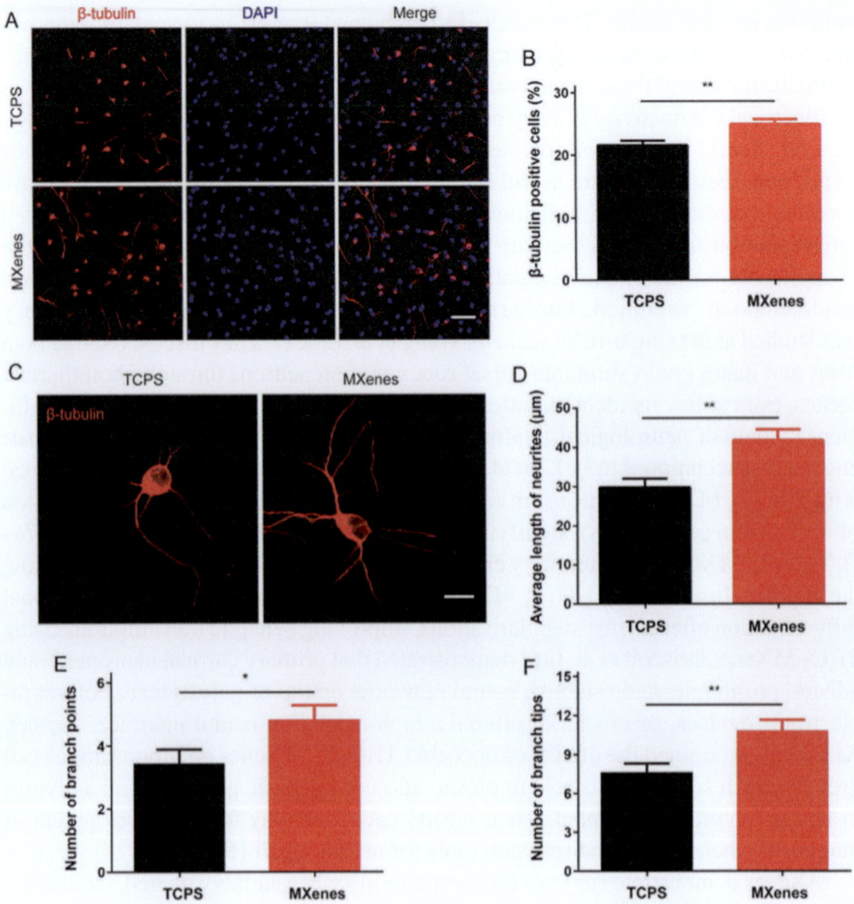

Fig. 7 $Ti_3C_2T_x$ MXene enhances the development and growth of NSCs. **a**, **b** Representative images of neurons derived from NSCs, stained with β-tubulin and DAPI at 7 DIV, with a scale bar of 50 μm (**a**), and the percentage of β-tubulin-positive cells on TCPS and MXenes (**b**). **c**–**f** Representative images of newly formed neurons grown on TCPS and MXenes, scale bar 10 μm (**c**), along with quantitative analysis of neurite length (**d**), branch points (**e**), and branch tips (**f**). Statistical significance is indicated by $*p < 0.05$ and $**p < 0.01$. Reproduced with permission [64]. Copyright 2022, Springer Nature. (Open access)

tissue regeneration [69]. Zhao et al. reported the formation of Ti_3C_2 MXene nanosheet incorporated transglutaminase (TG) enzyme-containing gallic acid-imbedded gelatin (GGA) hydrogel. The hydrogel supported the growth and differentiation of neural cells, and its electrical conductivity enabled the transmission of electrical signals, further enhancing neural regeneration [70].

Further, the MXene-based nanocarrier system can provide sustained release of nerve growth factors (NGFs), thereby promoting sustained neural differentiation and axonal growth over several weeks [71]. Additionally, MXenes' photothermal

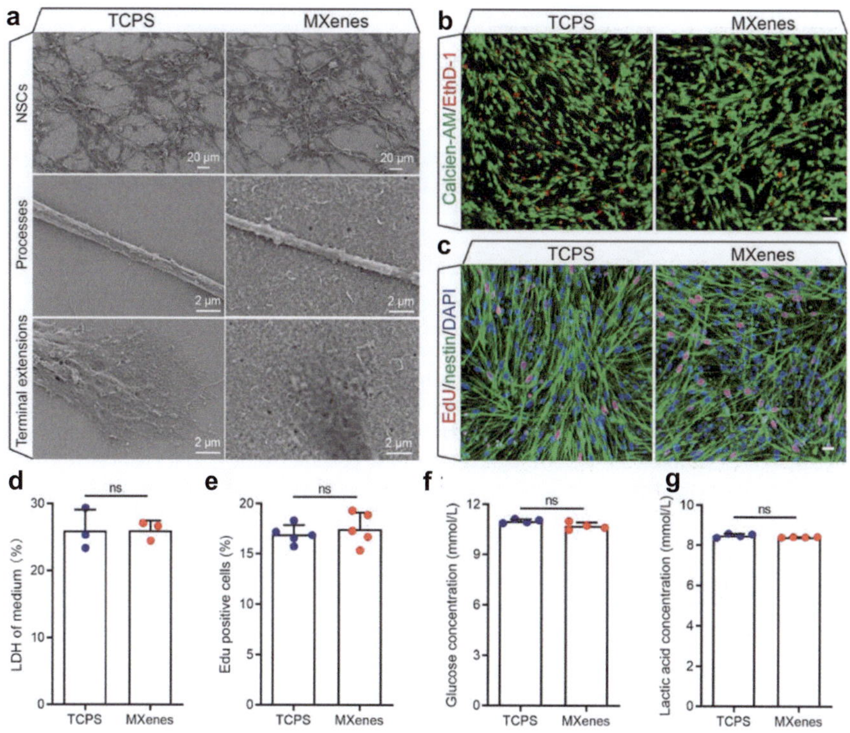

Fig. 8 Assessment of $Ti_3C_2T_x$ MXene film biocompatibility for NSC growth and proliferation. **a** Field-emission scanning electron microscopy images showing NSCs cultured on TCPS and $Ti_3C_2T_x$ MXene film. **b** Fluorescent live/dead staining of NSCs on TCPS and $Ti_3C_2T_x$ MXene film, with live cells stained green (cytoplasm) and dead cells stained red (nuclei). Scale bar: 50 μm. **c** Fluorescent images of NSCs after 3 days of culture, with EdU added on day 2. EdU is marked in red, nestin in green, and nuclei in blue. Scale bar: 20 μm. **d** LDH cytotoxicity levels in the medium after 3 days of NSC culture. **e** Histogram of the percentage of EdU-positive cells for samples cultured on $Ti_3C_2T_x$ MXene film and TCPS. **f**, **g** Glucose (**F**) and lactic acid (**G**) concentrations in the medium after NSCs were cultured for 3 days on $Ti_3C_2T_x$ MXene film and TCPS. Statistical analysis indicates no significant differences, as assessed by the Mann–Whitney test. Reproduced with permission [67]. Copyright 2022, Elsevier Ltd.

properties can provide an on-demand release of therapeutic agents, enabling precise control over the delivery process [72]. By incorporating MXenes into neural scaffolds, researchers observed improved survival of neurons in oxidative environments [73]. A 2024 study utilized an MXene-based conductive bioink to print nerve guidance conduits. These conduits provided a favorable environment for neural cell growth and guided axonal regeneration across the injury site. The ability to integrate MXenes into 3D bioprinting processes allows for the precise fabrication of patient-specific nerve repair constructs [73].

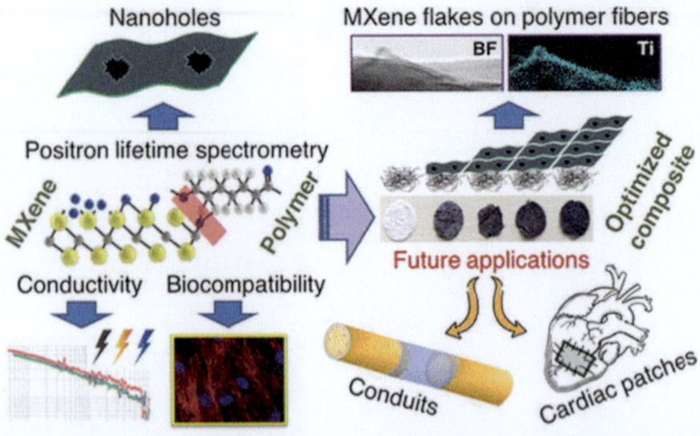

Fig. 9 $Ti_3C_2T_x$ MXene was layered onto electrospun PCL membranes, enhancing their porosity, conductivity, and structural properties. Conductive scaffolds demonstrated temperature-dependent defect behavior and long positron lifetimes. MXene coatings supported cell attachment, proliferation, and mild antibacterial effects, highlighting their potential as advanced conductive biomaterials for tissue engineering applications through combined structural, chemical, and biological advantages. Reproduced with permission [69]. Copyright 2023, American Chemical Society

3.3 Skin Tissue

MXene-based composites have demonstrated significant potential in wound healing due to their versatility and easily customizable surface properties. These materials can be tailored to exhibit desirable characteristics such as high porosity, biodegradability, and antibacterial effectiveness. When integrated into wound dressings, MXenes contribute to accelerating the healing process by facilitating cell movement, growth, and specialization. Moreover, they play a role in regulating the inflammatory response, which is an essential phase of wound repair and tissue regeneration.

A study conducted by Mao et al. [74] introduced innovative multifunctional hydrogels by combining regenerative $Ti_3C_2T_x$ MXene and bacterial cellulose (rBC). These hydrogels were designed to electrically influence cellular behavior, promoting effective wound healing when paired with external electrical stimulation (ES). Among the formulations, a hydrogel containing 2 wt% MXene (rBC/MXene) showcased an ideal balance of electrical conductivity, biocompatibility, mechanical strength, flexibility, and biodegradability. In vivo tests on rats with full-thickness wounds demonstrated that the rBC/MXene hydrogel significantly outperformed traditional Tegaderm film dressings in promoting wound healing. The combination of the hydrogel dressing with ES provided a powerful, synergistic strategy for enhancing the healing process (Fig. 10a–e).

MXenes show great potential for hemostasis, making them ideal for wound management and bleeding control. Their highly porous structure facilitates efficient absorption and entrapment of blood components, while their large surface area and interconnected pores promote rapid platelet adhesion and aggregation. The strong surface charge of MXenes further enhances platelet activation and clot

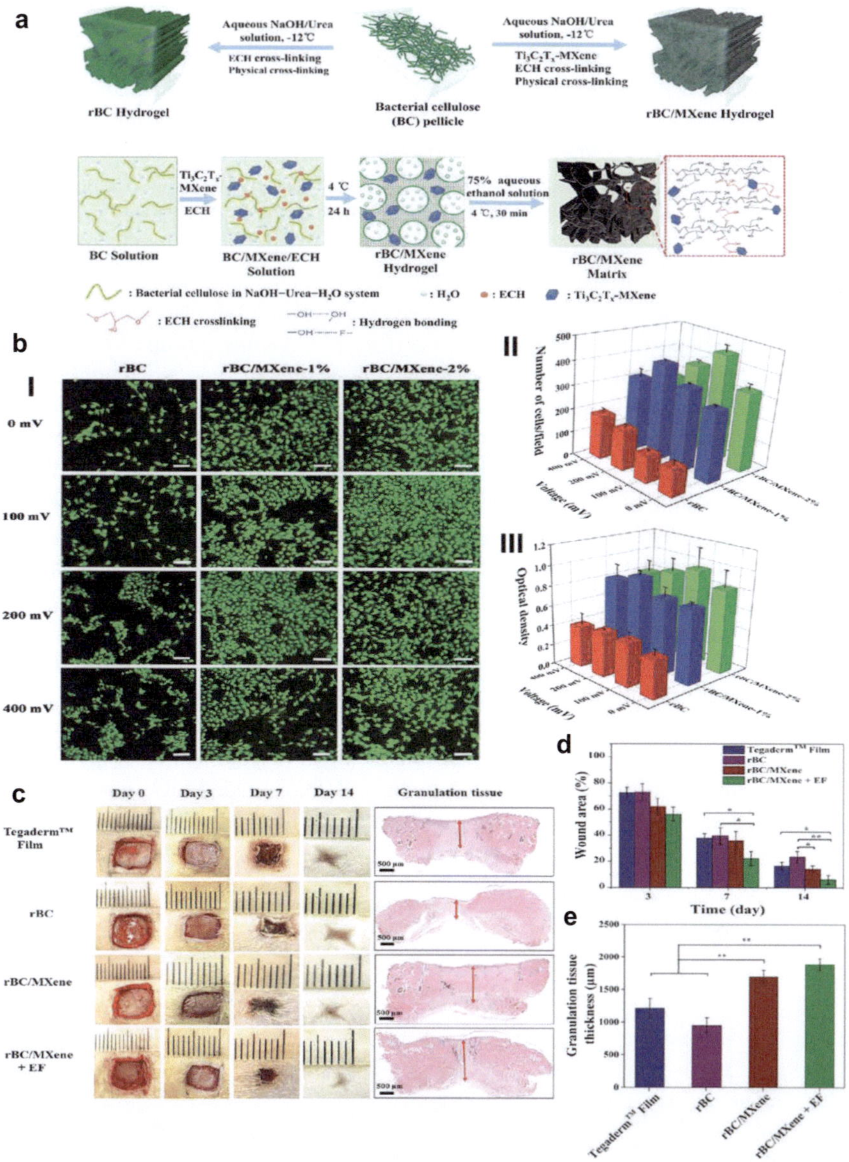

Fig. 10 **a** Illustration of the fabrication process for BC/MXene composite hydrogels. **b** (i) Confocal microscopy images showing live/dead staining of NIH_3T_3 cells after treatment with various formulations (scale bar = 100 μm). Quantitative analysis of cell number (ii) and viability (iii) after 3 days of treatment with different formulations. **c** Images of wound healing progression at days 0, 3, 7, and 14, highlighting granulation tissue (red arrows). **d** Quantitative assessment of wound area reduction over time and **e** granulation tissue thickness on day 14 for each treatment group, with significance levels indicated (*$p < 0.05$, **$p < 0.01$). Reproduced with permission [74]. Copyright 2020, Wiley Online Library

formation, aiding in stable clot development and coagulation. Additionally, surface modifications, such as functionalizing with carboxyl or amine groups, improve interactions with blood components, boosting their hemostatic performance. The team of Zhou et al. [51] developed HPEM scaffolds by combining poly(glycerol-ethylenimine), $Ti_3C_2T_x$ MXene@polydopamine (MXene@PDA) nanosheets, and oxidized hyaluronic acid. These scaffolds exhibited self-healing, electrical conductivity, tissue adhesion, antibacterial activity against MRSA, and rapid hemostasis. They supported skin cell proliferation with minimal toxicity and significantly accelerated MRSA-infected wound healing (96.31% closure rate). This was achieved through anti-inflammatory effects, enhanced cell growth, granulation tissue formation, collagen deposition, and angiogenesis (Fig. 11). Another group of researchers developed an antibacterial fiber membrane using electrospun polycaprolactone scaffolds combined with polydopamine-coated MXene/Ag_3PO_4 bioheterojunctions (MX@AgP bio-HJs) [75, 76]. Upon near-infrared (NIR) exposure, the membrane released Ag^+ ions, exhibiting strong antibacterial effects. Polydopamine facilitated the reduction of Ag^+ to Ag^0, enabling repeated ion release for sustained phototherapy. In vivo studies showed that the photoactivated nanofibrous membrane effectively eliminated bacteria, halted bleeding, and promoted wound healing through epithelialization, collagen deposition, and angiogenesis (Fig. 12a–d) [76].

Ti_3C_2 MXene-based flexible materials exhibit remarkable versatility, capable of photothermal antibacterial action, motion monitoring, and controlled drug release due to their excellent conductivity and photothermal properties. Jin et al. [77] developed a NIR laser-activated hybrid hydrogel (MNFs@V-H@DA) designed for regulating wound inflammation and enhancing revascularization (Fig. 13a–c). The hydrogel fabrication involved incorporating VEGF-loaded SiO_2 nanoparticles and MXene nanosheets into PLGA nanofibers, forming MNFs@V. These were

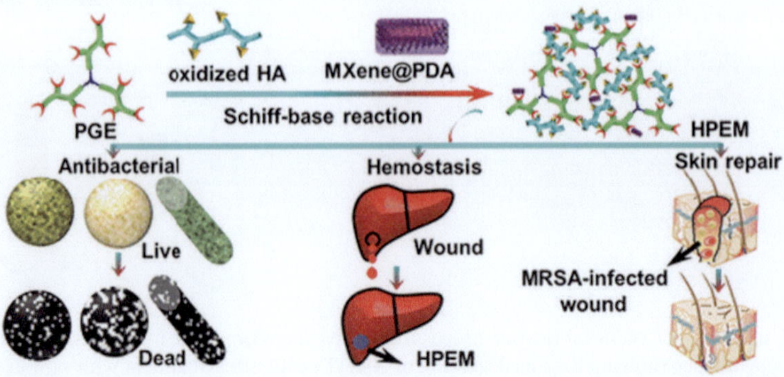

Fig. 11 Illustration depicting the development of multifunctional scaffolds incorporating $Ti_3C_2T_x$ MXene nanosheets. These scaffolds exhibit excellent conductivity, antibacterial properties, and hemostatic capabilities, effectively supporting the healing of wounds infected with multidrug-resistant bacteria. The design integrates advanced materials to simultaneously address infection, promote tissue repair, and enhance wound healing efficiency. Reproduced with permission [51]. Copyright 2021, American Chemical Society

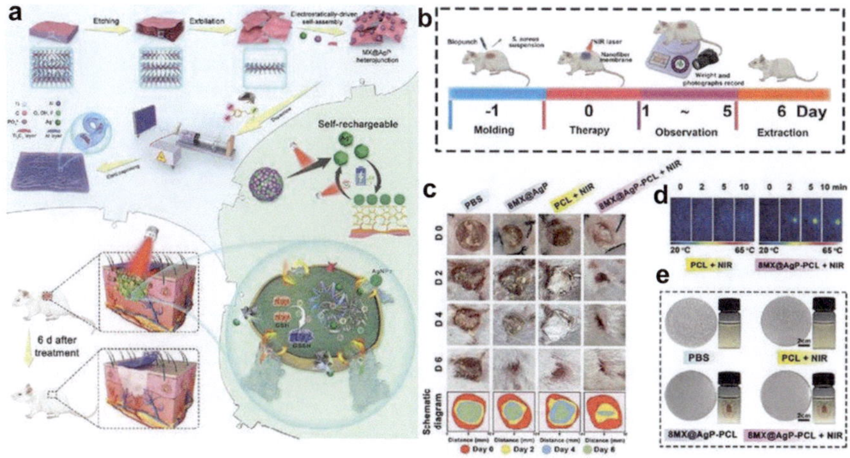

Fig. 12 **a** Diagram showing the fabrication and application process of MX@AgP-PCL membranes. **b** Overview of the in vivo experimental procedure. **c** Wound healing progression following treatments with various formulations. **d** NIR imaging of the membrane at different exposure durations. **e** Bacterial colony counts of S. aureus after exposure to different treatments. Reproduced with permission [76]. Copyright 2024, Royal Society of Chemistry. (Open Access)

then coated with dopamine-conjugated hyaluronic acid and diallyl trisulfide (DA), resulting in MNFs@V-H@DA hydrogel nanofibers. Upon application to wounds, the hydrogel allowed controlled VEGF release using NIR laser activation, while DA was gradually released, generating hydrogen sulfide (H_2S) to stimulate M2 macrophage formation. In vivo studies demonstrated that the hydrogel effectively promoted wound healing, reduced inflammation, and minimized scarring by regulating the immune response and encouraging vascularization. This multifunctional material shows promise as a smart dressing for scar-free healing and managing chronic wounds.

3.4 Cardiac Tissue

Designing and fabricating conductive cardiac patches to enhance the patch's electrical connection with the host tissue is one of the promising ways to increase heart tissue regeneration. In order to improve the electrophysiological characteristics of hydrogels and cryogels for use in cardiac tissue engineering, numerous conductive materials, such as MXenes [54], have been included. To create antioxidant/conductive hydrogels (E-MXene), Ren et al. [78] inserted MXene nanosheets (Ti_3C_2) into hydrogels made of temperature-sensitive extracellular matrix (ECM). As a result of its exceptional antioxidant and conductivity characteristics, E-MXene effectively reduced reactive oxygen species (ROS) in vitro and enhanced the synchronisation of

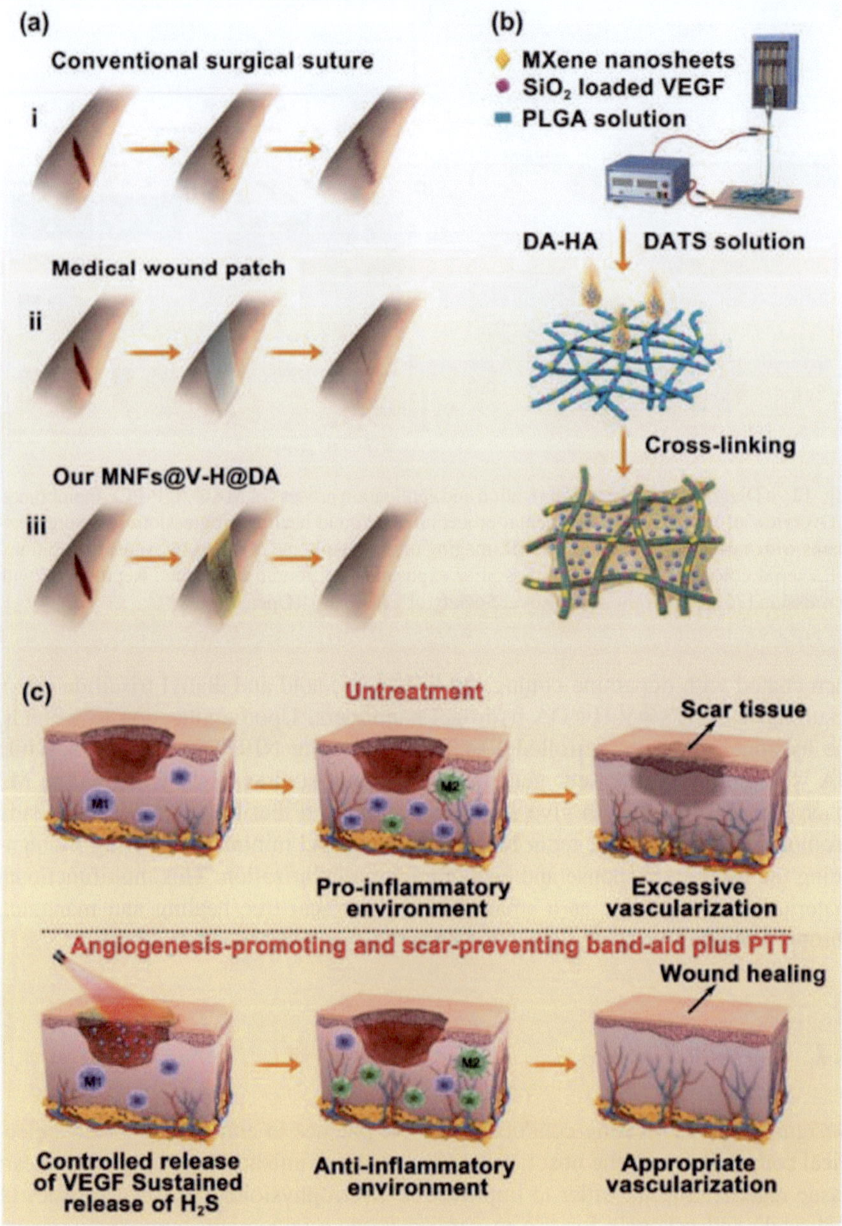

Fig. 13 Formulation and wound healing process of MNFs@V–H@DA. **a** Comparison of wound healing materials: (i) traditional surgical sutures, (ii) medical wound patches, and (iii) MNFs@V–H@DA. **b** Schematic representation of MNFs@V–H@DA fabrication through electrospinning and surface coating. **c** Illustration of the wound healing mechanism facilitated by MNFs@V–H@DA. Reproduced with permission [77]. Copyright 2022, Elsevier Ltd. (Open access)

calcium signaling in cardiomyocytes derived from human induced pluripotent stem cells (hiPSC-CMs). This finding offers novel theoretical principles and potential treatments for myocardial infarction (Fig. 14).

To engineer a cardiac patch, achieving proper alignment of cardiomyocytes (CMs) is crucial, as the functionality of cardiac tissue heavily relies on the linear organization of myofibrils, their bundling, and the elongated structure and interconnectivity of ventricular myocytes. In a pioneering study, Pinar Zorlutuna and her team demonstrated the use of aerosol jet-printed 2D titanium carbide ($Ti_3C_2T_x$) MXene-hydrogel composites to create human cardiac patches. These patches not only supported the alignment of cardiomyocytes derived from human-induced pluripotent stem cells (iCMs) but also highlighted the potential of $Ti_3C_2T_x$ MXene-PEG composite hydrogels as electroactive scaffolds for cardiac tissue repair [53].

Injectable shape-memory composite scaffolds have been developed by incorporating Ti_3C_2 MXene quantum dots into chitosan-based hydrogels for in vitro studies. These scaffolds demonstrated the ability to support cell survival and proliferation while facilitating electrical signal transmission between cells, making them potential candidates for tissue engineering applications [79]. Similarly, chitosan-based hydrogels containing Ti_3C_2 MXene, honey, and 0D fluorescent carbon dots have been fabricated for tissue engineering purposes. These hydrogels exhibited excellent

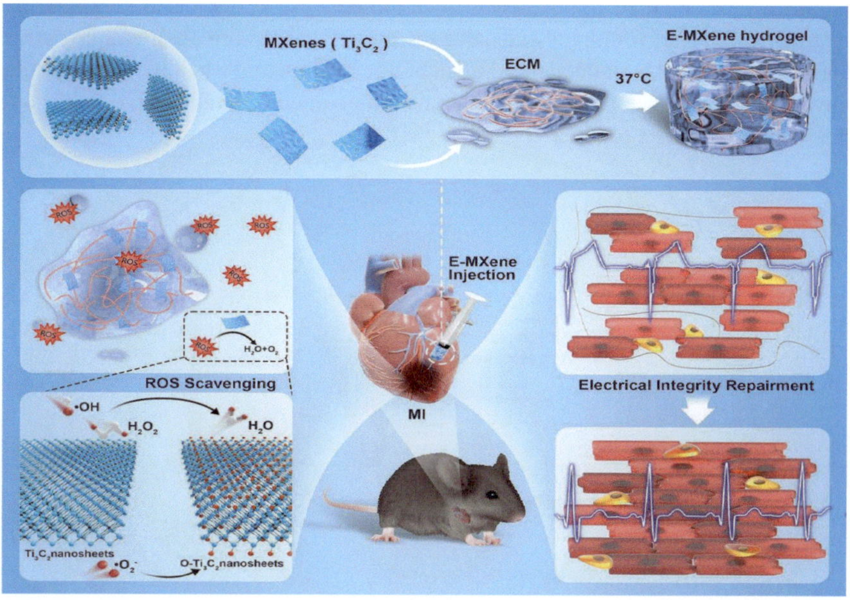

Fig. 14 Abstract depicting the role of an injectable MXene-based conductive hydrogel in facilitating myocardial infarction recovery through the neutralization of reactive oxygen species (ROS) and the restoration of myocardial electrical integrity. Reproduced with permission [78]. Copyright 2024, Elsevier Ltd.

compatibility with various stem cell types, along with anti-inflammatory and antibacterial properties, further highlighting their suitability for biomedical applications [80].

4 Applications of MXenes and Its Composites in Cancer Nanotherapy and Diagnosis

Nanomaterials have gained significant attention in biomedical research due to their potential applications in cancer therapy, diagnostics, sensors, and imaging. Their exceptional characteristics, such as a high surface area-to-volume ratio and quantum electronic effects, make them distinct from conventional bulk materials [81]. Among these, MXene nanomaterials have emerged as promising candidates for cancer theranostics. MXenes possess a range of exceptional physicochemical properties, including a high surface area, excellent electrical conductivity, magnetic behavior, low toxicity, luminescence, and notable biocompatibility, making them strong contenders for cancer therapy and theranostic applications. These 2D nanomaterials exhibit photothermal, photodynamic, and chemo-photothermal synergistic effects, offering significant potential for efficient and minimally invasive anti-tumor therapy. Their potential in targeted medication delivery systems and photothermal treatment (PTT) has prompted extensive research. Furthermore, their unique optical characteristics have facilitated their use in biological imaging and biological sensors, however, their efficient light-to-heat translation renders them very suitable and biocompatible agents for PTT (Fig. 15) [82].

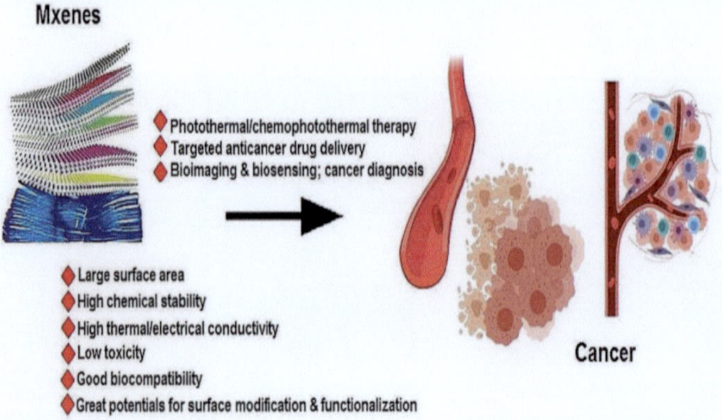

Fig. 15 Summary of MXenes in cancer detection and treatment. Reproduced with permission [82]. Copyright 2021, American Chemical Society

4.1 Application of MXenes in Photothermal Therapy

A scientific analysis of MXene-based photothermal tumour eradication utilising NIR-I and NIR-II biowindows, together with their function in various imaging-guided photothermal ablation techniques, underscores the effectiveness of NIR laser-induced phototherapy for addressing tumours in merely vascularised microenvironments. This method involves the introduction of photothermal agents or light-absorbing materials into tumour tissues, where the absorbed near-infrared laser energy is converted into localised heat, therefore successfully annihilating cancer cells [83]. Successful PTT necessitates that photothermal agents demonstrate robust light absorption in the NIR range, effective conversion of light energy to heat, superior biocompatibility, and the capability for rapid surface functionalisation. The application of NIR light as a remote external stimulus provides numerous advantages, including accurate spatial and temporal regulation of localised heating while reducing adverse side effects. To enhance tumour ablation, NIR irradiation must achieve a compromise between minimal tissue scattering and absorption for deep light penetration, while also assuring little self-heating of healthy tissues. Simultaneously, the photothermal agents must demonstrate superior light-to-heat conversion efficiency to maximize therapeutic performance.

A recent investigation demonstrated that Ta_4C_3-SP nanosheets exhibit exceptional photothermal conversion efficiency (PCE) of 44.7% under 808-nm laser irradiation, making them promising candidates for PTT [84] (Fig. 16a). These nanosheets possess a sheet-like structure with an average lateral dimension of approximately 100 nm and exhibit robust light absorption across a broad wavelength range (Fig. 16b). Under laser irradiation at a power density of 1.5 W cm^{-1}, the temperature can swiftly rise to about 55 °C within 5 min. Furthermore, Ta_4C_3-SP nanosheets demonstrate excellent thermal stability, maintaining their performance over five consecutive heating and cooling cycles (Fig. 16b). In the absence of laser exposure, cytotoxicity assessments revealed that Ta_4C_3-SP nanosheets had minimal impact on the viability of 4T1 cells, even at elevated concentrations of up to 400 μg ml^{-1}, highlighting their outstanding biocompatibility. The photothermal performance of the nanosheets is influenced by the laser power density, a relationship that was further validated through confocal fluorescence imaging following different treatments (Fig. 16c). When administered via intravenous (i.v.) or intratumoral (i.t.) routes in a mouse tumor model, the nanosheets rapidly accumulated in the tumor tissue. Under 808-nm laser irradiation for 6 min, the tumor temperature increased sharply, reaching approximately 60 °C for i.v. delivery and 68 °C for i.t. delivery from an initial 30 °C (Fig. 16d). This rapid heating effectively facilitated tumor ablation, as depicted in Fig. 16e. Additionally, images of 4T1 tumor-bearing mice taken 16 days post-treatment further confirmed the potent photothermal tumor eradication ability of Ta_4C_3-SP nanosheets (Fig. 16f).

Li et al. [85] established an innovative MXene-based photothermal platform that integrates heat shock protein (HSP) inhibition to enhance tumor cell sensitivity to heat stress, thereby improving PTT outcomes. This platform, referred to as Ti_3C_2@Qu, combines Ti_3C_2 nanosheets with their excellent photothermal conversion properties

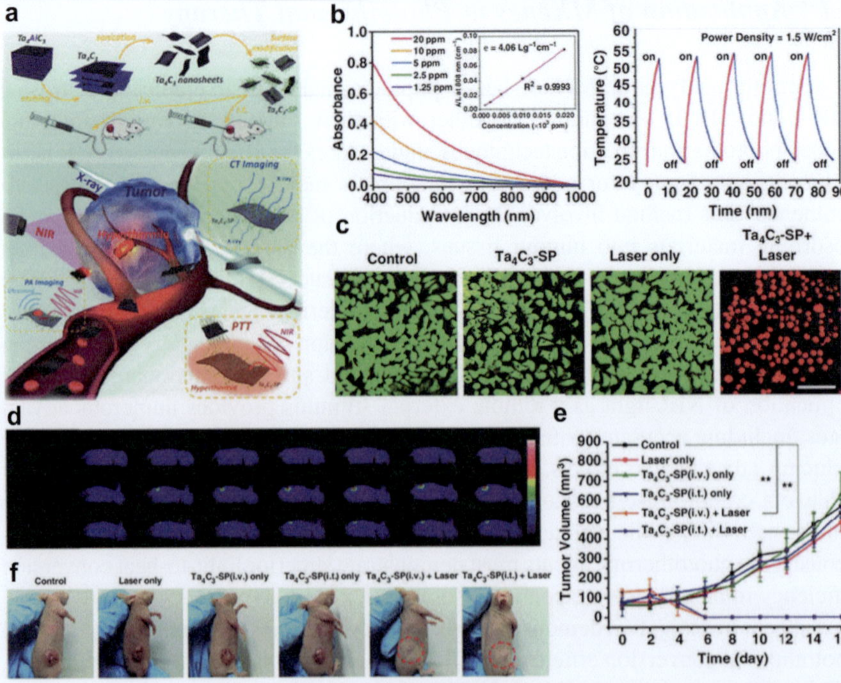

Fig. 16 PTT effects of Ta_4C_3-SP MXene nanosheets both in vitro and in vivo. **a** A schematic representation of Ta_4C_3-SP MXene nanosheets applied for PTT. **b** Absorbance spectra and photothermal stability of Ta_4C_3-SP MXene nanosheets after five heating and cooling cycles. **c** In vitro photothermal ablation of 4T1 cells following different treatments under 1.5 W cm^{-1}, confirmed by confocal fluorescence imaging (scale bar: 100 μm). **d** Infrared thermal imaging of tumor sites in 4T1 tumor-bearing mice from control, Ta_4C_3-SP (i.v.) + NIR laser, and Ta_4C_3-SP (i.t.) + NIR laser groups during laser irradiation at various time intervals. **e** Tumor growth progression following different treatments. **f** Images of tumor-bearing mice after undergoing PTT. Reproduced with permission [84]. Copyright 2020, De Gruyter. (Open access)

and quercetin (Qu), a known HSP70 inhibitor. Qu molecules are incorporated into the nanocomposite in a manner that allows controlled, pH-responsive release. Within the acidic tumor microenvironment, the Qu molecules are rapidly released, leading to the suppression of HSP70 levels in tumor cells. By depleting HSP70, the cells are left vulnerable, as their heat-resistant survival pathways are disrupted. This sensitization significantly amplifies the efficacy of hyperthermia-based treatment, resulting in enhanced PTT performance (Fig. 17).

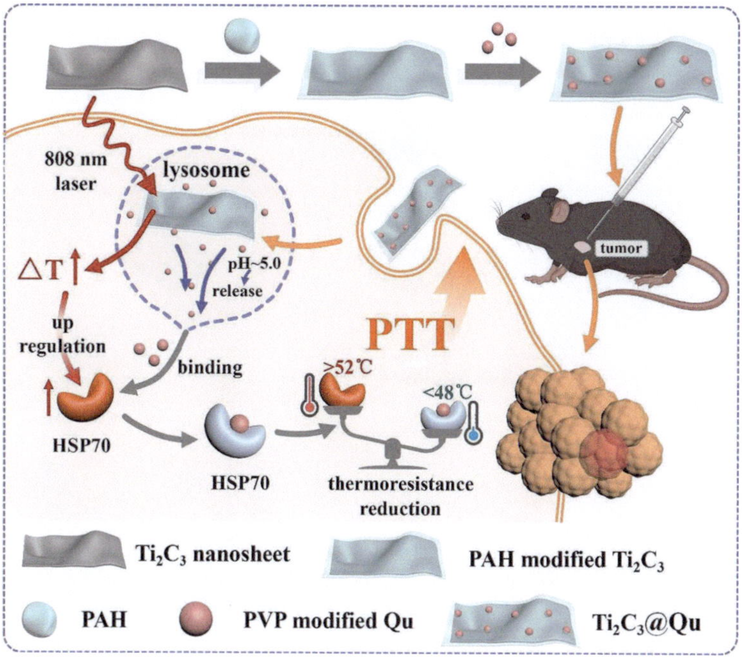

Fig. 17 The Ti_3C_2@Qu nanocomposites are formed by exploiting the outstanding photothermal conversion capacity of Ti_3C_2 nanosheets with quercetin (Qu), which serves as an inhibitor of heat shock protein 70 (HSP70). Qu molecules are incorporated onto the nanoplatform in a pH-sensitive, controlled release approach. The tumor's acidic environment induces the rapid release of Qu molecules, resulting in the reduction of HSP70 levels in tumour cells, hence removing their protection from the heat-resistant survival pathway and enhancing the efficiency of hyperthermia. Reproduced with permission [85]. Copyright 2024, Elsevier Ltd.

4.2 *Application of MXenes in Photodynamic Therapy*

Liu and their team [86] successfully synthesized ultrathin Ti_3C_2 MXene nanosheets (~100 nm) by introducing Al^{3+} as an additive to mitigate aluminum loss during preparation. These nanosheets demonstrated exceptional optical and therapeutic properties, including a high mass extinction coefficient of 28.6 Lg^{-1} cm^{-1} at 808 nm, excellent photothermal conversion efficiency (~58.3%), and significant generation of singlet oxygen ($1O_2$) under 808 nm laser exposure. To create a multifunctional therapeutic platform, the nanosheets were modified using a layer-by-layer approach with doxorubicin (DOX) and hyaluronic acid (HA), forming Ti_3C_2-DOX. Both in vitro and in vivo evaluations revealed that this platform offers improved biocompatibility, targeted tumor accumulation, and pH-responsive drug release. The combined photothermal, photodynamic, and chemotherapeutic effects of Ti_3C_2-DOX effectively induced cancer cell death and achieved substantial tumor suppression (as shown in Fig. 18).

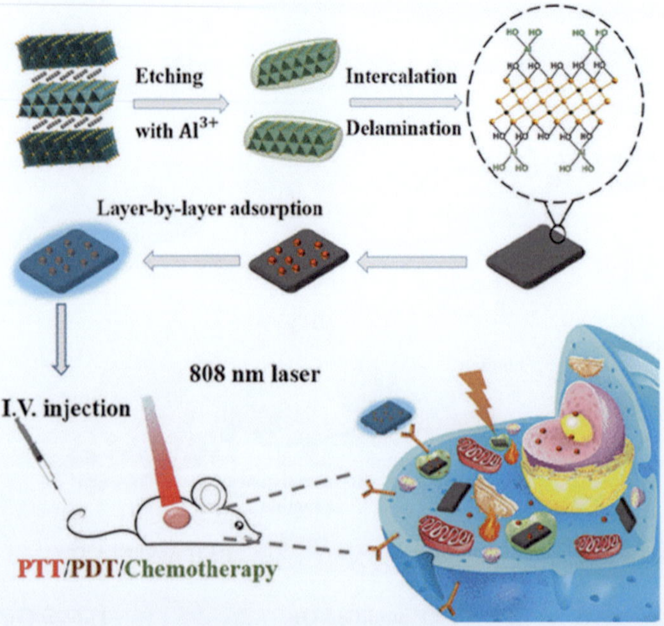

Fig. 18 Graphical abstract showcasing surface-modified Ti_3C_2 MXene nanosheets for targeted cancer therapy. These nanosheets integrate photothermal, photodynamic, and chemotherapy into a synergistic treatment platform, enhancing tumor-targeting efficiency. Functional modifications improve stability and controlled drug release, enabling precise therapeutic effects through heat generation, reactive oxygen species production, and localized drug delivery under laser activation. Reproduced with permission [86]. Copyright 2017, American Chemical Society

4.3 Application of MXenes in Enhancing Biocompatibility

Another research done by Lin et al. [87] presented a biocompatible 2D Ti_3C_2 nanosheet through HF etching of the bulk Ti_3AlC_2 MAX phase, followed by surface modification with SP molecules (Fig. 19a). The resulting Ti_3C_2-SP nanosheets feature an ultrathin structure, consisting of either monolayers or a few layers, and exhibit a notably high extinction coefficient (ε) of 25.2 l g^{-1} cm^{-1}. This value is significantly greater compared to graphene oxide (GO) nanosheets (3.6 l g^{-1} cm^{-1}) and gold nanorods (13.9 l g^{-1} cm^{-1}). Additionally, as depicted in Fig. 19b, the Ti_3C_2-SP nanosheets demonstrate broad-spectrum light absorption ranging from 400 to 900 nm, which suggests their excellent capability to absorb near-infrared (NIR) light. Upon exposure to NIR laser irradiation for 6 min, the temperature of a Ti_3C_2-SP nanosheet dispersion rapidly rises to 55 °C (Fig. 19c), confirming its efficient photothermal conversion capability. This photothermal effect was further evaluated in vitro using live/dead cell staining, where red fluorescence indicated dead cells and green fluorescence indicated live cells. The results showed significant hyperthermia-induced cell death, with most 4T1 cancer cells successfully ablated after treatment with Ti_3C_2-SP nanosheets and laser irradiation. The in vivo therapeutic potential of these nanosheets was also validated using a mouse xenograft tumor model. Infrared thermal imaging and tumor photographs (Fig. 19d and e) demonstrated the ability of Ti_3C_2-SP nanosheets to accumulate effectively at the tumor site after administration (20 mg kg^{-1} dose). Upon NIR laser exposure, a strong photothermal effect was observed in vivo. Tumor volume measurements recorded throughout the treatment confirmed a substantial reduction in tumor growth, indicating that Ti_3C_2-SP nanosheets exhibit effective photothermal tumor suppression under laser irradiation (Fig. 19f) [84].

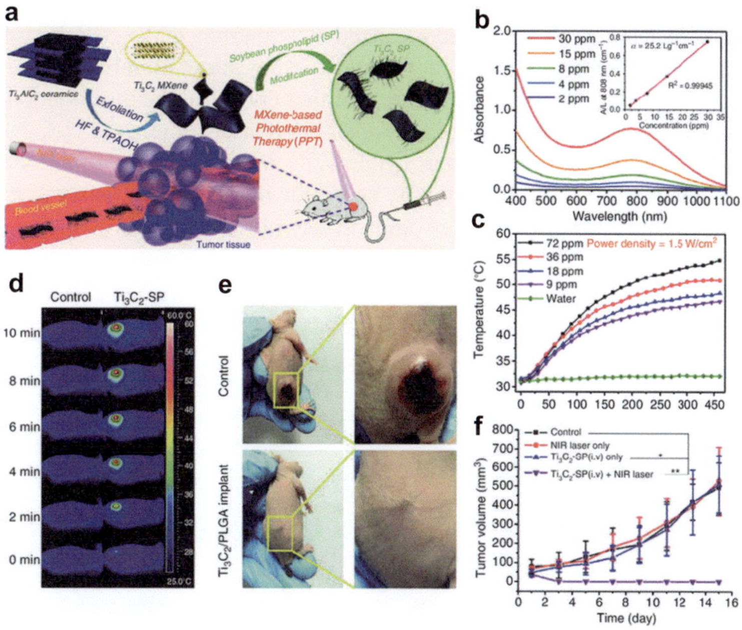

Fig. 19 Photothermal characteristics of Ti_3C_2-SP MXene nanosheets and their PTT effects in a mouse model. **a** Diagram showing tumor ablation using Ti_3C_2-SP MXene nanosheets for PTT. **b** Absorption spectra of Ti_3C_2-SP MXene measured from 400 to 1100 nm. **c** Temperature changes of Ti_3C_2-SP MXene under laser exposure. **d** Infrared thermal images of tumor-bearing nude mice treated with or without Ti_3C_2-SP MXene during laser irradiation. **e** Images of tumor-bearing mice following PTT. **f** Tumor growth inhibition observed after different treatment approaches. Reproduced with permission [84]. Copyright 2020, De Gruyter (Open access)

4.4 Application of MXenes in Sonodynamic Therapy

Zhang et al. [88] formed a dual-responsive nanoplatform for photoacoustic/photothermal imaging-guided therapy by synthesizing oxygen-deficient TiO_2-x nanoparticles in situ on 2D titanium carbide (Ti_3C_2) MXene nanosheets (Fig. 20). This platform was specifically designed to enhance near-infrared II (NIR-II) PTT and sonodynamic therapy (SDT) for tumors. The TiO_2-x nanoparticles, approximately 10 nm in size, exhibited a high surface area and a narrow bandgap of 2.65 eV, owing to their oxygen vacancy-rich structure. These properties allowed them to act as charge traps, minimizing charge recombination and boosting reactive oxygen species (ROS) generation under ultrasound (US) stimulation. Simultaneously, the Ti_3C_2 MXene nanosheets generated localized hyperthermia under NIR-II laser irradiation, improving tumor oxygenation by increasing intratumoral blood flow and alleviating hypoxia. This synergistic effect further enhanced the therapeutic impact of SDT. When subjected to combined US and NIR-II laser stimuli, the $Ti_3C_2@TiO_2$-x nanoplatform demonstrated potent cytotoxicity in vitro and achieved complete tumor elimination in vivo. Importantly, no tumor recurrence or systemic toxicity was observed, highlighting the safety and efficacy of the approach.

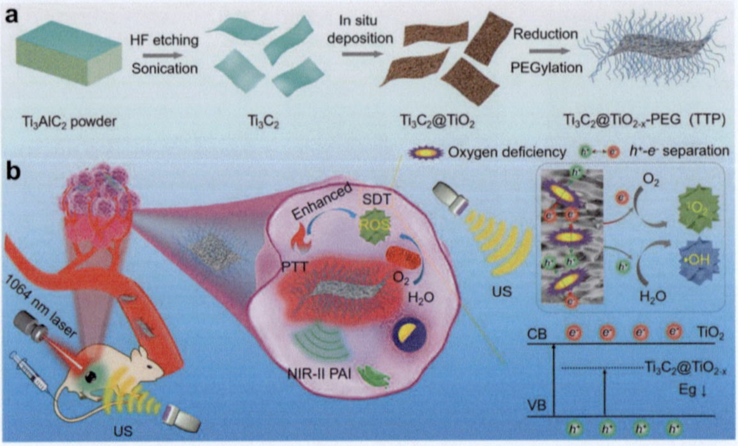

Fig. 20 Diagrammatic representation of **a** the in-situ fabrication of TTP nanohybrids and **b** their use in synergistic photothermal-enhanced sonodynamic therapy guided by bimodal PA/PT imaging in NIR-II biowindow. Reproduced with permission [88]. Copyright 2022, Springer nature. (Open access)

4.5 Application of MXenes for Chemotherapeutic Drug Delivery

MXenes can be loaded with chemotherapeutic drugs like doxorubicin (DOX) due to their large surface area. Surface functionalization enables targeted and pH-responsive drug release in the tumor microenvironment. $Ti_3C_2T_x$ MXene loaded with Au nanoparticles (Au NPs) was synthesized by Liu et al. [89] via in situ growth. In this study, thiol-polyethylene glycol aldehyde (SH-PEG-CHO) polymers, which contain both aldehyde (–CHO) and thiol (–SH) functional groups, were used to bind gold nanoparticles (AuNPs) through the thiol groups. This simple surface modification enabled the subsequent loading of the chemotherapeutic drug doxorubicin (DOX) onto the MXene@Au-PEG system through the formation of pH-sensitive Schiff base bonds, as depicted schematically in Fig. 21a–c. The resulting MXene@Au-PEG-DOX nanocomposite demonstrated dual-responsive drug release behavior: pH sensitivity enabled drug release in acidic environments, while near-infrared (NIR) laser irradiation triggered additional controlled drug release. Moreover, the incorporation of Au and PEG onto the MXene surface enhanced the system's photothermal stability and improved its biocompatibility. The formation of Schiff base bonds further ensured controlled drug release, contributing to enhanced targeted therapeutic effects.

4.6 Application of MXenes for Cancer Nanotheranotics

MXenes, known for their distinctive structure and versatile surface chemistry, have been utilized as a platform for the in situ growth of superparamagnetic Fe_3O_4 nanocrystals, enabling the development of superparamagnetic 2D Ti_3C_2-based composites for advanced cancer theranostics. These biocompatible composites demonstrated notable photothermal conversion efficiency (~48.6%), facilitating effective PTT for cancer cell destruction and tumor ablation, both in vitro and in vivo. Additionally, the composites exhibited excellent T_2 relaxivity (~394.2 $mM^{-1}\ s^{-1}$), significantly enhancing MRI contrast for precise tumor imaging. This multifunctional design highlights a promising approach for integrated cancer diagnosis and therapy.

In 2017, Dai and colleagues [35] developed composite nanosheets based on tantalum carbide (Ta_4C_3) MXenes for multimodal imaging-guided photothermal tumor treatment (Fig. 22). Using a surface redox reaction, manganese oxide nanoparticles (MnOx) were grown in situ on the Ta_4C_3 nanosheets, forming MnOx/Ta_4C_3 composites. The tantalum within the structure served as an efficient contrast agent for enhanced computed tomography (CT) imaging, while the MnOx component responded to the tumor microenvironment to provide T_1-weighted magnetic resonance imaging (MRI) contrast. Furthermore, the composite nanosheets exhibited

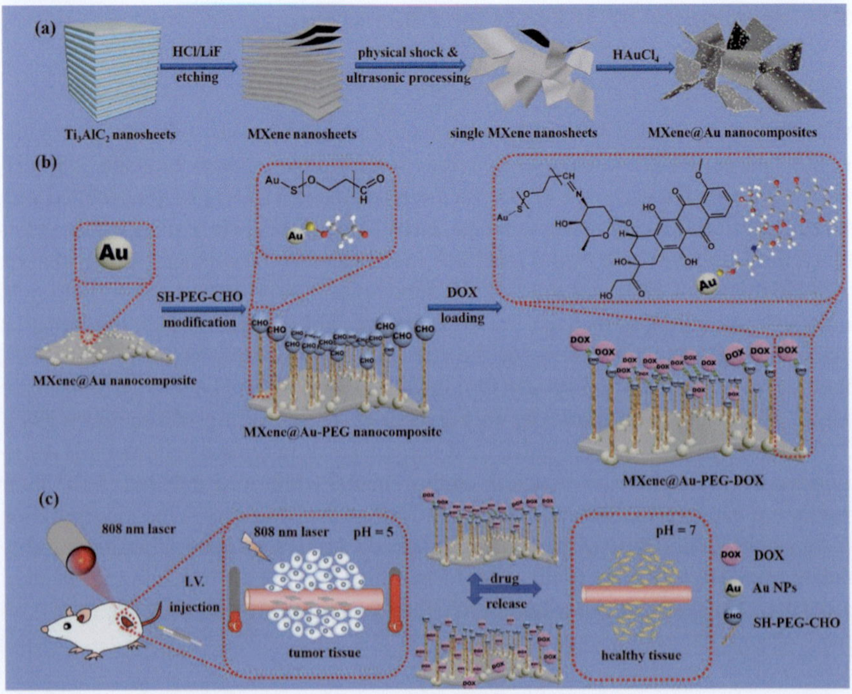

Fig. 21 Diagram illustrating the synthesis and application of MXene in cancer treatment. **a** The preparation method for MXene and MXene@Au nanosheets. **b** Surface modification of MXene@Au with thiol polyethylene glycol aldehyde (SH-PEG-CHO), followed by the loading of doxorubicin (DOX) on the surface. **c** Illustration of the MXene-based drug delivery system designed for near-infrared (NIR) laser-triggered and pH-sensitive drug release in tumor tissue. Reproduced with permission [89]. Copyright 2022, Elsevier Ltd.

excellent photothermal conversion capabilities, enabling contrast-enhanced photoacoustic imaging and achieving significant suppression of tumor growth through PTT.

A W-based 2D i-MXene for multimodal-imaging-guided cancer theranostics was developed by Zhou et al. [31]. Derived from the quaternary i-MAX phase $(W_{2/3}Y_{1/3})_2AlC$, $W_{1.33}C$ i-MXenes were synthesized by selectively removing Y and Al, resulting in structures with ordered basal-plane vacancies. Theoretical simulations revealed strong NIR absorption, which was experimentally validated through photothermal studies in both the NIR-I and NIR-II bio-windows. Modification with bovine serum albumin (BSA) improved the stability and performance of $W_{1.33}C$ i-MXenes, creating $W_{1.33}C$-BSA nanosheets. These nanosheets exhibited high photothermal conversion efficiency, pH-dependent biodegradability, and selective tumor accumulation, followed by rapid clearance with minimal systemic toxicity. They also demonstrated excellent imaging properties for CT, photoacoustic, and photothermal imaging, enabling non-invasive therapy monitoring. Mechanistic

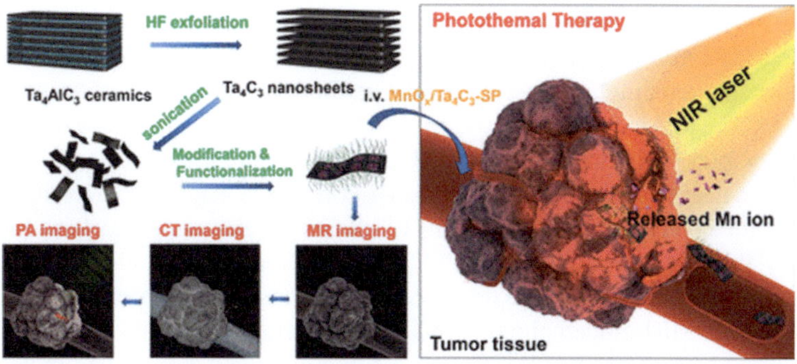

Fig. 22 Represents the development of Ta_4C_3 MXene-based composite nanosheets for imaging-guided photothermal tumor ablation. A redox reaction on the Ta_4C_3 surface enabled in-situ growth of manganese oxide (MnOx) nanoparticles where Ta_4C_3 component served as a contrast agent for computed tomography, while MnOx acted as a tumor-responsive agent for T1-weighted MRI. These MnOx/Ta_4C_3 composites enhanced photoacoustic imaging and effectively suppressed tumor growth through PTT. Reproduced with permission [35]. Copyright 2017, American Chemical Society

studies revealed that PTT with $W_{1.33}C$-BSA activated pathways leading to cell membrane rupture and tumor cell lysis. This work highlights $W_{1.33}C$ i-MXenes as a promising platform for efficient and biodegradable cancer Theranostics (Fig. 23).

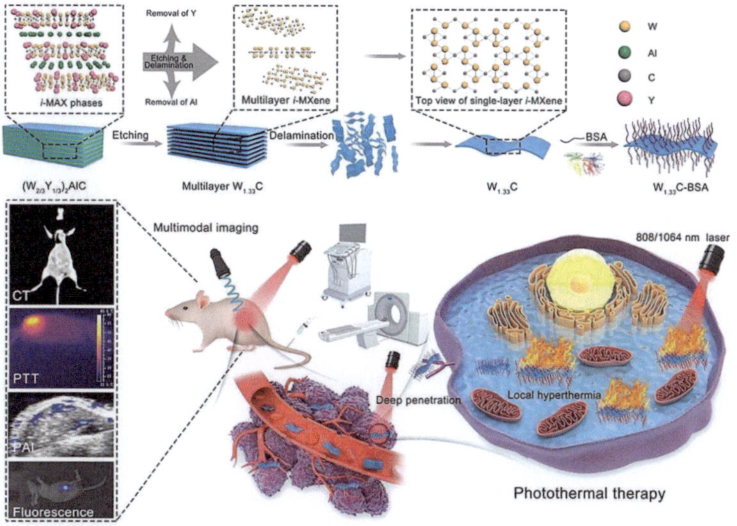

Fig. 23 A schematic representation depicting the fabrication of 2D $W_{1.33}C$-BSA nanosheets, designed as an advanced phototherapeutic agent for cancer treatment. These nanosheets are intended for use in multimodal imaging-guided therapy, integrating multiple imaging techniques to enhance precision in cancer treatment and monitoring therapeutic effectiveness. Reproduced with permission [31]. Copyright 2021, Wiley Online Library (Open access)

Mo_2C MXene nanospheres (~50 nm) have been developed as theranostic agents, capable of absorbing light across the entire NIR spectrum. Upon NIR irradiation, these nanospheres simultaneously induce hyperthermia and generate reactive oxygen species (ROS), making them effective for combined photothermal and photodynamic cancer therapies. This dual approach facilitates the elimination of cancer cells and the liquefactive necrosis of solid tumors. Furthermore, these nanospheres exhibit strong potential for in vivo photoacoustic and CT imaging, alongside excellent biocompatibility [90]. In a separate study [91], Ti_3C_2 MXene nanosheets functionalized with $NaErF_4$ nanoparticles were designed as multifunctional platforms for NIR-IIb (1530 nm) and MRI-guided photothermal cancer therapy using 808 nm laser excitation. These nanocomposites demonstrated a photothermal conversion efficiency of 43.62%, high photothermal stability, and a tumor inhibition rate of ~92.9%. Their T_2-weighted MRI capabilities were attributed to the magnetic properties of Er^{3+} ions, and no significant toxicity was observed at the administered dose (Fig. 24) [92]. Zhang et al. [88] fabricated photo- and sono-responsive nanoplatforms by decorating TiO_{2-x} nanoparticles (~10 nm) onto Ti_3C_2 MXenes. These composites enabled bimodal photoacoustic and photothermal imaging for guiding NIR-II photothermal-enhanced sonodynamic therapy. Enhanced ROS generation and localized hyperthermia contributed to effective tumor eradication in vivo, with no recurrence or systemic toxicity reported.

Liu et al. [34] developed an advanced nanoplatform that offers both therapeutic and diagnostic capabilities for breast cancer treatment. Their work focused on leveraging the unique properties of 2D MXenes, specifically tantalum carbide (Ta_4C_3), which were enhanced through surface modification with superparamagnetic iron oxide nanoparticles (SPIONPs). This resulted in the creation of Ta_4C_3-IONP-SPs composite nanosheets with superior biocompatibility and stability in physiological conditions. The nanoplatform demonstrated dual functionality by enabling PTT and serving as an imaging agent for techniques such as T_2-weighted magnetic resonance imaging (MRI) and contrast-enhanced computed tomography (CT). The high atomic number of tantalum facilitated effective CT imaging, while the SPIONPs acted as contrast agents for MRI (Fig. 25). The photothermal properties of the nanosheets contributed to complete tumor eradication with no recurrence, attributed to their efficient photothermal conversion. This innovative approach highlights the potential of MXene-based nanoplatforms for cancer theranostics, showcasing the effectiveness of combining novel MXene materials with advanced surface modification techniques for enhanced biomedical applications.

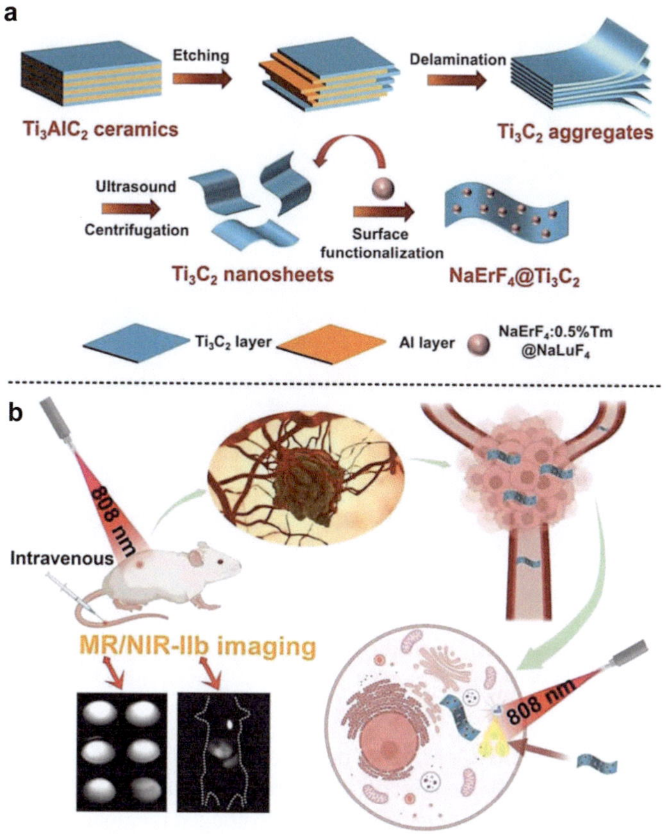

Fig. 24 **a** The fabrication of nanosystems based on $NaErF_4$@Ti_3C_2 MXene for dual cancer therapy and diagnosis. **b** MR/NIR-IIb imaging of tumours employing MXene nanocomposites and PTT. Reproduced with permission [92]. Copyright 2022, MDPI (Open access)

5 Challenges Associated with the Employment of MXene in Biomedicine

Despite extensive research on the synthesis and applications of MXenes, studies on their toxicity and biosafety, especially in biomedical contexts, remain limited. Key factors influencing MXene toxicity include chemical composition, size, solubility, morphology, aggregation, and surface properties [93]. Addressing these factors is crucial for advancing MXenes from in vitro to in vivo and ultimately to clinical applications. Comprehensive preclinical studies are needed to evaluate their efficacy, off-target effects, and interactions with biological systems during circulation [94]. Biodegradability is another critical factor in assessing MXene biosafety. For example, Nb_2C MXenes decompose within 24 h in the presence of human myeloperoxidase and

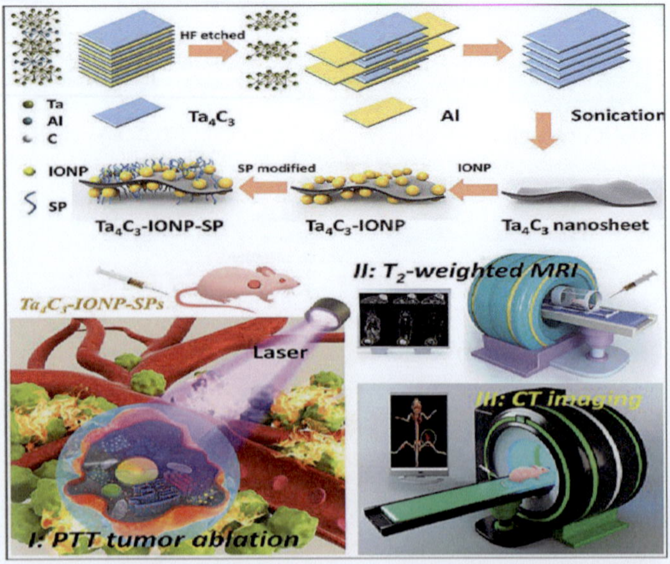

Fig. 25 Breast cancer photothermal ablation guided by dual-modal contrast-enhanced MRI/CT imaging: a schematic depiction of the fabrication process of Ta_4C_3-IONP-SPs composite nanosheets and their distinctive capabilities. Reproduced with permission [34]. Copyright 2018, Ivyspring International Publisher (Open access)

hydrogen peroxide [58], Ti_3C_2 MXenes degrade under enzymatic and ROS stimulation [95], and Mo_2C MXenes exhibit pH-dependent degradation [33]. While MXenes show stability in tumor environments, enabling prolonged therapeutic effects, understanding their biodegradability mechanisms requires further research. Optimizing synthesis methods, such as minimizing harsh etching processes, can improve MXene stability. However, physiological conditions, including ionic strength and enzymatic activity, can accelerate their biodegradation and clearance. Studies on MXene toxicity and biocompatibility suggest promising results. Ti_3C_2-soybean phospholipid nanocomposites showed no acute toxicity or organ damage, with ~18.7% excreted in urine and ~10.3% in feces within 48 h. Similarly, MnOx/Ti_3C_2 composites demonstrated normal vital signs and no toxic effects after intravenous administration. Nb_2C MXenes modified with polyvinyl pyrrolidone also showed high biocompatibility, with 20% excretion through urine and feces within 48 h and no histological abnormalities in major organs. These findings highlight the potential of MXenes as safe and effective theranostic agents, though further studies are essential to address the remaining challenges [96].

6 Future Directions

MXenes hold great potential for precision therapies by enabling ligand functionalization to target specific tumor markers, allowing highly accurate delivery of therapeutic agents while minimizing side effects on healthy tissues. This specificity is particularly valuable in treatments like chemotherapy or PTT, where collateral damage is a concern. However, extensive testing is required to confirm that functionalized MXenes maintain their therapeutic effectiveness without compromising safety. Their versatility extends to multimodal treatments, making them ideal for combination therapies such as chemo-PTT. To advance these approaches, studies must focus on pharmacokinetics, biodistribution, and tissue interactions in animal models to ensure effective tumor targeting and minimize systemic toxicity. In vivo research is crucial to transition promising in vitro results into clinical applications. Scaling up MXene production also presents challenges, particularly regarding environmental and safety concerns. Sustainable and cost-effective synthesis methods, such as ultrasonic exfoliation and green chemistry approaches, are being developed to reduce reliance on toxic reagents like hydrofluoric acid. These innovations aim to lower production costs, improve scalability, and address potential environmental and health risks, ensuring MXenes are viable for both industrial and biomedical applications.

7 Summary and Conclusion

MXenes and related composites are pioneering technologies in tissue engineering and cancer nanotheranostics due to their biocompatibility, flexibility, and physicochemical features. It has been shown to improve PTT, drug delivery platforms, antimicrobial materials, and tissue regeneration scaffolds, thereby revolutionising these fields. Nevertheless, therapeutic and biological potential of MXene-based materials remains unrealised despite these encouraging advances. There are a number of concerns that need to be addressed, including long-term biocompatibility, toxicity profiles, scalability of synthesis, and accurate functionalisation procedures. Furthermore, it is essential to have a molecular level understanding of the interactions that take place between MXenes and biological systems in order to maximise their effectiveness and minimise any unexpected side effects.

Future research should be focused on creating cost-effective and sustainable MXene composites using better surface engineering methodologies to improve targeting specificity, drug loading efficiency, and stability. When it comes to moving MXenes and their composites from the laboratory to clinical applications, interdisciplinary collaborations that integrate materials science, biology, and medicine will play a crucial role in advancing the field. Through the overcoming of these obstacles, MXene-based technologies have a tremendous amount of promise to reshape the landscape of tissue engineering and cancer nanotheranostics, hence paving the way for novel and efficient therapeutic approaches.

Acknowledgements The authors gratefully acknowledge the support provided by Centurion University of Technology and Management, Odisha, India, for carrying out the present research work.

Conflicts of Interest The authors (s) declare no conflict of interest.

References

1. Haleem, A., Javaid, M., Singh, R.P., Rab, S., Suman, R.: Applications of nanotechnology in medical field: a brief review. Global Health J. **7**(2), 70–77 (2023)
2. Baig, N.: Two-dimensional nanomaterials: a critical review of recent progress, properties, applications, and future directions. Compos. A Appl. Sci. Manuf. **165**, 107362 (2023)
3. Mishra, D.P., Acharya, B., Tripathy, S., Barik, B., Sahu, P.K.: An overview of the biosensing potential of organometallic compounds. Chem. Phys. Impact **28**, 100326 (2023)
4. Mishra, D.P., Sarangi, A.K.: Sensing application of organometallic compounds. In: Recent Advances in Organometallic Chemistry, pp. 217–237. Elsevier (2024)
5. Novoselov, K.S., Colombo, L., Gellert, P.R., Schwab, M.G., Kim, K.A.J.N.: A roadmap for graphene. Nature **490**(7419), 192–200 (2012)
6. Acun, A., Zhang, L., Bampoulis, P., Farmanbar, M.V., van Houselt, A., Rudenko, A.N., Lingenfelder, M., Brocks, G., Poelsema, B., Katsnelson, M.I., Zandvliet, H.J.W.: Germanene: the germanium analogue of graphene. J. Phys. Condens. Matter **27**(44), 443002 (2015)
7. Zhao, J., Liu, H., Yu, Z., Quhe, R., Zhou, S., Wang, Y., Liu, C.C., Zhong, H., Han, N., Lu, J., Yao, Y.: Rise of silicene: a competitive 2D material. Prog. Mater. Sci. **83**, 24–151 (2016)
8. Mannix, A.J., Zhou, X.F., Kiraly, B., Wood, J.D., Alducin, D., Myers, B.D., Liu, X., Fisher, B.L., Santiago, U., Guest, J.R., Yacaman, M.J.: Synthesis of borophenes: anisotropic, two-dimensional boron polymorphs. Science **350**(6267), 1513–1516 (2015)
9. Lin, H., Chen, Y., Shi, J.: Insights into 2D MXenes for versatile biomedical applications: current advances and challenges ahead. Adv. Sci. **5**(10), 1800518 (2018)
10. Karthikeyan, P., Elanchezhiyan, S.S., Preethi, J., Talukdar, K., Meenakshi, S., Park, C.M.: Two-dimensional (2D) $Ti_3C_2T_x$ MXene nanosheets with superior adsorption behavior for phosphate and nitrate ions from the aqueous environment. Ceram. Int. **47**(1), 732–739 (2021)
11. Kulkarni, M., Lalic, A., Balu, R., Zhang, H., Dutta, N.K., Roy Choudhury, N.: Hydrothermal synthesis of MXenes in alkali environment and development of MXene/PEDOT: PSS composite electrodes for supercapacitor applications. MRS Adv. **9**(17), 1310–1317 (2024)
12. Zhang, R., Sun, Y., Jiao, F., Li, L., Geng, D., Hu, W.: MXene-MoS_2 nanocomposites via chemical vapor deposition with enhanced electrocatalytic activity for hydrogen evolution. Nano Res. **16**(7), 8937–8944 (2023)
13. Protyai, M.I.H., Rashid, A.B.: A comprehensive overview of recent progress in MXene-based polymer composites: their fabrication processes, advanced applications, and prospects. Heliyon (2024)
14. Urbankowski, P., Anasori, B., Makaryan, T., Er, D., Kota, S., Walsh, P.L., Zhao, M., Shenoy, V.B., Barsoum, M.W., Gogotsi, Y.: Synthesis of two-dimensional titanium nitride Ti_4N_3 (MXene). Nanoscale **8**(22), 11385–11391 (2016)
15. Sun, W., Shah, S.A., Chen, Y., Tan, Z., Gao, H., Habib, T., Radovic, M., Green, M.J.: Electrochemical etching of Ti_2AlC to Ti_2CT_x (MXene) in low-concentration hydrochloric acid solution. J. Mater. Chem. A **5**(41), 21663–21668 (2017)

16. Verger, L., Xu, C., Natu, V., Cheng, H.M., Ren, W., Barsoum, M.W.: Overview of the synthesis of MXenes and other ultrathin 2D transition metal carbides and nitrides. Curr. Opin. Solid State Mater. Sci. **23**(3), 149–163 (2019)
17. Salim, O., Mahmoud, K.A., Pant, K.K., Joshi, R.K.: Introduction to MXenes: synthesis and characteristics. Mater. Today Chem. **14**, 100191 (2019)
18. Ronchi, R.M., Arantes, J.T., Santos, S.F.: Synthesis, structure, properties and applications of MXenes: current status and perspectives. Ceram. Int. **45**(15), 18167–18188 (2019)
19. Cheng, W., Zeng, X., Chen, H., Li, Z., Zeng, W., Mei, L., Zhao, Y.: Versatile polydopamine platforms: synthesis and promising applications for surface modification and advanced nanomedicine. ACS Nano **13**(8), 8537–8565 (2019)
20. Dong, L.M., Ye, C., Zheng, L.L., Gao, Z.F., Xia, F.: Two-dimensional metal carbides and nitrides (MXenes): preparation, property, and applications in cancer therapy. Nanophotonics **9**(8), 2125–2145 (2020)
21. Naguib, M., Kurtoglu, M., Presser, V., Lu, J., Niu, J., Heon, M., Hultman, L., Gogotsi, Y., Barsoum, M.W.: Two-dimensional nanocrystals produced by exfoliation of Ti_3AlC_2. MXenes, pp. 15–29. Jenny Stanford Publishing (2023)
22. Gao, L., Li, C., Huang, W., Mei, S., Lin, H., Ou, Q., Zhang, Y., Guo, J., Zhang, F., Xu, S., Zhang, H.: MXene/polymer membranes: synthesis, properties, and emerging applications. Chem. Mater. **32**(5), 1703–1747 (2020)
23. Rasool, K., Helal, M., Ali, A., Ren, C.E., Gogotsi, Y., Mahmoud, K.A.: Antibacterial activity of $Ti_3C_2T_x$ MXene. ACS Nano **10**(3), 3674–3684 (2016)
24. Zhu, X., Liu, P., Xue, T., Ge, Y., Ai, S., Sheng, Y., Wu, R., Xu, L., Tang, K., Wen, Y.: A novel graphene-like titanium carbide MXene/Au–Ag nanoshuttles bifunctional nanosensor for electrochemical and SERS intelligent analysis of ultra-trace carbendazim coupled with machine learning. Ceram. Int. **47**(1), 173–184 (2021)
25. Yin, J., Han, Q., Zhang, J., Liu, Y., Gan, X., Xie, K., Xie, L., Deng, Y.: MXene-based hydrogels endow polyetheretherketone with effective osteogenicity and combined treatment of osteosarcoma and bacterial infection. ACS Appl. Mater. Interfaces **12**(41), 45891–45903 (2020)
26. Yang, C., Luo, Y., Lin, H., Ge, M., Shi, J., Zhang, X.: Niobium carbide MXene augmented medical implant elicits bacterial infection elimination and tissue regeneration. ACS Nano **15**(1), 1086–1099 (2020)
27. Zhan, X., Si, C., Zhou, J., Sun, Z.: MXene and MXene-based composites: synthesis, properties and environment-related applications. Nanoscale Horiz. **5**(2), 235–258 (2020)
28. Han, M., Maleski, K., Shuck, C.E., Yang, Y., Glazar, J.T., Foucher, A.C., Hantanasirisakul, K., Sarycheva, A., Frey, N.C., May, S.J., Shenoy, V.B.: Tailoring electronic and optical properties of MXenes through forming solid solutions. J. Am. Chem. Soc. **142**(45), 19110–19118 (2020)
29. Gao, G., Ding, G., Li, J., Yao, K., Wu, M., Qian, M.: Monolayer MXenes: promising half-metals and spin gapless semiconductors. Nanoscale **8**(16), 8986–8994 (2016)
30. Liu, Y., Xiao, H., Goddard, W.A., III.: Schottky-barrier-free contacts with two-dimensional semiconductors by surface-engineered MXenes. J. Am. Chem. Soc. **138**(49), 15853–15856 (2016)
31. Zhou, B., Yin, H., Dong, C., Sun, L., Feng, W., Pu, Y., Han, X., Li, X., Du, D., Xu, H., Chen, Y.: Biodegradable and excretable 2D W1. 33Ci-MXene with vacancy ordering for theory-oriented cancer nanotheranostics in near-infrared biowindow. Adv. Sci. **8**(24), 2101043 (2021)
32. Liang, R., Li, Y., Huo, M., Lin, H., Chen, Y.: Triggering sequential catalytic fenton reaction on 2D MXenes for hyperthermia-augmented synergistic nanocatalytic cancer therapy. ACS Appl. Mater. Interfaces **11**(46), 42917–42931 (2019)
33. Feng, W., Wang, R., Zhou, Y., Ding, L., Gao, X., Zhou, B., Hu, P., Chen, Y.: Ultrathin molybdenum carbide MXene with fast biodegradability for highly efficient theory-oriented photonic tumor hyperthermia. Adv. Func. Mater. **29**(22), 1901942 (2019)
34. Liu, Z., Lin, H., Zhao, M., Dai, C., Zhang, S., Peng, W., Chen, Y.: 2D superparamagnetic tantalum carbide composite MXenes for efficient breast-cancer theranostics. Theranostics **8**(6), 1648 (2018)

35. Dai, C., Chen, Y., Jing, X., Xiang, L., Yang, D., Lin, H., Liu, Z., Han, X., Wu, R.: Two-dimensional tantalum carbide (MXenes) composite nanosheets for multiple imaging-guided photothermal tumor ablation. ACS Nano **11**(12), 12696–12712 (2017)
36. Tang, W., Dong, Z., Zhang, R., Yi, X., Yang, K., Jin, M., Yuan, C., Xiao, Z., Liu, Z., Cheng, L.: Multifunctional two-dimensional core–shell mxene@ gold nanocomposites for enhanced photo–radio combined therapy in the second biological window. ACS Nano **13**(1), 284–294 (2018)
37. Seyedin, S., Uzun, S., Levitt, A., Anasori, B., Dion, G., Gogotsi, Y., Razal, J.M.: MXene composite and coaxial fibers with high stretchability and conductivity for wearable strain sensing textiles. Adv. Func. Mater. **30**(12), 1910504 (2020)
38. Yin, J., Pan, S., Guo, X., Gao, Y., Zhu, D., Yang, Q., Gao, J., Zhang, C., Chen, Y.: Nb_2C MXene-functionalized scaffolds enables osteosarcoma phototherapy and angiogenesis/osteogenesis of bone defects. Nano-Micro Lett. **13**, 1–18 (2021)
39. Shokri, A., Ramezani, K., Jamalpour, M.R., Mohammadi, C., Vahdatinia, F., Irani, A.D., Sharifi, E., Haddadi, R., Jamshidi, S., Amirabad, L.M., Tajik, S.: In vivo efficacy of 3D-printed elastin–gelatin–hyaluronic acid scaffolds for regeneration of nasal septal cartilage defects. J. Biomed. Mater. Res. B Appl. Biomater. **110**(3), 614–624 (2022)
40. He, C., Yu, L., Yao, H., Chen, Y., Hao, Y.: Combinatorial photothermal 3D-printing scaffold and checkpoint blockade inhibits growth/metastasis of breast cancer to bone and accelerates osteogenesis. Adv. Func. Mater. **31**(10), 2006214 (2021)
41. He, C., Yu, L., Yao, H., Chen, Y., Hao, Y.: Combinatorial photothermal 3D-printing scaffold and checkpoint blockade inhibits growth/metastasis of breast cancer to bone and accelerates osteogenesis. Adv. Funct. Mater. **31**(10), 2006214 (2021)
42. Fu, Y., Zhang, J., Lin, H., Mo, A.: 2D titanium carbide (MXene) nanosheets and 1D hydroxyapatite nanowires into free standing nanocomposite membrane: in vitro and in vivo evaluations for bone regeneration. Mater. Sci. Eng. C **118**, 111367 (2021)
43. Yang, Q., Yin, H., Xu, T., Zhu, D., Yin, J., Chen, Y., Yu, X., Gao, J., Zhang, C., Chen, Y., Gao, Y.: Engineering 2D mesoporous Silica@ MXene-integrated 3D-printing scaffolds for combinatory osteosarcoma therapy and NO-augmented bone regeneration. Small **16**(14), 1906814 (2020)
44. Li, C., Chu, D., Jin, L., Tan, G., Li, Z.: Synergistic effect of the photothermal performance and osteogenic properties of MXene and hydroxyapatite nanoparticle composite nanofibers for osteogenic application. J. Biomed. Nanotechnol. **17**(10), 2014–2020 (2021)
45. Liao, M., Hu, Y., Zhang, Y., Wang, K., Fang, Q., Qi, Y., Shen, Y., Cheng, H., Fu, X., Tang, M., Sun, S.: 3D $Ti_3C_2T_x$ MXene–matrigel with electroacoustic stimulation to promote the growth of spiral ganglion neurons. ACS Nano **16**(10), 16744–16756 (2022)
46. Driscoll, N., Richardson, A.G., Maleski, K., Anasori, B., Adewole, O., Lelyukh, P., Escobedo, L., Cullen, D.K., Lucas, T.H., Gogotsi, Y., Vitale, F.: Two-dimensional Ti_3C_2 MXene for high-resolution neural interfaces. ACS Nano **12**(10), 10419–10429 (2018)
47. Zhang, Z., Gao, S., Hu, Y.N., Chen, X., Cheng, C., Fu, X.L., Zhang, S.S., Wang, X.L., Che, Y.W., Zhang, C., Chai, R.J.: $Ti_3C_2T_x$ MXene composite 3D hydrogel potentiates mTOR signaling to promote the generation of functional hair cells in cochlea organoids. Adv. Sci. **9**(32), 2203557 (2022)
48. Cai, J., Zhang, H., Hu, Y., Huang, Z., Wang, Y., Xia, Y., Chen, X., Guo, J., Cheng, H., Xia, L., Lu, W.: GelMA-MXene hydrogel nerve conduits with microgrooves for spinal cord injury repair. J. Nanobiotechnol. **20**(1), 460 (2022)
49. Li, Y., Han, M., Cai, Y., Jiang, B., Zhang, Y., Yuan, B., Zhou, F., Cao, C.: Muscle-inspired MXene/PVA hydrogel with high toughness and photothermal therapy for promoting bacteria-infected wound healing. Biomater. Sci. **10**(4), 1068–1082 (2022)
50. Rozmysłowska-Wojciechowska, A., Karwowska, E., Gloc, M., Woźniak, J., Petrus, M., Przybyszewski, B., Wojciechowski, T., Jastrzębska, A.M.: Controlling the porosity and biocidal properties of the chitosan-hyaluronate matrix hydrogel nanocomposites by the addition of 2D $Ti_3C_2T_x$ MXene. Materials **13**(20), 4587 (2020)
51. Zhou, L., Zheng, H., Liu, Z., Wang, S., Liu, Z., Chen, F., Zhang, H., Kong, J., Zhou, F., Zhang, Q.: Conductive antibacterial hemostatic multifunctional scaffolds based on $Ti_3C_2T_x$

MXene nanosheets for promoting multidrug-resistant bacteria-infected wound healing. ACS Nano **15**(2), 2468–2480 (2021)
52. Xu, X., Wang, S., Wu, H., Liu, Y., Xu, F., Zhao, J.: A multimodal antimicrobial platform based on MXene for treatment of wound infection. Colloids Surf. B **207**, 111979 (2021)
53. Basara, G., Saeidi-Javash, M., Ren, X., Bahcecioglu, G., Wyatt, B.C., Anasori, B., Zhang, Y., Zorlutuna, P.: Electrically conductive 3D printed $Ti_3C_2T_x$ MXene-PEG composite constructs for cardiac tissue engineering. Acta Biomater. **139**, 179–189 (2022)
54. Asaro, G.A., Solazzo, M., Suku, M., Spurling, D., Genoud, K., Gonzalez, J.G., Brien, F.J.O., Nicolosi, V., Monaghan, M.G.: MXene functionalized collagen biomaterials for cardiac tissue engineering driving iPSC-derived cardiomyocyte maturation. npj 2D Mater. Appl. **7**(1), 44 (2023)
55. Cui, Y., Li, H., Li, Y., Mao, L.: Novel insights into nanomaterials for immunomodulatory bone regeneration. Nanoscale Adv. **4**(2), 334–352 (2022)
56. Xu, X., Chen, X., Wang, H., Mei, X., Chen, B., Li, R., Qin, Y.: Balancing the toxicity, photothermal effect, and promotion of osteogenesis: photothermal scaffolds for malignant bone tumor therapy. Mater. Today Adv. **13**, 100209 (2022)
57. Pan, S., Yin, J., Yu, L., Zhang, C., Zhu, Y., Gao, Y., Chen, Y.: 2D MXene-integrated 3D-printing scaffolds for augmented osteosarcoma phototherapy and accelerated tissue reconstruction. Adv. Sci. **7**(2), 1901511 (2020)
58. Lin, H., Gao, S., Dai, C., Chen, Y., Shi, J.: A two-dimensional biodegradable niobium carbide (MXene) for photothermal tumor eradication in NIR-I and NIR-II biowindows. J. Am. Chem. Soc. **139**(45), 16235–16247 (2017)
59. Zhang, J., Fu, Y., Mo, A.: Multilayered titanium carbide MXene film for guided bone regeneration. Int. J. Nanomed., 10091–10103 (2019)
60. Mi, X., Su, Z., Fu, Y., Li, S., Mo, A.: 3D printing of Ti_3C_2-MXene-incorporated composite scaffolds for accelerated bone regeneration. Biomed. Mater. **17**(3), 035002 (2022)
61. Halim, A., Qu, K.Y., Zhang, X.F., Huang, N.P.: Recent advances in the application of two-dimensional nanomaterials for neural tissue engineering and regeneration. ACS Biomater. Sci. Eng. **7**(8), 3503–3529 (2021)
62. Liao, M., Cui, Q., Hu, Y., Xing, J., Wu, D., Zheng, S., Zhao, Y., Yu, Y., Sun, J., Chai, R.: Recent advances in the application of MXenes for neural tissue engineering and regeneration. Neural Regen. Res. **19**(2), 258–263 (2024)
63. Wang, X., Fan, X., Li, M., Zhu, W., Xue, J., Ye, F., Cheng, L.: Structure and electromagnetic properties of $Ti_3C_2T_x$ MXene derived from Ti_3AlC_2 with different microstructures. Ceram. Int. **47**(10), 13628–13634 (2021)
64. Li, Y., Hu, Y., Wei, H., Cao, W., Qi, Y., Zhou, S., Zhang, P., Li, H., Li, G.L., Chai, R.: Two-dimensional $Ti_3C_2T_x$ MXene promotes electrophysiological maturation of neural circuits. J. Nanobiotechnol. **20**(1), 398 (2022)
65. Xiao, M., Li, X., Pifferi, S., Pastore, B., Liu, Y., Lazzarino, M., Torre, V., Yang, X., Menini, A., Tang, M.: 2D MXene interfaces preserve the basal electrophysiology of targeted neural circuits. Nanoscale **14**(30), 10992–11002 (2022)
66. Gürbüz, B., Ciftci, F.: Bio-electric-electronics and tissue engineering applications of MXenes wearable materials: a review. Chem. Eng. J., 151230 (2024)
67. Guo, R., Xiao, M., Zhao, W., Zhou, S., Hu, Y., Liao, M., Wang, S., Yang, X., Chai, R., Tang, M.: 2D $Ti_3C_2T_x$ MXene couples electrical stimulation to promote proliferation and neural differentiation of neural stem cells. Acta Biomater. **139**, 105–117 (2022)
68. Li, X., Wang, S., Zheng, M., Ma, Z., Chen, Y., Deng, L., Guang, F., Khademolqorani, S., Banitaba, S.N., Osman, A.I.: Synergistic integration of MXene nanostructures into electrospun fibers for advanced biomedical engineering applications. Nanoscale Horiz. (2024)
69. Diedkova, K., Pogrebnjak, A.D., Kyrylenko, S., Smyrnova, K., Buranich, V.V., Horodek, P., Zukowski, P., Koltunowicz, T.N., Galaszkiewicz, P., Makashina, K., Bondariev, V.: Polycaprolactone–MXene nanofibrous scaffolds for tissue engineering. ACS Appl. Mater. Interfaces **15**(11), 14033–14047 (2023)

70. Zhao, J., Wang, T., Zhu, Y., Qin, H., Qian, J., Wang, Q., Zhang, P., Liu, P., Xiong, A., Li, N., Udduttula, A.: Enhanced osteogenic and ROS-scavenging MXene nanosheets incorporated gelatin-based nanocomposite hydrogels for critical-sized calvarial defect repair. Int. J. Biol. Macromol. **269**, 131914 (2024)
71. Nan, L.P., Lin, Z., Wang, F., Jin, X.H., Fang, J.Q., Xu, B., Liu, S.H., Zhang, F., Wu, Z., Zhou, Z.F., Chen, F.: $Ti_3C_2T_x$ MXene-coated electrospun PCL conduits for enhancing neurite regeneration and angiogenesis. Front. Bioeng. Biotechnol. **10**, 850650 (2022)
72. Han, X., Huang, J., Lin, H., Wang, Z., Li, P., Chen, Y.: 2D ultrathin MXene-based drug-delivery nanoplatform for synergistic photothermal ablation and chemotherapy of cancer. Adv. Healthcare Mater. **7**(9), 1701394 (2018)
73. Li, J., Hashemi, P., Liu, T., Dang, K.M., Brunk, M.G., Mu, X., Nia, A.S., Sacher, W.D., Feng, X., Poon, J.K.: 3D printed titanium carbide MXene-coated polycaprolactone scaffolds for guided neuronal growth and photothermal stimulation. Commun. Mater. **5**(1), 62 (2024)
74. Mao, L., Hu, S., Gao, Y., Wang, L., Zhao, W., Fu, L., Cheng, H., Xia, L., Xie, S., Ye, W., Shi, Z.: Biodegradable and electroactive regenerated bacterial cellulose/MXene ($Ti_3C_2T_x$) composite hydrogel as wound dressing for accelerating skin wound healing under electrical stimulation. Adv. Healthcare Mater. **9**(19), 2000872 (2020)
75. Yang, Y., Zhou, X., Chan, Y.K., Wang, Z., Li, L., Li, J., Liang, K., Deng, Y.: Photo-activated nanofibrous membrane with self-rechargeable antibacterial function for stubborn infected cutaneous regeneration. Small **18**(12), 2105988 (2022)
76. Zarepour, A., Rafati, N., Khosravi, A., Rabiee, N., Iravani, S., Zarrabi, A.: MXene-based composites in smart wound healing and dressings. Nanoscale Adv. (2024)
77. Jin, L., Guo, X., Gao, D., Liu, Y., Ni, J., Zhang, Z., Huang, Y., Xu, G., Yang, Z., Zhang, X., Jiang, X.: An NIR photothermal-responsive hybrid hydrogel for enhanced wound healing. Bioact. Mater. **16**, 162–172 (2022)
78. Ren, X., Xiao, M., Xu, Y., Wu, Y., Yang, J., Wang, Y., Hu, Y., Jiang, Z., Li, X., Shen, Z., Hu, S.: Injectable MXene conductive hydrogel improves myocardial infarction through scavenging ROS and repairing myocardium electrical integrity. Chem. Eng. J. **481**, 148791 (2024)
79. Rafieerad, A., Yan, W., Sequiera, G.L., Sareen, N., Abu-El-Rub, E., Moudgil, M., Dhingra, S.: Application of Ti_3C_2 MXene quantum dots for immunomodulation and regenerative medicine. Adv. Healthcare Mater. **8**(16), 1900569 (2019)
80. Rafieerad, A., Sequiera, G.L., Yan, W., Kaur, P., Amiri, A., Dhingra, S.: Sweet-MXene hydrogel with mixed-dimensional components for biomedical applications. J. Mech. Behav. Biomed. Mater. **101**, 103440 (2020)
81. Sahu, R., Jha, S., Pattanayak, S.P.: Therapeutic silencing of mTOR by systemically administered siRNA-loaded neutral liposomal nanoparticles inhibits DMBA-induced mammary carcinogenesis. Br. J. Cancer **127**(12), 2207–2219 (2022)
82. Iravani, S., Varma, R.S.: MXenes for cancer therapy and diagnosis: recent advances and current challenges. ACS Biomater. Sci. Eng. **7**(6), 1900–1913 (2021)
83. Mishra, D.P., Sahu, P.K., Acharya, B., Ranajit, S.K.: Harnessing synthetic strategies for enhanced gold nanoparticle functionality in cancer treatment. BioNanoScience, 1–22 (2024)
84. Huang, Z., Cui, X., Li, S., Wei, J., Li, P., Wang, Y., Lee, C.S.: Two-dimensional MXene-based materials for photothermal therapy. Nanophotonics **9**(8), 2233–2249 (2020)
85. Li, B., Fu, G., Liu, C., Lu, Y., Mi, Y., Yan, D., Wu, J., Dai, X., Cao, D., Liu, W., Liu, X.: Ti_2C_3 MXene-based nanocomposite as an intelligent nanoplatform for efficient mild hyperthermia treatment. J. Colloid Interface Sci. **665**, 389–398 (2024)
86. Liu, G., Zou, J., Tang, Q., Yang, X., Zhang, Y., Zhang, Q., Huang, W., Chen, P., Shao, J., Dong, X.: Surface modified Ti_3C_2 MXene nanosheets for tumor targeting photothermal/photodynamic/chemo synergistic therapy. ACS Appl. Mater. Interfaces **9**(46), 40077–40086 (2017)
87. Lin, H., Wang, X., Yu, L., Chen, Y., Shi, J.: Two-dimensional ultrathin MXene ceramic nanosheets for photothermal conversion. Nano Lett. **17**(1), 384–391 (2017)
88. Zhang, D.Y., Liu, H., Younis, M.R., Lei, S., Chen, Y., Huang, P., Lin, J.: In-situ TiO_2-x decoration of titanium carbide MXene for photo/sono-responsive antitumor theranostics. J. Nanobiotechnol. **20**(1), 53 (2022)

89. Liu, A., Liu, Y., Liu, G., Zhang, A., Cheng, Y., Li, Y., Zhang, L., Wang, L., Zhou, H., Liu, J., Wang, H.: Engineering of surface modified $Ti_3C_2T_x$ MXene based dually controlled drug release system for synergistic multitherapies of cancer. Chem. Eng. J. **448**, 137691 (2022)
90. Zhang, Q., Huang, W., Yang, C., Wang, F., Song, C., Gao, Y., Qiu, Y., Yan, M., Yang, B., Guo, C.: The theranostic nanoagent Mo_2C for multi-modal imaging-guided cancer synergistic phototherapy. Biomater. Sci. **7**(7), 2729–2739 (2019)
91. Pan, J., Zhang, M., Fu, G., Zhang, L., Yu, H., Yan, X., Liu, F., Sun, P., Jia, X., Liu, X., Lu, G.: Ti_3C_2 MXene nanosheets functionalized with NaErF4: 0.5% Tm@ $NaLuF_4$ nanoparticles for dual-modal near-infrared IIb/magnetic resonance imaging-guided tumor hyperthermia. ACS Appl. Nano Mater. **5**(6), 8142–8153 (2022)
92. Iravani, S., Varma, R.S.: MXenes in cancer nanotheranostics. Nanomaterials **12**(19), 3360 (2022)
93. Lim, G.P., Soon, C.F., Ma, N.L., Morsin, M., Nayan, N., Ahmad, M.K., Tee, K.S.: Cytotoxicity of MXene-based nanomaterials for biomedical applications: a mini review. Environ. Res. **201**, 111592 (2021)
94. Johnson, K.K., Koshy, P., Yang, J.L., Sorrell, C.C.: Preclinical cancer theranostics—from nanomaterials to clinic: the missing link. Adv. Func. Mater. **31**(43), 2104199 (2021)
95. Zhao, X., Wang, L.Y., Li, J.M., Peng, L.M., Tang, C.Y., Zha, X.J., Ke, K., Yang, M.B., Su, B.H., Yang, W.: Redox-mediated artificial non-enzymatic antioxidant MXene nanoplatforms for acute kidney injury alleviation. Adv. Sci. **8**(18), 2101498 (2021)
96. Chen, L., Dai, X., Feng, W., Chen, Y.: Biomedical applications of MXenes: from nanomedicine to biomaterials. Acc. Mater. Res. **3**(8), 785–798 (2022)

MIX
Papier aus verantwortungsvollen Quellen
Paper from responsible sources
FSC® C105338

If you have any concerns about our products,
you can contact us on
ProductSafety@springernature.com

In case Publisher is established outside the EU,
the EU authorized representative is:
Springer Nature Customer Service Center GmbH
Europaplatz 3, 69115 Heidelberg, Germany

Printed by Libri Plureos GmbH
in Hamburg, Germany